FLUID MECHANICS

FLUID MECHANICS

W. M. SWANSON

Washington University

HOLT, RINEHART AND WINSTON

New York Chicago San Francisco Atlanta
Dallas Montreal Toronto London Sydney

ACKNOWLEDGMENTS

Figure 5-7 is reproduced from *Fluid Meters, Their Theory and Application,* Fifth Edition, American Society of Mechanical Engineers (1959) by permission of the publisher.

Figure 10-18 is reproduced from "Friction Factor for Pipe Flow," by L. F. Moody, *Transactions,* American Society of Mechanical Engineers, 66, 671-684 (1944) and is used by permission of the publisher.

Figures 10-10, 10-12, 10-13, 10-16, 10-17, 10-26, 11-6, and 11-14 are from *Boundary-Layer Theory* by H. Schlichting. Copyright © 1968 by McGraw-Hill, Inc. Used by permission of McGraw-Hill Book Company.

Figure 11-8b is reproduced from J. G. Richards, *British Aircirculation Research Council Reports and Memoranda,* No. 1590, 1934. British Crown copyright. Reproduced by permission of the Controller of HBM Stationery Office.

Library of Congress Catalog Card Number: 69-20457

SBN: 03-081006-X

Printed in the United States of America

1 2 3 4 5 6 7 8 9

PREFACE

The motivation for writing another fluid mechanics text came primarily from teaching both undergraduate and graduate courses, and following students as they were confronted with problems after graduation. Details and interpretations not usually presented with an introduction, and passed over in intermediate or advanced texts, are included with the objective of providing a greater physical understanding to mathematical developments and analyses. Consequently, vector mathematics is used because of its direct physical interpretation and simplicity. The single vector equation for force and momentum is more directly interpretative (as well as simpler to write) than an array of partial differential equations.

Developments of the equations in the streamline coordinates are used to give a simple, direct interpretation of quantities such as rotation (or vorticity) and strain rate in terms of their descriptive (verbal) definitions.

Differential equations must be integrated to obtain most final results. It is therefore logical to develop integral field equations as the most general form and deduce differential equations from them. Therefore more rigorous treatment of control volume (both integral and differential) developments is presented than in other texts. The necessity of the *actuator disk* concept is shown in systems where work is done on the fluid by a moving part of the control surface.

This text is arbitrarily organized into four parts:

Part I. Introduction and Statics. The groundwork for fundamental methods of analysis is laid. Properties of fluids are first presented. More properties are considered than are actually used in the text; however, these additional properties are directly connected with the others and the interrelations are valuable for a general understanding of the behavior of properties, such as surface tension, vapor pressure, and energy of vaporization.

An introduction to field behavior, vector fields, and point functions are in Chapter 3.

Part II. Fluid Dynamics Fundamentals. The basic fundamental conservation principles for mass, energy, and momentum are developed for both integral and differential fluid volumes. Combinations of one-dimensional, differential, and integral methods are presented with the gain of additional information from possible comparisons. Also, informative physical insight is gained from comparisons of different viewpoints, for example, Lagrangian and Eulerian to give descriptive meaning to divergence, and so forth.

Dimensional Analysis is arbitrarily included in this Part as a method of analysis to which reference is made in following chapters.

Part III. Basic Applications. Applications to internal and external flows are introduced. Chapter 8 presents classical Couette and Poiseuille flows. An introduction to the Boundary Layer and Turbulence is presented in Chapters 9 and 10. Applications of Chapters 9 and 10 with experimental and semi-empirical correlation follow in Chapter 11.

Potential Flow (Chapter 13) does not appear until Part IV, Special Topics. Although the external flow solution is required for boundary-layer analyses, introductory boundary-layer analysis does not go much beyond the flat plate and, therefore, does not require coverage of potential flow. The inclusion of potential flow in an introductory course is sometimes questioned because two-dimensional potential flow is more efficiently handled with complex variable mathematics, which many students do not have at this level.

Part IV. Special Topics. Compressible Flow (Chapter 12) is probably the most general subject of this Part and may be covered with a one-semester course. Open channel flow is becoming of more general interest. In connection with process and biological problems, Unsteady Flow (Chapter 15) is a subject of increasing and lasting importance. An introduction to simple, one-dimensional magnetohydrodynamics appears in the last chapter, which will give undergraduate students a working understanding of such common devices as electromagnetic flow meters.

In many chapters, more material is included than can be covered in a one-semester course. The additional coverage was prompted by questions from students in teaching introductory fluid mechanics. The first two Parts, and a portion of Part III were used in a one-semester, junior-year course over a period of three years.

A suggested one-semester course would cover Parts I and II (excluding the starred sections) plus selected subjects from Parts III and IV to suit the instructor. More complete coverage of Parts I and II and most of the material in Parts III and IV may be used for a full-year course.

St. Louis, Mo. W. M. Swanson

CONTENTS

FLUID MECHANICS

PART I INTRODUCTION, FLUID PROPERTIES, AND STATICS

CHAPTER 1

INTRODUCTION

There are many fascinating areas of study in fluid mechanics. The one closest to everyone is the human circulatory system. Some of the first studies of fluid flow were motivated by the desire to understand this complicated system. Mechanically, the heart is a simple pulsating pump with check valves. The fluid is a complicated mixture of organic fluids and particles. The "piping" system is a multibranched, elastic tubing terminating in the capillaries of diameter of the same size as the red blood cells (erythrocytes). It is necessary to understand the flow behavior of this complicated system to control, remedy, and prevent diseases of the heart and other organs, stroke, and so on. The mathematical analysis of this complex fluid in an oscillating system of living passages is obviously extremely complicated. Experimental measurements of heart action, valve action, and blood flow are a challenge.

In the food and chemical process industries, a knowledge of fluid mechanics is necessary not only in the design of pumping and piping transport systems but also as a fundamental part of the process design. The products of many complicated organic reactions are determined by the concentrations of the reactants; therefore, complete mixing is necessary in a time less than that required for these reactions. Such rapid mixing is possible only in high-speed, high-contact, continuous-flow devices.

Fluid mechanics is the foundation for aerodynamics, gas dynamics, and magneto fluid dynamics. The design of turbines, rocket engines, VSTOL, and supersonic and conventional aircraft involves calculations of fluid flow fields, reactions forces, and so on. Fluidic control devices with no moving parts are in use that perform switching, computer, and control functions. It is then possible to have both the control and power devices operated pneumatically or hydraulically without a dual electric-fluid system. A knowledge of compressible flow and gas dynamics is required in the design and analysis of these machines and components. The study becomes more complex in determining the effects on vehicles reentering the earth's atmosphere at speeds of 10,000 to 50,000 miles per hour. The gas dynamics studies are further complicated by disassociation and ionization of the air molecules.

Problems of fluid behavior are of primary concern in the design of jet engines and rockets. Specific design areas are piping, pumping, cooling jackets and fuel tanks to prevent sloshing, fuel-oxidant mixing systems for optimum combustion, nozzles for maximum thrust, and vehicles for optimum lift-drag characteristics. Rocket propellant pumps must be designed to deliver 1000 pounds per second of liquid hydrogen, hydrogen fluoride, liquid oxygen, and so on, at pressures exceeding 1000 pounds per square inch. The properties of the fluids must be known to design such pumps (and their drive turbines) for optimum weight-efficiency considerations.

Other examples of interesting fluid flow studies are found in the field of power generation from conventional methods such as hydroelectric, steam and gas turbines, to newer ones involving magneto fluid dynamics. The surrounding environment and its behavior and control also present interesting studies in fluid mechanics. Meteorology, oceanography, and geology in the study of man's environment and the forces that shape the earth are fascinating subjects.

As physicists continue to discover new phenomena, engineers must learn and understand these phenomena in order to be able to apply them as well as those that already exist.

Fortunately, most known natural phenomena comply with certain fundamental laws—the laws of conservation of mass, energy, and momentum. The study of fluid mechanics involves (1) the study of these principles, (2) their formulation into manageable mathematical models, (3) manipulation of the mathematics to obtain certain predictions, and (4) the interpretation of the mathematical results to obtain physically meaningful information. The results are information used for design or to aid in understanding and predicting flow behavior.

The fundamental physical "laws" and methods of analysis used in the study of solid mechanics and thermodynamics are used in the analysis of fluid systems. These are:

1. Newton's laws for the motion of a fluid mass element, or the fluid passing through the confines of a machine.
2. The "conservation" principles and laws for rigid-body and thermodynamic systems.
3. Methods of analysis: free-body diagrams and open and closed systems of thermodynamics.

These natural laws are used with physical concepts and definitions to formulate physical statements regarding the behavior of a fluid under certain imposed conditions. These imposed conditions or restraints are in the form of initial or boundary conditions on the flow geometry that require the flow to conform to the physical restricting boundaries. These observations, concepts, and boundary conditions are then translated into symbolic mathematical language. These mathematical statements can be manipulated to provide information beyond that which is immediately obvious from the physical statements. Since this study involves forces and motion, it is both logical and advantageous to use vector mathematics, which is best suited for describing and dealing with these quantities.

The calculus allows predictions in time and space to be determined from a limited amount of initial given information. Using the integral calculus, rudimentary information can be summed to yield an overall or gross result. For example, pressure and shear stress distributions on an airplane wing can be integrated to yield its net lift and drag.

The results of mathematical predictions must be translated from the mathematical symbolic script back to the physical statement for the result to be of use. The final step is the most important in a physical or engineering sense, that is, the evaluation of the results in the light of the methods and assumptions used to obtain them. The value of the results as a new piece of information or as design information is determined by (1) the accuracy with which the original physical statement is specified, (2) the accuracy and validity of the mathematical translation, (3) the accuracy of the mathematical manipulations with respect to the assumptions required to obtain a solution to the mathematical problem (approximate methods of solution may be necessary), and (4) the limitations of the translation into the final physical statement. All these restrictions, assumptions, and limitations result in a qualitative and/or quantitative specification of error in the final result. Therefore, no engineering problem is complete without a statement specifying its limitations and accuracy! The limitations of an engineering design also involve economics and usefulness to mankind. The final step involves follow-up to evaluate the usefulness and economic success of the design solution.

Fluid mechanics, and its application to engineering and physical problems, is probably the study most subject to the engineering approach as outlined above. Only a small number of fluid flow problems possess exact solutions. Therefore, many simplifying assumptions and approximations are required to obtain any information at all, in most cases. With the advent of analog and high-speed digital computing techniques, it has become easier to obtain numerical solutions to some of the more difficult problems. However, all problem solutions obtained using these tools require an even more accurate engineering judgment in their set-up and evaluation.

The study of the physical phenomena by means of experimental investigation is an equally important area. Analytical and experimental methods are not two independent means of investigation; neither is complete without the other. An analytical solution to a problem is of little value unless it is verified by physical observation. It is also pointless to begin an experimental investigation without carrying the analysis as far as possible. The analysis provides background and insight into the variables of primary interest and significance, and possible direction to the experimental procedure. In some cases the main part of the information must come from experimental work, or else experimental verification of analytic results is necessary when it is not possible to determine analytically the limitations of accuracy or error in the analysis. In many cases full-scale experimental design tests are not possible or are prohibitive in cost, so scale-model tests must be made. Analytical evaluation and experience is then required in specifying the applicability of the results of the model study to the full-scale prototype.

Much of the development of engineering apparatus primarily involving fluid mechanics has been accomplished by a combined analytic plus experimental approach. Engineers and physicists should be familiar with, and be able to use, both methods of investigation. Because of the nature of the mathematical analyses, experimental confirmation of the validity of the analytical results is usually a necessity. As more effects (and terms) are included in the mathematical solutions of fluid problems, more complete solutions will become available; the degree of uncertainty of the correspondence between analytical results and actual flow behavior will decrease and the time and expense of conducting experiments will be eliminated. At the same time, new fluid phenomena will be discovered, requiring new and different types of analytic and experimental investigations. The field of magneto fluid dynamics is a current example wherein many mathematical solutions have been obtained for hypothetical problems, but these solutions have doubtful meaning because corresponding physical phenomena have not been observed.

The study of fluid mechanics involves investigation of the relations governing the effects and actions of fluid masses or the fluid within defined system boundaries. The above principles and laws must be applied to a mass element or region of a complete or continuous fluid field. Fluid mechanics is then characterized within the subject of continuum field mechanics. A point of differentiation between the study of fluid mechanics and solid mechanics is provided by the definitive nature of a *fluid*. In determining the behavior of a solid under imposed loads or stresses, the stress is proportional to the strain. In general, determination of the strain is not easy. In the case of a fluid the relation is even more complicated by the fact that the stress is proportional to the *rate* at which a fluid is deformed—the strain rate.

In an elementary course in fluid mechanics only simple cases are considered. Most of the emphasis is on what is termed a one-dimensional flow analysis of fluid flow systems and problems that can be put into this category. Some introductory material on two- and three-dimensional flows is also presented.

As a course of study, fluid mechanics is essentially a course in the physics of fluids involving applied mathematics: algebra, differential and integral calculus, differential equations, and vector mechanics.

CHAPTER 2

CONTENTS

FLUID PROPERTIES

2.1 INTRODUCTION

In the study of fluid mechanics with the objective of obtaining an understanding of physical flows or design information, a familiarity with the properties of fluids is necessary. There are two frames of reference for considering the behavior of fluid properties in a changing environment (for example, the variation in the viscosity of a fluid as a result of a variation of its temperature or pressure). From the macroscopic viewpoint, the overall behavior of the substance can be studied without regard for its atomic and molecular structure. In this case the fluid is considered as an amorphous *continuum.* This macroscopic frame of reference is the easiest and most logical for formulation of relationships between the forces and motion for a fluid whose properties are given. The microscopic- or molecular-scale consideration allows a deeper understanding of the behavior of material properties. In this case it is nearly impossible to formulate the physical behavior of the material with the physical mechanics into mathematical expressions that can be analyzed. Molecular considerations of the fluid materials help in understanding the reasons for the behavior of materials as observed in tests and experiments.

In order to formulate the physical principles of the mechanics of the motion into mathematical relationships, the macroscopic or continuum frame of reference will be used. To analyze the motion of a fluid such as water it is only necessary to know the overall properties and their behavior, not the atomic and molecular structure.

2.2 DEFINITION OF A FLUID

Macroscopically, a fluid is a substance that can sustain no static shear stress or, as a consequence of this definition, a substance that will deform continuously under any applied shear load. Other definitions, such as "a fluid is a substance that will take the shape of its container," are seen to be consequences of the more general definition given above. A fluid is also a substance that has no characteristic shape or extensive physical property such as crystalline structure.

Consider the definition of a fluid from a microstructure or molecular viewpoint: In the solid state the atoms and molecules are bonded into a crystalline lattice structure, and the only possible motions are vibrations and small displacements of the atoms within this structure. In a fluid there is no definite structure and the molecules are in an incessant state of random translational and rotational motion. Vibratory modes of motion of the atoms within the molecule are also possible.

Both the gas and the liquid state fall within the above definition of a fluid. These states are best defined from the microscopic viewpoint. The gas state is characterized by a relatively large distance between molecules, so the intermolecular forces of attraction are very small or negligible with respect to the inertia forces. The gas molecules are free to escape or expand indefinitely under conditions of reduced pressure or to fill an enclosing container. In the liquid state, the intermolecular cohesive bonding forces hold the liquid into a coherent amorphous mass, giving it the characteristic of a free surface. In the liquid state, a large force or pressure is required to compress the molecules closer together, so the liquid state is one of relatively fixed density under conditions of varied pressure.

A fluid may usually be considered, for most formulations and calculations, an amorphous *continuum*. This assumption is not valid if the distance between molecules becomes significant with respect to dimensions of the physical equipment. The mechanics must then be studied on a molecular basis, as in rarefied gas dynamics and aerodynamics.

2.3 DEFINITION OF A FIELD

A field is a region (either real or imaginary) within which physical or mathematical quantities can be either specified or measured at every point. As an example consider a subregion of the atmosphere at a particular instant. At every altitude, latitude, and longitude as measured from an established reference point, the pressure, density, and temperature of the air can be measured or calculated. A pressure (or temperature or density) field is then specified by giving the value of the pressure (or other quantity of interest) as a function of the three independent space coordinates measured from a specified reference point or *coordinate system origin.* The quantities of pressure, density, and temperature are scalar quantities; their distribution in space is termed a *scalar field.* Velocity is a vector quantity; its distribution in space is termed a *vector field.*

The properties at a given point in the field (or the entire field configuration) may also be functions of time, since the quantities may vary from one instant to another at any given, fixed point. Any physical field in a fluid flow problem (pressure, temperature, density, velocity, and so on) can be specified by giving the value of this quantity of interest as a function of three space coordinates and time. An example of a flow field is the fluid flowing between two adjacent blades of a turbine or a compressor. The cross section of this field is illustrated in Fig. 2-1. The lines between the blades represent the directions of the flow velocity. Such lines are called *streamlines*. If this flow section represents that between two inlet guide

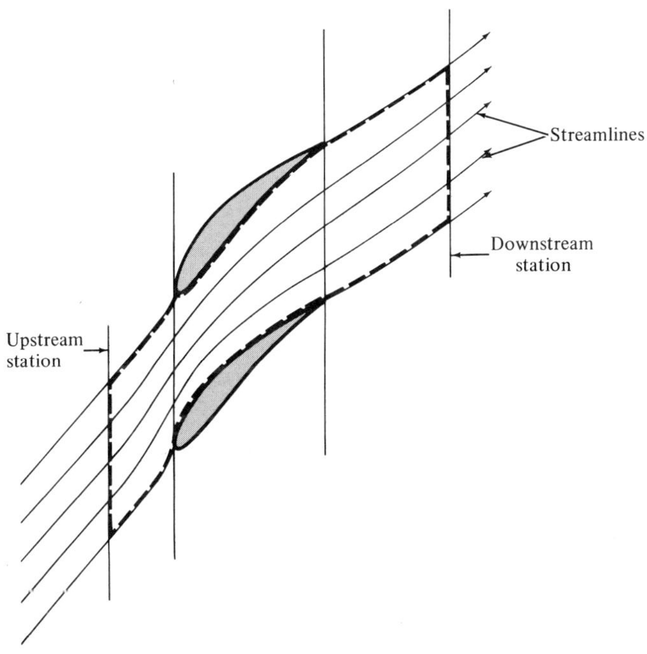

Fig. 2-1. Flow region between the adjacent blades of a compressor cascade as an example of a flow field. The pressure, density, velocity, and so on, are given or are to be determined at every point within the flow field. The field boundaries are indicated by the dashed line.

vanes, and if the flow pattern does not vary with time, the flow is referred to as *steady*. The velocity field, pressure field, and so on, for this flow are then seen to be invariant with respect to time, or the *flow fields* are termed *stationary*. The terms *steady flow* and *stationary field* are then equivalent.

If an orthogonal, rectilinear (cartesian) coordinate system with coordinates x, y, and z is selected for representing the location of a quantity of interest (for example, pressure), the pressure field is specified as $p(x, y, z)$.

The x,y,z coordinates imply a very specific coordinate system, and the pressure field would be specified in this system only when it is advantageous to do so (for example, the pressure field in a box). On the other hand, if the pressure field in a large subregion of the atmosphere is being considered, it may be advantageous to use the standard spherical polar coordinates of altitude, latitude, and longitude (r, φ, θ).

A space field may be designated independent of any specific coordinate system as $p(\vec{r}, t)$, where $\vec{r}$ is the radius or position vector from a reference "origin" to the point where the pressure is to be specified or measured within the field. The components of the vector $\vec{r}$ can be specified in any specific coordinate system. The single quantity $\vec{r}$ is then used in the mathematical manipulations without carrying all the system coordinates (for example, x, y, z; or r, φ, θ; and so on), through the manipulations. The choice of coordinate system is then made in the final calculations that best fits the geometry of the problem. It is not logical to make mathematical manipulations in an x,y,z coordinate system if a polar coordinate system is to be used to plot the field.

Two property classifications may be defined. Properties that depend on the amount (mass) of the substance are called *extensive.* Volume, energy, and momentum are examples. Properties that are independent of the amount are called *intensive.* Temperature, pressure, and viscosity are examples of intensive properties. Extensive properties may be made intensive by dividing the property by its mass. The result is a *specific* property. The volume divided by its mass becomes specific volume; energy divided by mass is specific energy.

The use of physical and mathematical quantities as field quantities will become more familiar in the specific examples and developments to follow.

2.4 PHYSICAL PROPERTIES OF A FLUID

Some fluid properties of importance in fluid mechanics studies are:

1. Density ρ, specific volume v_s, and specific gravity S
2. Molecular mass number (molecular "weight") $\mathscr{M}$
3. a. Viscosity or absolute viscosity μ
   b. Kinematic viscosity ν
4. Compressibility κ and bulk modulus or modulus of elasticity E
5. Thermal coefficient of expansion β
6. Vapor pressure p_v and latent heat of vaporization l_v
7. Surface tension σ
8. Specific "heats" c_p and c_v
9. Gas constant R and universal gas constant R_0
10. Thermal conductivity k

The above physical quantities can be measured experimentally; in addition, their behavior is a consequence of the molecular properties of the substance.

The above physical quantities are intrinsic to the substance (with the exception of R). Most intrinsic thermodynamic quantities can be expressed as functions of the two primary-state variables, temperature and pressure.

TEMPERATURE (T)

In a fluid the temperature is a direct indication of the kinetic energy of the continuous random motion of the molecules. Temperature is, therefore, an indication of the mean random speed, or of the inertial force capability of the random molecular motion. The intermolecular bonding or cohesive forces are electrostatic in nature and are essentially independent of the temperature. If a substance in a liquid state is heated (energy added), the state of molecular random motion will increase. At some temperature, the inertial forces of the molecules will become approximately equal to the forces of attraction between the molecules. When the molecular inertia is greater than the intermolecular attraction, the molecules will start to "break loose" from one another and expand into the gas or vapor state. A large amount of energy is required during this transition from the liquid to the gaseous state. This *energy of vaporization* must be added to overcome the molecular attraction forces. As expected, the energy of vaporization is a function of the temperature at which this boiling transition takes place.

PRESSURE (p)

Pressure is exerted by a gas on a surface primarily by the momentum exchange of the molecules as they collide with and rebound from the surface. In a liquid at rest, the hydrostatic pressure results from the gravity force acting on the liquid mass, a very large mass (or tall column) of gas is required to produce a significant pressure. The entire height of the atmosphere only produces a pressure of 1 atm. This same pressure increase is obtained at a depth of 34 ft of water.

DENSITY (ρ)

The mass density of a fluid is defined as mass per unit volume (designated herein as ρ). It has measurement units of pounds-mass (lb_m) per unit volume (ft^3) in the absolute English system of units, g/cm^3 in the cgs system, kg/m^0 in the mks system, or units as appropriate in other systems. (See Appendix A for a further discussion of unit systems.)

The density can be determined by finding the mass (m) of any given volume (V):

$$\rho \equiv \frac{m}{V}$$

In a compressible flow field, for example, a high-speed gas flow, the density can vary from point to point. The density at a point is then defined as the limit

$$\rho \equiv \lim_{\Delta V \to 0} \frac{\Delta m}{\Delta V}$$

where m is the mass of a small element of the fluid having a volume V. If this limiting process were actually carried out on the volume of a cube, of side dimension l, the results illustrated in Fig. 2-2 would be obtained. As

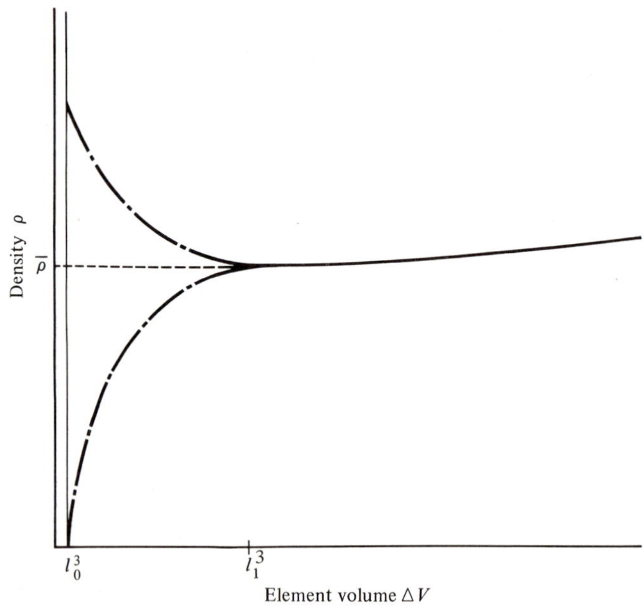

Fig. 2-2. Density at a point as a limit. l_0 represents the limit of the size of ΔV for continuum considerations. l_0 can be considered to represent the collision cross-sectional diameter, or the "size of the molecule."

the volume element gets extremely small, so that the sides of the cube have a length on the order of the distance between the molecules l_1, the time when the measurement is made then becomes important. Since the molecules of a fluid are in constant agitation, the volume element may contain no molecules, or several, at a particular instant. The density may then have any value within the spread indicated by the dashed lines. If the measurements are averaged over a period of time much greater than a characteristic period equal to the length l_1 divided by the velocity of the

TABLE 2–1

CHARACTERISTIC PROPERTIES OF SOME LIQUIDS AT 68°F (20°C) AND 1 ATM

	ρ, $\frac{lb_m}{ft^3}$	S[a]	$\mu \times 10^5$, $\frac{lb_f\text{-sec}}{ft^2}$	$E \times 10^{-4}$, psi	p_v, psia	σ, $\frac{lb_f}{ft}$	c_p, $\frac{Btu}{lb_m\text{-}°R}$	k, $\frac{Btu}{ft\text{-}hr\text{-}°R}$	$\beta \times 10^3$, $\frac{1}{°R}$
Benzene (C_6H_6)	54.9	0.879	1.36	16.1	1.45	0.0020	0.406	0.093	0.69
Carbon tetrachloride (CCl_4)	99.1	1.595	2.04	16	1.48	0.0018	0.201	0.0596	0.69
Castor oil	59.9	0.96	2060	21	—	0.0027	—	0.077	—
Ethyl alcohol (C_2H_5OH)	49.2	0.7893	2.51	19.5	0.85	0.0015	0.565	0.0966	0.62
Glycerine [$C_3H_5(OH)_3$]	78.6	1.260	3110	63	2×10^{-6}	0.0043	0.570	0.166	0.28
Glycol $(CH_2OH)_2$	69.5	1.116	44.8	—	—	0.0033	0.569	0.144	0.36
Mercury	845	13.546	3.24	380	2.8×10^{-5}	0.035	0.0333	5.02	0.101
Methyl alcohol (CH_3OH)	49.5	0.7928	—	—	1.25	0.00155	0.565	0.117	0.665
Tetrabromoethane $(CHBr_2)_2$	185.0	2.964	—	—	4.7	0.0034	0.225	—	—
Water	62.3[b]	0.9982	2.10	31.3	0.339	0.0050	0.9988	0.345	0.11

[a] Standard specific gravity of substance at 20°C referred to water at 4°C.

[b] The standard reference (maximum) density of water at 4°C is 62.43 lb_m/ft^3 or 1.000 g/ml.

molecules, the limiting value at the end of the dashed line is obtained. This limit is defined as the *density at a point*. The density of some fluids at various temperatures is presented in Tables 2-1 and 2-2.[1]

TABLE 2–2

CHARACTERISTIC PROPERTIES OF SOME GASES AT 68°F (20°C) AND 1 ATM

	ρ, $\frac{lb_m}{ft^3}$	$\mathcal{M}$	$\mu \times 10^5$, $\frac{lb_f\text{-sec}}{ft^2}$	c_p, $\frac{Btu}{lb_m\text{-}°R}$	γ	k, $\frac{Btu}{ft\text{-}hr\text{-}°F}$	NBP,[a] °C
Air	0.0752	28.97	0.0377	0.2402	1.40	0.0148	—
Carbon dioxide	0.115	44.01	0.0307	0.206	1.28	0.00929	−78.5
Helium	0.0104	4.003	0.0411	1.242	1.66	0.082	−268.9
Hydrogen	0.00522	2.016	0.0185	3.427	1.40	0.103	−258.2
Nitrogen	0.0728	23.016	0.0377	0.248	1.40	0.0147	−195.8
Oxygen	0.0830	32.000	0.0420	0.219	1.40	0.0151	−183

[a] Normal boiling point; boiling temperature at 1 atm pressure.

SPECIFIC VOLUME (v_s)

The specific volume of any substance is defined as the reciprocal of the density: $v_s \equiv 1/\rho$. The units of specific volume are volume per unit mass (ft^3/lb_m, m^3/kg, cm^3/g). Note that the units are volume per unit *mass*, and that in most tabulated material (for example, the steam tables) the units of specific volume are cubic feet per pound-mass and not cubic feet per pound-weight.

SPECIFIC GRAVITY (S)

The specific gravity of a substance is the ratio of its density to the density of water. It is, therefore, dimensionless. Specific gravity can be specified either as the ratio of the density of the mass of the substance at its temperature and pressure to the density of water at that same temperature and pressure, or to the maximum density of water at 1 atm. The maximum or standard reference density of water at 1 atm is 62.43 lb_m/ft^3 (1.0000 g/cm^3 by arbitrary definition) at 4°C or 39.2°F. Since the density of the substance and the reference fluid (water) both vary with temperature and pressure (and each in a different manner), the temperature and pressure of the substance and of the reference water should be given when accurate values of specific gravity are specified. The specific gravity referred to the

[1] For more data on properties of substances, see *International Critical Tables*, McGraw-Hill, Inc., New York, and *Handbook of Chemistry and Physics*, Chemical Rubber Publishing Co., Cleveland.

standard reference density of water at 4°C and 1 atm may be denoted as the *standard* specific gravity. Specific gravities of some liquids are given in Table 2-1.

MOLECULAR MASS NUMBER ($\mathcal{M}$)

There is no absolute law that determines what standard should be chosen as *the* unit mass.[2] The mass of any particle or system must be specified in units of arbitrarily determined unit systems of measurement. The relative masses of atoms of the elements are determined by their atomic numbers as they appear in the periodic table with oxygen arbitrarily given the number 16.000. Since molecules are definite combinations of atoms, molecular mass numbers are obtained by adding the atomic mass numbers according to the molecular formula. This molecular mass number is sometimes referred to as the *molecular weight*,[3] although *molecular mass number* is a more accurate term.

By definition the magnitude of $\mathcal{M}$ for any substance is proportional to the mass of a molecule of that substance. If m_i is the mass of a single molecule, nm_i is the mass of n molecules. A given number of molecules (say n_1) of a *given* substance (denoted by the subscript i) would have a mass m equal to n_1m_i. $\mathcal{M}_i$ (the *molar mass number* of the ith substance) is directly proportional to m_i by definition; therefore, $\mathcal{M}_i$ contains a definite number of molecules; that is, $\mathcal{M}_i = N_A m_i$, where N_A is this definite number, constant for all substances. Since this number is a "universal constant," it is referred to as *Avogadro's number* after the person who first determined and used this relationship. There are approximately 6.023×10^{23} molecules in a *gram-mole* and 2.73×10^{26} molecules in a *pound-mass-mole*. $\mathcal{M}$ is usually specified in the system of units being used, and a mass of units in that system is defined as a molar mass. For example, 32 lb_m of oxygen (O_2) is one molar mass, or 1 *mole*, of this substance. A kilogram-mole of O_2 would be 32 kg of O_2. Similarly, a mass-atom of O would be 16.000 lb_m (g or kg). The mass magnitude of the mole or molar mass is determined by the system of units specified.

FLUID DEFORMATION AND ABSOLUTE VISCOSITY(μ)

By definition, a fluid can sustain no static shear stress. However, a shear stress will result if a fluid volume is subjected to a continuous shear. Figure 2-3 indicates the simplest geometry in which a continuous shear stress can be applied to a fluid. A large plate is moved at a velocity U over a thin sheet of fluid of thickness b above a flat surface. The length and

[2] The mass of an electron (as a basic particle) might be a logical fundamental mass unit.

[3] Weight is a force (the gravitational attraction force between masses); therefore, molecular "weight" is a poor choice of terminology in this case, since an absolute unit of mass (independent of any gravitational system) is implied.

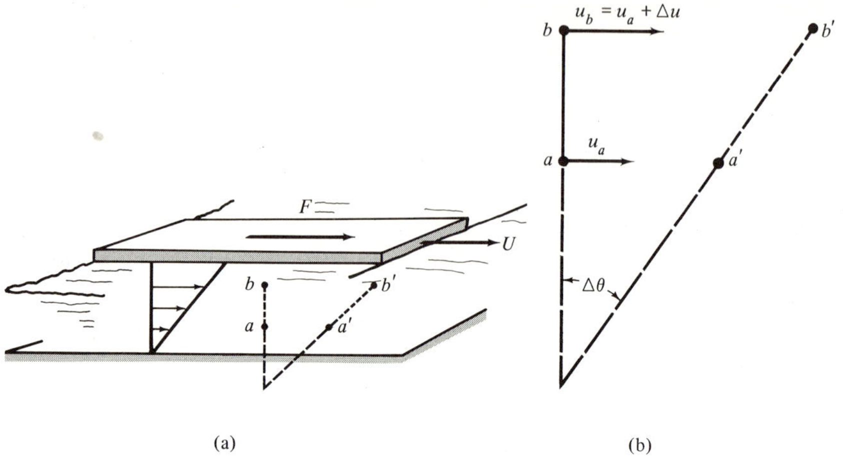

Fig. 2-3. Shear and velocity distribution in a semi-infinite flow.

width of the plate are much larger than b, so edge effects can be neglected. In the case of a solid, Hooke's law states that stress is proportional to strain. A similar relation exists for a fluid; shear stress is proportional to *rate of strain* for a linear or Newtonian fluid. The proportionality factor is the *viscosity.*

Shear stress in a solid is a result of the atoms of the substance being displaced from their neutral position in their attraction–repulsion field. Figure 2-4a shows a simple atomic distribution within a crystalline material. Under application of a shear stress, the distance between atoms along the horizontal and vertical directions is essentially unchanged. However, along one set of diagonals (for example, 1-5) the atoms are brought into closer proximity. A repulsion force then exists between 1 and 5, since the original position was one of neutral equilibrium. The distance is increased along perpendicular sets (for example, 2-4), producing an attractive force. This force system tends to return the object to its initial configuration, or to oppose the applied shearing torque. For very small displacements the intermolecular forces vary linearly with displacement between atoms so that the displacements are proportional to the applied stress. The internal atoms are in a state of force-balance, or equilibrium. For example, atom 5 has two equal and opposite tensile and two equal and opposite compressive forces acting on it. However, on the "surface" atoms are unbalanced with respect to the internal force system; the unbalanced component is just the external shear force. A microscopic explanation is then obtained for the cancellation of internal pressure and shear forces.

A shear stress is produced in a fluid by application of a dynamic strain, or a rate of strain. Consider, from a microscopic viewpoint, a perfect gas

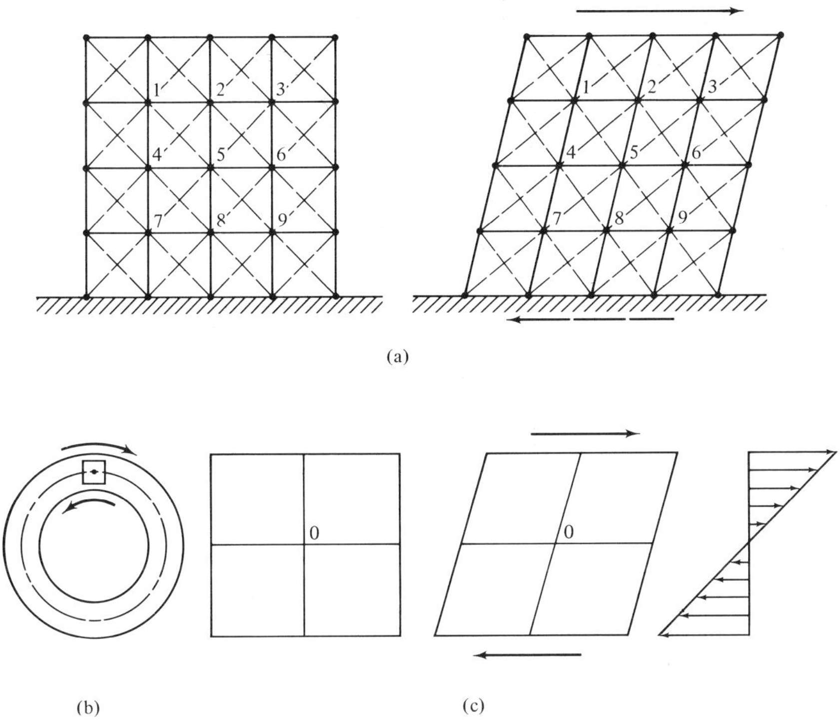

Fig. 2-4. (a) Lattice deformation in a simple crystalline solid; (b) Concentric cylinder fluid shearing device; (c) Deformation of a fluid element and the velocity distribution at point 0.

(no intermolecular forces) between the two cylinders of Fig. 2-4b. Application of a constant torque on one cylinder will result in a constant rotation of that cylinder with respect to the other. Assuming the cylinders are rotating in opposite directions, there will be a cylindrical surface about midway between the two with zero velocity. Consider a small fluid element, as shown in Fig. 2-4c, with this plane passing through its center. The fluid molecules are in a state of continuous random motion plus a "directed" tangential component produced by the relative angular motion of the cylinders. Molecules are being transported into and out of this small volume by the random motion. Those near the center plane have no directed velocity. Some molecules will travel upward because of their random motion. When they collide with those above that have a tangential directed velocity, they will be accelerated to the velocity there. If the state of motion is continuous, this momentum exchange process occurs continuously throughout the fluid to the outer cylinder, where the molecules collide with and exchange momentum with the cylinder wall. In the collisions at the wall, the surface imparts the momentum to the fluid such that the velocity of the fluid layer immediately adjacent to the wall has the velocity of that surface: condition of no slip at the wall (see Appendix C). The force at the wall is the one that continuously imparts tangentially directed velocity or momentum to the fluid. This tangential momentum exchange, brought about by the transverse random fluid motion component, continuously varies the directed velocity until the other surface is reached.

The *shear deformation* or *shear strain* illustrated in Figs. 2-3 and 2-4 is characterized by an angular deformation. The shear stress in a fluid is proportional to the strain rate.

Only the momentum exchange was considered in the above discussion considering the mechanism of viscous shear in a fluid. The viscous shear stress in a liquid is primarily a result of the intermolecular forces. The displacement effect on these forces that results in a viscous shear action is quite complicated. There are some theories in existence[4] that attempt to explain this action, but, as yet, none has been experimentally verified. The variation of liquid viscosity with temperature and pressure is quite complicated. The only reliable method for determining the viscosity of liquids is by experiment. Viscosities of some gases and liquids are presented in Fig. 2-5. These data can be used over a fairly wide range of pressures for both the gases (below the critical pressures) and liquids, since the absolute viscosity is not much affected by pressure.

With respect to the simplified flow representation of Fig. 2-3, the annular deformation and its rate can be determined by considering a line element *a-b* drawn in the fluid at an initial instant. At some time increment δt later, the fluid particles that initially made up line *a-b* are at the position *a′b′*. Point *a* has traveled the distance *a-a′* equal to the velocity at point *a* times δt, or $u_a\ \delta t$. Similarly, the distance *b-b′* is equal to $u_b\ \delta t$. The angular deformation is the angle $\delta\theta$. Let the line element $ab \equiv \Delta n$ and let $u_b = u_a + \Delta u$; then the distance *bb′* is greater than the distance *aa′* by the quantity $\Delta u\ \delta t$. For a very small period of time (δt) the angle $\delta\theta$ will be small and

$$\delta\theta \approx \frac{(u + \Delta u)\ \delta t - u\ \delta t}{\Delta n} = \frac{\Delta u\ \delta t}{\Delta n}$$

where n is the coordinate measured normal to the flow direction. The angular speed or angular deformation rate is $\dot{\theta} \equiv \delta\theta/\delta t = \Delta u/\Delta n$. The shear stress ($\tau$) is proportional to this angular velocity and the constant of proportionality is the viscosity (μ), or $\tau = \mu(\Delta u/\Delta n)$. Assuming that the flow is steady (there is no acceleration), consideration of Newton's second

[4] S. Glasstone, K. J. Laidler, and H. Eyring, *Theory of Rate Processes*, McGraw-Hill, Inc., New York, 1941.

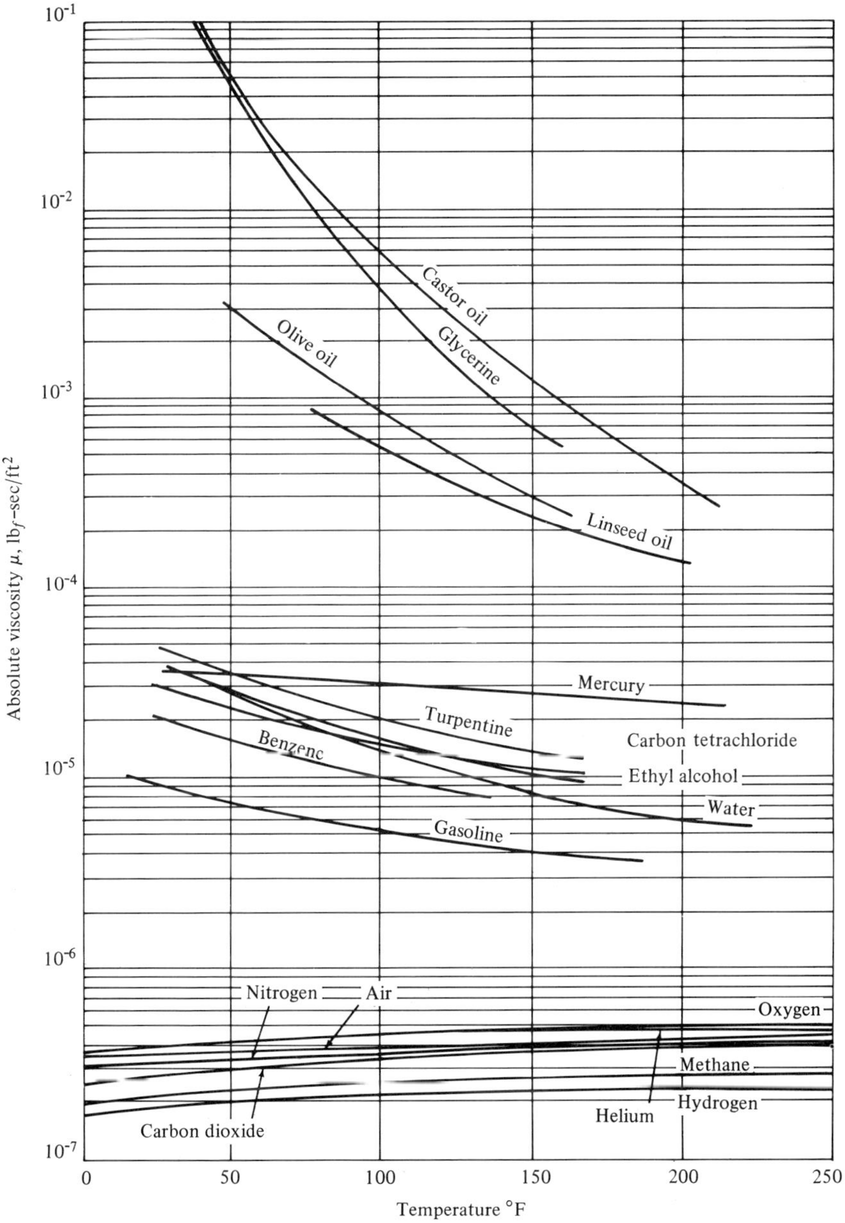

Fig. 2-5. Absolute viscosity as a function of temperature for various fluids at 1 atm.

law yields the result that the shear stress must be constant throughout the fluid. The velocity distribution, or velocity profile, within the fluid is then linear, or the angular deformation rate (strain rate) is constant everywhere in the fluid field and is U/b. The shear stress for this particular case is $\tau = \mu U/b$. The force required to move the plate at the constant uniform velocity U is just the shear stress times the area of $F = \tau A = \mu(UA/b)$, where A is the area of the plate. The input work rate (power) is $FU = \mu U^2 A/b$. The conservation of energy, or relation between work and energy, states that this power input goes into increasing the internal energy of the fluid.

In the flow between the two flat plates described above, it must be assumed that there is no externally applied pressure difference between any two points in the fluid. A differential pressure applied to the fluid across the ends of the plate would produce an additional flow.

The viscosity coefficient may be determined with an apparatus such as illustrated in Fig. 2-4b to produce a uniform shear within the liquid (for example, one cylinder rotating concentrically within another with a very close clearance between them). The shear stress and strain rate are determined from direct measurements and the viscosity is calculated. The details of such an apparatus are presented in Chapter 8.

Newtonian flow behavior is defined as one for which the viscosity is not a function of velocity gradient but may be a function of pressure and temperature. The viscosity relation can be determined for a wide range of strain rates (or velocity gradients) du/dn. The results of investigations on a number of substances are indicated in Fig. 2-6. An apparent viscosity can be determined for any of these substances as the ratio $\tau/(du/dn)$. By definition this ratio is constant for Newtonian flow behavior, but it can be considered a function of the strain rate du/dn for substances that exhibit *non-Newtonian* behavior. Air, water, most gases at normal or standard conditions, and many fluids of low viscosity exhibit Newtonian flow behavior. The type of fluids that exhibit non-Newtonian behavior are usually characterized by a very long chain, coil, or a platelet type of molecular structure. Suspensions or solutions in water or other solvents of substances with long chain or platelet molecules or particle geometries also exhibit this type of behavior. Blood, polymer solutions, and suspensions of graphite or clay are examples of such fluids. The tendency of these particles to line up in the plane of maximum stress has the effect of reducing the shear stress produced, thus reducing the effective or apparent viscosity. This type of behavior is categorized as viscoelastic, pseudoplastic, or "shear thinning."

Liquid metals are also reported to exhibit non-Newtonian flow characteristics. This behavior is probably explained by the fact that the viscous property in liquid metals is primarily due to intermolecular force and displacement action under strain. The atomic mass and density of liquid

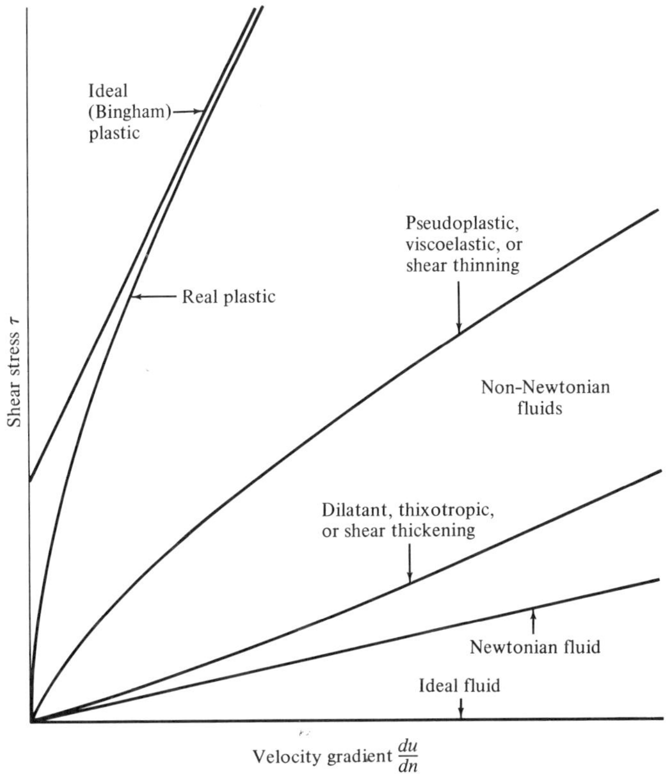

Fig. 2-6. Indication of the relative viscous behavior of several types of substances.

metals is quite large compared with fluids that have simple Newtonian viscosity behavior.

Dilatant or "shear-thickening" substances are also non-Newtonian but exhibit a behavior opposite to that described above. When it is attempted to shear, tear, or break them rapidly they are brittle, but will spread like a viscous liquid when placed on a flat surface. These substances are usually puttylike, with long, complicated, but weakly bonded molecules that break apart when strained rapidly and reassociate or polymerize when the external strain is removed. Such behavior is also exhibited by sols with asymmetric particle geometries with polymerizing tendencies and with asymmetric polar molecules that have electrolytic tendencies. "Thixotropic" is sometimes used to refer to "shear-thickening" behavior (see Fig. 2-5) for which the viscosity is time-dependent. In reality all non-Newtonian viscosity behavior is time-dependent; a relaxation phenomenon is present.

The viscosity of a gas has been primarily attributed to its molecular motion; indeed, it is a momentum exchange or "transfer" coefficient that is due to the momentum exchange between layers or laminae of fluid that have different flow velocities. Consider the gas moving in the vicinity of a solid boundary as indicated in Fig. 2-7, and an imaginary surface in the gas with the laminae of fluid immediately below the surface with a flow velocity u and the one above with $u + \Delta u$. The motion of a molecule can be imagined to consist of a random (in space and time) velocity $\tilde{v}$

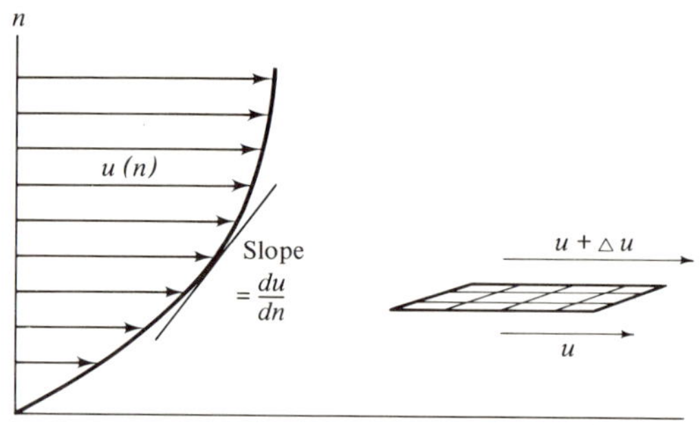

Fig. 2-7. Laminar viscous flow representation.

superimposed on the flow velocity. Considering only the vertical component of this random motion with respect to the mean motion, some molecules with momentum $m_i u$ in the flow, or x, direction will be transported across the surface into a region where the velocity is $u + \Delta u$. The slower molecules will collide and exchange x momentum with these molecules, tending to slow them down. The converse will happen with respect to molecules above the surface that pass downward through the surface. The details of momentum exchange analysis for a perfect gas are presented in Appendix C. The analysis yields the result that $\mu \propto \rho\tilde{v}\lambda$, where λ is the mean free path between collisions. The more densely the molecules are packed and the larger the effective collision cross section, the smaller λ becomes, or $\lambda \propto m_i/\rho A_c$, where A_c is the collision cross section. With this value of λ, $\mu \propto m_i\tilde{v}/A_c$, or the viscosity coefficient is proportional to the mean momentum, $m_i\tilde{v}$, and inversely proportional to the "size" of the molecule. This representation of viscosity (for a perfect gas) indicates that it is a *momentum exchange*, *momentum transfer*, or *momentum diffusion* coefficient. (This *momentum diffusivity* is actually the quantity $g_c\mu/\rho$, the *kinematic viscosity*, discussed in the next section.)

The viscosity of a perfect gas is also expected to be proportional to the square root of the gas temperature, since $\tilde{v} \propto \sqrt{T}$. For gases that approach the conditions assumed for the perfect gas, μ is found to be nearly proportional to $\sqrt{T}$. Perfect gas behavior is approached by real gases at pressures below the critical pressure and temperatures above the critical temperature. For air, the critical pressure is 37.2 atm (547 psia) and the critical temperature is 132.4°K (−327°F), so air behaves nearly as a perfect gas under normal atmospheric conditions.

Since conductive heat transfer is a molecular-energy exchange or transport process, the analysis for a heat-conduction coefficient (thermal conductivity, k) for a gas might be expected to be the same as for the viscosity coefficient. The thermal-energy exchange or *thermal-diffusion coefficient* is actually $k/\rho c_p$. If both of these coefficients were the same, the ratio $g_c \mu c_p/k$, defined as the *Prandtl number*, N_{Pr},[5] would be expected to be unity. Because of nonideal effects and a difference in the momentum transfer and energy excitation mechanisms when molecules collide, this ratio is usually different from unity and may be a function of pressure and temperature. This dimensionless ratio is of importance in heat-transfer phenomena in gases and liquids. For air at standard conditions, it has the value of 0.7, indicating a deviation from the ideal simplified behavior assumed. In the heating of a polyatomic perfect gas, the energy transferred on collision goes into the several possible modes of energy, while the momentum transfer does not depend on these modes. Therefore, the ratio of energy to momentum transfer coefficients would be expected to vary from any ideal gas behavior. In addition to the variation in k and c_p produced by the various energy modes, there is an effect due to the intermolecular force fields as the molecules become more complex and "heavier."

A *mass-diffusion coefficient* is important in mass-transport phenomena. The ratio of this coefficient to kinematic viscosity is known as the Schmidt number, $N_{\mathrm{Sc}} = \mathscr{D}/\nu$, where $\mathscr{D}$ is the mass or concentration diffusivity.

The units of the coefficient of viscosity (or absolute viscosity) are obtained from the defining relationship for the viscous shear stress,

$$\tau = \mu \frac{du}{dn}$$

The units of τ are those of stress, so the units of μ are Ft/L^2, or $\mathrm{lb_f}$-sec/ft^2 in the English system, N-sec/m^2 in the mks system, and dyne-sec/cm^2 in the cgs system. The unit of 1 dyne-sec/cm^2 is defined as a *poise*, after Poiseuille.[6] By comparing the size of the units involved, it can be determined that one $\mathrm{lb_f}$-sec/ft^2 is equivalent to 479 P; the conversion factor is 479 P/($\mathrm{lb_f}$-sec/ft^2). Other conversion factors are presented in Appendix D.

[5] A further discussion of this and other significant dimensionless numbers will be found in Chapter 7.

[6] Jean Louis Poiseuille (1799–1869), physician and experimental physiologist. His work on viscous fluid flow in tubes was done in connection with determining fluid resistance and heart pumping power required in the blood circulatory system.

KINEMATIC VISCOSITY (ν)

The quotient of viscosity with density appears quite often in fluid mechanics and therefore is used as a single quantity defined as the *kinematic viscosity*, ν. In the absolute units system,

$$\nu \equiv \frac{g_c \mu}{\rho}$$

and the units are ft²/sec (or m²/sec in mks and cm²/sec, or *stokes*, in cgs).

The kinematic viscosity of liquids is primarily a function of temperature because of their relative incompressibility. For gases ρ, and therefore ν, is a function of both p and T, so that the pressure dependence must be considered. Figure 2-8 presents the kinematic viscosities of some fluids as a function of temperature. For the gases, these values are for a pressure of 1 atm only. However, for pressures below the critical pressure, the kinematic viscosity may be approximated by multiplying the value from Fig. 2-8 by $\rho_a/\rho = p_a T/p T_a$ (where ρ_a is the standard atmospheric density).

COMPRESSIBILITY (κ)

Compressibility is the coefficient that gives the relative volume or density change resulting from an applied pressure. The compressibility of any substance is due primarily to two phenomena: the resistance of the molecular field to a change in the distance between molecules (Fig. 2-9), and the momentum of the molecules in random motion as they exchange momentum and apply an impulse to the surface with which they collide. The force–distance relation indicated in Fig. 2-9 (representative of any physical substance) indicates about an equal ability to resist both compressive and tensile forces for a small change in intermolecular distance from the equilibrium point e_1.

Since the relationship between intermolecular force and distance varies rapidly from attraction to equilibrium to repulsion for a very small change of intermolecular distance, volume or length changes will be small under relatively large tensions or compressions. Since the molecules of a solid are fixed in the crystal lattice except for a relatively small vibratory motion, only a very small amount of the ability of a solid to resist compressive or tensile forces is due to the motion of the molecules.[7]

In the gaseous state, the intermolecular forces of attraction are quite small. Such forces are neglected in the limiting consideration of a perfect gas, and any resistance to compression is due to the inertial force from changing the momentum of the molecules where the compression is

[7] Indeed, this random motion actually lessens the resistance of a solid to tension (and compression), as indicated by the increase in the yield stresses of crystalline solids as temperature is decreased (the random molecular vibration decreases).

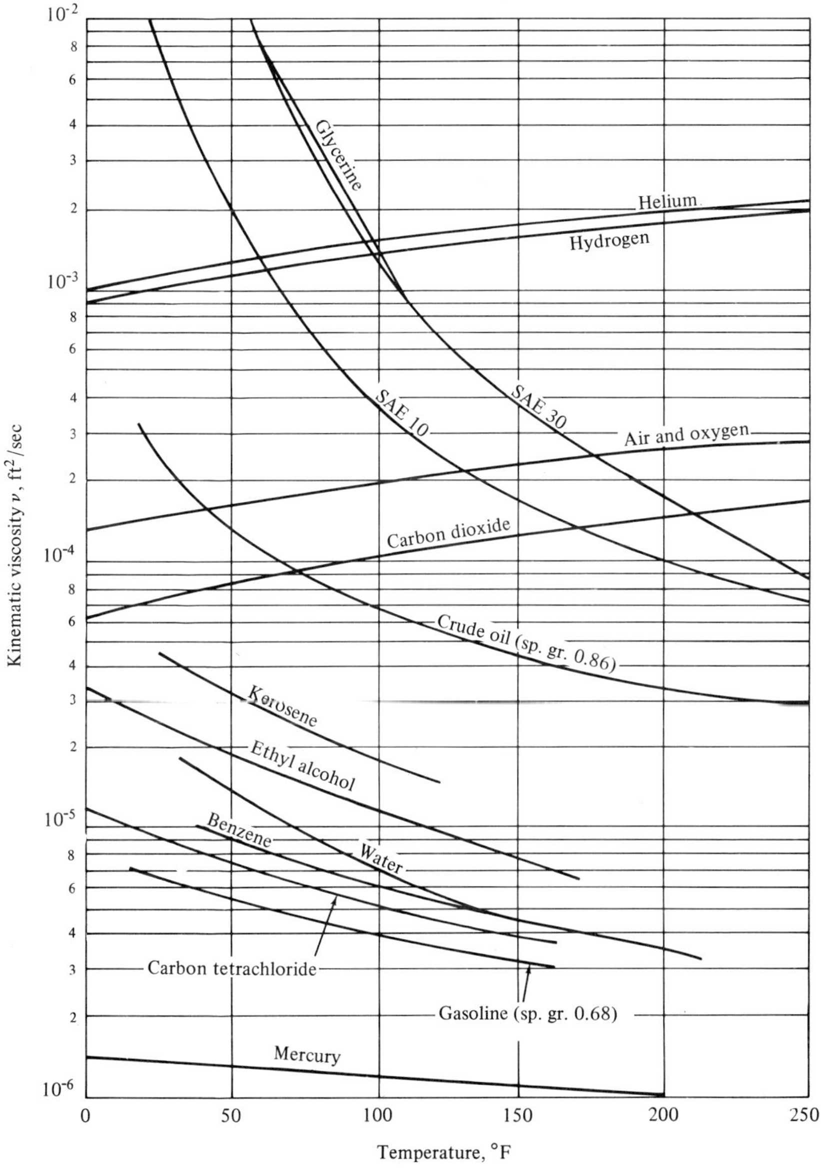

Fig. 2-8. Kinematic viscosity as a function of temperature for various fluids at 1 atm.

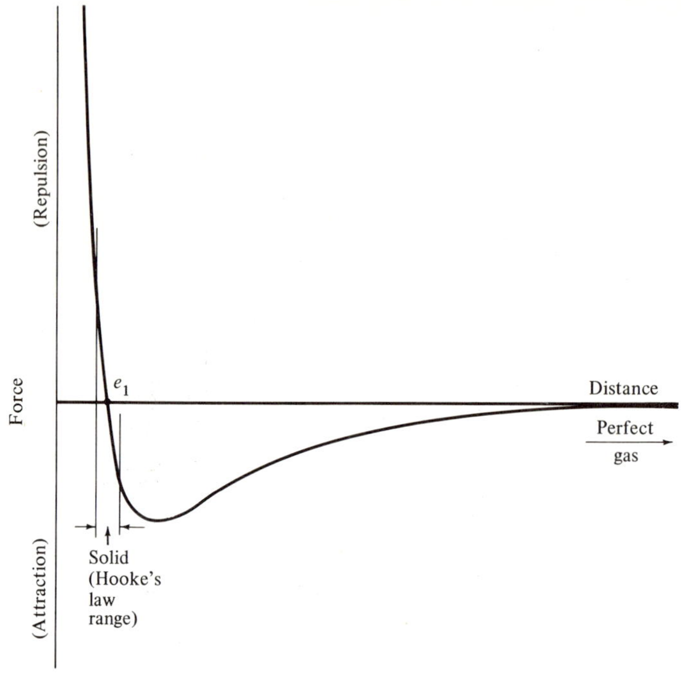

Fig. 2-9. Intermolecular force as a function of distance between molecules.

applied. The compressibility of a perfect gas is then a direct function of the pressure.

The compressibility of a substance is defined as the relative change of volume produced by a unit normal stress (pressure),

$$\kappa \equiv -\frac{dV/V}{dp}$$

The negative sign is introduced because pressure is defined as a compressive stress and the volume of most substances decreases under a compressive stress (dV is therefore negative when dp is positive); κ is therefore a positive quantity in most cases. The reciprocal of κ is defined as the bulk modulus, the stress produced by a unit volume (bulk) strain:

$$E \equiv -\frac{dp}{dV/V}$$

If the volume of the substance under consideration is replaced by its mass/density equivalent, $V = m/\rho$, then

$$\kappa = \frac{d\rho/\rho}{dp} \quad \text{and} \quad E = \frac{dp}{d\rho/\rho}$$

The actual numerical values of κ (or E) for any substance are seen to depend on the manner in which the stress (p) and strain (dV/V) are accomplished. If a substance is compressed isothermally (with heat transfer to maintain the temperature constant) a smaller stress (p) will be required to produce a given strain (dV) than would be required if the process had been accomplished without the energy removal (that is, an adiabatic process). Therefore, a better definition of the above quantities is

$$\kappa_T = \frac{1}{\rho}\left(\frac{\partial \rho}{\partial p}\right)_T \quad \text{or} \quad E_T = \rho\left(\frac{\partial p}{\partial \rho}\right)_T$$

$$\kappa_q = \frac{1}{\rho}\left(\frac{\partial \rho}{\partial p}\right)_q \quad \text{or} \quad E_q = \rho\left(\frac{\partial p}{\partial \rho}\right)_q$$

where the subscript T indicates an isothermal process and q indicates an adiabatic process.

As a simple example, the equation of state of a perfect gas is $p/\rho = RT$. The isothermal process equation between any two states would then be $p/\rho =$ const., giving $E_T = p$; therefore, the bulk modulus relating the change in compressive strain to change in applied normal stress is just equal to the compressive stress, or the pressure. Physically this result is logical; the only possible resistance that a perfect gas can offer to a compressive strain is the pressure, which is a consequence of the molecular motion. If the temperature (and thereby the motion) is kept constant, the isothermal modulus must be just the pressure.

The bulk modulus is the only type of modulus that is significant for a fluid since it is not possible to set up linearly oriented strains within a fluid volume at rest (as demonstrated in Chapter 3). The bulk modulus, or compressibility, in a substance determines its elasticity and the speed of the propagation of an elastic wave. These elastic or "sound" waves are propagated at a sufficient speed that there is not enough time for any significant heat transfer between particles, or energy dissipation. Therefore, the propagation process is very nearly isentropic (adiabatic and no dissipation). The relation giving the speed of such a wave (a) in an elastic medium is derived in Chapter 8 as

$$a = \sqrt{\frac{g_c E}{\rho}} = \sqrt{g_c\left(\frac{\partial p}{\partial \rho}\right)_s}$$

where the subscript s denotes constant entropy.

A plot of the pressure–volume relation for any given substance would appear as shown in Fig. 2-10. The value of E (or κ) is seen to depend on the slope measured at a given value of p and the volume V (or density ρ) and also on the temperature T; that is, $E(p, \rho, T)$. One of these variables can be eliminated using the thermodynamic "equation of state" that relates p, ρ, and T. In the case of a perfect gas, this relationship is quite

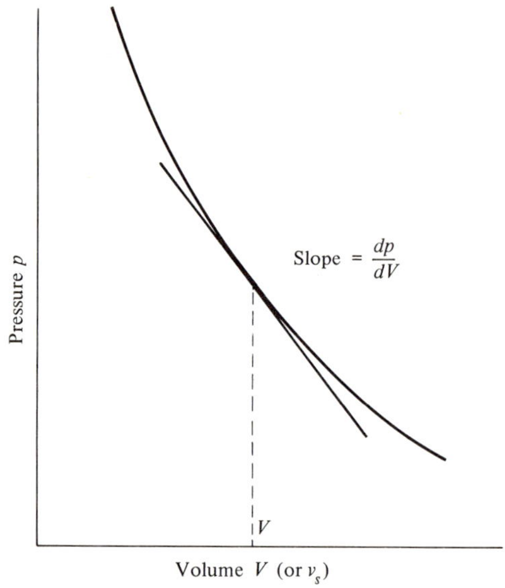

Fig. 2-10. Pressure–volume plot for an arbitrary, compressible substance.

simple. For a more complicated substance (for example, water) the relationship is more complex and must be determined experimentally and presented graphically to specify all states and phases of interest (for example, the Mollier diagram). For water the change in volume is quite small for pressures ranging from 1 to hundreds of atmospheres (that is, the curve shown in Fig. 2-10 is quite steep). An average value for the bulk modulus in the neighborhood of 1 atm and usual atmospheric temperatures (60°F) is 310,000 psi. The bulk modulus for various liquids is given in Table 2-1.

THERMAL CONDUCTIVITY (k)

Thermal conductivity is the coefficient relating the heat transfer to the temperature gradient. The thermal conductivity is of primary interest in fluid mechanics in determining the heat transfer between a boundary and a fluid or through the fluid itself. Fluid mechanics and thermodynamics are primarily related in the convective mode of heat transfer, which is aided by the fluid flow currents. The *thermal conductivity* also enters in compressible flow as a consequence of the conversion of work to internal energy in overcoming the friction forces. The thermal conductivity can be termed a transport property[8] that indicates the ability of the substance to transmit thermal energy internally by means of molecular collisions. By reasoning similar to that above, thermal conductivity is expected to be a function of temperature. (Further discussions on thermal conductivity and other thermodynamics properties of a fluid are given in Appendix C.)

THERMAL COEFFICIENT OF EXPANSION (β)

Volume thermal coefficient of expansion is defined as the relative volume change per unit temperature change or $\beta_V = (dV/V)/dT$. This coefficient is a "thermodynamic property" of the substance. It is then more properly written

$$\beta_V = \frac{1}{V}\left(\frac{\partial V}{\partial T}\right)_p \qquad \text{or} \qquad \beta_V = -\frac{1}{\rho}\left(\frac{\partial \rho}{\partial T}\right)_p$$

The quantity β is then a function of two independent thermodynamic properties, for example, temperature and pressure; that is, $\beta(T, p)$. For a perfect gas with a simple equation of state $p/\rho = RT$, β is just $1/T$. Some representative values of thermal expansion coefficient are indicated in Table 2-1.

VAPOR PRESSURE (p_v) AND LATENT HEAT OF VAPORIZATION (l_v)

The vapor pressure of a liquid is defined as the pressure at which vaporization or "boiling" occurs. It is a function of temperature and pressure. A vapor-pressure dependence on temperature is expected if the microscopic or molecular behavior of a fluid is considered as it is heated and passes from the liquid to the vapor state as discussed in Section 2.2. Consider the microscopic molecular behavior at a liquid-free surface in a closed container with a gas above the free surface. Because of the random state of translational molecular motion, some molecules with greater velocities than others will escape into the region above the free surface. If the entire system is in equilibrium (that is, if the free surface remains at a fixed level), some of molecules that are in the container above the free surface will have velocities toward the free surface and will become entrapped in the liquid state. An equilibrium state exists when the rate at which fluid molecules escape from the free surface is equal to the rate

[8] For a more comprehensive discussion and presentation of transport properties and their determination and specification see R. B. Bird, W. E. Stewart, and E. N. Lightfoot, *Transport Phenomena*, John Wiley & Sons, Inc., New York, 1960.

at which molecules return to it. If the gas above the free surface were initially air, an equilibrium mixture of air and molecules of the liquid in the vapor state would exist above the free surface. The pressure above the surface is made up of the gas partial pressure plus the partial pressure due to the vapor (according to Dalton's law of partial pressures). At equilibrium, the liquid and the gas above it are at the same temperature. The partial pressure of the vapor is just the vapor pressure of the liquid at the given temperature. For water, at a container pressure of 1 atm and a temperature of 70°F, the vapor pressure is 0.3631 psia (the partial pressure of the air is approximately 14.33 psia). If the fluid is heated and the top of the container is vented to maintain the total pressure at 1 atm, the higher-energy liquid molecules escape more readily. At 150°F the partial pressure due to the water vapor would be 3.718 psia, and the partial pressure due to the air is now only 10.98 psia. At 212°F the water-vapor partial pressure is 1 atm. If the temperature of the liquid is maintained at slightly more than 212°F, and if the pressure above the liquid is maintained at 1 atm, all the liquid will be vaporized.

The state in which the inertia forces of the molecules are larger than the intermolecular attraction forces is by definition a gas (or vapor). Having raised the temperature of the liquid to 212°F in a surrounding ambient pressure of 1 atm, the amount of energy required to overcome the molecular forces and transform the liquid into a vapor at this temperature is the *latent energy of vaporization* (l_v). For water at this temperature, l_v is 970.3 Btu/lb$_m$. Increasing the pressure above the free surface presents a greater resistance to the molecules attempting to escape from this surface; consequently, a greater temperature is required to vaporize the liquid. A relation then exists between pressure and temperature, with the vapor pressure increasing with an increase in temperature. The same reasoning would imply a relationship between energy of vaporization and temperature. This relationship is somewhat more complicated. The relationship between energy of vaporization and temperature is tabulated in the "Steam Tables."[9]

The same type of argument also holds for other phase changes. A schematic representation of the phase-change boundaries for water is shown in Fig. 2-11. The vapor pressure is seen to be a significant quantity in considering the mechanism of boiling. An interesting field of study in fluid mechanics is *cavitation*, which is a condition of local boiling as a consequence of a local reduced pressure.

The units of energy of vaporization are Btu/lb$_m$ in English-system heat units, or ft—lb$_f$/lb$_m$ in mechanical-energy units[10]; in mks units, J/kg.

[9] J. H. Keenan and F. G. Keyes, *Thermodynamic Properties of Steam*, John Wiley & Sons, New York, 1936.

[10] The conversion from thermal energy to mechanical energy is 778 ft-lb$_f$/Btu. No numerical conversion constant is needed in the mks system.

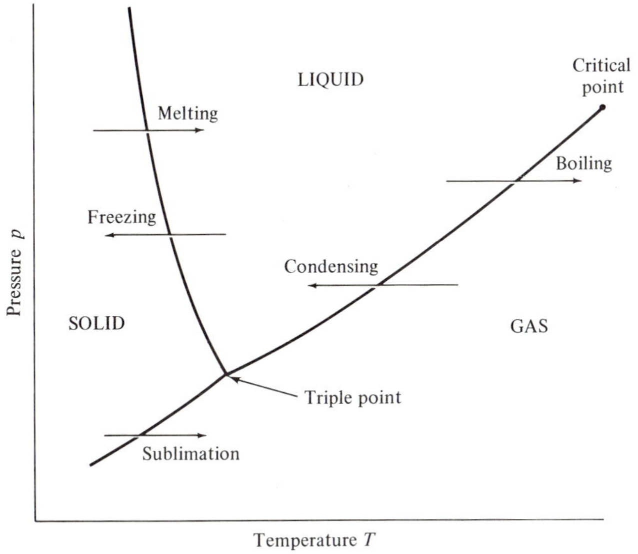

Fig. 2-11. Phase boundaries for water.

Values of vapor pressure for some representative liquids are given in Table 2-1.

SURFACE TENSION (σ)

A liquid film and liquid droplets are maintained by the intermolecular forces of attraction. A resultant *surface tension* maintains the film or the near-spherical shape of the droplet. A liquid in contact with a solid surface will gather into droplets if its *cohesive* molecular forces are greater than the *adhesive* forces between the surface molecules and the liquid molecules. It will spread if the adhesive forces are greater than the cohesive forces.

On considering the relationship between the intermolecular attractive forces and the molecular inertia, surface tension is expected to be a function of temperature. As the temperature of a liquid film is increased, the molecular motion increases with respect to the relatively fixed attractive force and tends to disrupt it. The liquid surface tension at the interface between the liquid and any gas is only slightly dependent on the gas, since the molecules in the gas are separated by a distance on the order of hundreds of times the distance between the liquid molecules. The forces between the gas or vapor and liquid molecules are negligible with respect to the intermolecular cohesive forces within the liquid. This relationship is an observed fact; it is noted that the surface tension of water with water vapor is the same as that of water with air as the surrounding gas.[11]

[11] This statement is true as long as the gas density or its molecular mass is relatively small.

The surface tension at the interface between two liquids or between a liquid and a solid depends on the ratio of the *cohesive* forces between the liquid molecules to the *adhesive* forces between the molecules of the liquid and solid or with the other liquid.

Capillarity is a surface-tension effect that depends upon the relative intermolecular attraction between different substances. Water will rise and form a concave meniscus in a small-diameter glass "capillary tube" (Fig. 2-12a), because the adhesive forces between the glass and water

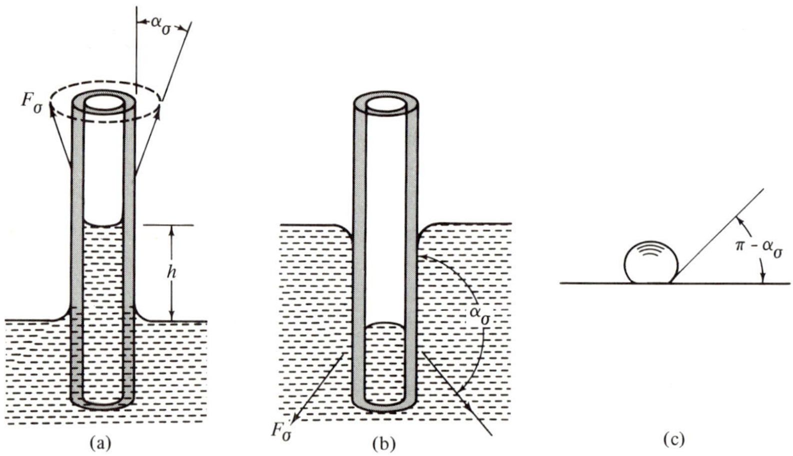

Fig. 2-12. Surface tension and capillarity: (a) Concave meniscus and capillary rise (example, water); (b) convex meniscus and capillary depression (example, mercury); (c) droplet on a "waxed" surface.

molecules are greater than the cohesive forces in the water. Conversely, the surface of mercury in a glass capillary will be depressed and forms a convex meniscus (Fig. 2-12b), because the cohesive forces in the mercury are greater than the cohesive forces between mercury and glass. The meniscus angles indicated are approximately 0° for water in contact with clean glass and about 130° between mercury and glass.

The phenomenon of "wetting" is also a consequence of the surface tension. A "wetting agent" added to water, or a surface treated with a wetting agent, changes the relationship between the cohesive and adhesive forces. The adhesive forces between the liquid and the surface predominate, so the liquid spreads out over the surface. Waxing a surface produces the opposite effect. The cohesive forces predominate, so that the liquid is drawn together into small droplets (Fig. 2-12c). Surface tension can be measured directly as fine film tension by using a thin wire, as indicated in

Fig. 2-13. The surface tension for the apparatus shown is determined as the net line-film force (the total force minus the wire weight) divided by the film length. A better idea of this force balance can be obtained by considering only a small film element immediately adjacent to the wire, as indicated in Fig. 2-13b. In determining σ the weight of the fluid film is neglected. (This approximation is satisfactory since the weight is an order of magnitude smaller than the surface tension force involved.) Surface tension can also be measured by considering a force equilibrium for the

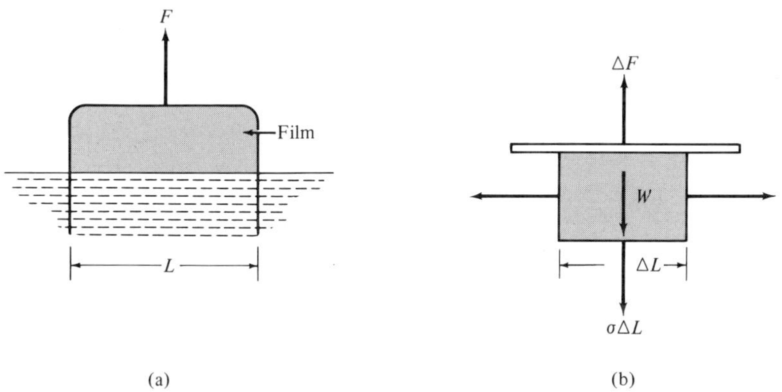

Fig. 2-13. Surface tension measurement by means of a wire loop.

fluid within the capillary tube above the external free surface (Fig. 2-12a). A free-body force balance on this mass gives

$$F_\sigma = \sigma\pi d \cos \alpha \qquad W = \frac{\rho g}{g_c}\frac{\pi d^2}{4}h$$

Therefore,

$$\sigma = \frac{\rho g d h}{4 g_c \cos \alpha}$$

By analyzing the force balance on a small droplet or bubble in equilibrium with its surroundings, the pressure may vary significantly from the surrounding pressure, owing to the surface tension acting at the surface interface. For the droplet indicated in Fig. 2-14, a force equilibrium exists between the surface tension force at a line drawn around the equator and the difference between the pressure force inside and outside of the droplet. This net pressure force can be determined as the difference in pressure times the projected area, as indicated. The condition of static equilibrium

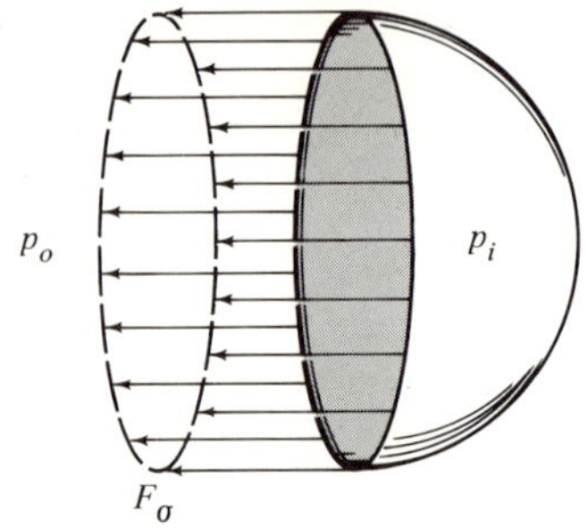

Fig. 2–14. Half of a spherical droplet showing equilibrium between differential pressure force and surface tension force.

then gives

$$(p_i - p_o)\frac{\pi d^2}{4} = \sigma\pi d \qquad \text{or} \qquad p_i - p_o = \frac{4\sigma}{d}$$

The units of surface tension are force per unit length (F/L) or energy per unit area ($FL/L^2 = F/L$). From Fig. 2-13, the work required to overcome the surface tension forces to lift the wire to the height H is FH, which is also the potential energy attained for this configuration. The potential energy per unit area for this configuration is $FH/LH = F/L$. Because of this equivalence between surface tension and potential energy, the term *surface energy* is sometimes used. The surface energy of a free surface also represents the potential energy possessed by the surface, which is equivalent to the work required to form the free surface.

Some values of surface tension are presented in Table 2-1. For more complete data, refer to the *International Critical Tables* and publications of the National Bureau of Standards.

SPECIFIC ENERGIES (c_p and c_v)

The *specific energy* of any substance is energy required to raise the temperature of a unit mass of that substance 1 degree at the initial or average temperature specified. The specific energy is generally a function of temperature. The specific energy of a fluid is of primary importance in compressible flow (gas dynamics), where energy-balance considerations involve the intrinsic or internal (molecular) energy as well as the kinetic energy of the mean flow motion. Heating a fluid increases its (1) random translational, (2) rotational, and (3) vibrational modes of molecular motion. The specific energy therefore depends on the mass of the molecule and its molecular structure. For a given substance the specific energy is a function primarily of temperature (which is an indication of the kinetic energy of the three modes of motion described) and, to a lesser extent, of pressure. In the simplest case of a billiard-ball perfect gas model, the specific energy is a constant.

At temperatures greater than the critical temperatues and pressures less than the critical pressure, some of the rotation and vibration modes of "real" gases are quite small compared to the translational modes, so the specific energy is determined primarily by the translational mean kinetic energy of the molecules. The pressure dependence is due to the density variation, which gives a variation in the intermolecular distance and therefore a variation in the potential energy.

The magnitude of the specific energy will depend on the heat-transfer process to the substance. If energy is added at constant volume, all the energy that is added merely goes into the three modes of kinetic (intrinsic or internal) energy of the molecules. The quantity of energy required to produce a 1-degree temperature rise in a unit mass held at constant volume is defined as the *specific energy at constant volume*, c_v. For a perfect gas, the only internal kinetic-energy mode is that due to the random, translational motion, which is directly proportional to the temperature. The specific energy at constant volume for a perfect gas is therefore a constant.

If the process by which the energy is added is a constant-pressure process, the substance usually expands against the ambient pressure, expending work on the surroundings. The energy added then goes into producing this work in addition to raising the temperature. It might therefore appear that the *specific energy at constant pressure* would depend upon the external pressure against which the expansion work is done. However, if the external pressure is high, the displacement will be relatively small, and conversely. The expansion or compression work will then be about constant independent of the pressure and, therefore, primarily a function of the temperature. For the idealized perfect gas the specific energy at constant pressure (c_p) is a constant. The *ratio* of specific energy at constant pressure to specific energy at constant volume frequently appears in various thermodynamic and fluid flow processes. This ratio of specific energies is denoted by the symbol gamma: $\gamma \equiv c_p/c_v$.

The general units of specific energy are energy per unit mass per 1-degree temperature rise. In the English system of units the energy can be specified in either of two units: thermal—Btu/lb_m—or mechanical—ft-lb_f/lb_m. Therefore, the units of specific energy are: thermal—Btu/lb_m-°R—or mechanical (or potential)—ft-lb_f/lb_m-°R. The temperature unit can be degrees Rankine, °R (absolute), or degrees Fahrenheit, °F (a temperature difference is implied and any reference or base temperature cancels). In the mks system, thermal, mechanical, and electrical energy all have the units J/kg-°K; in the cgs system the units are dyne-cm/g-°K (or ergs/g-°K). (In the mks system 1 J is equivalent to 1 N-m or 1 W-sec.) The mks system is seen to have the great advantage that all forms of work and energy (that is, thermal, mechanical, and electrical) have the same unit, the joule. There is an international trend toward the use of the mks system. Many rocket parameters (with respect to work and energy) are universally specified in joules and the power in watts (1 W is equivalent to 1 J/sec). More information on units and their use can be found in Appendix A.

GAS CONSTANT (R)

The *gas constant* is a proportionality constant between the temperature of a gas and the mean molecular kinetic energy. (See Appendix C for a more detailed discussion of this relationship.) Since the gas constant is a specific energy quantity, a relationship between the specific energy at constant pressure, the specific energy at constant volume, and the gas constant might be anticipated, that is, $f(c_p, c_v, R)$. In the case of the idealized perfect gas this relationship is $c_p - c_v = R$. The units of R are then obviously the same as c_p and c_v.

The relationship between a gas constant for specific gas (R) and the universal gas constant (R_0) is $R = R_0/\mathscr{M}$. The universal gas constant is then based on a standard molecular reference or the molecular mass. Since a molecular mass of any substance contains a fixed number of molecules (Avogadro's number), a gas constant per molecule may also be defined as R_0/N_A, which is the Boltzmann constant $\mathscr{k}$. This quantity represents the amount of energy required to increase the translational kinetic energy of the molecule by unit temperature rise. Numerical values of Avogadro's number, the Boltzmann constant, and the universal gas constant are given in Appendix C.

The equation of state for a perfect gas is $p/\rho = RT$. Since temperature appears singly and not as a difference, absolute temperature is, of course, implied. The general units for the gas constant are energy per unit mass per degree of absolute temperature. In the English system the units are ft-$\text{lb}_\text{f}/\text{lb}_\text{m}$-°R (mks: J/kg-°K; cgs: ergs/g-°K).

ELECTRICAL PROPERTIES

When the motion of a fluid is affected by electrical and/or magnetic forces (or force field), the properties of electrical conductivity, dielectric strength, magnetic permeability, and so on, are also of importance. The study of magnetohydrodynamics considers the interactions of electric and magnetic force fields with the motion of fluids (Chapter 16).

2.5 ILLUSTRATIVE EXAMPLE 2-1 : Superheat and Surface Tension

When a vapor bubble starts to form in boiling water, a small amount of superheat and accompanying overpressure inside the bubble (in excess of the surrounding pressure) is necessary for the bubble to grow. Calculate this overpressure and the corresponding superheat when the bubble is 0.001 in. in diameter and the water is at 1 atm pressure.

From the relation for the differential pressure that balances the surface-tension force,

$$\Delta p = \frac{4\sigma}{d} = \frac{(4)(0.004)}{0.001/12} = 192 \text{ lb}_f/\text{ft}^2, \quad \text{or} \quad 1.33 \text{ lb}_f/\text{in.}^2$$

From the steam tables, the saturation temperature at 1 atm is 212°F. At 16.0 psia, the temperature of the *saturated* steam within the bubble must be 216.3°F, or the water must be *apparently* superheated by 4.3°F for the bubbles to grow at this diameter. If the bubble diameter is less than 0.001 in., it would not be able to grow against the external pressure plus surface-tension forces at this temperature and external pressure.

2.6 ILLUSTRATIVE EXAMPLE 2-2: Surface-Tension Effect on Manometers

A reservoir type of water manometer uses tubing with a nominal 2-mm bore (Fig. 2-15). Near the zero setting, the bore is actually 1.9 mm. At

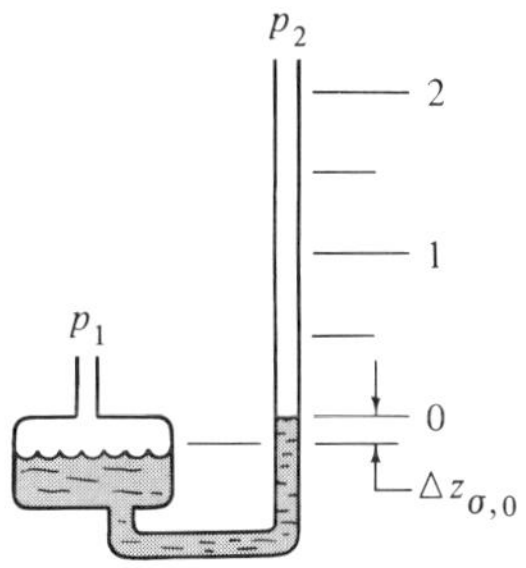

Fig. 2-15.

2 in. up the tube the bore is 2.1 mm. What are the absolute and relative errors when the manometer reads 2 in.? If the bore is 2.1 mm everywhere above the 1.5-in. mark, how does the relative reading vary with reading in the range above 1.5 in.? If the taper in the bore is linear from 0 to 1.5 in., specify a correction factor for the entire manometer range from 0 to 2.0 in.

Measured from the reservoir level, the capillary rise at zero reading is

$$z_{\sigma,0} = \frac{4g_c \cos\alpha}{\rho g d} = \frac{(4)(0.005)(25.4)(12)^2}{(62.4)(1.9)} = 0.618 \text{ in.}$$

No error is involved by this rise, since the zero is set on the initial meniscus.

At the 2-in. reading point, the capillary rise is

$$z_{\sigma,2} = (0.618)\frac{1.9}{2.1} = 0.560 \text{ in.}$$

The manometer then reads 0.058 in. *lower* than it would have if the diameter had been uniform. The reading was specified as 2 in. Therefore, the true value of the applied pressure head is 2.058 in. The correction is 0.058 in. (*Note:* Correction $=$ $-$error.) The indicated value (z) is the true value (z^0) plus the error, or error $=$ indicated $-$ true $= -0.58$ in. The relative error is $-0.058/2.058 = -0.028$ (or -2.8 percent). In the range above 1.5 in., the relative error is $-0.058/(z + 0.058) \approx -0.058/z$.

An expression can be derived for the error and relative error as a function of the known taper, or for the diameter as a function of the reading, z. The error at $z = 1.5$ in. (at the end of the taper) is -0.058 in., and the error varies linearly from zero at $z = 0$ to this point. Thus the error as a function of z (for range $z = 0$ to $z = 1.5$ in.) is

$$e = \frac{-0.058}{1.5} z = -0.0387z$$

The relative error (e') is the error (e) relative to the true value, z, or $e' = e/(z - e)$, where z is the indicated value and $z - e$ is the true value. The relative error between $z = 0$ and $z = 1.5$ is then $e' = -0.0387z/(1 + 0.0387)z = -0.0373$ (-3.7 percent). For $z > 1.5$ in., e is constant at -0.058.

The true value (z^0) can be obtained from the relations $z^0 = z - e$ and $e' = e/(z - e)$, giving $z^0 = z/(1 + e')$, or approximately $z^0 = z(1 - e')$. For example, the true reading at $z = 1$ in. would be $z^0 = 1/1 + (-0.0373) = 1.038$ in., or $z^0 \approx z(1 - e') = 1.037$ in.

The *reading resolution* of the instrument must also be considered in determining and making such corrections. With a vertical manometer a resolution of 0.01 in. is about the best that can be expected with the unaided eye. The above correction is greater than this value and is therefore significant. With a shallow, inclined draft gage, the resolution may approach 0.001 in. The surface-tension error (and correction) for a draft gage is also greater than for a vertical manometer. Consequently, when using a shallow-draft gage, precautions must be taken (1) to use a uniform bore tube, (2) to use a tube of sufficient diameter to minimize surface-tension effects, and/or (3) to use a fluid with low surface tension. In the case of water, dope may be used to decrease the surface tension.

Liquids with a strong adhesion to glass tubes (for example, water with $\sigma \approx 0.005$ and $\alpha = 0$) can produce a surface-tension hysteresis effect during small changes in head. The contact line tends to be maintained as a small change in pressure occurs. The curvature of the meniscus then changes and an erroneous reading is obtained.

2.7 ILLUSTRATIVE EXAMPLE 2-3: Bulb-Type Temperature Indicator

A temperature-gage actuator for an automobile consists of a bulb, a connecting tube, and a Bourdon-tube pressure gage (Fig. 2-16). The

Fig. 2-16.

bulb is immersed in the cooling-water jacket in the engine block. Its inside dimensions are 1/4 in. in diameter and 1 in. long. The tube is 3 ft long with an I.D. of 0.016 in. Both the tube and bulb are filled with air. The gage is to indicate temperature from 100° to 240°F. The system is sealed at 5 atm and 80°F. Find the pressure range of the gage for the temperature variation to be indicated. Assume that the bulb material (brass) has a *linear* coefficient of expansion of 1.05×10^{-5}/°F within the range 0 to 100°C and that the tube remains at the ambient temperature of about 100°F during the operation.

The pressure corresponding to the 100°F indication will be determined from

$$pV = mRT \qquad \text{or} \qquad \frac{p_0 V_0}{T_0} = \frac{p_1 V_1}{T_1}$$

since the contained mass (m) remains constant. The zero-subscript conditions are at 5 atm and 80°F. Solving for the pressure at 100°F gives

$$p_1 = p_0 \frac{T_1}{T_0} \frac{V_0}{V_1}$$

The volume change from V_0 to V_1 will be due to (1) the thermal expansion of the metal, and (2) the pressure-stress expansion of the metal. Since the relative volume expansion, $\Delta V/V$, is going to be quite small, the two effects can be considered as independent and therefore linearly additive.[12] The total internal volume is $V_0 = (\pi/4)(0.25)^2(1) + (\pi/4)(0.016)^2(36)$ (neglecting the volume of the Bourdon element), or

$$V_0 = 0.0491 \text{ in.}^3 \text{ (bulb)} + 0.00724 \text{ in.}^3 \text{ (tube)} = 0.0563 \text{ in.}^3$$

The increase in temperature will cause each of the linear internal dimensions of the tube to increase in proportion to β_L (linear coefficient). For the tube, $\beta_L = (\Delta l/l)/T$, or for the bulb diameter at T_0, $(d_1 - d_0)/d_0 = \beta_L \Delta T$, or $d_1 = (1 + \beta_L \Delta T)d_0$. The bulb volume, V_1, is then

$$V_1 = \frac{\pi d_1^2}{4} l_1 = \frac{\pi}{4}(1 + \beta_L \Delta T)^2 d_0^2(1 + \beta_L \Delta T)l_0$$

The quantity $\beta \Delta T$ is quite small (much less than unity), so $V_1 \approx (\pi/4)\, d_0^2 l_0(1 + 3\beta_L \Delta T)$, where quantities on the order of $(\beta \Delta T)^2$ and higher have been neglected. This result can be expressed as

$$V_1 = V_0(1 + \beta_V \Delta T) \qquad \text{or} \qquad V_1 - V_0 = V_0 \beta_V \Delta T$$

The volume coefficient of expansion β_V is $\beta_V = (\Delta V/V)/T$. The general conclusion is that, for any geometry,

$$\beta_V \approx 3\beta_L$$

as long as $\beta_V \Delta T$ is small (0.01 or less). The quantity appearing in the pressure (p) to be determined is $V_0/V_1 = (1 + \beta_V \Delta T)^{-1} = (1 + 3\beta_L \Delta T)^{-1}$. Since $\beta_L \Delta T$ is very small, expanding $(1 + \beta_L \Delta T)^{-1}$ by the binomial theorem (or Taylor expansion series[13]) gives

$$(1 + 3\beta_L \Delta T)^{-1} \approx 1 - 3\beta_L \Delta T$$

Neglecting the pressure-stress expansion for the present, the pressure p_1

[12] $$V_2 = V_1\left[1 + \left(\frac{\Delta V}{V}\right)_{(1)}\right]\left[1 + \left(\frac{\Delta V}{V}\right)_{(2)}\right] \approx V_1\left[1 + \left(\frac{\Delta V}{V}\right)_{(1)} + \left(\frac{\Delta V}{V}\right)_{(2)}\right]$$

[13] $$(1 + x)^n = 1 + nx + n(n - 1)\frac{x^2}{2!} + \cdots + r(r - 1)(r - 2)\cdots(1)\frac{x^r}{r!} + \cdots$$

where $x < 1$ and r designates the rth term. The number of terms that must be considered for a good approximation decreases as x becomes smaller. This is a useful expression where approximations and errors are involved or where answers are required to only two or three significant places.

given by[14]

$$p_1 = p_0 \frac{560}{540} [1 - 3\beta_L(100 - 80)] = p_0 \frac{560}{540} [1 - 0.00063]$$

Note that $3\beta_L \Delta T$ is $\ll 1$ or $\ll 0.01$. For this type of application, the thermal expansion of the metal is negligible. The relative pressure change

$$\frac{\Delta p}{p_0} = \frac{p_1 - p_0}{p_0} = \frac{p_1}{p_0} - 1 \approx \frac{\Delta T}{T_0} = \frac{20}{540} = 0.037 \text{ (3.7 percent)}$$

is then due primarily to the temperature change, and is small. The pressure-stress change and the relative change in volume would be proportional to $\Delta p/E$, where E is Young's elastic modulus (10^7 psi for brass), so the pressure-stress effect is definitely negligible. The initial operating design pressure for the Bourdon element is then $p_1 = (5)(1.037) = 5.185$ atm, or 76.2 psia (4.185 atm or 61.5 psig relative to a standard atmospheric pressure of 14.7 psia).

Neglecting the thermal expansion of the metal at an engine temperature of 240°F, the pressure would be

$$\frac{p_2}{p_1} = \frac{T_2}{T_1}\frac{V_1}{V_2} = \frac{700}{560}\frac{0.0491}{0.0563}[1 - 3(140)(1.05 \times 10^{-5})]$$

$$= \frac{700}{560}\frac{0.0491}{0.0563}[1 - 0.0044] = 1.09$$

Note that only the active portion of the volume (the bulb) expands from 100 to 240°F, since it was assumed that the tube remains at 100°F. The thermal expansion of the metal parts still produces less than 0.5 percent in the pressure change, again a negligible consideration in this application.

The maximum operation design pressure for the gage element is

$$p_2 = 1.09p_1 = 5.65 \text{ atm}$$

or 83.1 psia (4.65 atm-gage or 68.4 psig). (The pressure-stress variation on the metal bulb, and the corresponding correction on pressure, would again be much less than 0.5 percent.)

The gage-element pressure variation is only $5.65 - 5.185 = 0.465$ atm, or 6.9 psi for the temperature variation of 100 to 240°F. Note that the pressure range is only 10 percent of the maximum reading, a result that

[14] Note that the only possible and significant way to express temperature ratios is as ratios of *absolute* temperatures. However, temperature *differences* can be expressed in terms of absolute (°R or °K) or relative (°F or °C), since the base or reference value cancels. The same observation is true for any quantity.

forces more stringent quality requirements on the assembly (especially the Bourdon-tube element) than are desirable for a low-cost, mass-produced item. For these reasons, the above design would be rejected. Although it would be possible to load the air at 1 atm, the operating range would be only about 1.4 psi. Such a small pressure range would require a gage element much too delicate for the application.

The maximum altitude for automobile engine operation is 12,000 ft. The gage would then read high, ranging from 66.2 to 73.1 psi. Since the Bourdon-tube gage indicates differential or gage pressure, an error of $14.7 - 9.0 = 5.7$ psi is superimposed on the 6.9-psi range, with the result that the indication is in error by about 80 percent. Since the absolute error does not depend on the gage mechanism (only on the altitude at which the engine is to be operated), the relative error depends primarily on the initial pressure. A 10 percent altitude error is probably tolerable at 12,000 ft, or the active differential pressure must be increased to 57 psi from 6.9, or by a factor of about 8. The same factor must then be applied to the charge pressure, p_0, or the charging pressure would have to be about 600 psi.

Since the air-charge system is not acceptable, other fluids must be investigated. Water is one possibility. If the bulb were partially filled with water, with the air in the rest of the system at 1 atm, the pressure is then the sum of the partial pressures of the air and water at that temperature. The partial pressure of the liquid is a function only of temperature, and the volume of the air is constant. Therefore, the total pressure could be calculated. The relatively high freezing point is a significant disadvantage.

The designer must consider other fluids with more desirable behavior for this application. Note that the initial "solution" to this design problem (with respect to the working fluid) was specified before the design criteria were established, a very poor way to initiate a design (see Problems 2-65 to 2-67).

PROBLEMS

2-1 When full, a 5-gal tank contains what *mass* of water? Of air? (1 gal = 231 in.^{3})

2-2 What would be the *weight* of 5 gal of water on the earth? On the moon? (The moon's gravitational acceleration is 5.47 ft/sec^2.)

2-3 What is the *weight* of 1 ft^3 of water (at 20°C) on the moon, where the gravitational acceleration is 5.47 ft/sec^2?

2-4 What is the *weight* of 1 gal (231 in.^{3}) of mercury (a) on the earth? (b) on the moon?

2-5 Calculate the standard specific gravity for mercury at 150°F. What is the specific volume at this temperature?

2-6 What is the specific volume and specific gravity of (a) hydrogen, (b) air, (c) water, and (d) mercury at 2 atm and 68°F?

2-7 Calculate the density of air at 68°F and 1 atm (14.696 psia or 2116.2 psfa) assuming air to be a perfect gas, and compare the result with that given in Table 2-2.

2-8 A test crew weighs 1 lb of gasoline on a spring scale and runs a test on a gasoline–liquid oxygen engine and obtains an overall test efficiency of 30 percent, based on a heating value of 18,900 Btu/lb of fuel. The entire test rig is transported to the U.S. moon base, where the test is checked. The calculated efficiency is 175 percent! What happened? (Gasoline has a specific gravity of ⅔.)

2-9 What mass of air can be stored in a 1 ft^3 tank at 1000 psi and −80°F?

2-10 In the isothermal region of the atmosphere (the stratosphere) (36,152 to about 80,000 ft), the temperature is −69.7°F. The pressures at 36,152 and 80,000 ft are 3.28 and 0.41 psia, respectively. Assuming that air behaves as a perfect gas, calculate the densities at these altitudes and compare the results with the values given for the standard atmosphere in Appendix D.

2-11 An automobile tire having a volume of 2500 $in.^3$ is inflated to 25 psi at sea level (T = 50°F). What is the pressure in the tire when it is driven over Trail Ridge Pass, Colorado, elevation 12,122 ft?

2-12 If the tire in Problem 2-11 is driven out into the desert and reaches a temperature of 125°F, what is the pressure? What is the mass of air in the tire?

2-13 How many molecules of water are there in a droplet 0.001 in. in diameter? How many air molecules at 1 atm and 20°C?

2-14 For a fluid with a viscosity of 15 cP and a specific gravity of 0.86, determine the viscosity in lb_f-sec/ft^2 and the kinematic viscosity in stokes and in ft^2/sec.

2-15 Calculate the conversion constant from stokes to ft^2/sec and from centistokes to ft^2/sec.

2-16 Determine the absolute and kinematic viscosities of air, water, and mercury at 70°F and comment on the comparison.

2-17 What is the shear rate and angular deformation velocity for the lubricant in a 0.0015-in. radial clearance between a 1-in.-diameter shaft and concentric bearing surface? The lubricant viscosity is 200 cP.

2-18 What is the kinematic viscosity of air in stokes at 1 psia and 200°F?

2-19 Calculate the kinematic viscosity of air at 50,000 ft, assuming $\mu = C\sqrt{T}$ and that only the information from Table 2-2 is available. (T = −69°F at 50,000 ft.)

2-20 Derive an expression for determining kinematic viscosity when the absolute viscosity and specific gravity of a liquid are given.

2-21 What is the kinematic viscosity of air at 68°F and 1 atm? 2 atm? 100 atm?

2-22 What is the torque required to turn the shaft in a 2-in. journal bearing with a clearance of 0.002 in. and 4 in. in length at 900 rpm? The fluid is oil with a viscosity of 120 cP.

2-23 What is the friction torque per inch of length on a journal-bearing shaft 1 in. in diameter with an 0.0015-in. radial clearance running at 1800 rpm? The viscosity of the oil is 100 cP.

2-24 For the case of one plate sliding over another with a viscous fluid between (Fig. 2-3), determine the velocity profile for a fluid with a non-Newtonian viscosity behavior as given by $\mu = \mu_0(du/dn)^\alpha$, where μ_0 and α are constants.

2-25 Assuming air to behave as a perfect gas, plot viscosity versus temperature (absolute °R) for air according to the relation $\mu = C\sqrt{T}$. Determine the constant C from the data of Table 2-2. Also plot μ versus T from the data given in Appendix D and compare and comment on the results.

2-26 Using the relationship for the viscosity of air that $\mu = C\sqrt{T}$ and assuming air to behave as a perfect gas, determine an equation for $\nu = f(p, T)$; show that $\nu = C_2T^{3/2}/p$ and find C_2.

2-27 For a viscous fluid between two concentric cylinders of radii r_1 and r_2 with one rotating with respect to the other, consider Newton's laws and show that the shear stress at the inside cylinder must be different from that at the outside cylinder. At which cylinder is the shear stress greater?

2-28 Consider first the conditions that must apply from Newton's laws and derive the velocity profile for a viscous fluid in the clearance space between two concentric cylinders of radii r_1 and r_2, with the inside cylinder rotating with an angular speed ω rad/sec. What is the radial shear stress distribution in the fluid? What is the torque and power necessary to drive the shaft for a sleeve length L?

2-29 For a sleeve sliding axially along a concentric shaft with a viscous fluid between the two, find the velocity and shear stress profiles at a function of radius.

2-30 What pressure is necessary to compress a volume of water by 1/10 of 1 percent of its initial volume at 70°F?

2-31 With κ defined as $\kappa \equiv -(dV/V)/dp$, prove the validity of the relation $\kappa = (d\rho/\rho)/dp$.

2-32 Give a detailed explanation of why the adiabatic bulk modulus E_q should be greater than the isothermal modulus E_T.

2-33 The pressure in a hydraulic actuating cylinder on a truck reaches a peak load value of 5000 psi. If the loading vibrates the cylinder so that the pressure varies from 0 to 5000 psi, how much does the piston move due to compression of the liquid in the cylinder if its bulk modulus is 200,000 psi? The steel cylinder is 20 in. long, 5 in. in diameter, and ½ in. thick. (Neglect the elasticity of the cylinder. Note that if the wall thickness to I.D. ratio for the cylinder is greater than 0.1, the volume change due to the expansion of the steel cylinder would be less than 10 percent of the volume change of the fluid.)

2-34 The valve on the far end of a 100-ft control line is set to open at 1000 psi. If the line is 0.1 in. in diameter, and initially full at 1 atm, how much fluid ($E = 120{,}000$ psi) must be pumped before the valve opens?

2-35 Find the spring constant and natural sonic frequency of a 10-ft-long, ¼-in.-I.D. steel control line filled with hydraulic fluid with a modulus of 200,000 psi. (Assume that the steel tube is rigid.) ($S = 0.8$.)

2-36 The linear modulus E_l for steel is 3×10^7 psi. For the cylinder in Problem 2-33 show that the volume increase due to expansion of the cylinder is less than 10 percent of the fluid volume change.

2-37 If the process equation for an isentropic (reversible or idealized adiabatic) process for a perfect gas is $p/\rho^\gamma = \text{const.}$, show that $E_s = \gamma p$.

2-38 The equation $p/\rho^n = \text{const.}$ is used to represent a polytropic process in a perfect gas. What is the corresponding polytropic bulk modulus, E_n?

2-39 The Mach number (M) is defined as the speed of the object divided by the local speed of sound as determined by the local ambient conditions. What is the Mach number for an airplane traveling at 800 mph at sea level if the pressure is 1 atm and the temperature is (a) 0°F? (b) 100°F? (c) What is the Mach number at 40,000 ft where the pressure is 0.185 atm and the temperature is −70°F?

2-40 Determine the speed of sound in the following media at 68°F: air, water, alcohol, and mercury.

2-41 Starting with the definition for κ, derive the expression for the speed of sound in a perfect gas, assuming that the sound wave is propagated (a) isothermally, (b) isentropically, and (c) polytropically. (d) Which process would you choose as most closely representing the sound-wave-propagation process and why?

2-42 In the flow over a hydrofoil, at what local pressure will cavitation start if the water temperature is (a) 40°F? (b) 80°F? (Consult steam tables.)

2-43 Oil spreads when poured on water. Discuss the relations between adhesive and cohesive forces for the two liquids.

2-44 A U-tube manometer is constructed as shown. Determine the absolute and relative error when the tube diameters are 2.2 mm (left side) and 1.8 mm (right side).

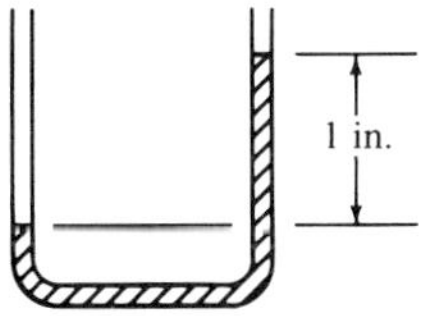

Fig. P. 2-44.

2-45 What is the overpressure (pressure over the surrounding ambient) in a water droplet 0.001 in. in diameter? What is it for an air bubble 0.001 in. in diameter in water?

2-46 A U-tube manometer is constructed with glass tubing with a nominal I.D. of 2 mm. If the I.D. of one leg is 2.2 mm and the other leg is 1.8 mm, what is the absolute error in reading when the fluid is (a) water? (b) mercury?

2-47 Determine the surface-tension capillary rise (h) in an inclined tube as a function of d and α for water and for mercury.

2-48 How does the surface energy of a droplet vary with the size of the droplet?

2-49 What is the net surface energy for 10,000 water droplets 0.001 in. in diameter? Comment on the significance of such an energy value.

2-50 Derive the following forms of the equation of state for a perfect gas starting with $p/\rho = RT$.

(a) $pV = mRT$.

(b) $pv_s = RT$.

(c) $p/\rho = (R_0/\mathscr{M}) - T$. (What are the general units of R_0?)

(d) $pV = nR_0T$ (where n is the number of moles in the volume V at p and T).

(e) $pV/T =$ const. (What are the conditions for this relationship to hold?)

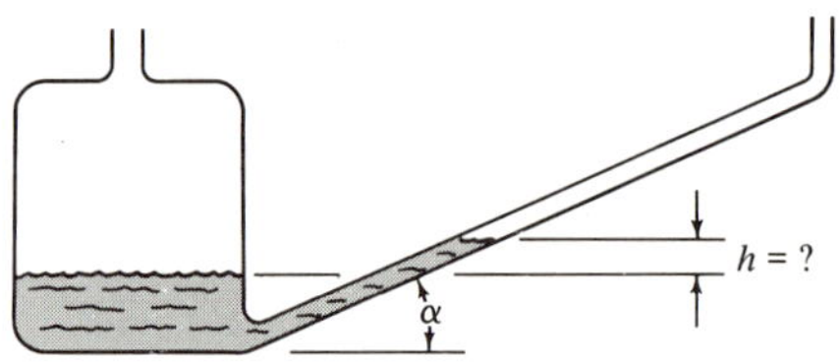

Fig. P. 2-50.

2-51 Calculate the variation of standard specific gravity of water if it is compressed from 1 to 1000 atm. What assumptions are necessary in making this calculation?

2-52 Calculate the standard specific gravity of alcohol at a pressure of 1000 psi. What is the percentage variation from the value at 1 atm?

2-53 On a day when the temperature is 110°F, what is the absolute and the relative (percentage) correction that must be applied to a barometer reading to account for the expansion of the mercury if the standard reference temperature is 20°C?

2-54 Specify an expression for a thermal coefficient of density variation and determine its specific value for air.

2-55 If a mercury manometer scale is calibrated at 60°F, what is the percentage error introduced when it is used at 150°F? (*Note:* The thermal-expansion coefficients given in Table 2-1 are *volume* coefficients. What is the relation between volume thermal-expansion coefficient and linear thermal-expansion coefficient?)

2-56 A mercury thermometer is constructed using a volume of 0.005 in.^{3} connected with a 0.010-in.-diameter tube 6 in. long. At 0°F the mercury is 1 in. from the bulb and the rest of the tube (closed at the top) contains air at 0.1 atm. What is the height the mercury will rise when the temperature is raised to 50°F? To 100°F? Neglect the glass expansion for the calculation, but discuss its effect.

2-57 The volume thermal expansion coefficient for mercury for the temperature range 0 to 100°C is given by the formula

$$\beta = \beta_0(1 + 0.182 \times 10^{-3}T + 1.1405 \times 10^{-6}T^2 - 0.539 \times 10^{-8}T^3)$$

where T is in °C. Calculate β at 20°C and compare the result with the value given in Table 2-1. Also calculate the percentage variation in β between 0 and 100°C.

2-58 A mass of liquid is subjected to a simultaneous change of pressure and temperature. Specify an expression for obtaining the relative density change.

2-59 How would you expect pressure level or pressure variations to affect β for a gas?

2-60 The average value of β for a perfect gas over a given temperature range (from T_1 to T_2) would be defined as

$$\bar{\beta} = \frac{1}{\Delta T}\int_{T_1}^{T_2} \beta(T)\, dT$$

Find $\bar{\beta}$ for air for the range $T_1 = 32°\text{F}$ to $T_2 = 212°\text{F}$ and compare the result with the value given in Problem 2-64. Comment on the comparison.

2-61 The thermal expansion coefficient for a gas at constant pressure would be

$$\beta - \frac{1}{V}\left(\frac{\partial V}{\partial T}\right)_p = -\frac{1}{\rho}\left(\frac{\partial \rho}{\partial T}\right)_p$$

Determine β for a perfect gas.

2-62 What is the value of β for air at 32°, 100°, 212°?

2-63 Plot β versus T as determined from the equation of Problem 2-61 for air between 0 and 250°F. (Check the results with those given in Table D-2 and comment.)

2-64 The handbook value of the expansion coefficient (change in volume per unit volume per degree centigrade) for air at a constant pressure of 1 atm, for a temperature range of 0 to 100°C, is given as 3.67×10^3/°C. Discuss the significance of this value. The corresponding value of $\Delta V/V$ per degree Fahrenheit would be $\beta = 2.04 \times 10^{-3}$/°F in the range 32 to 212°F.

2-65 What possible fluids would have desirable property behavior for the application described in Example 2-3? First consider and list the necessary operating characteristics that the indicating gage must have for this application, then consider the required fluid properties to match these gage characteristics.

2-66 A plot of vapor pressure as a function of temperature (saturation liquid–vapor curve) is presented for some fluids in Appendix D. Consider some of these fluids for charging the thermometer bulb of Example 2-3 with respect to the temperature and altitude variations required, and specify the fluid that you believe would give the best gage characteristics. Justify the choice.

2-67 If the thermometer bulb in Example 2-3 is charged with Freon at room temperature and saturation pressure, find the pressure range for the Bourdon-tube element corresponding to the operating range 100 to 240°F. What is the relative error produced by a change of ambient atmospheric pressure corresponding to an altitude change from sea level to 12,000 feet?

2-68 Why does the air cool, and why are updrafts formed, when it rains?

CHAPTER 3

CONTENTS

FLUID STATICS

3.1 INTRODUCTION

Fluids at rest, or in motion as a solid mass, are studied to determine hydrostatic forces and pressure distributions. Static forces in tanks, in hydraulic actuators, and on dams and ships are necessary in the design of structures and equipment. The study of the fluid statics of a pressure field also yields useful scalar-field formulations. The study and development of fluid mechanics involves the same procedures that are used for analyzing the behavior of solids. The behavior of a fluid at rest is more easily analyzed than that of a solid because no friction forces act on the fluid.

Only two types of forces can act on a fluid element or fluid mass: These are *surface forces* that are the result of surface stresses, and *body* or mass-acting *forces*, the magnitudes of which are proportional to the mass of fluid. The surface stresses are pressure (normal force per unit area) and shear (tangential force per unit area). An additional force can be provided by surface tension, the line stress at the edge of a surface element when two fluids or fluids and a solid are in contact, as discussed in Chapter 2. For a fluid in static equilibrium, pressure is the only surface force to be considered. Gravity and electrostatic force[1] are two examples of body forces.

A force balance will then involve both surface and body forces.

The mathematical relations to be derived for determining the variation of pressure, density, and so on, in a static fluid are those that express the condition for no motion. Consequently, the sum of all forces acting on the body must be in equilibrium. It is not necessary to consider torque equilibrium, since a body can be rotated only by applying a torque with tangential force, and it is not possible to apply a tangential force or stress to a fluid body at rest and have it remain at rest.

[1] An electrostatic force is produced by the action of an electrostatic field on a charged body of fluid. The charge capability of the fluid is a molecular, consequently a mass, property.

3.2 PRESSURE AS A POINT FUNCTION: Pressure Field

Since a static fluid can sustain no shear stress, a normal or pressure force is the only force applicable to a fluid at rest. Pressure as a stress is defined as a conservative (nondissipative) normal force per unit area.[2] Pressure within a fluid will be shown to be independent of orientation or direction of a fluid element. If it were possible to move a body through a closed path within a fluid field without the action of any viscous forces, then the work $\oint \vec{F}_p \cdot \vec{ds}$ evaluated around any closed path would be zero. This condition defines the pressure-force field $\vec{F}_p$ (and therefore the pressure field) as *conservative*.

To illustrate the idea that the pressure at a point within a fluid is independent of direction of orientation of any arbitrary surface within the fluid,[3] consider the small tetrahedron within a static fluid, as shown in Fig. 3-1. If this fluid element is at rest, the system consisting of the five forces shown in Fig. 3-1 is in equilibrium. *The pressure force on the fluid volume always acts inward toward the fluid.* Force is a vector quantity and pressure is a scalar quantity. Therefore, the *plane* area, which is by nature a vector quantity, determines the direction of the pressure *force.* A plane area or area element A is a vector quantity by virtue of its surface area (magnitude), and the only unique direction that can be assigned to it, its normal. The convention of positive direction is adopted as *outward* from the enclosed fluid volume. Since the pressure force acts *inward,*

$$\vec{F}_p = -p\vec{A}$$

The minus sign appears because of these assumed conventions. In the case where A is not a simple planar surface, it has no single unique direction, so the force must be determined from

$$\vec{F}_p = -\iint p\,\vec{dA}$$

The above equation is, in general, difficult to evaluate because of the difficulties of integrating a vector quantity. Therefore, it is necessary to work with the components of the vector in each of the coordinate directions.

[2] On the six surfaces of a coordinate fluid element cube in an arbitrary state of motion, there are three normal-force and six shear-force components. Since nine stress elements are required to determine completely the state of stress of the fluid element, stress is a tensor quantity. For a fluid at rest or moving without shear stresses (no *relative* fluid motion), the normal or pressure stress in the only stress possible. It will be shown that the magnitude of the pressure stress is independent of orientation, so the three components of pressure force at a point must be equal. The general stress tensor then reduces to the special case of a spherical tensor. The pressure-stress field is then isotropic and behaves as a scalar field.

[3] This observation was first made and proved by Blaise Pascal (1623–1662), a French philosopher and mathematician who had a hand in founding the calculus.

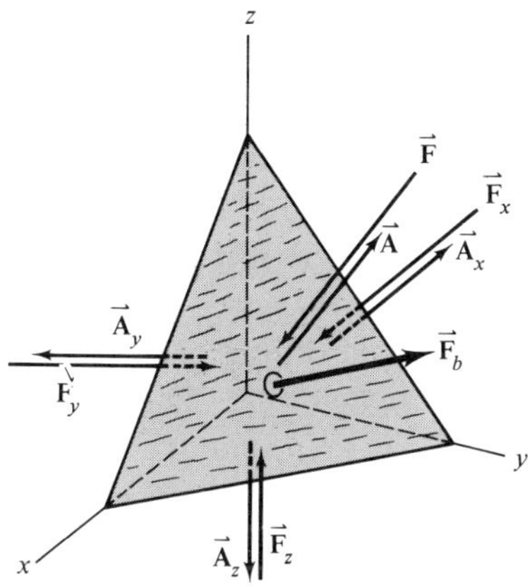

Fig. 3-1. Pressure forces acting on a fluid element in relative equilibrium with the surrounding fluid.

In the y direction, $\vec{F}_y = -p_y\vec{A}_y$ is the force acting on the left-hand face (where p_y denotes the pressure there). $\vec{A}_y$ may be written $-\hat{y}A_y$, where $\hat{y}$ denotes a unit vector in the direction of the positive y axis; then $\vec{F}_y = \hat{y}(p_yA_y)$, or the magnitude of the force on the left-hand face, is $F_y = p_yA_y$. F_y is positive, denoting that it acts in the direct of the positive y axis.

The y component of $\vec{\mathrm{F}}$ on the slant face [denoted as $(\vec{\mathrm{F}})_y$], may be dealt with in the same manner, maintaining the same sign and direction conventions. This component is denoted as $(\vec{\mathrm{F}})_y = -(p\vec{\mathrm{A}})_y$, and may be written $(\vec{\mathrm{F}})_y = \vec{\mathrm{F}}\cos(\vec{\mathrm{A}}, \hat{y})$, where $(\vec{\mathrm{A}}, \hat{y})$ denotes the angle between the vectors $\vec{\mathrm{A}}$ and $\hat{y}$ measured in the direction from $\vec{\mathrm{A}}$ into $\hat{y}$. In the illustration $(\vec{\mathrm{A}}, \hat{y})$ is a positive angle smaller than $90°$, so its cosine is positive. Using the value $-p\vec{\mathrm{A}}$ for $\vec{\mathrm{F}}$, $\mathrm{F}_y = -p\mathrm{A}\cos(A, y)$. It is now observed that $\mathrm{A}\cos(A, y) = A_y$, since it has the magnitude of A projected onto the x-z plane; then $(\mathrm{F})_y = -pA_y$. For static equilibrium in the y direction, summation of the y-component forces gives $p_yA_y - pA_y + \mathrm{F}_{by} = 0$. F_{by} is the y component of the body force $\vec{\mathrm{F}}_b$, which can be written $\vec{\mathrm{F}}_b = \vec{\mathrm{f}}_b\rho\,\Delta V$; $\vec{\mathrm{f}}_b$ is the body force per unit mass and ΔV is the element volume, $\Delta V = \Delta x\,\Delta y\,\Delta z/6$ or $\Delta V = A_y\,\Delta y/3$. Making these substitutions gives

$$p_yA_y \quad pA_y + \frac{f_{by}\rho A_y\,\Delta y}{3} = 0$$

In the limit as $\Delta V \to 0$ (which implies that Δx, Δy, and Δz must each $\to 0$), $p_y = p$. A similar process gives $p_x = p = p_z$, indicating that the pressure at a point is independent of direction and therefore is a scalar.

The pressure exerted by a perfect gas on a surface is produced by the "elastic rebounding" of the gas molecules from the surface. The impact and rebound produces an impulse and the impulses of the large number of impulses per second yields an integrated effect over a time period, or a force $F = (1/\tau)\int_\tau F\,dt = (1/\tau)\int_\tau d(mv)/g_c$. Since both the speed and direction of the molecular motion are random, the pressure effect has no preferred orientation. The pressure is then indicated to be a scalar quantity from microstructure considerations.

3.3 STATIC EQUILIBRIUM OF A FLUID

The conditions necessary for equilibrium of a fluid element are the same as those for a solid body. The static pressure force and the force due to the attraction of gravity are the only forces to be considered. The pressure field will be assumed to be constant with respect to time; that is, the pressure can vary from point to point in space but will not vary with time at any fixed point. The pressure is then expressed as a function of position in space as $p(\vec{r})$ or $p(x_i)$, where $\vec{r}$ is a vector location coordinate from an arbitrary reference coordinate system origin, and x_i implies the three coordinate components of $\vec{r}$ or the three coordinates of the point at which the pressure p exists (Section 2.4). Over a differential space distance, $d\vec{r}$, the difference in pressure is designated as dp.

The body force acting on a mass element Δm will be $\vec{f}_b\,\Delta m$, where $\vec{f}_b$ is the body force per unit mass (therefore specific force, or force density), and $\vec{f}_b(\vec{r})$ expresses the body force field. This force may vary from one point to another in space; therefore, it forms a *vector field.*

EQUILIBRIUM OF A FLUID ELEMENT

If a fluid element is at rest within a surrounding body of fluid, the sum of all forces on that element is zero. Consider first a complete body of fluid (perhaps the fluid within a container) under the influence of a net body force field $\vec{f}_b$ in Fig. 3-2. If the fluid is at equilibrium, then every element within the overall fluid mass must be at rest. A surface pressure force exists on each surface of the differential elemental cube. The element is in equilibrium under the action of the body and surface forces. If the total force system is in equilibrium, forces in each of the component directions must also be in equilibrium. The condition for static equilibrium in the y direction is $\sum \vec{F}_y = 0$. The pressure is $p(x, y, z)$ and $\vec{f}_b$ is $\vec{f}_b(x, y, z)$ *at the center of the element.* If a vector field is a function of position in space, each component of that vector field is also a function of position in space. The components are then considered as component fields, for example, $f_{b,y}(x, y, z)$. The body-force component in the y direction is $f_{b,y}\,\Delta m \equiv f_{b,y}\rho\,\Delta V$. The pressure force on the left-hand face is the force at the center

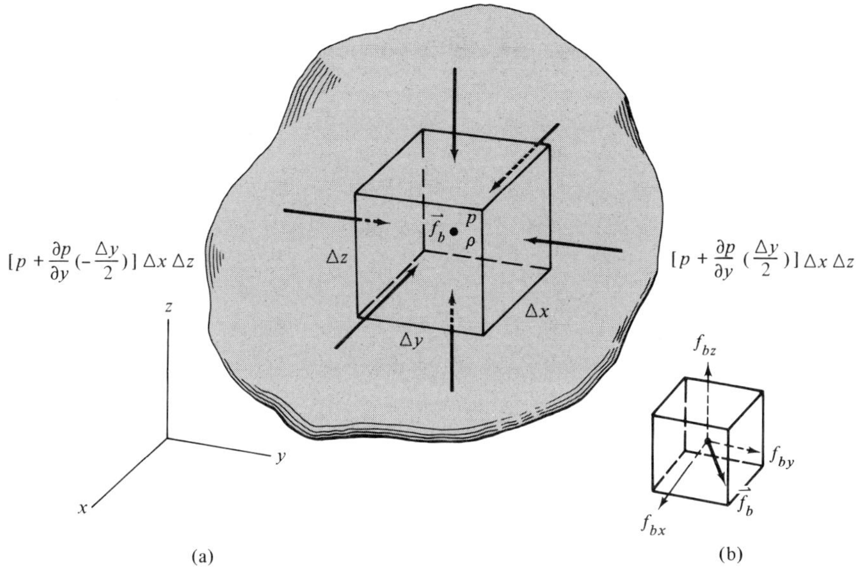

Fig. 3-2. Pressure and body forces acting on a fluid element. The field quantities (p, f_b, ρ, and so on) are specific at the reference point at the center of the element. The quantities on the faces are then determined by the first-order variations of these quantities from the center. The pressure forces are indicated in (a) and the body forces (which act through the mass center) are indicated in (b).

plus the rate of change of force in the y direction times the distance from the reference point (center) to the left-hand face,[4]

$$pA_y + \frac{\partial pA_y}{\partial y}\left(-\frac{\Delta y}{2}\right)$$

$A_y = \Delta x\,\Delta z$ does not change with y; therefore, this force is

$$+\left[p + \frac{\partial p}{\partial y}\left(-\frac{\Delta y}{2}\right)\right]\Delta x\,\Delta z$$

The pressure force on the right-hand face is

$$-\left[p + \frac{\partial p}{\partial y}\left(\frac{\Delta y}{2}\right)\right]\Delta x\,\Delta z$$

[4] Note that the sign (or direction of variation) of the quantity $\partial p/\partial y$ is not known unless $p(x, y, z)$ is known. Since $\partial p/\partial y$ is not yet determinable, it is implied as positive. The direction of travel from the center point to the left-hand face is negative, so the negative sign properly belongs with this distance, $\Delta y/2$.

The condition for equilibrium for these forces is $\sum F_y = 0$, or

$$+f_{by}\rho\,\Delta x\,\Delta y\,\Delta z + \left(p - \frac{\partial p}{\partial y}\frac{\Delta y}{2}\right)\Delta x\,\Delta z - \left(p + \frac{\partial p}{\partial y}\frac{\Delta y}{2}\right)\Delta x\,\Delta z = 0$$

or

$$\left(\rho f_{by} - \frac{\partial p}{\partial y}\right)\Delta V = 0$$

Since the mass element is arbitrary (any ΔV in the control volume of Fig. 3-2a) the y-component differential equation is

$$\rho f_{by} - \frac{\partial p}{\partial y} = 0 \tag{3-1}$$

The component differential equations in the x and z directions can be similarly determined as

$$\rho f_{bx} - \frac{\partial p}{\partial x} = 0 \qquad \text{and} \qquad \rho f_{bz} - \frac{\partial p}{\partial z} = 0 \tag{3-2}$$

A complete vector equation can be derived by multiplying each component equation by the corresponding unit vector in that direction and adding the component effects or component equations to give

$$\rho(\hat{x}f_{bx} + \hat{y}f_{by} + \hat{z}f_{bz}) - \left(\hat{x}\frac{\partial p}{\partial x} + \hat{y}\frac{\partial p}{\partial y} + \hat{z}\frac{\partial p}{\partial z}\right) = 0 \tag{3-3}$$

or

$$\rho\hat{x}_i f_{bi} - \hat{x}_i\frac{\partial p}{\partial x_i} = 0 \tag{3-4}$$

where $i = 1, 2, 3$ for the three component directions.[5] The vector quantity formed by the three partial derivative components as indicated in Fig. 3-3

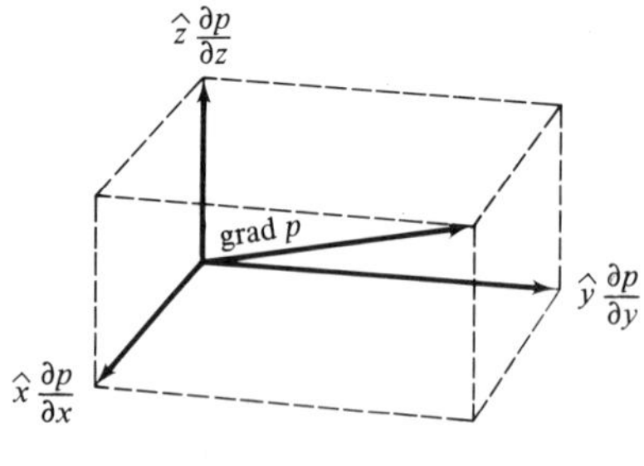

Fig. 3-3. Pressure gradient at a point showing the decomposition of grad p into component vectors along the x, y, and z axes.

[5] Equation (3-4) is a subscript-notation abbreviation of (3-3). A subscript notation convention exists whereby a repeated subscript on a single term implies that that term is to be summed over all values of the subscript, that is, for each of the three component directions.

gives the maximum rate of change of p at the point in question. This maximum rate of change is defined as the *gradient* of the quantity. The vector equation is then

$$\rho \vec{f}_b - \text{grad}\, p = 0 \tag{3-5}$$

The gradient of p may be written

$$\text{grad}\, p \equiv \left(\hat{x}_i \frac{\partial}{\partial x_i}\right)(p) \tag{3-6}$$

where $\hat{x}_i(\partial/\partial x_i)$ is a vector operator referred to as *del* and denoted by ∇; that is, grad $p \equiv \nabla p$. This same operator is also involved in other types of operations. Applied to a scalar, it produces a gradient of that scalar.

The gradient as a maximum derivative can be seen with the aid of Fig. 3-4. Consider the identity $\vec{dr} \cdot \text{grad}\, p = dp$, which can be proved by taking the product in the form

$$(\hat{x}_i\, dx_i) \cdot \left(\hat{x}_i \frac{\partial}{\partial x_i}\right) = dx_i \frac{\partial}{\partial x_i} \tag{3-7}$$

operating on p. With respect to Fig. 3-4, grad p depends only on the point at which it is measured. The pressure difference Δp (or dp) between the two surfaces is the same, independent of the points between which it is measured, and is therefore independent of $\vec{\Delta r}$ (or $\vec{dr}$). Two possible $\vec{\Delta r}$'s are indicated from point 1. The minimum $\vec{\Delta r}$ occurs when it is perpendicular to the surface $p =$ const. or $\vec{\Delta r}_{\min}$ is parallel to ∇p. Therefore, ∇p is perpendicular to $p =$ const.; ∇p is the maximum or normal spacial rate of change. It can also be interpreted as

$$\nabla p = \hat{n} \frac{dp}{dn}$$

where $\hat{n}$ is the normal direction to the surface $p =$ const. Since the surfaces are farther apart at point 1 than at point 2, the pressure is changing faster with respect to distance at 2 than at 1. The magnitude of the gradient is then greater at 2 than at 1.

The variation in position location, $\vec{dr}$, can be identified with a differential line segment of a path in space as designated by the space curve s; that is, $\vec{dr} = \vec{ds}$ (Fig. 3-4c). If s is a streamline, $dp = \nabla p \cdot \vec{ds}$ is the differential pressure along this streamline.

Equation (3-5) is the vector form of the differential equation expressing the condition of static equilibrium for a fluid element or fluid field. To obtain the specific pressure field required to balance a certain body force field, it is necessary to integrate this equation. This integration to determine

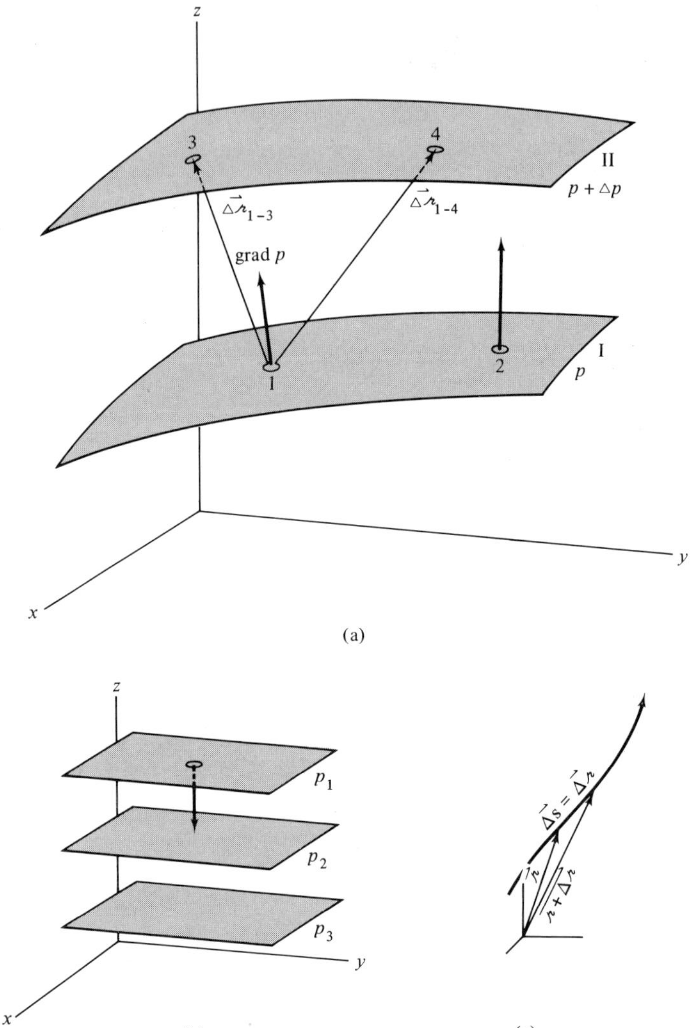

Fig. 3-4. Relation between the gradient of a scalar quantity (pressure) and surfaces on which the quantity is a constant : (a) general pressure field ; (b) earth's gravitational field ; (c) relation between line increment and position vector increment.

p requires that the $\vec{f}_b$ field and the ρ field be known as functions of the space coordinates (for example, x, y, z).

As an example, if $\vec{f}_b$ is the force due only to the earth's gravitational field acting on the fluid, this force acts vertically downward (toward the center of the earth). Specifying the $+z$ axis vertically upward (the altitude), the body force per unit mass acting on the fluid is then $\vec{f}_b = \hat{z} f_{bz}$ only; $f_{b,x}$ and $f_{b,y}$ are zero. The three component equations of (3-5) are then

$$\frac{\partial p}{\partial x_1} = 0$$

$$\frac{\partial p}{\partial x_2} = 0$$

$$\frac{\partial p}{\partial x_3} = \frac{\partial p}{\partial z} = \rho f_{bz}$$

where x_1 and x_2 are any two orthogonal coordinates (for example, x and y in a cartesian coordinate system).[6] In this particular case, where the rates of change of p in these two component directions (partial derivatives) are zero, the pressure is not a function of these two coordinates but is a function of z only. Therefore, the last of the above equations is a total differential equation. A gravitational force can be specified in terms of an equivalent acceleration $\vec{g}$,

$$\vec{f}_b = \frac{\vec{g}}{g_c} \qquad \text{or} \qquad \vec{f}_b = -\hat{z}\frac{g_e}{g_c}$$

where the negative sign indicates that the earth's gravitational force field ($\vec{g}_e$) acts in a direction opposite to that of the altitude coordinate z. These conditions give[7]

$$\frac{dp}{dz} = -\frac{\rho g}{g_c} \tag{3-8}$$

To integrate this relationship requires further inspection. If the fluid is compressible (air, for example), ρ is not a constant. From the equation of state, ρ is a function of p and T: A new variable T is introduced. Separating the variables as indicated gives

$$\frac{dp}{\rho(p, T)} = -\frac{g(z)}{g_c}\,dz \tag{3-9}$$

The strength of the earth's gravitational field varies inversely with the distance from the center of the earth (neglecting variations with latitude

[6] Because the earth is nearly spherical, a spherical polar coordinate system (r, φ, θ) would be a more logical choice. The above system can be so considered with x_1 or z defined as $r - r_0$, $dx_2 \equiv r\,d\varphi$, and $dx_3 \equiv r \sin\varphi\, d\theta$.

[7] The quantity $\rho g/g_c$, the gravity field force on a unit mass, is sometimes denoted as γ, the "specific weight" of the fluid. It is seen to be a variable quantity, depending on the fluid density and the magnitude of the gravitational field, g. In this text, γ is used to denote the specific energy ratio c_p/c_v only.

and longitude). Therefore, g is generally a function of z. However, g can be assumed to be about constant within the extent of the atmosphere. To evaluate (3-9) further, a relation for T in terms of p and ρ, or of z, is required. Some special cases of this equation will be considered in Section 3.5.

3.4 DERIVATION OF THE VECTOR INTEGRAL RELATION FOR A STATIC FLUID

Equation (3-5) for the pressure variation in a static fluid was obtained for a differential element of fluid. In order to obtain a "working formula," such a relation must be integrated. It is also possible to derive directly an integral relation for the overall fluid region of interest. Consider the fluid region indicated in Fig. 3-5, where the differential element previously

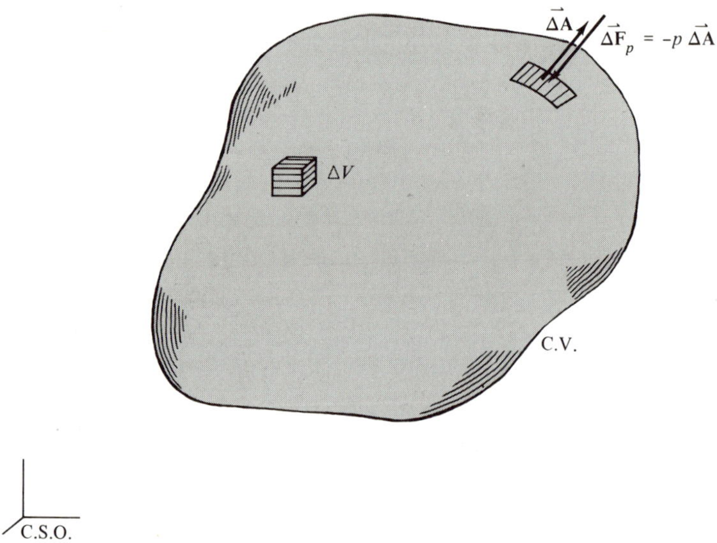

Fig. 3-5. Fluid region or control volume in static equilibrium with the surroundings.

considered is also shown. This particular indefinite-looking region might be the fuel within a collapsible fuel tank.

The pressure force exerted by the tank surface on the fluid is

$$\vec{F}_p = \oiint -p\, d\vec{A} \tag{3-10}$$

The symbol $\oiint$ implies integration over the entire enclosing surface of the fluid region.

The gravitational body force acting on the fluid region is

$$\vec{F}_g = \oiiint \frac{\rho \vec{g}}{g_c}\, dV \tag{3-11}$$

where $\oiiint$ implies integration over the entire fluid *control volume*.[8] The term *control volume* is given to a fluid region under consideration to which the principles of the mechanics or the conservation principles are applied.

The condition for static equilibrium of a fluid within the tank is that summation of the forces must be zero:

$$\sum \vec{F} = -\oiint p\, \vec{dA} + \oiiint \rho \vec{f}_b\, dV = 0 \tag{3-12}$$

An interesting result is obtained by comparing the results of the derivations of the differential equation (3-5) and the integral equation (3-12). The two results were derived for the same basic principles. The differential equation (3-5) in the form $\nabla p\, dV = \rho \vec{f}_b\, dV$ can be integrated over the control volume (assuming that the p and $\vec{g}$ fields are well-behaved, that is, finite, continuous, and single-valued) to give

$$-\oiiint \nabla p\, dV + \oiiint \rho \vec{f}_b\, dV = 0 \tag{3-5a}$$

which may be compared directly with (3-12) in the form

$$-\oiint p\, \vec{dA} + \oiiint \rho \vec{f}_b\, dV = 0 \tag{3-12a}$$

This comparison implies the mathematical equivalence

$$\oiint p\, \vec{dA} \equiv \oiiint \nabla p\, dV \tag{3-13}$$

This result (3-13) will be termed the *gradient theorem*. It is seen to be a very useful relation if, for example, the integral relationship (3-12) were derived first and it were desired to deduce a differential equation from it. It is not possible to deduce or write immediately a differential equation from (3-12), since both area and volume integrals are involved. Use of the integral identity relation (3-13) allows the area integral to be transformed

[8] The designation $\iiint$ is sufficient for a volume integration. The redundant designation $\oiiint$ is used here to designate a control volume integral, $\iiint_{\text{c.v.}}$.

into a volume integral, so that all terms of (3-12) can be written under the same integral sign:

$$\oiiint \left(-\nabla p + \frac{\rho g}{g_b} \right) dV = 0 \tag{3-14}$$

from which the differential equation

$$\nabla p = \frac{\rho \vec{g}}{g_b} \tag{3-15}$$

can be deduced.

Consider that the derivation procedure started with the differential element analysis. The process of integration implies adding all the pressure forces on each element to obtain the net effect on the whole control volume. At the interface between two adjacent elements, the pressure forces are equal and opposite and therefore cancel. This cancellation effect continues until the external boundary is reached. Adding all the forces on each element therefore results in obtaining just the pressure force at the external surface, or, mathematically,

$$\left.\oiiint \left[\underbrace{\frac{\partial p}{\partial x} dx}_{p \text{ AT ELEMENT FACE}} \underbrace{(\hat{x} dy\, dz)}_{dA_x} + \frac{\partial p}{\partial y} dy(\hat{y} dx\, dz) + \frac{\partial p}{\partial z} dz(\hat{z} dx\, dy) \right] \right\} \text{SUMMATION OVER ALL ELEMENTS}$$

$$\equiv \oiiint \text{grad } p\, dV = \underbrace{\oiiint p\, \overrightarrow{dA}}_{\text{NET SURFACE EFFECT}} \tag{3-16}$$

3.5 STATIC EQUILIBRIUM OF A FLUID IN THE EARTH'S GRAVITATIONAL FIELD

Since it can be shown that the earth's gravitational field does not vary appreciably within the limits of the atmosphere, g can be taken as g_0 (32.174 ft/sec^2 or approximately 32.2 ft/sec^2). The static equilibrium condition (3-8) is then

$$\frac{dp}{dz} = -\frac{\rho g_0}{g_z} \tag{3-17}$$

Some special cases and solutions of this equation will be considered:

CASE I

CONSTANT DENSITY; CONSTANT ρ

This case gives directly

$$p_2 - p_1 = -\frac{\rho g_0}{g_c}(z_2 - z_1) \tag{3-18}$$

An arbitrary altitude datum may be chosen as $z_1 = 0$, giving

$$p = p_0 - \frac{\rho g_0}{g_c} z \tag{3-19}$$

The limitations on the relation are that (1) variations in x, y, z are small enough that variations in g are negligible, and (2) density variations are negligible for the range of pressure or altitude encountered.

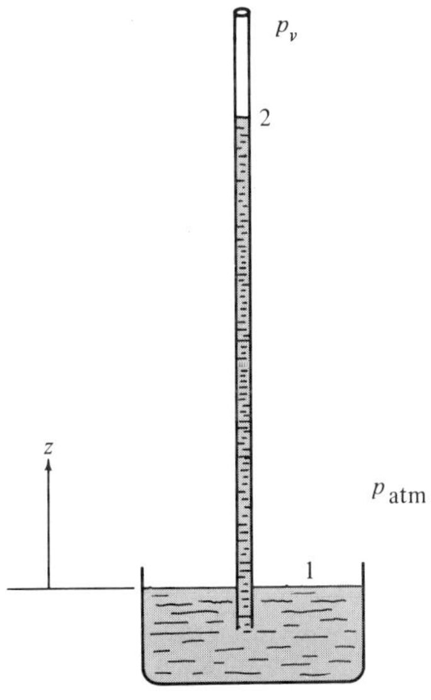

Fig. 3-6. Torricelli tube; mercury barometer.

Consider a mercury barometer (Torricelli tube)[9] as represented in Fig. 3-6 as an application of this relationship. The pressure at z_1 is atmospheric, p_a; the pressure in the space at the top of the tube can only be the vapor pressure of mercury, p_v, which is determined by the ambient

[9] Evangelista Torricelli (1608–1647), Italian. Performed important analytical and experimental investigations in astromechanics, fluid mechanics, and ballistics.

temperature, T_a. The static equilibrium equation applied to the mercury column above the free surface altitude gives

$$p_v - p_a = -\frac{\rho g_0}{g_c}(z_2 - z_1) \tag{3-20}$$

Calling $z_2 - z_1$ the height of the mercury column, H, the local atmospheric pressure is

$$p_a = \frac{\rho g_0}{g_c} H + p_v \tag{3-21}$$

The height H and the compressibility of mercury are sufficiently small that ρ, the density of the mercury, may be considered to be independent of H, or p_a. However, ρ will be a function of T_a. The vapor pressure is also a function of T_a and must be determined if an accurate determination of P_a is desired. Although g is essentially constant at g_0 over the distance H, it is a function of latitude, longitude, and altitude. The local value of g may be determined from U.S. Coast and Geodetic Survey or other tables. To determine local atmospheric pressure accurately, the correct values of ρ and p_v must be determined in addition to an accurate measurement of H. Gases may be considered as incompressible in some cases (see Problem 3–32).

Case II

COMPRESSIBLE FLUID; CONSTANT g

If the fluid is compressible, two additional equations among p, ρ, and T must be specified. Still using the barometer as an example, atmospheric pressure is the pressure resulting from the weight (force of attraction between the earth and air molecules) of the column of air above the free surface of the mercury pool. Air may be assumed to be a perfect gas with the equation of state

$$\frac{p}{\rho} = \frac{R_0}{\mathscr{M}} T \tag{3-22a}$$

or

$$\frac{p}{\rho} = RT \tag{3-22b}$$

where R_0 is the universal gas constant, $\mathscr{M}$ the average molecular mass number for air (28.9), and R the specific or "engineering" gas constant ($R = 53.3$ ft-lb$_f$/lb$_m$-°R). Since another variable, T, has been added, another relationship involving T is needed.

CASE IIA

ISOTHERMAL FLUID; CONSTANT g

If it is assumed that the fluid is at constant temperature (for example, an isothermal atmosphere), (3-17) may be written

$$\frac{dp}{\rho} = -\frac{g_0}{g_c}\, dz$$

This equation may be put into the following form, using (3-22b):

$$RT \int_{p_0}^{p} \frac{dp}{p} = -\frac{g}{g_c} \int_{z_0}^{z} dz \tag{3-23}$$

Integration of (3-23) gives

$$p = p_0 \exp\left(\frac{-g_0 z}{g_c RT}\right) \tag{3-24}$$

indicating that the pressure in a compressible, isothermal, perfect gas at rest decreases exponentially with altitude.

CASE IIB

ADIABATIC FLUID; CONSTANT g

Again assume a perfect gas, but the adiabatic "process" relation implies no heat transfer between layers of the atmosphere. The barotropic[10] relation for this isentropic process is

$$\frac{p}{\rho^\gamma} = \frac{p_0}{\rho_0^\gamma} = \text{const.} \tag{3-25}$$

where $\gamma \equiv c_p/c_v$. For this case, integration of (3-17) gives

$$\frac{(p_0)^{1/\gamma}}{\rho_0} \int_{p_0}^{p} p^{-1/\gamma}\, dp = -\frac{g_0}{g_c} \int_{0}^{z} dz \tag{3-26}$$

which, when evaluated, gives

$$p = p_0\left(1 - \frac{(\gamma - 1)g_0 z}{\gamma g_c RT_0}\right)^{\gamma/(\gamma - 1)} \tag{3-27}$$

or p decreases with a $\gamma/(\gamma - 1)$ power function of z.

[10] The relation $p(\rho)$ or $\rho(p)$ is termed *barotropic.*

CASE IIc

CONSTANT "LAPSE-RATE" ATMOSPHERE; CONSTANT g

It has been determined experimentally that the temperature in the atmosphere varies inversely with altitude up to an altitude of about 35,000 ft. This observation is expressed as $dT/dz = -\lambda$, or

$$T = T_0 - \lambda z \tag{3-28}$$

where λ is the lapse rate and has a value of $\lambda = 0.003566°\text{R/ft}$ (or °F/ft). Again assuming a perfect gas and substituting (3-17) into (3-28) gives

$$\int_{p_0}^{p} \frac{dp}{p} = -\frac{g}{Rg_c}\int_0^z \frac{dz}{T_0 - \lambda z} \tag{3-29}$$

Integration of (3-29) yields

$$\frac{p}{p_0} = \left(1 - \frac{\lambda z}{T_0}\right)^{g/g_c\lambda R} \tag{3-30}$$

The atmosphere from 36,000 ft to about 80,000 ft is nearly isothermal at −69.7°F. Check calculations with the above formulas in the appropriate ranges show very good agreement with standard atmospheric tables.[11]

In cases where changes in g are important for navigational purposes, the variations with longitude and latitude are of the same order as the variations in altitude. Tabulated values must then be used with numerical procedures to predict possible pressure variations. Even then, meteorological variations usually overshadow geodetic variations.

3.6 STABILITY AND EQUILIBRIUM OF THE ATMOSPHERE

When a refrigerator door is opened, the cold air "spills" onto the floor. Therefore, it is deduced that the atmosphere would be in a state of stable equilibrium if the temperature increases. If a small cold air mass were displaced upward into the warm air surrounding it, it would tend to "sink" to its initial equilibrium state. This argument is equally valid for compressible and incompressible fluids. However, as a mass of gas (air) is moved from one region to another of different pressure, its temperature decreases (or increases) as it expands (or is compressed) by the surrounding fluid pressure field. The density of the gas may vary significantly with

[11] Manual of the ICAO Standard Atmosphere; *NACA Report 1235* (also *NACA Technical Note 3182*). R. A. Minzner, W. S. Ripley, and T. P. Condron, *U.S. Extension of the ICAO Standard Atmosphere*, U.S. Department of Commerce, Weather Bureau, Washington, D.C., 1958.

altitude because of variations in atmospheric pressure and temperature. Therefore, the equilibrium condition involves pressure as well as temperature considerations. To investigate the equilibrium conditions for the atmosphere, the equilibrium process to be considered must first be well-defined. If the atmosphere is in equilibrium at rest, an equilibrium force and weight balance will exist, as indicated in Fig. 3-7a. If this particular

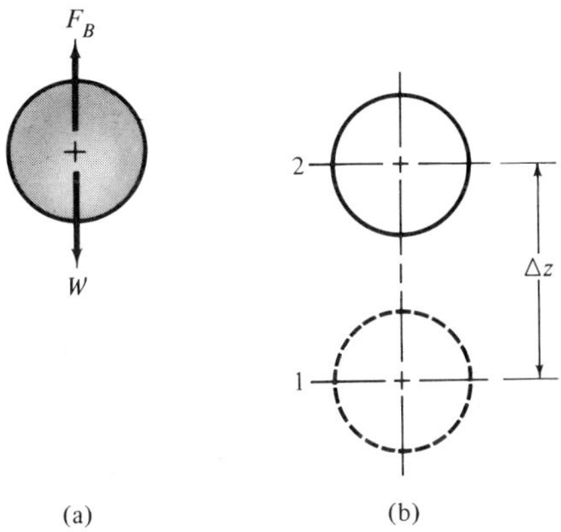

Fig. 3-7. (a) An element of the atmosphere in equilibrium with its surroundings; (b) an element displaced a distance Δz.

element is surrounded by an imaginary surface[12] and displaced a distance Δz, the following conditions can be defined for stability of the element:

1. Stable: If a small displacement is given to a system initially in stable equilibrium, and the system tends to return to its initial state, the initial condition is said to be one of stable equilibrium. The potential energy of the system is increased by the displacement.
2. Neutral: If the system remains at equilibrium in the displaced state such that no net forces tend to return it to its initial state, or to drive it toward another state, the initial state is said to be one of neutral equilibrium. Its potential energy is not changed.
3. Unstable: If the state of the system after the small displacement has been made is one where the acting forces tend to further displace it to a condition different from the initial equilibrium state, the initial state is said to be one of unstable equilibrium. Displacement tends to

[12] The surface will allow adiabatic expansion to a volume at the new pressure. Since the conductivity of air is small, heat transfer is neglected. The conditions are similar to those that exist in the atmosphere, where a mass of air is convected to a different altitude and, like a limp balloon, expands (or is compressed) according to the pressure variation. These assumed conditions are disturbed by atmospheric turbulence and mixing but can be considered to hold initially for the stability determinations given here.

"upset" the system from its initial state. Its potential energy is decreased by the displacement.

If a mass of air initially located at an altitude z_1 where the pressure, density, and temperature are p_1, ρ_1, and T_1 (Fig. 3-7b) is displaced upward to a point where the surrounding conditions are p_{2a}, ρ_{2a}, and T_{2a} (where the subscript a refers to the surrounding or ambient atmospheric conditions), the gas will expand adiabatically to $p_{2s} = p_{2a}$, ρ_{2s}, and T_{2s}. The subscript s is used to imply the slow, adiabatic, or isentropic expansion of the gas volume; the air can be assumed to be a perfect gas for both the surrounding atmosphere and the displaced mass element. The mass element expands according to the isentropic process relationship p/ρ^γ = const. The surrounding atmosphere is assumed to behave arbitrarily according to the polytropic process p/ρ^n = const., where n is any arbitrary number, possible variations of which will be considered later. For example, $n = 1$ implies an isothermal process and $n = \gamma$ implies an isentropic process.[13]

The buoyancy force and weight in the displaced condition are then

$$F_{B2} = \frac{\rho_{2a} g}{g_c} V_2 \qquad \text{and} \qquad W_2 = \frac{\rho_{2s} g}{g_c} V_2$$

or
$$(W - F_B) = \frac{p_2 g V_2}{g_c R}\left(\frac{1}{T_{2s}} - \frac{1}{T_{2a}}\right) = \frac{p_2 g V_2}{g_c R T_1}\left[\left(\frac{p_1}{p_2}\right)^{(\gamma-1)/\gamma} - \left(\frac{p_1}{p_2}\right)^{(n-1)/n}\right] \tag{3-31}$$

The stability conditions are then

$$\begin{aligned} (W - F_B)_2 &> 0 \qquad \text{stable } n < \gamma \\ (W - F_B)_2 &= 0 \qquad \text{neutral } n = \gamma \\ (W - F_B)_2 &< 0 \qquad \text{unstable } n > \gamma \end{aligned} \tag{3-32}$$

In the above example, $p_1/p_2 > 1$. Therefore, if $W - F_B$ is to be positive (or greater than zero), $\gamma > n$ or $n < \gamma$.

Comparison of the pressure variations for the adiabatic, polytropic, and constant lapse-rate atmosphere indicates that an equivalent lapse-rate relation can be determined for each type of variation; that is,

$$\lambda_n = \frac{n-1}{n}\frac{g}{g_c R} \text{ (°F/ft)} \qquad \text{and} \qquad \lambda_s = \frac{\gamma - 1}{\gamma}\frac{g}{g_c R} \tag{3-33}$$

[13] If $n = \gamma$, a condition of neutral stability is obtained, since conditions for the displaced mass and the surrounding atmosphere are identical.

or, conversely, if λ is given, an equivalent polytropic exponent can be determined. For example, if $\lambda = 0.00357°F/ft$ for the observed "standard" atmosphere, then $n_\lambda = 1.24$, indicating a stable atmosphere.

The standard atmosphere is represented in Fig. 3-8. The lower atmosphere is stable. The isothermal portion with $n = 1$ is also stable. For the upper atmosphere, where T increases with altitude, $n < 1$, which also indicates stability.

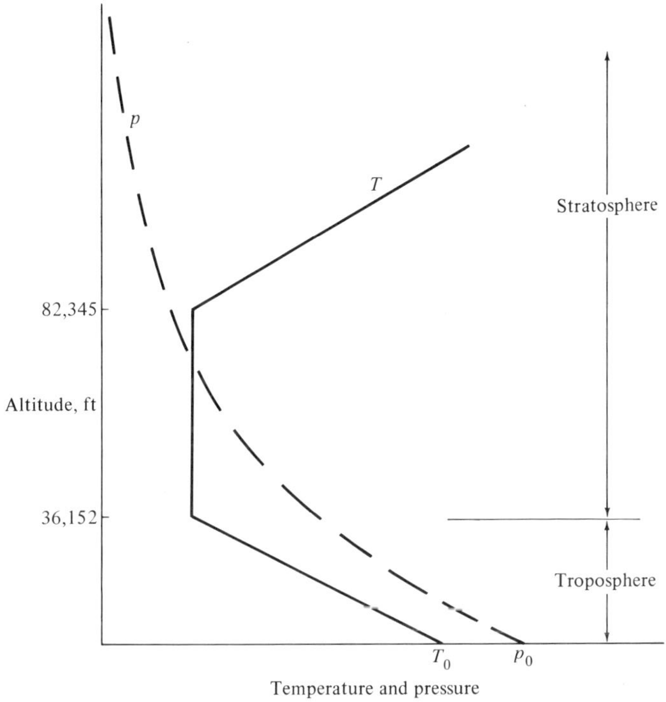

Fig. 3-8. Representation of standard atmospheric properties as a function of altitude. The density ρ_a as a function of altitude is similar to p_a.

Instabilities such as thunderstorms, winds, and so on, are caused by nonuniformities in geography. Land and water masses, mountains, and so on, have different heating and cooling rates under the influence of cyclic heating and cooling periods from day to night. These and other effects produce density and pressure variations yielding wind and storm currents. The earth's rotation also produces the "westerlies" and rotational currents that mix with polar and equatorial air masses. Density variations in sea-water masses produced by differences in geographical temperature combined with sloping ocean floors give rise to ocean currents in a similar fashion.

3.7 PRESSURE MEASUREMENT

The relation developed for the static equilibrium of a fluid under the influence of the earth's gravitational force field was

$$\frac{dp}{dz} = -\frac{\rho g}{g_c}$$

where z is the vertical coordinate. For liquids, it is sometimes advantageous to deal with a coordinate depth vertically downward from some reference

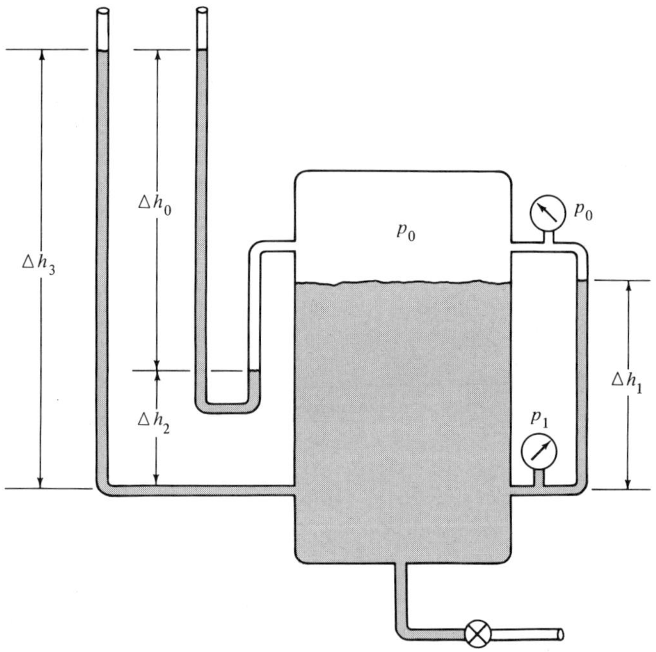

Fig. 3-9. Pressure in a liquid showing the dependence on depth.

point or free surface. If this coordinate is designated as h, the relation between the altitude z and depth h is $dz = -dh$. The hydrostatic variation of pressure with depth is then given by the relation

$$\frac{dp}{dh} = \frac{\rho g}{g_c} \tag{3-34}$$

With ρ = const., (3-34) gives

$$p = \frac{\rho g}{g_c} h + \text{const.}$$

or

$$p - p_0 = \frac{\rho g}{g_c}(h - h_0) \tag{3-35}$$

where $p_0(h_0)$ is a known reference pressure at the depth h_0. This relation is historically known as Pascal's principle of hydrostatic pressure variation, stating that pressure varies directly with depth.

Figure 3-9 illustrates a liquid container with a gas or vapor at a pressure $p = p_0$ in the space above the liquid free surface. Since the pressure variation within the gas volume is negligible (because of the small density of the gas relative to that of the liquid), the free surface pressure can be taken as p_0. The pressure p_1 at the depth h is then

$$p_1 = p_0 + \frac{\rho g}{g_c}\Delta h_1 \tag{3-36}$$

PRESSURE GAGES

The gages measuring p_0 and p_1 can measure either absolute or gage pressure, depending on how they are constructed. *Absolute pressure* is measured with respect to *absolute zero. Gage pressure* is measured with respect to the *local ambient* atmospheric pressure. Most Bourdon or expanding-element gages indicate gage pressure, since they deflect in proportion to the difference between the external (local ambient) and internal applied pressure force. The difference between absolute pressure and the pressure designated as "gage pressure" is then the local ambient atmospheric pressure:

$$p_{\text{gage}} = p_{\text{abs}} - p_{\text{loc atm}}$$

An ordinary Bourdon gage is illustrated in Fig. 3-10a. This type of gage can be converted to measure differential pressure by sealing the case and applying the second pressure to the case, as illustrated in Fig. 3-10b. The lower pressure (p_2) is usually connected to the case, since the entire case must withstand this pressure. By completely evacuating the case, the pressure (p_1) will be measured with respect to absolute zero; thus the gage can be made to read absolute pressure.

An aneroid gage (or aneroid barometer) is a small diaphragm gage constructed as indicated in Fig. 3-10c. The region under the diaphragm is evacuated to indicate atmospheric pressure. This type of gage can be used to measure absolute pressure if it is enclosed and the pressure to be determined is applied to the case. Aneroid and accurate absolute-pressure gages are usually equipped with a mechanism that automatically corrects for differential thermal expansion of the gage parts.

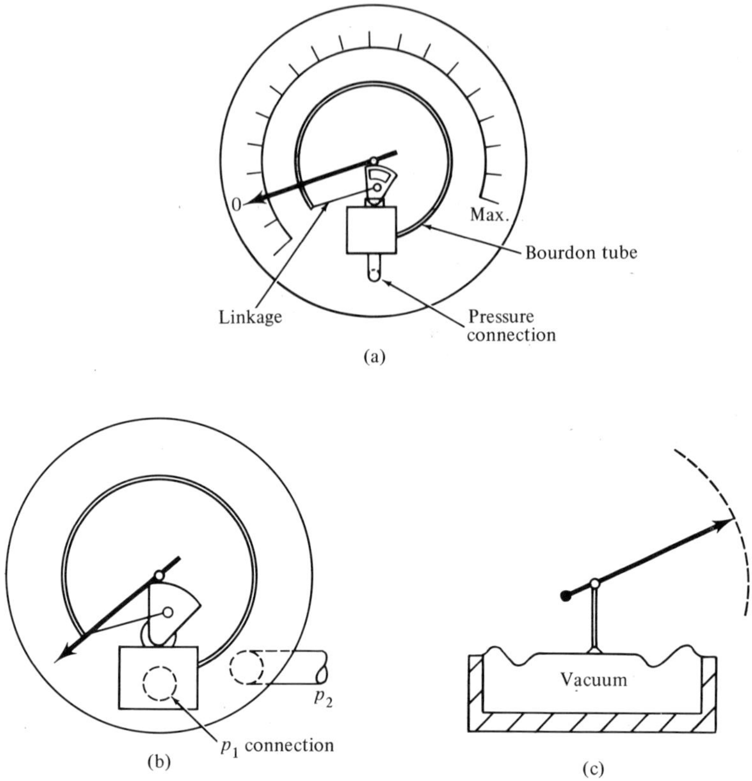

Fig. 3-10. (a) Bourdon-type pressure gage; (b) Bourdon pressure gage for measuring differential pressure; (c) aneroid gage indicating absolute pressure.

If the pressure gages in Fig. 3-10 are the "ordinary" type that indicate gage pressure, then the absolute pressure corresponding to the pressure p_0 indicated on Fig. 3-9 is

$$p_{0,\text{abs}} = p_{0,\text{gage}} + p_{\text{atm}}$$

The atmospheric pressure varies with geographic position and weather conditions. The absolute pressure can be determined if the local atmospheric pressure or "local barometer" reading is known. The barometer is usually specified in inches of mercury reduced to "standard" sea-level conditions (that is, at standard gravity and temperature), so the local atmospheric pressure is given by

$$p_{\text{atm}} = \frac{\rho_{\text{Hg}} g_0}{g_c} \frac{H_{\text{bar}}}{12} \tag{3-37}$$

(g_0 is the standard equatorial reference value of g, the earth's gravitation, and is numerically equal to g). The specific gravity of mercury is 13.60. A barometer reading of 29.92 in. Hg will give the local atmospheric pressure as

$$p_{\text{atm}} = \frac{(13.60)(62.43)(29.92)}{12} = 2116.5 \text{ lb}_f/\text{ft}^2$$

or $p_{\text{atm}} = 14.696 \text{ lb}_f/\text{in.}^2$ (or approximately 14.7 psia[14]) is the standard atmospheric pressure.

Pressures below zero gage (that is, negative gage pressure) are designated "vacuum" with units of lb/in.^{2} vacuum. A usual unit of vacuum is the magnitude of the pressure expressed as an equivalent inches-water-vacuum [or millimeters-mercury-vacuum, mm Hg, or torr (for Torricelli)]. The "vacuum" pressure is a gage measurement and therefore not an absolute quantity. Figure 3-11 illustrates these basic designations of pressure measurement.

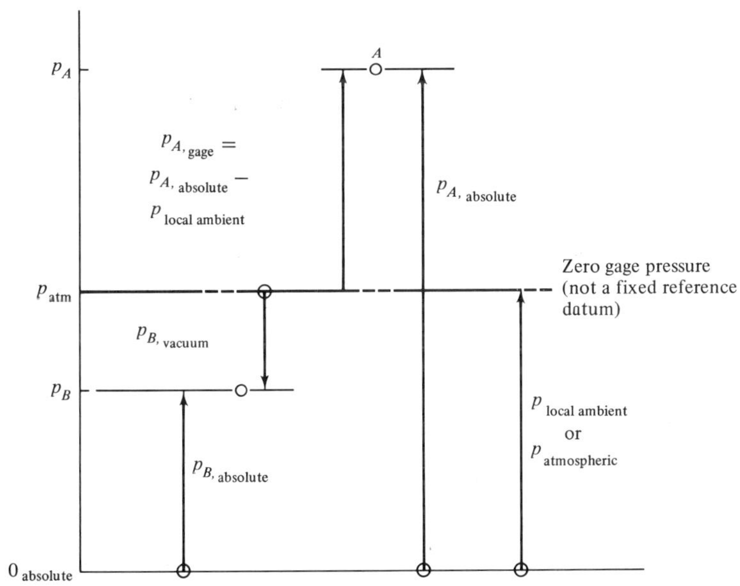

Fig. 3-11. Pressure measurement base references.

MANOMETERS

Manometer tubes that indicate pressure as a *differential head*, h, of manometer fluid are also shown connected to the tank of Fig. 3-9. The

[14] It is usual to specify gage pressure as "psi" (unspecified) and *absolute pressure* as "psia," where the "absolute" is specifically indicated.

fluid shown in both manometers of Fig. 3-9 is the same as that in the tank. The pressure on the right leg of the manometer connected to the gas volume above the liquid is p_0, and the pressure on the open end is p_a. Therefore,

$$p_{0,\mathrm{abs}} - p_a = \frac{\rho g}{g_c}\,\Delta h_0 \tag{3-38}$$

or

$$p_{0,\mathrm{abs}} - p_a = p_{0,\mathrm{gage}} = p_0 = \frac{\rho g}{g_c}\,\Delta h_0$$

A manometer then indicates differential pressure and an open manometer indicates gage pressure. The gage pressure is $p_1 = (\rho g/g_c)\,\Delta h_2$ and the differential pressure is

$$p_1 - p_0 = \frac{\rho g}{g_c}(h_3 - h_0) = \frac{\rho g}{g_c}\,\Delta h_1$$

It is emphasized that the hydrostatic pressure is a function of the fluid density, gravity, and the depth only and is independent of the shape of the container. This fact is evident from (1) the derivation, (2) the fact that p is a scalar, and (3) consideration of the physical setup shown in Fig. 3-12.

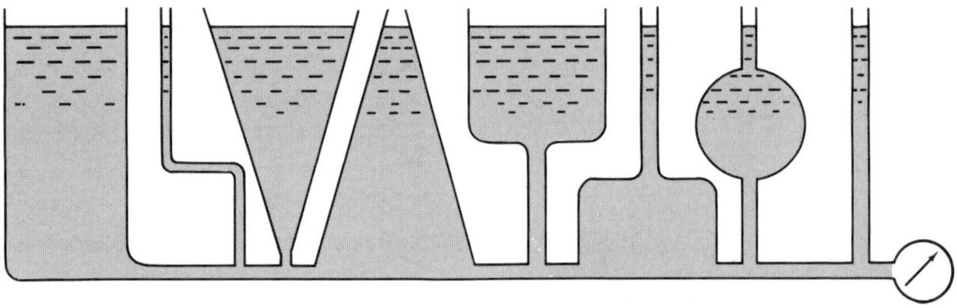

Fig. 3-12. Illustration of Pascal's principle that pressure is dependent on depth only and is independent of container geometry for static liquids.

Figure 3-13 illustrates a simple manometer used to determine differential pressure between two pipes or vessels. The pressure at the top of the left column of manometer fluid is the pressure in the pipe, p_1, plus the hydrostatic pressure increase from the pipe to the top of this column, that is, $p_1 + (\rho_1 g/g_c)h_a$. The pressure at the lower level of the manometer fluid column is this pressure plus the additional hydrostatic pressure $(\rho_m g/g_c)h_m$, or the pressure here is

$$p_1 + \frac{\rho_a g}{g_c}\,h_a + \frac{\rho_m g}{g_c}\,h_m$$

This pressure is the same as that at the interface between the manometer fluid m and the fluid 2 in the right-hand-column system, since it is at the

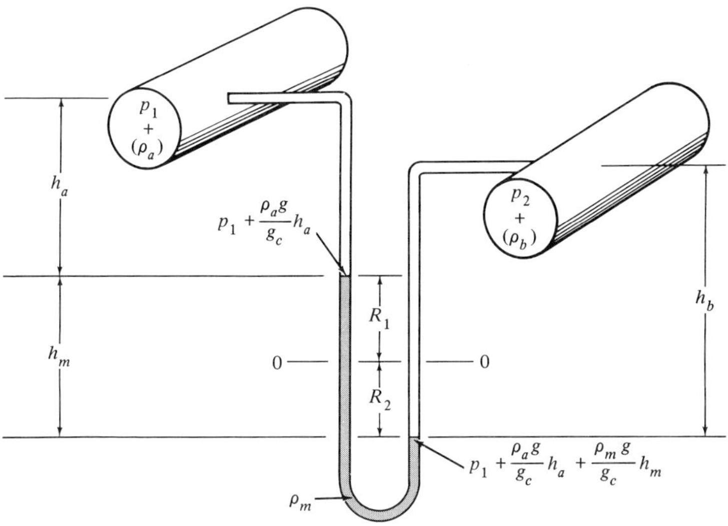

Fig. 3-13. Simple manometer between pipes or tanks.

same level in the same fluid. From this point to the center of the pipe 2, the hydrostatic pressure *decreases* by the amount $(\rho_2 g/g_c)h_b$, or the pressure at pipe 2 is

$$p_2 = p_1 + \frac{\rho_a g}{g_c} h_a + \frac{\rho_m g}{g_c} h_m - \frac{\rho_b g}{g_c} h_b \tag{3-39}$$

If one pressure (p_1 or p_2) is known, the other pressure can be determined by this procedure, or the differential pressure ($p_2 - p_1$) can be determined as a function of the geometry (h's) and fluid properties. (*Note procedure:* Start from one point as a reference and work through the system to the other point, where the second pressure exists. This method is the most direct and accurate.)

In most cases where this type of differential manometer is used, the zero pressure difference, or initial level, is 0 – 0 as indicated, and the elevations of the pipe centerlines 1 and 2 are known with respect to this level. The procedure for determining the pressure difference as a function of the level differences, fluid properties, and half-deflection reading R is exactly as outlined above.

In the case of liquids, the specific gravity is sometimes specified rather than the density. The specific gravity is defined as the ratio of the density of the fluid to the density of water at the reference standard conditions of 4°C (39.2°F) and 1 atm ($\rho_{std} = 62.43\ \text{lb}_m/\text{ft}^3$). If the specific gravity is denoted by S,

$$S \equiv \frac{\rho(T, p)}{\rho_{H_2O}(4°\text{C}, 1\ \text{atm})}$$

The specific gravity is seen to be a variable that is primarily a function of the temperature (since the compressibility of most liquids is small). The densities of water and mercury are tabulated as functions of temperature in Tables D-3 and D-4. In the case where the fluid's specific gravity is given in place of the density, the relation given above for the differential pressure

$$p_2 - p_1 = \frac{\rho_0 g}{g_c}(S_1 h_a + S_m h_m - S_2 h_b) \tag{3-40}$$

where ρ_0 is the standard reference density of water.

Manometers are used as reference calibration standards for measuring pressures or differential pressures, since they are direct reading and require no calibration if the density or specific gravity of the fluid is known. A wide range of *sensitivity* (deflection obtained for a given applied differential pressure) is available through choice of manometer fluids. A range of specific gravities of about 0.8 for alcohol to 13.6 for mercury can be used. Some commercial manometer fluids have been developed to have low thermal-expansion coefficients, so that their specific gravity is nearly constant over the usual range of ambient temperatures at which manometers are used.

Where it is desired to measure very small differential pressures, a low-density manometer fluid is necessary. For wind-tunnel work, small air differential pressures are usually measured with alcohol inclined manometers or micromanometers, examples of which are shown in Fig. 3-14.

PRESSURE TRANSDUCERS

When it is necessary to record a large amount of pressure data for subsequent processing, it is advantageous to use pressure transducers. These instruments are diaphragm-type gages with an electronic transducer element, a strain gage, capacitance, or inductance pickup that senses the diaphragm deflection produced by the applied pressure and "transduces" the signal into a voltage output. This electrical signal may be transmitted directly to analog or digital data-reduction computers or stored for subsequent operation. It may be used in on-line process control.

For good response to high-frequency pressure variations (of the order of 100 cycles/sec and up), ceramic or crystal materials exhibiting a *piezoelectric* effect may be used directly or in conjunction with a diaphragm (used as a "pressure signal amplifier"). Some of these transducer-type gages are illustrated in Fig. 3-15.

3.8 FORCES ON PLANE SUBMERGED SURFACES

Dams, tanks, bulkheads, ship plates, liquid- and solid-propellant rocket tanks, cases, engines, and so on, are examples of submerged surfaces where pressure forces are required for design information. In most of these examples the surfaces are not plane but curved. However, by first considering the forces on plane surfaces, the principles and results derived in the previous sections can be used to determine the forces on curved surfaces.

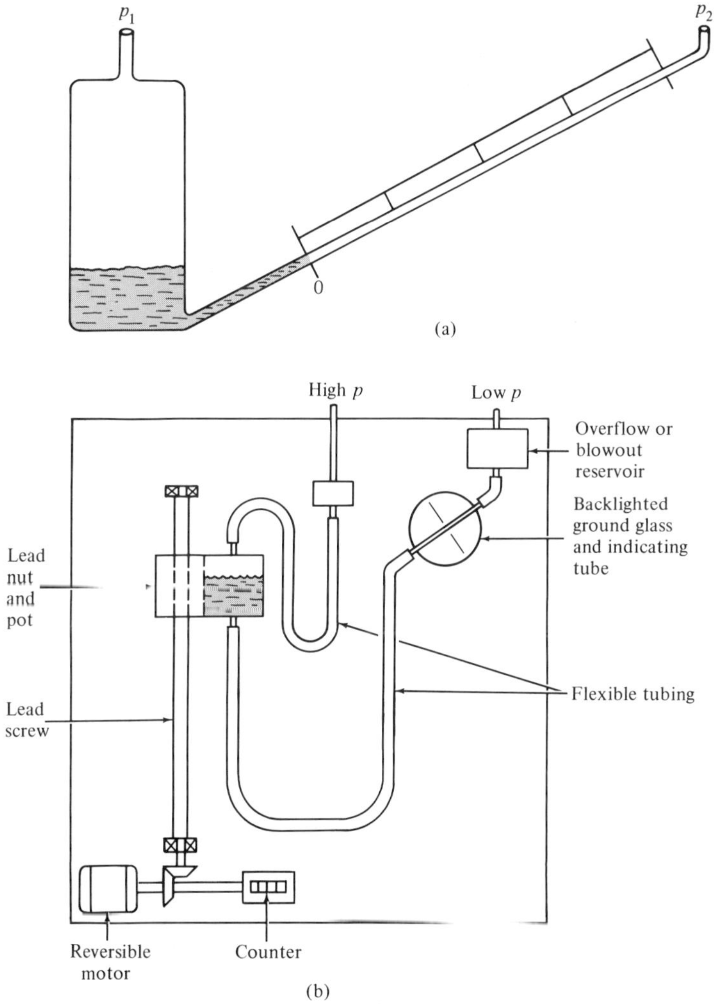

Fig. 3-14. (a) Simple inclined manometer; (b) direct-reading micromanometer. Gear ratio, lead screw pitch, and counter combination are designed to give readout directly in thousandths of an inch of manometer fluid.

The simplest case is a horizontal surface where the net force is due to the uniform pressure loading. The force is then just the hydrostatic pressure (as determined by the depth of the surface below the free liquid surface or reference pressure level) times the area. When the surface is not horizontal

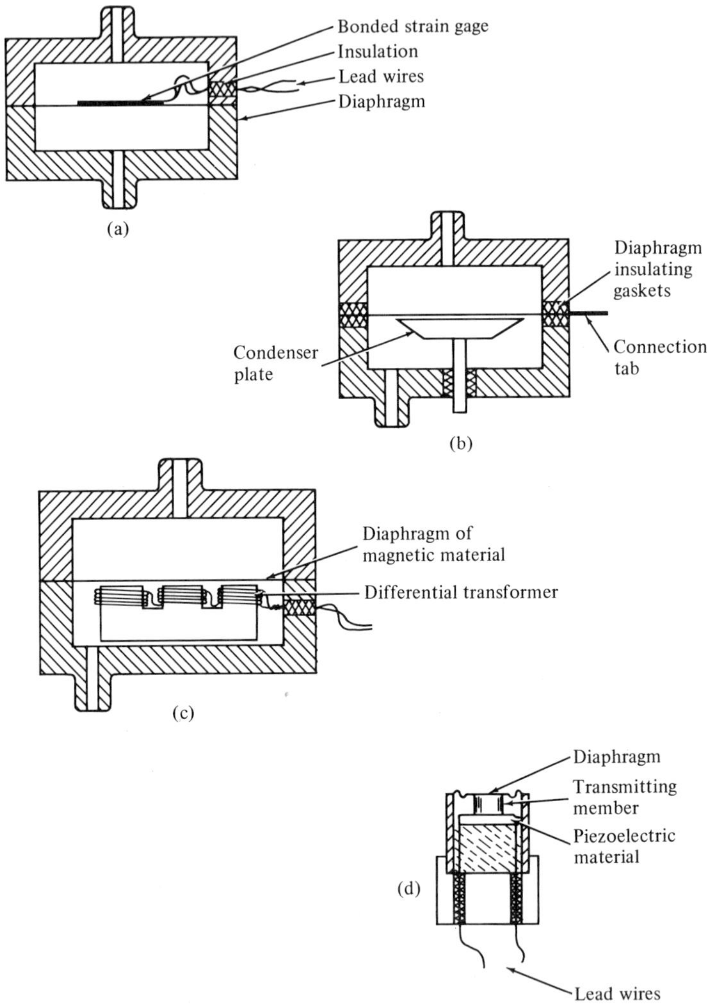

Fig. 3-15. Pressure transducers: (a) bonded strain-gage type; (b) variable-capacitance type; (c) variable-inductance type; (d) piezoelectric type. In the piezoelectric type, the diaphragm and transmitting member are used to protect the piezoelectric material (crystal, ceramic, and so on) from damage. The top and bottom of the piezoelectric material are coated with a conducting paint or surface to which the lead wires are connected.

but is inclined, the pressure loading is nonuniform but is a simple (linear) function of depth, so the net pressure force is easily determined.

The methods of statics can be applied to determine the magnitude and line of action of the resultant force and force components as well as the moments of the forces. As in any problem involving the determination of forces and their lines of action, it is mandatory to set up a coordinate system including a defined coordinate system origin, and to be consistent with respect to all coordinate directions and directions of forces or force components.

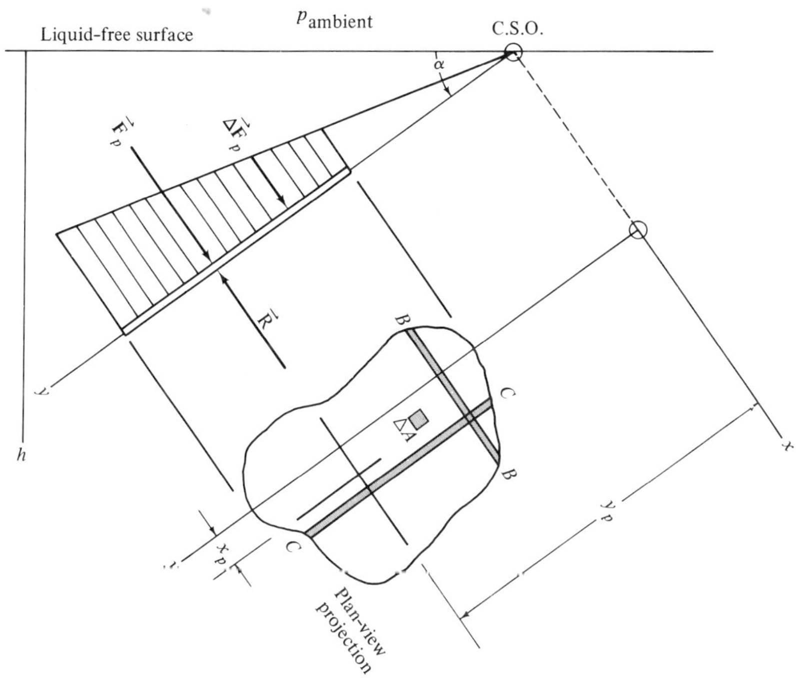

Fig. 3-16. Representation for forces on an inclined surface.

An arbitrary plane surface inclined at an angle α to the free surface (or some reference, constant-pressure surface) is represented in Fig. 3-16. On a small element of area, two things are known; first, the pressure on the area element ΔA is

$$p = \frac{\rho g}{g_c} h + p_a \tag{3-41}$$

where p_a is the reference pressure at the free surface. (If the free surface is open to the atmosphere, $p_a = 0$ gage.) Since the ambient pressure field, p_a, is uniform (that is, constant everywhere) it will contribute no net force (unless the problem specifically requires the pressure force with respect to an absolute zero reference). Second, the pressure force acts perpendicular to the area or area element, as shown. These two conditions determine the direction and magnitude of the force on any element ΔA.

The pressure loading on a horizontal area element (B-B) is uniform because the depth is constant along B-B. On the other hand, loading on C-C increases linearly with depth (h), as indicated on the slant projection. The resultant pressure force, F_p, is obtained by integrating dF_p over the area

$$\underset{\text{on fluid}}{F_p} = \int dF_p = -\iint p\, dA \quad \text{or} \quad \underset{\text{on surf.}}{F_p} = \iint p\, dA \tag{3-42}$$

Since the direction is known, the magnitude is determined with

$$p = \frac{\rho g}{g_c} h$$

as

$$F_p = \frac{\rho g}{g_c} \iint h\, dA \tag{3-43}$$

(where variations in density and gravity over the area have been neglected).

The choice of the coordinate system to be used in specifying the area, dA, depends on the shape of the area. It should be chosen so that the area boundary can be expressed in the most simple manner (for example, plane polar coordinates for a circular area). The relation between the depth h, the slant coordinate y, and the constant slant angle α is $h = y \sin \alpha$. The force is then

$$F = \frac{\rho g}{g_c} \iint y \sin \alpha\, dA = \frac{\rho g}{g_c} \sin \alpha \iint y\, dA \tag{3-44}$$

If the area and dA are not specified in the reference x-y coordinate system, a transformation of the coordinate y is necessary. The quantity $\iint y\, dA$ specifies the centroid of the area *with respect to the initial reference x-y coordinate system*; that is,

$$\bar{y}A = \iint y\, dA \quad \text{or} \quad \bar{y} \equiv \frac{\iint y\, dA}{A} \tag{3-45}$$

Using this identity,

$$F = \frac{\rho g}{g_c} (\bar{y} \sin \alpha) A \tag{3-46}$$

With $\bar{y} \sin \alpha \equiv \bar{h}$, the net pressure force on one side of the surface is equal to the pressure ($\bar{p}$) at the area centroid times the area: $F = \bar{p}A$ [or $F = (\bar{p} + p_a)A$ if the reference ambient pressure, p_a, is considered]. Since p_a is ambient, a force equal and opposite to $p_a A$ will be produced on the other side of the surface. However, if a different reference pressure exists on the other side, the force $p_a A$ must be considered.

The handbook result $F = \bar{p}A$ is not recommended for use. The foolproof method is to (1) start with $F = \iint p\, dA$, (2) specifically set up the coordinate systems needed, and (3) work through the details of the problem. This method eliminates the possibility of errors from ambiguous definitions of such quantities as $p = (\rho g/g_c)h$. The h (and y) are measured with respect to the specific, indicated reference (free surface) coordinate system, or from an initial point where there is a known pressure.

The line of action of the resultant pressure force is determined by taking moments about any arbitrary axis and determining a line of action of a resultant reaction R that must balance the distributed pressure force. For example, taking moments about the x axis to determine the y coordinate of the resultant pressure force y_p gives (with $R = F$ in magnitude)

$$Fy_p = \iint py\, dA = \frac{\rho g}{g_c} \iint y^2 \sin \alpha\, dA = \frac{\rho g}{g_c} \sin \alpha \iint y^2\, dA$$

or

$$y_p = \frac{1}{F} \frac{\rho g}{g_c} \sin \alpha \iint y^2\, dA \tag{3-47}$$

With (3-44) for F,

$$y_p = \frac{\iint y^2\, dA}{\iint y\, dA} \tag{3-48}$$

The expression for the determination of x_p is

$$x_p = \frac{\iint px\, dA}{F} = \frac{\iint xy\, dA}{\iint y\, dA} \tag{3-49}$$

The derivation of handbook formulas for x_p and y_p similar to that for $\bar{y}$ is left as an exercise (Problem 3-23).

Nonsimple or composite areas must be divided into elements that can be analyzed by the above methods. The resultant forces and moments are then combined to obtain the net force and its equivalent line of action.

3.9 FORCES ON CURVED SURFACES

Figure 3-17 represents the plate of a ship's hull as an example of an arbitrary curved surface. The net force and its line of action must be known

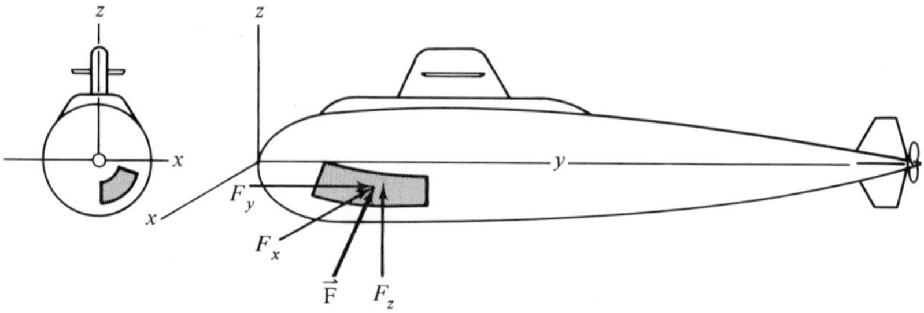

Fig. 3-17. Hull plate illustrating the pressure force on a curved surface.

to design the bulkheads and ribbing to withstand the maximum-depth hydrostatic force. The same principles used to determine the force and its line of action on a plane surface are applicable here: The pressure force acts everywhere normal to the surface, and the pressure is determined by the hydrostatic depth. The pressure force acting on the surface boundary can be considered as a field (point) function, determined by the hydrostatic depth field with the free surface as the reference plane. The total pressure force on the surface is again just

$$\underset{\text{on surf.}}{\vec{F}_p} = +\iint p\, d\vec{A}$$

The above vector equation represents three force components in each of three independent coordinate directions. These force components are evaluated independently, and the three mutually perpendicular components added vectorially to obtain the net vector force. Its direction is also determined by the magnitudes of the components.

Taking the above hull plate as an example and specifying a coordinate system for the example hull plate of Fig. 3-17 as indicated (x beamwise or athwartships, y lengthwise, and z vertically), the net force can be considered with respect to its three components, F_x, F_y, and F_z,

$$\vec{\text{F}} = \hat{\text{x}}F_x + \hat{\text{y}}F_y + \hat{\text{z}}F_z \tag{3-50}$$

or

$$\vec{\text{F}} = \iint p\, \text{dA} = \iint p(\hat{\text{x}}\, dA_x + \hat{\text{y}}\, dA_y + \hat{\text{z}}\, dA_z)$$

where dA_x, and so on, are the components of the vector $\overrightarrow{\text{dA}}(x, y, z)$. Therefore,

$$F_x = \iint p\, dA_x \qquad F_y = \iint p\, dA_y \qquad F_z = \iint p\, dA_z \tag{3-51}$$

But now dA_x is the projection of $\overrightarrow{dA}$ in the x direction, or on the y-z plane. F_x is then

$$F_x = \iint \frac{\rho g}{g_c} h \, dA_x \tag{3-52}$$

which is the pressure force on the plane-area projection projected onto the vertical y-z plane. The other horizontal component, F_y, is just the pressure force on the plane-area projection projected onto the vertical x-z plane.

For the vertical component

$$F_z = \iint \frac{\rho g}{g_c} h \, dA_z \tag{3-53}$$

where $\iint h \, dA_z$ is the *equivalent volume of fluid above the surface.* The vertical component is then the weight of the fluid in the volume above the surface.

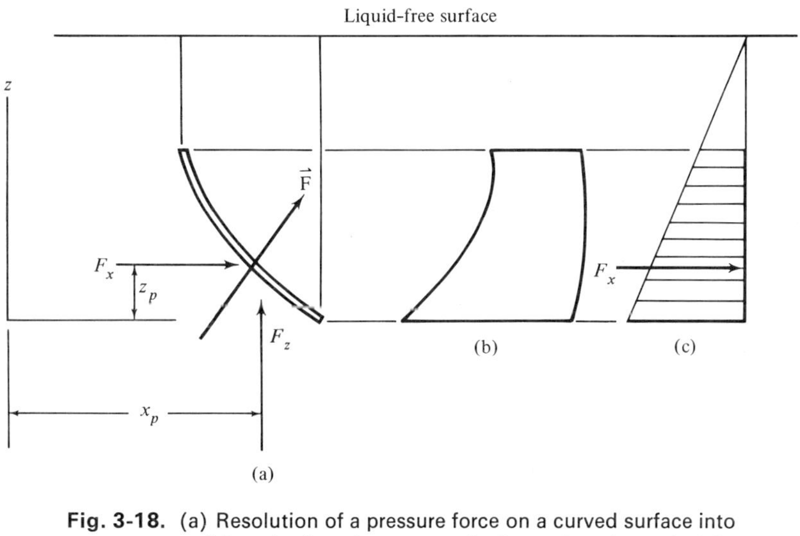

Fig. 3-18. (a) Resolution of a pressure force on a curved surface into components ; (b) projection view perpendicular to that shown in (a) ; (c) vertical projection of the surface area.

To determine the line of action of the force on a curved surface, consider the geometrically more simple case of a surface curved in one plane only, Fig. 3-18. F_x is the pressure force on the vertical projection of the surface. The area projected onto the plane of the page is shown in Fig. 3-18b and the pressure-loading diagram is shown in Fig. 3-18c. F_z is the weight of the liquid above the actual surface. The line of action is located at the point (x_p, z_p) determined in exactly the same manner as for the plane surface; x_p is also seen to be at the centroid of the fluid volume above the surface. The angle at which the force acts is arctan (F_z/F_x).

This method can be generalized to the arbitrary curved surface. Three coordinate distances locating the center of pressure with respect to the specified coordinate system are obtained, and the three components of the force give three angles, which determine the line of action. For such a surface, the most difficult problem is obtaining the area projection. In the case of a ship, this is best done graphically. The centroids and moments are then determined graphically to yield the force components and the coordinates of the center of pressure.

3.10 BUOYANCY AND STABILITY OF FLOATING BODIES

Archimedes[15] first discovered the principle that a body floating or immersed in a fluid experiences a buoyant force equal to the weight of displaced fluid. A floating body then displaces its weight in a fluid. These observations are easily verified by the principles of statics previously developed.

Consider a body submerged in a fluid body as shown in Fig. 3-19 and a vertical cylinder with a small cross section ΔA.

The vertical component of the force on the lower area element is

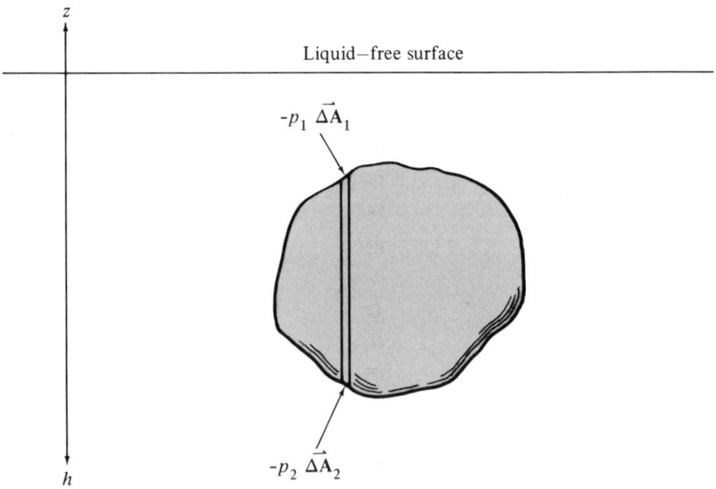

Fig. 3-19. Equilibrium of a submerged body.

[15] Archimedes (287–212 B.C.), born in Sicily, postulated the basic theorems of hydrostatic buoyancy and flotation. He also invented the screw water pump. He worked with and perfected instruments of war and was killed when the Romans sieged and sacked Syracuse.

$p_2 \Delta A_{2,H}$, where $\Delta A_{2,H}$ is the horizontal projection of ΔA_2. The force on the upper area element is $p_1 \Delta A_{1,H}$. If ΔA_1 and ΔA_2 are end areas of the same prismatic element as indicated, $\Delta A_{1,H} = \Delta A_{2,H}$ and the vertical force component of $p_2 \Delta A_2$ and $p_1 \Delta A_1$ is

$$F_z = \iint \frac{\rho g}{g_c} (h_2 - h_1)\, dA_H = \frac{\rho g}{g_c} V_D \tag{3-54}$$

where V_D is the displaced volume of the fluid and the quantity $(\rho g/g_c)V_D$ is the weight of this displaced volume. The pressure at both ends of a horizontal prismatic element is the same; therefore, the horizontal force components cancel each other. If the body is lighter than an equivalent volume of fluid, it will float, displacing a weight of liquid equal to its weight. If the body is heavier than its volume in fluid, it will sink. If a submerged body has weight initially equal to that of its displacement volume at a given initial depth, its stability depends upon its compressibility relative to that of the liquid. If it is relatively more compressible than the fluid (larger κ, smaller E), then it is in a state of unstable equilibrium. If displaced upward, it will expand more than the surrounding liquid and will rise to the surface. If displaced downward, it will "collapse" and displace a weight of liquid less than its own weight, since its mass (and weight) are constant.

3.11 ROTATIONAL STABILITY

The buoyant force was shown to act through the centroid of the displaced fluid. The weight (or gravity force) acts through the center of mass or "center of gravity" (c.g.) of the body. If the line of action of these two forces is colinear, the body is in equilibrium. Its equilibrium condition for a submerged body may be stable or unstable, as indicated in Fig. 3-20.

For the case shown in Fig. 3-20a, the condition of stability is determined by rotating the body slightly, as in Fig. 3-20b. The weight still acts vertically through the c.g. while the buoyant force acts through the centroid of the displacement volume. For the case shown, a righting moment (equal to Wr) is produced which tends to return the body to its initial position. The body is therefore in stable equilibrium.

From the above statements, the rotational stability of a submerged body is determined by the location of the c.g. with respect to the centroid of its displacement volume as follows:

Stable: c.g. located below centroid
Neutral: c.g. located at centroid
Unstable: c.g. located above centroid

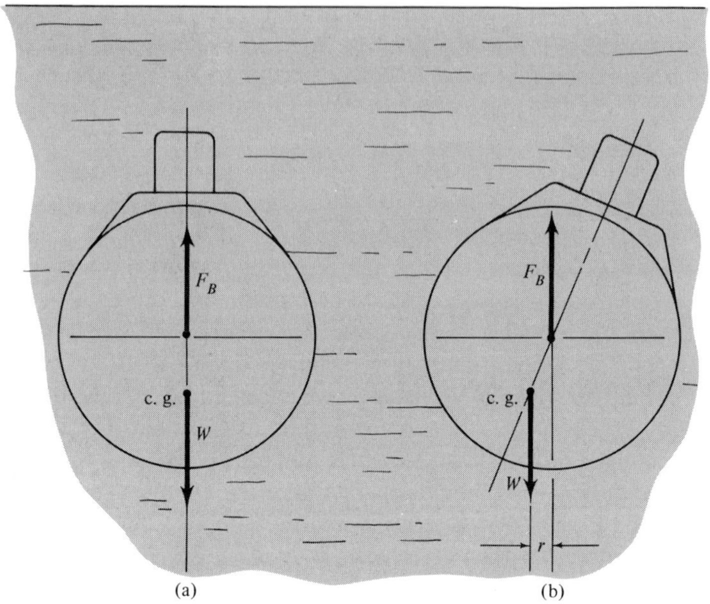

Fig. 3-20. Stability of a submerged body.

A free unstable body will rotate to a position of stability.

A balloon is essentially a body submerged in a fluid (gas), so its stability criteria are the same. However, if the balloon is large enough that the pressure varies significantly from top to bottom, the buoyant force acts through the center of mass of the displaced air, which may be slightly below the volume centroid.

The rotational stability condition for a floating body is more complicated. Surprisingly, the c.g. can be above the displacement volume centroid and the floating body may be stable, neutral, or unstable, depending on the geometry of the body.

In the case of a ship cross section as shown in Fig. 3-21, the location of the *center of buoyancy* (the centroid of the displaced fluid volume and the point through which the buoyant force acts) shifts as the ship rolls. The stability criterion then depends on whether a righting moment is produced when this center of buoyancy is displaced. In the rolled position, the buoyant forces passes through the displacement volume centroid, which has shifted from B to B'. This ship is seen to be stable, even though its c.g. (G) is above the centroid of the displacement volume. The stability criterion for a floating body is: The vertical line through B' (buoyant force line of action) must intersect the ship's vertical centerplane (at a point M) above the c.g., G. The distance MG is called the *metacentric height*. The location of point M is a function of the roll angle θ, but is defined for small angles of

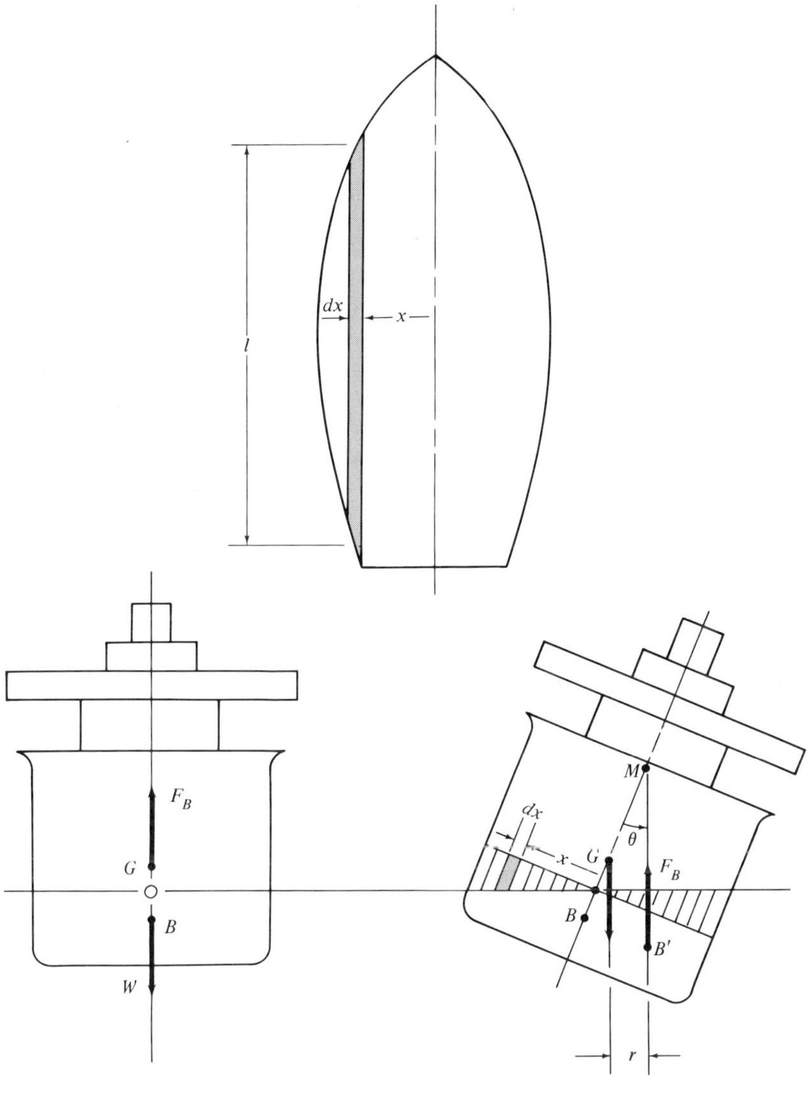

Fig. 3-21. Rotational stability of a floating body.

roll (that is, as θ approaches zero). The metacentric height MG is then determined by locating B' for a small roll angle θ and knowing the location of the c.g., G. The shift B-B' is easily determined by noting that this shift is produced by lifting the left-hand wedge between the initial waterline and the "rolled" waterline and submerging a similar wedge on the right side. The net unbalance in submerged volume is then equivalent to the volume of the wedges. The displacement moment is determined by taking the moments of these wedges about O. This moment must also be identically equivalent to the buoyant force F_B times its displacement from B to B':

$$\overline{BB'} \cdot F_B \equiv \int \frac{\rho g}{g_c} [(x\theta)l\, dx]x \tag{3-55}$$

where $[(x\theta)l\, dx]$ is the displacement volume element and $l\, dx$ is the area of the element of waterline-area element $dA_{\text{w.l.}} = l\, dx$; $\overline{BB'}$ = displacement moment/F_B. With $F_b = (\rho g/g_c)V_D$,

$$BB' = \frac{\int x\theta\, dA_{\text{w.l.}}}{V_D} = \frac{\theta I_{\text{w.l.}}}{V_D} \tag{3-56}$$

where $I_{\text{w.l.}}$ is the waterline-plan-form-area moment of inertia and V_D is the displacement volume. Also, taking $\overline{BB'} \approx \overline{MB}\theta$ gives

$$\overline{MB} = \frac{I_{\text{w.l.}}}{V_D} \tag{3-57}$$

The stability criterion is that $\overline{MG}$ must be greater than zero. If G is above B as indicated in Fig. 3-21, this condition is equivalent to

$$\overline{MB} - \overline{GB} > 0 \tag{3-58}$$

Assuming that the locations of G and B are known from the ship geometry and loading, then $\overline{GB}$ is known. The stability criterion is then

$$\frac{I_{\text{w.l.}}}{V_D} - \overline{GB} > 0 \tag{3-59}$$

for G above B. For the case where G is below B,

$$\overline{MG} - \overline{MB} + \overline{GB} = \frac{I_{\text{w.l.}}}{V_D} + \overline{GB} > 0 \tag{3-60}$$

The righting moment is Wr, and $r \approx \overline{MG}\theta$. If the ship is displaced through a small angle θ, it will roll-oscillate about O. The roll frequency can be determined by considering the ship as a pendulum, the motion of which is determined from the relation between torque and angular acceleration,

$$T = \frac{J}{g_c}\ddot{\theta} \tag{3-61}$$

where J is the polar moment of inertia of the waterline area. The torque is $T \equiv -W\overline{MG}\theta$, giving

$$\ddot{\theta} + \frac{W\overline{MG}g_c}{J}\theta = 0 \tag{3-62}$$

The negative sign appears because the torque (as a righting moment) has a direction opposite to that of the angular displacement θ. For small roll angles, the metacentric height $\overline{MG}$ is approximately constant, as determined above. Equation (3-62) represents an undamped, oscillating pendulum (or a simple harmonic oscillator) with an angular frequency

$$\omega = \sqrt{\frac{W\overline{MG}g_c}{J}} \tag{3-63}$$

or with a roll period of

$$\tau_r = 2\pi\sqrt{\frac{J}{W\overline{MG}g_c}} \tag{3-64}$$

The actual roll period is different from that given by (3-64) because

1. The relations derived above are for very small angles of roll.
2. The damping of the water has been neglected.
3. The effect of the displaced water mass as the ship rolls gives another mass that must somehow be considered in determining the torque. This last effect is referred to as a "virtual mass" or additional "apparent mass," the details of which must be left for a more advanced course of study.

3.12 UNIFORM ACCELERATION OF A FLUID

A mass of fluid may be accelerated uniformly (constant magnitude and direction) so that no relative motion of one part of the fluid with respect to another exists. This problem is then relatively simple, since there will be no shear forces on or in the fluid.

The effects of a constant rectilinear acceleration, $\vec{a}$, can be determined by considering a mass of fluid subjected to this uniform acceleration field and the earth's gravitational field. First, a coordinate system is chosen with z vertical, and the acceleration vector $\vec{a}$ in the x-z plane as shown in Fig. 3-22. The forces on a differential element are then considered and Newton's laws applied to determine the relation between the force field and the acceleration. The forces on the element are due to the pressure on the

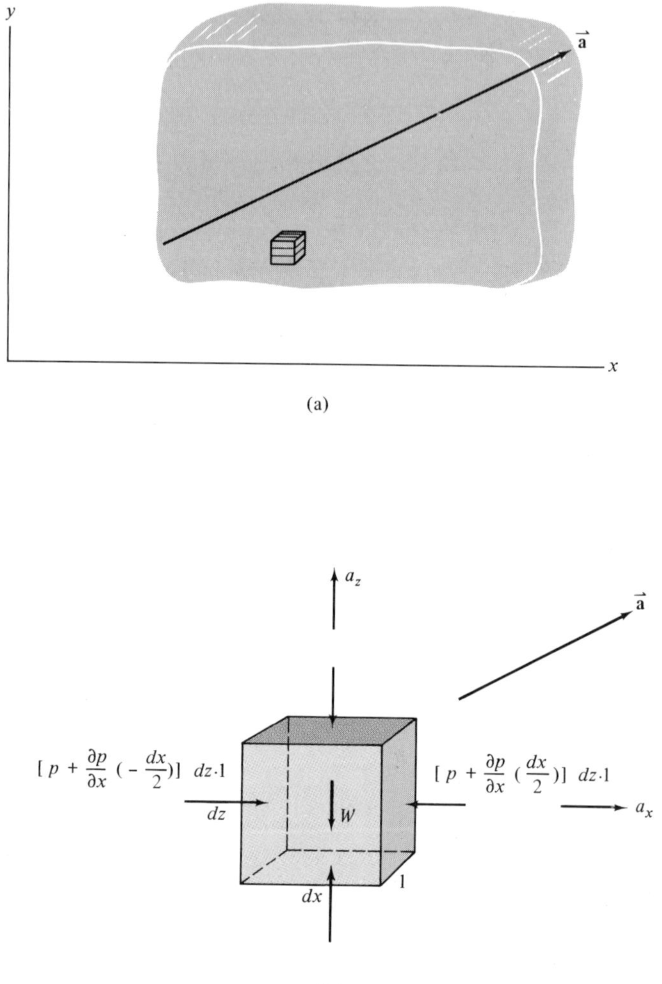

Fig. 3-22. Forces on a uniformly accelerated body of fluid.

surfaces and the weight of the element. The acceleration will affect the pressure variation within the fluid, so it will be desired to determine this pressure field.

Newton's second law states that the external forces acting *on* a given mass produces an acceleration of the mass proportional to the net applied force, or, in the form of an equation,

$$\sum \vec{F} = \frac{m}{g_c}\vec{a}$$

This is a vector equation; therefore, the direction of $\vec{a}$ is the same as that of the resultant vector force, $\sum \vec{F}$.

The gravity force acts vertically downward and the pressure force acts normal to the fluid surface. There are then two independent component equations,

$$F_x = \frac{m}{g_c} a_x \qquad \text{and} \qquad F_z = \frac{m}{g_c} a_z$$

The component forces and accelerations are indicated in Fig. 3-22b. The pressure at the center of the element is taken as the reference pressure p, which is a function of x and z, $p(x, z)$. Since there are only two significant directions, the problem will be set in these two coordinate directions. The element is of unit width in the direction perpendicular to the x-z plane.

The net force in the x direction is

$$F_x = \left[p + \frac{\partial p}{\partial x}\left(-\frac{dx}{2}\right)\right] dz - \left(p + \frac{\partial p}{\partial x}\frac{dx}{2}\right) dz \equiv -\frac{\partial p}{\partial x}\, dx\, dz \tag{3-65}$$

which is proportional to ma_x (the mass of the element, $\rho\, dx\, dz$, and the x component of the acceleration, a_x),

$$-\frac{\partial p}{\partial x}\, dx\, dz = \frac{\rho\, dx\, dz}{g_c} a_x \tag{3-66}$$

In the z direction the force of gravity also acts, so

$$F_z \equiv \left[p + \frac{\partial p}{\partial z}\left(-\frac{dz}{2}\right)\right] dx - \left(p + \frac{\partial p}{\partial z}\frac{dz}{2}\right) dx - \frac{\rho g}{g_c}\, dx\, dz = \frac{\rho}{g_c}\, dx\, dz\, a_z$$

or

$$-\left(\frac{\partial p}{\partial z} + \frac{\rho g}{g_c}\right) dx\, dz = \frac{\rho}{g_c}\, dx\, dz\, a_z \tag{3-67}$$

Solving (3-66) and (3-67) for the pressure-gradient components in the x and z directions gives

$$\frac{\partial p}{\partial x} = -\frac{\rho a_x}{g_c}$$

$$\frac{\partial p}{\partial z} = -\frac{\rho}{g_c}(a_z + g) \tag{3-68}$$

The differential equation for the total derivative or variation of p from one point in the fluid to another is

$$dp(x, z) = \frac{\partial p}{\partial x}\,dx + \frac{\partial p}{\partial z}\,dz$$

or

$$dp = -\frac{\rho a_x}{g_c}\,dx - \frac{\rho}{g_c}(a_z + g)\,dz \tag{3-69}$$

The equation for the gradient of the pressure field is

$$\text{grad}\,p = \hat{\mathbf{x}}\frac{\partial p}{\partial x} + \hat{\mathbf{z}}\frac{\partial p}{\partial z} = \hat{\mathbf{x}}\left(-\frac{\rho a_x}{g_c}\right) + \hat{\mathbf{z}}\left[-\frac{\rho}{g_c}(a_z + g)\right] \tag{3-70}$$

Equation (3-69) is a simple differential equation with constant coefficients and can therefore be integrated immediately. The indefinite integral is

$$p(x, z) = -\frac{\rho a_x}{g_c} - \frac{\rho}{g_c}(a_z + g)z + \text{const.} \tag{3-71}$$

If a value of p is known at some reference point (x_0, z_0), then the constant can be evaluated and (3-71) gives the pressure field. The pressure difference between any two points is obtained by evaluating the definite integral of (3-69) between these points, or

$$p_2 - p_1 = -\frac{\rho a_x}{g_c}(x_2 - x_1) - \frac{\rho}{g_c}(a_z + g)(z_2 - z_1) \tag{3-72}$$

Surfaces of constant pressure (or the free surface, if one exists) are determined for the condition that $p = \text{const.}$, or $dp = 0$. The slope of the free surface is then determined from (3-69) by setting $dp = 0$ and solving for the slope:

$$\left(\frac{dz}{dx}\right)_p = -\frac{a_x}{a_z + g} \tag{3-73}$$

Note that these results hold only for the coordinate system specified in Fig. 3-22, where the z axis is vertically upward and the x axis is horizontal.

3.13 ILLUSTRATIVE EXAMPLE 3-1: Uniform Acceleration

A container of liquids sitting on a table in an airplane is accelerated with components $a_x = -5$ ft/sec^2 and $a_z = -15$ ft/sec^2 when the airplane hits a downdraft. See Fig. 3-23.

(a) Find the equilibrium slope of the free surface.

(b) Find the pressure at the bottom of the container before and during the acceleration. Neglect the transient condition.

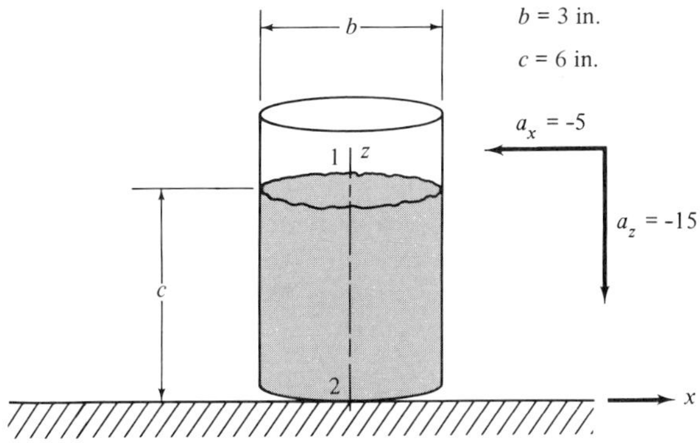

Fig. 3-23.

First the coordinate system is specified and the acceleration components are indicated. The equilibrium slope of the free surface during the downdraft is

$$\frac{dz}{dx} = -\frac{-5}{-15 + 32.2} = \frac{5}{71.2} \qquad \text{or} \qquad \alpha = +16.2^\circ$$

The pressure at 2 at the bottom of the glass before the downdraft is

$$p_2 = \frac{\rho g \, \Delta h}{g_c} = \frac{62.4}{144}\frac{6}{12} = 0.217 \text{ psi}$$

During the downdraft it is

$$p_2 = -\frac{\rho}{g_c}(-15 + 32.2)(0 - c) = \frac{\rho c}{g_c}(17.2) = 0.116 \text{ psi}$$

Note the absolute necessity of (1) definitely specifying the location and direction of the coordinate system, and (2) the consistent use of coordinate directions for acceleration components and distance.

3.14 CONSTANT ANGULAR ROTATION

If a fluid mass in a container is rotated about a fixed axis with a constant angular velocity $(\vec{\omega})$, an equilibrium condition will result with no relative motion of one element of fluid with respect to another. The shear forces are therefore zero. In this type of motion it is unusual to have an a_x or a_z acceleration component in addition to the centripetal acceleration component. Note that although $\vec{\omega}$ is constant, the acceleration is a function of r, with a magnitude of $r\omega^2$ and a direction radially inward toward and perpendicular to the axis of rotation (the ω axis). Expressed as a vector, $\vec{a} = -\hat{r}r\omega^2$. The coordinate system has the z axis in the direction of the rotation vector $\vec{\omega}$, and the radial coordinate perpendicular to the z axis, as indicated in Fig. 3-24. With the coordinate system and acceleration fields specified, the rotational equilibrium condition for a fluid element will be used to determine the pressure distribution within the rotating fluid mass. The fluid element is chosen to conform to the cylindrical coordinate system with r, θ, and z coordinates. With no acceleration in the z direction,

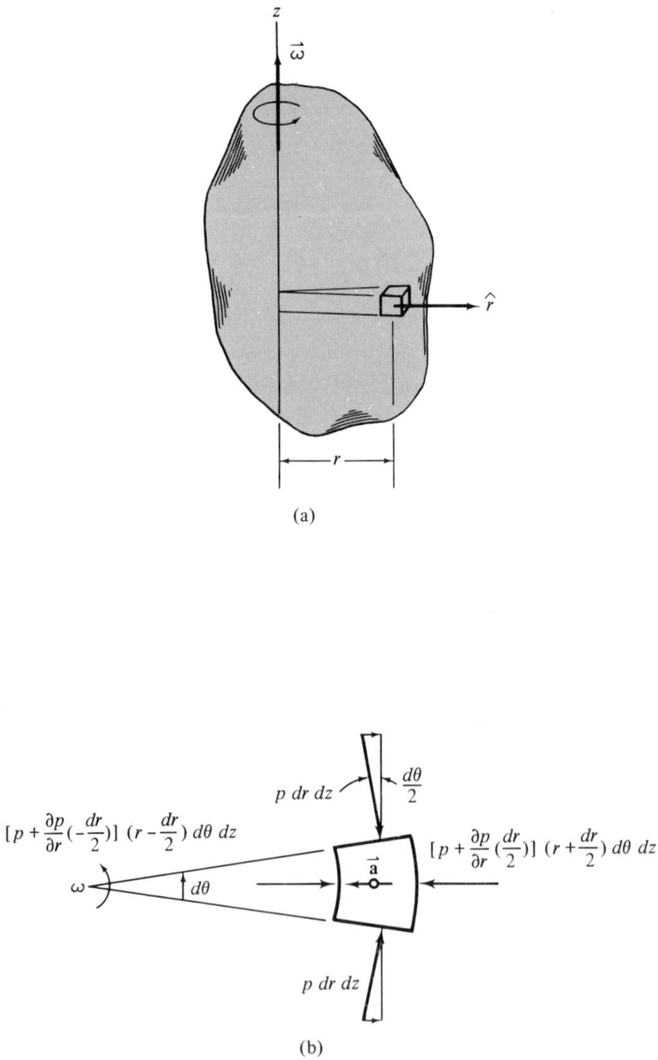

Fig. 3-24. Forces on an element of fluid in "solid-body" rotation.

the z component of the pressure gradient is the same as for a fluid at rest,

$$\frac{\partial p}{\partial z} = -\frac{\rho g}{g_c}$$

To obtain the radial component of the pressure gradient (or the pressure variation in the radial direction), an element is taken with a depth dz and the plan form indicated in Fig. 3-24b. The pressure forces on the sides of the fluid element have radial components $p(dr\, dz)(d\theta/2)$ [where $\sin(d\theta/2) \approx d\theta/2$, since $d\theta/2$ is quite small]. Summing all the radial force components and dividing by the volume of the element yields the radial-pressure-gradient component,

$$\frac{\partial p}{\partial r} = \frac{\rho}{g_c} r\omega^2 \tag{3-74}$$

The differential equation for the pressure field within a fluid rotating at a constant angular velocity is then given by

$$dp(r, z) = \frac{\partial p}{\partial z} dz + \frac{\partial p}{\partial r} dr = \frac{\rho r\omega^2}{g_c} dr - \frac{\rho g}{g_c} dz \tag{3-75}$$

which can be integrated immediately to give

$$p = \frac{\rho r^2\omega^2}{2g_c} - \frac{\rho g z}{g_c} + \text{const.} = \frac{\rho}{g_c}\left(\frac{v_\theta^2}{2} - gz\right) + \text{const.} \tag{3-76}$$

where $v_\theta = r\omega$ is the tangential velocity at a radius r.

The equations of the constant pressure surfaces are given by

$$\left(\frac{dz}{dr}\right)_p = \frac{r\omega^2}{g}$$

or

$$z = \frac{r^2\omega^2}{2g} + \text{const.} \tag{3-77}$$

These constant-pressure surfaces for a spinning container of liquid with no relative fluid motion are parabolas of revolution.

This result suggests a method for making parabolic mirrors. A form can be constructed into which a plastic monomer is poured. The form and monomer are rotated at a constant speed until an equilibrium parabolic surface is obtained. The catalyst can then be sprayed on the surface and the plastic polymerized or "set" in the shape of a perfect parabola. If the plastic polymerizes slowly, the catalyst can be mixed before pouring into the form. This surface can then be aluminized.

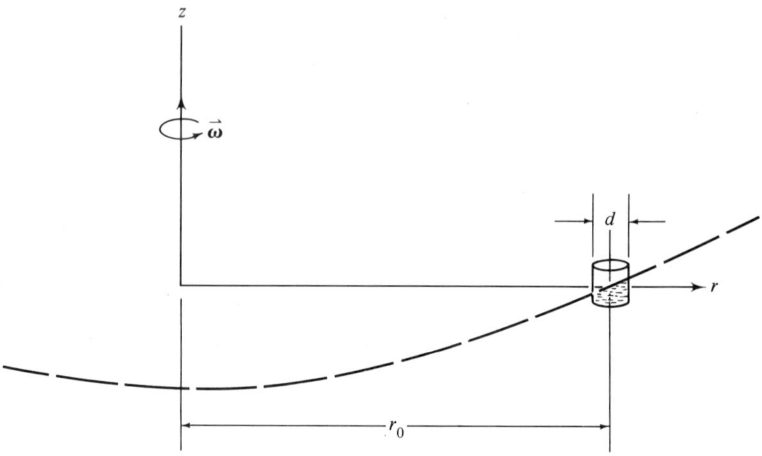

Fig. 3-25. Rotation of a small body of fluid located far from the axis of rotation.

A special case exists for the rotation of a small volume of fluid about an axis that is a large distance from the volume, as illustrated in Fig. 3-25. If $d \ll r_0$, the problem can be considered as of the first type, with $a_x = -r\omega^2$ and $a_z = 0$. The constant-pressure surfaces (and free surface) have a slope of

$$\frac{dz}{dx} = \frac{r_0\omega^2}{g}$$

This slope line is a small straight-line approximation to the parabola of revolution passing through the container with its vertex at the axis of revolution.

3.15 SUMMARY

1. Pressure is a scalar quantity, and is therefore independent of orientation or direction. Pressure force, however, is a vector. Its direction of action depends on the geometry of the area on which the pressure acts. The pressure force on a fluid surface is given by

$$\vec{\mathrm{F}}_p = \iint -p\, d\vec{\mathrm{A}}$$

2. The summation of *all* forces acting on the fluid must be zero if the fluid is in a state of static equilibrium. Specifically there must be a balance between surface (pressure) and body (gravitational-type) forces.

3. In the process of summing all surface forces on differential elements within a body of fluid, the surface forces on adjacent elements cancel, yielding only the net pressure force on the area surrounding the entire fluid volume (control volume) in question. An equation expressing this physical observation is stated mathematically by the gradient theorem

$$\oiiint \text{grad}\, p\, dV = \oiint p\, d\vec{A}$$

4. The pressure field in a static fluid in the earth's gravity field is a function primarily of altitude. For a fluid with negligible compressibility the pressure is seen to be a function of depth only (Pascal's principle).
5. Hydrostatic forces on surfaces or bodies are most directly determined from the definition of pressure force,

$$\vec{F}_p = \iint_{\text{on surf.}} p\, d\vec{A}$$

In the case of curved surfaces the components of F_p are determined by judiciously selecting coordinate (and component) directions that best fit the geometry of the surface or body.
6. A fluid under the influence of a uniform accelerational field (constant vector acceleration), or rotating with no relative motion between any regions of the fluid, can be treated as a solid body. The pressure field is determined by applying Newton's laws of motion to this "solid body" of fluid.

PROBLEMS

3-1 Perform the integration indicated in (3-29) to obtain the result of (3-30).

3-2 Using (3-30) with standard sea-level values of $p_0 = 14.7$ psia and $T_0 = 59°F$, find T/T_0, p/p_0, and ρ/ρ_0, and T, p, and ρ at the tropopause (36,152-ft altitude). Compare the results with those given in Table D-1.

3-3 Assuming the temperature to remain constant at the tropopause temperature for z greater than 36,152 ft, calculate the p, ρ, and T at 50,000 ft and compare with the results of Table D-1.

3-4 A Bourdon-tube pressure gage on a fuel tank reads 10 psi. If the local barometer reads 29.00 in. Hg, what is the absolute pressure in the tank?

3-5 Determine a, the difference in the levels.

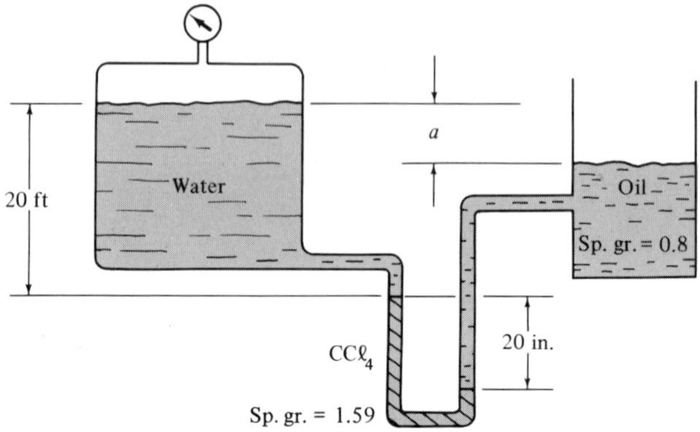

Fig. P3-5.

3-6 A mercury manometer connected to a vacuum system reads 1.52 in. Hg. What is the absolute pressure in the system in $lb_f/in.^2$ and in feet of water head? What is the gage pressure in psi? It may be necessary to make one or more assumptions to work this problem. State all assumptions made.

3-7 A pump is installed in a pipeline as shown with a manometer to measure the differential pressure across the pump. If the manometer is installed one floor below the pump (that is, 10 ft below), find the pressure rise across the pump when the manometer indicates 12 in. ($= h_m$ as shown) if the specific gravity of the manometer fluid is 2.94.

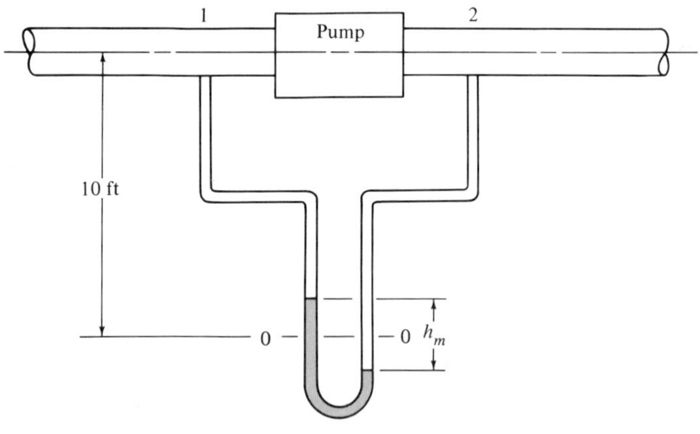

Fig. P3-7.

3-8 A manometer containing carbon tetrachloride is used to measure the pressure difference across a vertical pump as shown. Find the pressure difference (psi) between points 1 and 2 in terms of $z_2 - z_1$, the specific gravity of CCl_4 (S_c), the specific gravity of the fluid being pumped (S_p), and the manometer reading R.

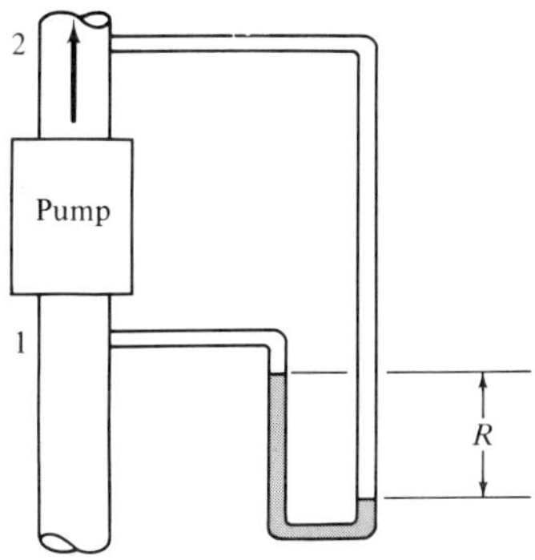

Fig. P3-8, 9.

3-9 If $z_2 - z_1 = 4$ ft, the fluid is water, and the manometer reading is 10 in., find $p_2 - p_1$ in psi.

3-10 A gasoline tank is buried in the ground with the piping system shown. For the particular levels indicated, determine p_1, p_2, p_3, and p_4 in terms

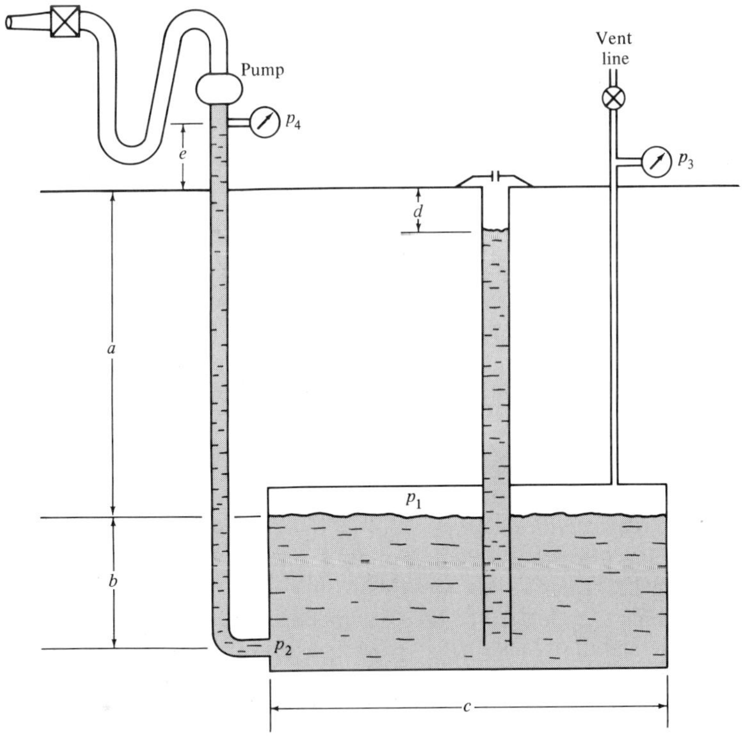

Fig. P3-10, 11, 12.

of the specific gravity of the gasoline and the dimensions indicated (a, b, c, d, and e). (See the figure.)

3-11 If $a = 6$ ft, $b = 2$ ft, $c = 8$ ft, $d = 2$ ft, and $e = 3$ ft, find the pressures p_1 through p_4. (The specific gravity of gasoline is approximately 0.70.)

3-12 How high above the ground level could the fuel be drawn by a vacuum pump (or be siphoned) on a day when the temperature is 90°F? (Consider the gasoline to be essentially octane, C_8H_{18}.)

3-13 For the inclined manometer ("draft gage") shown in Fig. 3-14a, find the "sensitivity" (or "gain") of the manometer in inches of scale reading per inch of applied water differential ($h_1 - h_2$) in terms of the inside diameter of the reservoir d_1, the tube i.d. d_2, the angle of inclination of the tube θ, and the specific gravity of the manometer fluid. What factors have the largest effect on maximizing this sensitivity?

3-14 If the reservoir diameter is 1 in. and the tube diameter is 5 mm, find the inclination angle for the draft gage (Fig. 3-14a) to show a 10-in. scale deflection for 1 in. of applied water head differential when a manometer oil with $S = 0.832$ is used.

3-15 Find the net force and moment on the butterfly valve shown. Work this problem by setting up the necessary integrals for force and moment. Sketch the pressure-loading diagrams for each side of the valve.

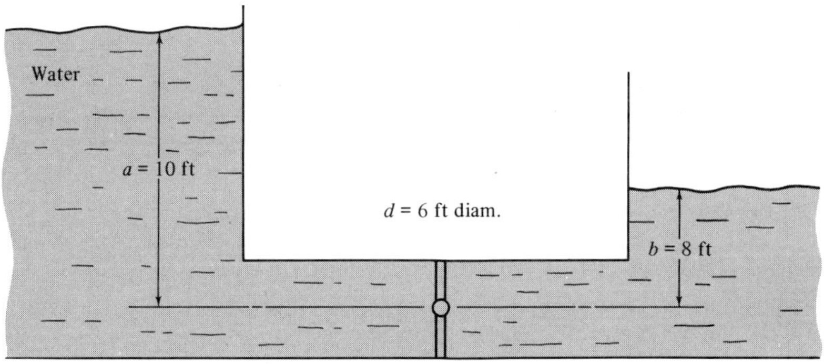

Fig. P3-15.

3-16 A barrel is anchored in a lake as shown. Find the water-pressure force on the bottom end.

3-17 Find the force on the dike check valve shown. The flow passage is semicircular with a diameter of 4 ft.

3-18 Find the force on the bulkhead as a function of a, h, and the specific gravity, S.

3-19 A simple U tube is to be used as an accelerometer in an automobile. If the acceleration is 10 ft/sec^2, find the deflection if the fluid is carbon tetrachloride.

3-20 Find the sensitivity of the U-tube accelerometer in inches of deflection reading R per g of acceleration a. What is the primary factor affecting the sensitivity? Explain.

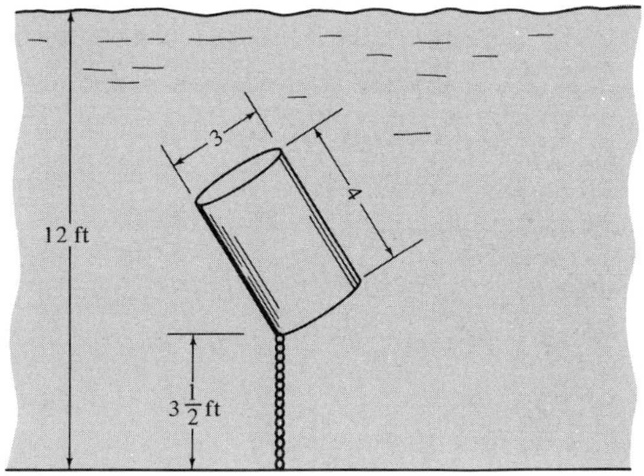

Fig. P3-16.

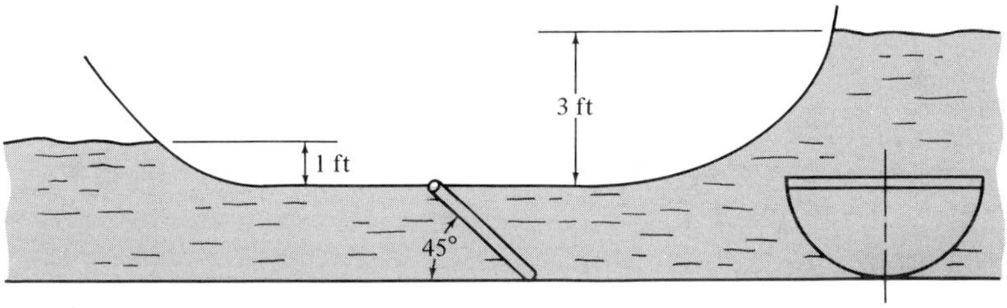

Fig. P3-17.

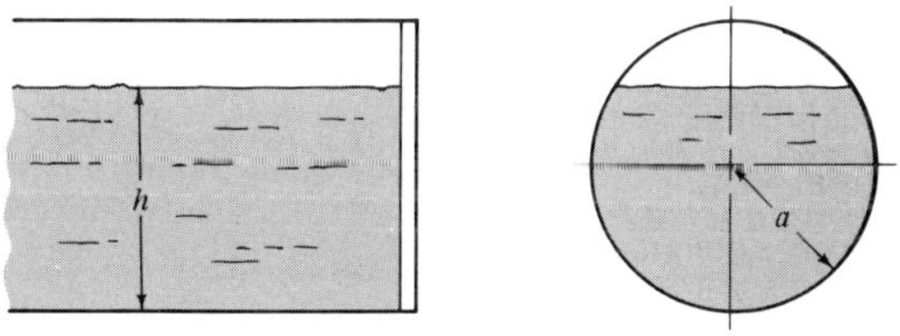

Fig. P3-18.

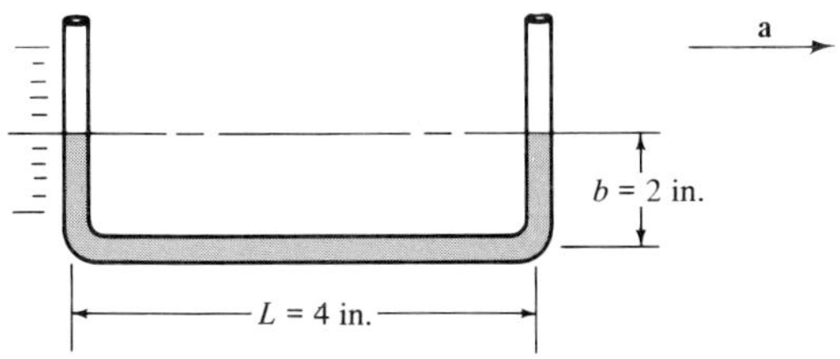

Fig. P3-19, 20.

3-21 A parabolic plastic mirror is to be "cast" on a phonograph turntable running at $33\frac{1}{3}$ rpm. Find the focal length that will be obtained. If the mirror diameter is 10 in., what is its f number (ratio of diameter/focal length)? The specific gravity of the plastic is 1.00.

3-22 Find an expression giving the focal length of a spun-cast plastic parabolic mirror as a function of rpm and specific gravity.

3-23 (a) Show that the coordinates of the center of pressure can be given by $x_p = I_{xy}/\bar{y}A$ and $y_p = I_{zx}/\bar{y}A$, where I_{xy} is the moment of inertia of the area about the x axis and I_{xy} is the product of inertia with respect to the x and y axes.

(b) Show that the above relations can be written

$$x_p = \frac{I_{\bar{x}\bar{y}}}{\bar{y}A} + \bar{x} \qquad \text{and} \qquad y_p = \frac{I_{\bar{x}\bar{x}}}{\bar{y}A} + \bar{y}$$

where $\bar{x}$ and $\bar{y}$ are the coordinates of the centroid of the plane area with respect to the coordinate axes x and y.

(c) In general, what statement can be made regarding the location of y_p with respect to y (assuming the x axis is on the free surface)?

3-24 A sluice gate is held in place by a bar attached to the top of the gate at A and B and two pins at C and D that fit into the sluice bottom. Find the forces on the retainers at A, B, C, and D.

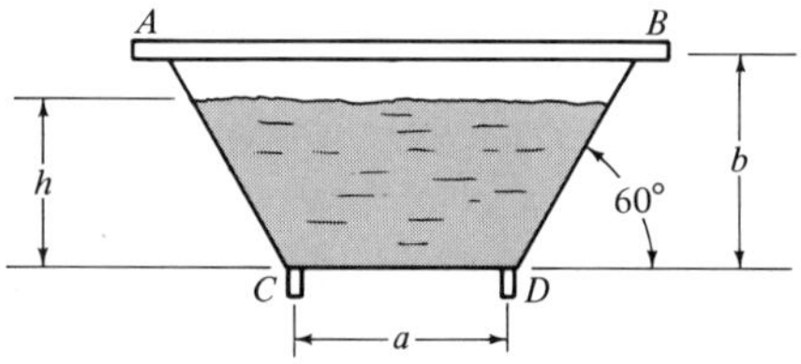

Fig. P3-24.

3-25 A round tank gate of radius a and weight W is pivoted about the seal point A on a spillway. Find the force and moment required to hold it in place as a function of a, b, h, and W.

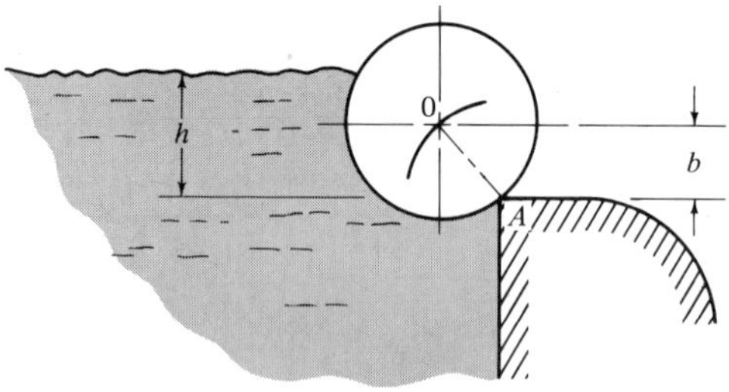

Fig. P3-25.

3-26 A quadrant spillway gate of weight W is pivoted at A as shown. What is the torque required to hold the gate in place as a function of h, a, b, and W?

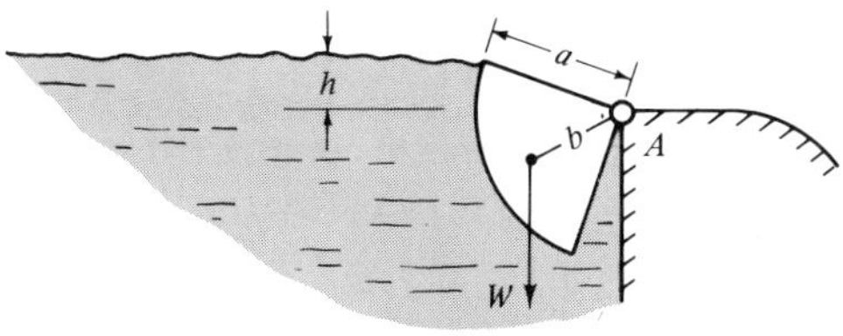

Fig. P3-26.

3-27 A limp balloon is to be used for a high-altitude investigation. The initial mass of the helium must be sufficient to lift the package. Since g is about constant, the weight is about constant. Assume the takeoff lift is 10 percent greater than the weight.

(a) What is the initial takeoff acceleration?
(b) What affects the rise rate?
(c) Will the rise rate increase or decrease?
(d) What is the excess lift percentage at 50,000 ft?

Assume that the balloon is designed to become taut at 100,000 ft, that skin expansion is negligible, and that gas is dumped to maintain the taut (but equilibrium) condition of constant volume.

(e) What is the maximum altitude possible?

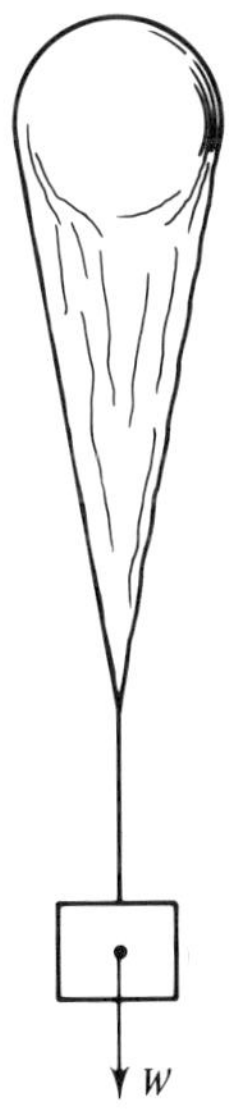

Fig. P3-27.

3-28 Discuss the stability of the ocean. What are the fluid properties that determine this stability? What is the basic stability condition?

3-29 On top of Mt. Whitney, a plaque gives the altitude as 14,495.866 ft. Calculate the mean temperature, pressure, and density there, assuming a constant-lapse-rate atmosphere ($\lambda = 0.00357°F/ft$).

3-30 Calculate the temperature, pressure, and density ratios at a jet-cruise altitude of 30,000 ft referenced to sea level (a) based on a constant lapse rate ($\lambda = 0.00357°\text{F/ft}$), (b) for an adiabatic atmosphere ($\gamma = 1.40$), and (c) for an isothermal atmosphere. (d) Tabulate the results of (a) through (c) and calculate the percentage error based on the values given in Table D-1.

3-31 Calculate the pressure and density at an altitude of 60,000 ft for a constant-lapse-rate atmosphere from 0 to 36,152 ft ($T_0 = 59°\text{F}$) and isothermal to 60,000 ft. If an airplane cabin is pressurized to an equivalent altitude of 7000 ft, what is the stress in the $\frac{1}{6}$-in.-thick aluminum skin of a 28-ft-diameter fuselage (neglect framework strengthening).

3-32 Calculate the altitude at which the density variation from sea level is 0.1 percent and 0.5 percent.

PART II FLUID DYNAMICS FUNDAMENTALS

All studies in fluid mechanics are founded upon the basic physical laws of mass, energy, and momentum conservation developed in Chapters 4, 5, and 6, respectively. The characteristics of fluid flow fields required to develop the mathematical statements of these laws are presented in Chapter 4.

Two viewpoints are possible for starting analyses of the motion of a fluid. The *Lagrangian*[1] approach develops the basic equations for a specified mass of fluid in a manner similar to the dynamic analysis of a mass (or the center of mass of a system of masses) in rigid-body dynamics. This fluid mass may be a differential-sized element or a large integral mass. Energy may be transferred to or from the fluid and work done on or by it. This fixed-mass representation is analogous to the closed system in thermodynamics.

The Eulerian[2] approach considers the instantaneous mass, energy, and momentum of the fluid flowing through a defined *control volume* fixed with respect to a designated coordinate system. Integral equations are derived for a control volume enclosed by a complete piece of apparatus, or differential equations for the flow through a control volume of differential size. This open-control-volume representation is analogous to the *open system* of thermodynamics.

The basic equations are more easily and simply derived with the Lagrangian approach; however, the boundary conditions are more easily and directly specified in the Eulerian representation. Solution of physical

[1] Joseph Louis Lagrange (1736–1813), Italian–French. Succeeded Euler as the world's leading mathematician. He introduced the concepts of stream function and potential function into hydrodynamics, but, perhaps of equal importance, he realized the limitations of idealized or potential flow representation. The "Lagrangian" method of formulation of the equations of motion with respect to a fixed defined mass was actually first considered by Euler.

[2] Leonhard Euler (1707–1783), Swiss mathematician of the Bernoulli school and one of the founders of classical hydrodynamics. He also formulated much of the basic analytical approach to the study of mechanics of motion and of deformable bodies. Euler was extremely prolific in his works in mathematics and mechanics, and also other lines; he is reported to have fathered somewhere between 13 and 21 children.

boundary-value problems is usually easier with the Eulerian method because description of the manner in which the Lagrangian mass deforms to conform to physical boundaries is generally quite complicated.

It is frequently necessary to formulate relations for control volumes with deforming boundaries (such as a deflating toy balloon in free flight). Problems with moving or deforming control volume boundaries, with flow across part of the control surfaces, must be treated as special cases.

Since mechanical work is the result of a force and displacement, no work is possible if the walls of the control volume are fixed so that no displacement can occur. Therefore, the production of mechanical work requires a moving boundary.

CHAPTER 4

CONTENTS

FLUID MOTION

4.1 FLOW PATTERNS AND FLOW FIELDS

A flow field is any region in space where a fluid flow exists. An obvious physical condition is that any flow region or subregion must be completely occupied by a fluid to be considered as a flow field.

Some examples of flow fields are:

1. The flow in a pipe. The fluid within the confining boundaries of the pipe walls, or any part thereof, can be considered as a flow field to which the "laws," principles, and equations of motion can be applied.
2. Any part of the atmosphere.
3. The fluid region surrounding any moving vehicle, missile, and so on, or part thereof.

An example of a flow field was previously illustrated in Fig. 2-1.

Flow field characteristics were defined in Chapter 2 which aid in applying the available physical principles for the purpose of obtaining more information about the flow pressure distributions, consequent net pressure forces on bodies in flow fields, and so on.

If the flow and fluid properties are constant at a fixed point (that is, if the configuration of the flow field does not vary with respect to time), the flow was described as steady and the field stationary. The mathematical condition for steady flow is that the variations of any quantity at a fixed point in space are zero or

$$\frac{\partial(q)}{\partial t} = 0$$

where q is any quantity. This partial derivative denotes variation with respect to time only, keeping all other independent variables (for example, space coordinates) constant.

As an example of an unsteady flow, consider the smoke rising from a chimney on a gusty day. At some instant a velocity field can be determined in a region above the chimney by taking a short time-exposure photograph and noting the direction and distance traveled by the smoke particle during the exposure period. If a similar velocity-field determination is made at a different time, the field will be entirely different.

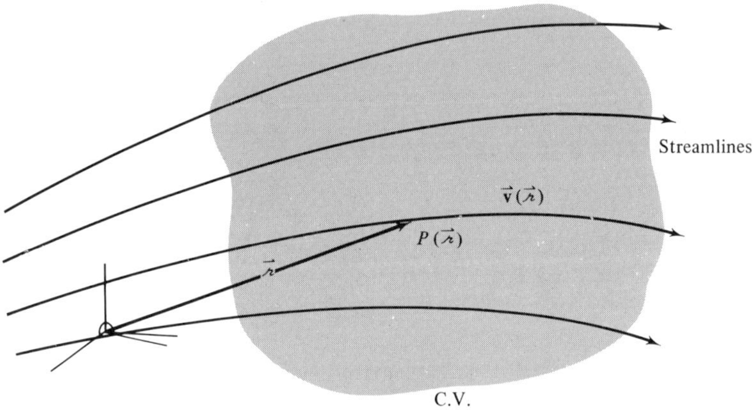

Fig. 4-1. Representation of a flow field.

A flow field can be represented by indicating the *streamline* pattern as in Fig. 4-1. A streamline is a line that is everywhere tangent to the direction of the fluid motion at any instant. It can be seen that there exists only one unique streamline through any given point in a flow field at a given time. The flow and fluid properties can be specified, calculated, or measured at any point P located and specified by the distance $\vec{r}$ from an arbitrarily defined coordinate system origin.

The location of P is most simply determined by specifying the radius vector $\vec{r}$ from the coordinate system origin to the point P, or $P(\vec{r})$. The radius vector $\vec{r}$ does not imply any specific coordinate system, which is an advantage of using vector notation. This radius or position vector is given in some of the more common coordinate systems as[1]

Cartesian: $\vec{r}(x, y, z) = \hat{x}x + \hat{y}y + \hat{z}z$
Cylindrical polar: $\vec{r}(r, \theta, z) = \hat{r}r + \hat{z}z$
Spherical polar: $\vec{r}(r, \varphi, \theta) = \hat{r}r$

[1] In the simple cartesian coordinate system, the unit vectors are constant in both magnitude and direction. In curvilinear coordinate systems, the unit vectors are constant in magnitude, but their directions may vary so that $\hat{x}_i$ are a function of the other coordinates, that is, $\hat{x}_i(x_j)$; see Appendix B.

These coordinate systems and corresponding position vectors are indicated in Fig. 4-2. The relationship between the three systems is indicated by the reference system axes shown in brackets (for example, [x]).

The field quantities of interest, such as velocity, pressure, and so on, may be generally specified as function of space $(\vec{r})$ and time (t):

$$\vec{v}(\vec{r}, t), \quad p(\vec{r}, t), \quad \text{and so on}$$

The flow about a cylinder would logically be specified by the use of cylindrical polar coordinates. The pressure field would then be given as $p(r, \theta)$ for a steady flow. There is no variation of any property with z. All flow sections perpendicular to the z axis are identical. A single plane perpendicular to the z axis can therefore represent the entire flow field, and a plane polar coordinate system is then sufficient for this representation.

Vector quantities may be represented by their components in the directions of the coordinate axes as shown in Fig. 4-3. The velocity $\vec{v}$ is a function of both coordinates in a two-dimensional flow so that the components must also be a function of both coordinates; that is,

$$\vec{v}(x, y) = \hat{x}u(x, y) + \hat{y}v(x, y)$$

The dotted portion of the field represented in Fig. 4-4 represents the flow pattern between two adjacent compressor blades. The flow is shown as diffusing (velocity decreasing, pressure increasing), indicating a compression in the direction of flow.

There are some unsteady flows that can be reduced to steady flow (or a stationary flow field) by merely transforming the frame of reference of the observer. Consider an airplane flying in a straight path at constant speed past a fixed observer. An observed region in the atmosphere fixed in space and defined as a fluid field through which the plane flies will experience an unsteady motion; the velocity, pressure, and so on, will vary with time. On the other hand, if the observer and his frame of reference move with the same velocity as the airplane, the flow conditions at every point in the field will apparently not vary with time, and the flow will appear to be steady and the field stationary.

The same result of reducing the flow fluid to a steady one can be obtained by first considering the airplane flying past at its flight velocity $-U_f$. A uniform flow field with velocity $+U_f$ is then superimposed, which effectively "stops" the airplane motion and causes the air to flow past it with a speed U_f. By this simple "transformation" of adding a uniform field to the flow, the nonstationary flow field is transformed to a stationary one. The end result is seen to be equivalent to "flying" the airplane (or a model) in a wind tunnel.

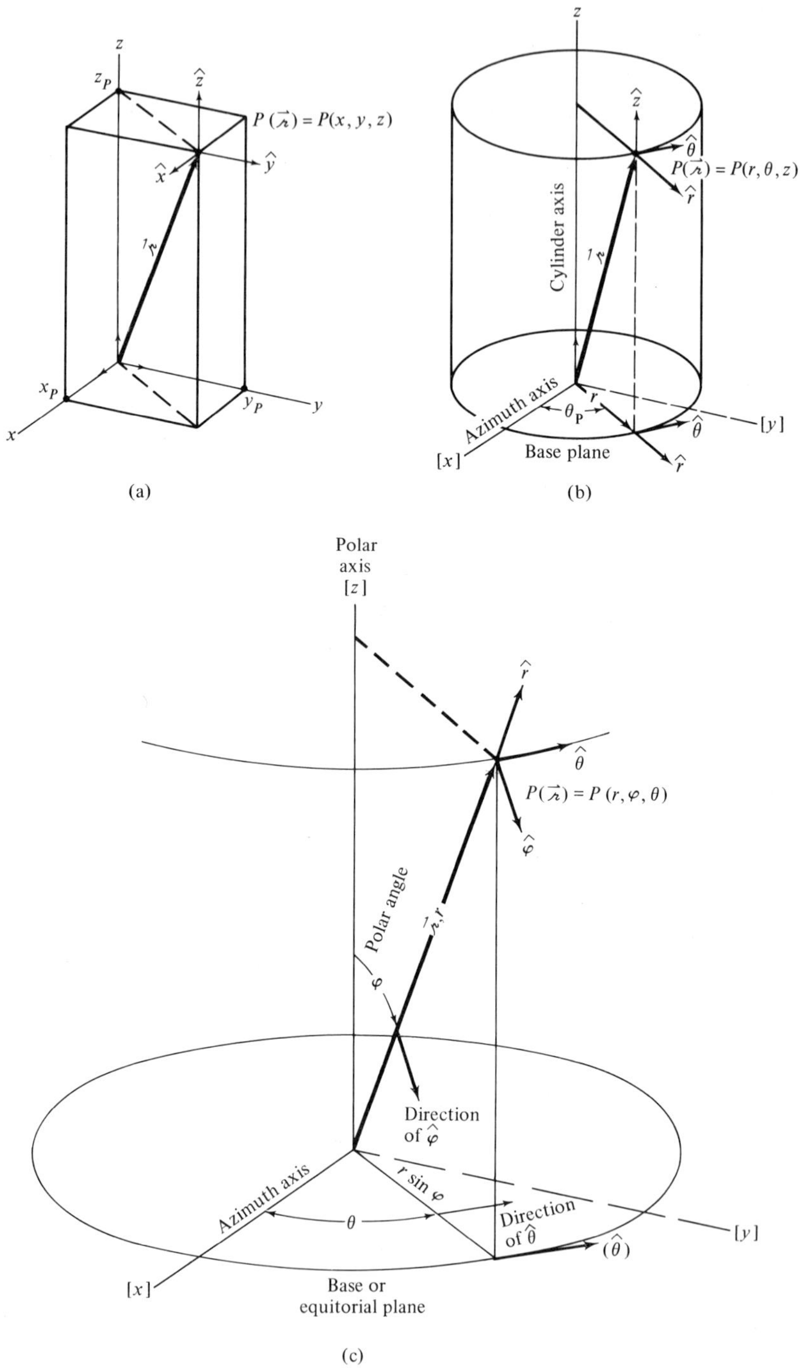

Fig. 4-2. Coordinate system representations showing coordinates and relations between coordinates and unit vectors: (a) cartesian; (b) cylindrical polar; and (c) spherical polar coordinate systems.

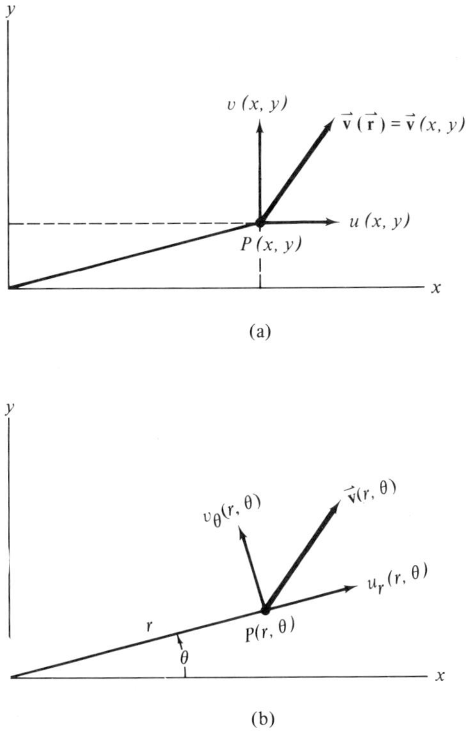

Fig. 4-3. Representation of components of a vector in (a) cartesian and (b) plane polar coordinate systems.

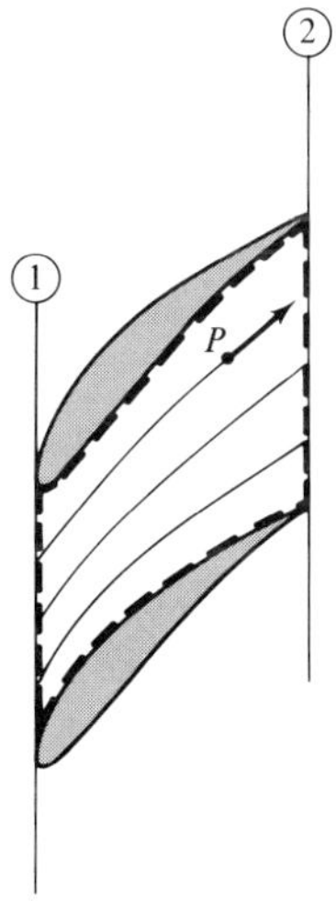

Fig. 4-4. Flow field in one cylindrical section between two compressor blades. If the blades are moving, all the flow properties and quantities (p, v, ρ, T, and so on) at P are varying with time. If the blades are stationary (compressor stator) and the inlet and outlet conditions are not time-varying, the flow is steady or the field is stationary.

It is emphasized that all field quantities must be considered in making this transformation. It was specifically stated that the plane was flying with a *constant* (vector) *velocity*. If it were accelerating or maneuvering, the velocity and pressure would vary within the field even though the reference coordinate system and observer were attached to the airplane.

It is obvious that only certain flow patterns can be reduced to a condition of steady flow. For the smoke exhausting from the chimney, no such reduction or transformation is possible, since this flow field is nonuniformly nonstationary. Other restrictions on such transformation will be considered in Chapter 6.

4.2 LAMINAR AND TURBULENT FLOW

The slow flow of a viscous fluid is characterized by a smooth flow of one layer or lamina of fluid over another. Fluid mixing occurs by random molecular motion only. Such flow is termed *laminar*. Viscous shear resistance in laminar flow is due to the molecular momentum exchange and the intermolecular forces described in Section 2.8. Dye or smoke tracers injected into such a flow appear as long, undisturbed streaklines[2] because the random molecular cross mixing in such flows is extremely slow.

The flow pattern associated with the rapid flow of a low-viscosity fluid is quite different. Injected tracers are almost immediately mixed throughout the flow by random currents superimposed on the mean through-flow. This random eddying motion is called turbulence.

The laminar and turbulent behavior of fluid flow was investigated by Osborn Reynolds[3] in the classical Reynolds experiment. The apparatus is illustrated in Fig. 4-5a. The water in the tank is first allowed to become perfectly still (by letting it sit until all convection currents have ceased). If the discharge valve (A) is open to allow only a very small flow, the dye injected into the flow at B will form a well-defined streak down the tube. As the valve A is opened further, a flow rate will be obtained where the dye streak will start to wave and then break up at the downstream end, as illustrated in Fig. 4-5b. If the flow is increased further, the transition to the slightly wavy unstable flow point moves upstream, as will the complete breakup point, and the dye mixing will become more violent.

The method by which these turbulent currents originate, propagate, are damped out, and re-form is still not completely understood. This random

[2] A streakline is then defined as a line made up of all particles that have passed through a given fixed "dye" point. A streakline is a "contamination" line.

[3] Osborn Rèynolds (1842–1912). His correct analyses and evaluations of phenomena observed during experimental laboratory investigations led to important formulations about laminar, turbulent, and transient flow.

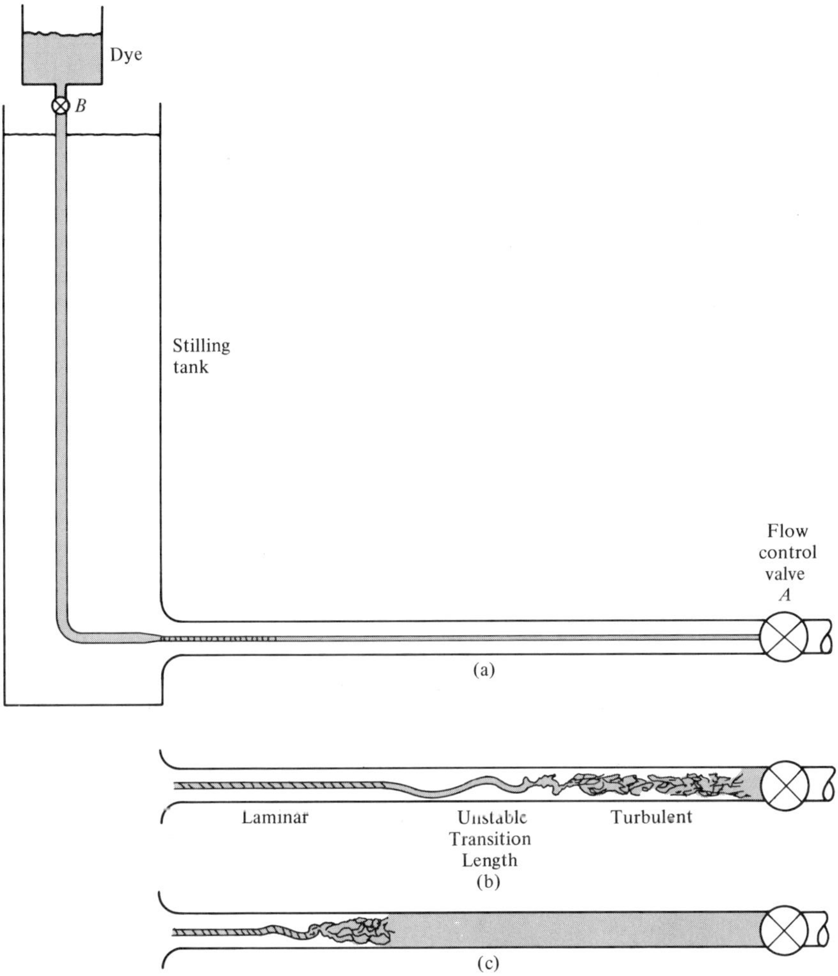

Fig. 4-5. Reynolds apparatus for the visualization of flow regimes in a tube: (a) laminar flow, (b) flow with transition to turbulent flow, (c) same as (b) but with velocity (Reynolds number) higher than that for the case shown in (b).

eddy motion is much more effective than the molecular mode in transporting mass and momentum in parallel with, and transverse to, the mean flow motion. The existence and degree of turbulence depend on (1) the energy state of the mean fluid flow, (2) the ability of the fluid to damp out the random turbulent motion through viscous dissipation, and (3) the damping effect due to the confining solid boundaries. Since no fluid can pass through a solid boundary, the normal components of all motion are zero at the boundaries. The random turbulent motion is thereby damped out at the boundary. However, turbulence apparently originates in the high-shear-stress, nearly laminar flow layer immediately adjacent to these same solid boundaries.

The study of turbulence and turbulent shear flow has presented a difficult problem for quite some time and continues to be a topic for advanced fluid flow study and research.

REYNOLDS NUMBER

As stated above, the transition to turbulent flow and the degree of turbulence intensity depends on (1) the kinetic energy of the mean flow, which is an indication of the ability of the flow to supply the energy that makes up the random turbulent motion; (2) the dissipative effects of the viscous forces[4]; and (3) the physical confinement of the fluid. If these effects are combined to indicate a comparison by considering (1) the kinetic energy of a fluid mass, (2) the viscous work being done on the mass, and (3) the boundaries within which the flow occurs, a quantity can be found that indicates whether the flow is *laminar* or *turbulent* and the degree of turbulence.

Consider the rate of kinetic-energy propagation through an area normal to the pipe axis at any station along the pipe, and the rate at which work is done in overcoming the shear stresses, or the ratio

$$\frac{\text{KINETIC-ENERGY RATE}}{\text{SHEAR STRESS WORK RATE}}$$

This quotient is dimensionless because of the equivalence between work and energy conversion. The kinetic-energy rate is

$$\frac{\dot{m}\mathrm{v}^2}{2g_c}$$

where $\dot{m}$ is the mass flow rate, which depends on the velocity, density, and area, and is $\dot{m} = \rho A\mathrm{v}$. The shear stress work rate is the shear force times the velocity, which is $\tau A\mathrm{v}$. From the previous discussion of shear stress, a representative value for τ can be taken as $\mu\mathrm{v}/L$, where all the quantities

[4] Although the viscous forces damp or dissipate the turbulent eddy motion, it is of interest that if the viscosity were identically zero, there would be no mechanism by which this random eddy turbulence could be originated within a flow. This apparent paradox is perhaps the underlying reason for the observed behavior of turbulent flow. The eddies are dissipated by the turbulence, but the velocity gradients caused by the viscous shear produce a *dynamic instability* within the fluid, which breaks down into a turbulent eddy formation.

are not specifically defined flow field quantities but are quantities characteristic of the entire flow. The characteristic velocity v is taken as the mean velocity obtained by dividing the mass flow rate by the product of the area and density at that area station. The ratio then becomes

$$\frac{\rho A \mathrm{v}^3}{(\mu \mathrm{v}/L)Ag_c} = \frac{\rho \mathrm{v} L}{\mu g_c} = N_R$$

where the factor ½ has been dropped because only characteristic quantities are of significance. This dimensionless ratio, which characterizes the flow, is known as the *Reynolds number*, after Reynolds. In terms of the kinematic viscosity ν,

$$N_R = \frac{\mathrm{v}L}{\nu}$$

In the case of flow through a round tube or pipe, the only possible unique characteristic dimension L is the diameter or radius. The diameter is conventionally used. Because of the arbitrariness in choosing characteristic quantities, the *magnitude* of the Reynolds number has no particular significance. Since it depends on the specific characteristic quantities and on the geometry, these quantities should be specified when a value of Reynolds number is given. For example: "The lower critical[5] Reynolds number for flow in a *round pipe* based on pipe *diameter* and *mean through-flow* velocity is 2000." It is of particular interest that the fluid need not be specified, for the same reason that the magnitude of the velocity is not specified: These quantities are *intrinsic*, or are *independent of the specific size* of the system.

The Reynolds number can also be obtained as a ratio of momentum rate to viscous force:

$$\frac{\text{MOMENTUM RATE}}{\text{VISCOUS FORCE}} = \frac{m\mathrm{v}/g_c}{\tau A} = \frac{\rho A \mathrm{v}^2}{\mu(\mathrm{v}/L)Ag_c} = \frac{\rho \mathrm{v} L}{\mu g_c} = N_R$$

The momentum transport rate characterizes the inertia of the fluid flow and therefore the effectiveness by which turbulence can be generated in the flow. The significance of the Reynolds number will be considered in later chapters.

[5] "Critical" Reynolds number will be discussed in Chapter 9. The lower critical Reynolds number for pipe flow is that below which turbulence will be damped out by viscous action.

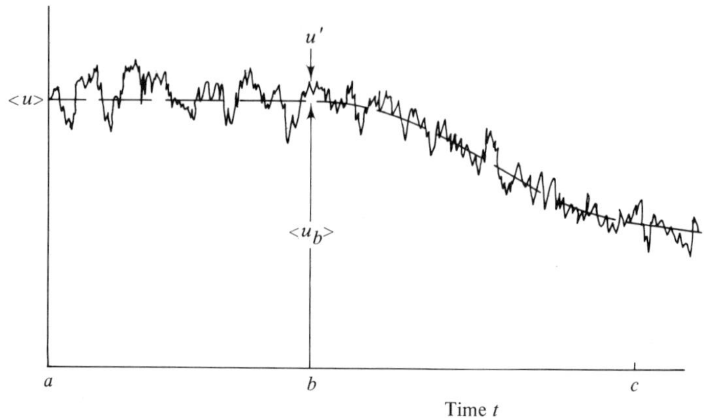

Fig. 4-6. Time history of velocity at a fixed point in a turbulent flow field. During the interval a-b, the time-average (or time-mean) velocity is $\langle u \rangle$. The velocity u in the x direction at b can be considered to be made up of $\langle u \rangle$ plus u' at the time t_b.

If an instrument for measuring instantaneous velocity (such as a hot-wire anemometer[6]) were placed in a turbulent stream, an oscillograph recording of the velocity signal as shown in Fig. 4-6 would be obtained. This trace represents the time history of the velocity at a fixed point in the flow field or $u(t$ only). During the period a-b the time-mean velocity $\langle u \rangle$ is seen to be constant. This mean velocity is defined as

$$\langle u \rangle \equiv \frac{1}{\Delta t} \int_{\Delta t} u(t)\, dt$$

where Δt is an averaging period that is at least an order of magnitude longer than the largest period of the random fluctuations. During the period a-b the mean velocity $\langle u \rangle$ is said to be "steady." However, the mean velocity varies with time during the period b-c and is therefore unsteady.

If the point at which the above record was made is investigated more thoroughly, the flow velocity can be considered to be made up of the mean velocity $\langle \vec{u} \rangle$ plus a random vector component $\vec{v}'$ due to the turbulence, as indicated in Fig. 4-7. This component is random in (1) time, (2) direction, and (3) magnitude. Several possibilities for $\vec{v}'$ are indicated.

In turbulent flow it is sometimes advantageous and possible to consider only the mean flow (ignoring turbulent contributions to stresses) when

[6] An extremely small (typically 1-mm-long, 0.005-mm-diameter) heated wire. The wire resistance changes as it is cooled in the airstream. An electronic circuit senses the resistance change and gives a signal that is a function of the instantaneous velocity.

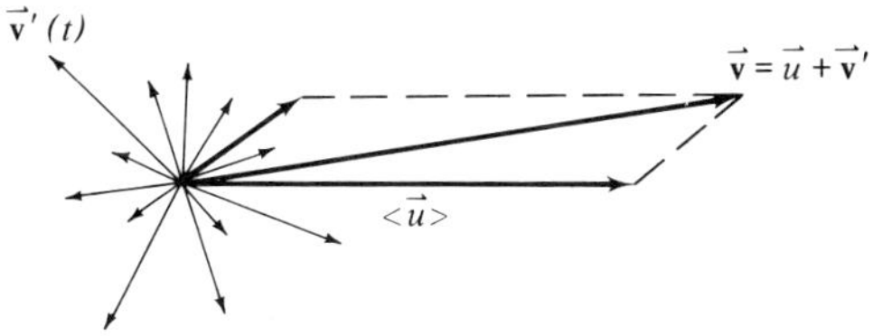

Fig. 4-7. Velocity at a fixed point in a flow field. The velocity can be considered to be a combination of the mean velocity $\langle u \rangle$ and a random fluctuating component $\vec{v}'(t)$. At one particular instant, the velocity is given by $\vec{v}$, as shown.

determining viscous and pressure forces and their relationship with the velocity field. In this case, the details of the turbulent motion are not explicitly included but appear implicitly in the shear stress term as an overall effect.

4.3 IDEAL FLOW—IDEAL FLUID

It is sometimes possible to neglect the viscous forces altogether to determine an approximate or partial evaluation of a fluid flow field and its consequences (for example, determination of the pressure forces). A hypotheoretical fluid that would conform to this behavior would be one with zero viscosity and is defined as an *ideal fluid*. Obviously no real fluid can conform to such a definition. However, Prandtl[7] observed that for the flow around streamlined bodies (for example, airfoils) of a fluid with relatively small viscosity, the region in which the viscosity affected the flow was confined to an extremely thin layer (*boundary layer*) immediately adjacent to the solid boundaries. The fluid in the flow field outside this boundary layer can therefore be considered to behave essentially as an *ideal fluid*. The viscous boundary-layer effect can sometimes be neglected to determine the velocity and pressure fields around the body when the layer is quite thin with respect to the thickness or characteristic dimension of the body. These conditions and their consequences and results will be discussed in further detail in Chapters 7 and 14.

Whereas an *ideal* fluid exhibits nonviscous behavior, the term *perfect* gas connotes gas behavior conforming to the perfect gas law $p/\rho = RT$, and negligible intermolecular forces. Such a "perfect gas" is indeed an

[7] Ludwig Prandtl (1875–1953), German, the founder of present-day fluid mechanics. Introduced the concept of the physical existence of the boundary layer, probably the single most important factor governing fluid flow phenomena. Prandtl's greatest attribute was (like Reynolds) the ability of understanding both the physical phenomena he studied and the ability to formulate these into simple mathematical models that could be manipulated and analyzed.

idealized gas. However, a perfect gas does possess a viscosity. An ideal, perfect gas is then a hypothetical perfect gas with zero viscosity. This convention facilitates discussion of these two independent idealizations without going into specific details each time.

4.4 STREAMLINES AND STREAMTUBES

The concepts of streamlines and streamtubes are helpful in considering the derivation of equations of motion from the physical principles involved. *A streamline is defined as a line drawn everywhere tangent to the velocity vector in a flow field.* As a consequence of this definition: (1) only one unique streamline can exist through a given point in a flow field, (2) two streamlines cannot cross, and (3) there can be no flow across streamlines. In nonsteady flow, the streamline pattern or field changes with time and a fluid particle on one streamline at one instant may be on another streamline at a later time. In the case of steady flow, fluid particles follow streamlines; therefore, streamlines and particle *path lines*[8] are identical. In turbulent flows the concept of the streamline is applied to the mean flow as discussed in Section 4.1.

A *streamtube* is a tube having a surface made up of streamlines. The cross-sectional area of the streamtube is small enough that the variation of any flow field property (p, ρ, v, and so on) over this area is negligible.

EQUATIONS OF A STREAMLINE

An equation (or set of equations) that represents the streamlines in space with respect to some reference coordinate system can be determined. Considering the streamline shown in Fig. 4-8a, the velocity $\vec{v}$ at P is collinear with the streamline element $\vec{\delta s}$ (which follows from the definition of the streamline); also, at P, $\vec{\delta s} = \vec{v}\,\delta t$. The three components of $\vec{\delta s}$ and $\vec{v}$ in cartesian and cylindrical coordinates are shown in Fig. 4-8b and c. The velocity components in the cylindrical coordinate system (Fig. 4-8d) are radial (u_r), tangential (v_θ), and axial (w_z). The unit vectors $\hat{x}$, $\hat{y}$, $\hat{z}$ and $\hat{r}$, $\hat{\theta}$, $\hat{z}$ are indicated in Fig. 4-2b and c, respectively. It is noted that the vectors $\vec{\delta s}$ and $\vec{v}$ can be expressed in terms of their components in the example x,y,z coordinate system as

$$\vec{\delta s} = \hat{x}\,\delta x + \hat{y}\,\delta y + \hat{z}\,\delta z \qquad \text{and} \qquad \vec{v} = \hat{x}u + \hat{y}v + \hat{z}w$$

so that the equation $\vec{\delta s} = \vec{v}\,\delta t$ can be written

$$\hat{x}\,\delta x + \hat{y}\,\delta y + \hat{z}\,\delta z = (\hat{x}u + \hat{y}v + \hat{z}w)\,\delta t$$

[8] A *path line* is the line traced out by a single particle in the flow. Streamlines, streaklines, and path lines are seen to be all different for unsteady flow and coincident for steady flow.

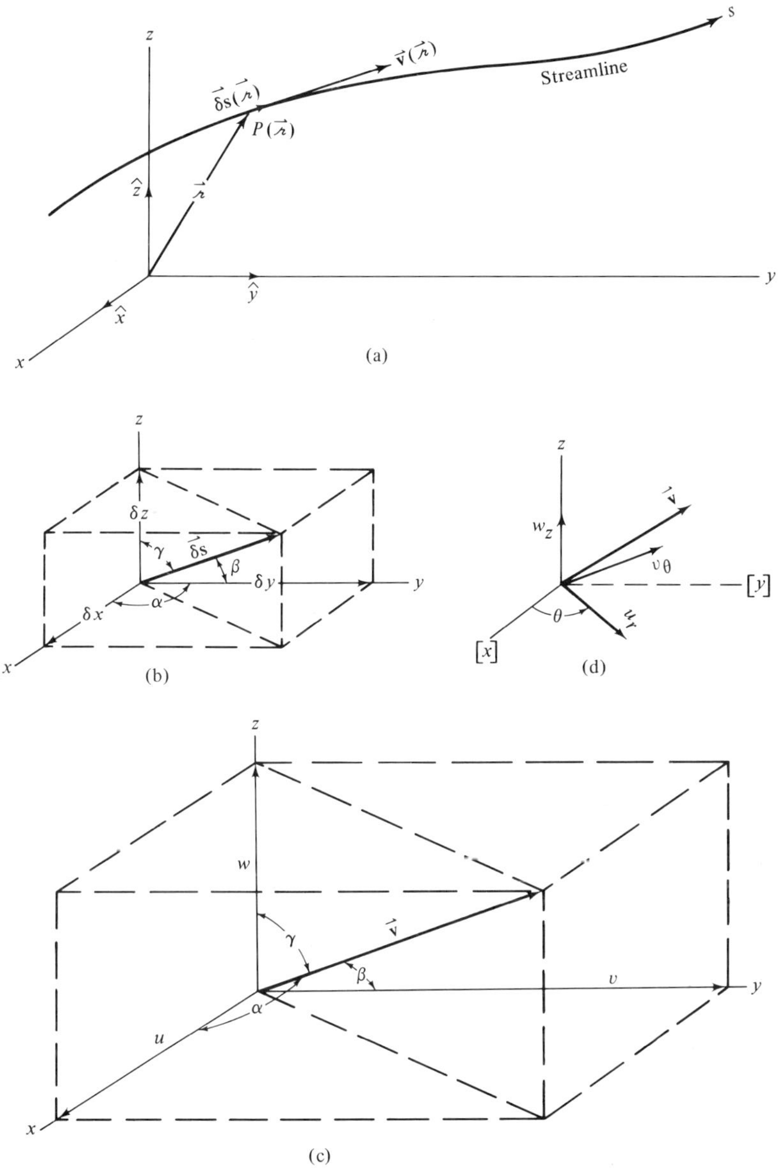

Fig. 4-8. Velocity and a line element of a streamline showing the local condition that $\vec{v}$ is parallel to $\vec{\delta s}$ and the decomposition of the vectors into their coordinate components.

Since each coordinate direction is independent, this equation can be considered as three independent component equations,

$$\delta x = u\,\delta t$$

$$\delta y = v\,\delta t$$

$$\delta z = w\,\delta t$$

The quantity δt is the same in each equation. In the limit with δs, δx, and so on, considered as differential quantities, this condition gives

$$\frac{dx}{u} = \frac{dy}{v} = \frac{dz}{w} \tag{4-1}$$

which represents *two* independent equations for the component slopes of the streamline. If the velocity field is known, these equations can be integrated to give the equations of the streamline, for example,

$$\frac{dy}{dx} = \frac{y(x, y, z)}{u(x, y, z)} \qquad \frac{dz}{dx} = \frac{w(x, y, z)}{u(v, y, z)} \tag{4-2}$$

It is noted that the direction cosines of the streamline element (and velocity vector) with respect to the specified coordinate system are given by

$$\cos\alpha = \frac{dx}{ds} \qquad \cos\beta = \frac{dy}{ds} \qquad \cos\gamma = \frac{dz}{ds} \tag{4-3}$$

4.5 CONTROL VOLUME

Several preliminary steps are necessary when setting up equations that mathematically describe the physical behavior of a flow field:

1. Locate and specify a coordinate system origin in the physical flow field.
2. Choose a coordinate system that best matches the characteristics of the physical boundaries of the problem.
3. Specify the flow region (that is, the flow field) to be considered.

In any physical or engineering problem in mechanics it is highly advisable to make a sketch of the problem that includes the above quantities. The specific flow region to be considered and for which the mathematical equations are to be written is defined as a *control volume*. It may be a large flow field for which the overall or integral relationships are to be determined, or it may be a differential volume element in which case differential equations or relationships are determined. *In any case, the control volume (C.V.) must be specified so that it contains only and wholly fluid with no "machine" or other bodies included.* This restriction is a consequence of the physical principles and "laws" to be applied. For example, Newton's second law correctly applied to a fluid C.V. states that the rate of change of momentum *of the fluid within the control volume* is a consequence of all the forces applied *on the fluid* within the control volume. Where pump or turbine vanes or other moving or stationary objects project into a fluid, the control volume boundaries must be drawn around them. Forces at such *moving* surfaces provide the means by which mechanical work is done *on* the fluid.[9]

4.6 CONSERVATION OF MASS AND THE CONTINUITY EQUATION

The principle of *conservation of mass applied to a flowing fluid* results in a mathematical expression referred to as the *continuity equation.* Stated physically, the mass that flows into an open system (or control volume) flows out or is partially stored. For the fixed control volume, it is obvious that partial storage cannot continue indefinitely, so the flow must be unsteady. For incompressible flow, any fluid storage is impossible in a fixed control volume. (By definition there are no initial pockets or voids in a flow field or control volume.)

Consider the process of gas compression within a cylinder with a leaky valve (Fig. 4-9) as a physical illustration of an unsteady-flow problem. The volume above the piston cannot be defined as a control volume, because the piston moves through it, or else its lower boundary must move up with the piston. The displacement volume that remains above the piston at the end of its stroke (as outlined in Fig. 4-9a and reproduced in Fig. 4-9b) is the only fluid control volume that can be considered as fixed in space. As the piston moves up, gas crosses the lower control volume boundary; part of this gas will be stored and part will leak out through the valve. Since the gas *inflow rate*, the *storage rate*, and *leakage* or *outflow rate* vary with time, any equation written for this conservation principle must be a *rate equation* expressing the inflow, outflow, and storage rates at an arbitrary instant:

$$\text{MASS INFLOW RATE} \xrightarrow[\text{INTO}]{\text{GOES}} \text{STORAGE RATE} + \text{OUTFLOW RATE}$$

Mathematically it is preferable to deal with the *net* outflow rate, which is just the outflow rate minus the inflow rate. Transposing the inflow

[9] This control volume definition appears overrestricted. A method of relaxing this restriction is given in Chapter 5 by introducing the concept of an actuator disc.

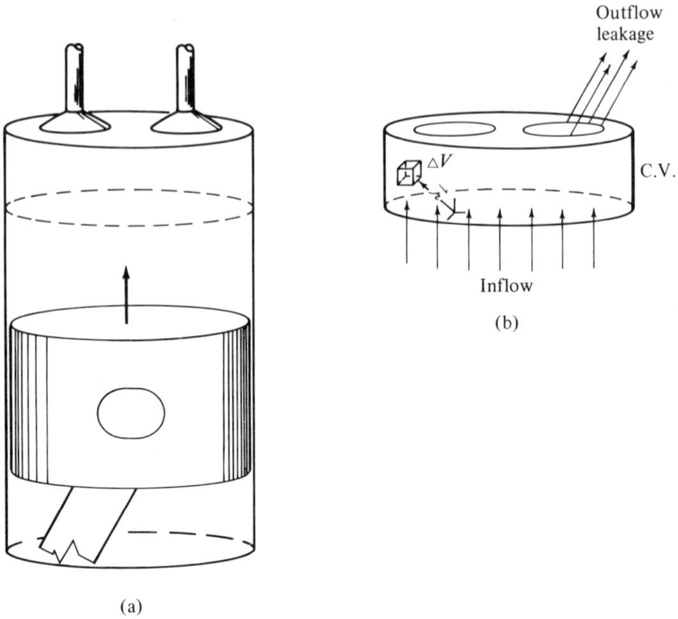

Fig. 4-9. Eulerian control volume for the unsteady flow problem of the compression of a gas within a cylinder with a leaking exhaust valve. Part (b) illustrates the outline of the fluid control volume that satisfies the condition of a fixed-in-space control volume.

quantity, the above statement becomes

$$\text{STORAGE RATE} + \text{NET OUTFLOW RATE} = 0 \tag{4-4}$$

Equation (4-4) is the statement of conservation of mass applied to the general case of unsteady, compressible fluid flow.

At any instant the mass of any differential element within the control volume is just $\rho\,\Delta V$, where the density of the element depends on the instant at which it is determined and on the position within the control volume where it is measured, or $\rho(\vec{r}, t)$. The total mass within the control volume at any instant is obtained by summing or integrating over all the mass elements within the C.V.,

$$\iiint_{\text{C.V.}} \rho(\vec{r}, t)\, dV = \text{TOTAL MASS IN C.V.}$$

The *mass storage rate* must be identical to the rate of change of the mass within the control volume with respect to time,[10] or, stated mathematically,

[10] A *positive storage* rate corresponds to a net mass *increase* rate within the control volume.

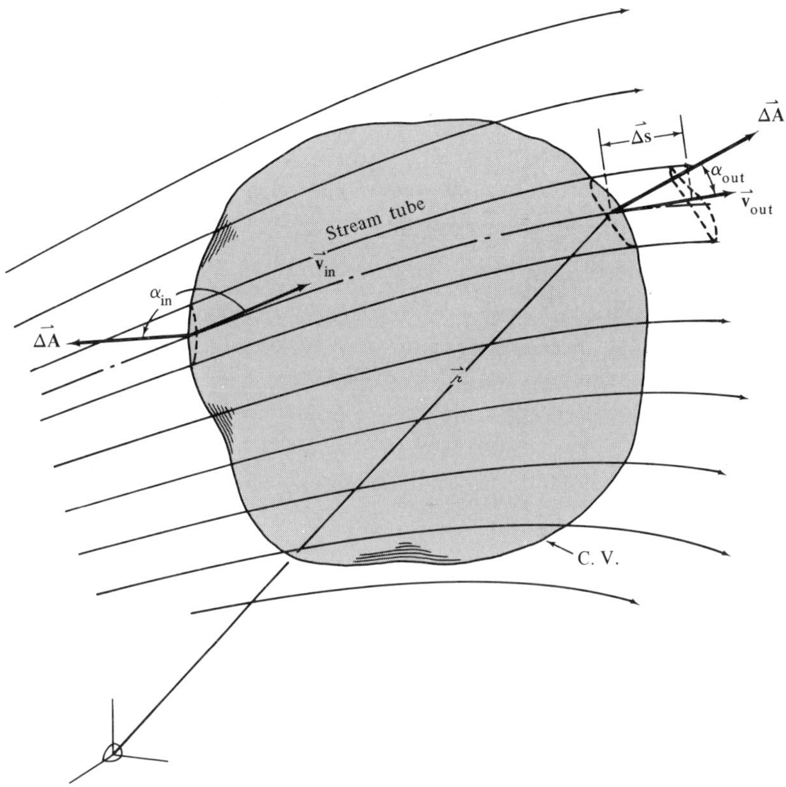

Fig. 4-10. General or arbitrary control volume showing a streamtube and the relationship between inflow and outflow velocities and area elements

$$\frac{\partial}{\partial t}\left[\iiint_{\text{C.V.}} \rho(\vec{r}, t)\, dV\right] = \text{MASS STORAGE RATE}$$

In order to determine an expression for a mass outflow rate, consider the more general representation for the flow field as illustrated in Fig. 4-10. In a small time period δt,[11] an amount of fluid within the volume element indicated will leave the system through the streamtube. This volume of fluid is $\vec{\delta s}\cdot\vec{\Delta A}$,[12] where $\vec{\delta s}$ is measured along the streamline. Therefore,

$$\vec{\delta s} = \vec{v}\,\delta t$$

or

$$\vec{\delta s}\cdot\vec{\Delta A} = \vec{v}\cdot\vec{\Delta A}\,\delta t$$

[11] Since time and space coordinates are independent, their variations are independent. To emphasize this independence, δ will be used to denote a small time increment or a quantity variation occurring at δt, and Δ will be used with elemental space dimensions.

[12] δs is the distance moved by a fluid particle with velocity $\vec{v}$ in $\vec{\delta t}$, $\delta s = \vec{v}\,\delta t$.

and the mass outflow is

$$\rho\, \overrightarrow{\delta s} \cdot \overrightarrow{\Delta A} = \rho \vec{v} \cdot \overrightarrow{\Delta A}\, \delta t$$

The outflow *rate* through the streamtube is obtained by dividing by δt and taking the limit as δt approaches zero. The volume outflow rate is then $\vec{v} \cdot \overrightarrow{dA}$ and the mass outflow rate is $\rho \vec{v} \cdot \overrightarrow{dA}$. Note that because $\overrightarrow{\Delta A}$ is measured outward from the C.V., the volume element $\overrightarrow{\delta s} \cdot \overrightarrow{\Delta A}$ and therefore the quantity $\vec{v} \cdot \overrightarrow{dA}$ is negative at the inflow side of the streamtube because the angle between $\vec{v}$ (or $\overrightarrow{\delta s}$) and $\overrightarrow{\Delta A}$ is always greater than 90° here. Therefore, the summation $\sum \rho \vec{v} \cdot \overrightarrow{\Delta A}$ over the inlet and outlet ends of the streamtube gives the *net* or *total outflow rate*. By definition there can be no flow across the sides of the streamtube. If the net outflow is then summed over all streamtubes, the net outflow rate for the entire control volume is

$$\oiint_{A_{\text{C.V.}}} \rho \vec{v} \cdot \overrightarrow{dA} = \text{NET OUTFLOW RATE}$$

where $\oiint_{A_{\text{C.V.}}}$ symbolizes integration over the entire area enclosing the control volume. The complete continuity equation is then

$$\frac{\partial}{\partial t} \iiint_{\text{C.V.}} \rho\, dV + \oiint_{A_{\text{C.V.}}} \rho \vec{v} \cdot \overrightarrow{dA} = 0 \tag{4-5}$$

For the Eulerian representation, the control volume boundary is defined as fixed in space; its location and shape do not change with respect to time. Since the *value* of an integral is a function only of its limits, and since these do not vary with time, the differentiation with respect to time may be placed inside the integral sign, so that (4-5) becomes[13]

$$\iiint \frac{\partial \rho}{\partial t}\, dV + \oiint \rho \vec{v} \cdot \overrightarrow{dA} = 0 \tag{4-6}$$

This equation is almost the most general equation of continuity for a fluid flow field in a fixed control volume. The only restrictions are (1) those stated with respect to choice and specifications of the control volume, (2) the condition that the system is nonrelativistic (that is, the conversion of mass to energy is accomplished by a negligible change in mass if the

[13] For example, refer to the material on the Leibnitz rule for differentiation under the integral sign in an advanced calculus text. The C.V. and $A_{\text{C.V.}}$ integral designations will be dropped. $\iiint$ and $\oiint$ are understood to be control volume and control surface integrals.

fluid flow system is one wherein energy conversion is occurring), and (3) there are no *sources of mass* within the control volume. The concept of sources (and sinks) will be considered in Chapter 14.

CASE I

STEADY FLOW

If the flow is steady (the field is stationary) so that ρ ($\vec{r}$ only), $\partial\rho/\partial t = 0$ and the continuity equation becomes

$$\oint\!\!\oint \rho\vec{\mathrm{V}}\cdot\vec{\mathrm{dA}} = 0 \tag{4-7}$$

which states that the net mass outflow (sum of all in- and outflows) $\equiv 0$, since there can be no storage within the control volume, or that the outflow

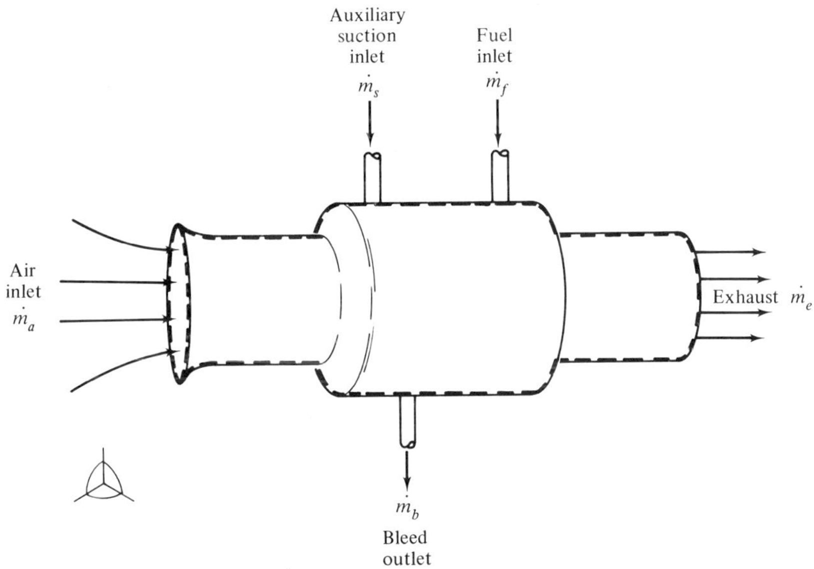

Fig. 4-11. Gas turbine as an example of a control volume with several flow paths.

rate must be equal to the inflow rate. If this result is written in the form of a scalar equation in terms of the inflows and outflows at the various branches of a system as indicated in Fig. 4-11, the statement of continuity becomes

$$\sum_{\text{net out}} \dot{m} = \sum_{\text{out}} \dot{m} - \sum_{\text{in}} \dot{m} = 0 \tag{4-8}$$

or, for the example illustrated,

$$\dot{m}_a + \dot{m}_s + \dot{m}_f = \dot{m}_b + \dot{m}_e$$

where all $\dot{m}$'s are scalar quantities, without regard for the sign indicating in- or outflow (the units are m/t or lb_m/sec, kg/sec, g/sec, and so on). Because the velocity is not perpendicular to all the elements at the inlet area A_a, and the density and velocity magnitude possibly vary over the inlet area, the inlet mass flow is given by

$$\dot{m}_a \equiv \left| \iint_{A_a} \rho \vec{\text{v}} \cdot \vec{\text{dA}} \right|$$

where the integral in general would have to be evaluated in detail in the form

$$\dot{m}_a = \int_{\theta=0}^{2\pi} \int_{r=0}^{a} \rho(r, \theta)\text{v}(r, \theta) \cos(\vec{\text{v}}, \vec{\text{dA}}) r \, dr \, d\theta$$

where $\cos(\vec{\text{v}}, \vec{\text{dA}})$ specifically means the cosine between the velocity vector and the area-element vector measured *from* $\vec{\text{v}}$ *toward* $\vec{\text{dA}}$. In Fig. 4-11 it is the angle between $\vec{\text{v}}$ and the "inlet axis." This angle would have to be expressed for each area element as a function of r and θ to evaluate the above integral.

Case II

INCOMPRESSIBLE FLOW

If the fluid and other conditions are such that compressibility effects are negligible, the density is constant and $\partial\rho/\partial t = 0$. Since the density is constant everywhere within the control volume flow field, ρ can be removed from the integration procedure and canceled, leaving

$$\oiint \vec{\text{v}} \cdot \vec{\text{dA}} = 0 \tag{4-9}$$

In the same manner that the mass flow rate at any station was defined in case I, the instantaneous volume flow rate at any station in the system at any instant is $Q = \left| \iint \vec{\text{v}} \cdot \vec{\text{dA}} \right|$. (The units of Q are L^3/t; ft^3/sec, m^3/sec, cm^3/sec, or possibly gallons per minute, designated gpm.)

CASE III

STEADY, INCOMPRESSIBLE FLOW

For this case the continuity equation reduces to

$$\oiint \vec{v} \cdot d\vec{A} = 0$$

Again defining $Q = \left|\iint \vec{v} \cdot d\vec{A}\right|$ at any station in the flow system, the continuity equation in the form of (4-9) can be interpreted as $Q_{out} - Q_{in} = 0$. Therefore $Q = \text{const.}$ at any station. The difference between this case and case II is that Q at any station is a function of time and therefore is not constant.

AVERAGE VELOCITY: MEAN FLOW VELOCITY

The *average or mean through-flow velocity* $\bar{v}$ at any flow section is defined in the same manner as any other average quantity where a variation exists with respect to an independent variable,

$$\bar{v} = \frac{1}{A}\iint_A \vec{v} \cdot d\vec{A} \equiv \frac{Q}{A} \qquad \text{or} \qquad Q \equiv A\bar{v} \tag{4-10}$$

4.7 ILLUSTRATIVE EXAMPLE 4-1: Nozzle

An inlet adapter is constructed for a static test of a jet engine on a thrust stand. The inlet design is obtained from the specifications for

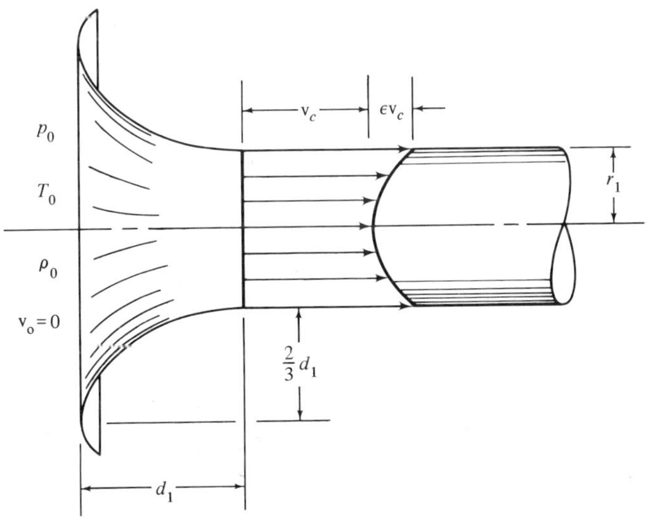

Fig. M4-1.

A.S.M.E. flow nozzles (A.S.M.E., *Fluid Meters: Their Theory and Application*, 5th ed., 1959), which specifies a 3:2 axis ratio elliptical profile, as shown. This nozzle construction is too abrupt and produces the approximate parabolic velocity profile indicated at the end of the contraction section. This profile is symmetrical about the centerline, with a 20 percent velocity variation from the center to the walls, and can be represented by

$$\mathrm{v}(r) = \mathrm{v}_c\left[1 + \epsilon\left(\frac{r}{r_1}\right)^2\right]$$

The centerline velocity is 180 ft/sec. The ambient pressure p_0 is 14.8 psia and the ambient temperature is $T_0 = 100°\text{F}$. Assuming that the flow from the surroundings to the end of this inlet is essentially incompressible, determine the mass flow rate ($\dot{m}$) and the average velocity $\bar{\mathrm{v}}$ at the end of the inlet contraction nozzle section; $d_1 = 2r_1 = 18$ in., $\epsilon = 0.2$.

$$\mathrm{v} = \mathrm{v}_c\left[1 + \left(\frac{r}{r_1}\right)^2\right]$$

$$\dot{m} = \left|\oint\!\!\oint_{A_1} \rho\vec{\mathrm{v}}\cdot\vec{\mathrm{dA}}\right|$$

$$= \rho_1 \int_{r=0}^{r_1} \mathrm{v}_c\left[1 + \epsilon\left(\frac{r}{r_1}\right)^2\right]2\pi r\, dr$$

$$= \pi r_1^2\rho_1\mathrm{v}_c \int_{r=0}^{r_1} \left[1 + \epsilon\left(\frac{r}{r_1}\right)^2\right] d\left(\frac{r}{r_1}\right)^2$$

Let $(r/r_1)^2 \equiv \xi$ and $A_1 = \pi r_1^2$:

$$\dot{m} = \rho_1 A_1 \mathrm{v}_c \int_0^1 (1 + \epsilon\xi)\, d\xi = \rho_1 A_1 \mathrm{v}_c\left(\xi + \frac{\epsilon\xi^2}{2}\right)\Bigg|_0^1$$

$$= \left(1 + \frac{\epsilon}{2}\right)\rho_1 A_1 \mathrm{v}_c$$

For $p_0 = 14.8$ psia and $T_0 = 100°\text{F}$, the density can be calculated as $\rho_0 \approx \rho_1 = 0.0714\ \text{lb}_\text{m}/\text{ft}^3$. With $r_1 = 9$ in. (or ¾ ft) and $\epsilon = 0.2$, the mass flow rate is

$$\dot{m} = (1.1)(0.0714)\pi\left(\frac{9}{16}\right)(180) = 25\ \text{lb}_\text{m}/\text{sec}$$

The average velocity $\bar{v}$ is determined from

$$\dot{m} = \rho_1 A_1 \bar{v}_1 = \rho_1 A_1 v_c\left(1 + \frac{\epsilon}{2}\right)$$

giving

$$\frac{\bar{v}}{v_c} = 1 + \frac{\epsilon}{2} = 1.1 \qquad \bar{v} = 198 \text{ ft/sec}$$

4.8 ONE-DIMENSIONAL FLOW: Uniform Flow

If the velocity is constant in magnitude and direction and if all other flow properties are constant over an area normal to the flow direction, the flow is said to be *uniform over that area.* If the flow properties are constant everywhere in a field, the *field* is said to be *uniform.* The flow in a wind-tunnel test section is an example of a uniform flow over an area and a uniform field. Every attempt is made to obtain a uniform flow field in the test section to "model" flight conditions.

A flow field where all flow properties can be represented as a function of only one space dimension is called one-dimensional flow. The flow in the ducting system represented in Fig. 4-12 is one example of a one-dimensional

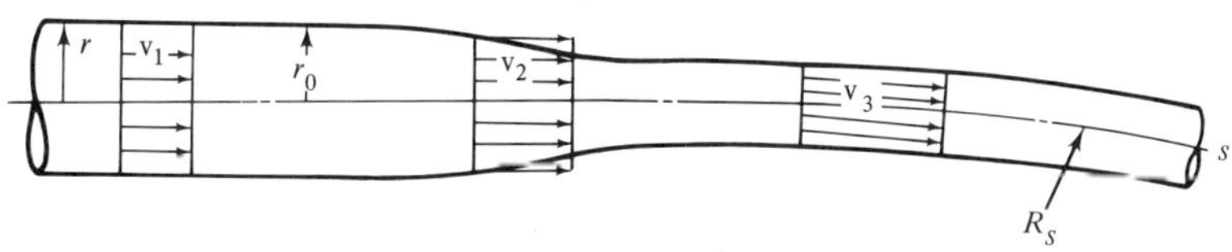

Fig. 4-12. Representation of an ideal (nonviscous) one-dimensional flow in a duct.

flow. The velocity at any station is constant. It does not depend on the normal coordinate n (measured perpendicular to the streamline coordinate s), but it does vary in the direction of flow (s direction). The pressure, density, temperature, viscosity, and so on, might also vary in the s direction, but not in the n direction. These quantities can then be represented as functions of s only, and the flow is then termed one-dimensional, with s as the "one dimension." Variations of velocity (consequently, pressure, density, and so on) in the n direction would make this flow two- or three-dimensional. However, if the area variation and curvature are small, the n variations of v, p, ρ, and so on, are also small. Specification of a quantity as "small" implies a comparison (that is, small with respect to *what*?). The comparison of significance in the above conditions is $R_s/d_0 \gg 1$; the radius of curvature R_s of any streamline must be much greater than the duct diameter d_0 for the flow to be considered as one-dimensional. This restriction is a consequence of the fact that a normal pressure *gradient* is necessary to curve the flow direction according to R_s. The pressure difference across the duct section depends on the width d_0 of the duct.

The example of Fig. 4-12 has been indicated to be physically impossible for a real flow because the condition of zero velocity at the wall has not been maintained. (Such a flow would be hypothetically possible for an *ideal* fluid.) For a real flow with the necessary condition of zero velocity at the wall, a strictly one-dimensional flow is seen to be impossible (neglecting the trivial case of no flow at all), since the velocity must vary in the n direction.

A flow such as that shown in Fig. 4-13 is termed *quasi-one-dimensional* because the conditions of small variation of area and radius curvature are maintained. The velocity varies in the n direction and this variation[14]

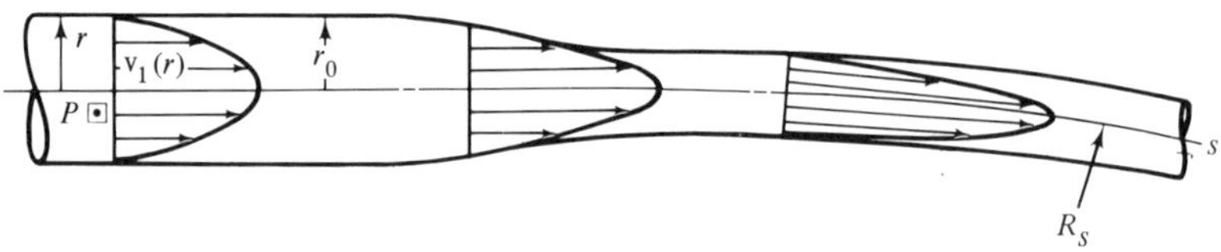

Fig. 4-13. Representation of the quasi-one-dimensional flow of a real fluid in a duct.

$[\partial \mathrm{v}/\partial n$ or $(\partial \mathrm{v}/\mathrm{v}_0)/(\partial n/d_0)]$ is seen to be quite large near the wall. Such a flow is quasi-one-dimensional if the velocity profiles are the same *shape* at every station. For example, if they are all parabolic, these profiles could all be represented mathematically by

$$\frac{\mathrm{v}(r)}{\mathrm{v}_0} = 1 - \left(\frac{r}{r_0}\right)^2$$

where the functional relationship $\mathrm{v}(r)/\mathrm{v}_0 = f(r/r_0)$ is the same everywhere.

For small streamwise-area variation $(dr_0/ds \ll 1)$ and small curvature $(R_s/d_0 \ll 1)$, the normal variation of pressure $\partial p/\partial n$ will be negligible $(\ll \partial p/\partial s)$, even though the velocity varies from zero to a maximum in the normal direction at any station (or s). Consider the small element of fluid at P with the velocity $\mathrm{v}(r)$ indicated in Fig. 4-13. Newton's first law states that a particle at rest will remain at rest unless acted upon by an external force and that a particle in motion will remain in a state of *uniform* motion

[14] It is sometimes preferable to present such variations as dimensionless quantities in making comparisons. Here, for example, $\partial \mathrm{v}/\partial n > 1$ would have no significance, because "1" is intended to be "unity" and therefore dimensionless. However, $\partial(\mathrm{v}/\mathrm{v}_0)/\partial(n/d_0)$ is a relative comparison of significant quantities, so the equivalent statement $(d_0/\mathrm{v}_0)(\partial \mathrm{v}/\partial n) > 1$ has a significant meaning. (v_0 is the centerline velocity.)

(constant speed and direction) unless acted on by an external force, in which case its motion will change according to Newton's second law. At station 1, the fluid particle (or element) must be traveling in a straight line according to the velocity v at that radius [v(r)]. Since its motion is not changing here, there can be no unbalanced external forces on the element in the normal direction. The pressure on the top must then be the same as that on the bottom of the element, or $\partial p/\partial n = 0$. For a straight duct section this condition can be proved analytically by setting up Newton's second law for the fluid element and noting that the normal acceleration a_n is zero. The pressure, and consequently the density, will not vary in the n direction across any section perpendicular to the flow.

At the station where the area changes, or the duct curves, there is a change in motion in the normal direction that produces a change in pressure in this direction. However, if the rate of change of area and curvature are small, this pressure variation will also be small, and the flow can be assumed to be quasi-one-dimensional. Since all real flows conform to the condition of zero velocity at the wall, the adjective "quasi" is sometimes dropped, and flows where the rate of area change with s (or dr_0/ds) and the curvature (or R_s/d_0) are small are referred to as *one-dimensional.*

4.9 DIFFERENTIAL EQUATION OF CONTINUITY

The overall or integral approach was used in the above development of the continuity equation from the principle of conservation of mass. A differential equation may be derived by applying the same principles to a differential-sized C.V. *in a specific coordinate system.*

Consider the element illustrated in Fig. 4-14a and b. The starting point for determination of the various rates of change or derivatives of the quantities will again be taken at the center of the volume ΔV so as to have a fixed reference point to avoid ambiguity, and to eliminate the necessity of having to consider second-order differentials in the final steps. First, the rate of change of the mass within the differential control volume with respect to time is

$$\frac{\partial}{\partial t}(\rho\, dV) \equiv \frac{\partial}{\partial t}(\rho\, dx\, dy\, dz) \equiv \frac{\partial \rho}{\partial t}\, dx\, dy\, dz$$

since x, y, z, and t are independent coordinates. The *net* outflow is just $\sum$ outflows $-$ $\sum$ inflows through all the surfaces with the velocity components assumed positive, as indicated. The y component of mass flow through the center face due to the propagation of mass in the y direction by the velocity field is just $\rho v\, dx\, dz$, as shown in Fig. 4-14c. The quantities ρ and v vary in the y direction, so the mass flow out through the right-hand

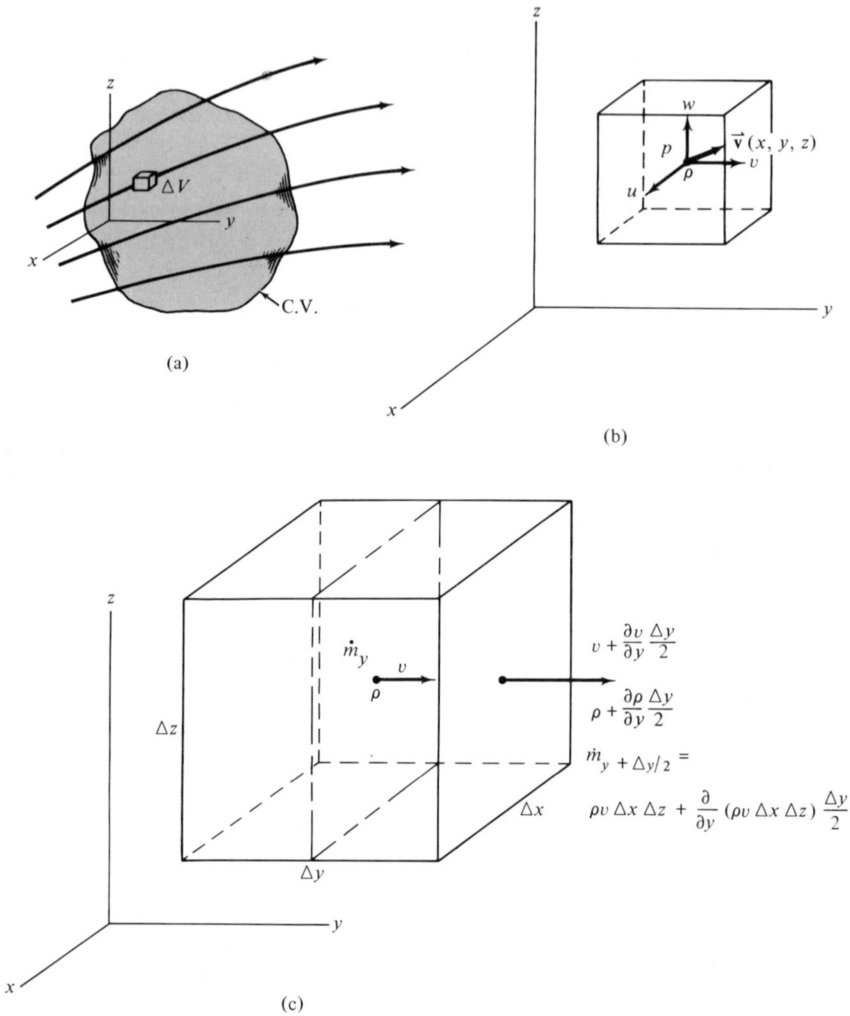

Fig. 4-14. Representation of a volume element of a control volume in a flow field to which the principle of conservation of mass is applied. The element is sufficiently small that the values of the variable quantities (v, ρ, p, and so on) can be considered to be the average values over any *face* of the element, and these values can be considered to be the values at the center of that face [for example, the y face as shown in (c)].

face is the mass flow that existed at the center face, $\rho v(dx\,dz)$, plus the variation from the center to the right-hand face, or

$$\rho v(\Delta x\,\Delta z) + \frac{\partial}{\partial y}[\rho v(\Delta x\,\Delta z)]\frac{\Delta y}{2} \equiv \rho v\,\Delta x\,\Delta z + \frac{1}{2}\frac{\partial \rho v}{\partial y}\Delta x\,\Delta y\,\Delta z$$

The $(\partial\,\Delta x/\partial y)$ and $(\partial\,\Delta x/\partial z)$ are zero since x, y, and z are *independent* coordinates. The mass flow *in* through the left face is

$$\rho v\,\Delta x\,\Delta z + \frac{\partial}{\partial y}[\rho v(\Delta x\,\Delta z)]\left(-\frac{\Delta y}{2}\right) \equiv \rho v\,\Delta x\,\Delta z - \frac{1}{2}\frac{\partial \rho v}{\partial y}\Delta x\,\Delta y\,\Delta z$$

or the net mass efflux from the element through the y faces[15] (surface area $\Delta x\,\Delta z$) is the outflow minus the inflow, or

$$\frac{\partial \rho v}{\partial y}\Delta x\,\Delta y\,\Delta z$$

Similarly, the net outflows through the x and z faces are determined as

$$\frac{\partial \rho u}{\partial x}\Delta x\,\Delta y\,\Delta z \qquad \text{and} \qquad \frac{\partial \rho w}{\partial z}\Delta x\,\Delta y\,\Delta z$$

Since the sum of the storage rate and net outflow rate must be equal to zero,

$$\frac{\partial \rho}{\partial t}\Delta V + \left(\frac{\partial \rho u}{\partial x} + \frac{\partial \rho v}{\partial y} + \frac{\partial \rho w}{\partial z}\right)\Delta V = 0$$

If the net effect for all elements within this volume is added, or the above expression is integrated over the overall control volume (C.V.),

$$\oiiint \frac{\partial \rho}{\partial t}\,dV + \oiiint \left(\frac{\partial \rho u}{\partial x} + \frac{\partial \rho v}{\partial y} + \frac{\partial \rho w}{\partial z}\right) dV = 0 \tag{4-11}$$

or

$$\left(\frac{\partial \rho}{\partial t} + \frac{\partial \rho u}{\partial x} + \frac{\partial \rho v}{\partial y} + \frac{\partial \rho w}{\partial z}\right) dV = 0 \tag{4-12}$$

From either (4-11) or (4-12) the differential equation of continuity is determined as

$$\frac{\partial \rho}{\partial t} + \frac{\partial \rho u}{\partial x} + \frac{\partial \rho v}{\partial y} + \frac{\partial \rho w}{\partial z} = 0 \tag{4-13}$$

where ρ, u, v, and w (the dependent variables) are functions of x, y, z, and t (the independent variables). Corresponding to cases I, II, and III above, this equation would reduce to the following forms.

[15] The face is designated by the direction of its area vector.

CASE I

STEADY FLOW

$$\frac{\partial \rho u}{\partial x} + \frac{\partial \rho v}{\partial y} + \frac{\partial \rho w}{\partial z} = 0$$

where ρ, u, v, and w are functions of x, y, and z, or expressed symbolically,

$$\begin{pmatrix} \rho \\ u \\ v \\ w \end{pmatrix}(x, y, z) \tag{4-14}$$

CASE II

INCOMPRESSIBLE FLOW

$$\frac{\partial u}{\partial x} + \frac{\partial v}{\partial y} + \frac{\partial w}{\partial z} = 0 \tag{4-15}$$

where

$$\begin{pmatrix} u \\ v \\ w \end{pmatrix}(x, y, z, t)$$

CASE III

STEADY, INCOMPRESSIBLE FLOW

$$\frac{\partial u}{\partial x} + \frac{\partial v}{\partial y} + \frac{\partial w}{\partial z} = 0 \tag{4-15}$$

where

$$\begin{pmatrix} u \\ v \\ w \end{pmatrix}(x, y, z)$$

It is emphasized that the above differential equations have been derived specifically for an orthogonal, rectilinear (cartesian) coordinate system (x,y,z coordinates). The form of the equations is entirely different for any other type of coordinate system. The differential equation of continuity in plane polar and cylindrical polar coordinate systems is presented as the results of Problems 4-11 and 4-12. The method of development is, of course, identical. A volume element is set up with sides of elemental length as indicated by the coordinate system, and the storage rate within the volume and the net outflow rate through all the surfaces are determined.

DIVERGENCE THEOREM

Comparison of (4-6) and (4-11) indicates that they are both statements of the same principle with identical first (time-variation) terms. Equivalence of the second terms is then established:

$$\oiint \rho \vec{V} \cdot d\vec{A} \equiv \iiint \left(\frac{\partial \rho u}{\partial x} + \frac{\partial \rho v}{\partial y} + \frac{\partial \rho w}{\partial z} \right) dV \tag{4-16}$$

Since the left side of this apparent identity is expressed as a vector combination, it is implied that the right side is a coordinatewise expansion of a vector operation involving derivatives. Considered physically, both of these terms were obtained as the *net outflow* of mass from the control volume. In the case of the differential expression, the net outflow from one element was specifically derived and the net effect for all such elements was added (integrated) for the entire volume to yield the expression on the right side of (4-16). It can be seen that this term would yield just the left term if the actual process of adding the effects for all elements within the control volume is considered as illustrated in Fig. 4-15. Starting with one element and considering the adjacent elements, the outflow from one is the inflow to the adjacent element. The effects of the addition or integration cancel at the interfaces. This process continues until the boundary is

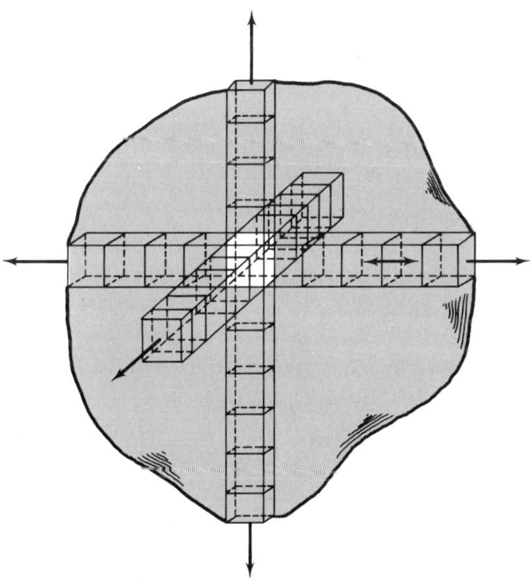

Fig. 4-15. Illustration of the principle that the net outflow through the control volume surface is equivalent to the sum of the outflows from each of the elemental cubes that make up the volume within the control volume.

reached, leaving only the net outflow through the control volume surface. The same is true with respect to every coordinate direction and every element.

In both cases the process is one of taking the net efflux of a quantity. Net outflux may also be connoted by *divergence*. By considering the divergence of flow from all the volume elements within the control volume, the net outflow through the area is then obtained. This statement is a translation of the vector identity of (4-16). The quantity in parentheses on the right side of (4-16) is then just the divergence of $\rho\vec{v}$, the mass flux per unit volume in the limit as the volume size approaches zero,[16] or, written symbolically, div $(\rho\vec{v})$ or $\nabla\cdot(\rho\vec{v})$. The *divergence* is then a vector operator defined for any coordinate system as

$$\operatorname{div}(\ ) \equiv \nabla\cdot(\ ) \equiv \sum_i \hat{x}_i \frac{\partial}{\partial x_i}\cdot(\ ) \tag{4-17}$$

where $\hat{x}_i$ and dx_i or Δx_i are unit vectors and characteristic coordinate differential or elemental lengths, respectively. In cartesian coordinates the divergence operator is

$$\operatorname{div}(\ ) \equiv \nabla\cdot(\ ) = \left(\hat{x}\frac{\partial}{\partial x} + \hat{y}\frac{\partial}{\partial y} + \hat{z}\frac{\partial}{\partial z}\right)\cdot(\ )$$

In cylindrical polar coordinates, the "∇ operator" is

$$\hat{r}\frac{\partial}{\partial r} + \theta\frac{\partial}{r\,\partial\theta} + \hat{z}\frac{\partial}{\partial z}$$

Note that the dot-product-multiplication indication is part of the definition of "divergence" or "div" and that both this vector multiplication and a differentiation process are involved. Although the density is a scalar and does not enter into the vector multiplication, it is still affected by the differentiation because density does vary in space (and therefore with respect to the coordinates).

The identity relation

$$\oiint (\ )\cdot d\vec{A} = \oiiint \operatorname{div}(\ )\,dV \equiv \oiiint \nabla\cdot(\ )\,dV \tag{4-18}$$

is known as the *Gauss divergence theorem*. It can be applied to a vector quantity (for example, $\vec{v}$) or the product of a scalar and vector (for example,

[16]
$$\operatorname{div}(\rho\vec{v}) \equiv \lim_{\substack{\Delta V\to 0\\(\Delta A\to 0)}} \frac{1}{\Delta V}\iint (\rho\vec{v})\cdot d\vec{A}$$

$\overrightarrow{\rho \mathrm{v}}$), or to higher-order tensor quantities. The above method of establishing this identity from a physical principle is an "indicative" proof of this theorem.

The expansion of the vector operation indicated by div $(\overrightarrow{\rho \mathrm{v}})$ is, for example,

$$\operatorname{div}(\overrightarrow{\rho \mathrm{v}}) \equiv \nabla \cdot (\overrightarrow{\rho \mathrm{v}}) = \left(\hat{\mathrm{x}} \frac{\partial}{\partial x} + \hat{\mathrm{y}} \frac{\partial}{\partial y} + \hat{\mathrm{z}} \frac{\partial}{\partial z}\right) \cdot [\hat{\mathrm{x}}(\rho u) + \hat{\mathrm{y}}(\rho v) + \hat{\mathrm{z}}(\rho w)]$$

Derivatives of the unit vectors are implied in the products

$$x_i \frac{\partial}{\partial x_i} \cdot [\hat{\mathrm{x}}_j(\rho u_j)] = \hat{\mathrm{x}}_i \cdot \hat{\mathrm{x}}_j \frac{\partial \rho u_j}{\partial x_i} + \rho u_j \hat{\mathrm{x}}_i \cdot \frac{\partial x_j}{\partial x_i}$$

for which $\hat{\mathrm{x}}_i \cdot \hat{\mathrm{x}}_j$ has the value unity for $i = j$ and zero for $i \neq j$,[17] and the second term is zero for a rectilinear, orthogonal (cartesian) coordinate system. In a curvilinear coordinate system (for example, polar coordinate system) the unit vectors are of constant (unity) magnitude but vary in direction, so their derivatives are not necessarily zero. See Appendix B for further discussion of the derivatives of unit vectors.

If the Gauss divergence theorem were known when the integral form of the continuity equation

$$\oiiint \frac{\partial \rho}{\partial t} dV + \oiint (\overrightarrow{\rho \mathrm{v}}) \cdot \overrightarrow{\mathrm{dA}} = 0 \tag{4-6}$$

was developed, the volume integral could be substituted for the area integral, giving

$$\oiiint \left[\frac{\partial \rho}{\partial t} + \operatorname{div}(\overrightarrow{\rho \mathrm{v}})\right] dV = 0$$

from which the differential equation

$$\frac{\partial \rho}{\partial t} + \operatorname{div}(\overrightarrow{\rho \mathrm{v}}) = 0 \tag{4-19a}$$

could be deduced immediately.

Noting that div is a vector *differential operator*, it might be expected that div $(\overrightarrow{\rho \mathrm{v}})$ could be expanded in a manner similar to that for the expansion of the derivative of a product. This is the case, but the details of the

[17] This particular double-index product is designated $\delta_{ij} = \hat{\mathrm{x}}_i \cdot \hat{\mathrm{x}}_j$ and is referred to as the "Kronecker delta."

expansion are not immediately obvious. However, writing div $(\rho\vec{v})$ as

$$\operatorname{div}(\rho\vec{v}) = \hat{x}_i \frac{\partial}{\partial x_i}\cdot(\rho\hat{x}_j u_j)$$

indicates the manner in which the expansion is carried out. For example, in the simple case of a two-dimensional (x, y) flow,

$$\operatorname{div}(\rho\vec{v}) = \left(\hat{x}\frac{\partial}{\partial x} + y\frac{\partial}{\partial y}\right)\cdot(\hat{x}\rho u + \hat{y}\rho v)$$

An equivalent form of the above expansion is

$$\hat{x}\frac{\partial}{\partial x}\cdot(\hat{x}\rho u + \hat{y}\rho v) + \hat{y}\frac{\partial}{\partial y}\cdot(\hat{x}\rho u + \hat{y}\rho v)$$

$$= \hat{x}\frac{\partial}{\partial x}\cdot(\hat{x}\rho u) + \hat{x}\frac{\partial}{\partial x}\cdot(\hat{y}\rho v) + \hat{y}\frac{\partial}{\partial y}\cdot(\hat{x}\rho u) + \hat{y}\frac{\partial}{\partial y}\cdot(\hat{y}\rho v)$$

where derivatives of the product of three quantities (ρ, unit vectors, and velocity components) must now be expanded. However, the unit vectors are constant in a cartesian coordinate system and therefore have no derivatives, leaving only terms of the form

$$\hat{x}\frac{\partial}{\partial x}\cdot(\hat{x}\rho u) = \frac{\partial \rho u}{\partial x} = \rho\frac{\partial u}{\partial x} + u\frac{\partial \rho}{\partial x}$$

The only possible vector product operation is the scalar (dot) product of the two unit vectors.

Expanding all such products, the final result is

$$\rho\frac{\partial u_i}{\partial x_i} + u_i\frac{\partial \rho}{\partial x_i}$$

which can be shown to be the expression for

$$\rho \operatorname{div}\vec{v} + \vec{v}\cdot\operatorname{grad}\rho$$

Using this expansion for the divergence of $\rho\vec{v}$ in (4-19a) gives

$$\frac{\partial \rho}{\partial t} + \vec{v}\cdot\operatorname{grad}\rho + \rho\operatorname{div}\vec{v} = 0 \tag{4-19b}$$

for the continuity equation.

Considering the first two terms of (4-19b) and the fact that the density is a field function of time [that is, $\rho(\vec{r}, t)$ or $\rho(x_i, t)$], the expression for $d\rho/dt$ can be expanded to give

$$\frac{d\rho(x_i, t)}{dt} = \frac{\partial \rho}{\partial t} + \frac{\partial \rho}{\partial x_i}\frac{dx_i}{dt}$$

If $\vec{dr} \equiv \vec{ds}$ along a streamline of the flow field,

$$\frac{dx_i}{dt} = u_i$$

which gives

$$\frac{d\rho}{dt} \equiv \frac{\partial \rho}{\partial t} + u_i \frac{\partial \rho}{\partial x_i}$$

which is identical to

$$\frac{d\rho}{dt} \equiv \frac{\partial \rho}{\partial t} + \vec{v} \cdot \text{grad } \rho$$

In the Lagrangian sense of traveling with the particle, the derivative is just $d\rho/dt$. This expression then gives the equivalence of the total derivative in the two forms of reference. This total derivative is also referred to as the "substantial" or "material" derivative of a quantity going along with or following its motion.

This particular type of *total derivative* involving the velocity in a flow field is termed the Eulerian or total derivative (where total implies both time and *convective* or transport rates of change). It is the rate of change at a point (that is, the time rate of change) plus the space variation rate (derivative). The density at a particular fixed reference point in space may change with time (nonstationary density field) and, at a particular fixed instant (time not changing), can vary from one point to another (that is, a density gradient exists at this point in the field).

An alternative form of the differential equation of continuity is then

$$\frac{D\rho}{Dt} + \rho \, \text{div } \vec{v} = 0 \tag{4-20}$$

where the notation D/Dt is used to emphasize this particular type of total derivative in the Eulerian sense. Since ρ is a function of x_i and t, $d\rho/dt$ would produce exactly the same result as $D\rho/Dt$. Therefore, the $D\rho/Dt$ notation is, as stated, used only to emphasize this derivative with respect to time and space as a time variation plus a convective variation by the flow field.

4.10 LAGRANGIAN DEVELOPMENT OF THE CONTINUITY EQUATION

The Lagrangian viewpoint considers a definite fixed mass of fluid as it flows through some physical system boundary. The control volume is then attached to the mass, so the control volume boundary and the physical system or "machine" boundary will not generally be the same. It is then necessary to consider the fixed mass of fluid as it flows through the system. The equation for the conservation of mass for a fixed mass of fluid m (in a nonrelativistic flow) is then simply

$$\frac{dm}{dt} = 0$$

with the derivative taken in its usual, literal mathematical sense. Since it is possible for the density to vary within the control volume that encompasses the mass at any instant, the mass of the fluid under consideration is just $m = \oiiint \rho \, dV$; the "continuity equation" then becomes

$$\frac{d}{dt} \oiiint \rho \, dV = 0 \tag{4-21}$$

If only the initial and final states in a physical system are required, integrating (4-21) over the time period for the mass to pass through the system gives

$$m_2 = m_1 = \text{const.}$$

or, for a multibranched physical system,

$$\sum_{\text{out}} m = \sum_{\text{in}} m$$

This equation and its development are much simpler in concept than the equation of continuity in the Euler frame of reference. However, the Lagrangian equation is much more difficult to apply because it is now necessary to describe the method by which the initially defined mass control volume deforms and conforms to the boundaries of the physical system as it passes through them.

Another interesting conclusion can be obtained from comparing the Lagrangian equation for a small mass element Δm with the Eulerian differential equation (4-19) or (4-20). The total derivative with respect to time in the Lagrangian sense is

$$\frac{d(\Delta m)}{dt} = \Delta V \frac{d\rho}{dt} + \rho \frac{d(\Delta V)}{dt}$$

which must be zero for mass conservation. The relative mass rate of change is obtained by dividing this expression by Δm:

$$\frac{1}{\Delta m}\frac{d(\Delta m)}{dt} = \frac{1}{\rho}\frac{d\rho}{dt} + \frac{1}{\Delta V}\frac{d(\Delta V)}{dt} = 0$$

Dividing (4-20) by ρ gives

$$\frac{1}{\rho}\frac{D\rho}{Dt} + \operatorname{div}\vec{v} = 0 \tag{4-20a}$$

A comparison of these two equations for the same physical principle indicates that

$$\underset{\text{LAGRANGIAN}}{\frac{d\rho}{dt}} = \underset{\text{EULERIAN}}{\frac{D\rho}{Dt}}$$

as previously given. It is also indicated that

$$\operatorname{div}\vec{v} \equiv \lim_{\Delta V \to 0} \frac{1}{\Delta V}\frac{d(\Delta V)}{dt}$$

which states that the divergence as the outflux through a differential volume in the Eulerian sense is equivalent to the relative rate of volume expansion in the Lagrangian sense (in the limit as the volume approaches zero).

4.11 CONSERVATION OF SPECIES

In some cases it is necessary to make a detailed accounting of the constituents of the substance that flows through a machine or apparatus (control volume). For example, when fuel and oxidant enter a combustion engine, two separate constituents are flowing into the control volume. However, exhaust products flow out of the exhaust ports (plus possible excess fuel, oxidant, or inert components that do not enter into the combustion). Therefore, neither fuel nor oxidant is conserved because a chemical reaction has taken place. What then is conserved? A microscopic evaluation of the inlet and exhaust streams reveals that the same atoms that went in come out.

The combustion reaction, or any chemical reaction, involves an energy conversion between the chemical potential of the reactants and the internal energy with possible work production. The atoms are not significantly changed and the reaction products are nothing more than different molecular combinations of the same atoms that constitute the reactants.

In these reactions the atoms' tally must balance on both sides of the bookkeeping ledger that constitutes the conservation-of-mass equation. Either the actual masses of each specific species of atom could be considered as being conserved, or the atoms could be counted and the number of each must be conserved.

Another possibility is to use the atomic mass, which involves both the number and the mass of the atomic species. The number of mass atoms of the products must be the same as the number of mass atoms of the reactants.

In using the Eulerian representation to formulate an equation for the conservation of energy, mass per unit volume or mass density was employed. In accounting for the number of atoms (or moles), a number density (number of atoms or moles per unit volume) could be used, but it is difficult to formulate the mass storage rate and mass transport rate in terms of this number density. In considering chemical reactions (for example, combustion) it is easier to adopt the Lagrangian viewpoint. The fact that nothing can happen to the atoms in such a reaction can be stated simply mathematically as

$$\sum_{r=1}^{\nu} \frac{dn_r}{dt} = 0 \tag{4-22}$$

where n_r is the number of atoms of the r species in the mass under consideration, which contains ν different types of atoms. Also, each specific type (r) of atom must also be "conserved," so that

$$\frac{dn_r}{dt} = 0 \tag{4-23}$$

which states the conservation of mass for each species.

The conservation of mass as stated in (4-22) could be written

$$\frac{d}{dt} \sum_{r=1}^{\nu} n_r m_r^0 = 0 \tag{4-24}$$

Since the total mass m could be expressed as $m = \sum n_r m_r^0$ (where m_r^0 is the mass of one atom of the r species), (4-24) can be expanded in the form

$$\sum \left(m_r^0 \frac{dn_r}{dt} + n_r \frac{dm_r^0}{dt} \right) = 0 \tag{4-25}$$

(since the $\sum$ and d/dt operations are independent). But the mass of each atom (m_r^0) is assumed to remain constant or

$$\frac{dm_r^0}{dt} = 0 \qquad \text{leaving } \frac{dn_r}{dt} = 0$$

Each of these conservation principles is seen to be actually independent—almost. When a detailed microbalance account is attempted for a chemical-reaction energy conversion (or any physical-chemical energy conversion) the question always arises: Where does the energy actually come from? Interestingly enough, by Einstein's principle (or equation) relating energy and mass conversion, the energy converted from the chemical structure into internal or intrinsic energy can be traced to a variation of the mass of the reactants (but not the number of atoms or basic particles), so that $dm/dt \neq 0$. For an ordinary type of hydrocarbon combustion reaction, the relative change of mass ($\Delta m/m$) is only 1 part in 10^7, which is immeasurable on the scale with which the reactants and products can be measured. Thus the relativistic effect for these types of reactions is negligible. On the other hand, the mass change is not negligible for atomic and nuclear reactions (fusion, fission, and so on).

Rather than deal with the numbers of atoms (since these can only be inferred from mass measurements anyway), the number of moles of each species is used, giving

$$\frac{dN_r}{dt} = 0 \qquad \text{or} \qquad \frac{d}{dt}\sum_{r=1}^{\nu} N_r = 0 \tag{4-26}$$

where N_r is the number of "mass atoms" of the rth species. These equations are (of course) for a fixed total mass of substance (that is, with respect to the Lagrangian frame of reference) and for a system in which the relativistic effect is negligible.

★4.12 NONSTATIONARY, DEFORMABLE CONTROL VOLUME: Modified Euler Frame of Reference

A control volume of fixed shape, fixed in space, was considered in the above developments. However, this restriction is not possible to achieve in some cases. In the example of the piston moving within the cylinder, it may not be possible to choose the clearance volume as the control volume because of certain other restrictions on the problem.[18] Other examples are an emptying tank of liquid, a collapsible gas tank, and an emptying balloon. The last example is probably the most general. Consider the balloon moving through space due to its own jet propulsion (Fig. 4-16). The most obvious choice of control volume surface is the balloon skin and the plane across the outlet area. The choice of coordinate system origin is difficult, because there can be relative motion between any two points on the control surface as it deforms. A coordinate system origin must be

[18] In attempting to work this problem, the C.V. must be defined as the varying volume within the cylinder (bottom area adjacent to top of piston).

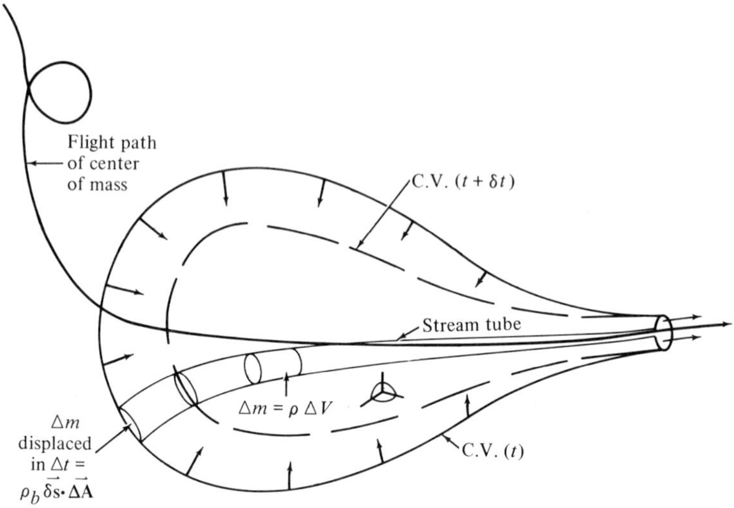

Fig. 4-16. Collapsing balloon as an example of a nonstationary control volume.

specified, because positions and velocities must be measured with respect to a definite reference position (and orientation). The only sure way of making such measurements in this case is with respect to a space-fixed coordinate system origin. Other possible locations for relative or moving coordinate system origins would be the instantaneous center of mass of the fluid within the balloon and the center of the mouth area. The rate statement of the conservation of mass applied directly to this case is

$$\begin{bmatrix}\text{RATE OF DECREASE}\\ \text{OF MASS IN C.V.}\end{bmatrix} = \begin{bmatrix}\text{RATE OF OUTFLOW OF MASS}\\ \text{ACROSS C.V. SURFACE}\end{bmatrix}$$

or

$$-\frac{\partial}{\partial t}\oiiint \rho\, dV = \iint_{A_e} \rho \vec{v} \cdot \vec{dA}$$

where A_e is the open exhaust or exit area and the C.V. is now a function of time. The first term (which represents the time rate of change of mass within the control volume at any instant) is actually the result of two effects: The mass of any given (fixed) volume element ($\Delta m = \rho\, dV$) within the control volume is changing as a consequence of the density variation, and the mass within the balloon skin is changing because of the displacement of mass at the skin by its motion. This displacement mass is seen to be just $\rho_b \vec{\delta s} \cdot \vec{\Delta A}$, where ρ_b is the density at the skin boundary or at the ΔA position of the control volume. The rate of mass displacement for all streamtubes is obtained by dividing by δt and taking the limit as δt

approaches zero. Subsequent integration for all streamtubes yields

$$\iint_{A_b(t)} \rho_b \vec{v}_b \cdot \vec{dA}$$

where the subscript b indicates values at the control volume boundary. The rate of mass decrease is then equivalent to the sum of the two terms

$$\oiiint \frac{\partial \rho}{\partial t} \, dV + \iint_{A_b} \rho_b \vec{v}_b \cdot \vec{dA}$$

or the continuity equation for the nonstationary control volume is given by

$$\oiiint \frac{\partial \rho}{\partial t} \, dV + \iint_{A_b} \rho_b \vec{v}_b \cdot \vec{dA} + \iint_{A_e} \rho_e \vec{v}_e \cdot \vec{dA} = 0 \tag{4-27}$$

It is seen that $A_b + A_e = A_{\text{C.V.}}$, so that (4-27) could be written

$$\oiiint \frac{\partial \rho}{\partial t} \, dV + \oiint \rho \vec{v} \cdot \vec{dA} = 0 \tag{4-28}$$

where it is understood that at the open portion of the control volume the velocity is the exhaust velocity, and on the rest of the control surface (skin) the velocity is the skin or the control volume displacement velocity. It is apparent that the control volume and $A_{\text{C.V.}}$ are functions of time.

For the particular case of negligible density variation, the velocity $\vec{v}$ can be taken as either absolute velocity with respect to some space-fixed coordinate system or with respect to a point moving *with* a balloon along its flight path because the area integration covers the entire control volume. In either case, the balloon velocity $\vec{v}_A$ (constant over the entire control volume at any instant) produces no effect, because the quantity $\rho \vec{v}_A \cdot \vec{dA} = 0$ (that is, the integral encloses a volume: therefore, for every area component in one direction there is an equal and opposite component).

★4.13 ILLUSTRATIVE EXAMPLE 4-2: Nonstationary Control Volume

The emptying of a rocket-fuel tank in flight (Fig. 4-17) is a simple example that indicates the significance of the above quantities. The fluid control volume is taken coincident with the tank walls, the upper liquid-free surface, and the arbitrarily designated tank exit plane. Since the tank walls are assumed to be rigid, the coordinate system and its origin can be taken in the tank, as shown, because there is no problem in specifying the motion of the liquid with respect to this coordinate system or in specifying

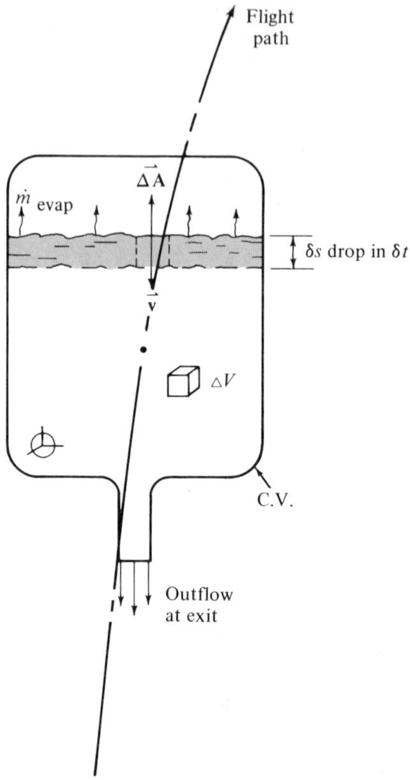

Fig. 4-17. Emptying of a rocket fuel tank as an example of a non-stationary, time-varying control volume.

the motion (translation and rotation) of the coordinate system with respect to a point fixed in space if the rocket path is known. In a small time period δt: (1) mass $\rho\ \vec{\delta s}_e \cdot \vec{\Delta A}_e$ will flow out of the exit connection, (2) a small amount δm_v will evaporate from the free surface, and (3) the free surface will drop by an amount $\vec{\delta s}_{fs}$. If the liquid is assumed to be incompressible, there can be no net mass storage within any control volume element. Taking into account the loss of liquid by evaporation, the continuity relation gives

$$\oiint \rho \vec{V} \cdot \vec{dA} + \dot{m}_v = 0 \tag{4-29}$$

$$\iint_{A_e} \rho \vec{V} \cdot \vec{dA} + \iint_{A_{fs}} \rho \vec{V} \cdot \vec{dA} + \dot{m}_v = 0$$

Equation (4-29) is most easily interpreted by writing it in the form

$$-\iint_{A_{fs}} \rho\vec{v}\cdot\vec{dA} = \iint_{A_e} \rho\vec{v}\cdot\vec{dA} + \dot{m}_v$$

which states that the rate of drop at the free surface is due to the outflow rate plus the evaporation rate. The quantity $-\iint_{A_{fs}} \rho\vec{v}\cdot\vec{dA}$ is positive, because the correspondence between the vector $\vec{v}$ and $\vec{dA}$ produces another negative sign.

4.14 GENERAL TRANSPORT EQUATIONS

An interesting generalization of the method used to derive the integral-continuity and conservation-of-mass equations for a fixed control volume is possible. A verbal equation for the change of any property or quantity X (vector or scalar) as it moves through a flow field can be formulated with respect to the control volume as

$$\begin{bmatrix}\text{TOTAL NET RATE OF}\\ \text{CHANGE OF } X \text{ IN C.V.}\end{bmatrix} = \begin{bmatrix}\text{NET ACTION PRODUCED ON THE FLUID IN}\\ \text{THE C.V. AS } X \text{ VARIES WITHIN THE C.V.}\end{bmatrix}$$

If the control volume is fixed in space (Eulerian frame of reference),

$$\begin{bmatrix}\text{NET RATE OF CHANGE}\\ \text{OF } X \text{ IN C.V.}\end{bmatrix} \equiv \begin{bmatrix}\text{RATE OF INCREASE}\\ \text{OF } X \text{ WITHIN C.V.}\end{bmatrix} + \begin{bmatrix}\text{NET OUTFLOW TRANSPORT}\\ \text{RATE OF } X \text{ THROUGH C.V.}\\ \text{SURFACE}\end{bmatrix}$$

If the control volume is fixed in a given mass of the fluid as it flows through a system (Lagrangian frame of reference), the transport term is not present. The "net action" is a cause or consequence of the variation of the quantity X that is produced as X passes through the field or as the X field changes.

With respect to the Eulerian viewpoint, the law of conservation of mass applied to a control volume states that

$$\begin{bmatrix}\text{RATE OF INCREASE OF}\\ \text{MASS WITHIN C.V.}\end{bmatrix} + \begin{bmatrix}\text{NET OUTFLOW TRANSPORT}\\ \text{RATE OF MASS THROUGH C.V.}\end{bmatrix} = \begin{bmatrix}\text{NET SOURCE OF MASS}\\ \text{CREATION (RATE)}\\ \text{WITHIN C.V.}\end{bmatrix} + \begin{bmatrix}\text{NET RATE OF MASS}\\ \text{CONVERSION FROM ENERGY}\end{bmatrix}$$

The *action* is the *net* creation rate of mass within the control volume from relativistic energy conversion or from hypothetical sources.

The law of conservation of *energy*, including the possibility of conversion of energy to work, would be

$$\begin{bmatrix}\text{RATE OF INCREASE}\\ \text{OF ENERGY WITHIN}\\ \text{THE C.V.}\end{bmatrix} + \begin{bmatrix}\text{NET OUTFLOW TRANSPORT}\\ \text{RATE OF ENERGY THROUGH}\\ \text{THE C.V.}\end{bmatrix}$$

$$= \begin{bmatrix}\text{RATE AT WHICH WORK IS}\\ \text{DONE ON THE FLUID AS}\\ \text{IT PASSES THROUGH THE C.V.}\end{bmatrix}$$

$$+ \begin{bmatrix}\text{RATE OF ENERGY}\\ \text{CREATION}\\ \text{(for example, FROM MASS)}\end{bmatrix} + \begin{bmatrix}\text{ENERGY}\\ \text{ADDED BY}\\ \text{HEAT TRANSFER}\end{bmatrix}$$

The "action" on the fluid in the C.V. that produces a change in the energy of the fluid is the net conversion of work to energy, plus the possible conversion from mass to energy, plus the energy added through the C.V. boundary as heat.

Newton's second law relating the rate of change of momentum to the forces acting on the fluid would be stated

$$\begin{bmatrix}\text{RATE OF INCREASE OF}\\ \text{MOMENTUM WITHIN}\\ \text{THE C.V.}\end{bmatrix} + \begin{bmatrix}\text{NET OUTFLOW TRANSPORT}\\ \text{RATE OF MOMENTUM}\\ \text{THROUGH THE C.V.}\end{bmatrix}$$

$$= \begin{bmatrix}\text{ALL FORCES}\\ \text{ACTING } \textit{on} \text{ THE FLUID}\\ \text{IN THE C.V.}\end{bmatrix}$$

The action on the fluid in the C.V. that produces the change in momentum of the fluid is the *net force on* the fluid.

The above principles of work and energy and force and momentum will be investigated in Chapters 5 and 6 in the manner implied here to yield useful mathematical relations. Although the above three examples of the general variations of any quantity X are of primary interest in fluid mechanics, the application of this principle is not limited to these three examples.

4.15 SUMMARY

1. A vector statement of the conservation of mass for a flowing fluid can be written as the integral equation

$$\oiiint \frac{\partial \rho}{\partial t}\, dV + \oiint \rho \vec{\mathrm{v}} \cdot \vec{\mathrm{dA}} = 0$$

or

$$\oiiint \left[\frac{\partial \rho}{\partial t} + \operatorname{div}(\rho \vec{\mathrm{v}})\right] dV = 0$$

or as a differential equation

$$\frac{\partial \rho}{\partial t} + \operatorname{div}(\rho \bar{\mathrm{v}}) = 0$$

where the Gauss divergence theorem has been employed. These general relations can be used in the solution of any problem by considering the specific conditions of the problem and by stating the vectors and vector operators in the form that best suits the coordinate system as determined by the shape of the physical system.

2. Because of the literal statements of the physical laws to be applied, a fluid control volume must be defined to contain only and entirely fluid. The C.V. may surround solid or machine elements or voids that form part of the C.V. boundary and can exert an action (for example, force, energy exchange, and work) on the fluid boundary.
3. In a literally one-dimensional flow field (duct flow), no property may vary over any cross-sectional area perpendicular to the flow field. Such a flow is hypothetical, since the condition of zero velocity at a solid boundary cannot be satisfied.
4. In a quasi-one-dimensional flow, no significant property variation may exist over any cross-sectional area perpendicular to the flow direction, with the exception of the velocity distribution, which must satisfy the condition of zero velocity at the wall.

PROBLEMS

4-1 Show that the equations of the streamlines can be derived from the vector property describing the condition that streamlines are everywhere parallel to the velocity.

4-2 From the vector equation describing the condition that the velocity is parallel to the streamline at any point on the streamline ($\vec{\mathrm{v}} \times \vec{\mathrm{ds}} = 0$), derive the coordinate differential equation of the streamline in cylindrical polar coordinates similar to (4-1).

4-3 For a two-dimensional flow, the equations of the streamlines are given by (4-1) and (4-2). In Chapter 14, a stream function is derived such that $u = \partial \Psi / \partial y$ and $v = -\partial \Psi / \partial x$. With these substitutions for u and v, show that the equation of a streamline is $\Psi = \text{const.}$

4-4 Using the relationship $\vec{v} \times \vec{ds} = 0$ together with the equation for a streamline $\Psi(r, \theta) =$ const. in plane polar coordinates, find the velocity components u_r and v_θ in terms of the stream function Ψ. (See also Problem 4-3.)

4-5 In Fig. 14-13, the spacing between the cylinder and the first streamline above it at the top is about one half relative to the spacing at infinity. What is the relative velocity (with respect to U_∞) at the top of the cylinder?

4-6 In Fig. 14-14, the spacing between the top of the cylinder and the first streamline above it is about 0.32, and the spacing at the bottom is 0.84. Find the relative velocities at the top and bottom of the cylinder, and their ratio.

4-7 Calculate (a) the Reynolds number for air and for water flowing at 0.5 ft^3/sec through a 2-in.-diameter duct at 20°C and 1 atm, and (b) N_R for air at these conditions.

4-8 For the laminar flow of a fluid in a round pipe or tube, the velocity profile is given by $v/v_0 = 1 - (r/r_0)^2$, where the velocity v_0 is the centerline velocity. Find the volume flow rate and the average velocity as functions of v_0 and r_0.

4-9 For the laminar flow of corn syrup of a viscosity of 100 P and $S = 1.13$ at 5 lb_m/min through a 1-cm-diameter tube, find $\bar{v}$, v_0, and N_R.

4-10 What is the Reynolds number for a fast-pitched baseball?

4-11 Calculate the Reynolds number for flow about a 3/8-in. power line in a 40-mph wind.

4-12 Calculate the Reynolds number per foot ($N_R/D = v/\nu$) for (a) a velocity of 100 ft/sec on a hot (100°F) humid (80 percent relative humidity) day, and (b) a cold (30°F) dry (negligible humidity) day.

4-13 The flow discharge from a diffuser may have a nearly triangular velocity profile. Calculate $\bar{v}/v_0$ for this type of discharge from a circular diffuser.

4-14 Find $\bar{v}/v_0$ for the flow in an annulus between two concentric tubes of radii r_1 and r_2 for which the profile is approximately parabolic.

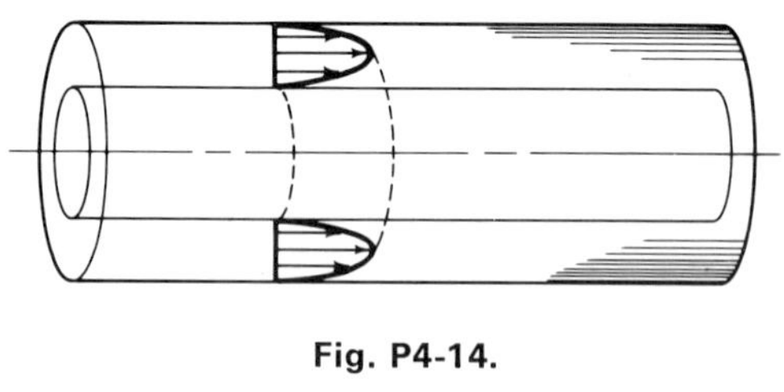

Fig. P4-14.

4-15 If the velocity distribution in a round duct is given by $v/v_0 = 1 - (r/r_0)^n$, find the volume flow rate and the average velocity as functions of v_0 and n.

4-16 For a range of Reynolds number between 10^4 and 10^5, the velocity profile for turbulent pipe flow may be closely approximated by $v(y)/v_0 = (y/r_0)^{1/7}$ (where $y \equiv r - r_0$ is the perpendicular distance from the wall and v_0 is the centerline velocity at $r = r_0$). Find $\overline{v}/v_0$.

4-17 Calculate $\overline{v}$ for flow in a pipe at $N_R = 50{,}000$, where the measured centerline velocity is 50 ft/sec.

4-18 For Reynolds numbers outside the range given in Problem 4-16, the velocity profile may be approximated by $v(y)/v_0 = (y/r_0)^{1/n}$. Find $\overline{v}/v_0$ as a function of n and plot the result for $5 < n < 10$.

4-19 The velocity distribution at the discharge from the blade row is given in the figure. Find the volume flow rate and the average velocity per blade passage. Assume that the velocity is uniform in the direction perpendicular to the plane illustrated.

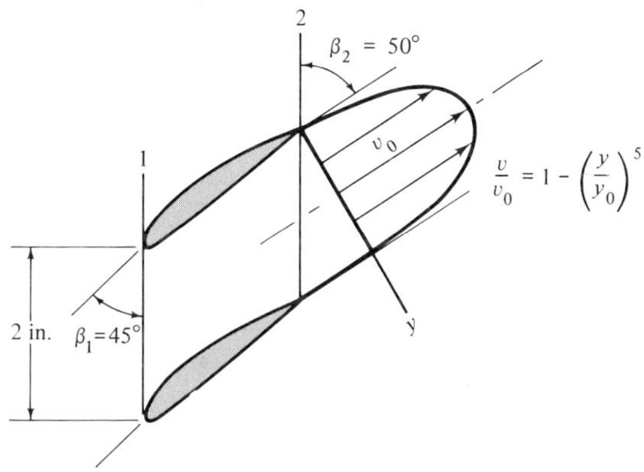

Fig. P4-19.

4-20 Derive the differential equation for the one-dimensional flow of a gas in a variable-area ducting system:

$$\frac{d\rho}{\rho} + \frac{dA}{A} + \frac{dv}{v} = 0$$

4-21 A gas is flowing in a piping system. At one station the pressure, temperature, and area are p_1, T_1, and A_1 and the velocity is v_1; downstream, the values are p_2, T_2, and A_2. Find (a) the mass flow rate, (b) v_2, and (c) the volume flow rates at 1 and 2 in terms of the given quantities.

4-22 In Problem 4-21, if $p_1 = 15$ psia, $T_1 = 80°$F, $A_1 = 5$ in.^{2}, $v_1 = 400$ ft/sec, and $p_2 = 10$ psia, $T_2 = 40°$F, and $A_2 = 3.5$ in.^{2} find (a) the mass flow rate; (b) v_2; (c) the volume flow rate, Q_1 and Q_2; and (d) the Mach numbers, M_1 and M_2, for air.

4-23 A scuba (self-contained underwater breathing apparatus) tank 15 cm in diameter and ½ m long is charged to a pressure of 100 atm. How long can this tank be used at a depth of 150 m if the breathing rate is nominally 10 liters/min?

4-24 Determine the percentage mixture of helium and oxygen to be used in an underwater breathing apparatus at a depth of 100 m if the mass of oxygen is the same as that breathed at 1 atm. What mass and volume must be stored at 200 atm to sustain normal breathing for a period of 2 hr?

4-25 A tank with a volume V_0 is charged to a pressure p. The temperature of the gas is T_0. If the tank is bled down through a valve, find the density, pressure, and temperature as functions of p_0, T_0, and the valve opening during the period that the tank/ambient pressure ratio is large enough to maintain sonic velocity at the valve. The tank is insulated. (Neglect acceleration effects.)

4-26 For a volume element in cylindrical polar coordinates, consider the *net* mass outflow and derive the continuity equation in this coordinate system in a manner similar to that used to derive (4-13). From the results, deduce the expanded form for the divergence of a vector in cylindrical polar coordinates:

$$\operatorname{div} \vec{\mathrm{v}}\,(r, \theta, z) = \frac{\partial u_r}{\partial r} + \frac{u_r}{r} + \frac{\partial v_\theta}{r\,\partial \theta} + \frac{\partial w_z}{\partial z}$$

where $\mathrm{v} = \hat{\mathrm{r}} u_r + \hat{\theta} v_\theta + \hat{\mathrm{z}} w_z$.

4-27 Expand the operation indicated by

$$\operatorname{div} \vec{\mathrm{v}}\,(r, \theta, z) = \nabla \cdot \vec{\mathrm{v}} = \left(\hat{\mathrm{r}} \frac{\partial}{\partial r} + \hat{\theta} \frac{\partial}{r\,\partial \theta} + \hat{\mathrm{z}} \frac{\partial}{\partial z}\right) \cdot (\hat{\mathrm{r}} u_r + \hat{\theta} v_\theta + \hat{\mathrm{z}} w_z)$$

and compare the results with those given in Problem 4-26.

4-28 A reciprocating pump and pulsation damper (capacitance) are set up as shown. The output from the pump at 1 is given by $Q = AU_{01}(1 - \cos \omega t)$ and at 2 by $Q = AU_{02}(1 - \epsilon \cos \omega t)$. Find the time-average flow rate

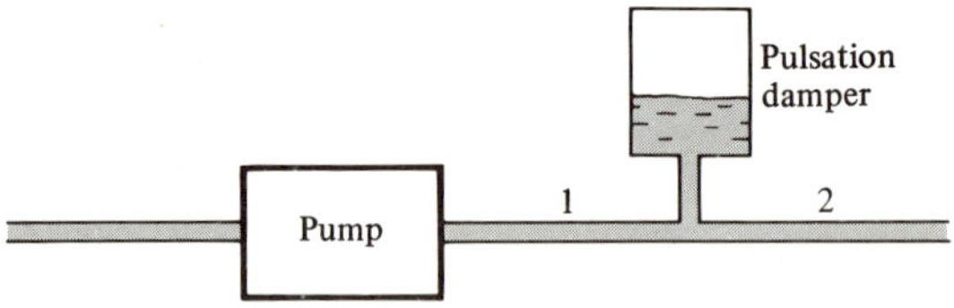

Fig. P4-28.

Q and the maximum variation about this mean value at each station. What is the relative decrease in fluctuation about the mean flow at 2 relative to that at 1 (the "pulsation attenuation"). Sketch the flow rates as functions of time (ϵ is < 1).

4-29 Describe how a time-averaged velocity could be obtained at time *a* during the interval *b-c* of Fig. 4-6.

4-30 In the derivation procedure leading to (4-13), the mass flow rate through the right face of Fig. 4-14 was determined in terms of the increment in mass flow rate from the reference value at the center of the element. Show that the same result can be obtained by considering the density and velocity separately at the right face and determining the mass flow rate there and similarly at the left face, as indicated on Fig. 4-14c.

4-31 The exhaust from a conical rocket nozzle is essentially a conical flow, as illustrated. If the exit cone angle is 2α with radius r_0, find (a) the mass flow if ρ_0 and v_0 are constant, and (b) the ratio of mean exit speed to gas speed v/v_e.

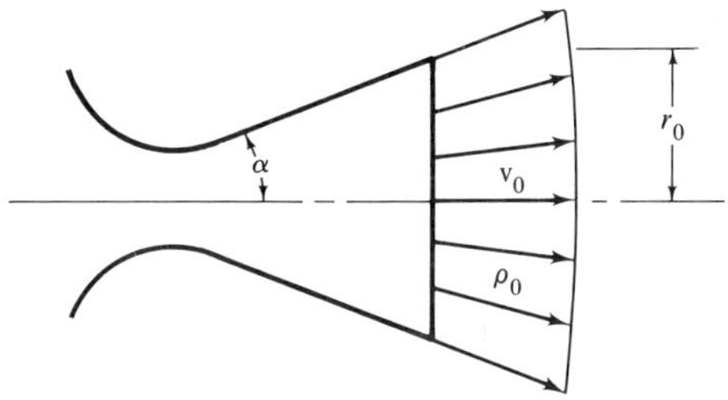

Fig. P4-31.

4-32 For the rocket nozzle of Problem 4-31, what is the mass flow rate for an exhaust speed of $v_0 = 5000$ ft/sec, molecular mass number $\mathscr{M} = 30$, $\gamma = 1.2$, $p_e = 10$ psia, $T_e = 1000°$F, $A_e = 1$ ft^2, and $\alpha = 15°$.

4-33 A velocity traverse of the exhaust of a jet engine was experimentally plotted and gave the profile shown. This radial profile indicates a nearly uniform core of radius with velocity v_c and a parabolic velocity distribution decreasing to zero at $r = r_0$. The density was nearly constant across the exit plane. (a) Determine the expression for the mass flow rate in terms of ρ, v_c, r_c, and r_0. (b) What is the ratio v/v_c?

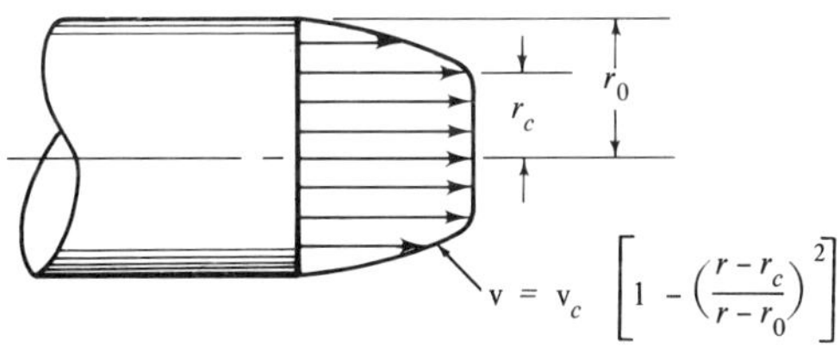

Fig. P4-33.

4-34 For the rocket nozzle of Problem 4-33, if $p_e = 1$ atm, $T_e = 1000°$F, and $M_e = 0.8$, find the mass flow rate for $r_0 = 18$ in., $r_c = 12$ in., assuming the exhaust gas is air.

4-35 For the cylinder, piston, and leaking valve of Fig. 4-9, calculate the rate of change of density within the cylinder at the instant that the piston is moving at 2000 ft/min (at midstroke). The leakage area is a slit around the valve 0.020 in. wide and 0.6 in. long. The bore and stroke are both 4.0 in. and the volume compression ratio is 8. Assume that the gas (air) pressure and temperature at this instant are 250 psi and 500°F and that the leakage velocity is sonic. Also assume that the density within the cylinder at any instant is about constant. Actually this assumption is quite invalid in the immediate neighborhood of the leak, but therefore over only a small portion of the entire fluid volume; thus it causes only a small error for the gas within the cylinder. The density of the escaping gas is $0.63\rho_0$ and its temperature is $T_0/1.2$.

4-36 An end-burning, solid-propellant rocket engine is shown. The propellant burns uniformly over the exposed grain area at a constant rate $\dot{r}_b$ (units of velocity). The exit conditions, ρ_e, V_e, and therefore $\dot{m}_e$ are constant. Write an integral continuity equation for this engine. (Consider a gas control volume at any instant and the variations occurring in a time δt.)

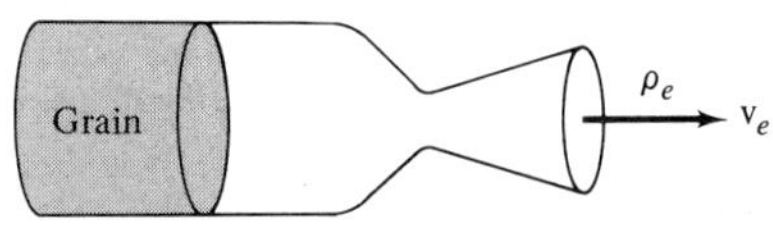

Fig. P4-36.

CHAPTER 5

CONTENTS

CONSERVATION OF ENERGY AND AN ENERGY EQUATION FOR A FLOWING FLUID

For processes in a closed system with static end states, it is a physical law that energy is conserved. As an instantaneous-rate statement for an open or closed system, energy conversion occurs at a rate determined by the work done on, and heat transfer to, the substance.

The principle of conservation of energy applied to a system is used primarily to determine work and energy balances for the system or the rate of energy conversion to useful power, as in the conversion of available energy to shaft output in a turbine. In an energy-conversion balance it is necessary to consider all forms of energy that are transformed within the system and all forms of work done on the working fluid (or substance) while it is in the system. Since fluid systems usually involve the *flow* of a substance, it is advantageous to consider the *rates* of energy change in the system.

Energy must be specified or measured with respect to a *space-fixed coordinate system*. If the coordinate system were moving, the reference datum from which *potentials* are measured would shift. The potential energy measured in a moving coordinate system would be different from that measured in a fixed system. The *kinetic* energy of a uniform reference field motion (u) is not canceled when calculating the difference in kinetic energies:

$$\frac{(v_1 - u)^2}{2g_c} - \frac{(v_2 - u)^2}{2g_c} \neq \frac{v_1^2 - v_2^2}{2g_c}$$

where v_1 and v_2 are the *absolute* velocities (in a space-fixed coordinate system).

When it is necessary to develop the energy equation within a moving frame of reference, the transformation must be made from a frame that is initially space-fixed (see Section 6.18).

All forms of energy must be considered in the work–energy balance for a system. Some of these forms of energy contain basic quantities found in other basic principles. One example is the kinetic energy of a mass of fluid that is proportional to the square of the *velocity*, a quantity also found in the conservation of mass and in the principle relating force and momentum. The mathematical equation stating the physical concept of conservation of energy may be combined with other equations to obtain information where only a limited amount of initial information is known. Mathematical statements of the principles of conservation of mass and momentum may be formulated with the energy equation as a set of simultaneous equations to determine "a solution" to a problem. The more carefully the equations are formulated, the more accurate is the solution. Also, the problem-solution accuracy is determined by the degree of validity of the assumptions and approximations necessary to formulate and solve the problem.

If enough initial information is known, the energy equation or a simple form thereof is sufficient to give a fairly accurate answer to a flow problem. For example, the Bernoulli equation (5-25) will yield a close approximation for the exhaust velocity of a nozzle if the upstream pressure and velocity, along with the exhaust pressure and nozzle geometry, are known. In other cases, additional relations among the variables may be required when not enough information is available to yield a solution to a single equation.

In relativistic mechanics the principles of conservation of energy and conservation of mass are not independent but are related by the Einstein equation,

$$\Delta\mathscr{E} = \frac{\Delta mc^2}{g_c}$$

where $\Delta\mathscr{E}$ is the change in energy produced by conversion of mass Δm and c is the velocity of light. The following material will be developed for nonrelativistic systems.

The above principle of conservation of energy can be developed either for a unit of the fluid mass as it flows through the system, or with respect to a control volume, as was previously done in the development of the continuity equation. The control volume development will be carried through here.

5.1 WORK AND ENERGY

The types of energy and work to be considered for a system can be determined by means of the example of a gas turbine engine, as illustrated in Fig. 5-1. In dealing with energy into or out of a system, it is evident that different forms of energy must be considered. One form that can be

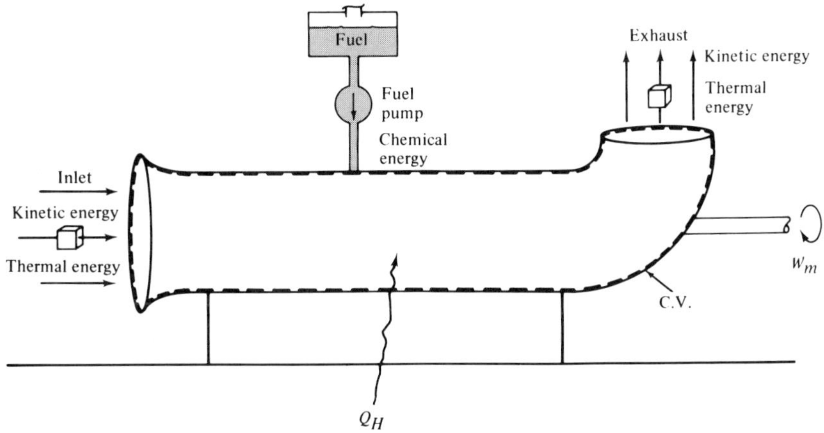

Fig. 5-1. Gas turbine engine as an example of a "system" or "control volume" (C.V.) on which an energy balance is to be made. The control volume boundary must also circumscribe any moving or stationary nonfluid elements within the case.

termed "mass associated" is the intrinsic energy of a mass that is a consequence of its static and dynamic *states*. Examples of this type of energy are:

1. *Kinetic Energy.* The macroscopic or ordered field motion kinetic energy of a small mass element Δm is $\Delta m \mathrm{v}^2/2g_c$. If the flow is turbulent, as described in Section 4.4 (Figs. 4-6 and 4-7), the time mean of v^2 must be taken:

$$\langle \mathrm{v}^2 \rangle \equiv \frac{1}{T}\int_T \mathrm{v}^2\, dt = \frac{1}{T}\int_T (\bar{\mathrm{v}} + \mathrm{v}')^2\, dt = \bar{\mathrm{v}}^2 + \langle \mathrm{v}'^2 \rangle = \bar{\mathrm{v}}^2\left[1 + \frac{\langle \mathrm{v}'^2 \rangle}{\bar{\mathrm{v}}^2}\right] \tag{5-1}$$

The *turbulence intensity*[1] $\sqrt{\langle \mathrm{v}'^2 \rangle / \bar{\mathrm{v}}^2}$ is usually small, on the order of 0.05, so that its square is quite small.

2. *Internal Intrinsic Thermal Energy* ($\tilde{u}$). From the macroscopic viewpoint the internal energy ($\tilde{u}$) can be considered as potential energy.

[1] In turbulence literature, the mean, local velocity components are usually denoted by U, V, W, and the root-mean-square (rms) turbulence components by u, v, and w. The relative, component turbulence intensities with a local mean velocity U in the x direction only would then be u/U, v/U, and w/U; the total intensity would be $\sqrt{u^2 + v^2 + w^2}/U$. This total intensity is $\sqrt{\langle \mathrm{v}'^2 \rangle}/U$, which will arbitrarily be simplified to v/U in Chapter 10.

From the microscopic viewpoint, it is both kinetic energy of the random motion of the molecules and potential energy of the intermolecular field forces.

3. *Chemical Energy* (e_{chem}). The chemical energy (e_{chem}) may also be considered as a form of internal potential energy. The energy that can be released through a chemical reaction is determined by the difference in chemical potential that a substance undergoes in a system.

4. *Electric Charge Potential Energy* (e_e). The charge potential is associated with a discrete mass element. The charge potential (e_e) enters into the energy balance in two ways: as it is associated with chemical reactions, and by an interaction with magnetic fields to produce a force and therefore work.

5. *Magnetic Potential.* Magnetic potential is a consequence of the *polar moment* of polarized molecules.

6. *Molecular*, *Atomic*, *Nuclear*, and *Subnuclear Energies* (e_n). These energies include those not included in 2, 3, and 4, such as radiation potential.

7. *Other Forms of Energy* ($e_{\text{etc.}}$). In general, all types of energy must be considered with respect to the possibility of energy conversion into work or other forms of energy.

If energy is examined on a strictly microscopic basis, all the forms of energy listed above can be traced to just two independent, identifiable forms: the kinetic energy of the independent units (molecules, atoms, and subatomic particles) and the interparticle force fields, or potentials. The *random-particle-motion* kinetic energy (plus the interparticle potential energy) is the quantity referred to as internal or intrinsic energy. The *ordered*, or *macroscopic*, *field motion* is the ordered kinetic energy.

Heat is an *energy-transfer rate* that cannot be associated with a fluid element mass. The energy transferred to the working fluid across the solid boundaries of a heat exchanger in a nuclear-energy-powered steam or gas turbine engine is an example of *heat transfer*. This heat increases the internal energy of the fluid, which is then converted into work by the turbine. In the case of a gas turbine engine, the casing is probably hotter than the surroundings, so some heat transfer occurs from the working fluid to the surroundings through the metal walls of the engine. In both cases, the energy is transferred not by *direct convection* of the mass but through an intermediate substance, the walls of the heat exchanger and engine casing.

Since the principle of conservation of energy allows the transformation of energy into work, the types of work must be specified. These are: (1) mechanical or shaft work done on the fluid, (2) work done on the fluid in overcoming friction forces, (3) work done on the fluid by pressure forces at the system boundaries, and (4) work done by mass-acting (gravitational-type) body forces. Electrical work can be defined in a manner similar to the pressure work done on a fluid by its surroundings. Electrically charged atoms or molecules are convected by the flow field. If a magnetic field is present, work will be done on (or by) the fluid as a consequence of its motion and the magnetomotive force. The subject of magneto fluid dynamics is treated in Chapter 16.

5.2 AN ENERGY-RATE, WORK-RATE EQUATION

The principle of conservation of energy can be stated for the fluid within a *control volume* by considering the following energy changes and work occurring during a small time period δt:

1. The energy stored within the system $\equiv$ the increase of the energy of the fluid in the control volume.
2. The *net* energy transported into the control volume by the mass flow.
3. The energy increase as a consequence of the heat transfer to the fluid in the control volume.
4. The energy increase as a consequence of the work done on the fluid within the control volume:

$$\overset{(1)}{\begin{bmatrix}\text{INCREASE}\\ \text{OF ENERGY}\\ \text{WITHIN C.V.}\end{bmatrix}} = \overset{(2)}{\begin{bmatrix}\text{NET ENERGY}\\ \text{TRANSPORTED}\\ \text{INTO THE C.V.}\end{bmatrix}} + \overset{(3)}{\begin{bmatrix}\text{ENERGY ADDED}\\ \text{TO THE C.V. BY}\\ \text{HEAT TRANSFER}\end{bmatrix}} + \overset{(4)}{\begin{bmatrix}\text{ENERGY INCREASE BY}\\ \text{WORK DONE ON THE}\\ \text{FLUID IN THE C.V.}\end{bmatrix}} \tag{5-2}$$

Note that the statement specifically applies to the *energy of the fluid in the C.V.* and the *work done on the fluid in the C.V.* Consequently, *the C.V. must be constituted only and entirely of fluid.* The work is applied at the fluid boundaries by surface forces, and throughout the C.V. by body- or mass-acting forces.

Note that the energy in transit or rate quantities of heat transfer $\dot{Q}_H$ and work rate $\dot{W}$ exist only for a dynamic system (motion implied). They are, therefore, instantaneous *integral*[2] rate quantities associated with the *process* by which the energy conversion occurs. No fluid *state* of work (or work rate) or heat transfer exists, since these quantities are defined only for *dynamic processes.*

[2] No differential amount of work or heat is definable.

Since most of the quantities involved in the statement (5-2) are in the form of transport or transient quantities, it is an advantage to divide through by the quantity δt and form a rate statement:

$$\begin{bmatrix}\text{RATE OF ENERGY}\\ \text{INCREASE IN C.V.}\end{bmatrix} = \begin{bmatrix}\text{NET INFLOW RATE}\\ \text{OF ENERGY INTO C.V.}\end{bmatrix} + \begin{bmatrix}\text{HEAT TRANSFER}\\ \text{TO C.V.}\end{bmatrix} + \begin{bmatrix}\text{WORK RATE ON}\\ \text{FLUID IN C.V.}\end{bmatrix} \quad (5\text{-}3a)$$

Transposing the energy influx term to the left side gives

$$\begin{bmatrix}\text{RATE OF ENERGY}\\ \text{INCREASE IN C.V.}\end{bmatrix} + \begin{bmatrix}\text{NET } \textit{outflow} \text{ RATE}\\ \text{OF ENERGY FROM C.V.}\end{bmatrix} = \begin{bmatrix}\text{HEAT TRANSFER}\\ \text{TO C.V.}\end{bmatrix} + \begin{bmatrix}\text{WORK RATE ON}\\ \text{FLUID IN C.V.}\end{bmatrix} \quad (5\text{-}3b)$$

As stated above, a space-fixed Eulerian frame of reference is necessary, with fixed, nonvarying control volume boundaries.[3] Consider the energy-rate and work-rate terms in the above statements as applied to the generalized control volume of Fig. 5-2. First, the energy transport rate *out* of the system by the mass flow can be determined by considering the energy of a quantity of mass δm that has flowed out through the control volume boundary at 1 in the period δt. The energy in this volume is the energy per unit mass multiplied by the mass in the volume element, or[4]

$$e\,\delta m = e\rho\,\delta V = e\rho\,\overrightarrow{\delta \mathrm{s}}\cdot\overrightarrow{\Delta \mathrm{A}} \quad (5\text{-}4)$$

where e is the summation of all mass associated or "*intrinsic*" energies. A *positive* outflow quantity at 1 is indicated by the relation between the vectors $\overrightarrow{\delta s}$ and $\overrightarrow{\Delta A}$. The energy outflow *rate* through this streamtube is obtained by dividing by δt, and since $\overrightarrow{\delta s}/\delta t = \vec{\mathrm{v}}$, the energy outflow rate is

$$e(\rho\vec{\mathrm{v}}\cdot\overrightarrow{\Delta \mathrm{A}}) \qquad \text{or} \qquad e(\rho\vec{\mathrm{v}}\cdot\overrightarrow{\mathrm{dA}}) \quad (5\text{-}5)$$

[3] It is interesting to note that the basic reference point or coordinate system origin is usually taken as "earth-fixed." With respect to most small-scale earth-bound engineering problems, this choice violates the conditions with respect to the energy, since the system would possess some kinetic energy as a consequence of the earth's rotation and translation within the universe. Such effects are considered at the end of Chapter 6. The earth's motion has been considered on a small-scale engineering problem in a paper by G. S. Benton, The Effect of the Earth's Rotation on Laminar Flow in Pipes, *J. Appl. Mech.*, **23**, 123–127 (1956).

[4] Note that the δ quantity is used to denote a change occurring in δt, while a Δ quantity denotes an elemental quantity connected with the metric size as expressed in terms of coordinate system lengths, or the general coordinate element length Δx_i.

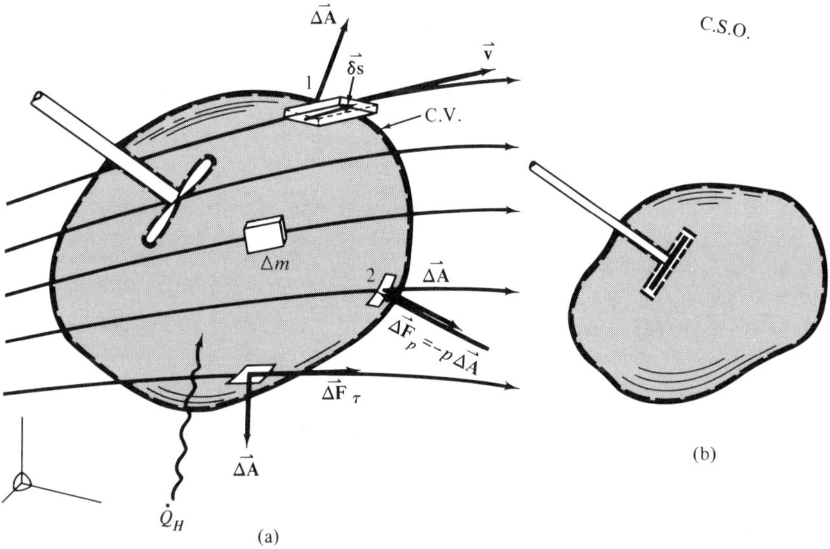

Fig. 5-2 (a) Generalized control volume to which the principle of conservation of energy is to be applied. The C.V. is fixed in space with the exception of the moving propeller shaft which does work on the fluid in the C.V. (b) Reduction of (a) to an entirely fixed C.V. representation by replacing the periodic motion of the propeller blades by a hypothetical "actuator disc" which produces the work by an imaginary steady flow process at its interface with the fluid. The control volume boundary is indicated by the heavy dashed line.

in the limit for a differential-sized streamtube.[5] The net outflow of energy transported through the control volume surface is determined by summing the net outflows for all streamtubes. This net energy efflux can be expressed as the integral over the control volume area $\oiint d\vec{A} \cdot \rho \vec{v} e$, indicating the mass transport rate of e, or $\oiint (e)\rho\vec{v}\cdot d\vec{A}$. The $\vec{v}\cdot d\vec{A}$ product accounts for the influx as a negative efflux.

HEAT AND WORK

The heat-energy input rate is by nature an integral-type term and therefore can only be written $\dot{Q}_H$ (general units of heat energy/time: Btu/sec, J/sec, or any equivalent unit). Similarly, the specific manner by which the mechanical work is produced will not be determined in detail.

[5] Although $\rho\,\delta V = \delta m$ in the expression above, the limiting condition of $\delta m \to dm$ does not have physical significance, because dm mathematically signifies a differential *change* in mass but no true change in mass is allowed in the nonrelativistic type of system assumed. Similarly, the mass flow rate or transport of mass (δm) that occurs in time δt is $\delta m/\delta t \equiv \dot{m}$. In a rigorous mathematical sense, $\dot{m}$ cannot be taken as dm/dt, since this implies a change in the mass of an object with respect to time.

The mechanical work rate (power) is specified as a net-effect or integral-type term as $\dot{W}_m$ (general units of work/time: ft-lb$_f$/sec, J/sec = W, dyne-cm/sec). The detailed determination of energy conversion to work consists of several subjects of study in itself; the study and design of turbines, pumps, and compressors, or the field of turbomachinery, are examples.

The work done on the fluid in moving it through the control surface boundaries is supplied by the external pressure forces. The mathematical definition of this pressure work is

$$W_p = \int \vec{\mathrm{F}}_p \cdot \vec{\mathrm{ds}}$$

The work can be evaluated by considering the pressure force.

The pressure force exerted *on* the fluid control surface element $\overrightarrow{\Delta \mathrm{A}}$ at 2 in Fig. 5-2a by the surroundings is $\overrightarrow{\Delta \mathrm{F}_p} = -p\,\overrightarrow{\Delta \mathrm{A}}$. The work corresponding to this $\overrightarrow{\Delta \mathrm{F}_p}$ is

$$\Delta W = \overrightarrow{\Delta \mathrm{F}_p} \cdot \overrightarrow{\delta \mathrm{s}}$$

The work rate (power) is

$$\dot{W} = \frac{\Delta W}{\delta t} = \overrightarrow{\Delta \mathrm{F}_p} \cdot \frac{\overrightarrow{\delta \mathrm{s}}}{\delta t} = \overrightarrow{\Delta \mathrm{F}_p} \cdot \vec{\mathrm{v}} = -p\,\overrightarrow{\Delta \mathrm{A}} \cdot \vec{\mathrm{v}}$$

The pressure work rate on the entire surface of the C.V. is then

$$\dot{W}_p = -\oiint p\,\vec{\mathrm{dA}} \cdot \vec{\mathrm{v}} = -\oiint \vec{\mathrm{dA}} \cdot (p\vec{\mathrm{v}}) \tag{5-6}$$

which represents the net power applied to the fluid C.V. by the pressure forces at the C.V. surface. The fact that the work at the inlet of the streamtube is of opposite sign is again accounted for by the $\vec{\mathrm{v}} \cdot \vec{\mathrm{dA}}$ product.

In general, the pressure can vary continuously *within* the control volume (nonuniform pressure field). There is no net work effect from this internal pressure field even if fluid is in motion. The pressure forces at the interfaces between mass elements of the C.V. are equal and opposite and therefore cancel, as was the case for the static fluid. No net work is then done by the pressure field *wholly within the fluid* that does not touch the control volume boundary. External work *by* the surroundings would be performed by pressure forces exerted by an object such as the propeller or turbine wheel shown in Fig. 5-2. However, these surfaces must be considered as part of the control volume boundary and not as contained with the fluid C.V.

The work done on the fluid by the shear forces: The internal shear stress forces occur in equal and opposite pairs on adjacent fluid elements and cancel, as do the internal pressure forces; therefore, only the work done at the control surface is relevant. Noting that the shear force acts perpendicular to the surface-area-element vector implies a more complicated relation between the shear stress and area than was derived for the pressure force. The evaluation of the shear work rate (or power required from the fluid to overcome the shear forces) involves a dot or "scalar" product of the shear stress tensor with the velocity (which will yield a vector, since a tensor is essentially a second-order vector) and then the scalar product of this result with the area element.[6]

It must be noted that mechanical (shaft-type) work can only be produced by a force exerted on the fluid by a work-producing actuator such as a piston surface, the surfaces of pump and turbine impellers, or other solid surface adjacent to the fluid. Since this work-production requires displacement of the actuator, the C.V. cannot possibly be "fixed in space," nor can the flow be steady in the neighborhood of the actuator. The forces producing the mechanical work are just the pressure and shear forces exerted on the fluid by the moving actuator surfaces. The flow across the inlet and outlet areas of the C.V. may be steady, as in the case of the machine of Fig. 5-1, so that it is convenient to consider the C.V. in two parts: one part surrounds the actuator, and the other part encloses the fixed, solid surfaces and cross-flow surfaces. Since the actuator motion is cyclic (or periodic) when observed from a fixed point in space, the cyclic forces may be integrated over a number of cycles to calculate the time-mean, or steady average, work. This work is actually applied as a steady work rate or power at the input shaft. The effect of the nonsteady, cyclic-work-producing actuator is simulated by a hypothetical time-averaged-steady-flow device, and the entire overall C.V. is then treated as a steady-state device with the work input by this *actuator disc*. The pressure and shear stress work at the moving actuator surfaces is

$$\left\langle \oiint \vec{dA}\cdot(-p + \tau)\cdot\vec{v} \right\rangle \equiv \langle \dot{W}_m \rangle$$

where p, τ, and $\vec{v}$ are functions of time. Integration is over the actuator surface area and averaged with respect to time. Time averaging is denoted by the pointed brackets. This procedure is advantageous for machine control volumes, where the flow and control surfaces are otherwise steady (for example, Fig. 5-1). Since this part of the control surface and volume has been removed from the pressure and shear work rate C.V. integrals, the notation $\oiint$ must now be considered as applying to the remaining part of the C.V. and crossing the shaft intersection.

[6] $\dot{W}_\tau = \int \vec{dF}_\tau\cdot\vec{v} = \oiint \vec{dA}\cdot\tau\cdot\vec{v}$, where τ is the shear or viscous stress tensor.

Transient problems involving unsteady actuator motion (where the period is varying with time as in a start-up condition) are complicated unsteady flow problems and require specific study. Transients in simple flow cases are treated in Chapter 15.

Most mass-acting, gravitational-type body forces are conservative; they can be specified as gradients of a *force potential.* Such a potential also represents the potential energy of the force field. The relation between potential energy, work, and force is

$$\Delta\text{P.E.} = W = \int \vec{F}\cdot d\vec{s} = \int \vec{F}\cdot d\vec{r}$$

where $\vec{\Delta s}$ is the path-element distance equal to $\vec{\Delta r}$, a coordinate distance along a path s.

The above relation applies to *conservative* force fields, where the work done is equal to the change in potential energy. There is no conversion of work to unavailable forms of energy (for example, internal thermal energy) in overcoming friction forces. A full conversion of the potential energy to work and a complete reversible reconversion are possible.

The force $\vec{F}$ is considered as positive acting *on* the body or fluid mass. Consequently, work *on* the fluid will be positive.[7]

If $\vec{f}_b$ is the body force per unit mass, the change in potential energy Ω per unit mass is

$$\Omega_1 - \Omega_2 = \int_1^2 \vec{f}_b\cdot d\vec{s} \qquad \text{and} \qquad -\Delta\Omega = \vec{f}_b\cdot\vec{\Delta s}$$

for the work over a small distance $\vec{\Delta s}$ or $\vec{\Delta r}$. This relation can be expressed interpretively as

$$\vec{f}_b = -\frac{\Delta\Omega}{\vec{\Delta r}}$$

or in the limit as a differential equation,

$$\vec{f}_b = -\frac{d\Omega}{d\vec{r}}$$

[7] This sign convention is in direct agreement with Newton's laws of motion (Chapter 6), which deal with the change of the momentum of a body or substance resulting from the forces acting *on* it. Unfortunately, some thermodynamics texts consider work done *by* the substance as positive, which leads to a sign confusion when energy and momentum equations are used simultaneously.

which is the vector derivative of a scalar, or its *gradient* (as in Chapter 3)

$$\vec{f}_b = -\text{grad}\ \Omega = -\nabla\Omega$$

The body-force work ΔW_b *on* a fluid mass element $\rho\,\Delta V$ by the conservative force field $\vec{f}_b$ is

$$\Delta W_b \equiv \vec{f}_b \cdot \vec{\delta s}\rho\,\Delta V$$

where $\vec{\delta s}$ (or $\vec{\delta r}$) is the path traversed in δt. The work rate for the fluid in the C.V. is

$$\dot{W}_b = \oiiint \vec{f}_b \cdot \vec{v}\rho\, dV = -\oiiint \nabla\Omega \cdot \vec{v}\rho\, dV$$

The work done on a unit mass in moving it along a path from 1 to 2 is

$$w_{1-2} = \int_1^2 \vec{f}_b \cdot \vec{ds} = -\int_1^2 \Delta\Omega \cdot \vec{ds} = -\int \vec{dr} \cdot \nabla\Omega = -\int_1^2 d\Omega = \Omega_1 - \Omega_2$$

which states that this work is equivalent to the change in potential energy of the mass. If the path is closed, or as $2 \rightarrow 1$,

$$w = -\oint d\Omega \equiv 0$$

or the work done in moving a mass around a closed path in a conservative force field is zero, which is the defining condition for the field to be conservative.

Gravitational fields can be expressed in terms of an equivalent acceleration $\vec{g}$ as

$$\vec{f}_b = \frac{\vec{g}}{g_c}$$

The gravitational potential (geopotential) is obtained from

$$\vec{f}_b = -\nabla\Omega = \frac{\vec{g}}{g_c}$$

Multiplying by $d\mathbf{r}$ gives

$$d\Omega = -\frac{\vec{g} \cdot \vec{dr}}{g_c}$$

or the geopotential is

$$\Omega = -\int \frac{\vec{g} \cdot \vec{dr}}{g_c}$$

For the earth's gravitational field, where g acts radially inward, $\vec{g} = -\hat{z}g$ (z is the positive altitude direction). Therefore,

$$\Omega = +\int \frac{g\,dz}{g_c}$$

In terms of $\vec{g}$, the body-force work is

$$\dot{W}_b = \iiint \frac{\vec{g} \cdot \vec{V}}{g_c} \rho \, dV$$

An energy conversion could take place through the interaction of the moving charge with a magnetic field (as in the case of a magnetohydrodynamic power generator or accelerator, cyclotron, linear accelerator, and so on). A work term $\dot{W}_{\text{em}}$ (electromagnetic work rate) must then also be included.

The same argument is valid for any other type of energy. Therefore, the term $\dot{W}_{\text{etc.}}$ is added to account for any other possible work conversion that may occur.

AN ENERGY-RATE EQUATION

The rate of energy storage within a system is due to the storage of mass associated or intrinsic energy. The energy within the system at any time is then the sum of all energies associated with all mass elements Δm, or $\iiint (e)\rho \, dV$. The storage rate is identical to the time rate of change of this total energy,

$$\frac{\partial}{\partial t} \iiint (e)\rho \, dV$$

Since ρ and e are the only possible time-varying quantities in the Eulerian representation, this term can be rewritten

$$\iiint \frac{\partial \rho e}{\partial t} \, dV$$

With all the work and energy terms described above now translated into mathematical quantities, the conservation of energy statement (5-3b) can be written as a rate balance,

$$\underbrace{\oiiint \frac{\partial e\rho}{\partial t}\, dV + \oiint e\rho\vec{\mathrm{V}}\cdot\vec{dA}}_{\text{NET RATE OF CHANGE OF INTRINSIC ENERGY OF FLUID IN C.V. AT ANY INSTANT}}$$

$$= \underbrace{-\oiint \rho\vec{\mathrm{V}}\cdot d\vec{\mathrm{A}} + \dot{W}_\tau + \langle \dot{W}_m \rangle + \dot{W}_{\mathrm{em}} + \dot{W}_{\mathrm{body}}}_{\text{WORK DONE ON FLUID IN C.V. BY SURROUNDINGS}} + \underbrace{\dot{Q}_H}_{\text{H.T. TO FLUID IN C.V.}} \qquad (5\text{-}7)$$

The pressure-work-rate term can be transposed and the area-integral terms combined to give a general, nonrelativistic form of the energy equation,

$$\oiiint \frac{\partial e\rho}{\partial t}\, dV + \oiint \left(e + \frac{p}{\rho}\right)\rho\vec{\mathrm{V}}\cdot d\vec{\mathrm{A}} = \sum \dot{W} + \dot{Q}_H \qquad (5\text{-}8)$$

Equation (5-8) may be compared with the "first law of thermodynamics," usually written for the static end states of a *closed system* as

$$\underbrace{\Delta E}_{\text{CHANGE IN INTERNAL ENERGY}} = \underbrace{\Delta E_W}_{\text{ENERGY INCREASE BY WORK DONE ON SUBSTANCE}} + \underbrace{\Delta E_Q}_{\text{ENERGY INCREASE BY HEAT TRANSFER TO SUBSTANCE}} \qquad (5\text{-}9)$$

This comparison emphasizes the fact that the most general statement of a law such as the conservation of energy is a verbal statement. The translation to an equation depends on the terms to be included and the method by which the "system" is described. Here the "system" is specified as a *fixed-in-space control volume*, or as a *fixed open system*. The energy terms to be included are arbitrary and dependent on the physical situation to be evaluated, the information available, and the information desired.

The term e is understood to include all types of energy listed above as "mass-associated" or "intrinsic":

$$e \equiv e_{\mathrm{chem}} + \frac{\mathrm{V}^2}{2g_c} + \tilde{u} + e_e + e_n + e_{\mathrm{etc.}} \qquad (5\text{-}10)$$

where $v^2/2g_c$ = kinetic energy of mean flow motion
$\tilde{u}$ = thermal internal energy
e_e = electrostatic potential energy
e_n = nuclear and atomic energy (of type not included in other specific terms)
$e_{etc.}$ = any other type of energy not specifically included in above terms

(The general units of e are energy per unit mass, specifically ft-lb_f/lb_m, J/kg, dyne-cm/g, and so on.)

For the purpose of the following developments, the e_{chem}, e_e, e_n, and $e_{etc.}$ will be considered as combined in the term $e_{etc.}$ with a similar combination for the $\dot{W}$ (work rate) terms. These "etc." energy terms will be completely dropped when analyzing simple fluid flow systems, giving specifically

$$\oiiint \frac{\partial e\rho}{\partial t}\,dV + \oiint \left[\frac{v^2}{2g_c} + \tilde{u} + \frac{p}{\rho}\right]\rho\vec{v}\cdot d\vec{A}$$
$$= -\oiiint \nabla\Omega\cdot\vec{v}\rho\,dV + \dot{W}_\tau + \langle\dot{W}_m\rangle + \dot{Q}_H \quad (5\text{-}11)$$

If (5-11) is to have a physical significance, all terms must have the same general units of energy rate (or, equivalently, work rate or power). Since work and energy may be most generally case (and easily) specified as F-L (ft-lb_f, N-m ≡ J, dyne-cm ≡ ergs, and so on), the mechanical equivalent of "heat"-type terms must be used. For example, the heat-transfer term must be $\dot{Q}_H J$, where $\dot{Q}_H$ is the heat-transfer rate in heat-transfer units (for example, Btu/sec) and J is the conversion from mechanical units to heat (J = 778 ft-lb_f/Btu). The same situation exists for the internal energy term $\tilde{u}$. (Note that no such dual unit exist in the mks system; both types of energy are expressed as joules and power is in watts.) Two courses of action are now possible: (1) to use the equation as it is, or (2) to apply the Gauss divergence theorem to the area-integral term so that the integral terms can be combined as a single volume integral. The latter procedure gives

$$\oiiint \left\{\frac{\partial e\rho}{\partial t} + \operatorname{div}\left[\left(\frac{v^2}{2g_c} + \tilde{u} + \frac{p}{\rho} + \Omega\right)\rho\vec{v}\right]\right\}dV = \dot{W}_\tau + \langle\dot{W}_m\rangle + \dot{Q}_H \quad (5\text{-}12)$$

CASE I

STEADY FLOW

If the flow is steady, the time-derivative terms are zero. If the C.V. boundaries are fixed (invariant with time), there is literally no possibility of any mechanical work; however, the time-mean effect of this periodic work can be considered as a quasi-steady work $\langle \dot{W}_m \rangle$ if the *period* or *frequency* of the C.V. boundary motion is constant, as discussed previously.

A simplification of notation is possible by noting that $\tilde{u} + (p/\rho) \equiv \tilde{h}$, the enthalpy per unit mass (or specific enthalpy) and, further, that $\tilde{h} + (\mathrm{v}^2/2g_c) \equiv \tilde{h}_0$, the stagnation enthalpy. Equation (5-12) then becomes

$$\oiiint \operatorname{div} [(\tilde{h}_0 + \Omega)(\rho\vec{\mathrm{v}})]\, dV = \dot{W}_\tau + \langle \dot{W}_m \rangle + \dot{Q}_H \tag{5-13a}$$

The Gauss divergence theorem can be applied to (5-13a) to give[8]

$$\oiint [(\tilde{h}_0 + \Omega)(\rho\vec{\mathrm{v}})] \cdot \vec{\mathrm{dA}} = \dot{W}_\tau + \langle \dot{W}_m \rangle + \dot{Q}_H \tag{5-13b}$$

If the flow is *quasi-one-dimensional* (as described in Section 4.9) a modification of the energy relation is possible. If the control volume surface area integral is broken up into the surface integrals over the inlet and outlet flow paths, and the body force is assumed to be caused by gravity only, (5-13b) can be rewritten

$$\sum_{\substack{\text{OUTFLOW}\\ \text{AREAS}}} \iint \left(\tilde{h}_0 + \frac{gz}{g_c}\right) \rho\vec{\mathrm{v}} \cdot \vec{\mathrm{dA}} + \sum_{\substack{\text{INFLOW}\\ \text{AREAS}}} \iint \left(\tilde{h}_0 + \frac{gz}{g_c}\right) \rho\vec{\mathrm{v}} \cdot \vec{\mathrm{dA}} = \dot{W}_\tau + \langle \dot{W}_m \rangle + \dot{Q}_H \tag{5-14}$$

The quantities that appear as linear terms may be replaced by average values over the outflow or inflow surface areas if the integrand quantity is nearly uniform over the area A. For example,

$$\iint_1 \tilde{h}\rho\vec{\mathrm{v}} \cdot \vec{\mathrm{dA}} = \dot{m}_1 \bar{\tilde{h}}_1$$

where $\bar{\tilde{h}}_1$ is the average value of $\tilde{h}$ over the area A at station 1, as indicated in Fig. 5-3. In this example it must be assumed that the product $\rho\mathrm{v}$ is also nearly uniform over the area A and that $\tilde{h}$ varies only slightly, as indicated in Fig. 5-3. In general, for a flow to be quasi-one-dimensional, it is necessary that the flow (1) has small curvature, (2) is turbulent, and (3) is nearly adiabatic. The requirement of turbulence will yield a nearly uniform velocity profile. Negligible heat transfer plus turbulence will give a nearly uniform temperature ($\tilde{u}$ and $\tilde{h}$) profile. In addition, the pressure, and consequently the density, will be about constant over the area if the curvature is small. T and ρ variations are only significant over areas where heat

[8] Alternatively, the gradient theorem could be applied to the Ω term of (5-11) to give the same result directly.

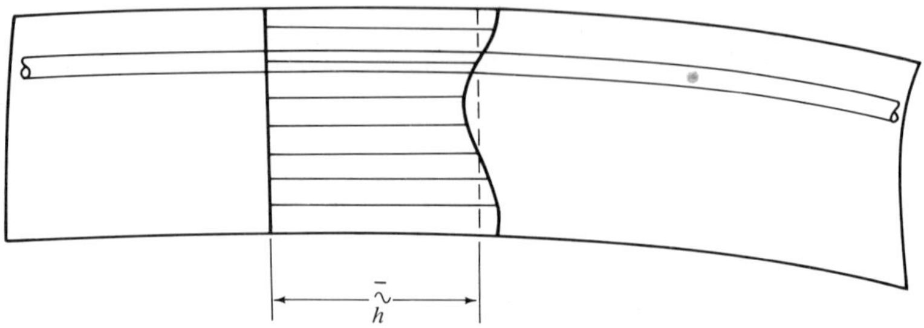

Fig. 5-3. Distribution and average value of enthalpy at a station or section in a flow field. The variation of $\tilde{h}$ from the average value $\bar{\tilde{h}}$ is assumed to be small.

transfer is on the order of the other terms in the equation, for example, $\dot{Q}_H$ and $\dot{m}\,\Delta\tilde{h}$, in (5-14).

The kinetic-energy term is definitely not linear: The variation of velocity with the variable of integration A is cubic. The kinetic-energy term is then

$$\dot{\text{K.E.}} = \iint_{A_1} \frac{\rho \mathrm{v}^2}{2g_c} \vec{\mathrm{v}} \cdot d\vec{\mathrm{A}} = \iint_{A_1} \frac{\rho \mathrm{v}^3}{2g_c}\, dA$$

where the surface area A_1 is chosen perpendicular to the flow direction. The *average velocity* at an area station is defined as

$$\bar{\mathrm{v}} \equiv \frac{\dot{m}}{\rho A} \equiv \frac{\iint \rho \vec{\mathrm{v}} \cdot d\vec{\mathrm{A}}}{\rho A} \equiv \frac{\iint \vec{\mathrm{v}} \cdot d\vec{\mathrm{A}}}{A} \equiv \frac{Q}{A}$$

where Q is the volume flow rate (ft^3/sec, m^3/sec, and so on) and the density is assumed to be constant over the area of integration.[9] The kinetic energy cannot be written directly in terms of the average velocity,

$$\frac{\rho(\bar{\mathrm{v}})^3 A}{2g_c} = \frac{\dot{m}(\bar{\mathrm{v}})^2}{2g_c} \neq \frac{\rho}{2g_c} \iint \mathrm{v}^2 \vec{\mathrm{v}} \cdot d\vec{\mathrm{A}}$$

For the kinetic energy to be specified in terms of the average velocity, a correction factor for the above inequality is required. Such a "*kinetic-energy correction factor*" α can be defined as

$$\frac{\alpha m(\bar{\mathrm{v}})^2}{2g_c} = \frac{\alpha \rho(\bar{\mathrm{v}})^3 A}{2g_c} = \frac{\rho}{2g_c} \iint_A \mathrm{v}^2 \vec{\mathrm{v}} \cdot d\vec{\mathrm{A}}$$

[9] Note that Q is used to represent the volumetric flow rate; $\dot{Q}_H$ represents the heat-transfer rate, as is indicated explicitly by the subscript H.

or

$$\alpha \equiv \frac{1}{(\bar{\mathrm{v}})^3 A} \iint \mathrm{v}^2 \vec{\mathrm{v}} \cdot d\vec{\mathrm{A}}$$

If the surface area is everywhere perpendicular to the flow across the area A,

$$\alpha = \frac{1}{(\bar{\mathrm{v}})^3 A} \iint \mathrm{v}^3 \, dA = \frac{\overline{\mathrm{v}^3}}{\bar{\mathrm{v}}^3} \tag{5-15}$$

That is, α is the ratio of the mean value of v^3 to the cube of the average value of the velocity.

Using these average values, the energy equation can be written for a quasi-one-dimensional flow as

$$\sum_{\substack{\text{OUTFLOW}\\ \text{AREAS}}} \dot{m}\left(\frac{\alpha \mathrm{v}^2}{2g_c} + \tilde{h} + \frac{gz}{g_c}\right) - \sum_{\substack{\text{INFLOW}\\ \text{AREAS}}} \dot{m}\left(\frac{\alpha \mathrm{v}^2}{2g_c} + \tilde{h} + \frac{gz}{g_c}\right) = \dot{W}_\tau + \langle \dot{W}_m \rangle + \dot{Q}_H \tag{5-16}$$

The advantage of (5-16) over (5-13) is that the integration procedure can be eliminated if the mass flow rate, velocity profiles (or α's), and average velocities are known.

Two factors are now of interest:

1. At the solid boundaries of the control surface, the velocity is zero, so that the work rate $(\vec{F}_\tau \cdot \vec{\mathrm{v}})$ is zero. The shear work term can then be dropped by a proper choice of control volume. If the inlet and outlet areas are perpendicular to the inflow and outflow directions, there can be essentially no shear stress on these areas.[10] Although the shear work term may not enter explicitly into the energy equation, by choice of the control volume, the shear stresses within the system are responsible for a conversion of energy and pressure availability into unavailable internal energy. The work and energy converted in overcoming the friction forces goes into the internal or random kinetic energy of motion of the fluid molecules (and into energy to rearrange the intermolecular force fields). These work and energy

[10] A shear stress force can exist on a surface perpendicular to the flow direction in a compressible flow where the velocity change is extremely rapid in the flow direction. However, such rapid rates produce significant shears only in regions such as those in the center plane of a shock wave. This type of shear stress can be eliminated from consideration by choosing the control volume to be to one side or the other of such a shock. This perpendicular shear action is also responsible for the internal damping of purely oscillatory motions. However, in most types of flows, the friction damping at solid surfaces is much greater than such internal vibration damping, so this internal friction can be neglected.

forms can again be redistributed in the energy balance of the fluid as it continues through the system.

2. There is no "*energy-loss*" term in the equation. Indeed, the principle of conservation of energy states that energy cannot be "lost," only converted from one form to another or into transient energy forms (heat and work).

For a simple, single flow path system (flow field control volume), $\dot{m}_{\text{in}} = \dot{m}_{\text{out}} = \dot{m} = \text{const}$. Equation (5-16) may be divided through by $\dot{m}$ to give

$$\left(\frac{\alpha \text{v}^2}{2g_c} + \tilde{h} + \frac{gz}{g_c}\right)_{\text{out}} - \left(\frac{\alpha \text{v}^2}{2g_c} + \tilde{h} + \frac{gz}{g_c}\right)_{\text{in}} = \langle w_m \rangle + \tilde{q} \qquad \text{(5-16a)}$$

where $\langle w_m \rangle \equiv \langle \dot{W}_m \rangle / \dot{m}$ and $\tilde{q} \equiv \dot{Q}_H / \dot{m}$ (work and energy per unit mass of fluid: ft-lb_f/lb_m, J/kg, dyne-cm/gm, and so on) and the C.V. boundary has been chosen to give $\dot{W}_\tau = 0$, $w_\tau = 0$, as described above.

Case II

STEADY, ONE-DIMENSIONAL FLOW

If the flow is literally one-dimensional, the p, ρ, v, $\tilde{h}$, and so on, will not vary across any inlet or outflow station, for all the quantities within the area integral will be independent of the specific area element (dA) at that station. If the total control volume area integral is broken down into the inlet and outlet flow path surface areas where the flow enters and leaves, (5-16) can be written for these inlet and outlet flow stations as

$$\sum_{\substack{\text{OUTFLOW}\\ \text{AREAS}}} \left[\left(\frac{\text{v}^2}{2g_c} + \tilde{h} + \frac{gz}{g_c}\right) \iint \rho \vec{\text{v}} \cdot d\vec{\text{A}}\right] + \sum_{\substack{\text{INFLOW}\\ \text{AREAS}}} \left[\left(\frac{\text{v}^2}{2g_c} + \tilde{h} + \frac{gz}{g_c}\right) \iint \rho \vec{\text{v}} \cdot d\vec{\text{A}}\right] = 0 \qquad \text{(5-17)}$$

The quantity $\dot{W}_\tau$ cannot exist for a literally one-dimensional flow, because there can be no significant velocity variation at any cross section within the flow boundaries. If the flow were viscous, this condition would imply an infinite shear stress at the wall, which is not a physically allowable result. Therefore, "one-dimensional" flows can only be *ideal* or frictionless flows. *Parts* of some flow fields sufficiently far from closed solid boundaries (for example, the flow field and in a wind tunnel) can be considered frictionless, and therefore ideal. Similarly, any significant heat transfer cannot be considered in a strictly one-dimensional compressible flow field because of the temperature gradients that would be produced normal to the boundary or normal to the flow direction. Even for flow in a straight duct, normal density gradients resulting from the normal temperature variations would also produce normal velocity gradients. In an ideal frictionless flow field, there is also no means by which mechanical work $\langle \dot{W}_m \rangle$ can be initiated.[11]

The quantity $\iint \rho \vec{\mathrm{v}} \cdot \vec{\mathrm{dA}}$ is $\dot{m}_{\text{out}}$ over an outflow area and $-\dot{m}_{\text{in}}$ over an inflow path (where $\dot{m}$ is the magnitude of the mass flow rate). Equation (5-17) then becomes

$$\sum_{\text{OUT}} \dot{m}\left(\frac{\mathrm{v}^2}{2g_c} + \tilde{h} + \frac{gz}{g_c}\right) - \sum_{\text{IN}} \dot{m}\left(\frac{\mathrm{v}^2}{2g_c} + \tilde{h} + \frac{gz}{g_c}\right) = 0 \tag{5-18}$$

If the device is a simple, single-flow-path piece of equipment (such as a ducting or piping system where friction can be neglected), (5-18) can be divided through by $\dot{m}$ to give

$$\left(\frac{\mathrm{v}^2}{2g_c} + \tilde{h} + \frac{gz}{g_c}\right)_1 = \left(\frac{\mathrm{v}^2}{2g_c} + \tilde{h} + \frac{gz}{g_c}\right)_2 \tag{5-19}$$

where 1 implies the inlet and 2 the outflow station.

Case III

STEADY, QUASI-ONE-DIMENSIONAL, INCOMPRESSIBLE FLOW

If the flow is also incompressible, (5-16a) is not changed. The only additional, but implicit, condition is that $\rho_1 = \rho_2$. However, the energy balance is different with respect to the internal friction work terms. Any conversion of available work and energy in overcoming friction forces can only go into internal energy ($\tilde{u}$), which then can appear only as an increase in the temperature of the fluid. In an incompressible fluid there exists no mechanism by which this internal energy can be converted back into any type of useful energy or work, so it is termed a *loss of availability*. This quantity is sometimes given the name energy dissipation or energy loss. The same reasoning indicates that any heat transfer to an incompressible fluid cannot be used to perform work of any kind, but can only go into increasing the internal energy. In a *compressible flow*, a heat transfer (or internal energy increase) can produce work by subsequent fluid expansion or change in kinetic energy. The molecular potential energy and random kinetic energy can be converted to work by *expansion* in a *compressible* flow, but (by definition) no such expansion can be obtained in an *incompressible* flow. The useful or available energy and work converted to unavailable energy in overcoming the shear forces in compressible flow can

[11] The situation is similar to trying to push a wagon while standing on a frictionless surface.

be determined, but development of this "dissipation function" is a subject of more advanced study.

TOTAL INTERNAL ENERGY AND TOTAL ENTHALPY

For the purpose of defining some often-used quantities it is of interest to reconsider the more general form for the intrinsic energy term plus the pressure work terms as they would appear in detail in the integrands of (5-11) or (5-12):

$$\tilde{u} + \frac{p}{\rho} + e_{\text{chem}} + e_e + e_{\text{n}} + e_{\text{etc.}} + \frac{\text{v}^2}{2g_c} \tag{5-20}$$

If the mean or ordered kinetic-energy term is excluded and all the intrinsic energy terms are grouped and defined as the total internal energy I_{tot},

$$\tilde{u} + e_{\text{chem}} + e_e + e_n + e_{\text{etc.}} \equiv I_{\text{tot}}$$

Using the definitions for enthalpy $\tilde{h} \equiv \tilde{u} + (p/\rho)$ and stagnation enthalpy $\tilde{h}_0 \equiv \tilde{h} + (\text{v}^2/2g_c)$, (5-20) can be written

$$I_{\text{tot}} + \frac{p}{\rho} + \frac{\text{v}^2}{2g_c} \equiv \tilde{h}_{\text{tot}}$$

where $\tilde{h}_{\text{tot}}$ is the *total enthalpy*. This total enthalpy (measured at any point in the fluid field) gives a measure of the *total* capability or *availability* of a unit mass of fluid for performing useful work. The total enthalpy (and each of its components) is a *point* or *state function* or variable.

In a flow process where changes in all *e*'s of expression (5-20) are negligible, the total enthalpy is just the *stagnation enthalpy*,

$$\tilde{h}_{\text{tot}} = u + \frac{p}{\rho} + \frac{\text{v}^2}{2g_c} \equiv \tilde{h}_0 \tag{5-21}$$

where the zero subscript explicitly implies the value of the enthalpy that would be obtained by bringing the fluid to rest (zero velocity) adiabatically at a particular position in the flow field. It is a physically observed law that any thermodynamic quantity can be expressed or determined as a function of any two other *independent* thermodynamic quantities.[12] The enthalpy is a thermodynamic quantity that can be expressed or measured as a function of two primary quantities, pressure and temperature, or

[12] It is important to note that a thermodynamic quantity is a function of *two* other independent *thermodynamic* quantities only when other nonthermodynamic actions are not present. For example, density (ρ) is a function only of pressure and temperature, *only* in the absence of other actions that could have an effect on the density such as magnetic fields, electric fields, and so on. See, for example, Section 2.4.

$\tilde{h}(p, T)$. The stagnation enthalpy can then be determined by measuring the pressure and temperature of a flowing fluid that is brought to rest adiabatically (that is, without heat transfer from one fluid element to another). The pressure and temperature at a stagnation point are then defined as the stagnation values, p_0 and T_0 (the zero subscript explicitly denoting zero velocity). Note that this definition of the stagnation quantities follows directly from the energy equation, since the process of bringing the flow to rest adiabatically implies no heat transfer ($\tilde{q} = 0$) and assumes no work (other than the pressure work) done by the fluid ($w_m = 0$, $w_\tau = 0$, and so on). Under these conditions the energy equation gives [from (5-19)]

$$\left(\tilde{h} + \frac{\mathrm{V}^2}{2g_c}\right)_1 = \tilde{h}_0$$

where station 1 is on the streamline that is brought to rest. The stagnation enthalpy $\tilde{h}_0$ may be calculated from measurements of p and T at a point in a flow field where the flow is brought to rest adiabatically. For adiabatic duct flow, $\tilde{h}_0$ is seen to be the reservoir enthalpy (real or imaginary). These measurements or interpretations then indicate that $\tilde{h}_0$ is a useful reference quantity. Sometimes it is useful to employ the stagnation values of p, T, $\tilde{h}$, and so on, even though these values may not exist anywhere in the fluid. They are then the hypothetical values that would be obtained at any point if the fluid were adiabatically brought to rest there. (For additional considerations of the energy equation for some special conditions and interpretations see Problems 5-1 and 5-2.)

5.3 DERIVATION OF BERNOULLI'S EQUATION FROM THE ENERGY EQUATION

A form of the energy equation was originally determined by Daniel Bernoulli[13] for the specific conditions of (1) steady, (2) one-dimensional or streamtube, (3) isentropic (no viscous forces or shear work and no heat transfer), (4) incompressible flow with (5) no mechanical work:

$$\frac{\mathrm{V}^2}{2g_c} + \frac{p}{\rho} + \frac{gz}{g_c} = \text{const.}$$

[13] Daniel Bernoulli (1700–1782) was born in Holland of Swiss parents. Along with Euler, d'Alembert, and Clairaut, he "founded" the field of hydrodynamics. "Bernoulli's equation" as he formulated it bears little resemblance to the forms in use. The principle is similar, but Bernoulli arrived at the result through the combined processes of observation, induction, and deduction.

For the steady flow in a duct under the above conditions, the energy equation is

$$\oiint \left(\frac{\mathrm{v}^2}{2g_c} + \tilde{h} + \frac{gz}{g_c}\right)\rho\vec{\mathrm{v}}\cdot d\vec{\mathrm{A}} = 0 \tag{5-22}$$

In addition, if the control volume is a streamtube, the quantities within the integral are constant over the inflow and outflow areas, so (5-22) reduces to

$$\left(\frac{\mathrm{v}^2}{2g_c} + \tilde{h} + \frac{gz}{g_c}\right)_{\mathrm{out}} = \left(\frac{\mathrm{v}^2}{2g_c} + \tilde{h} + \frac{gz}{g_c}\right)_{\mathrm{in}}$$

or

$$\Delta \frac{\mathrm{v}^2}{2g_c} + \Delta\tilde{h} + \Delta\frac{gz}{g_c} = 0 \tag{5-23}$$

Now $\Delta\tilde{h} = \Delta\tilde{u} + \Delta(p/\rho)$ and from the thermodynamic considerations of the reversible heat transfer to ($\tilde{q}_{\mathrm{rev}}$), or work process on a mass of fluid as it flows through the streamtube,

$$\tilde{q}_{\mathrm{rev}} = T\,\Delta\tilde{s} = \Delta\tilde{u} + p\,dv_s = \Delta\tilde{h} - \frac{dp}{\rho}$$

(where $\tilde{s}$ denotes the entropy). Assumption 3 above gives $\tilde{q}_{\mathrm{rev}} = 0$ (or $\Delta\tilde{s} = 0$), or

$$\Delta h = \int_{\mathrm{in}}^{\mathrm{out}} \frac{dp}{\rho}$$

so (5-23) becomes[14]

$$\Delta\frac{\mathrm{v}^2}{2g_c} + \int\frac{dp}{\rho} + \Delta\frac{gz}{g_c} = 0 \tag{5-24}$$

If the flow is incompressible, the density ρ is constant and (5-24) becomes

$$\Delta\left(\frac{\mathrm{v}^2}{2g_c} + \frac{p}{\rho} + \frac{gz}{g_c}\right) = 0$$

or

$$\frac{\mathrm{v}^2}{2g_c} + \frac{p}{\rho} + \frac{gz}{g_c} = \text{const.} = B \tag{5-25}$$

In some cases, the viscous work term ($\dot{W}_\tau$) in (5-13) can be dropped by judicious choice of control surface, as described above. An example is the flow in a nozzle, where the kinetic energy is increased as a result of the

[14] This form of the equation is sometimes incorrectly referred to as the "compressible Bernoulli equation." More correctly, (5-24), it is the integrated Euler equation for steady, one-dimensional flow.

work done by the pressure forces. The conversion from pressure work to kinetic energy (and possible conversion between gravitational potential energy and kinetic energy if the nozzle is not horizontal) produces large variations in these terms. If the pressure difference across the nozzle (and Δz) is known, the kinetic energy can be determined by assuming that the irreversible conversion to internal energy $\tilde{u}$ is negligible. The consequent error will be on the order of 1 or 2 percent. This assumption is possible only because a nozzle is usually short, and the variation in pressure (and its conversion from force potential into kinetic energy) is then large with respect to the change in pressure required to overcome the shear forces.

In Chapter 6 it will be shown that (5-24) can also be derived from the principle of force and momentum if the same restrictions (1, 2, 3, and 5) are imposed. The important conclusion is that, *under these restrictions, the two fundamental principles yield identical equations* and therefore give no new information (or independent equations) that could be used to solve problems in fluid mechanics.

For the quasi-one-dimensional flow of an incompressible fluid in a ducting system, it is interesting to observe that Bernoulli's equation might be written

$$\frac{\alpha \mathrm{v}^2}{2g_c} + \frac{p}{\rho} + \frac{gz}{g_c} = \text{const.} + B$$

where the kinetic-energy correction factor (α) is included to account for the velocity variation (from zero at the walls to some maximum) within the duct. Such a modification is not logical. Including α implies that a friction effect is only partially included. The actual effect of friction in producing a change in $\tilde{u}$ is not included. Although it was shown that the friction work term ($\dot{W}_\tau \approx \vec{F}_\tau \cdot \vec{\mathrm{v}}$) would be omitted if the control surface were taken as coincident with the walls, assumption 3 used in the derivation of the Bernoulli equation is violated: If friction is not completely neglected, the change in entropy ($\Delta\tilde{s}$) is not zero even for adiabatic flow, and the assumption that $\Delta\tilde{h} = \Delta p/\rho$ is no longer valid. Therefore, the α correction to Bernoulli's equation is only a half-correction and is not worthwhile.

It is also interesting to note that the quantity $p + \rho gz/g_c$ is the equivalent *manometric* or static pressure that would be indicated by a simple manometer connection (that is, *piezometer tap*) on the side of a pipe or duct containing an incompressible flow (Fig. 5-4). For a straight duct, this pressure "head" would be the same for a manometer tube connected to the top or bottom (or any intermediate location) of the duct, since only the static p is measured. The quantity $z + g_c p/\rho g$ is termed the *manometric* or *static head*. If the manometric pressure is denoted by $p_m \equiv p + \rho gz/g_c$,

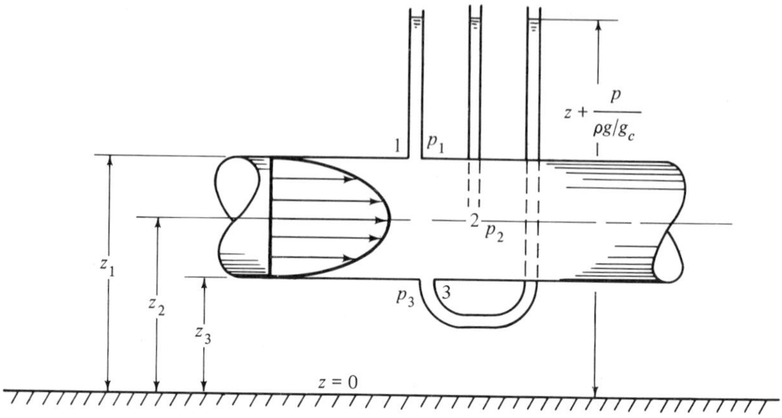

Fig. 5-4. Representation of the "manometric," or "static," "head."

then Bernoulli's equation can be written

$$\frac{\mathrm{v}^2}{2g_c} + \frac{p_m}{\rho} = \text{const.}$$

or, defining $z_m = z + g_c p/\rho g$,

$$\frac{\mathrm{v}^2}{2} + g z_m = \text{const.}$$

★5.4 LAGRANGIAN DEVELOPMENT OF THE ENERGY EQUATION

The Lagrangian viewpoint varies from the Eulerian in that the control volume is drawn around a *fixed mass* of fluid, and changes in energy of this fixed mass are investigated as it passes through a physical system. The boundaries of the control volume and the physical system are therefore different, and they cannot usually be chosen to coincide, which is one disadvantage of this method of representation. The control volume mass will deform as it passes through the system. Its energy will change and the surrounding system will do work on it.

In stating the principle of conservation of energy, an element of fluid (such as indicated at the engine intake of Fig. 5-1) would be considered as an element within a larger mass of fluid entering the system. The work on this mass and the variation of its energy as it passes through the system would be determined and an equation written expressing the conservation principle for this mass. As the defined mass control volume passes through the system, the change in intrinsic energy would be

$$\frac{d}{dt}\iiint e\rho\, dV = \frac{dE}{dt}$$

where

$$E = \iiint e\rho\, dV$$

is the total intrinsic energy at any time. This rate of change of E must be just equal to the rate at which work is done on the fluid mass plus the heat transfer to the mass. The entire rate of change of intrinsic energy of the fluid mass as it passes through the system is then dE/dt. The pressure work on the mass as it passes through the system is again $\iint_A p\vec{v}\cdot d\vec{A}$, where A is the area around the mass.

The energy equation is then

$$\underbrace{\frac{dE}{dt}}_{\text{RATE OF INCREASE OF ENERGY}} = \underbrace{\dot{W}}_{\text{WORK RATE ON FLUID}} + \underbrace{\dot{Q}_H}_{\text{HEAT TRANSFER TO FLUID}} \tag{5-26}$$

Integrating with respect to time over the period required for the fluid to pass through the system gives[15]

$$\Delta E = \Delta E_W + \Delta E_Q \equiv W + Q$$

If the flow is steady and only a unit mass is considered as it passes through a one-dimensional system (so that the energy and work can be considered as the same for all representative mass units or elements), then

$$\Delta e = \Delta e_w + \Delta e_q \equiv w + \tilde{q}$$

states the principle of conservation of energy for that unit mass element. The intrinsic energy terms in e and the work terms are the same as described previously, so that the energy equation for a "nearly one-dimensional" flow (that is, $\alpha \sim 1.0$) could be written

$$\Delta\left(\tilde{u} + \frac{p}{\rho} + \frac{gz}{g_c} + \frac{\alpha v^2}{2g_c}\right) = \langle w_m\rangle + \tilde{q} \tag{5-27}$$

[15] W represents the change in internal energy as a consequence of work done on the fluid during the process. No static state of work (or Q) is definable.

which is the same result as previously obtained and presented in (5-16a) *for the conditions of steady, quasi-one-dimensional flow*. Note that the results are the same only for these restrictive conditions, and α now implies an averaging integration of kinetic energy over all particles in the fixed mass being carried through the system.

The development of the Lagrangian flow equations for the more general case of nonsteady flow is much more difficult for the following reasons: (1) following a given mass through a system requires a knowledge of the manner in which this mass deforms as its energy changes and work is performed on it; (2) the system boundaries must also be prescribed as mathematical boundary conditions; and (3) a relation must be formulated to express the correspondence of the mass boundaries and the system boundaries as the mass flows through the system.

5.5 SUMMARY

The most general statement of the principle of conservation of energy is a verbal statement. For a flowing fluid, this principle is best formulated from a statement expressing the work *rate* (power) and energy rates:

$$\oiiint \frac{\partial e\rho}{\partial t}\, dV + \oiint \left(e + \frac{p}{\rho}\right)\rho\vec{\mathrm{v}}\cdot d\vec{\mathrm{A}} = \sum \dot{W} + \dot{Q}_H \qquad (5\text{-}8)$$

where e contains all the forms of intrinsic or mass-associated energy. If the flow is steady and only gravitational potential, thermodynamic and fluid kinetic-energy terms are considered, the energy equation can be written (from (5-12) with the divergence theorem)

$$\oiint \left(\frac{\mathrm{v}^2}{2g_c} + \tilde{u} + \frac{p}{\rho} + \Omega\right)\rho\vec{\mathrm{v}}\cdot d\vec{\mathrm{A}} = \dot{W}_\tau + \langle \dot{W}_m \rangle + \dot{Q}_H \qquad (5\text{-}12a)$$

By an appropriate choice of the control volume boundary, $\dot{W}_\tau$ can be made zero.

The energy equation for (1) steady, (2) one-dimensional (or streamtube or streamline), (3) isentropic flow with (4) no mechanical work involved, is

$$\frac{\mathrm{v}^2}{2g_c} + \int \frac{dp}{\rho} + \Omega = \text{const.} \qquad (5\text{-}24a)$$

For the flow of a liquid (negligible compressibility or $\rho \approx$ const.) Bernoulli's equation is obtained:

$$\frac{\mathrm{v}^2}{2g_c} + \frac{p}{\rho} + \Omega = \text{const.}$$

For the case in which g is the earth's gravitational field g_e, Ω becomes gz/g_c.

Consideration or interpretation of availability is not possible without simultaneous consideration of the principles of energy and force and momentum (or the equations derived therefrom). The subject of "losses" for an incompressible flow will be treated in Chapter 6.

5.6 ILLUSTRATIVE EXAMPLE 5-1: Gas Turbine

A "twin-spool" gas turbine engine is constructed as illustrated in Fig. 5-5. One turbine is used to drive the compressor while a separate turbine (in series with the first) supplies the output shaft power. A test on a turbine rated at nominally 3000-kW output yielded the following data:

Ambient conditions: $T_0 = 70°\text{F}$, $p_0 = 14.5$ psia
Inlet flow rate: $\dot{m}_{\text{air in}} = 40\ \text{lb}_\text{m}/\text{sec}$
Compressor outlet (station 2): $T_2 = 1100°\text{R}$, $\text{v}_2 = 200$ ft/sec, $p_2 = 159.4$ psia
Fuel rate: $\dot{m}_f = 0.8\ \text{lb}_\text{m}/\text{sec}$ (enthalpy of combustion of fuel, 18,000 Btu/lb_m; combustion efficiency, $\eta_B = 0.95$)
Turbine outlet conditions: $T_4 = 1370°\text{R}$, $p_4 = p_0 = 14.5$ psia [kinetic-energy change, $(\text{v}_4^2 - \text{v}_3^2)/2g_c$, across the turbine is negligible]

It is desired to find (a) the compressor power, (b) the total turbine horsepower and the output shaft power, and (c) the operating efficiency at these conditions.

Assuming steady, one-dimensional flow and that the working fluid behaves as a perfect gas and can be considered as air throughout, an

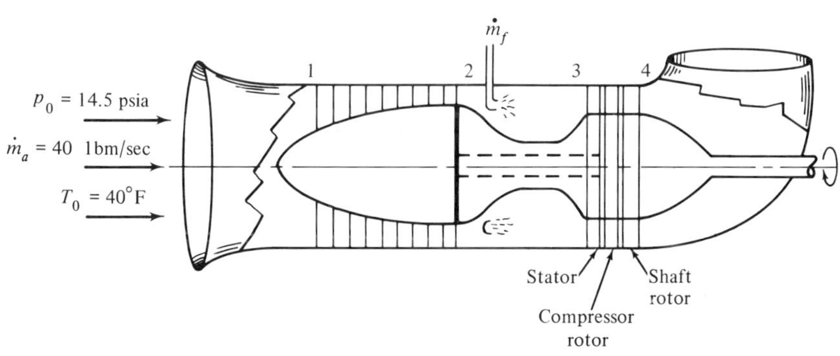

Fig. 5-5. "Twin-spool" gas turbine engine. One "spool" consists of a turbine stage and the compressor it drives; the second "spool" is the output turbine stage and shaft.

energy balance (or application of the energy equation) for adiabatic flow across the compression gives

$$\dot{m}_2\left(\frac{\mathrm{V}_2^2}{2g_c} + \tilde{h}_2\right) - \dot{m}_1\left(\frac{\mathrm{V}_1^2}{2g_c} + \tilde{h}_1\right) = \dot{W}_C$$

where $\dot{W}_C$ is the work done *on* the fluid *by* the compressor. The control volume is taken as the inlet opening, the walls of the compressor and the exit plane at the compressor discharge (station 2). Assume (1) that the air behaves as a perfect gas with constant specific energies ($c_p = 0.24$ Btu/lb$_m$), (2) that $\dot{m}_1 = \dot{m}_2 = \dot{m}_a =$ constant is the airflow rate, and (3) that $\tilde{h} = c_pT$. The stagnation temperature at 1 is then T_0, since the temperature of the air at rest is T_0. The work done by the compressor on the fluid is

$$\dot{W}_{\text{compressor on fluid}} = \dot{m}\left(c_pT_2 + \frac{\mathrm{V}_2^2}{2g_c} - c_pT_0\right) = \dot{m}\left(c_p(T_2 - T_0) + \frac{\mathrm{V}_2^2}{2g_c}\right)$$

$$= 40[10.64 \times 10^4 + 622] \text{ ft-lb}_f/\text{sec}$$

$$\text{HP}_C = 7800 \text{ hp}$$

The kinetic-energy term (622) is relatively small. The total compressor input to the air is 7800 hp, which must be supplied by the first turbine. The useful or available energy input between 2 and 3 is also obtained by consideration of the energy equation. There is no work done on the fluid within the burner-chamber control volume, and this part of the engine is assumed to be insulated, so the heat transfer to the surroundings is again negligible. The energy equation then gives

$$(\dot{m}_3h_{03}) - (\dot{m}_2h_{02} + \dot{m}_fe_{f,\text{chem}}) = 0$$

where $\dot{m}_3 = \dot{m}_a + \dot{m}_f = 40.8$ lb$_m$/sec.

Noting that the f/a ratio is only 0.02 (or about 1/7 of the stoichiometric value) indicates that the fluid at 3 is about 15 percent combustion gases and 85 percent air; therefore, as a first approximation, it will be assumed that the fluid is all air.[16] The energy available from the fuel combustion is

$$\dot{m}_fe_{\text{chem}} = \dot{m}_f\eta_B\,\Delta h_B = (0.8)(0.95)(18{,}000) = 13{,}680 \text{ Btu/sec}$$

[16] At the higher temperatures, the c_p of air will start to vary. Also the assumption that the mixture is all air at 3 is in error. More exact calculations can be made using the actual enthalpy values for air and the mixture as determined from tables such as J. H. Keenan and J. Kaye, *Gas Tables*, John Wiley & Sons, Inc., New York, 1948.

Or the turbine inlet enthalpy is

$$\tilde{h}_{03} = \frac{\dot{m}_a}{\dot{m}_3} h_{02} + \frac{\dot{m}_f}{\dot{m}_3} e_{\text{chem}} = \frac{1}{1.02} [\tilde{h}_{02} + 0.02(13{,}680)] \text{ Btu/lb}_\text{m}$$

or
$$\tilde{h}_{03} = \frac{(0.24)(778)(1100) + 622 + (273.6)(778)}{1.02}$$
$$= 4.175 \times 10^5 \text{ ft-lb}_\text{f}/\text{lb}_\text{m}$$

Neglecting the change in kinetic energy across the turbine the work done *on the fluid by the turbine* is

$$\dot{W}_T = \dot{m}_4\tilde{h}_{04} - \dot{m}_3\tilde{h}_{03} = 40.8[(0.24)(778)(1370) - 4.175 \times 10^5] \text{ ft-lb}_\text{f}/\text{sec}$$

or
$$\dot{W}_T = -11{,}770 \text{ hp}$$

for both turbines. The negative sign indicates negative work *on* the fluid, or work *by* the fluid on the turbine. The compressor requires 7800 hp, leaving 3970 hp (or 2960 kW) for the shaft turbine.

The overall efficiency is the shaft output divided by the fuel energy input, or

$$\eta_{oa} = \frac{(3970)(550)}{(0.08)(18{,}000)(778)} = 0.195$$

or 19.5 percent.

It is interesting to note that the turbines must develop nearly 12,000 hp to obtain a useful output of about 4000 hp. It is equally interesting to observe that if the compressor and its turbine were 100 percent efficient and there were no losses anywhere in the system and no power were taken from the output shaft, then, if the fuel were cut off, the turbine shaft would continue to deliver 8000 hp to the compressor, which would continue to do work on the air, using the 8000 hp to increase the air pressure. The air would then expand through the turbine, providing the 8000 hp to drive the turbine. This "circulating power" is analogous to the power required to compress the gas in a piston engine, which would be regained on expansion if there were no losses and no combustion occurred.

5.7 ILLUSTRATIVE EXAMPLE 5-2: Flow Through a Venturi Meter

A Venturi meter[17] is used to indicate volume or mass flow rate as a function of the differential pressure resulting from the velocity variation

[17] Giovanni Battista Venturi (1746–1822), Italian. Physicist, ordained priest, and Secretary of the Italian Legation in Paris during the French Republic. Developed the Venturi profile in studying ways to eliminate eddy formation at the entrance section to pipes.

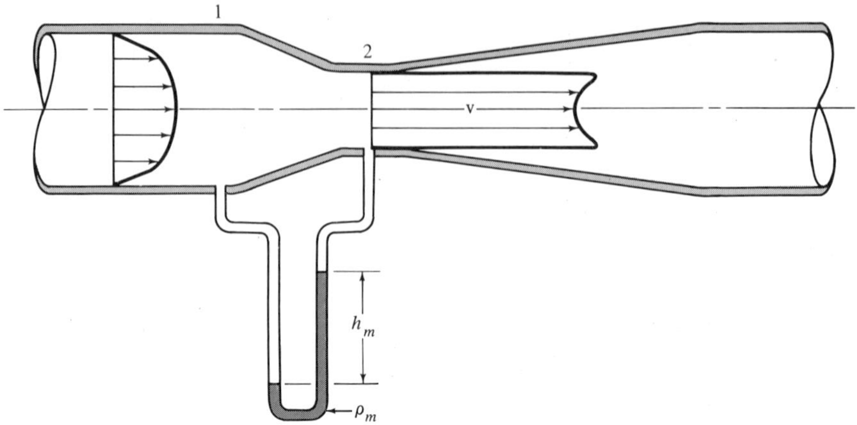

Fig. 5-6. Venturi-meter schematic.

between two flow sections of different cross-sectional area as illustrated in Fig. 5-6.

To determine the mass flow rate in the system under conditions of steady flow, Bernoulli's equation can be used to calculate an approximate velocity. Writing Bernoulli's equation between points 1 and 2 gives

$$\frac{v_1^2}{2g_c} + \frac{p_1}{\rho} + \frac{gz_1}{g_c} = \frac{v_2^2}{2g_c} + \frac{p_2}{\rho} + \frac{gz_2}{g_c}$$

Application of this equation to the above system implies the following assumptions:

(1) steady flow,
(2) incompressible flow ($\rho_1 = \rho_2$),
(3) uniform velocity profiles at 1 and 2 ($\alpha_1 = \alpha_2 = 1$),
(4) no losses (friction neglected), and
(5) no heat transfer to the fluid between 1 and 2.

The assumption of constant density is satisfactory for all liquids. For liquid flow measurements made with a Venturi meter, the density variation would not cause an error greater than 0.05 percent in the velocity or mass flow rate. For air, the error would be less than 0.5 percent for v_2 less than 120 ft/sec. Because of the nozzle shape, the flow at the walls is initially decelerated in the neighborhood of 1, where the flow curvature is concave, and is accelerated in the neighborhood of 2, where the curvature is convex. These local deceleration and acceleration effects are proportional to the velocity as well as the curvature (as will be shown in Chapter 6), so that the profile at 2 is influenced more by the acceleration than by the deceleration effect. The overall result is to produce a more pointed or parabolic profile at 1 than at an upstream station and then a dented profile (as shown) at 2. Since the flow is turbulent in most cases where a Venturi meter is used, the profile at 1 is nearly uniform, and the kinetic-energy correction factor α_1 does not vary appreciably from 1.0. If the nozzle curvature is small (if the nozzle length is greater than d_1), α_2 will also be of the order of 1.0.

Because of the complicated behavior of the flow in the neighborhood of the walls, the actual friction loss would be extremely difficult to calculate and even then the validity of the calculations would be questionable. The amount of heat transferred to (or from) the surroundings is usually quite small compared to the rate at which the internal energy is transported past 1 or 2, so the assumption of adiabatic flow is valid.

The mass flow rate is usually measured with respect to point 2 (that is, $\dot{m} = \rho_2 A_2 \mathrm{v}_2$) because the diameter of the meter at 2 is more accurately known or specified, since it is constructed as part of the meter, whereas d_1 may be the nominal pipe diameter. Using the continuity equation as

$$\dot{m} = \rho_1 A_1 \mathrm{v}_1 = \rho_2 A_2 \mathrm{v}_2 \qquad \text{and} \qquad \rho_1 = \rho_2 = \rho$$

and solving for v_2 from Bernoulli's equation, assuming the Venturi is horizontal ($z_1 = z_2$), gives

$$\mathrm{v}_2' = \frac{\sqrt{2g_c[(p_1 - p_2)/\rho]}}{\sqrt{1 - (A_2/A_1)^2}}$$

The notation v_2' is used to imply a hypothetical (or "ideal") velocity determined subject to the above approximations. If the effects of the approximations are lumped into a single correction factor, the velocity coefficient C_v, then $\bar{\mathrm{v}}_2 = C_v \mathrm{v}_2'$ is the average velocity at 2. The mass flow rate is

$$\dot{m} = \frac{C_v \rho_1 A_2}{\sqrt{1 - (A_2/A_1)^2}} \sqrt{\frac{2g_c\,\Delta p}{\rho_1}}$$

Expressing Δp as a function of the manometer head h_m and manometer fluid specific gravity, S_m,

$$\Delta p = \frac{\rho_0 g}{g_c}(S_m - 1)h_m$$

where ρ_0 is the standard density of water. The mass flow rate is then

$$\dot{m} = \frac{C_v \rho_1 A_2}{\sqrt{1 - (A_2/A_1)^2}} \sqrt{2\rho_0 g h_m(S_m - 1)}$$

As described previously, the α's and the friction behavior are a function primarily of Reynolds number. The flow profile behavior was described in terms of the curvature, which is a function of A_2/A_1 (or d_2/d_1) and the nozzle L/d ratio. The behavior of the velocity coefficient C_v, as a function of Reynolds number N_R and size, is shown in Fig. 5-7 for a series of Venturi

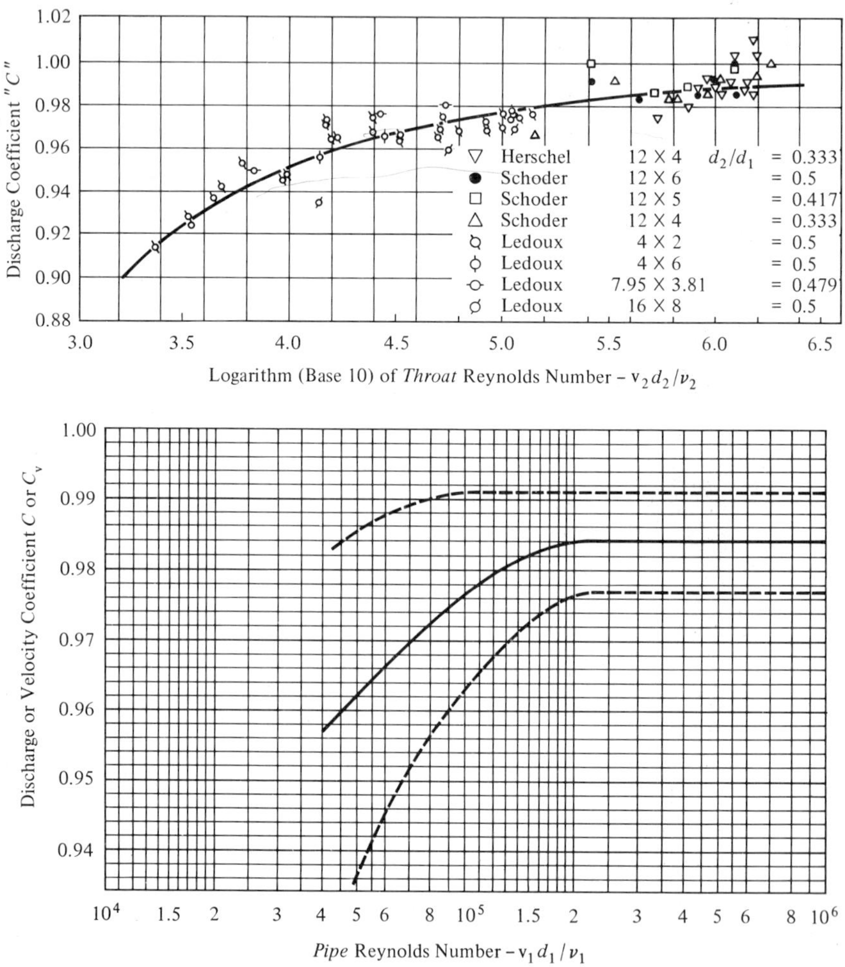

Fig. 5-7. (a) Experimentally determined discharge coefficients for several Venturi meters. (b) Discharge coefficients of Herschel-type Venturi tubes as a function of the pipe Reynolds number. Note that the Reynolds number for part (a) is based on the velocity and diameter at the throat, while the Reynolds number for part (b) is based on the pipe velocity and diameter. The range of values shown in (b) is for diameter ratios (d_2/d_1) from 0.25 to 0.75 and for pipe sizes 2 in. and larger. (Reproduced from A.S.M.E., Fluid Meters, Their Theory and Application, 1959.)

meters. The factor $1/\sqrt{1-(A_2/A_1)^2} = 1/\sqrt{1-(d_2/d_1)^4}$ is termed the *velocity of approach factor*. More information on the Venturi, its use, and standardization can be found in the reference cited in the legend of Fig. 5-7.

The only way of determining the velocity of discharge coefficient for a Venturi meter is by calculating C_v for measured values of mass flow rate and Δp or h_m. The coefficient will vary for different methods of construction and pressure tap installation. Variations caused by some of these differences are illustrated in Fig. 5-7.

If the mass flow rate for a given installation is to be determined to accuracies better than 0.5 percent, it is necessary to calibrate the meter in its operating configuration. Accurate calibrations can be made using a "weigh" tank for liquids, or by using a previously calibrated meter for gases. C_v can also be determined for gases by means of a velocity traverse obtained with a Pitot tube (Section 5.8).

When changes in density cannot be neglected, the relation for mass flow rate as a function of p_1, p_2, and the upstream temperature T_1, is considerably more complicated. Compressible flow through ducts and nozzles will be treated in Chapter 12.

Orifice plates and flow nozzles will be considered in Chapter 6, since relatively large availability losses occur with these flow-measuring devices.

5.8 ILLUSTRATIVE EXAMPLE 5-3: The Pitot Tube

A Pitot-static tube[18] allows the determination of the local velocity in a flow field by indicating the local static and stagnation pressures from which the velocity can be calculated.

A schematic sketch of a Pitot tube is shown in Fig. 5-8. The fluid is brought to rest at the point 0, and the pressure here is termed the *stagnation pressure*. The fluid in the immediate vicinity of the nose of the Pitot tube accelerates around the nose and then gradually decelerates (to a velocity about equal to the free stream velocity v_1) as the flow passes back along the sides and bottom of the tube. Along the top, however, the flow will again decelerate and stagnate on the stem, as indicated in Fig. 5-8b. At some location (position 1) the average pressure around the circumference of the tube will be about equal to the free stream pressure p_1. A slot (or several holes drilled around the circumference at this point) will then allow the average pressure at 1 to be determined. One possible proportional design using a spherical tip is indicated in Fig. 5-8. This design was determined

[18] Henri de Pitot (1695–1771), French. Educated as a mathematician but worked extensively as a hydraulic engineer. The first Pitot tube was a glass stagnation tube with a 90° band at one end; the vertical part of the tube indicated static pressure. Pitot did not apply Bernoulli's equation to obtain the correct velocity, but observed that the liquid rise in the tube was a function of the velocity.

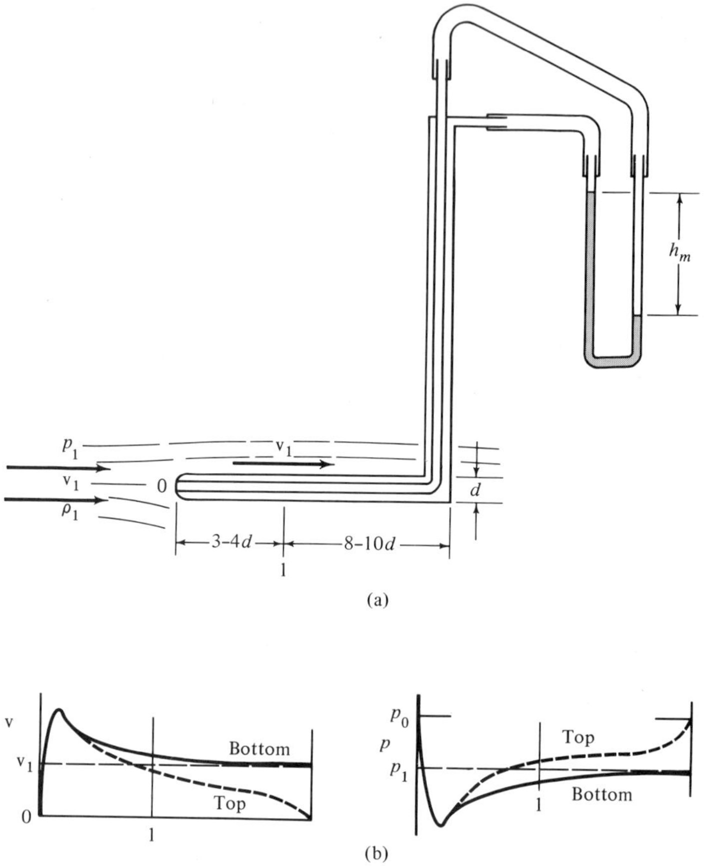

Fig. 5-8. (a) Schematic representation of a Pitot-static tube. The stagnation pressure is measured at 0 and the static pressure at 1. (b) Representative velocity and pressure distributions along the tube.

by Prandtl, and such Pitot tubes are termed Prandtl Pitot tubes or just Prandtl tubes.

Applying Bernoulli's formula between the points 0 and 1 on the Pitot tube gives

$$v_1 = C_v \sqrt{\frac{2g_c(p_0 - p_1)}{\rho}}$$

for an incompressible fluid, C_v is a velocity coefficient to account for the invalidity of the assumptions made in deriving Bernoulli's equation. Although friction exists along the tube, the friction effect on the variation of pressure perpendicular to the flow direction is negligible (as will be proved in Chapter 9). Therefore, the pressure indicated by the static pressure taps at 1 is the static pressure in the free stream flow at this location.[19] It is very difficult to imagine any significant heat transfer between the Pitot tube and the fluid in the immediate neighborhood of the tube. For a gas flowing at low enough speeds to be considered incompressible, local temperature variations and any significant heat transfer over a flow path length on the order of that of the tube are unlikely, so all assumptions made in deriving Bernoulli's equation are satisfied. Consequently, a Pitot tube of the Prandtl (or similar) design will have a velocity coefficient, C_v, of 1.00.

If the Pitot tube is used to measure high-speed gas flows such that the gas can no longer be considered to be incompressible, there is no simple method by which a correction factor can be used with Bernoulli's equation to yield the velocity. The pressure varies with velocity and is measured at points of zero and high velocity. With the gas state a function of both pressure and temperature, significant variations of density with velocity would be expected. These variations are considered in Chapter 12.

PROBLEMS

5-1 Discuss the significance of the p/ρ term as it occurs in the energy equation. Show that if no mechanical work is done on a system and the control volume boundaries are chosen to coincide with the system walls and inflow and outflow station areas, the energy equation for steady flow can be written

$$\dot{Q}_H = \oiint (\tilde{h}_0)\rho\vec{v}\cdot\vec{dA}$$

and that this form can also be written

$$\tilde{q} = \tilde{h}_{0_{out}} - \tilde{h}_{0_{in}}$$

Write out explicitly the terms $\tilde{h}_0$.

5-2 Show that

$$\tilde{q}_{in} = \Delta\tilde{h}_0 + w_{out}$$

for a steady flow system with the control volume boundaries as described in Problem 5-1. List and discuss the assumptions necessary for this equation to be valid. Define and describe the type of work terms that would be included in "w_{out}."

[19] If the flow streamlines are curved, the length (from the nose to the static taps) must be much smaller than the radius of curvature of the neighboring flow.

5-3 A viscous liquid is flowing in a 3-in.-diameter pipe. The Reynolds number is subcritical, so the flow is laminar with a parabolic profile given by

$$\mathrm{v}(r) = \mathrm{v}_0\left[1 - \left(\frac{r}{r_0}\right)^2\right]$$

(where r_0 is the pipe radius). If the mass flow is 1030 $\mathrm{lb_m/min}$ and the density of the liquid is 70 $\mathrm{lb_m/ft^3}$: (a) sketch the velocity distribution, and (b) find the average velocity, (c) the kinetic-energy transport rate, (d) the kinetic-energy correction factor, and (e) the maximum velocity. (f) What is the horsepower represented by the kinetic-energy transport rate? (g) What is the minimum viscosity that the liquid can have for the flow to be laminar?

5-4 Find (a) the kinetic-energy transport rate and (b) the kinetic-energy correction factor for the flow at the end of the contraction nozzle of Example 4-1. (c) What is the horsepower required to accelerate the flow to the state at the end of the nozzle? (d) Where did this power come from?

5-5 Express (5-18) and (5-19) explicitly for an *incompressible* flow and for the other assumptions under which these equations were specified.

5-6 Show that the potential energy of the earth's gravitational field can be obtained as $\Omega_0 = gz/g_c$. What are the units in the ft-$\mathrm{lb_f}$-$\mathrm{lb_m}$-sec, mks, and cgs systems?

5-7 Explain why a differential equation formed by differentiating (5-16a) (with $\alpha \approx 1.0$) to get $(\mathrm{v}\,d\mathrm{v}/g_c) + d\tilde{h} + (g\,dz/g_c) = d\langle w_m\rangle + d\tilde{q}_H$ does not yield a result that can be accurately interpreted. How would such quantities as $d\tilde{q}_H$ and $d\langle w_m\rangle$ be defined or interpreted?

5-8 Can (5-16a) or (5-19) be applied to the flow along a single streamline? Explain.

5-9 Can (5-16a) or (5-19) be applied to the flow through a streamtube? Explain.

5-10 Can a differential or derivative (d/ds, where ds is the distance in the stream direction) form of the energy equation properly be written for the case of (a) a one-dimensional adiabatic flow in a system with no mechanical work? (b) A quasi-one-dimensional adiabatic flow with no work?

5-11 Following (5-17) it was stated that heat transfer is not possible if a compressible flow is to be one-dimensional. Is it possible to have one-dimensional incompressible flow and also have heat transfer? Explain.

5-12 How does the stagnation pressure vary along an insulated duct? What is the reason or cause for a variation in stagnation pressure for this case?

5-13 Assuming steady, one-dimensional flow, find $d\mathrm{v}/ds$ and $\mathrm{v}(d\mathrm{v}/ds)$ as a function of s, v_1, d_1, and θ in the diffusing section of circular cross section shown. Plot $\mathrm{v}(d\mathrm{v}/ds)$ as a function of s for $\theta = 8°$ (since $\theta/2$ is small, assume that $\sin \theta/2 \approx \tan \theta/2 \approx \theta/2$). Make the plot nondimensional by plotting $[v(dv/ds)]/(v_1^2/d_1)$ versus s/d_1.

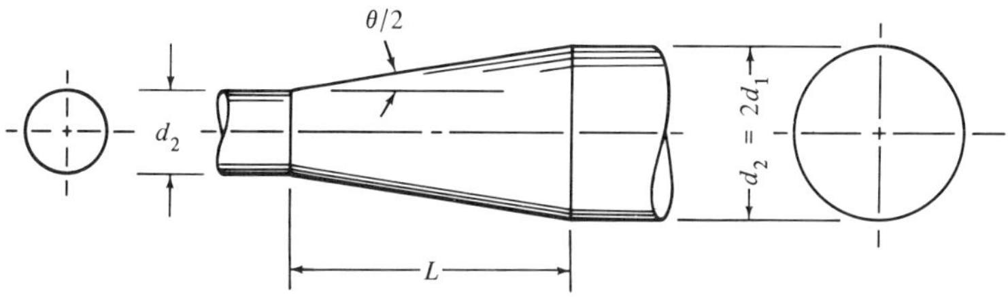

Fig. P5-13.

5-14 Find the relationship between d and s that will provide a uniform acceleration from 1 to 2 in a nozzle of circular cross section as illustrated. Assume steady, one-dimensional flow.

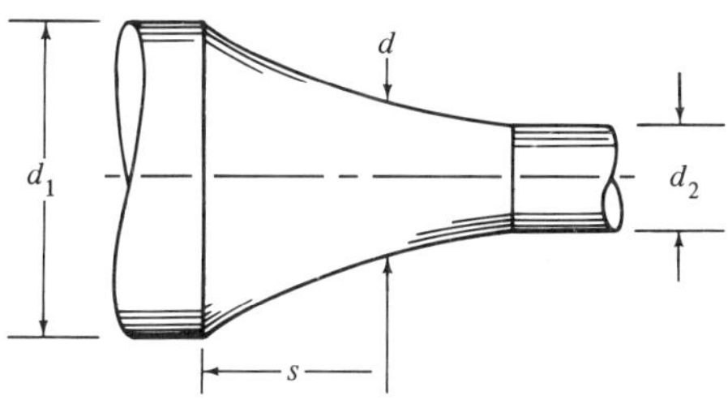

Fig. P5-14.

5-15 Find the diameter d as a function of s that will produce a constant-pressure gradient dp/ds between 1 and 2 for the nozzle illustrated in Problem 5-14.

5-16 A liquid enters a duct of circular cross section with a uniform velocity distribution as shown. The flow is laminar, so the profile becomes parabolic some distance downstream. The pressure difference between 1 and 2 is due to two effects: the pressure force required to overcome the

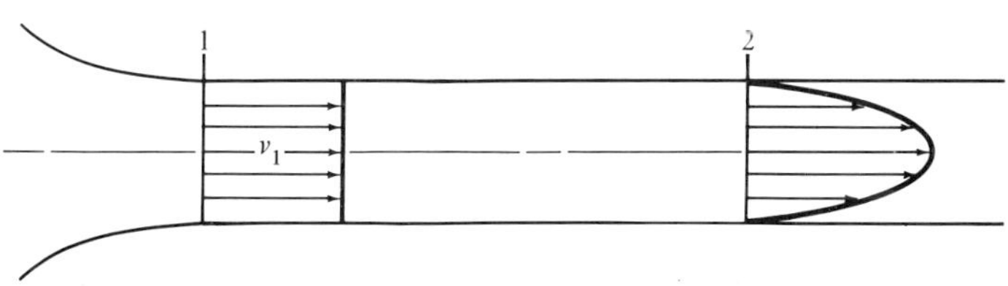

Fig. P5-16.

friction forces and the pressure work involved in changing the kinetic energy from 1 to 2. Find only the pressure difference due to the kinetic-energy change.

5-17 For turbulent flow in a straight pipe, tube, or duct, the velocity profile closely fits the relation

$$\frac{\mathrm{v}}{\mathrm{v}_0} = \left(\frac{y}{r_0}\right)^{1/m}$$

where v_0 is the centerline velocity (at $r = 0$), r_0 is the radius, and $y = r_0 - r$ is the coordinate measured from the wall. The value of m is between 6 and 10 and is a function of the Reynolds number. (a) Find the value of the kinetic-energy correction factor for this profile; that is, find α as a function of m. (b) Determine the value of α for $m = 7$.

5-18 A liquid flows out of the bottom of a tank from a rounded opening of diameter d_0 at a constant flow rate. (a) Find the diameter of the falling fluid column as a function of distance measured from the opening [find $d(x)$ neglecting viscous and surface tension effects]. (b) Discuss the effects that viscosity and surface tension have on the results.

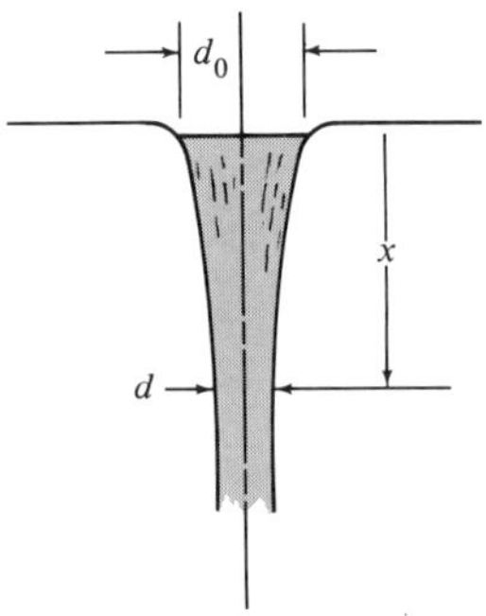

Fig. P5-18.

5-19 Water is flowing in a channel of constant width at a depth d_1 at section 1, as shown. What is the depth at section 2 if friction effects are neglected? Comment on the results.

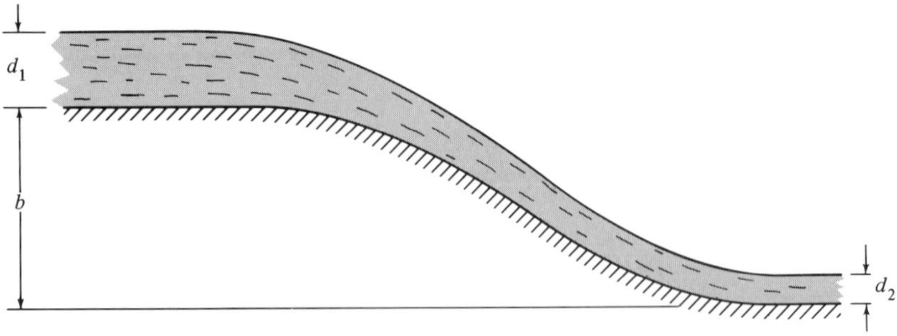

Fig. P5-19.

5-20 A blower is required to cool an electronic assembly. One hundred standard cubic feet per minute (scfm = volume rate at 1 atm pressure and 68°F) of air is required. The blower discharge is 2 in. in diameter. (a) Find the "air horsepower" input needed to deliver this air into the system, where the pressure is essentially atmospheric. (b) If the blower efficiency is 40 percent, find the motor-drive horsepower required. (c) A consideration of the energy balance on the blower will indicate that the difference in motor horsepower and air horsepower is "losses" in the form of (1) friction in the air flow, which appears as an increase in air temperature, and (2) bearing friction, which will probably appear as an increase in air temperature by heat transfer from the bearings. Since the blower is to provide convective cooling of the electronic components, it is necessary to determine the temperature rise across the blower. Is this ΔT significant?

5-21 A small blower takes in air from the ambient atmosphere and discharges it into a chamber (where the pressure is essentially atmospheric) through a square opening 4 in. long and 3 in. high. The velocity does not vary significantly over the 4-in. length but varies linearly from 150 ft/sec at the bottom to 80 ft/sec at the top of the 3-in. height. (a) Find the input power to the blower if the blower efficiency is 40 percent. (b) Find the power input for the above discharge rate if the "pressure head" in the chamber is 2 in. of water.

5-22 An axial-flow water pump picks up water from an open canal and discharges it into an irrigation ditch 5 ft above the canal level, as shown. If the required discharge rate is 1000 gpm and the pump efficiency is 65 percent, what is the approximate motor horsepower required? [Consider the application of the energy equation between points 1 and 2 and a corresponding fluid (flow field) control volume.]

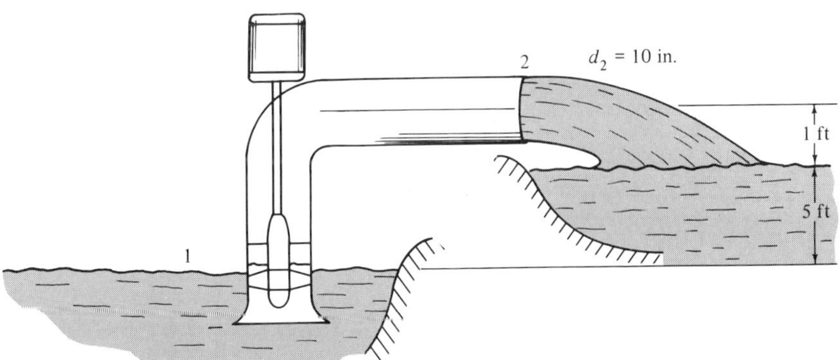

Fig. P5-22.

5-23 A radial flow pump test is run to determine the pump operating characteristics. The following data are obtained from the test setup shown on the sketch: $p_1 = -2$ psi, $p_2 = 125$ psi (at a point 5 ft above the gage

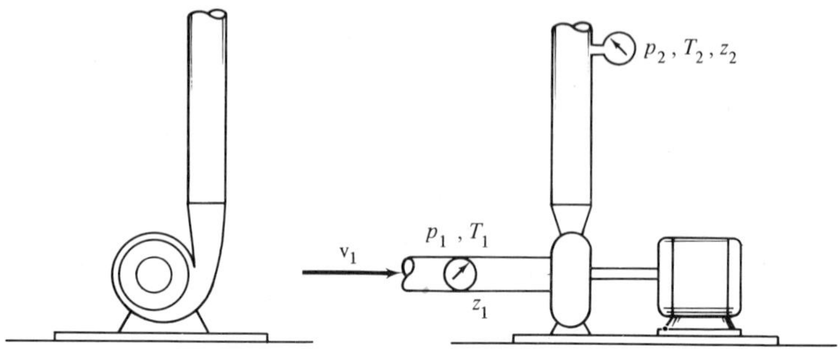

Fig. P5-23.

reading p_1), $T_1 = 54°F$, and a differential thermocouple reads 0.21°F between points 1 and 2. A discharge meter in the circuit indicates a flow rate of 485 gpm. The suction piping is 4 in. in diameter and the discharge pipe is 3 in. in diameter. Calculate (a) the input horsepower to the pump, (b) the pump efficiency, and (c) the proportion of the input that goes into (1) pressure work on the water, (2) change in the kinetic energy, (3) potential energy between points 1 and 2, and (4) thermal internal energy.

5-24 For the setup shown in Problem 5-23, the flow rate is 100 gpm (water) and the Δp is 150 psi. At these operating conditions, the characteristics supplied with the pump indicate an efficiency of 65 percent. (a) What is the horsepower input into the water? (b) What is the required horsepower output of the motor? (c) What is the useful power conversion by the pump? (d) What is the temperature rise from 1 to 2?

5-25 A turbine test setup as shown yields the data indicated. Assume that the tank is large enough that the velocity therein is negligible. Also assume that the density is constant from the tank to 1 and from 2 to the exit. (a) Calculate the pressure at 1 and 2. (b) Determine the turbine power. (c) Compute the torque if the turbine rotates at 10,000 rpm.

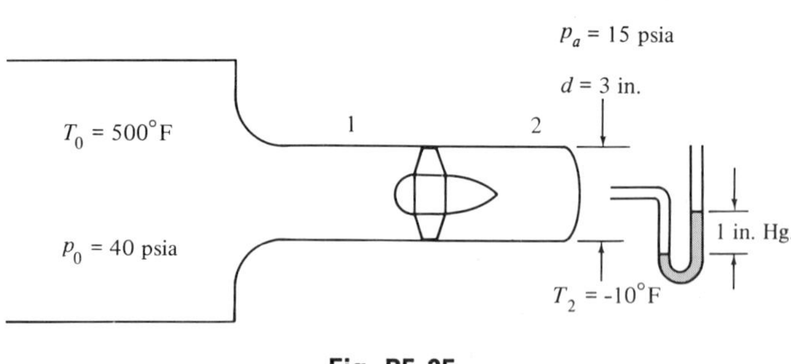

Fig. P5-25.

5-26 A jet issues from a nozzle as shown. Find v_2 and d_2 in terms of v_1, d_1, and θ_1. Point 2 is located at the top of the trajectory, as indicated.

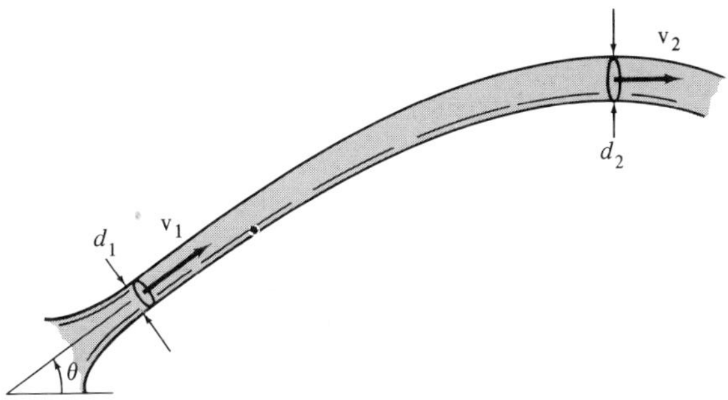

Fig. P5-26.

5-27 A pump is to be used to transfer water from a reservoir tank as shown. The required flow rate is 900 gpm. The availability loss in the pipe is determined from the relationship $\Delta p_L/\rho = f(L/D)(v^2/2g_c)$, where f is the friction factor for the pipe, L the length, and D the diameter. For this particular pipe $f = 0.02$. For an overall pump efficiency of 0.70, find the drive motor horsepower. What would be the easiest, most practical, and most economical method of reducing the power required to accomplish the above requirements (50-ft lift and 900 gpm)?

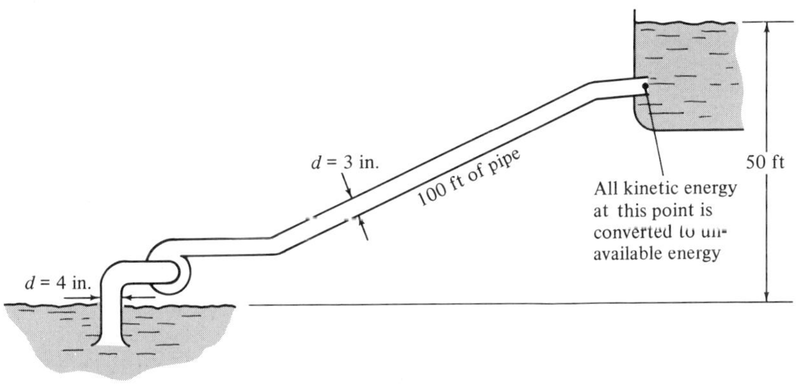

Fig. P5-27.

5-28 The header on a fire boat is supplied with a pressure of 150 psi. (a) What is the maximum possible jet velocity of the hose jets? (List and

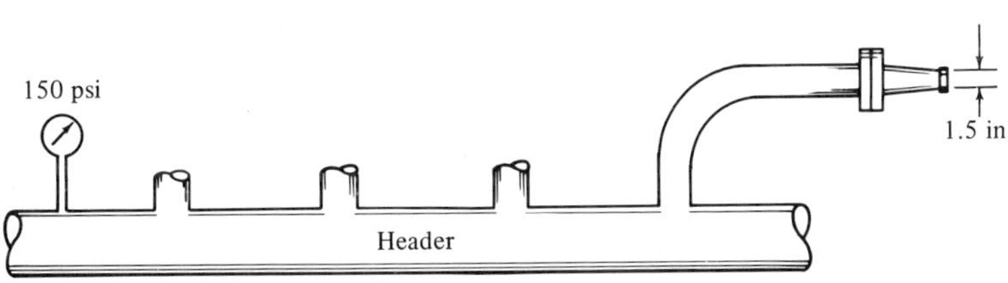

Fig. P5-28.

evaluate the assumptions made to obtain an answer.) (b) If the nozzle diameters are 1.5 in., what is the maximum flow rate Q (gpm) the pumps must deliver if five hoses are connected to each pump header? (c) What is the maximum height to which the water jets could reach?

5-29 A liquid-fuel rocket engine burns RP-1 fuel (a kerosene–gasoline–fuel-oil mixture) with liquid oxygen in the combustion chamber after these components are pumped in from the tanks. The fuel and oxidant rates are 80 lb_m/sec and 265 lb_m/sec. The fuel rate is 5.4 percent greater than the amount for complete combustion with the oxygen, so the combustion products will present a reducing rather than oxidizing atmosphere to the combustion chamber and nozzle walls. The chemical energy released on combustion is 18,500 Btu/lb_m of fuel. If 80 percent of the energy released in combustion is converted into kinetic energy at the nozzle exhaust, calculate the exhaust velocity. List and discuss the forms of work, energy, and so on, that make up the other 20 percent.

5-30 A Venturi meter is used to indicate the flow rate in a sloping pipeline as shown. Find an expression for the mass flow rate $\dot{m}$ or Q as a function of d_1, d_2, ρ_f, ρ_m, R, x, α, and meter coefficient C_v. Comment on the results.

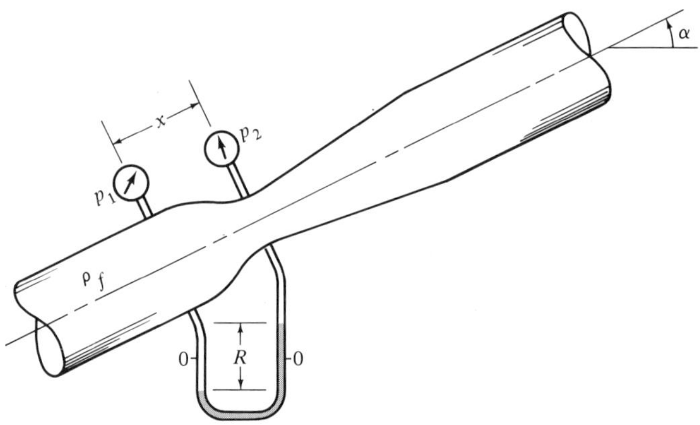

Fig. P5-30.

5-31 In Problem 5-30, find $\dot{m}$ or Q as a function of d_1, d_2, ρ_f, p_1, p_2, x, α, and the meter coefficient C_v.

5-32 In Problem 5-30, find the mass flow rate $\dot{m}$ or volume flow Q as a function of the variables indicated in the figure if the pressure difference is measured with the use of a pressure transducer. What care should be taken when connecting the transducer?

5-33 Find the pressure in the tank when $p_1 = 20$ psi, $p_2 = 5$ psi, and the fluid is water. Neglect losses.

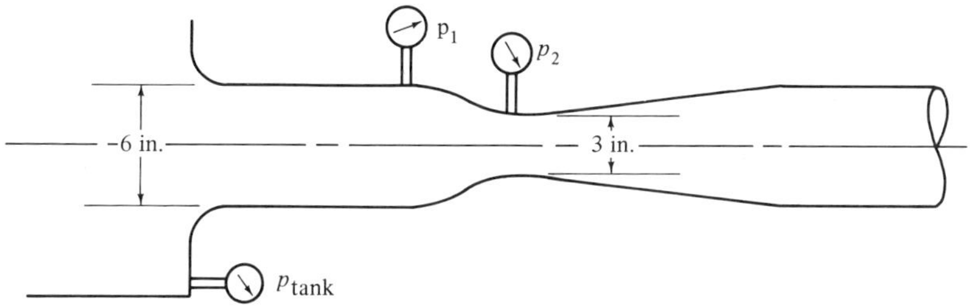

Fig. P5-33.

5-34 Water is siphoned from reservoir A to reservoir B as shown. If the losses are neglected, find the flow rate through a pipe of diameter d. What is the maximum height h for which water will continue to flow through the siphon?

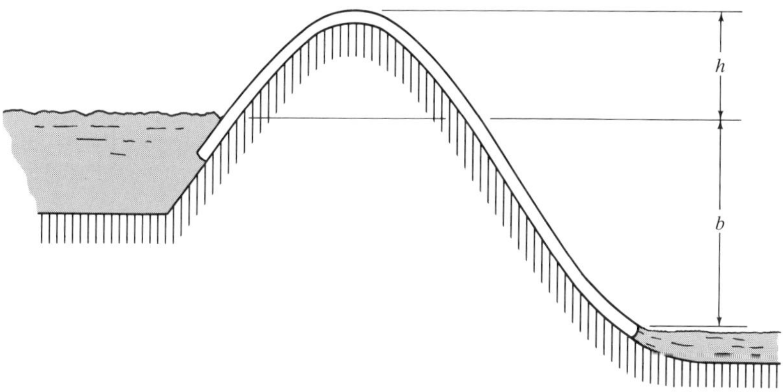

Fig. P5-34.

5-35 A Venturi (or flow nozzle) with an area ratio of 2 has a parabolic velocity distribution at the upstream measuring position and a uniform velocity distribution at the throat. Neglect friction effects and find the velocity coefficient C_v for this Venturi.

5-36 (a) A Venturi (or flow nozzle) with an area ratio of 2 has parabolic velocity distributions at both the upstream and throat measuring positions. Neglect friction effects and find the velocity coefficient C_v for this Venturi. (b) Comment on the results.

5-37 A Venturi (or flow nozzle) with an area ratio of 2 has turbulent flow upstream with a kinetic-energy correction factor, $\alpha_1 = 1.05$ and $\alpha_2 = 1.1$, due to a velocity profile of the type shown in Fig. 5-5. Neglect friction effects and find the velocity coefficient C_v for this Venturi.

5-38 Find a general expression for a Venturi or flow nozzle coefficient, $C_v(\alpha_1, \alpha_2, A_2/A_1)$. Neglect friction effects.

5-39 Assuming one-dimensional flow, what is the maximum flow rate of water per throat area (A_2) that can be put through a Venturi meter (or flow nozzle) with a diameter ratio of 2 if the upstream pressure is constant at 1 atm and the temperature is 70°F? Assume that $C_v = 0.97$. What is the maximum permissible value of the throat velocity v_2? What phenomenon determines the limiting velocity in this case?

5-40 Determine the velocity as a function of R_1, R_2, ρ_m, ρ_f, and a in the figure shown. (Consider the flow along the streamline s.) What is the static pressure (absolute) at point 1?

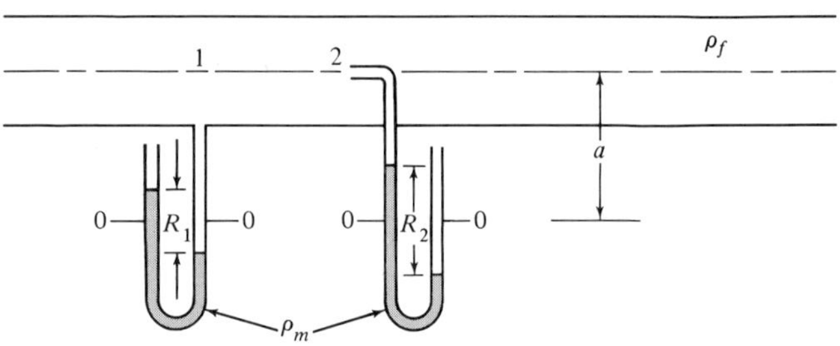

Fig. P5-40.

5-41 Find the difference in the level in the mercury in the manometer in the figure shown, and show on a sketch which leg is higher. Neglect friction effects.

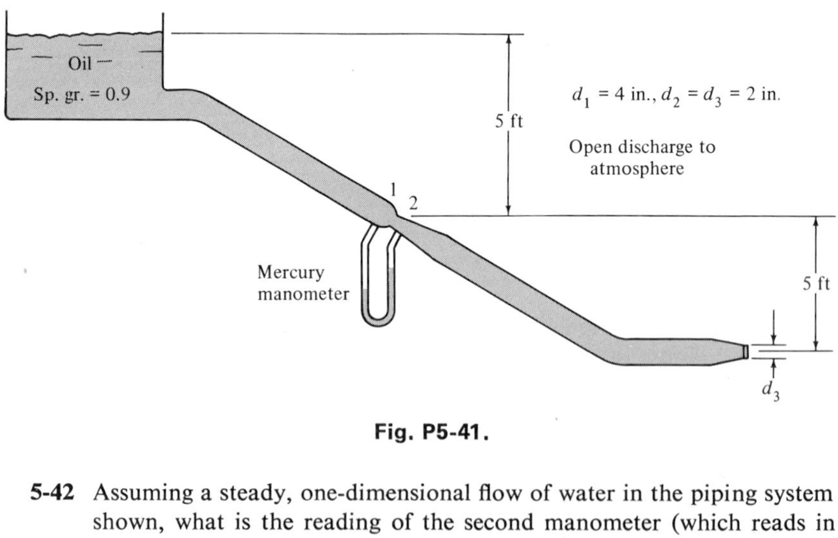

Fig. P5-41.

5-42 Assuming a steady, one-dimensional flow of water in the piping system shown, what is the reading of the second manometer (which reads in inches of mercury) if the flow rate is 20 ft³/sec? $A_1/A_2 = 2$.

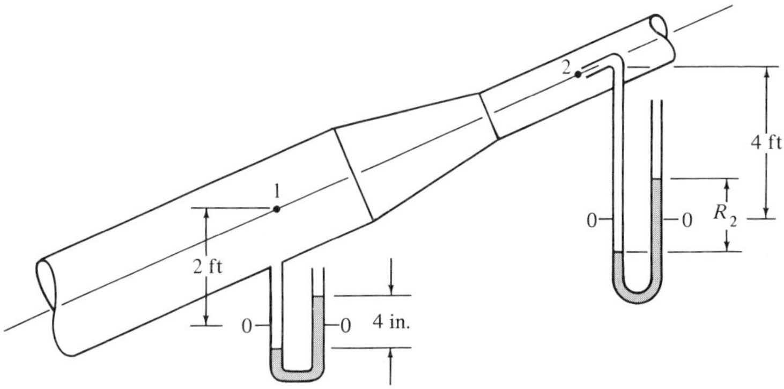

Fig. P5-42.

5-43 The velocity of water flowing in a pipe is measured with a Pitot-static tube as shown. Find the velocity and sketch the velocity distribution in the pipe.

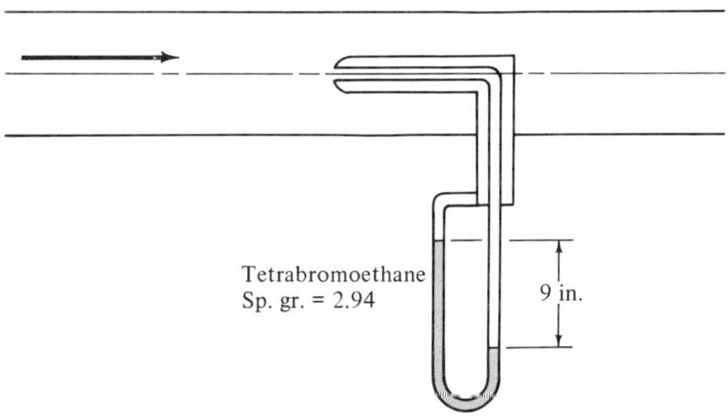

Fig. P5-43.

5-44 A Pitot-static tube is used to measure the flow rate in the pipe as shown. Assume the diameter of the Pitot-static tube to be negligible if compared to the diameter of the pipe and assume the Pitot-tube coefficient to be

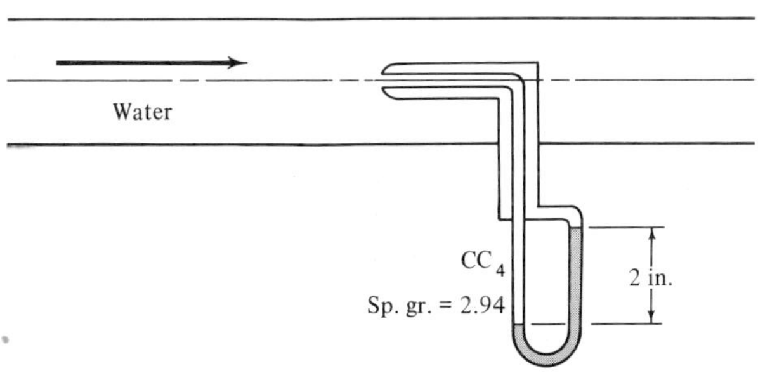

Fig. P5-44.

1.00. (a) Find the velocity at the centerline of the pipe. (b) Find the volume flow rate, if possible.

5-45 The flow rate in a pipe is to be determined by a single Pitot-static tube located on the centerline, which will indicate v_1. Find the flow coefficient C_f that must be applied to the indicated centerline velocity so that the mass flow rate can be determined from $m = \rho A\bar{v} = \rho A C_f v_1$ if the pipe flow Reynolds number is known to be subcritical.

5-46 For the setup in Problem 5-45, find C_f for a turbulent flow with the velocity profile given by $v/v_0 = (y/r_0)^{1/7}$, where $y \equiv r - r_0$ is the distance measured from the wall.

5-47 The flow rate in a round pipe or tube is to be determined from a Pitot-tube velocity traverse. Readings are to be taken at four radii (including the center), as shown. The locations are at the centers of equal annular areas, so the flow rate can be determined by the simple equation

$$Q = \frac{A}{4}\sqrt{\frac{2g_c}{\rho}}\left(\sqrt{\Delta p_1} + \sqrt{\Delta p_2} + \sqrt{\Delta p_3} + \sqrt{\Delta p_4}\right)$$

which can be abbreviated

$$Q = \frac{A}{4}\sqrt{\frac{2g_c}{\rho}}\sum_{i=1}^{4}\sqrt{\Delta p_i} \qquad (A = \pi R^2)$$

Find the locations as functions of R; that is, find r_i/R. Note that the Pitot-tube locations are in the *area* centers; that is,

$$r_i = \sqrt{\frac{a_i^2 + a_{i+1}^2}{2}} \qquad (i \geq 2)$$

where $\pi(a_i^2 + a_{i+1}^2)$ is the annular area over which the average flow rate is determined by the Pitot-tube velocity indication, and therefore $r_1 = 0$.

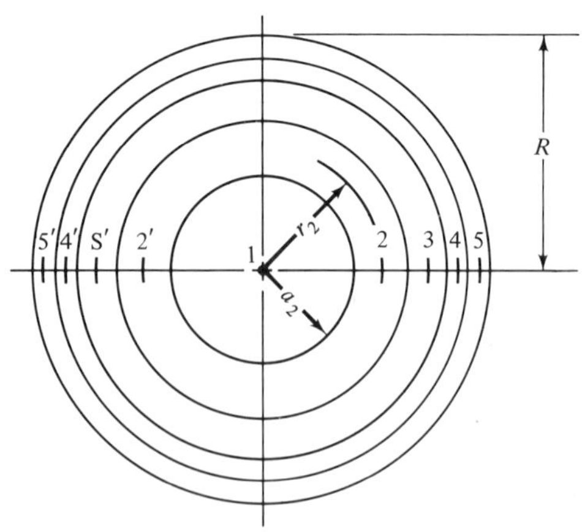

Fig. P5-47.

5-48 Find the radial locations for a Pitot-static tube velocity traverse in order that the flow rate can be determined from the simple summation

$$Q = \frac{A}{n}\sqrt{\frac{2g_c}{\rho}}\sum_{i=1}^{n}\sqrt{\Delta p_i} \qquad (i = 1, 2, \ldots, n)$$

where Δp_i is the Pitot-tube pressure differential and n readings on n radii are to be taken. Note that the locations are in the area centers; that is, $r_i = \sqrt{(a_i^2 + a_{i+1}^2)/2}$, where $\pi(a_i^2 + a_{i+1}^2)$ is the annular area over which the average flow rate is determined by the Pitot tube; therefore, $r_1 = 0$. Find r_i/R, where r_i is the ith radial location and R is the tube or pipe radius.

Answer: $\dfrac{r_i}{R} = \dfrac{2i - 1}{2n} \qquad i = 2, 3, \ldots, n$

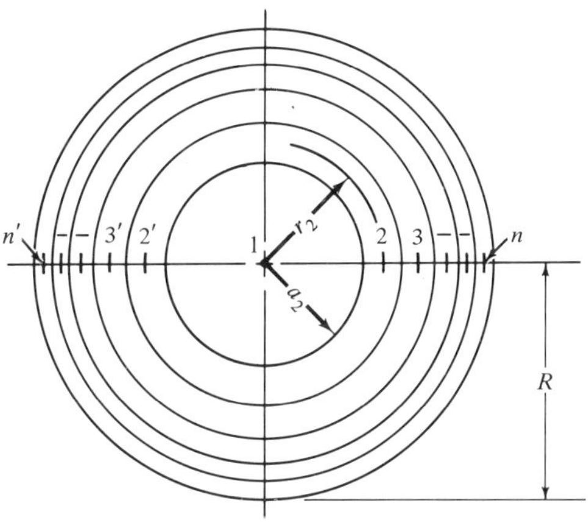

Fig. P5-48, 49.

5-49 For the method and results obtained from Problem 5-48, determine the radii for Pitot-tube locations. If the velocity profile is parabolic, $Q = A\mathrm{v}_0/2$, where $\mathrm{v}_0 = 2\bar{\mathrm{v}}$ is the centerline velocity. Calculate the Pitot-tube pressure-differential reading (in terms of ρ and v_0) at each location. Apply the results of the above problem to obtain the flow rate (Q) and check with the known value given. What is the percentage error produced by summing the readings from the traverse? What is the cause of this error? Would the error be larger or smaller for a turbulent profile: Explain.

Answer: Error = 3.75 percent

5-50 The method described for determining flow rates from velocity traverse data at specified locations as determined in the Problems 5-47, 5-48, and 5-49, are often used where results are desired to an accuracy of ±5 percent. Describe a method whereby a more accurate determination of flow rate can be made from velocity traverse data of an arbitrary profile that cannot be described by a mathematical equation. Consider the following mathematical equivalents:

$$Q = \iint_A \mathrm{v}\, dA = \int_0^R \mathrm{v} 2\pi r\, dr = \int_0^{R^2} \mathrm{v}\pi\, dr^2$$

$$= \int_0^1 \mathrm{v}\pi R^2\, d\left(\frac{r}{R}\right)^2 = A \int_0^1 \mathrm{v}\, d\left(\frac{r}{R}\right)^2$$

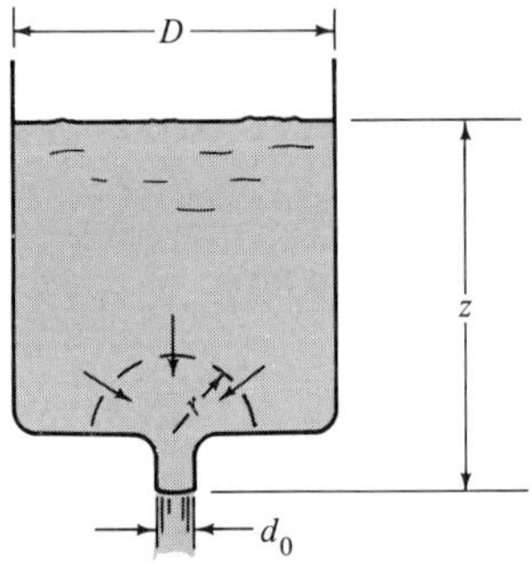

Fig. P5-51.

5-51 A fluid of density ρ flows out of the bottom of an open tank as shown. The tank diameter D is much greater than the opening diameter d_0. Find the volume flow rate Q as a function of time assuming that the flow can be considered at "quasi-steady"; that is, the acceleration $\partial \mathrm{v}/\partial t$ at the surface is very small with respect to the convective acceleration of the fluid as it enters the opening.

5-52 Show that for an incompressible flow process, the specific heats are equal; that is, show that $c_p = c_v = c$.

CHAPTER 6

CONTENTS

FORCE AND MOMENTUM:
The Equations of Motion

6.1 NEWTON'S LAWS OF MOTION

The physical principles represented in Newton's laws relating force and motion can be applied to a fluid system or control volume to obtain forces on the defined fluid mass. For example, the combustion pressure force in a rocket engine accelerates the gas out of the rocket nozzle; the *reaction* to this force is the propulsive thrust that accelerates the rocket. Changing the direction of a fluid in a turbine results in a force on the turbine blades. The blades of a pump apply a force to the fluid that changes its momentum or velocity. The velocity change implies a change in kinetic energy, which is converted to available work in the form of pressure.

The formulation of Newton's laws into mathematical statements allows the forces on or motion of a fluid to be predicted.

Newton's laws state that (1) a body will remain in a *uniform state of motion* unless acted upon by a force; (2) if acted upon by force, the *rate of change of* the momentum of the body will be directly proportional to the acting force; and (3) to every applied force there is an equal and opposite reaction. These natural laws deduced from natural physical observation apply to a single body or particle, a system of bodies, and to a continuous distribution or continuum of a substance. Newton's laws may be applied to a fluid (1) to predict its motion under the influence of a force (or force distribution or stress), or (2) to calculate the forces produced by the motion of the fluid through a given, confining control volume geometry.

The quantity characterizing *the motion* of a body is its momentum, which is defined as the product of the mass of the body with its velocity, $m\vec{v}$; it is a vector quantity. A uniform state of motion therefore implies a constant vector velocity of a fixed or defined mass. For example, a fluid jet would travel in a straight line at a constant speed (constant velocity) unless acted on by a force or system of forces. Gravity and friction will then tend to change its direction and speed in such a manner that the rate of change of momentum is proportional to all the forces acting on the jet. The reaction to the applied external friction force on the jet is the fluid friction force the jet exerts on the surrounding air. The reaction to the gravitational force is the inertial force of the changing motion. To obtain a desired result it may be necessary to use other information, such as that obtained from the principles of mass or energy conservation, information regarding the behavior of the fluid, the physical boundary restrictions on its motion, and so on.

For application of Newton's laws to a fluid, the first obvious problem is the definition of the "fluid body." To obtain useful results from this principle, the "body" of fluid and the forces acting on it must be defined very carefully.

6.2 PRINCIPLE OF FORCE AND MOMENTUM

The force–momentum principle yields a direct relation between the forces on a fluid mass element or system and its motion. This principle (or its resulting equations) may be used to determine the motion where the external forces are known. If the motion is known beforehand, the forces may be determined, for example, as design load information on a piece of equipment. In cases where insufficient information is available, the resulting equations for this and other principles must be combined to determine the required result.

As a principle involving force and momentum, Newton's second law states that the sum of all forces acting on a body produces a change in its motion such that the rate of change or momentum is directly proportional to the acting forces, or

$$\begin{bmatrix}\text{ALL FORCES}\\ \text{ACTING } on \text{ A}\\ \text{OF A SUBSTANCE}\\ \text{BODY OR MASS}\end{bmatrix} \propto \begin{bmatrix}\text{RATE OF CHANGE}\\ \text{OF MOMENTUM OF THE}\\ \text{BODY OR MASS}\end{bmatrix} \tag{6-1}$$

Writing the above principle as an equation gives

$$\vec{\mathrm{F}} = \frac{1}{g_c}\frac{d}{dt}(\mathrm{m}\vec{\mathrm{v}}) \tag{6-2}$$

where d/dt is the *total* time rate of change or total time derivative.[1]

It is interesting to note that a *constant vector* motion or field can be superimposed on (6-2) without producing any net change in effect as far as the forces are concerned. This result is obvious if the principle itself is considered for such a *constant motion.* If the motion (that is, velocity magnitude and direction) is constant, the force on the system for such a motion must be (in a nonrelativistic system)

$$\vec{\mathrm{F}} = \frac{1}{g_c}\frac{d}{dt}(m\vec{\mathrm{v}}_{\mathrm{ref}}) = \frac{d}{dt}(\text{const.}) = 0 \tag{6-3}$$

so the constant motion can be added to both sides of (6-2) to give

$$\vec{\mathrm{F}} = \frac{1}{g_c}\frac{d}{dt}[m(\vec{\mathrm{v}} \pm \vec{\mathrm{v}}_{\mathrm{ref}})] \tag{6-4}$$

where $\vec{\mathrm{v}}_{\mathrm{ref}}$ represents a *uniform*[2] velocity field; the forces are identical in both relationships of (6-3) and (6-4). A field of motion for which the velocity $\vec{\mathrm{v}}_{\mathrm{ref}}$ as above is everywhere constant (or zero) is termed an *inertial field.*

There are two viewpoints or frames of reference in which the motion may be considered:

1. The *Lagrangian* viewpoint considers the physical principles and their equations applied to a specifically defined mass of fluid. The forces

[1] The factor g_c is necessitated by the choice of mass units and the relation implied between force and mass units by this equation. If the mass is measured in units of $\mathrm{lb_m}$ and the velocity is ft/sec, $g_c = 32.1740\ \mathrm{lb_m\text{-}ft/lb_f\text{-}sec^2}$. In the mks system $g_c = 1\ \mathrm{kg\text{-}m/N\text{-}sec^2}$ and in the cgs system $g_c = 1\ \mathrm{g\text{-}cm/dyne\text{-}sec^2}$. See Appendix A.

[2] Note again the corresponding of definitions of *constant*, *uniform*, *steady*, and *stationary*. A constant motion implies a uniform field where the motion is steady or the field stationary.

on and motion of the mass are investigated as it is propagated by the velocity field through the confining boundaries of a piece of equipment.

2. The *Eulerian* viewpoint considers the flow field within a defined geometry (the control volume). The forces and motion are investigated for the fluid mass within this control surface at any instant.

In either of these methods, the mass or control volume may be a large integral amount for which integral equations are derived. For a differential-sized element, differential equations are obtained.

The Lagrangian approach is of primary use where general characteristics of fluid motion are desired. The Lagrangian method yields simpler equation forms; however, difficulty is encountered in specifying the motion of the mass as it passes through a given physical system.

6.3 MOTION ALONG A STREAMLINE: Equations of Euler and Bernoulli

The equations describing the motion of a differential fluid element (of fixed or constant mass) as it moves along a streamline will be developed first. Since the principle stating the action and effect of force and momentum is a vector relation by nature, it can be stated mathematically as a single vector equation or by three independent component equations.

It is necessary to specify the forces acting on the fluid. As listed in Chapter 5, the two basic types of forces are body (or mass-acting) forces and surface forces. Specifically, only gravity, pressure, and shear forces will be considered in the following development.

STREAMLINE OR NATURAL COORDINATES

The distance along a streamline s can be defined as a local *coordinate* along which the fluid and flow properties can be measured or determined (Fig. 6-1). The local normal to the streamline coordinate can be defined as an independent coordinate direction. This normal coordinate will be the *principal normal coordinate* n if it is further defined as being measured collinear with, but in opposite direction to, the *principal radius of curvature* of the streamline. A third local coordinate direction can be defined as the *conormal* (or binormal) direction c (normal to both s and n). The local unit vectors $\hat{s}$ and $\hat{n}$ (or the local coordinate elemental lengths Δs and Δn) also define a plane in which the principal radius of curvature lies. It therefore follows that the motion at the point under consideration is in this plane (or that the radius of curvature in the c direction is infinite).

Note that the entire flow field can be determined if the s field is specified.

That is, the flow field is specified by giving the equations of the streamlines in space (as functions of position-location coordinates and time). The field velocity, pressure, and so on, may then be calculated from the s field and certain initial and boundary conditions specifying the confining boundaries and where and how the flow originated. The component equations of motion will be considered with respect to each of these directions:

$$\sum F_s = \frac{1}{g_c}\frac{d}{dt}(m\text{v})_s$$
$$\sum F_n = \frac{1}{g_c}\frac{d}{dt}(m\text{v})_n \tag{6-5}$$

The s coordinate is defined as everywhere tangent to v, or $\vec{\text{v}} = \hat{s}\text{v}_s$ only, so v_n and v_c must be identically zero (Fig. 6-1). However, the velocity can vary also in the n direction at any point in the field, as indicated in Fig. 6-1, where Δs, $\vec{\text{v}}$, R_s, and $\hat{n}$ are all in the plane of the page. At P_1 the velocity

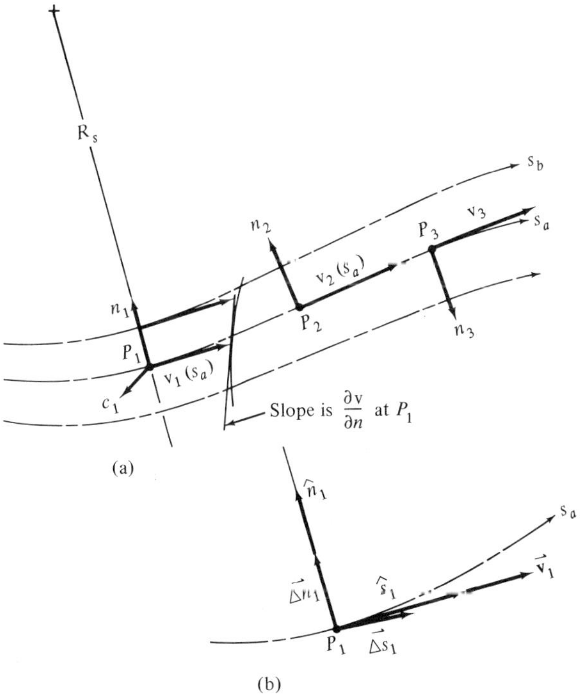

Fig. 6-1. Representation of the relation between the streamlines and the streamline (or natural) coordinate system showing the correspondence among the unit vectors, velocity, and velocity gradient.

is $v_1(s_1, n_1, t)$, indicating that the velocity can vary along the streamline and normal to it (and can also vary with time at the position P_1, which is considered to be fixed in space). At P_2 the velocity is $v_2(s_2, n_2, t)$, which is different in both magnitude and direction. In a general three-dimensional flow, s_b is not in the same plane as s_a. Also, the velocity distribution shown (along with the *s*-*n* plane) at P_3 would not be expected to be in the same plane as that at P_1. Since the field may be unsteady, all the field quantities (including the directions of the vector quantities) at the point P_1 would be different at another instant. This situation can be described by saying that v, *s*, *n*, and so on, are all functions of a fixed-in-space coordinate system and of time; that is, $\vec{v}(\vec{r}, t)$, for example, $\vec{v}(x, y, z, t)$, $\hat{s}(x, y, z, t)$, and so on, or

$$\begin{bmatrix} \vec{v} \\ \hat{s} \\ \hat{n} \\ p \\ \rho \\ \vdots \end{bmatrix} (\vec{r}, t)$$

where $\vec{r}$ is the position coordinate locating the points *P* with respect to the fixed-coordinate-system origin.[3] Note that the velocity cannot vary in the *c* direction at a given coordinate point P_1 because *n* is defined as the principal normal coordinate here. Therefore, the *c*-component equation of motion reduces to a simple hydrostatic force balance.

A small mass element is shown at P_1 in Fig. 6-2 with all quantities indicated as positive, including the rates of change (for example, the variation of $\vec{v}$ with *n*). By convention, all quantities and rates of change of quantities (for example, $\partial v/\partial n$) are assumed to be positive in the positive coordinate directions. The quantities p, ρ, $\vec{v}$, τ, and so on, which are known, or are to be determined at any arbitrary point along the streamline, are indicated at the center of the element. The pressure shear and gravity field forces are shown acting *on* the element.

For the following development, the body force is assumed to be produced by a conservative body force field; therefore, the body force can be determined as the gradient of a body force potential as described in Chapter 5. Because there is no variation of velocity in the *c* direction by definition, there will be no variation in the *c* direction ($\vec{v} = \hat{s}v$ only).

[3] This matrix notation implies that each quantity in the column is a function of all variables in the row.

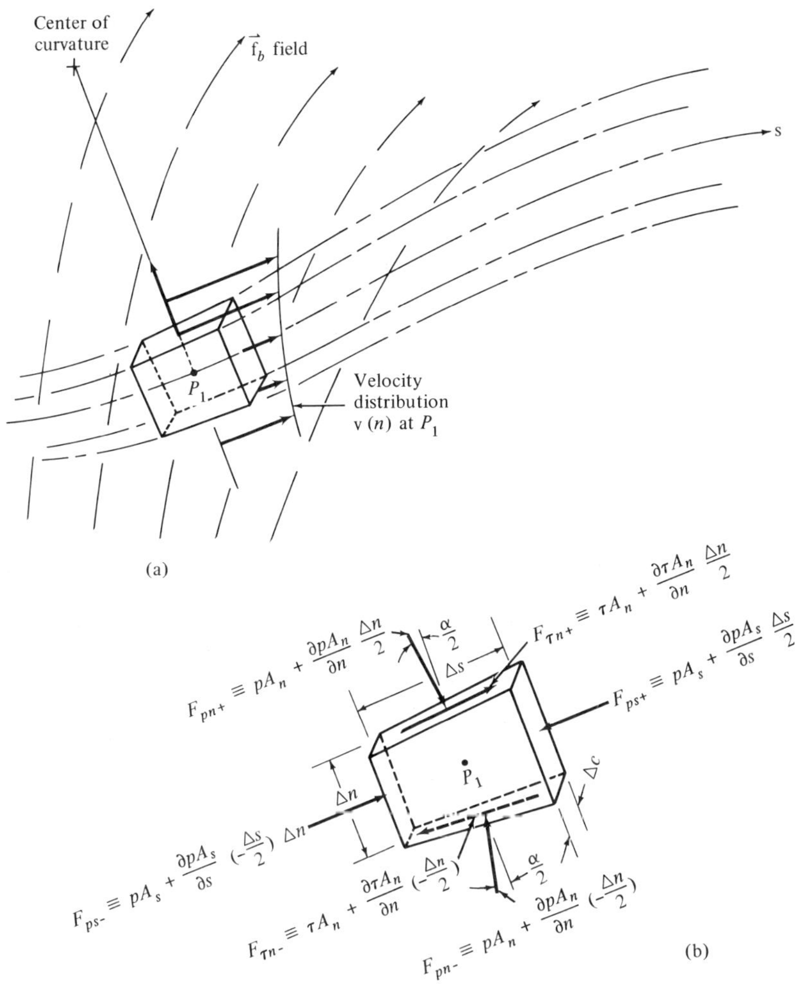

Fig. 6-2. (a) A fluid element in a flow field; (b) Force components acting on the element. The dimensions of the element measured through its center are Δs, Δn, and Δc. A_n, A_s, v, p, ρ, f_b, and τ are the values measured at P_1.

Therefore, there is no shear stress force on the front and back faces (n faces) of the element. The friction forces on the top and bottom surfaces are a direct consequence of the assumed velocity distribution. That is, the velocity on the top is greater than at the center; therefore, the shear force on the fluid element at the top surface is a drag on the element in the direction of the velocity as indicated in Fig. 6-2b.

At a particular point indicated, the equation of motion in the s direction (6-5) gives (with the mass of the element constant and equal to $\rho\,\Delta V = \rho\,\Delta s\,\Delta n\,\Delta c$)

$$\sum F_s = (F_{ps-} - F_{ps+}) + (F_{pn+} + F_{pn-})\sin\frac{\alpha}{2} + (F_{\tau n+} - F_{\tau n-})\cos\frac{\alpha}{2} + f_{bs}\rho\,\Delta V = \frac{\rho\,\Delta V}{g_c}\,a_s \tag{6-6}$$

where $\vec{f}_b$ is the body force per unit mass and a_s is the s component of the acceleration. For a streamtube-sized element, α will be quite small, so that $\sin(\alpha/2) \approx \Delta n/2\,\Delta s$ and $\cos(\alpha/2) \approx 1$. Making the substitutions indicated in Fig. 6-2b in (6-6) and dividing through by the mass of the element gives

$$-\frac{1}{\rho}\frac{\partial p}{\partial s} + \frac{1}{\rho}\frac{\partial \tau}{\partial n} + f_{b,s} = \frac{1}{g_c}\left(\frac{d\mathrm{v}}{dt}\right)_s \tag{6-7}$$

The units of the terms in this equation are force and momentum per unit mass ($\mathrm{lb_f/lb_m}$, N/kg, dynes/g, and so on).

Since $\vec{\mathrm{v}}$ (or v) is the velocity of the mass element as it travels along s, $\mathrm{v} = ds/dt$ and $d\mathrm{v}/dt = d^2s/dt^2$. Mathematically this statement is indicated by $\mathrm{v}(s, n, t)$, and also $s(t)$, $n(t)$, and $c(t)$.

In the above s-component equation of motion, only the variation of motion (or velocity) in the s direction is needed (as indicated by the subscript s on the acceleration term). Therefore,

$$\left(\frac{d\mathrm{v}(s, n, t)}{dt}\right)_s = \left(\frac{\partial \mathrm{v}}{\partial t}\right)_s + \left(\frac{\partial \mathrm{v}}{\partial s}\frac{ds}{dt}\right)_s \qquad \text{and} \qquad \frac{ds}{dt} = \mathrm{v}$$

The s subscript on the quantities on the right side is superfluous, since only s-directed components of velocity are possible. The term $\partial\mathrm{v}/\partial t$ is the variation of the velocity with time at a space-fixed point or the time-dependent acceleration. The term $\mathrm{v}(\partial\mathrm{v}/\partial s)$ is the variation of v from one point to another along s (or with respect to s) at any given instant (t fixed) and is termed the spatial variation of v with respect to s or the s component of the *convective acceleration*.[4] The convective term $(d\mathrm{v}/dt)_s$ can be represented by the mathematical limiting process

$$\left(\frac{d\mathrm{v}}{dt}\right)_s = \lim_{\delta t \to 0} \frac{(\mathrm{v} + \delta\mathrm{v})_s - \mathrm{v}_s}{\delta t} = \lim_{\delta t \to 0} \frac{(\delta\mathrm{v})_s}{\delta t}$$

[4] The term "*convective* acceleration" is seen to imply the acceleration of the fluid element as it is transported or *convected* from one point in the fluid to another by the velocity field.

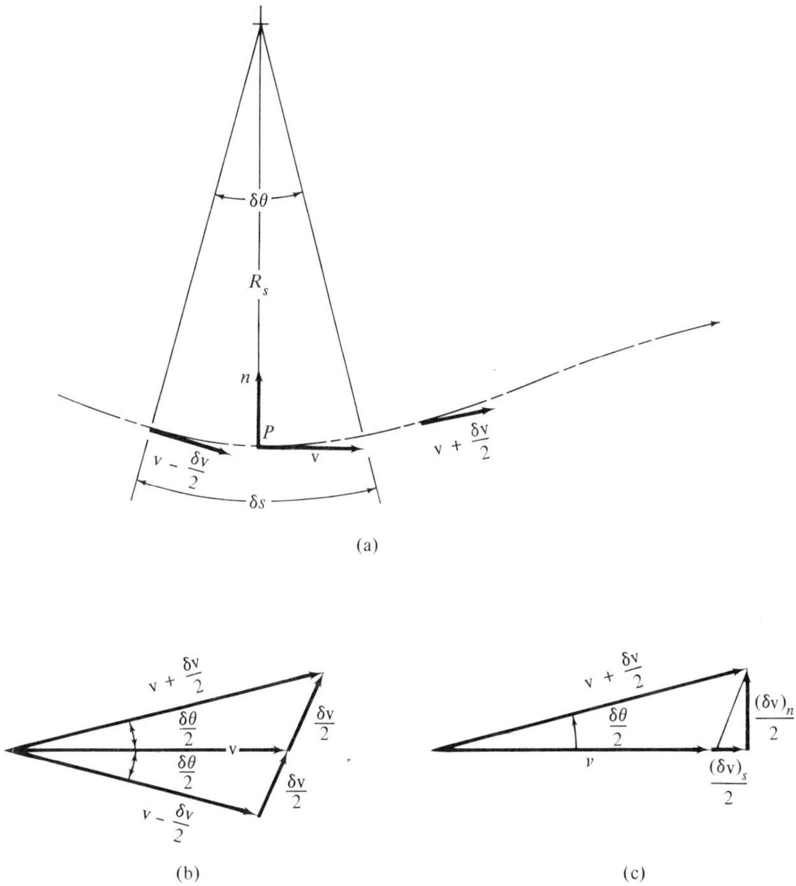

Fig. 6-3. Velocity variation along a streamline showing the streamwise velocity variation $(\delta v)_s$ and the normal variation $(\delta v)_n$ at a point *P*.

Since the variation of v along *s* is being determined, δt is the time required for the element to be propagated a distance $\delta s = v\, \delta t$ (or $\delta t = \delta s/v$); therefore,[5]

$$\lim_{\delta t \to 0} \frac{(\delta v)_s}{\delta t} = \lim_{\delta t \to 0} \frac{(\delta v)_s}{\delta s / v} = \lim_{\delta s \to 0} v \frac{(\delta v)_s}{\delta s} = v \frac{\partial v}{\partial s}$$

The velocity components appearing in the above expressions are indicated in Fig. 6-3. Note that even though $v_n = 0$, $(\delta v)_n$, which is the

[5] The two types of delta, Δ and δ, have been used to indicate the small size of the element ($\Delta s\, \Delta n\, \Delta c$), and the small change in the quantities produced as the element is propagated a small distance $\delta s = v\, \delta t$ along the streamline s_a in the small time period δt. These two delta designations are therefore independent but can be considered to be of the same order (that is, $\delta s \sim \Delta s$).

variation of v in the n direction as the fluid element progresses along the streamline, is not zero because of the curvature of the streamline.[6] The acceleration term expresses explicitly the acceleration at a fixed point in space with respect to time plus the *convective* acceleration along the streamline. This total derivative of the velocity dv/dt is termed the *substantial* or *material* derivative. It is the time rate of change of a quantity (here the velocity) following the motion of an element (or point) of the substances or material. A more general expression for this derivative is given in Sections 6.8 and 6.9.

With the above-determined values for the acceleration, the streamline component of the equation of motion becomes

$$-\frac{1}{\rho}\frac{\partial p}{\partial s} + \frac{1}{\rho}\frac{\partial \tau}{\partial n} - \frac{\partial \Omega}{\partial s} = \frac{1}{g_c}\left(\frac{\partial \mathrm{v}}{\partial t} + \mathrm{v}\frac{\partial \mathrm{v}}{\partial s}\right) \tag{6-8}$$

The relation between the conservative body force per unit mass *on the fluid* and the body-force potential is $f_b = -\nabla\Omega$ on fluid, where $\Omega = \int \vec{g}\cdot d\vec{r}/g_c$ for an equivalent accelerational field $(\vec{g})$, as in Chapter 5.

6.4 QUASI-ONE-DIMENSIONAL FLOW

A quasi-one-dimensional flow has been defined as having the velocity variation across a flow section as the only variation of significance. In connection with the pressure gradient perpendicular to the streamlines $\partial p/\partial n$, it will be shown that for this quantity to be negligible, the radius of curvature of the flow boundaries (R_s) must be much larger than the channel diameter (or $D_H/R_s \ll 1$). In Chapter 5 it was indicated that the condition for negligible temperature variation was that the heat transfer was small ($\tilde{q}$ smaller than other significant terms in the energy equation). If the variations of T and p in the n direction are small, the density variation will be small ($\Delta\rho/\rho \ll 1$).

A slight variation in the development of (6-8) is permissible for quasi-one-dimensional flows. Rather than using the local or point-value shear stress, which requires quite complex shear stress–strain rate relationships, it is sometimes an advantage to consider the average shear stress acting on the entire *peripheral* area of the element. The shear-force term would then be $\bar{\tau}\,\Delta A_{\text{per}}$ in the "resistive" direction, opposite to the motion. ΔA_{per} is the peripheral area, which is $2(\Delta s)(\Delta n + \Delta c)$ for the element in Fig. 6-2. When this shear-force term is divided by the mass of the element,

[6] Mathematically, this statement is equivalent to saying that although a function is zero at some point (or everywhere along a line), its derivative(s) is (are) generally not zero there.

$$\frac{\bar{\tau}\,\Delta A_{\text{per}}}{\rho\,\Delta V} = \frac{\bar{\tau}\,\Delta s 2(\Delta n + \Delta c)}{\rho\,\Delta s\,\Delta n\,\Delta \text{v}} = \frac{\bar{\tau} 2(\Delta n + \Delta c)}{\rho\,\Delta n\,\Delta c} \tag{6-9}$$

The quantity $2(\Delta n + \Delta c)$ is seen to be the periphery of the element and $\Delta n\,\Delta c$ is the cross-sectional area. Four times the cross-sectional area divided by the periphery is defined as the *hydraulic diameter*,

$$D_H = 4\,\frac{A_{\text{cross}}}{PER} \tag{6-10}$$

(For a round tube, the hydraulic diameter D_H is just the diameter.) With this definition of hydraulic diameter, the modified equation for the s direction becomes

$$-\frac{1}{\rho}\frac{\partial p}{\partial s} - 4\frac{\bar{\tau}}{\rho}\frac{1}{D_H} - \frac{\partial \Omega}{\partial s} = \frac{1}{g_c}\left(\frac{\partial \text{v}}{\partial t} + \text{v}\frac{\partial \text{v}}{\partial s}\right) \tag{6-11}$$

If it is desired at this point to consider the motion along one single streamtube, the above equation can be multiplied by ds and integrated with respect to the s direction only. Transposing all the terms and multiplying by ds gives

$$\frac{1}{g_c}\frac{\partial \text{v}}{\partial t}\,ds + \frac{1}{g_c}\frac{\partial(\text{v}^2/2)}{\partial s}\,ds + \frac{1}{\rho}\frac{\partial p}{\partial s}\,ds + \frac{4\bar{\tau}}{\rho}\frac{ds}{D_H} + \frac{\partial \Omega}{\partial s}\,ds = 0 \tag{6-12}$$

for an equivalent acceleration-type gravitational field.

If $\vec{g} = -\hat{z}g$ is the earth's gravity acceleration field (assumed constant) and z is the altitude coordinate, $\Omega - gz/g_c$. Making this substitution and integrating the above expression along the streamtube as indicated gives

$$\frac{1}{g_c}\int \frac{\partial \text{v}}{\partial t}\,ds + \frac{\text{v}^2}{2g_c} + \int \frac{dp}{\rho} + \overline{\frac{4\tau L}{\rho D_H}} + \frac{gz}{g_c} = E(n, t) \tag{6-13}$$

where an average value of $\overline{4\tau/\rho D_H}$ over the length L is used in the shear stress term.[7] E is not a function of c, since only a hydrostatic balance exists in the c direction.

An analytical determination of this shear stress term is difficult. However it can be estimated fairly closely from an evaluation of the flow conditions (tube roughness, geometry, Reynolds number, and some experience). If

[7] L in (6-13) is the distance $s_2 - s_1$ measured along the streamline. The term $\overline{4\tau/\rho D_H}$ would be defined by the calculation

$$\overline{\frac{4\tau}{\rho D_H}} \equiv \frac{4}{L}\int_0^L \frac{\bar{\tau}}{\rho D_H}\,ds$$

the duct diameter is constant, this estimation is not difficult. The specific determination of friction forces is treated in Chapter 7.

It is very important to note the significance of the above procedure in obtaining (6-13). First, the differential equation describing the forces and motion of a fluid element was multiplied by ds and integrated. Since $\int \vec{F} \cdot \vec{ds}$ represents work, there results an equation in terms of work and energy per unit mass. Second, only the s direction is considered in the above development; consequently, an integration function of the other independent variables (n, c, t) not specifically evaluated as integration limits during the integration process is obtained rather than a "constant of integration." If it is explicitly stated that the results of the procedure are to apply only to one single streamtube, then, for this streamtube, E is a function of t only.

For these quasi-one-dimensional flow conditions, an entire machine control volume as illustrated in Fig. 6-4 can be considered as a quasi-one-dimensional "streamtube," with average values at flow sections 1 and 2. These average values can be used in place of the streamtube values in (6-11) through (6-13). If a work element exists as illustrated in Fig. 6-4 and its motion is periodic (the flow is otherwise steady), the time mean of the pressure and shear work terms for all the streamtubes influenced by the work element will yield the mechanical work $\langle \dot{W}_m \rangle$ identical to that of

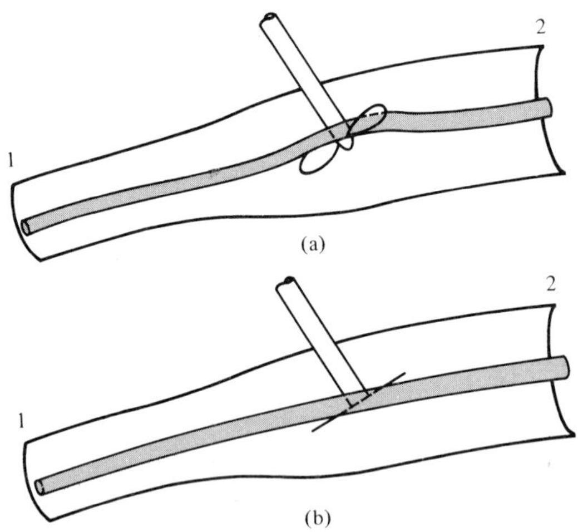

Fig. 6-4. (a) Quasi-one-dimensional machine control volume with periodic or quasi-steady flow in the neighborhood of the work element (propeller), and steady flow at the inflow and outflow stations 1 and 2; (b) equivalent steady-flow control volume with the periodic-flow-work element replaced by an equivalent hypothetical steady-flow actuator disc.

Chapter 5. Treating the entire machine C.V. as a streamtube and integrating over its length gives

$$\frac{1}{2g_cA_2}\iint_{A_2} \mathrm{v}^2\, dA - \frac{1}{2g_cA_1}\iint_{A_1} \mathrm{v}^2\, dA + \int_1^2 \frac{dp}{\rho} + \overline{\frac{4\tau_0 L}{\rho D_H}} + \frac{gz_2}{g_c} - \frac{gz_1}{g_c} = \langle w_m \rangle \tag{6-14}$$

where τ_0 represents the average shear stress at the stationary walls of the C.V. [The units of (6-14) are work and energy per unit mass.]

If it is desired to specify the momentum in terms of the average flow velocity $\bar{\mathrm{v}} \equiv Q/A$, it is noted that

$$\overline{\mathrm{v}^2}A = \iint \mathrm{v}^2\, dA \neq (\bar{\mathrm{v}})^2 A$$

since the integrand v^2 is a nonlinear function of the variable of integration A (for example, of r if $dA = 2\pi r\, dr$ for a circular cross section). If the plane of the area A is perpendicular to the mean or center streamline direction of the duct at the station under investigation, the momentum flux can be expressed as a function of the average velocity by introducing a correction factor β to account for the variation of velocity over the area,

$$\beta \equiv \frac{\iint \mathrm{v}^2\, dA}{\bar{\mathrm{v}}^2 A} \equiv \frac{1}{A}\iint \left(\frac{\mathrm{v}}{\bar{\mathrm{v}}}\right)^2 dA \equiv \frac{\overline{\mathrm{v}^2}}{\bar{\mathrm{v}}^2}$$

For fully developed laminar flow in a pipe, $\beta = 4/3$. For fully developed turbulent flow β has a value between 1.0 and 1.2. Note that the minimum value for β is 1.00 for a uniform velocity distribution over the area. With the use of β, (6-14) can be written

$$\frac{\beta_2 \bar{\mathrm{v}}_2^2}{2g_c} + \int_1^2 \frac{dp}{\rho} + \frac{(\overline{gz})_2}{g_c} = \frac{\beta_1 \bar{\mathrm{v}}_1^2}{2g_c} + \frac{(\overline{gz})_1}{g_c} + \langle w_m \rangle - \overline{\frac{4\tau_0 L}{\rho D_H}} \tag{6-15}$$

for steady flow. It is noted that β can be identically unity only for the condition $\tau = 0$ (that is, for frictionless, irrotational flow).[8]

6.5 FRICTIONLESS FLOW

For flow where friction forces are negligible (or the viscosity can be considered to be zero), (6-13) reduces to

$$\frac{1}{g_c}\int_s \frac{\partial \mathrm{v}}{\partial t}\, ds + \frac{\beta \bar{\mathrm{v}}^2}{2g_c} + \int_s \frac{dp}{\rho} + \frac{gz}{g_c} = E(n, t) \tag{6-16}$$

[8] The converse is not true, however. The condition that $\tau = 0$ does not necessarily imply that β is unity (Section 6.8).

where $$\begin{bmatrix} p \\ \rho \\ \mathrm{v} \\ z \end{bmatrix}(s, n, t)$$

In order to integrate the remaining terms in this equation, $\mathrm{v}(s, t)$ and $p(\rho)$ or $\rho(p)$ must be known for the flow along the streamline. The $p(\rho)$ relation defines the process that occurs as the flow traverses the streamline path. (Specifically, this process is a "work process," since it involves the pressure–force work term.) For example, for the isothermal flow of a perfect gas, $p/\rho = \text{const.}$; if there is no heat transfer or friction, the *isentropic* (reversible, adiabatic) flow process is described by $p/\rho^\gamma = \text{const.}$, where $\gamma \equiv c_p/c_\mathrm{v}$.

Equation (6-16) is sometimes termed the "unsteady Bernoulli equation" (although Bernoulli did not obtain the equation in this form). For a frictionless (ideal) flow, E is usually a constant for all streamlines at any instant (or is a function only of t).

STEADY, FRICTIONLESS FLOW

If the flow is also steady, there is no time dependency and the streamline (and entire flow field) is fixed in space, giving

$$\frac{\bar{\mathrm{v}}^2}{2g_c} + \int \frac{dp}{\rho} + \frac{gz}{g_c} = E(n) \tag{6-17}$$

where E may be function of n but is constant along a specified streamline.

6.6 STEADY, FRICTIONLESS INCOMPRESSIBLE FLOW: The Bernoulli Equation

The simplest process relation is obtained for an incompressible (constant-density) flow, which (with $\beta = 1$) gives the Bernoulli equation

$$\frac{\mathrm{v}^2}{2g_c} + \frac{p}{\rho} + \frac{gz}{g_c} = E(n) \tag{6-18}$$

Its use is subject to the conditions under which it was derived: (1) steady flow, (2) frictionless flow (ideal flow), (3) incompressible flow, (4) flow along a streamline, and (5) only pressure and gravity forces considered. The range of applicability of this equation is actually greater than expected from a literal interpretation of these restrictions. Possible modifications can be made as follows.

1. *Steady Flow and Quasi-Steady Flow.* If the flow changes slowly enough that the unsteady term of (6-13) or (6-16) is small (for example, with respect to the term $v^2/2g_c$), the temporal acceleration term $\partial v/\partial t$ may be neglected. Therefore, the above equation can be considered to apply *at any instant*, but each term is a function of time, so E is also a function of time. An example of this particular case is the starting of a turbocompressor, where timewise velocity changes are relatively small as compared to the acceleration due to flow curvature in the machine. The inertial force associated with accelerating the fluid at any single point within the control volume is then small compared to the deviational force (associated with the convective acceleration) exerted as the fluid direction is changed in the flow passages.

2. *Frictionless Flow.* For a real flow, (6-18) can be applied to a flow field that is everywhere outside the boundary-layer region. (The boundary layer was defined as the region in which the shear stress or shear forces were not negligible.) For cases in which the total length of flow path is small (for example, in a nozzle), the friction forces *may be* small compared to the other forces (terms) in the equation and may therefore be neglected. In this case, it is necessary to make an estimation of the order of magnitude of the friction forces.

3. *Incompressible Flow.* Equation (6-16) can be applied to compressible flows if the variations in the force and inertia terms ($v^2/2g_c$) do not produce a significant compression (or expansion) of the fluid. The "compressibility effect" is indicated by the Mach number and specific heat ratio γ. For most gases, compressibility effects are negligible up to a Mach number of about 0.2 (that is, the relative change in density $\Delta\rho/\rho$ is about 2 percent for a Mach-number variation from 0 to 0.2). The Mach number is defined as the ratio of the fluid (or flight) speed to the speed of sound. The speed of sound in air at sea-level conditions is about 1100 ft/sec, so that air behaves nearly as an incompressible fluid for speeds up to about 250 ft/sec. The Mach number also indicates the ratio of the compression or compressibility forces to the inertia forces acting in the flow, which implies its significance as a compressibility parameter. Compressibility will be discussed further in Chapter 12.

4. *Flow Along One Streamline, or One-Dimensional Flow.* If the flow starts from a tank or reservoir wherein the initial energy can be determined as $E = p_0/\rho + gz_0/g_c$ (the velocity in such a tank or reservoir is assumed to be negligible with respect to the velocity in the rest of the system), then all streamlines start with the same energy. If the flow is *nearly frictionless*, all streamlines will maintain the same total energy E. Note that this type of flow would correspond to the definition of *one-dimensional* flow but not quasi-one-dimensional flow. An example of this type of flow is illustrated in Fig. 6-5.

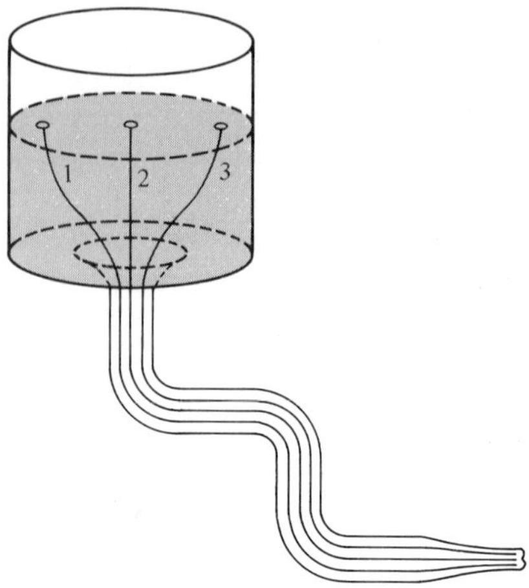

Fig. 6-5. Streamlines in a frictionless flow field. All streamlines (1, 2, 3) have the same total energy, which is constant throughout the field at any instant.

5. *Only Pressure and Gravity Forces Acting.* Any other forces that might act on the fluid must be negligible with respect to the pressure and gravity forces.

Under the condition of no flow or no motion, (6-7), or any of the subsequent forms derived, reduces to the simple equations for a static fluid presented in Chapter 3.

INTERPRETATION OF TERMS IN THE BERNOULLI EQUATION

For an incompressible flow, the pressure variations in the flow along a streamline are due to two independent effects: that produced by the action of gravity, or the *hydrostatic pressure variation,* and that produced by the change in motion or the inertia action, that is, the *dynamic pressure variation.* If the pressure is explicitly expressed as the sum of these two effects,

$$\frac{p_h}{\rho} + \frac{p_d}{\rho} + \frac{\mathrm{v}^2}{2g_c} + \frac{gz}{g_c} = E \tag{6-19}$$

since the hydrostatic pressure variation is a function only of the hydrostatic head z, and p_h varies according to the relation

$$p_h + \frac{\rho g}{g_c} z = \text{const.} \qquad \text{or} \qquad \frac{p_h}{\rho} + \frac{gz}{g_c} = \text{const.}$$

These terms may be included in the constant E to give

$$\frac{p_d}{\rho} + \frac{v^2}{2g_c} = E - \left(\frac{p_h}{\rho} + \frac{gz}{g_c}\right) = E' = \text{const.} \tag{6-20}$$

In determining the lift force on a body traveling through a fluid (or the equivalent problem of a body immersed in a fluid stream), (6-20) will give the pressure distribution on the body resulting from its motion. If the motion is known, p_d can be determined everywhere, and the component of this dynamic pressure force perpendicular to the direction of travel can be determined and integrated to yield the lift due to the dynamic forces. The hydrostatic pressure variation could also be determined to obtain the "hydrostatic lift," but this part of the lift is just the component of the *buoyant force*, which has already been determined as the "weight" of the displaced fluid volume, so it is sometimes easier to keep the two types of force (and pressure) separated.

HYDRAULICS INTERPRETATION

In civil engineering and hydraulics, which deal primarily with incompressible fluids, the Bernoulli equation is usually considered in the form

$$\frac{g_c p}{g\rho} + \frac{v^2}{2g} + z = \text{const.} = H \tag{6-21}$$

where the units are "feet of fluid head" and H is termed the "total head."[9] The terms in the equation are then designated the pressure head, the velocity head, and the hydrostatic head. These quantities are illustrated in Fig. 6-6 for an *ideal (frictionless) flow*. The pressure at the free surface of the reservoir and at all tubes is zero gage. The velocity head is then $H - (p/\gamma_h + z)$, as shown. The last tube shown is a stagnation (or impact) pressure probe. The velocity at its opening in the fluid stream is zero, so the height of its free surface above the datum is just equal to H or it is at the same level as the reservoir. At the Venturi throat, the velocity is higher than at any other point in the pipe, so the velocity head $v^2/2g$ is greater than in the pipe. The pressure at points 1, 2, and 3 can be determined from (6-1). For example, the pressure p_1 is

$$p_1 = \frac{\rho g}{g_c}\left(H - z_c - \frac{v_1^2}{2g}\right) = \gamma_h\left(H - z_c - \frac{v_1^2}{2g}\right)$$

The variation of the terms in the hydraulic Bernoulli equation and their relative significances as "heads" can be shown graphically on a drawing

[9] The quantity $\rho g/g_c$ is sometimes replaced by the quantity γ_h, the specific weight (not to be confused with the specific heat ratio γ).

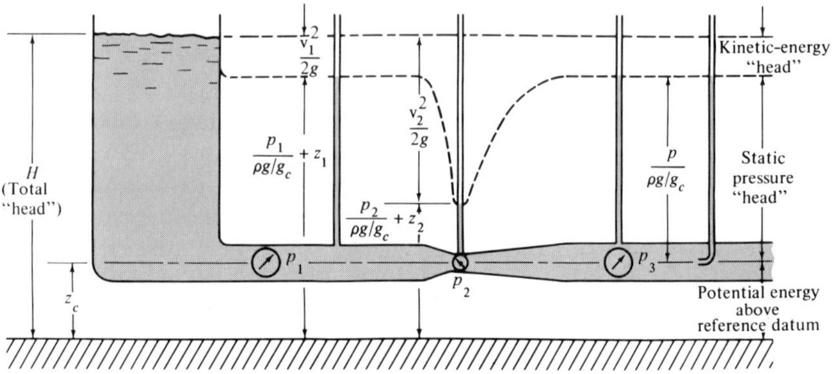

Fig. 6-6. Hydraulic or "manometric-head" interpretation of the terms in the Bernoulli (and energy) equation. The flow in the system illustrated is assumed to be ideal, so there is no loss in the total available energy which is represented by the constant "total head" H.

of the system as shown in Fig. 6-6. The level of the static pressure manometers, which indicates the quantity $p/\gamma_h + z$, is termed the "hydraulic grade line." Since friction forces have been neglected, the quantity H is the total available head at any station along the flow system and is called the total energy line. This graphical method is valuable in laying out piping systems, hydroelectric supply systems, and so on. In an open-channel system, the hydraulic grade line is the free surface of the flow (the pressure p is everywhere the local ambient pressure).

6.7 COMBINED EQUATIONS OF MOTION FOR STEADY, INCOMPRESSIBLE, QUASI-ONE-DIMENSIONAL FLOW

Three basic principles applying to the motion of a fluid have been examined. The equations derived from these principles can be combined to obtain other, perhaps more useful, relations. A particularly useful relation can be obtained for flows considered to be incompressible and quasi-one-dimensional within the modified restrictions listed with the incompressible Bernoulli equation.[10]

For a steady, incompressible, quasi-one-dimensional flow the following relationships have been developed:

[10] The following development is for a single-duct type of flow system. However, it can be generalized to multibranched series-parallel networks by a set of relations directly analogous to Kirchhoff's laws for electrical networks.

Continuity: $\dot{m}_1 = \dot{m}_2$ (6-22a)

Energy:[11]
$$\left(\tilde{u} + \frac{p}{\rho} + \frac{\alpha\bar{\mathrm{v}}^2}{2g_c} + \frac{gz}{g_c}\right)_2 = \left(\tilde{u} + \frac{p}{\rho} + \frac{\alpha\bar{\mathrm{v}}^2}{2g_c} + \frac{gz}{g_c}\right)_1 + \langle w_m \rangle + \tilde{q} \tag{6-22b}$$

Momentum:[12]
$$\left(\frac{p}{\rho} + \frac{\beta\bar{\mathrm{v}}^2}{2g_c} + \frac{gz}{g_c}\right)_2 = \left(\frac{p}{\rho} + \frac{\beta\bar{\mathrm{v}}^2}{2g_c} + \frac{gz}{g_c}\right)_1 + \langle w_m \rangle - \frac{\overline{4\tau_0 L}}{\rho D_H} \tag{6-22c}$$

The continuity equation has already been combined with the energy equation to obtain the form presented here. The subscripts 1 and 2 refer to any stations along the duct, pipe, or machine-type control volume of Fig. 6-4. L is the length of the ducting system in which units of fluid machinery involving the mechanical work term $\langle w_m \rangle$ may be included. These equations must be combined for identical flow conditions if the result is to have any significance.

Combining the energy and momentum equations and assuming that $\alpha \approx \beta$ gives

$$\tilde{u}_1 + \tilde{q} + \frac{\overline{4\tau_0 L}}{\rho D_H} = \tilde{u}_2 \tag{6-23a}$$

or

$$\Delta\tilde{u} = \tilde{q} + \frac{\overline{4\tau_0 L}}{\rho D_H} = \Delta\tilde{u}_q + \Delta\tilde{u}_\tau \tag{6-23b}$$

which states that the internal energy $\Delta\tilde{u}$ is increased by an external heat transfer ($\Delta\tilde{u}_q$) and/or by the conversion of the work done by the fluid in overcoming the friction forces ($\Delta\tilde{u}_\tau$). The loss of availability for a process (for example, the flow from one point in a system to another) is also defined as the irreversibility, Ir. From (6-23b), this irreversibility is

$$\Delta\tilde{u} - \tilde{q} = \frac{\overline{4\tau_0 L}}{\rho D_H} = \mathrm{Ir}$$

for an incompressible flow.

Another restriction or approximation on the development of (6-23) concerns the velocity profiles. Since the friction term is included, the flow cannot be *truly* one-dimensional (the velocity must be zero at the walls); consequently, the correction factors α and β are not unity, and they cannot be identical ($\alpha \neq \beta$). Therefore only quasi-one-dimensional flows with nearly uniform velocity profiles (that is, turbulent flow) having values of α

[11] From (5-16).
[12] From (6-15) for ρ = const.

and β near unity can be represented by these equations. If the profiles are approximately uniform, as indicated in Fig. 6-7, the values of α and β are approximately equal to unity. The error caused by assuming $\alpha \approx 1 \approx \beta$ is then small. For laminar flow it is interesting to note that α and β are usually of little importance because the kinetic-energy term (or momentum) is usually small with respect to the other terms in (6-22).

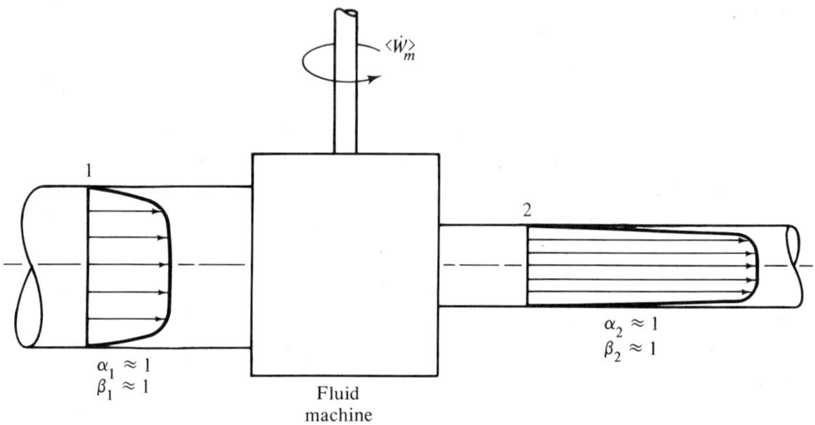

Fig. 6-7. Flow in a system in which the inlet and outlet velocity profiles are sufficiently uniform that the kinetic energy and momentum rates at stations 1 and 2 can be determined in terms of the average velocities.

The result (6-23) is valid only for an incompressible flow. Internal energy or heat transferred is *unavailable* for performing useful work or for conversion to other forms of energy (kinetic energy and so on) that could subsequently be converted into useful work. Two conclusions are then important: The $\Delta\tilde{u}_q = \tilde{q}$ does not enter into any consideration except determining the temperature of the fluid, and the quantity $\Delta\tilde{u}_\tau = \overline{4\tau_0 L/\rho D_H}$ represents a conversion of energy available for performing useful work into unavailable energy or a *dissipation of available energy* within the flowing fluid. This term is then an *energy-dissipation* term. However, long ago this loss of *available* energy was called an "*energy loss*" and misuse of this nomenclature remains. The loss of available energy (which is actually a consequence of the work required to overcome friction forces) appears as a change in the internal energy of the fluid. This dissipation of available energy (the friction work term) is replaced by the term $\Delta p_\tau/\rho$ or $\Delta p_{\text{loss}}/\rho$, which indicates the loss in availability of the fluid to accomplish work at the system boundaries and specifically implies the pressure work required to overcome the friction forces. This work is expended by the pressure forces in overcoming the shear forces at the fluid boundaries; that is, $\overline{4\tau_0 L/\rho D_H} \equiv \Delta\tilde{u}_\tau \equiv \Delta p_\tau/\rho$. The change in internal energy can then be written

$$\Delta\tilde{u} = \tilde{u}_2 - \tilde{u}_1 = \tilde{q} + \frac{\Delta p_\tau}{\rho} \tag{6-24}$$

Substituting this result into the original energy equation, (6-22b), gives

$$\left(\frac{p}{\rho} + \frac{\mathrm{v}^2}{2g_c} + \frac{gz}{g_c}\right)_1 - \frac{\Delta p_\tau}{\rho} + \langle w_m \rangle = \left(\frac{p}{\rho} + \frac{\mathrm{v}^2}{2g_c} + \frac{gz}{g_c}\right)_2 \tag{6-25}$$

which is an approximate energy equation for *steady, incompressible, quasi-one-dimensional flow*. Note that (6-23b) gives a method for determining the temperature difference in a fluid flow, since $\Delta\tilde{u} = c_v\,\Delta T$, where c_v is assumed to be constant. For a liquid flow where the compressibility is negligible, the specific heat can be taken as the average value for the expected temperature range (T_1 through T_2). If the equation is being applied to a gas in the range of negligible compressibility effects, and if the heat transfer $\tilde{q}$ is zero or small, the gas can usually be treated as a perfect gas with a constant c_v. Equation (6-25) states that the availability (available work and energy) of the fluid for energy conversion at station 2 is equal to the availability at station 1, minus the loss in availability incurred during the flow from 1 to 2, plus the mechanical work done on the fluid between 1 and 2.

6.8 COMPLETE EQUATIONS OF MOTION IN STREAMLINE COORDINATES: Euler's Equations

The s-component equation for a differential volume element is (6-8). The n-component of Newton's second-law equation of motion is (referring to Fig. 6-2)

$$(F_{pn-} - F_{pn+})\cos\frac{\alpha}{2} + f_{bn}\rho\,\Delta V = \frac{\rho\,\Delta V}{g_c}\,a_n \tag{6-26a}$$

Equations (6-26a), (6-6), and (6-7) are only approximate in the shear stress terms. The $\partial\tau/\partial n$ term is the only term of significant magnitude for the case of quasi-one-dimensional flow. With substitution of the pressure force quantities from Fig. 6-2, and with $\cos\alpha/2 \approx 1$, (6-26a) becomes

$$-\frac{1}{\rho}\frac{\partial p}{\partial n} - \frac{\partial\Omega}{\partial n} = \frac{1}{g_c}\left(\frac{d\mathrm{v}}{dt}\right)_n \tag{6-26b}$$

where pressure and gravity forces are the only external forces considered to be acting on the element in the n direction and where $a_n = (d\mathrm{v}/dt)_n$ is the normal component of acceleration. For pressure and gravity supplying the only acting forces in the c-component direction, the equation of motion in this direction is

$$-\frac{1}{\rho}\frac{\partial p}{\partial c} - \frac{\partial \Omega}{\partial c} = 0$$

The n-acceleration component $a_n = (d\text{v}/dt)_n$ is again the rate of change of the n component of the velocity at P (Fig. 6-3) with respect to time (which is zero, since $\text{v}_n = 0$) plus the n rate of change of v as the fluid moves along the streamline from P, or

$$a_n = \left(\frac{d\text{v}}{dt}\right)_n = \left[\lim_{\delta t\to 0} \frac{(\text{v} + \delta\text{v}) - \text{v}}{\delta t}\right]_n = \lim_{\delta t\to 0} \frac{(\delta\text{v})_n}{\delta t} \tag{6-27a}$$

The component $(\delta\text{v})_n$ is shown on Fig. 6-3c. From the similar triangles in Fig. 6-3,

$$\frac{(\delta\text{v})_n}{\text{v}} \approx \frac{\delta s}{R_s} \qquad \text{or} \qquad (\delta\text{v})_n \approx \frac{\text{v}\,\delta s}{R_s} \tag{6-27b}$$

which with $\delta s = \text{v}\,\delta t$ gives

$$a_n = \lim_{\delta t\to 0} \frac{\text{v}^2\,\delta t}{R_s\,\delta t} = \frac{\text{v}^2}{R_s} \tag{6-27c}$$

The set of s, n, and c equations of motion is then

$$\begin{aligned} s:&\quad \frac{1}{g_c}\left(\frac{\partial \text{v}}{\partial t} + \text{v}\frac{\partial \text{v}}{\partial s}\right) = -\frac{1}{\rho}\frac{\partial p}{\partial s} + \frac{1}{\rho}\frac{\partial \tau}{\partial n} - \frac{\partial \Omega}{\partial s} \\ n:&\quad \frac{1}{g_c}\frac{\text{v}^2}{R_s} = -\frac{1}{\rho}\frac{\partial p}{\partial n} - \frac{\partial \Omega}{\partial n} \\ c:&\quad 0 = -\frac{1}{\rho}\frac{\partial p}{\partial c} - \frac{\partial \Omega}{\partial c} \end{aligned} \tag{6-28}$$

A vector equation can be obtained from (6-28) by adding and subtracting the quantity $\text{v}(\partial\text{v}/\partial n)$ from the n-component equation; multiplying the s-component equation by $\hat{\text{s}}$, the n-component equation by $\hat{\text{n}}$, and the c-component equation by $\hat{\text{c}}$; and also noting that $\text{v}\,d\text{v} = d(\text{v}^2/2)$:

$$\begin{aligned} \frac{1}{g_c}\left(\hat{\text{s}}\frac{\partial \text{v}}{\partial t} + \hat{\text{s}}\frac{\partial \text{v}^2/2}{\partial s}\right) &= -\frac{\hat{\text{s}}}{\rho}\frac{\partial p}{\partial s} - \hat{\text{s}}\frac{\partial \Omega}{\partial s} + \frac{\hat{\text{s}}}{\rho}\frac{\partial \tau}{\partial n} \\ \frac{1}{g_c}\left(\hat{\text{n}}\frac{\partial \text{v}^2/2}{\partial n} - \hat{\text{n}}\text{v}\frac{\partial \text{v}}{\partial n} + \hat{\text{n}}\frac{\text{v}^2}{R_s}\right) &= \frac{\hat{\text{n}}}{\rho}\frac{\partial p}{\partial n} - \hat{\text{n}}\frac{\partial \Omega}{\partial n} \\ 0 &= \frac{\hat{\text{c}}}{\rho}\frac{\partial p}{\partial c} - \hat{\text{c}}\frac{\partial \Omega}{\partial c} \end{aligned} \tag{6-29}$$

In general, the gradient of a quantity (for example, X) is this s,n,c coordinate system would be

$$\text{grad } X = \hat{s}\,\frac{\partial X}{\partial x} + \hat{n}\,\frac{\partial X}{\partial n} + \hat{c}\,\frac{\partial X}{\partial c}$$

Adding the above vector-component equations then gives[13]

$$\frac{1}{g_c}\left[\frac{\partial \vec{v}}{\partial t} + \nabla\left(\frac{v^2}{2}\right) + \hat{n}v\left(\frac{v}{R_s} - \frac{\partial v}{\partial n}\right)\right] = -\frac{\nabla p}{\rho} - \nabla\Omega + \frac{\hat{s}}{\rho}\frac{\partial \tau}{\partial n} \tag{6-30}$$

It is interesting to consider the physical significance of the various terms of (6-30). The left-hand terms are obtained from the acceleration of the element. The right-hand terms are the forces acting on the element. If (1) the shear stress term were zero, (2) the second term on the left side is also taken as zero for the present, and (3) if the flow is considered to be steady, then (6-30) reduces to

$$\nabla\,\frac{v^2}{2g_c} + \frac{\nabla p}{\rho} + \nabla\Omega = 0 \tag{6-31}$$

Multiplying (6-31) by any arbitrary differential distance $\vec{d\imath}$ and integrating gives the result [where the previously shown identity $\vec{d\imath} \cdot \nabla(\ \ ) = d(\ \ )$ is used],

$$\frac{v^2}{2g_c} + \int \frac{dp}{\rho} + \frac{gz}{g_c} = \text{const.} \tag{6-32}$$

(Ω has been taken as the earth's gravitational potential). Since a constant of integration has been specified instead of $E(n)$ (as was obtained previously by considering only the s-component equation), it is implied that this result is related to the restricting conditions, whereby the shear stress terms and velocity term in the n direction (the term with the $\hat{n}$ in front of it) were dropped. It is therefore of interest to examine these terms for their physical significance and to correlate the above implied conclusions.

ROTATION AND SHEAR

The quantity $v/R_s - \partial v/\partial n$ in (6-30) will be shown to be the *net rotation rate* of the element. Since rotation implies a rotational axis direction, *rotation is a vector quantity.*

As the fluid element translates along the streamline, its shape and orientation will change according to the forces applied to it. The change of

[13] The fact that there is no c component of $v^2/2$ does not affect the generality of the procedure, since v is the magnitude of the vector velocity $\vec{v}$, and $\vec{v} = \hat{s}v$.

shape is caused by two types of forces: (1) shear forces will produce an angular deformation of the element, and (2) the element will expand (or contract) due to the application of normal forces. In addition, a rotating torque will produce a rotation of the element. In the natural or streamline coordinate system there is no motion in the conormal c direction at a point or in its immediate neighborhood. The above motion and deformation can then be investigated only in the s-n plane, which again indicates the advantage of the use of streamline coordinates: *Only two directions are significant*, so that the investigation is essentially two-dimensional at any point in the field.

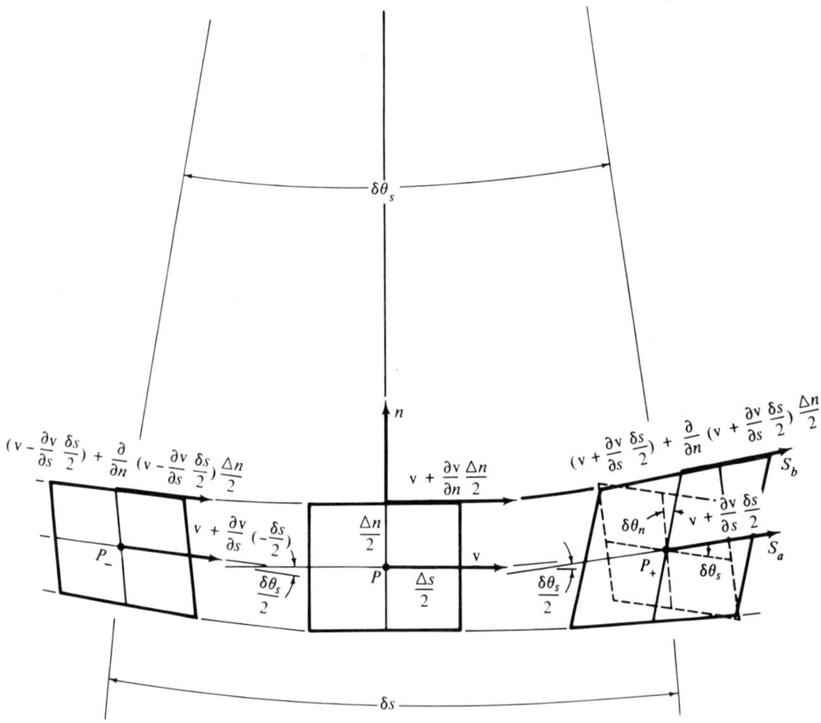

(a)

Fig. 6-8 (a). Deformation of a fluid element as it is propagated along a streamline. The reference shape (at the center position, P) is taken as approximately rectangular. The initial shape is shown superimposed on the final shape to show the manner in which the element is deformed in the flow field.

Deformation of the fluid element can be considered to result from the two effects mentioned above; a volume expansion or "dilatation" without apparent shear deformation as produced by the normal stress, and a shear deformation without dilatation as caused by the shear stresses. Figure 6-8a indicates the deformation and rotation as the element moves along in the neighborhood of the point P. Figure 6-8b shows the deformed shape and rotated orientation of the element superimposed on its initial shape; that is, the translation has been subtracted out of total change = translation + rotation + deformation undergone by the element. This net deformed shape can be shown to be equivalent to the sum of a net volume change or "dilatation," a shear deformation, and a rotation.

The force systems producing the overall and component deviations of the original element shape are shown in Fig. 6-9. Considering the case of dilatation alone, the conservation of mass yields one dynamic condition on the motion of the element from the continuity equation: the size (volume) change is inversely proportional to the density change at any instant, since the mass of the element $\Delta m = \rho\, \Delta V$ is constant (see Chapter 4).

Relative shear strain is an *angular deformation* that is defined as the rate of decrease of the angle φ indicated in Fig. 6-8c. That is, shear strain = $-\delta\varphi$ (where $\delta\varphi$ is the change in φ occurring in a period δt). The shear strain rate is then $-\dot{\varphi}$. From Fig. 6-8c, $-\delta\varphi$ is seen to be equivalent to 2γ; therefore, $-\dot{\varphi} = 2\dot{\gamma}$.

The instantaneous *rotation* of the element as it undergoes a small displacement from the initial position at P is indicated by the angles ϵ. A comparison of the total deformation produced in Fig. 6-8b with the separate component changes shown in Fig. 6-8c indicates that

$$\delta\theta_s = \epsilon + \gamma$$

$$\delta\theta_n = \epsilon - \gamma$$

The size of the element ($\Delta s\, \Delta n\, \Delta c$) is very small, as is the time δt in which the deformation occurs; therefore, the angles $\delta\theta_s$ and $\delta\theta_n$ can be determined by considering the motion of the element as represented in Fig. 6-8. The angle $\delta\theta_n$ is equal to the product of the radius and the angle.[14]

[14] The vector relation between these quantities is

$$\overrightarrow{\delta r} = \overrightarrow{\delta\theta} \times \overrightarrow{\Delta n}$$

The vector direction of an angle is perpendicular to the plane in which the angle is measured and is positive in a direction determined by the right-hand rule (Appendix B) in the same manner as the direction of an angular velocity vector is determined. If this relation is divided by δt,

$$\frac{\overrightarrow{\delta r}}{\delta t} = \frac{\overrightarrow{\delta\theta}}{\delta t} \times \overrightarrow{\Delta n} \qquad \text{or} \qquad \vec{\mathrm{v}} = \vec{\omega} \times \overrightarrow{\Delta n}$$

The deformation of the fluid element as shown in Fig. 6-8a from its initial shape to its final shape can be shown to be

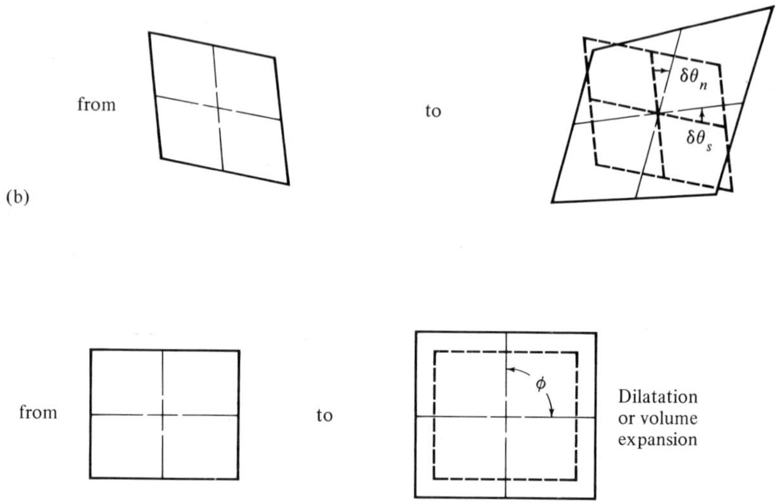

The element deviation of Fig. 6-8b is approximately equivalent to the following:

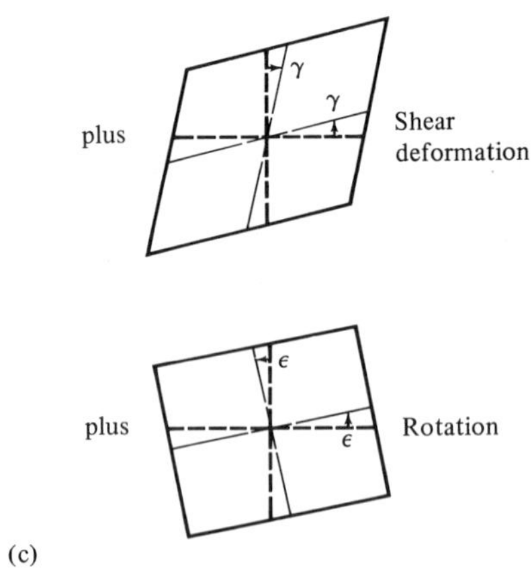

Fig. 6-8 (b) and (c).

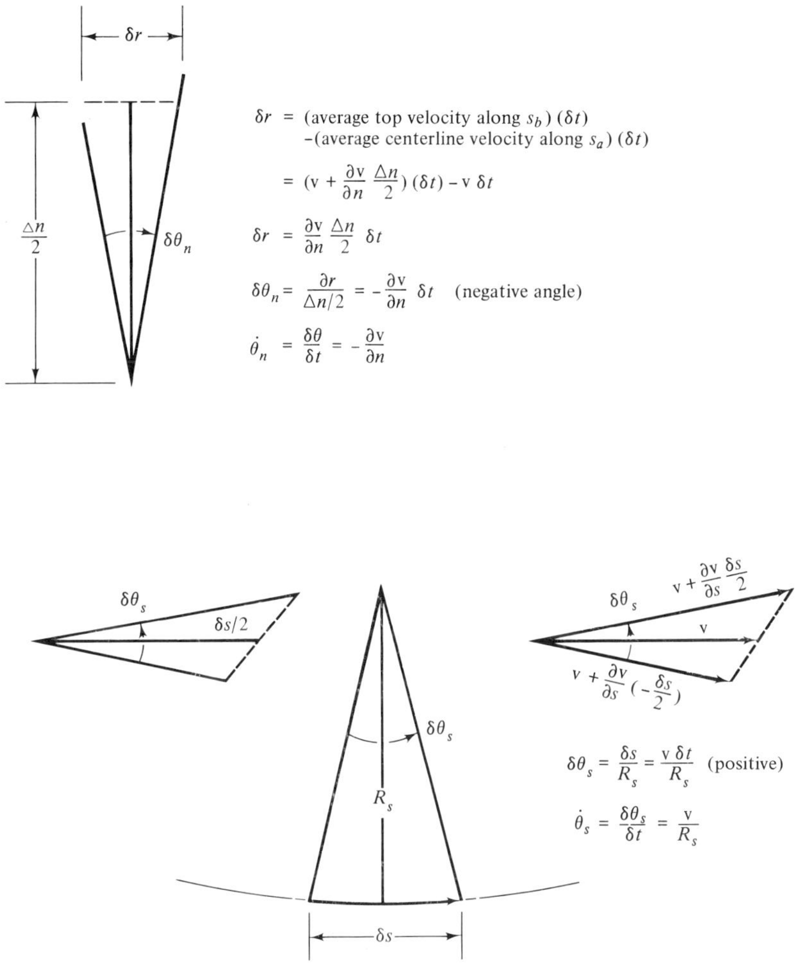

Fig. 6-8 (d).

Therefore, Fig. 6-8d indicates that

$$\delta r = \frac{\Delta n}{2}\, \delta\theta_n$$

This relation becomes more accurate as the displacement gets smaller, that is, as the points P_- and P_+ more closely approach P in Fig. 6-8a. From Fig. 6-8d, this displacement δr is equal to the average velocity at the top of the element (on s_b) minus the average centerline velocity (on s_a) times δt:

$$\delta r = \left[\left(\mathrm{v} + \frac{\partial \mathrm{v}}{\partial n}\frac{\Delta n}{2}\right) - \mathrm{v}\right] \delta t = \frac{\partial \mathrm{v}}{\partial n}\frac{\Delta n}{2}\, \delta t$$

Equating these expressions for δr then gives the angle $\delta\theta_n$ in terms of the normal velocity gradient

$$\delta\theta_n = -\frac{\partial \mathrm{v}}{\partial n}\,\delta t$$

(the minus sign indicates a clockwise or negative angle). The angle $\delta\theta_s$ is most easily determined from the relationships among the similar triangles shown in Fig. 6-8d:

$$\delta\theta_s = \frac{\delta s}{R_s} = \frac{\mathrm{v}\,\delta t}{R_s}$$

The force components producing the motion and deformation indicated in Fig. 6-8 are shown in (a),

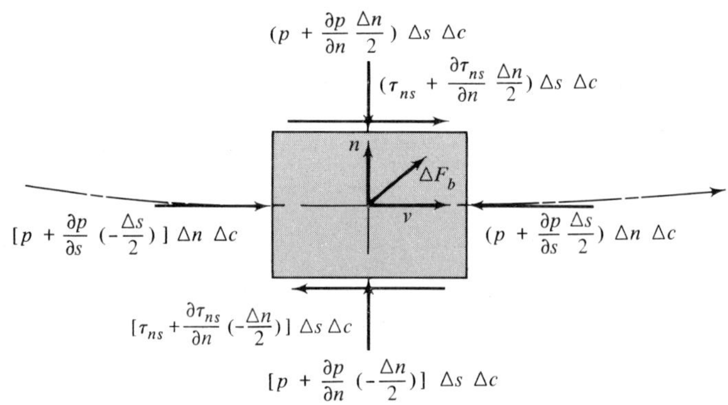

(a)

which is equivalent to (b), to which the system $\tau_{ns}\,\Delta n\,\Delta c\downarrow\uparrow\tau_{ns}\Delta n\,\Delta c$ can be added to the right and left sides of the element to give (c).

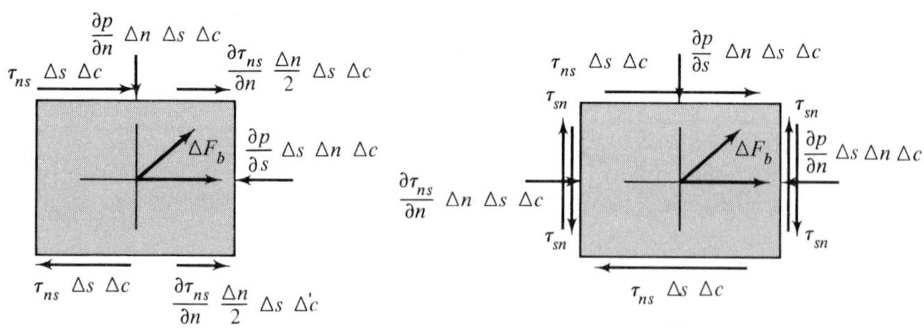

(b) (c)

Fig. 6-9. Relationships among the force components and deviations (deformations and rotation) of the fluid element.

The foregoing can be considered the sum of each of following separate force systems:

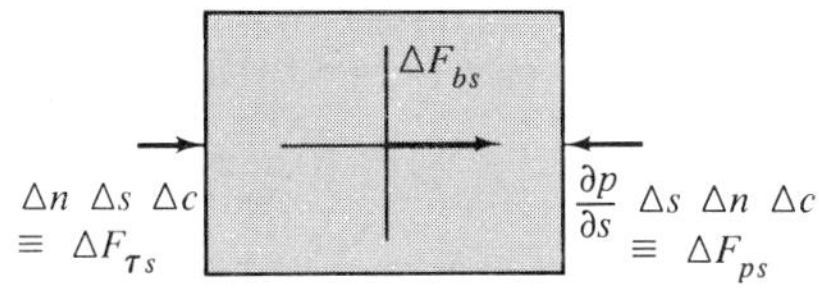

(e)

These n-directed forces are in equilibrium with the dynamic inertia force

$$\frac{\rho \mathrm{v}^2}{g_c R_s}\ \Delta s\ \Delta n\ \Delta c$$

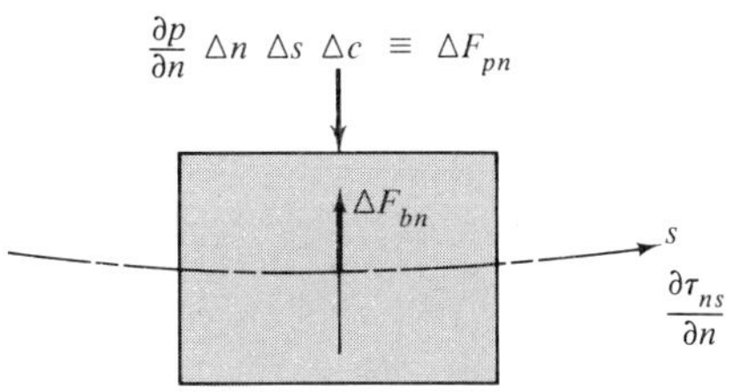

(d)

These s-directed forces are in dynamic equilibrium with

$$\frac{\rho}{g_c}\left[\left(\frac{\partial \mathrm{v}}{\partial t}\right)_s + \mathrm{v}\frac{\partial \mathrm{v}}{\partial s}\right]\Delta s\ \Delta n\ \Delta c$$

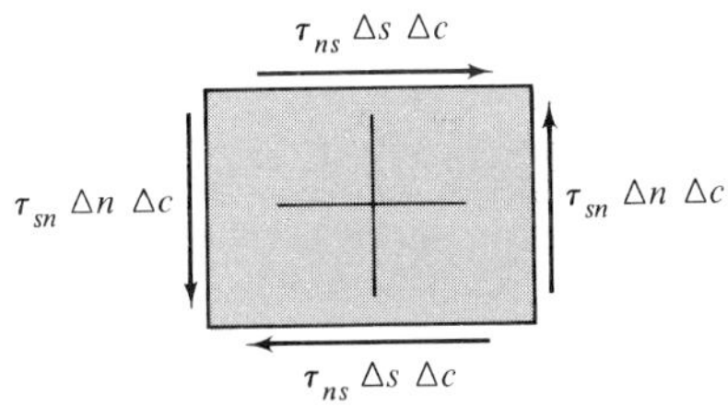

(f)

This shear force system is balanced, but produces the shear deformation of Fig. 6-8c.

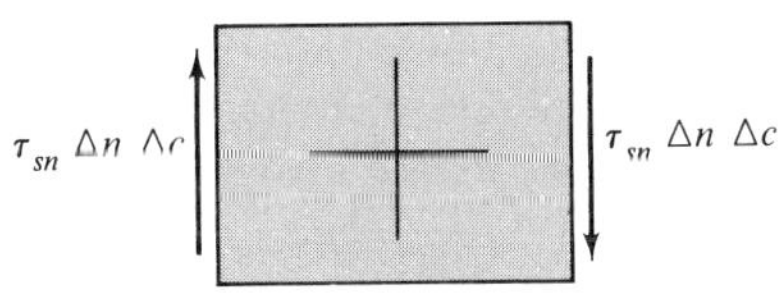

(g)

These forces are in equilibrium. The couple produces the rotation indicated in Fig. 6-8c.

Fig. 6-9 (continued).

In fluid mechanics, the *rates* of shear strain and angular rotation are the quantities of primary importance in the stress–strain relations. The *shear strain rate* is then characterized by $\dot{\gamma}$ (the angular deformation rate), which is defined as the rate of decrease of the angle φ, Fig. 6-8c; that is, $2\dot{\gamma} = -\dot{\varphi}$. The *rotation rate* is characterized by $\dot{\epsilon}$. If $\delta\theta_s$ and $\delta\theta_n$ are divided by δt and the resultant angular displacement rates (or rotations) are denoted by $\dot{\theta}_s$ and $\dot{\theta}_n$,

$$\dot{\theta}_s = \frac{\delta\theta_s}{\delta t} = \frac{\mathrm{v}}{R_s}$$

$$\dot{\theta}_n = \frac{\delta\theta_n}{\delta t} = -\frac{\partial \mathrm{v}}{\partial n}$$

or as vectors

$$\vec{\dot{\theta}}_s = \hat{c}\,\frac{v}{R_s} \qquad \vec{\dot{\theta}}_n = -\hat{c}\,\frac{\partial v}{\partial n}$$

In terms of $\dot{\gamma}$ and $\dot{\epsilon}$,

$$\dot{\theta}_s = \dot{\epsilon} + \dot{\gamma}$$

$$\dot{\theta}_n = \dot{\epsilon} - \dot{\gamma}$$

or

$$2\dot{\gamma} = \dot{\theta}_s - \dot{\theta}_n$$

$$2\dot{\epsilon} = \dot{\theta}_s + \dot{\theta}_n \equiv 2\bar{\omega}$$

The shear strain rate is then[15]

$$\dot{\gamma} = \frac{1}{2}\left(\frac{\partial v}{\partial n} + \frac{v}{R_s}\right) \tag{6-33a}$$

and the *average* rotation rate is

$$\vec{\bar{\omega}} = \hat{c}\bar{\omega} = \hat{c}\,\frac{\dot{\theta}_s + \dot{\theta}_n}{2} = \frac{\hat{c}}{2}\left(\frac{v}{R_s} - \frac{\partial v}{\partial n}\right) \tag{6-33b}$$

The quantity $v/R_s - \partial v/\partial n$ is then equal to twice the average rotation rate ($2\bar{\omega}$). The vector quantity $\vec{\omega}$ is then *defined* as the average of the rotation rates of two mutually perpendicular (for example, coordinate) axes of an element about the third axis, in the limit as the size of the element approaches a point. This average rotation rate is a function of the velocity field and is dependent on the rates of change of the velocities in the component directions. Therefore, it has the characteristics of a form of derivative of the velocity field. As a vector, this particular *net rotation rate* is directed perpendicular to the plane of the motion (perpendicular to the *s*-*n* plane, or in the $\hat{c}$ direction).[16] This type of operation involving a vector differentiation of a velocity (or vector) field to yield another vector field, which is the net rotation rate of the initial field, is termed a *curl* operation, yielding the curl of the initial (velocity) field:

[15] A vector designation is not given with $\dot{\gamma}$, since the state of strain (or the strain rate) of an element is a tensor quantity. Note that the correspondence of the n and R_s directions gives $dn = -dR_s$, so

$$\dot{\gamma} = -\frac{1}{2}\left(\frac{\partial v}{\partial R_s} - \frac{v}{R_s}\right) = -\frac{R_s}{2}\frac{\partial v/R_s}{\partial R_s}$$

[16] In three dimensions, the three components of $\vec{\omega}$ would be the vector sum of the rotations of each of the two mutually perpendicular axes about each coordinate rotation axis.

$$\vec{\omega} = \tfrac{1}{2}\,\text{curl}\,\vec{\text{v}} = \tfrac{1}{2}\,\nabla \times \vec{\text{v}} \tag{6-34}$$

where the ∇ operator is the same operator originally defined in Chapter 3 and curl signifies a net rotation or rotation rate. The term in which $\text{v}/R_s - \partial\text{v}/\partial n$ appears in (6-30) can now be written $\hat{\text{n}}\text{v}|\text{curl}\,\vec{\text{v}}|$ by considering $\vec{\text{v}} \times \text{curl}\,\vec{\text{v}} = \vec{\text{v}} \times (\nabla \times \vec{\text{v}}) = (\hat{\text{s}}\text{v}) \times \hat{\text{c}}\,\text{curl}\,\vec{\text{v}} = -\hat{\text{n}}\text{v}|\text{curl}\,\vec{\text{v}}|$, since $\hat{\text{s}} \times \hat{\text{c}} = -\hat{\text{n}}$. The shear stress terms can be obtained from the divergence of the shear stress in tensor form, div τ or $\nabla \cdot \tau$. With these substitutions, (6-30) becomes

$$\frac{1}{g_c}\left[\frac{\partial \text{v}}{\partial t} + \text{grad}\,\frac{\text{v}^2}{2} - \vec{\text{v}} \times (\text{curl}\,\vec{\text{v}})\right] = -\frac{\text{grad}\,p}{\rho} + \text{grad}\,\Omega - \frac{1}{\rho}\,\text{div}\,\tau \tag{6-35}$$

It is difficult to interpret physically the shear stress term as a function of the angular deformation rate as indicated in Fig. 6-8. This complication arises because of the tensor character of the shear stress and shear strain rate.

In general, the tangential shear stress may have different components on each of the six faces of a fluid mass element. These shear stresses can be denoted by τ_{ij}, where the subscript i denotes the *face* on which the stress acts and the subscript j denotes the *direction* of the shear stress force. For example, the shear stress in the above development would be represented as τ_{ns}. This shear stress acts on the top (and bottom) faces of the element which are characterized by the vector direction ($\hat{\text{n}}$) of these face areas ($\Delta s\,\Delta c$) and the shear stress force ($\tau\,\Delta s\,\Delta n$) acts in the s direction. If the flow is incompressible, the shear stress–strain rate relationship is $\tau_{ns} = 2\mu\dot{\gamma}_{ns}$. The strain rate, $\dot{\gamma}_{ns} = \dot{\gamma}$, is given in (6-33a). If (6-33a) is inserted into (6-35),

$$\frac{1}{g_c}\left[\frac{\partial \vec{\text{v}}}{\partial t} + \nabla\frac{\text{v}^2}{2} - \vec{\text{v}} \times (\nabla \times \vec{\text{v}})\right] = -\frac{\nabla p}{\rho} - \nabla\Omega + \frac{\hat{\text{s}}\,\partial}{\rho\,\partial n}\left[\mu\left(\frac{\partial \text{v}}{\partial n} + \frac{\text{v}}{R_s}\right)\right] \tag{6-35a}$$

A more complete analysis (with μ = const.) shows the last term to be $\mu\,\nabla \cdot \nabla\vec{\text{v}} = \mu\,\nabla^2\text{v}$ (where $\nabla \cdot \nabla \equiv \nabla^2$ is the Laplace operator), so (6-35a) can also be written

$$\frac{1}{g_c}\left[\frac{\partial \vec{\text{v}}}{\partial t} + \nabla\frac{\text{v}^2}{2} - \vec{\text{v}} \times (\nabla \times \vec{\text{v}})\right] = -\frac{\nabla p}{\rho} - \nabla\Omega + \frac{\nu}{g_c}\nabla^2\text{v} \tag{6-35b}$$

This equation of motion involving the acceleration of a fluid element under the influence of pressure, conservative field, and viscous shear forces is known as the *Navier–Stokes equation.*[17]

It is now important to note that the shear deformation and the rotation can only be produced by application of tangential or shear forces. It then follows immediately that a rotation cannot exist in an ideal fluid, where no mechanism exists for producing tangential stress on an element.[18] If the fluid is ideal, a strain rate may be set up in the flow field, but no *shear stress* can exist because the viscosity is zero by definition of an *ideal* fluid. A direct physical correlation then exists between shear stress (or strain rate) and rotation. If a fluid is ideal (no viscosity or negligible shear strain rate), there is no way for a tangential torque (shear force) to be applied to the fluid. Therefore, if the fluid is irrotational at some position in the field, it must be irrotational for the rest of the time that this particular fluid element is in that field. These conclusions are a direct result of the conditions on the fluid motion (Newton's laws) and may also be proved from purely mathematical considerations. The condition of zero rotation then implies that $\vec{\omega} = 0$ or curl $\vec{v} = 0$.

The equation of motion for a frictionless flow with no initial rotation is then

$$\frac{1}{g_c}\left(\frac{\partial \vec{v}}{\partial t} + \text{grad}\,\frac{v^2}{2}\right) + \frac{1}{\rho}\,\text{grad}\,p + \text{grad}\,\Omega = 0 \tag{6-36}$$

Using the vector identity that $\vec{v} \times \nabla \times \vec{v} = \nabla(v^2/2) + \vec{v}\cdot\nabla\vec{v}$, (6-35) may be written in the alternative form

$$\frac{1}{g_c}\left(\frac{\partial \vec{v}}{\partial t} + \vec{v}\cdot\nabla\vec{v}\right) = -\frac{\nabla p}{\rho} - \nabla\Omega + \frac{\nabla\cdot\tau}{\rho} \tag{6-35c}$$

The left side gives the local timewise acceleration plus the convective acceleration (or the total acceleration $\vec{a} = d\vec{v}/dt$), which allows (6-35c) to be written

$$\frac{\vec{a}}{g_c} = \frac{-\nabla p}{\rho} - \nabla\Omega + \frac{\nabla\cdot\tau}{\rho} \tag{6-35d}$$

[17] Louis Marie Henri Navier (1785–1836), French engineer. The viscous force was actually obtained by Navier from an analysis of intermolecular forces and represented a "rearrangement" coefficient for these forces. It was Jean Claude Barre de St. Vennant (1797–1886) who correctly determined the relation between stress and strain rate. George Gabriel Stokes (1819–1903), a British mathematician born in Ireland, derived the Navier–Stokes equation(s) directly by considering the viscous stresses acting on a fluid.

[18] If the fluid had an initial rotation, the angular motion or *angular momentum* could not change, because no torque could be applied to change it.

For a frictionless flow, (6-35d) becomes

$$\frac{\vec{a}}{g_c} = \frac{-\nabla p}{\rho} - \nabla\Omega \tag{6-35e}$$

which is *Euler's equation* in vector form. This vector equation can be translated to any coordinate system desired to obtain the component equations of motion in that coordinate system. The procedure for such a translation is purely a mathematical formality (see Appendix B).

An alternative method of derivation is to set up a fluid element in the desired coordinate system and work through the detailed derivation of the component equations in exactly the same manner as was done above. In curvilinear coordinate systems this type of detailed development becomes more tedious. An example of the added complication was shown in Chapter 3 for the determination of the pressure field for a rotating mass of fluid with no relative motion within the fluid. The radial pressure gradient term involved pressure–force components on four faces, but the end result was simple in form. That it must be simple is obvious from examination of the final vector result.

The gradient of a scalar X is just

$$\text{grad } X \equiv \hat{x}_1 \frac{\partial X}{\partial x_1} + \hat{x}_2 \frac{\partial X}{\partial x_2} + \hat{x}_3 \frac{\partial X}{\partial x_3} \equiv \nabla X$$

where dx_1, dx_2, and dx_3 are line-element lengths in each of the three coordinate directions. For example, in a cylindrical polar coordinate system they are dr, $r\,d\theta$, and dz (and the unit vectors in these three directions are $\hat{r}$, $\hat{\theta}$, and $\hat{z}$). If (6-36) is multiplied through by $\vec{d\imath}$ and integrated [again noting that $\vec{d\imath} \cdot \nabla(\ \ ) \equiv d(\ \ )$],

$$\frac{1}{g_c}\int \frac{\partial \mathrm{v}}{\partial t}\cdot\vec{d\imath} + \frac{\mathrm{v}^2}{2g_c} + \int \frac{dp}{\rho} + \frac{gz}{g_c} = E(t) \tag{6-37}$$

where the body-force potential Ω has been taken as the earth's gravitational potential field. Another important condition is obtained by noting that $E(t)$ is a function of time only for every point in the field.[19] Therefore, an irrotational flow field is also a constant-energy field at any given instant (that is, all streamlines have the same energy). If the flow is steady, the first term is zero and E is a constant. Equation (6-37) then reduces to (6-32).

The above derivation has evolved a more complete development of the Bernoulli equation. The restrictions on its various forms have a broader

[19] The integration was with respect to the space coordinates only and therefore does not involve the other independent coordinate, t.

interpretation. For example, the condition of one-dimensional flow also implies irrotational flow. A quasi-one-dimensional flow is usually not frictionless. A correlation is then obtained among the terms one-dimensional, irrotational, and frictionless (for flows emitted from a constant-energy reservoir). For a steady, irrotational, frictionless, incompressible flow, the vector equation of motion becomes

$$\text{grad}\left(\frac{\text{v}^2}{2g_c} + \frac{p}{\rho} + \frac{gz}{g_c}\right) = 0$$

which can be multiplied by any differential distance $\vec{d\imath}$ and integrated throughout the entire flow field to give *Bernoulli's equation*,

$$\frac{\text{v}^2}{2g_c} + \frac{p}{\rho} + \frac{gz}{g_c} = E \tag{6-38}$$

The total energy E (energy per unit mass of fluid) is the same everywhere in the flow field.

IRROTATIONAL FLOW

If the rotation is zero, the quantity $\partial\text{v}/\partial n - \text{v}/R_s = 0$. With $dn = -dR_s$,

$$\frac{\partial \text{v}}{\partial R_s} + \frac{\text{v}}{R_s} = 0$$

The variation of v with radius of curvature at any point in an irrotational flow field is obtained by integrating the above equation, giving

$$\int \frac{d\text{v}}{\text{v}} = -\int \frac{dR}{R} + c_1 \qquad \text{or} \qquad \text{v}R = c \tag{6-39}$$

(where the subscript s on the radius of curvature has been dropped for convenience).

With (6-39), a qualitative evaluation of the flow of an ideal fluid around a bend then indicates the hyperbolic velocity variation indicated in Fig. 6-10. The flow, which is initially uniform across the duct, tends to produce this profile as it goes through the bend. The flow must readjust to again become uniform downstream of the elbow. Although an ideal fluid has been considered, the same conditions can apply to a "high" Reynolds-number flow, where the velocity profile is nearly uniform. If the duct radius is short, an interaction between the production of this irrotational flow and the boundary-layer effect produces a flow separation near points A and B, as will be described in Chapter 9.

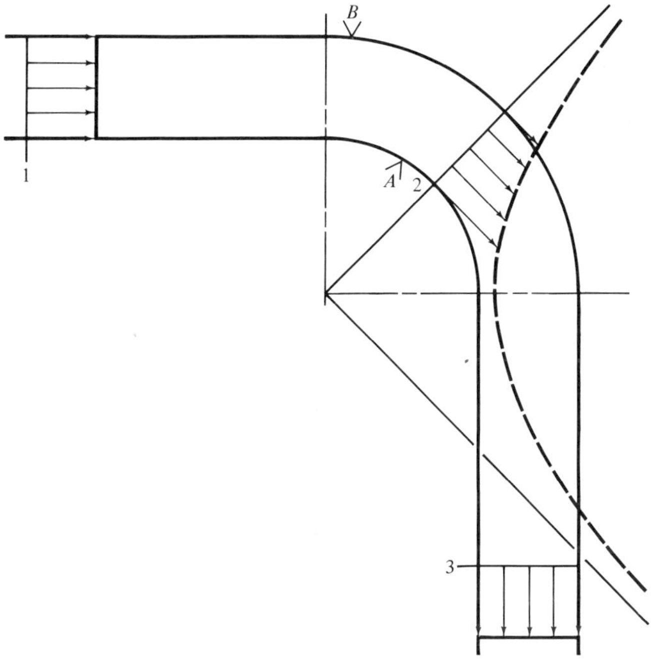

Fig. 6-10. Effect of curvature on the flow of an ideal fluid around a bend.

The pressure variation near the midpoint in the bend can be determined from Bernoulli's equation. If the flow is incompressible and steady,

$$p_1 + \frac{\rho \mathrm{v}_1^2}{2g_c} = p_0 = p_2 + \frac{\rho \mathrm{v}_2^2}{2g_c} \tag{6-40}$$

or

$$p_2(R) = p_0 - \frac{\rho}{2g_c}\left(\frac{c}{R}\right)^2$$

and the pressure decreases rapidly as the radius of curvature decreases (toward the inside of the curve).

Tornadoes derive much of their destructive power from the fact that outside a viscous "core" the fluid motion behaves as an "irrotational vortex." The pressure near the outside, where the air velocity is low, is nearly the local atmospheric pressure. Near the core where the velocity is quite high (on the order of 100 mph, or greater, even though the twister propagates along at only 15 to 25 mph) the pressure may fall to about 30 $\mathrm{lb_f/ft^2}$ below the local static ambient pressure. This 30-psf difference may be enough to lift the roof off a house, blow the windows out, or even cause the house to explode. The conclusion here is that the windows should be opened when a twister approaches. The consequence that provides the force to lift the whole house will be considered in Chapter 9, where boundary-layer interactions producing so-called secondary flows are considered.

ROTATIONAL FLOW

Most flows originate from a source (or reservoir) where all streamlines have the same initial energy (Fig. 6-5). If the flow is essentially irrotational (outside the boundary-layer region), it is difficult to imagine how a non-viscous or ideal *rotational* flow field could exist. One possible method of generating such a flow is indicated in Fig. 6-11a. For such an ideal ($\tau = 0$) steady flow field with rotation, the s-component equation of motion, (6-28), can be written

$$\frac{\mathrm{v}}{g_c}\frac{\partial \mathrm{v}}{\partial s} + \frac{1}{\rho}\frac{\partial p}{\partial s} + \frac{\partial (gz)}{g_c\,\partial s} = 0 \tag{6-41}$$

or

$$\frac{\mathrm{v}^2}{2g_c} + \int \frac{dp}{\rho} + \frac{gz}{g_c} = E(n)$$

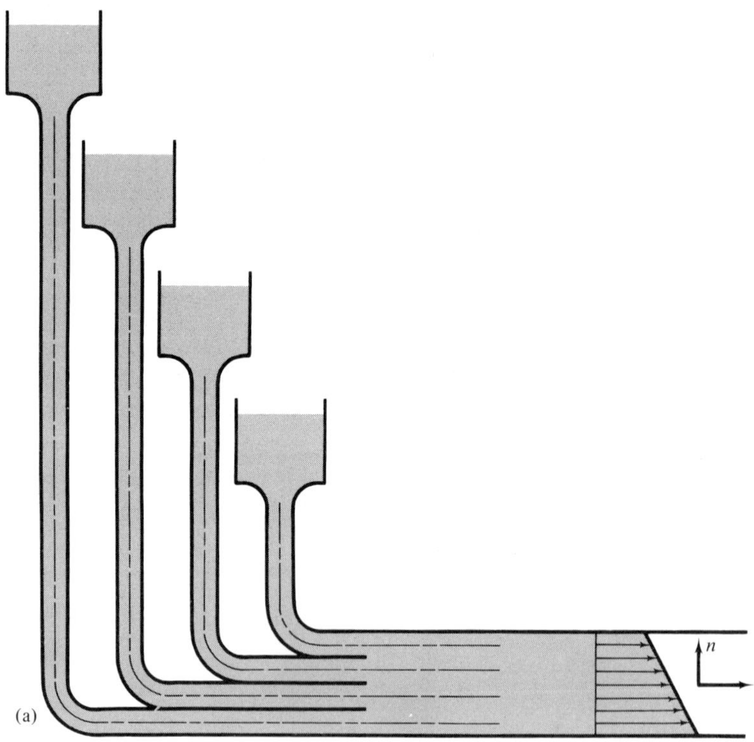

(a)

Fig. 6-11 (a). Example of a possible means by which a rotational flow in an ideal fluid could be generated from an initial nonuniform distribution of total energy. The profile shown indicates the shape of the velocity and the total energy (E) distribution in the normal (n) direction.

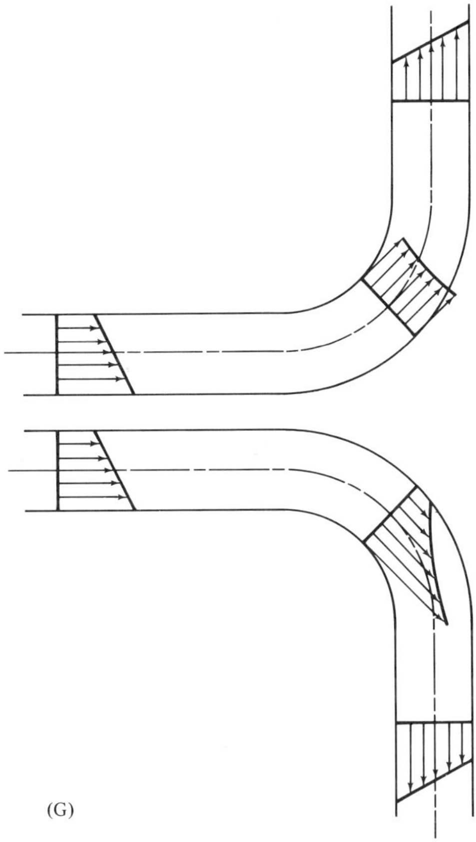

(G)

Fig. 6-11 (b). Effect of irrotational flow around bends on an initial non-uniform velocity distribution such as represented in Fig. 6-11a.

The n-component equation is

$$\frac{v^2}{g_c R} + \frac{1}{\rho}\frac{\partial p}{\partial n} + \frac{\partial(gz)}{g_c\, \partial n} = 0 \qquad (6\text{-}42)$$

Taking the partial derivative of (6-41) with respect to n gives

$$\frac{v}{g_c}\frac{\partial v}{\partial n} + \frac{1}{\rho}\frac{\partial p}{\partial n} + \frac{\partial(gz)}{g_c\, \partial n} = \frac{\partial E}{\partial n}$$

and combining with (6-42) gives

$$\frac{v}{g_c}\left(\frac{\partial v}{\partial n} - \frac{v}{R}\right) = \frac{\partial E}{\partial n} \qquad \text{or} \qquad \vec{v} \times (\text{curl}\, \vec{v}) = \hat{n} g_c \frac{\partial E}{\partial n} = \hat{n} g \frac{\partial H}{\partial n} \qquad (6\text{-}43)$$

The variation of the total energy E from one streamline to another is then directly proportional to $\vec{v} \times \text{curl}\, \vec{v}$ (or $\vec{v} \times 2\vec{\omega}$), as indicated in Fig. 6-11a.

Therefore, if the flow is frictionless and *irrotational*, the energy of all streamlines is the same throughout the field. If the flow is frictionless but *rotational*, the variation of energy from one streamline to another is given by (6-43). Furthermore, if the flow is frictionless, rotational, and steady, the variation of rotation or vorticity is given by (6-43). If the rotational flow of Fig. 6-11 goes around the bend as shown in Fig. 6-11b (top), the velocity profile will tend to straighten in the bend.

6.9 AN INTEGRAL EQUATION OF MOTION

It was stated that the problem of a fluid in motion can be considered with respect to a control volume fixed in space (Eulerian frame of reference) or fixed in the fluid (Lagrangian frame of reference). The Eulerian method is the more direct with which to deal, since the control volume boundaries can be chosen to coincide with the physical boundaries of the fluid system under consideration.

The integral equation for the flow of a fluid through a specified space-fixed control volume will now be determined by the Eulerian approach previously used in the derivation of the continuity and energy equations. The quantity stored and transported is now just the momentum $m\vec{v}$ of the mass.

Newton's second law relating the impulse and momentum applied to the fluid within a flow field control volume must be stated as

$$\begin{bmatrix}\text{THE INCREASE IN}\\ \text{MOMENTUM OF}\\ \text{THE FLUID IN}\\ \text{THE C.V. IN } \delta t\end{bmatrix} \xrightarrow[\text{QUENCE OF}]{\text{IS A CONSE-}} \begin{bmatrix}\text{THE NET MOMENTUM}\\ \text{TRANSPORTED BY THE}\\ \text{FLOW FIELD INTO}\\ \text{THE C.V. IN } \delta t\end{bmatrix} + \begin{bmatrix}\text{SUMMATION OF ALL}\\ \text{IMPULSES ACTING ON}\\ \text{THE FLUID IN THE}\\ \text{C.V. DURING } \delta t\end{bmatrix} \tag{6-44a}$$

The momentum transport term may be transposed and written as a positive net momentum transport *out* of the C.V. Dividing by δt, (6-44a) may be written as a rate statement,

$$\begin{bmatrix}\text{THE RATE OF INCREASE}\\ \text{OF MOMENTUM OF THE}\\ \text{FLUID IN THE C.V.}\end{bmatrix} + \begin{bmatrix}\text{THE NET RATE OF}\\ \text{MOMENTUM TRANSPORT}\\ \text{OUT OF THE C.V.}\end{bmatrix} = \begin{bmatrix}\text{SUMMATION OF ALL}\\ \text{FORCES ACTING ON}\\ \text{THE FLUID IN C.V.}\end{bmatrix} \tag{6-44b}$$

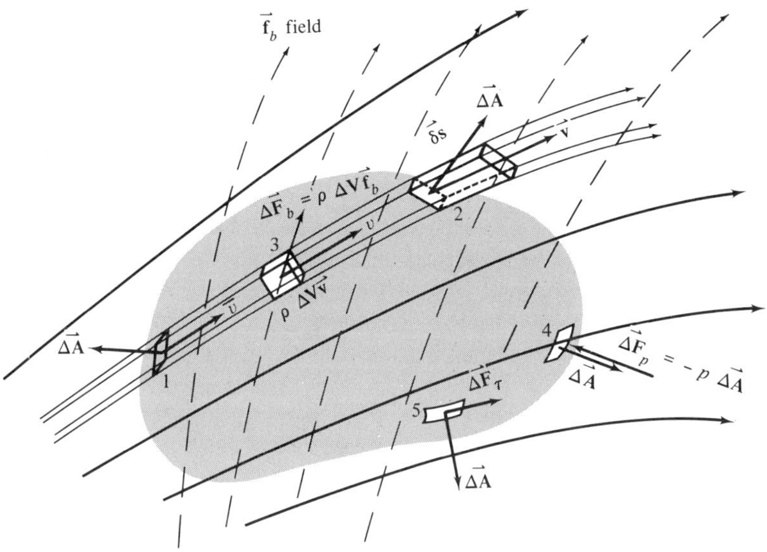

Fig. 6-12. General control volume showing the momentum transport across the control volume surface through a streamtube (in at 1, out at 2), the momentum of a small element within the control volume (at 3), the body force on a mass element (at 3), the pressure force acting on a surface element area (at 4), and the shear force acting on a surface element area (at 5).

The impulse–momentum, or force–momentum rate principle is stated as applying to *the fluid in the C.V.* Therefore, the C.V. must contain only and entirely fluid.

The momentum of a mass element Δm inside the control volume at point 3 of Fig. 6-12 is $\Delta m\vec{v}$. The momentum inside the C.V. at any instant is the sum of the momentum over all mass elements at a given instant, which is

$$\oiiint \vec{v}\rho \, dV$$

The rate of increase of the momentum of the fluid within the control volume (also the time rate of change or net storage rate of the momentum) is

$$\frac{\partial}{\partial t} \oiiint \vec{v}\rho \, dV$$

When the control volume is of fixed shape and either fixed in space or moving with a constant motion (velocity), the time differentiation can be placed under the integral sign.

The outflow rate of momentum through the streamtube as the flow leaves the control volume at 2 is the momentum carried out by the flow in δt divided by δt. In the time period δt, a mass $\Delta m = \rho\ \overrightarrow{\delta s} \cdot \overrightarrow{dA}$ flows out through the streamtube at point 2. This mass has a momentum $\Delta m\vec{v} = (\rho\ \overrightarrow{\delta s} \cdot \overrightarrow{dA})(\vec{v})$. The rate of outflow at 2 is then $(\rho\vec{v} \cdot \overrightarrow{\Delta A})\vec{v}$. Summing all inflows and outflows gives the net transport of momentum out of the control volume:

$$\oiint \vec{v}(\rho\vec{v} \cdot \overrightarrow{dA})$$

When the control volume is moving with a constant velocity (inertial frame of reference), the quantity $\vec{v} \cdot \overrightarrow{dA}$ must be evaluated for the flow velocity relative to the C.V., since it represents the transport rate at which the momentum of a unit mass is being carried across the control volume boundary. Since the momentum can be determined with respect to either the absolute or relative motion (inertial system only), it is advantageous to consider the control volume moving with the physical system.

The principle of momentum in the form of an equation in the Eulerian frame of reference is then

$$\sum_{\substack{\text{ON FLUID} \\ \text{IN C.V.}}} \vec{F} = \frac{1}{g_c} \iiint \frac{\partial \rho\vec{v}}{\partial t}\, dV + \frac{1}{g_c} \oiint \vec{v}(\rho\vec{v} \cdot \overrightarrow{dA}) \tag{6-45}$$

Before specifying the forces in detail, a few special cases of this principle will be considered:

Case I

STEADY FLOW

If the flow is steady, the momentum equation is simply

$$\sum \vec{F} = \frac{1}{g_c} \oiint \vec{v}(\rho\vec{v} \cdot \overrightarrow{dA}) \tag{6-46}$$

or expanded in three coordinate-component directions,

$$\sum F_1 = \frac{1}{g_c} \oiint u_1(\rho\vec{v} \cdot \overrightarrow{dA})$$

$$\sum F_2 = \frac{1}{g_c} \oiint u_2(\rho\vec{v} \cdot \overrightarrow{dA}) \tag{6-46a}$$

$$\sum F_3 = \frac{1}{g_c} \oiint u_3(\rho\vec{v} \cdot \overrightarrow{dA})$$

where 1, 2, and 3 refer to the three respective coordinates (for example, in a cartesian coordinate system $u_1 = u$, $u_2 = v$, $u_3 = w$, $F_1 = F_x$, and so on).

The surface area integral can be broken into the inflow and outflow paths, and (6-46) can be written in the form

$$\sum \vec{F} = \frac{1}{g_c} \iint_{\substack{\text{OUTFLOW}\\\text{AREAS}}} \vec{v}(\rho\vec{v}\cdot\vec{dA}) + \frac{1}{g_c} \iint_{\substack{\text{INFLOW}\\\text{AREAS}}} \vec{v}(\rho\vec{v}\cdot\vec{dA}) \tag{6-46b}$$

which states that the sum of all external forces acting on the fluid in the C.V. to the momentum flux out minus the momentum flux in through the surface of the control volume. [Note that $(\vec{v}\cdot\vec{dA})_{\text{in}}$ is negative.]

Case IIa

STEADY QUASI-ONE-DIMENSIONAL FLOW

When velocity is not uniform over the inflow and outflow areas, but where the flow can be considered to be quasi-one-dimensional as discussed in Sections 5.2 and 6.4, (6-46) can be written

$$\sum \vec{F} = \frac{1}{g_c} \iint_{\substack{\text{OUTFLOW}\\\text{AREAS}}} \vec{v}\rho v \, dA \cos(\vec{v}, \vec{dA}) + \frac{1}{g_c} \iint_{\substack{\text{INFLOW}\\\text{AREAS}}} \vec{v}\rho v \, \vec{dA} \cos(\vec{v}, \vec{dA}) \tag{6-47}$$

If the flow areas can be taken as planar and perpendicular to the direction of the flow (perpendicular to $\vec{v}$), then the magnitude of the momentum terms is

$$\iint_A \rho v^2 \, dA$$

According to the definition of quasi-one-dimensional flow, the density variation over the area A is negligible, so the density can be taken outside of the integral sign. If it is desired to express this momentum transport rate across the area boundary A in terms of the average velocity $\bar{v}$ (or in terms of $\rho\bar{v}^2 A$), the momentum correction factor β can be employed as in Section 6.4. With this substitution, the momentum equation can be written

$$\sum \vec{F} = \frac{1}{g_c} \sum_{\text{out}} \beta\dot{m}(\vec{\bar{v}}) - \frac{1}{g_c} \sum_{\text{in}} \beta\dot{m}(\vec{\bar{v}}) \tag{6-48}$$

where the direction of the mean velocity vector $\vec{\bar{v}}$ is taken as the direction of the mean streamline or the centerline streamline of the quasi-one-dimensional duct flow. If the system contains only one flow path,

$$\sum \vec{F} = \frac{\dot{m}}{g_c}(\beta_2\vec{V}_2 - \beta_1\vec{V}_1) \tag{6-48a}$$

Case IIb

STEADY ONE-DIMENSIONAL FLOW

If the flow is truly one-dimensional, so that the density and velocity do not vary over the inflow and outflow stations of the system, (6-46) can be written

$$\sum \vec{F} = \frac{1}{g_c} \sum_{\substack{\text{OUTFLOW}\\ \text{AREAS}}} \dot{m}\vec{V} - \frac{1}{g_c} \sum_{\substack{\text{INFLOW}\\ \text{AREAS}}} \dot{m}\vec{V} \tag{6-49}$$

If there is only one flow path, $\dot{m}_{\text{out}} = \dot{m}_{\text{in}}$ or

$$\sum \vec{F} = \frac{\dot{m}}{g_c}(\vec{V}_2 - \vec{V}_1) \tag{6-49a}$$

where 1 and 2 refer to the inlet and outlet positions, respectively.

SPECIFICATION OF FORCES

The forces that can act on the fluid within the control volume are the same as those described in Chapter 5. However, surface pressure and shear forces, and a body-acting gravity force, will be the only forces considered in the following developments. The manner in which these forces are treated is best described by an example.

6.10 ILLUSTRATIVE EXAMPLE 6-1: Forces on an Elbow

It is desired to determine the forces on a 4:1 horizontal reducing elbow to find the strength of the anchoring fixture necessary to hold it in place under the given conditions of maximum liquid flow rate ($Q_{\max}$) and upstream pressure p_1. The elbow turns through an angle α and discharges into a tank whose side is perpendicular to the inlet plane as shown in Fig. 6-13. The exit area is then elliptical. The flow is steady and the liquid can be assumed to be incompressible. Since no further information is given, the flow would have to be assumed to be one-dimensional (or $\beta \approx 1$). If the length and diameter of the upstream pipe were known, the Reynolds number at 1 could be determined and the shape of the velocity

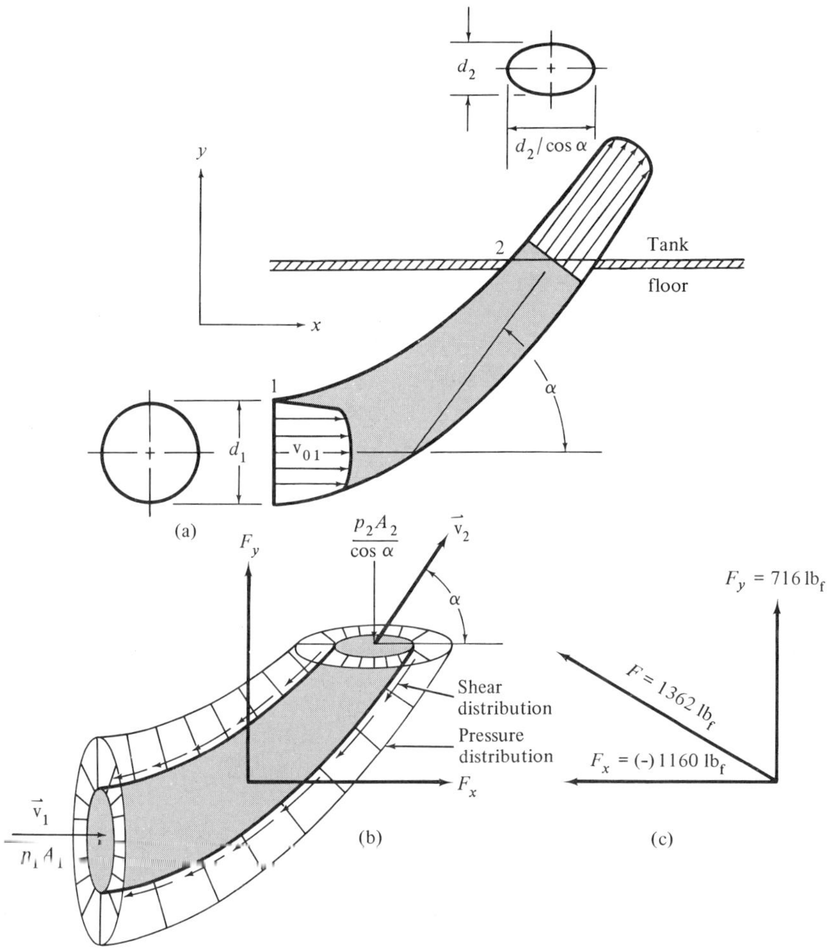

Fig. 6-13. Reducing elbow as an example of the application of the relationship between force and momentum. The pressure and shear distributions are shown in (b). The values of the net force components acting on the fluid in the control volume for this example are given in (c).

profile estimated. For a first approximation, however, assume one-dimensional flow, or $\beta_1 = \beta_2 = 1$. (Note that even if $\beta_2 = \beta_1$, the *effect* of the nonuniform distribution does not cancel.)

First, an x-y coordinate system is chosen in the plane of the elbow as indicated with one axis (preferably the x axis) in the direction of the inlet flow. The control volume is defined by the surface between the elbow walls and the fluid and the end areas at 1 and 2 as indicated in Fig. 6-13b. All forces that can combine to produce the change in momentum from v_1 to

v_2 must be considered. These forces are (1) the pressure forces at the end areas, whose difference tends to accelerate the magnitude of the velocity of the fluid from $\vec{v}_1$ to $\vec{v}_2$; (2) the distributed pressure forces exerted by the elbow walls on the fluid and producing a change in direction from 1 to 2; (3) the shear forces opposing the motion of the fluid; and (4) the force of gravity (or the weight), which will be neglected. The forces desired are the resultants of the distributions described in items 2 and 3. These forces might possibly be determined by solving for the three-dimensional flow pattern, calculating the pressure and shear fields, and integrating these over the elbow surface area. This procedure is nearly impossible. Moreover, by use of the momentum equation, the desired forces can be determined directly.

The two component equations for the control volume are

$$F_x = \frac{\dot{m}}{g_c}(v_{2x} - v_{1x})$$

$$F_y = \frac{\dot{m}}{g_c}(v_{2y} - v_{1y})$$

The sum of all forces in the x direction is just the pressure force p_1A_1 on the left end of the elbow and the unknown force F_x, which is the resultant of the pressure and shear forces exerted by the pipe on the fluid. F_x is then the x component of the distributed shear force acting on the fluid along the interface between the fluid and the pipe wall. Since both the magnitude and direction of this force are completely unknown (and it is generally not advisable to try to guess its direction), its direction is arbitrarily taken as *positive in the direction of the positive x axis.* If it is determined to be negative by the calculations, then it acts in a direction opposite to that originally assumed. (Note that it really does not make any difference which direction is originally chosen; choosing it positive in the direction of the positive x axis makes the bookkeeping easier, more accurate, and easier to interpret.)

The distributed pressure and shear forces that combine to give F_x and F_y are indicated on Fig. 6-13b. Since the velocity increases from 1 to 2, the pressure will decrease in this direction giving the pressure distribution of the form shown. Also the shear stress will increase from 1 to 2, because the velocity gradient at the wall becomes larger with the increasing mean velocity in the flow direction.

The given quantities are specifically p_1, A_1 (or d_1), and $A_2 = A_1/4$ perpendicular to the flow; the fluid (or fluid density ρ); the angle α; and the flow rate Q. It is desirable to try to determine the final results explicitly in terms of these quantities in order to minimize both calculation time and error. In the x direction,

$$F_x + p_1 A_1 = \frac{\dot{m}}{g_c}(\text{v}_2 \cos\alpha - \text{v}_1)$$

or

$$F_x = \frac{\dot{m}}{g_c}\text{v}_2 \cos\alpha - \left(\frac{\dot{m}}{g_c}\text{v}_1 + p_1 A_1\right)$$

F_x is completely determined by the given information: $\dot{m} = \rho Q$ and $\text{v}_1 = Q/A_1$. The quantity $pA + \dot{m}\text{v}/g_c$ is sometimes termed the impulse function $\mathscr{F}$, as defined by the above equation [although some authors define the impulse function as $\mathscr{F}/A = p + (\dot{m}\text{v}/g_c A) = p + (\rho\text{v}^2/g_c)$]. If the plane of the areas at both stations were perpendicular to the flow directions, it can be seen that the result $F_x = \mathscr{F}_2 - \mathscr{F}_1$ would be obtained.

In the y direction,

$$F_y - \frac{p_2 A_2}{\cos\alpha} = \frac{\dot{m}}{g_c}\text{v}_2$$

The only quantity that cannot be determined directly in terms of the given information is p_2. However, another form of the momentum equation is available from (6-22c):

$$\left(\frac{p}{\rho} + \frac{\text{v}^2}{2g_c} + \frac{gz}{g_c}\right)_1 - \frac{\overline{4\tau L}}{\rho D_H} = \left(\frac{p}{\rho} + \frac{\text{v}^2}{2g_c} + \frac{gz}{g_c}\right)_2$$

All quantities except p_2 and those in the friction term are known from given information. The friction term can be estimated if the surface roughness condition of the elbow is known and an average Reynolds number is determined. Since sufficient background is not yet available for such an estimate, the shear force will be assumed to be negligible with respect to the pressure forces or the inertia at this time. (The shear force term is usually significant, that is, $\sim$1 percent, for fairly long lengths of pipe, for example, $L/D > 10$.) Neglecting shear forces and assuming β_1 and β_2 to be nearly unity (or if ideal flow is assumed for a first approximation, $\tau = 0$ and $\beta_1 = \beta_2 = 1$),

$$p_2 = p_1 + \frac{\rho}{2g_c}(\text{v}_1^2 - \text{v}_2^2) = p_1 + \frac{\rho Q^2}{2g_c}\left(\frac{1}{A_1^2} - \frac{1}{A_2^2}\right)$$

$$= p_1 + \frac{\rho Q^2}{2g_c A_1^2}\left[1 - \left(\frac{A_1}{A_2}\right)^2\right]$$

$$F_y = \frac{A_2/A_1}{\cos\alpha}\left\{p_1 A_1 - \frac{\rho Q^2}{2g_c A_1}\left[\left(\frac{A_1}{A_2}\right)^2 - 1\right]\right\} + \frac{\rho Q^2}{2g_c A_1}\frac{A_1}{A_2}$$

in terms of given information.

As a numerical example, for $p_1 = 10$ psi (1440 psf), $Q = 4000$ gpm (8.92 cfs) of water ($\rho = 62.4\ \text{lb}_\text{m}/\text{ft}^3$), $A_1 = 1\ \text{ft}^2$, and $\alpha = 45°$, $F_x = -1160\ \text{lb}_\text{f}$, and $F_y = 716\ \text{lb}_\text{f}$. (As a side calculation, $p_2 = 2.0$ psi gage, or 16.7 psia.)

It is now of interest to inspect the effects that produce the magnitudes and directions of the force components. The x-force component F_x results from the pressure and shear-force distributions on the fluid as shown. The x component of the pressure-force distribution is negative, since the curved flow produces a pressure gradient in the direction of the principal normal, or toward the inside wall. This force is the x component of the turning force required to change the direction of the flow and push it around the corner. The shear-force component was neglected. F_x is then expected to be in the negative direction, as was obtained. A similar analysis indicates that F_y would be expected to be positive. The resultant reaction produced *by* the fluid *on* the elbow wall is equal and opposite to $\vec{F}$. The force that the mounting would have to exert to hold the elbow in place would in turn be equal and opposite to the reaction, or identical to $\vec{F}$. If p_1, p_2, v_1, and the velocity distribution functions were measured or known, the direct application of either (6-48a) or (6-49a) (as applicable) would yield F_x and F_y, the net pressure distribution force plus the shear distribution force components on the element. Note that in the above example the shear was assumed to be zero because only p_1 was given, making it necessary to calculate p_2. This calculation of p_2 was facilitated by assuming $\tau = 0$.

6.11 ILLUSTRATIVE EXAMPLE 6-2: Static Thrust Stand Firing of a Rocket

A rocket with its propellant supply tanks, lines, and turbopump fuel-pressurization system is illustrated in Fig. 6-14a. A simplified line sketch of the essential flow paths is shown in Fig. 6-14b. The fuel and liquid-oxygen tank pressures ($p_{f,T}$ and $p_{\text{ox},T}$) are the vapor pressures as determined by the respective temperatures.

The fuel is pumped to a pressure greater than the chamber pressure by an amount equal to the cooling-jacket pressure drop. The energy transferred to the fuel in this *regenerative cooling* scheme is available for net expansion work in the nozzle. The oxidant is not used for cooling because of severe oxidizing problems, particularly at high engine temperatures. The pumps also provide the pressure for a high injection kinetic energy of the propellant streams which are distributed in many fine spray streams for good mixing and rapid combustion in the chamber.

The turbine may operate on combustion gases generated from the main propellants, from a separate solid- or liquid-fuel gas generator, or on bleed

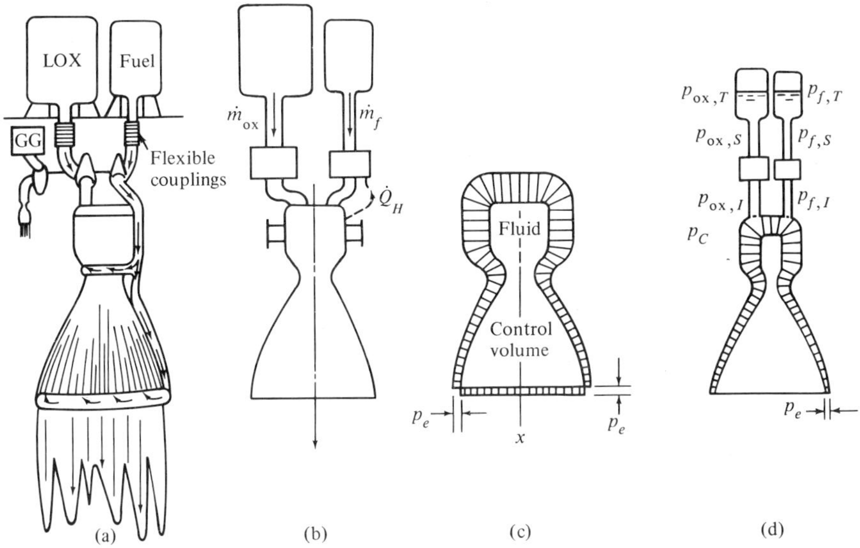

Fig. 6-14. (a) Sketch of a rocket engine mounted on a thrust stand showing details of propellant (fuel and oxidant) system; (b) sketch showing only basic elements for force–momentum analysis; (c) fluid control volume; (d) engine surfaces on which pressure distribution forces act to produce net thrust force. Local ambient pressure acts over entire external surface. This resultant ambient pressure force is just $+p_aA_e$.

gas from the main combustion chamber. The turbine discharge is exhausted through a small thrust nozzle to maximize overall thrust efficiency. The dashed line across the chamber inlets of Fig. 6-14d represents the injector nozzles, across which there is a pressure drop $p_I - p_c$. It is left as an exercise to show that this pressure drop does not enter into the determination of the thrust force. In general, all the internal pressure drops in the propellant supply systems are balanced.

The pressure forces acting on the fluid control volume are shown in Fig. 6-14c. The x component of the force–momentum relation for steady-state operation applied to the fluid control volume of Fig. 6-14c is proportional to the momentum flux out of the engine,

$$\underset{\substack{\text{ON FLUID}\\ \text{IN C.V.}}}{\sum F_x} \equiv \underset{\substack{\text{PRESSURE}\\ \text{EXERTED}\\ \text{BY WALLS}\\ \text{ON FLUID}}}{F} - p_eA_e = \frac{\dot{m}\mathrm{v}_e}{g_c}$$

where the subscript e refers to conditions at the nozzle exit plane. The net thrust *on the engine* is produced partially by the pressure force exerted by the fluid on the engine walls, as shown in Fig. 6-14d. This pressure force is equal and opposite to the pressure distribution force on the fluid control volume by the walls. The net unbalanced pressure acting on the engine fuel supply system is just the chamber pressure p_c, so the reaction to the propellant supply pressure force on the fluid is identically balanced by a pressure force from the walls and other parts connected to the engine (pumps, piping, and so on). This statement would also be true for any type of propellant feed system (for example, pressurized tanks). The ambient pressure p_a also exerts an additional pressure force *on the engine* (and its associated parts). This pressure force acts in the $+x$ direction and is just the ambient pressure times the net x-projected area, which is the nozzle exit area A_e. The forces acting on the engine in the $+x$ direction are then

$$F_{\substack{\text{NET ON}\\\text{WALLS OF}\\\text{ENGINE}}} = F_{\substack{\text{PRESSURE}\\\text{EXERTED}\\\text{BY WALLS}\\\text{ON FLUID}}} + p_a A_e$$

The *thrust force F_T on the engine* is defined as the net force on the engine in the $-x$ direction, or

$$F_T = -F_{\substack{\text{PRESSURE}\\\text{EXERTED}\\\text{BY FLUID}\\\text{ON WALLS}}} - p_a A_e = F_{\substack{\text{PRESSURE}\\\text{EXERTED}\\\text{BY WALLS}\\\text{ON FLUID}}} - p_a A_e$$

Substituting the pressure-force term from the above momentum equation into this defining relation for the thrust gives

$$F_T = \frac{\dot{m}\mathrm{v}_e}{g_c} + (p_e - p_a)A_e \tag{6-50}$$

Note that in the case of subsonic flow, it would not be possible for p_e to be different from p_a (unless there were considerable curvature of the streamlines at the nozzle exit); however, because of the characteristics of supersonic flow (to be considered in detail in Chapter 12), it is possible for this pressure difference to exist. The exhaust pressure, p_e, can be determined as a function of the nozzle exit design Mach number (or exit area/throat area ratio) and the chamber pressure; that is,

$$\frac{p_e}{p_c} = f_1(M_e) \qquad \text{or} \qquad \frac{p_e}{p_c} = f_2\frac{A_e}{A_t}$$

where p_c is the chamber pressure, A_t is the throat area, and M_e is the exit Mach number.

6.12 ILLUSTRATIVE EXAMPLE 6-3: Orifice Meter and Flow Nozzle

A simple, inexpensive flow-rate meter can be constructed by putting a constricting orifice in a duct or pipe line as illustrated in Fig. 6-15. As with the Venturi meter (Example 5-2), the velocity change through the orifice produces a pressure differential. The continuity, energy, and momentum relations can be applied to the flow field between the two points, where the pressures are measured to yield a relationship for the velocity and mass flow rate (or volume flow rate) as functions of the

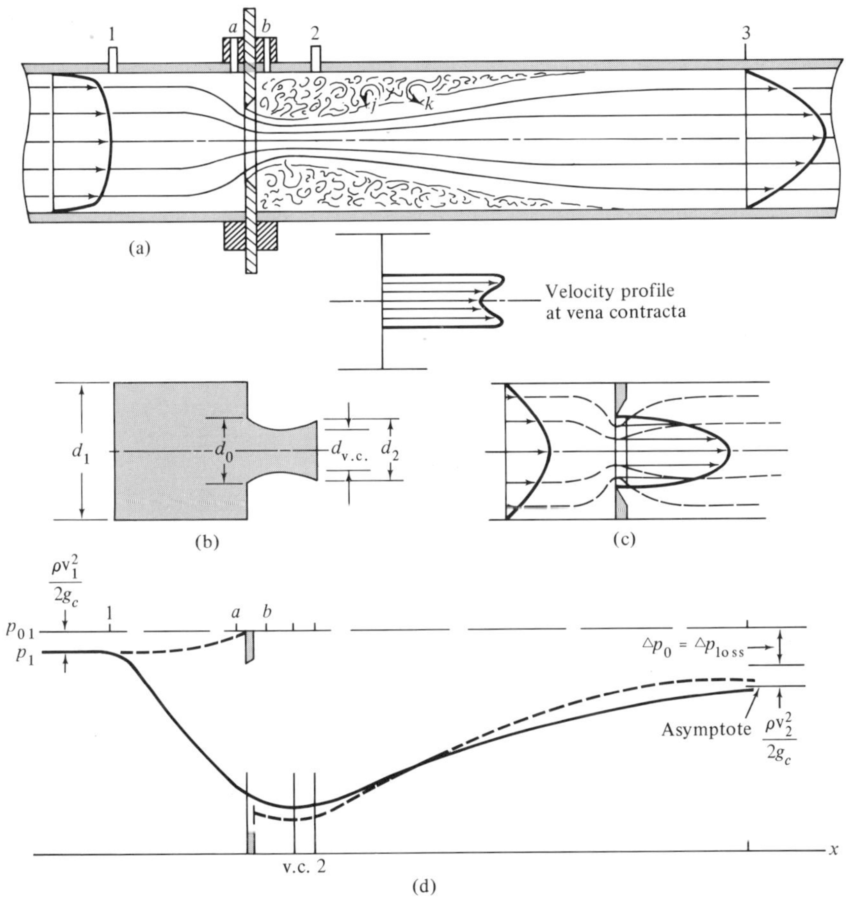

Fig. 6-15. Schematic representation of flow through an orifice: (a) section through flow showing streamlines and velocity profiles; (b) control volume; (c) diagram of streamlines and velocity profiles for low Reynolds number (on the order of 1); (d) pressure distributions for flow shown in part (a).

pressure differential, fluid properties, geometry, and certain calibration coefficients. These coefficients will be shown to be functions of the geometry and Reynolds number for incompressible flows. Measurement of quantities in compressible flows will be considered in Chapter 12.

The mass flow rate across any area perpendicular to the flow of Fig. 6-15 is $\dot{m} = \rho A \text{v}$ for steady flow. Because the orifice plate is the most accurately specified and machined element of the meter and because its diameter is stamped on a tab on the plate, the $\dot{m}$ (or Q, volume flow rate) is usually determined with respect to the flow through this area; that is, $\dot{m} = \rho A_o \text{v}_o$ (where the subscript o refers to quantities at the orifice diameter). With $\rho =$ const. for incompressible flow, the fact that the mass flow rate is constant gives the useful relation

$$A_1\text{v}_1 = A_o\text{v}_o = (A\text{v})_{\text{v.c.}} = A_2\text{v}_2 = \iint \vec{\text{v}}\cdot d\vec{A} = \text{const.}$$

where the subscript v.c. refers to the values at the minimum flow area or *vena contracta* section. This vena contracta occurs a short distance downstream of the orifice and is formed as a consequence of the inward momentum component of the flow as it enters the orifice.

The fluid control volume is shown in Fig. 6-15b. It will first be assumed that the flow in this control volume is one-dimensional, and therefore irrotational. This assumption appears invalid considering the requirements for one-dimensional and quasi-one-dimensional flow as stated in Chapters 4 and 5 and this chapter. The consequences of this assumption will be evaluated when considering application of the final results.

Note that the control volume includes only the main portion of the flow downstream of the orifice. The flow field surrounding this main through-flow jet is a random, turbulent eddy zone in which the time-average velocity $\langle \text{v} \rangle$ in any direction, defined as

$$\langle \text{v} \rangle \equiv \frac{1}{t_2 - t_1}\int_{t_1}^{t_2} \text{v}\, dt$$

is zero. The time period $t_2 - t_1$ must necessarily be taken as much larger than the characteristic period of the eddy turbulence in this wake zone.[20]

Because of the difficulty in determining the pressure forces on the plate and in this separated or "free" region of the jet, application of the momentum equation would prove difficult and would pose more problems than could be solved.

In addition to the assumptions of steady, incompressible flow, it will also be assumed that the flow is adiabatic and one-dimensional. Under all

[20] The mechanism of turbulent flow will be discussed in Chapter 10.

these assumptions, the energy equation reduces to Bernoulli's equation, (5-25). If the centerline is horizontal,

$$\frac{v_1^2}{2g_c} + \frac{p_1}{\rho} = \frac{v_2^2}{2g_c} + \frac{p_2}{\rho}$$

where v_2 is the velocity in the through-flow stream at the section where p_2 is measured.

The position and area ($A_{v.c.}$) of the vena contracta will vary as a function of Reynolds number and d_0/d_1 as described below, but the vena contracta will be located somewhere in the neighborhood of 2. This variation will be assumed to be negligible, giving $A_{v.c.} \cong A_2$ and $v_{v.c.} \cong v_2$. All velocities indicated are assumed to be average velocities (according to the assumption of one-dimensional flow). From the continuity equation and above assumptions,

$$v_2 = v_{v.c.} = \frac{A_0}{A_{v.c.}} v_0 \cong \frac{A_0}{A_2} v_0 = \frac{v_0}{C_c}$$

The area ratio of the vena contracta to the area of the orifice $A_{v.c.}/A_o$ is defined as the *contraction ratio* C_c, which will be shown to be a function of the Reynolds number. Rearranging Bernoulli's equation gives

$$v_2^2 - v_1^2 = v_2^2\left[1 - \left(\frac{v_1}{v_2}\right)^2\right] = 2g_c\frac{p_1 - p_2}{\rho}$$

or, in terms of v_0 and C_o,

$$v_0' = \frac{C_c}{\sqrt{1 - C_c^2(A_0/A_1)^2}}\sqrt{2g_c\frac{p_1 - p_2}{\rho}} \qquad (6\text{-}51)$$

The prime denotes v_0 as an *ideal* value of the velocity determined under all the above restrictions: (1) one-dimensional, (2) frictionless, (3) adiabatic, (4) incompressible, (5) negligible gravity or body forces, and (6) the vena contracta located at 2.

The actual velocity for steady, adiabatic incompressible flow can be written

$$v_0 = C_v v_0'$$

where C_v is a *velocity correction coefficient* to account for assumptions 1, 2, and 6. The effects of the invalidity of these assumptions are a function of the Reynolds number N_R, and the area ratio or diameter ratio d_0/d_1 so that C_v (and likewise C_c) are functions of these parameters only. With this definition of C_v, the mass flow rate can be determined as

$$\dot{m} = \rho A_0 \mathrm{v}_0 = \frac{\rho C_c C_v A_0}{\sqrt{1 - C_c^2 (A_0/A_1)^2}} \sqrt{2g_c \frac{p_1 - p_2}{\rho}}$$

or

$$\dot{m} = \frac{\rho C_d A_0}{\sqrt{1 - C_c^2 (A_0/A_1)^2}} \sqrt{2g_c \frac{\Delta p}{\rho}} \tag{6-52}$$

Since C_c and C_v are both functions of the same parameters, their product is lumped into a single *discharge coefficient* C_d for orifice meters. The radical in the denominator is also a function only of the area ratio (or diameter ratio) and Reynolds number. The reciprocal of this factor is defined as the velocity of approach factor, C_a. The product $C_d C_a$ is sometimes used as an overall discharge coefficient designated by C (no subscript), so the mass flow becomes

$$m = \rho C A_0 \sqrt{2g_c \frac{\Delta p}{\rho}} \tag{6-53}$$

Representative values of discharge coefficient C are given as a function of Reynolds number for various diameter ratios in Fig. 6-16. It must be

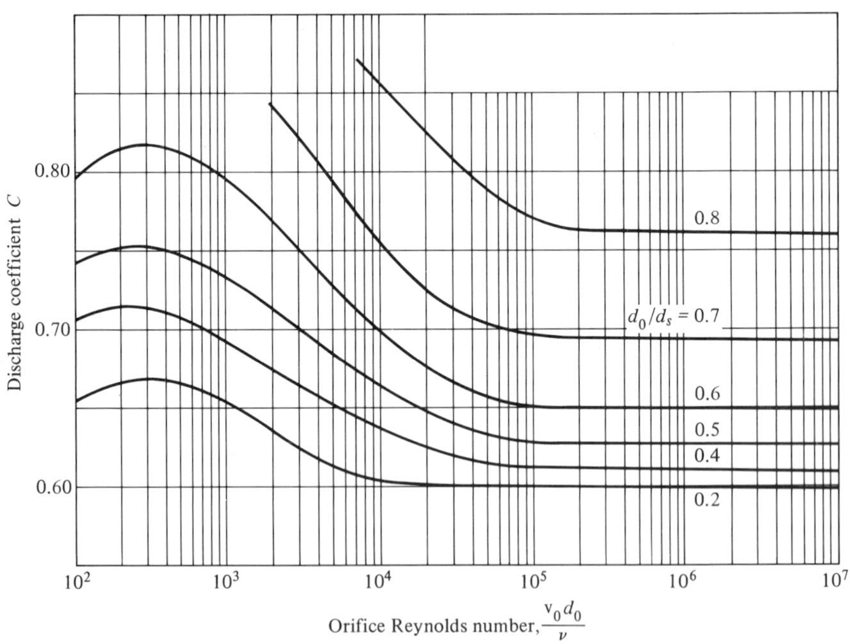

Fig. 6-16. Overall discharge coefficients for orifice meters as a function of orifice Reynolds number and diameter ratio.

emphasized that such plots are accurate and usable only under the specific conditions of overall similarity of geometry (shape and construction of orifice plate, placement of pressure taps, and so on).

There are several possible combinations for tap locations. The most common combinations are *1 pipe diameter upstream* and *½ diameter downstream* of the orifice plate, or *flange taps* nominally placed 1 in. upstream and 1 in. downstream of the orifice. A *vena contracta tap* placement (1 pipe diameter upstream and at the minimum pressure location downstream) is another possibility. Since the position of the vena contracta varies with Reynolds number, the downstream location is somewhat arbitrary if the meter is to be used for different flow rates. From Fig. 6-15, the maximum pressure difference occurs with the upstream tap at the flange and the downstream tap at the vena contracta. Further descriptions of orifice geometries, tap locations, and calibration data for several combinations are given in ASME, *Fluid Meters, Their Theory and Application*, 1959 (and in *NACA TM 962*). Representative calibration data for a square-edged orifice are given in Fig. 6-16.

The Reynolds number of the flow field illustrated in Fig. 6-15a is relatively large, implying that the inertia forces are large with respect to the viscous forces. For the flow to pass smoothly around the plate edges as shown in Fig. 6-15c would require that the available pressure forces (or pressure gradient) in a direction perpendicular to the flow be large with respect to the inertia forces along the streamlines immediately adjacent to the plate. To determine the behavior of the normal pressure gradient required to push the flow around the corner, the relations between the streamlines and normal gradients can be examined. Assuming steady flow and that the gravity and friction forces are negligible, the equations for the streamlines and normal pressure gradients (6-28) are given by

$$\frac{\partial p}{\partial s} = -\frac{\rho}{2g_c}\frac{\partial \mathrm{v}^2}{\partial s}$$

$$\frac{\partial p}{\partial n} = -\frac{\rho}{g_c}\frac{\mathrm{v}^2}{R_s}$$

Over a short distance along s, the first of these equations can be approximated as

$$(\Delta p)_s \propto -\Delta(\mathrm{v}^2) \qquad \text{or} \qquad p_s \propto -\mathrm{v}^2$$

Similarly, the second equation gives

$$\frac{(\Delta p)_n}{\Delta n} \propto -\frac{\mathrm{v}^2}{R_s} \propto \frac{p_s}{R_s}$$

As the inertia force increases, p_s or $(\Delta p)_s$ decreases, the available normal pressure gradient decreases. This result can also be regarded as specifying the minimum R_s that a flow streamline can navigate. If the R_s of the wall boundary geometry is smaller than this limiting R_s, the flow can no longer remain attached to the walls, so it *separates*, as shown.

The Reynolds-number dependence can be explained only by considering the friction forces and their effects. At the very low Reynolds numbers the flow is similar to that depicted in Fig. 6-15c, where the wall friction slows the fluid layers to produce the velocity profiles shown. The turbulent mixing in turbulent flow produces a more uniform initial velocity distribution as shown in Fig. 6-15a. The field curvature from 1 to the orifice accelerates the curved streamlines more than the center streamline producing profiles as indicated at the orifice or vena contracta. The velocity in the region of the orifice lip (or maximum turning region) is then greater for turbulent flow than for viscous flow. The velocities in the outer streamlines of the viscous flow are significantly less than the centerline velocity. Consequently, the limiting radius of curvature is smaller for the viscous flow. The transition from laminar (viscous) flow to turbulent flow takes place as the Reynolds number increases, proceeding from the one illustrated in Fig. 6-15c to that in Fig. 6-15a. The vena contracta also becomes smaller and moves farther downstream as the Reynolds number increases. The effects described above also depend upon the ratio d_0/d_1. As $d_0 \to d_1$, the gradients decrease so that the flow separates less readily, that is, at a higher Reynolds number. At some large value of Reynolds number (about 10^7) and for d_0/d_1 about 0.55, the flow-field geometry changes only a negligible amount for any further increase in Reynolds number above this *upper critical Reynolds number*.

The pressure distributions along the centerline and along the wall are shown in Fig. 6-15d. The flow stagnates in the corner near a, so the pressure there approaches the upstream stagnation pressure $(p \to p_{01})$. The center streamline velocity is accelerated as the flow approaches the orifice; consequently, the pressure decreases along this streamline to the point of maximum velocity at the vena contracta. In the neighborhood of the orifice, the curvature of the streamline bounding the jet produces a greater velocity and therefore a lower pressure at this boundary than at the centerline. With no mean directed motion (kinetic energy or momentum) in the wake region behind the plate, the pressure measured at the wall is essentially the pressure at the jet boundary. Downstream of the vena contracta, the jet boundary streamline has an inflection point, where the curvature becomes opposite to that in the neighborhood of the vena contracta. The effects are then reversed, and the boundary streamline decelerates more than the center streamline producing the velocity distribution shown at 3 in Fig. 6-15a. The pressure at the jet boundary is then greater than the centerline pressure, as indicated. As the flow progresses downstream, the normal pressure gradient disappears in the straight pipe flow and the velocity distributions revert to a fully developed profile similar to the one shown upstream.

A significant loss of availability occurs with an orifice meter, as indicated by the decrease in stagnation pressure p_0. The mechanism by which this "loss" occurs is interesting. If d_0/d_1 is ½, the ratio of the maximum kinetic energy at the vena contracta to the approach (and therefore the far-downstream) kinetic energy is about 20:1. The wake is then a region of zero mean motion bounded by a high-kinetic-energy turbulent stream. The shear boundary between the jet and wake region is unstable;[21] consequently, this boundary is randomly ragged, and there is a continuous exchange of fluid between the two regions as mass momentum and energy are exchanged by means of turbulent eddies. A small element of fluid with considerable kinetic energy will be transported from the main stream or jet boundary into the wake at *j*, which is due to the random fluctuations. The kinetic energy of this mass of fluid will then be converted ("dissipated") into internal energy that is unavailable for further use in an incompressible flow. As the element leaves the jet, it has a momentum component in the direction of the flow corresponding to its jet boundary velocity. This fluid element is brought to rest (zero mean velocity) by a pressure force acting opposite to the flow direction, or a positive pressure gradient exists to accomplish this result. This pressure increase accompanies that in the diffusing flow stream. The direct friction force that brings the fluid element to rest is equal and opposite to the pressure gradient. Simultaneously, elements such as the one at *k* are leaving the wake region and entering the main jet. The force required to accelerate such elements in the flow direction is supplied by a negative pressure gradient, which decreases the rate of pressure increase possible from divergence of the jet from $A_{\text{v.c.}}$ to A_3. Consequently, the pressure does not increase as much as is possible in an efficient diffuser (Chapter 7). The work required to accelerate these *k* elements to a velocity corresponding to the kinetic energy of the main stream must be supplied from this stream. Since $A_1 = A_3$ and $\text{v}_1 = \text{v}_3$, the loss of availability appears as a pressure decrease from 1 to 3. The above loss-mechanism description applies to any separated wake flow.

FLOW NOZZLE

Another device used for flow metering by measurement of a differential pressure is the flow nozzle schematically illustrated in Fig. 6-17. Several combinations of tap placement are again possible. The most commonly used is 1 pipe diameter upstream and ½ pipe diameter downstream. The

[21] The velocity profile from the wake into the jet centerline has two inflection points, which are regions of inherent instability in such a flow.

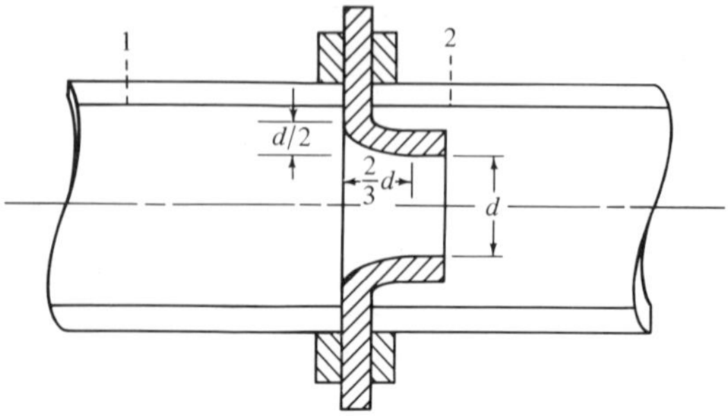

Fig. 6-17. Sketch of a flow nozzle. The ASME nozzle contour is a $\frac{2}{3}d$ major axis by $\frac{1}{2}d$ minor axis ellipse (d = nozzle discharge diameter).

standardized construction geometries and calibration data are given for flow nozzles in the ASME reference cited previously.

The Venturi is the most efficient of the three differential flow-rate meters (Venturi, orifice, and flow nozzle). The pressure loss for the Venturi is about 10 to 15 percent of the meter differential, 60 to 80 percent for the orifice, and 50 to 75 percent for the flow nozzle. (The exact loss depends primarily on the area ratio, Reynolds number, and construction details.)

6.13 GENERAL VECTOR EQUATION

A detailed evaluation of the forces in the equation of motion leads to a degree of complication usually considered beyond the scope of an introductory course. The following material is presented as an outline of the procedure by which the general equations for the arbitrary motion of a fluid are determined and used. Determination of the stress–strain relationship is outlined in Appendix B.

If the forces to be considered are limited to (1) conservative (or gravitational-type) body forces derivable from a potential, (2) pressure forces, and (3) shear forces, a complete vector equation (or set of component equations) can be derived for a control volume. The general case of a moving and deforming control volume will be considered first and then specialized to the case of a control volume fixed in a space.

Consider the control volume to be a deformable physical boundary (such as a balloon or deformable gas tank) moving through space with a motion determined by the net forces acting on it at any instant. The equations of motion can be written with respect to the fluid mass within this control volume at any instant. This general fluid control volume is represented in Fig. 6-12. The body force acting on any element of mass within this volume is $\overrightarrow{\Delta F_b}$, the pressure force on any surface area element is $\overrightarrow{\Delta F_p}$, and the shear force on any surface area element is $\overrightarrow{\Delta F_\tau}$, as shown.

The body force on a mass element $\rho\,\Delta V$ is $\vec{f}_b\rho\,\Delta V$, where $\vec{f}_b$ is the body force per unit mass, if the body forces are derivable from a potential, $\vec{f}_b = -\nabla\Omega$, where Ω is the body-force potential per unit mass of fluid. The total body force is

$$\oiiint \vec{f}_b\rho\,dV = -\oiiint \nabla\Omega\rho\,dV$$

where $\vec{f}_b$ or Ω and ρ are generally functions of both time and absolute space position with respect to a space-fixed coordinate system.

The surface forces are most directly represented by the mathematical form of notation that specifically indicates their nature, that is, the surface stress tensor. This surface stress tensor will in general contain the normal pressure stress p, which was shown to be simply a scalar.[22] It is therefore an advantage to remove the pressure from the complete stress tensor and denote the remaining tensor by τ, which now represents only the viscous shear stresses. The pressure force on the control volume at any instant is then

$$\oiint -p\,d\vec{A}$$

where p and the control surface ($A_{\text{C.V.}}$) are functions of time and position. The viscous shear stress can be shown to be represented by

$$\oiint d\vec{A}\cdot\tau$$

where the dyadic dot product of a tensor and a vector is generally not commutative; it is commutative where the tensor is symmetric. As used here, τ is the stress tensor written in dyadic form (see Appendix B). Since the dyadic form of stress tensor behaves as a second-order vector, the dot product with a vector will yield a vector. The sum of all external forces acting on the control volume at any instant is then

$$\sum \bar{F} = \oiiint \vec{f}_b\rho\,dV - \oiint p\,d\vec{A} + \oiint \tau\cdot d\vec{A}$$

The *internal* pressure and shear forces on adjacent mass elements occur in equal and opposite pairs, and therefore cancel. The integral momentum

[22] As a spherical tensor, the pressure has all its mutually perpendicular or diagonal components equal; therefore, there is no preferred orientation or axis to this tensor.

equation can then be expressed as

$$\oiiint \vec{f}_b \rho \, dV - \oiint p \, d\vec{A} + \oiint \tau \cdot d\vec{A}$$

$$= \frac{1}{g_c} \frac{\partial}{\partial t} \oiiint \vec{v} \rho \, dV + \frac{1}{g_c} \oiint \vec{v}(\rho \vec{v} \cdot d\vec{A}) \quad (6\text{-}54)$$

If the physical system is such that the control volume can be chosen as fixed in space (or as the control volume in an inertial system), the only change in (6-54) is that the partial time derivative can be placed under the integral sign in the temporal acceleration term. The limits of integration, C.V. and $A_{\text{C.V.}}$, are then fixed in space (or move with the uniform motion of the inertial system) and are no longer functions of time.

The difficulty in further expanding or using (6-54) arises first in determining the viscous shear stress field. It has already been established that the stress is proportional to the strain rate; therefore, the viscous shear stress tensor field is proportional to the strain rate tensor. The strain rate tensor involving the shear deformation rate (or angular displacement rate) and dilatation deformation rate ("divergence") is therefore a function of the velocity derivatives. The details of the mathematical manipulations involved and the determination of the relation between shear stress and strain rate are usually left to a more advanced course.

If the control volume can be considered to be fixed in space, (6-54) becomes

$$\frac{1}{g_c} \oiiint \frac{\partial \rho \vec{v}}{\partial t} \, dV + \oiint \vec{v}(\rho \vec{v} \cdot d\vec{A}) + \oiiint \nabla \Omega \rho \, dV + \oiint p \, d\vec{A} - \oiint \tau \cdot d\vec{A} = 0 \quad (6\text{-}55)$$

which contains volume and area integral terms with integrands in the form of scalars, mixed vector (or dyadic) products, and tensors. The scalar (zero-order tensor) and vector (first-order tensor) terms have already been handled by vector integral transformations such as the Gauss divergence theorem. This transformation theorem can also be applied to dyadic (second-order tensor) terms, yielding an equation with all volume integral terms:

$$\oiiint \left\{ \frac{1}{g_c} \frac{\partial \rho \vec{v}}{\partial t} + \frac{1}{g_c} \operatorname{div} [(\vec{v})(\rho \vec{v})] + \rho \nabla \Omega + \nabla p - \operatorname{div} \tau \right\} dV = 0 \quad (6\text{-}55a)$$

from which the vector differential equation

$$\frac{1}{g_c} \frac{\partial \rho \vec{v}}{\partial t} + \operatorname{div} [(\vec{v})(\rho \vec{v})] + \rho \nabla \Omega + \nabla p - \operatorname{div} \tau = 0 \quad (6\text{-}56)$$

can be deduced. The combination of parentheses and brackets in the div $[(\vec{v})(\rho\vec{v})]$ indicates the order in which the vector differentiation (divergence) is to be accomplished. The divergence of this product can be expanded in the same manner as for the derivative of a product. Note that this expansion must yield vector quantities. The first two terms of (6-56) can be expanded as follows:

$$\rho \frac{\partial \vec{v}}{\partial t} + \vec{v}\frac{\partial \rho}{\partial t} + \vec{v}[\text{div}\,(\rho\vec{v})] + \rho\vec{v}\cdot\text{div}\,\vec{v}$$

The two center terms are $\vec{v}$ multiplied by the continuity equation (4-19a) and therefore drop, leaving

$$\frac{\rho}{g_c}\left(\frac{\partial \vec{v}}{\partial t} + \vec{v}\cdot\nabla\vec{v}\right) + \rho\,\nabla\Omega + \nabla p - \text{div}\,\tau = 0 \tag{6-57a}$$

which can be divided by ρ/g_c to give

$$\frac{\partial \vec{v}}{\partial t} + \vec{v}\cdot\nabla\vec{v} + g_c\,\nabla\Omega + \frac{g_c\,\nabla p}{\rho} = \frac{g_c\,\nabla\cdot\tau}{\rho} \tag{6-57b}$$

Equation (6-57) must be essentially the same as the vector equation (6-30) or (6-35) derived by considering the flow in streamline coordinates. The shear term $\hat{s}(\partial\tau/\partial n)$ in Section 6.3 is not the complete result that would be obtained from $\nabla\cdot\tau(s, n, c)$, but only its major component for flows where the radius of curvature is large compared to the boundary dimensions.

The scalar (dot) product of the vector velocity $\vec{v}$ with this force–momentum equation will give an equation in work rate and energy rate. The resulting equation can be combined with an energy differential equation deduced from an integral approach (as above, or developed for a differential element) with the stress work expressed using the stress tensor. The combination of these equations then produces an equation similar to (6-23a) for the rate of energy "dissipation" or transformation to an unavailable form. A dissipation function analogous to the $\Delta p_{\tau}/\rho$ quantity is obtained. Although greatly simplified by the vector and tensor notation, the details are somewhat tedious and the results are difficult to use in general cases because of problems involved in obtaining enough information to determine the stress (or strain rate) tensor completely. For most physical cases, it is easier and more direct to use the vector equation derived in (6-57), where the shear stress is left to be determined by the particular conditions of the physical problem and certain simplifying approximations are possible.

Equation (6-57) can be applied to either laminar or turbulent flow. If the flow is turbulent, the velocity is space-variant and random in time.

However, the turbulence can be treated in either of two ways. The velocity can be considered as the mean directed flow velocity with the random dissipative turbulent motion considered as an inertial viscous dissipation produced by the eddy momentum transport. The second or microscopic interpretation is to consider the velocity as the instantaneous combined mean flow plus random turbulent motion (Fig. 4-7) with shear stress as the actual instantaneous viscous shear produced in the fluid. The viscous dissipation is then evaluated from the random eddy motion as eddies are continually formed and dissipated. Because of the mathematical complexity in formulating the random motion, the second method is not used. Chapter 10 will treat turbulence formulation by the first method.

STRESS–STRAIN RATE RELATIONSHIPS

The tangential viscous shear stress τ_{ij} on an element of a homogeneous, isotropic fluid obeying a simple linear (Stokesian) stress–strain rate relationship is

$$\tau_{ij} = 2\mu\dot{\gamma}_{ij}$$

The definition of the shear stress components on a fluid element depends on the orientation of the element (or on the coordinate system orientation chosen). Consider a fluid element under the influence of only normal stresses, as shown in Fig. 6-18. These normal stresses are indicated as viscous stresses, even though they are not tangent to the element as it is shown. If *the same stress system* were considered for a fluid element defined in another coordinate system diagonal to (or rotated with respect to) the original *s-n* system indicated, such as the coordinate system oriented along the sides of the dashed diagonal element, the original τ_{nn} and τ_{ss} systems can be resolved into the diagonal, or pure shear stress system indicated in Fig. 6-18b. (With no loss of generality, the sides Δs, Δn, and Δc have all been taken as equal in length for simplicity.) It is of interest to note that a tangential shear stress is produced only if $\tau_{ss} \neq \tau_{nn}$. The ability of the τ_{ss}, τ_{nn} system to produce a shear strain can be seen by considering the deformation of the original element as indicated in Fig. 6-18c, where an angular deformation of the diagonally oriented element is observed. Noting that a tangential shear stress can be produced only if $\tau_{ss} \neq \tau_{nn}$, this shear stress is $\tau_d = -\mu\dot{\varphi} = 2\mu\dot{\gamma} = \tau_{ss} - \tau_{nn}$. The deformation can be considered to be made up of the shear deformation without relative volume change, plus the relative volume change (dilatation) as indicated in Fig. 6-18c. The angular or shear deformation is a shear in the diagonal system giving $\tau_{ss} - \tau_{nn} = 2\mu\dot{\gamma}$ in the *s-n* coordinate system. The interaction of the relative molecular motion with the flow velocity gradients (as described in Section 2.4) was shown to produce the viscous stress in a gas. Relative field velocities also exist in a purely dilatational flow, so a viscous

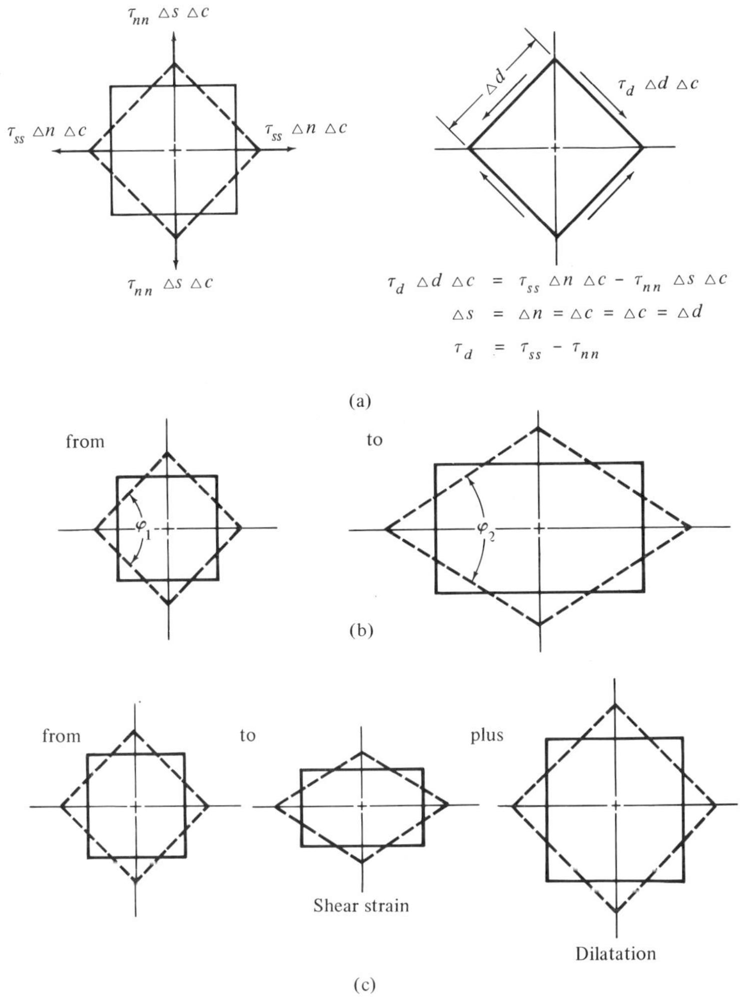

Fig. 6-18. Production of shear stress by a system of unequal normal stresses. The stress system (a) produces a deformation (b), which can be considered equivalent to (c).

molecular momentum action will result. In addition to this simple molecular viscous action, a second type of nonequilibrium (in the thermodynamic sense) normal or dilatational stress is produced, which is η div $\vec{v}$, where η is the second viscosity coefficient. This type of stress is significant only where the product η div $\vec{v}$ is significant. η is a function of the intermolecular force fields and therefore is small for gases (in states removed from the critical point). Div $\vec{v}$ may be relatively large for rapidly diverging fields (blast and shock waves), but the product is usually small for most flow fields, where pressure and inertia variations are not negligible. The damping of acoustical motions is the exceptional case where these terms are negligible, or are on the order of $\eta \operatorname{div} \vec{v}$. In liquids, the intermolecular force fields are large and η is significant (with respect to μ); however, $\operatorname{div} \vec{v}$ is not usually large for motions of liquid flow fields. The propagation and attenuation of sound or acoustic waves in liquids is a field of interest and importance, however, so this second viscosity coefficient and its behavior are worthy of further study in the area of the behavior of materials. Since the term is generally of little significance in the study of flow fields with appreciable mean directed motion, the product $\eta \operatorname{div} \vec{v}$ is usually neglected.

A general development of the strain rate expression and the stress–strain rate relation is best carried out using tensor methods. Attempted simplified geometric and trigonometric developments involve too many unconvincing arguments, assumptions, and difficult to prove hypotheses. The development using tensors allows a more direct physical interpretation of the mathematical statement and procedures. For a simple linear stress–strain rate (constant viscosity), this development yields

$$
\begin{aligned}
\tau_{ss} &= 2\mu \frac{\partial \mathrm{v}}{\partial s} + (\eta - \tfrac{2}{3}\mu) \operatorname{div} \vec{\mathrm{v}} \\
\tau_{nn} &= 2\mu \frac{\mathrm{v}}{R_s} + (\eta - \tfrac{2}{3}\mu) \operatorname{div} \vec{\mathrm{v}} \\
\tau_{cc} &= (\eta - \tfrac{2}{3}\mu) \operatorname{div} \vec{\mathrm{v}} \\
\tau_{sn} &= \tau_{ns} = \mu\left(\frac{\partial \mathrm{v}}{\partial n} + \frac{\mathrm{v}}{R_s}\right)
\end{aligned}
\tag{6-58}
$$

in the natural coordinate system. For a cartesian system,

$$
\begin{aligned}
\tau_{xx} &= 2\mu \frac{\partial u}{\partial x} + (\eta - \tfrac{2}{3}\mu) \operatorname{div} \vec{\mathrm{v}} \\
\tau_{yy} &= 2\mu \frac{\partial \mathrm{v}}{\partial y} + (\eta - \tfrac{2}{3}\mu) \operatorname{div} \vec{\mathrm{v}} \\
\tau_{zz} &= 2\mu \frac{\partial w}{\partial z} + (\eta - \tfrac{2}{3}\mu) \operatorname{div} \vec{\mathrm{v}} \\
\tau_{xy} &= \mu\left(\frac{\partial u}{\partial y} + \frac{\partial v}{\partial x}\right) \\
\tau_{yz} &= \mu\left(\frac{\partial v}{\partial z} + \frac{\partial w}{\partial y}\right) \\
\tau_{zx} &= \mu\left(\frac{\partial w}{\partial x} + \frac{\partial u}{\partial z}\right)
\end{aligned}
\tag{6-58a}
$$

It is interesting to note that the sum of the normal viscous shear stress components gives[23]

$$\tau_{ii} = 3\eta \operatorname{div} \vec{v}$$

which contains no μ term. A mean, normal viscous shear stress is defined as

$$\bar{\tau} \equiv \frac{\tau_{ii}}{3}$$

The *total* normal stress is then

$$\sigma_{ii} = -p + \tau_{ii}$$

If the second viscous effect is negligible (that is, $\eta \operatorname{div} \vec{v} = 0$), the mean normal stress is

$$\bar{\sigma} \equiv \frac{\sigma_{ii}}{3} = -p$$

or just the thermostatic pressure used in the developments presented at the beginning of the chapter.

The stress components in any general orthogonal coordinate system can be written

$$\tau_{ij} = 2\mu\dot{\gamma}_{ij} + \delta_{ij}(\eta - \tfrac{2}{3}\mu) \operatorname{div} \vec{v} \tag{6-59}$$

The equation of motion in vector form is then

$$\frac{\vec{a}}{g_c} = -\frac{\nabla p}{\rho} + \vec{f}_b + \frac{1}{\rho} \operatorname{div} \tau \tag{6-57c}$$

where τ is the viscous stress tensor expressed in dyadic form $(\hat{x}_i \tau_{ij} \hat{x}_j)$. This vector equation can be developed into a set of coordinate equations by carrying out the required mathematical manipulations in terms of specific coordinates and derivatives in that coordinate system. Note that use of the shear stress–strain rate relationship (6-59) implies a direct or linear relation, whereas the stress can be a tensor function of the strain rate for general rheological studies and investigations.[24]

Equations (6-57a, b, c and d) are termed the Navier–Stokes equations for incompressible Newtonian flow and are presented here in cartesian

[23] The repeated subscript conventionally implies summation, for example, $\tau_{ii} \equiv \sum \tau_{ii} = \tau_{xx} + \tau_{yy} + \tau_{zz}$.

[24] Rheology is the study of material flow and deformation behavior.

coordinates without derivation as an example of their form. Methods of deriving these equations are given in the references at the end of the chapter. As in the vector development given above, these are the component momentum equations with the continuity equation inserted. Using the shear stress terms from (6-58a), expansion of (6-57b) in cartesian coordinates yields

$$\frac{\rho}{g_c}\left(\frac{\partial u}{\partial t} + u\frac{\partial u}{\partial x} + v\frac{\partial u}{\partial y} + w\frac{\partial u}{\partial z}\right) = \rho f_{bx} - \frac{\partial p}{\partial x} + \mu\left(\frac{\partial^2 u}{\partial x^2} + \frac{\partial^2 u}{\partial y^2} + \frac{\partial^2 u}{\partial z^2}\right)$$

$$\frac{\rho}{g_c}\left(\frac{\partial v}{\partial t} + u\frac{\partial v}{\partial x} + v\frac{\partial v}{\partial y} + w\frac{\partial v}{\partial z}\right) = \rho f_{by} - \frac{\partial p}{\partial y} + \mu\left(\frac{\partial^2 v}{\partial x^2} + \frac{\partial^2 v}{\partial y^2} + \frac{\partial^2 v}{\partial z^2}\right)$$

$$\frac{\rho}{g_c}\left(\frac{\partial w}{\partial t} + u\frac{\partial w}{\partial x} + v\frac{\partial w}{\partial y} + w\frac{\partial w}{\partial z}\right) = \rho f_{bz} - \frac{\partial p}{\partial z} + \mu\left(\frac{\partial^2 w}{\partial x^2} + \frac{\partial^2 w}{\partial y^2} + \frac{\partial^2 w}{\partial z^2}\right) \tag{6-57d}$$

It may be noted that the viscous terms can be abbreviated $\mu\nabla^2 u$, and so on.

6.14 ILLUSTRATIVE EXAMPLE 6-4: Moving Vanes

As an illustration of some important points mentioned in the derivation of the force, rate of momentum equation, consider the hypothetical example of an open vane deflecting a jet of liquid as shown in Fig. 6-19. The vane is moving along the track at a constant velocity $\vec{u}$ and the incident jet has a velocity $\vec{v}_1$. It is desired to determine the force exerted by the liquid on the vane as a function of the vane velocity.

Since the vane is moving, no control volume can be constructed which will be "fixed in space" for the fluid in contact with the vane. However, since the vane is moving with a *constant velocity*, a uniform field with a velocity $-\vec{u}$ can be superimposed on the flow field, which will have no effect on the forces acting on either the fluid or the vane. (This control volume will be in an "inertial" system.)

The direction of the jet as it leaves the vane is determined by the vane angle at the discharge station. The jet configuration of Fig. 6-19 shows the jet as it leaves *relative to the moving vane*. Velocities relative to the moving vane will be denoted by $\vec{w}$. If $\vec{v}$ represents the absolute velocity, the relative velocity is then determined by the vector relationships obtained from geometric considerations:

$$\vec{w} = \vec{v} - \vec{u} \qquad \text{or} \qquad \vec{v} = \vec{u} + \vec{w}$$

The second relation is the easiest to remember: It states that the absolute velocity (with respect to a space-fixed coordinate system origin) is equal to the vane (or moving reference) velocity plus the velocity of the fluid relative to the vane. In the coordinate system frame of reference moving with the constant velocity $\vec{u}$, the motion of the fluid relative to the vane is shown in Fig. 6-19b.

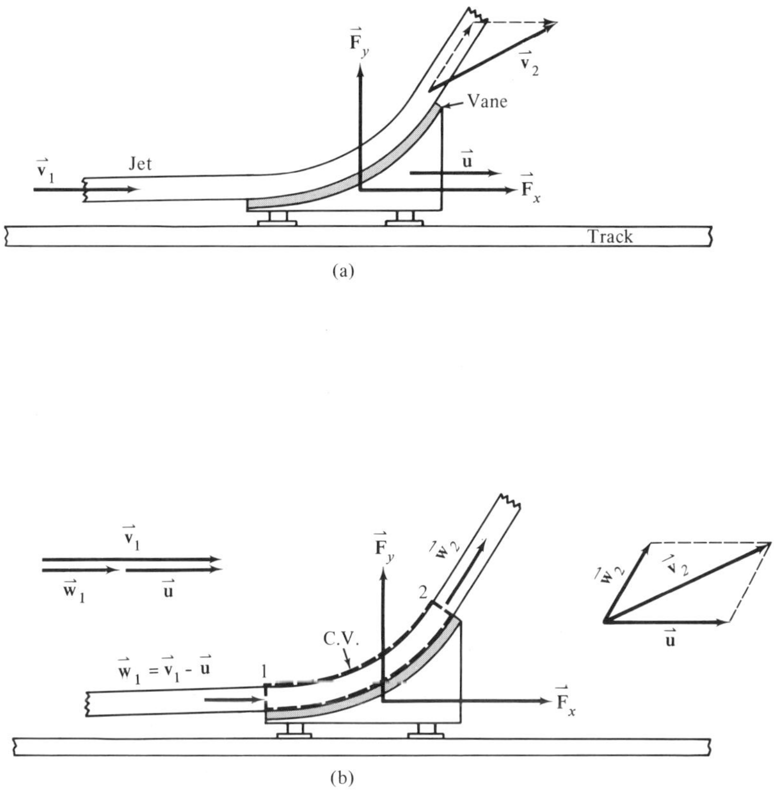

Fig. 6-19. (a) Continuous jet deflected by a vane moves along a track. The motion of the fluid relative to a controlled volume fixed in the vane is shown in (b).

The control volume can be drawn about a fluid volume as indicated. The flow through this control volume is now steady in the relative frame of reference. Since the vane angle determines the direction of $\vec{w}_2$, it is seen to be difficult to illustrate the direction that $\vec{v}_2$ actually takes in Fig. 6-19a without making the absolute jet velocity $(\vec{v}_2)$ appear discontinuous in direction at the vane discharge position.

The principle of force and rate of change of momentum for the stationary control volume of Fig. 6-19b can be written (assuming the velocity distributions at 1 and 2 to be nearly uniform such that $\beta_1 \approx 1 \approx \beta_2$)

$$\sum_{\substack{\text{ON}\\ \text{FLUID}}} \vec{F} = \frac{\dot{m}}{g_c}(\vec{w}_2 - \vec{w}_1) \tag{6-60}$$

In the original development of this integral force–momentum equation, it was specifically stated that $\dot{m}$ is the mass flow rate across the control volume boundary at the station indicated, or $\dot{m} = \rho_1 A_1 w_1 = \rho_2 A_2 w_2$. *The mass flow rate across the control volume surface is then determined by the relative velocity* $\vec{w}$ and not the absolute velocity, because the control volume has been established with respect to the moving vane.

To complete an approximate solution to this problem, the friction forces are assumed negligible compared to the deflecting force and the variation of the gravity force between 1 and 2 is neglected. The liquid is considered to be incompressible, so all of these assumptions allow Bernoulli's equation to be applied between 1 and 2. Bernoulli's equation as derived from the momentum principle can be applied in the relative (inertial) frame of reference using the relative velocities:

$$\frac{w_1^2}{2g_c} + \frac{p_1}{\rho} = \frac{w_2^2}{2g_c} + \frac{p_2}{\rho}$$

Since the jet is "free" (or open to the atmosphere) and since the curvature of the flow at stations 1 and 2 is zero, then $p_1 = p_2$, so that $w_1 = w_2$ (in magnitude only!).

Equation (6-60) can be written

$$\sum_{\substack{\text{VANE ON}\\ \text{FLUID}}} \vec{F} = \frac{\rho A_1 w_1}{g_c}(\vec{w}_2 - \vec{w}_1) = \frac{\rho A_1 w_1}{g_c}[(\vec{v}_2 - \vec{u}) - (\vec{v}_1 - \vec{u})] \tag{6-61}$$

or

$$\vec{F} = \frac{\rho A_1 w_1}{g_c}(\vec{v}_2 - \vec{v}_1)$$

again emphasizing that the force depends on the mass flow rate through the actual fluid control volume and the difference in either the absolute velocities or the relative velocities (for an inertial system only).

6.15 TORQUE AND ANGULAR MOMENTUM

It is frequently necessary to determine the torque exerted on the fluid within a system or control volume. For example, the power required to pump a fluid with a given pump impeller is determined by the torque the

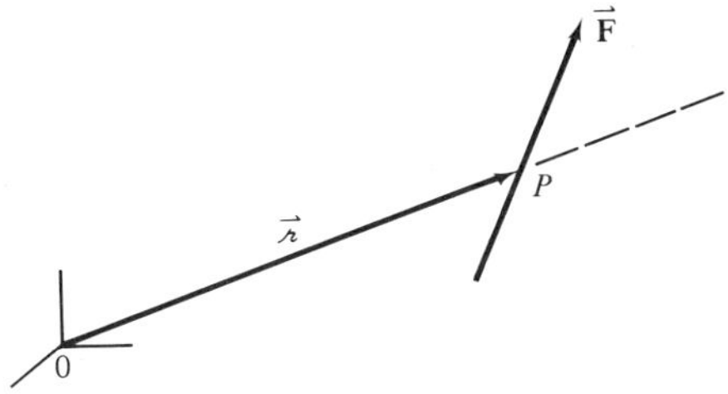

Fig. 6-20. Torque about a point O produced by a force $\vec{F}$ at the point P.

pump blades exert on the fluid and by the rotational speed of the pump. The power output of a turbine is determined by the torque the fluid imparts to the turbine wheel. It is therefore desirable to develop a direct expression for determining the torque on a fluid flowing through a defined control volume. This relation can be obtained directly from the momentum principle in the same manner in which the equation relating force and momentum was derived.

The torque of a force about a given point O is $\vec{r} \times \vec{F}$, as indicated in Fig. 6-20. This torque is obviously a vector quantity with a direction perpendicular to the plane formed by $\vec{r}$ and $\vec{F}$.

The torques due to various forces acting on fluid elements within a control volume are indicated in Fig. 6-21. The force $\overrightarrow{\Delta F}$ shown acting on the mass

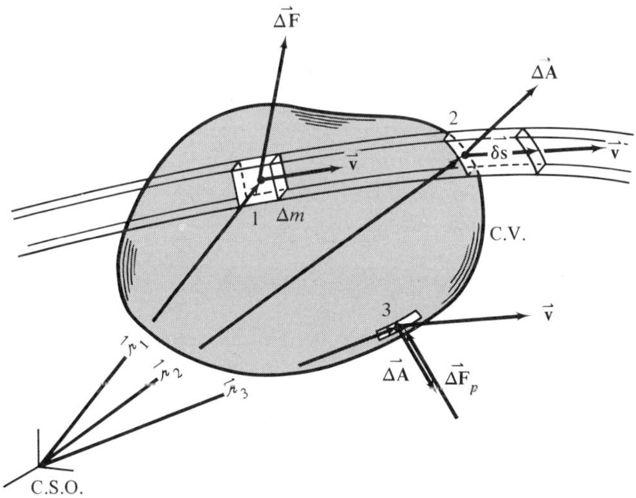

Fig. 6-21. Torques about a reference point (chosen as the coordinate system origin) by forces on element masses and areas of the control volume.

element Δm is the *net* resultant of all forces (gravity, and all other body forces, and pressure, shear, and other surface forces). The torque produced by this force at a distance from the point O (chosen as the coordinate system origin) is $\overrightarrow{\Delta T} = \vec{r} \times \overrightarrow{\Delta F}$. If the torques of all such forces acting on the fluid in the control volume are summed up over the entire control volume, the *net total* torque about O will be represented by $\sum \vec{T}$, just as the sum of all forces acting on the control volume was represented by $\sum \vec{F}$.

If the moment of all forces acting on a fluid element in the control volume is $\vec{r} \times \overrightarrow{\Delta F}$, the *moment of the momentum* or the *angular momentum* of the element with respect to the fixed coordinate system origin is $\vec{r} \times \Delta m\vec{v} = \vec{r} \times \vec{v}\,\Delta m = \vec{r} \times \vec{v}\rho\,\Delta V$ (since Δm and, therefore, ρ and ΔV are scalar quantities). The rate of change of angular momentum of all the fluid within the control volume at any instant is then

$$\frac{\partial}{\partial t} \oiiint \vec{r} \times \vec{v}\rho\, dV = \oiiint \frac{\partial \rho \vec{r} \times \vec{v}}{\partial t}\, dV$$

where the time derivative has been placed under the integral because the control volume and coordinate system origin are fixed in space.[25]

As in the case of the force–momentum relationship, the rate at which angular momentum is transported through the control volume surface must also be considered. The rate at which momentum is being carried out through the control volume at 2 is $(\vec{v})(\rho\vec{v}\cdot\vec{dA})$, as before. The moment of this momentum rate summed over the entire control surface is $\oiint (\vec{r} \times \vec{v})(\rho\vec{v}\cdot\vec{dA})$. The sum of all torques acting on the fluid in the control volume can be related to the rate of change of angular momentum by

$$\sum \vec{T} = \frac{1}{g_c} \oiiint \frac{\partial \rho \vec{r} \times \vec{v}}{\partial t}\, dV + \frac{1}{g_c} \oiint (\vec{r} \times \vec{v})(\rho\vec{v}\cdot\vec{dA}) \qquad (6\text{-}62)$$

STEADY, QUASI-ONE-DIMENSIONAL FLOW

If the flow is steady and considered quasi-one-dimensional, (6-62) can be written

$$\sum \vec{T} = \sum_{\substack{\text{OUTFLOW}\\ \text{PATHS}}} \frac{\dot{m}}{g_c}(\beta\, \vec{r} \times \vec{v}) - \sum_{\substack{\text{INFLOW}\\ \text{PATHS}}} \frac{\dot{m}}{g_c}(\beta\, \vec{r} \times \vec{v}) \qquad (6\text{-}63)$$

[25] Note that no inertial system, frame of reference, or coordinate system can be defined for the case where angular velocities are possible because of curvilinear accelerations (for example, v^2/R_s) that would exist.

In addition, if β is approximately unity and the control volume has only one through flow path, then

$$\vec{T} = \frac{\dot{m}}{g_c}[(\vec{r} \times \vec{v})_2 - (\vec{r} \times \vec{v})_1] \tag{6-64}$$

where 1 and 2 are the inflow and outflow stations, respectively. This equation states that the sum of all the torques on the fluid in a control volume is equal to the angular momentum rate out minus the angular momentum rate in. It is important to note that $\dot{m}$ is the mass flow rate across the control volume boundaries. For the inertial case in which the fluid is passing through a rotating-blade system as in an axial flow pump or turbine wheel, (6-63) or (6-64) gives the relationship between the torque and the inlet and outlet angular momentum or the inlet and outlet velocities and the geometry (that is, $\vec{r}_1$ and $\vec{r}_2$).[26]

If both sides of (6-64) are scalarly multiplied by the angular velocity $\vec{\omega}$, the power $\vec{T}\cdot\vec{\omega}$ is obtained:

$$P = \vec{T}\cdot\vec{\omega} = \frac{\dot{m}}{g_c}[(\vec{\omega}\cdot\vec{r} \times \vec{v})_2 - (\vec{\omega}\cdot\vec{r} \times \vec{v})_1] \tag{6-65}$$

If $\vec{\omega}$ is the angular velocity about the wheel axis and the flow is steady, then $\vec{\omega}$ is constant for all points in the flow field. It is now noticed that the dot and cross products in the triple scalar product are interchangeable, $\vec{\omega}\cdot\vec{r} \times \vec{v} = \vec{\omega} \times \vec{r}\cdot\vec{v}$. With $\vec{\omega} \times \vec{r} = \vec{u}$ (where $\vec{u}$ is the tangential velocity of the system at a distance $\vec{r}$ from the axis of rotation), the velocity $\vec{u}$ is then the tangential velocity of the wheel at that point. This substitution in the above equation gives

$$P = \vec{T}\cdot\vec{\omega} = \frac{\dot{m}}{g_c}[(\vec{u}\cdot\vec{v})_2 - (\vec{u}\cdot\vec{v})_1)] \tag{6-66}$$

Now $\vec{u}\cdot\vec{v}$ can be interpreted as the tangential velocity u times the tangential or u-component of v (v_u), giving

$$P = \frac{\dot{m}}{g_c}[(uv_u)_2 - (uv_u)_1] \tag{6-67}$$

where P is the power that could be obtained by passing a mass flow rate $\dot{m}$ through a turbine wheel, where it is deflected such that its angular momentum changes by the amount $\Delta(uv_u)$. Notice that an effect is produced

[26] The flow in a radial flow pump or turbine is necessarily *noninertial* because of the angular (centripetal) acceleration field $\vec{\omega} \times (\vec{\omega} \times \vec{r})$. However, in Sections 6.19 and 6.20, (6-65) through (6-57) are shown to be valid also for radial-flow machines.

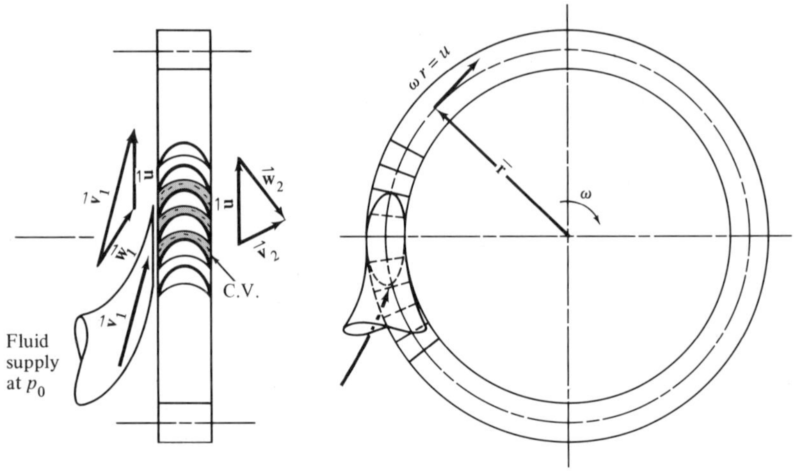

Fig. 6-22. Fluid jet passing through a turbine wheel. The flow in the control volume can be considered quasi-steady with the fluid passing through the channels between the moving vanes. Work is done as a consequence of the motion of the vanes and the force produced by them on the fluid in the control volume.

by both a change in the direction of the absolute velocity as it enters and leaves the control volume and by a change in the radius at which it enters and leaves. For the particular case of a purely "axial-flow" turbine, as illustrated in Fig. 6-22, the radius at which the flow enters and leaves is constant, so (6-64) and (6-67) can be written

$$\sum \vec{r} \times \vec{F} = \sum \vec{T} = \frac{\dot{m}}{g_c} \vec{r} \times (\vec{v}_2 - \vec{v}_1) \tag{6-64a}$$

$$P = \frac{\dot{m}}{g_c} u(v_{u2} - v_{u1}) \tag{6-67a}$$

Since the blade lengths are small compared to the average radius $\bar{r}$, the quantities in these equations can be evaluated with respect to this mean radius. The actual measured entering and leaving velocities give the $\sum \vec{T}$ or P for a turbine as the entire torque or power applied *to* the fluid in a control volume by the surroundings (the turbine wheel).[27] For example, $\vec{T}$ and P are negative quantities, implying that there is a power output *by* the fluid to the surroundings ("shaft output"). Also $\sum \vec{T}$ represents the sum of all torques from all forces within the control volume to produce

[27] One of the reasons for defining the control volume as made up only of fluid is now apparent: The turbine wheel and its blades are a part of the surroundings that supply the force or torque *to the fluid* to change its momentum.

the actual measured change in momentum (or angular momentum). Since some of these forces are friction forces, the input fluid power (or the energy change across the control volume) must be greater than the useful power output by the amount of work required to overcome the friction forces. This power is then "dissipated" or transformed into unavailable forms of energy and is partially responsible for the fluid mechanic efficiency of a pump or turbine being less than 100 percent. Other losses of availability in the energy-conversion process arise from fluid friction "losses" in the inlet and outlet passages, bearing friction, and so on. The total availability is represented by the kinetic energy, internal thermal energy, and the work available from the pressure forces; the sum of these terms is the total input enthalpy. The design of a turbomachine then requires the application of the physical principles of mass, energy, and momentum conservation along with consideration of the detailed system geometries possible within the control volume, which change the angular momentum and the net effects of adding energy to the fluid (pump) or converting the available energy to useful work (turbine).

★6.16 NONINERTIAL SYSTEMS

The equations of motion were previously developed only for inertial systems where the coordinate system (with respect to which the positions and motions were measured) was at rest or moving with a constant velocity. It is sometimes easier or necessary to have the coordinate system moving with the body geometry. For example, the physical principles are most easily applied to the flow in a rocket engine by choosing the coordinate system and its axes fixed with the rocket. This axis system may then have translational and rotational velocities and accelerations. The inertia forces associated with these motions must then be considered when formulating the equations of motion of the fluid flowing in this noninertial control volume.

Figure 6-23a indicates an element of fluid moving along the streamline s at a particular instant t. The motion of this element and the field streamlines is measured with respect to a coordinate system fixed in the moving body (for example, the rocket mentioned above). If the fluid element is in a hydro or pneumatic control system, which may be continually varying with time, the flow field relative to the rocket would be nonstationary.

In Fig. 6-23 the coordinate system and its origin O are fixed with the body. The coordinate system with origin designated by O' is space-fixed. The distance $\vec{r}'$, velocities $\vec{v}_f$, and accelerations $\vec{a}_f$ (the trajectory, flight velocity, and acceleration of O) are then absolute quantities. The moving coordinate system origin is located with respect to the absolute coordinate

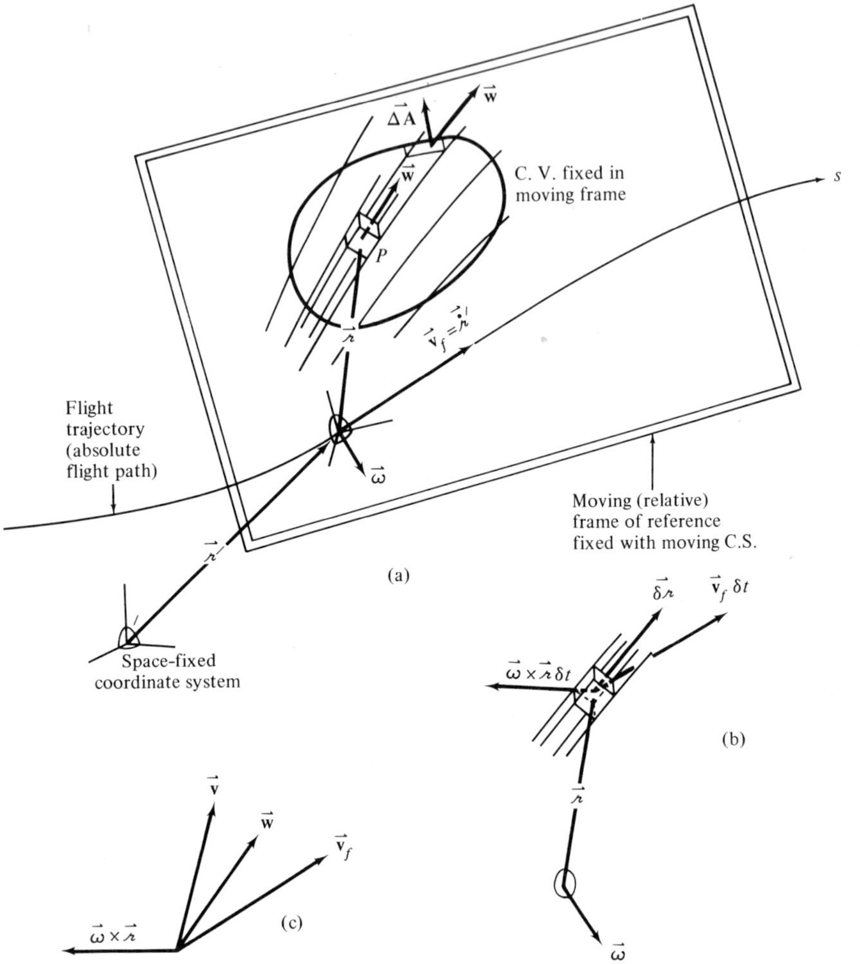

Fig. 6-23. Relationship between relative and absolute frames of reference and coordinate systems. (a) Relative frame fixed with relative coordinate system; frame translates with velocity $\vec{v}_f$, accelerates with acceleration $\vec{a}_f$, and rotates with angular velocity $\vec{\omega}$. (b) Displacement components during time period δt. (c) Relationship among velocity components: $\vec{v} = \vec{w} + \vec{v}_f + \vec{\omega} \times \vec{r}$.

system origin by the location vector $\vec{r}'$. The location of the fluid element within the control volume is determined by $\vec{r}$. The control volume is considered to be *nondeformable* and *fixed with respect to the moving coordinate system O* (relative Eulerian viewpoint). The control volume, body geometry, and moving coordinate system may therefore have a velocity $\vec{v}_f$, an acceleration $\vec{a}_f$, an angular velocity $\vec{\omega}$, and an angular acceleration $\vec{\alpha}$ relative to O'. The moving coordinate system and control volume are considered to be in the moving frame of reference indicated in Fig. 6-23a. At a later time, this entire reference frame will have moved along the flight path, as determined by $\vec{v}_f$, and will have rotated through an angle determined by the rotational velocity $\vec{\omega}$.

If the fluid element travels the distance $\vec{\delta r}$ in δt, as indicated in Fig. 6-23b, the velocity $\vec{w}$ measured with respect to O is[28]

$$\vec{w} = \lim_{\delta t \to 0} \frac{\vec{\delta r}}{\delta t} = \frac{\vec{dr}}{dt}$$

In this small time δt, the entire frame of reference will have translated a distance $v_f \, \delta t$ and rotated through an angle $\omega \, \delta t$. The total or absolute (vector) displacement of the element is then the sum of the three displacements: $\vec{\delta r} + \vec{v}_f \, \delta t + \vec{\omega} \times \vec{r} \, \delta t$. The absolute velocity of the element at P with respect to the fixed coordinate system O' is

$$\vec{v} = \vec{w} + \vec{v}_f + \vec{\omega} \times \vec{r} \tag{6-68}$$

which is obtained by adding the various motion vectors of the element as indicated in Fig. 6-23c.

From the geometry of Fig. 6-23a, the derivative d'/dt (where d'/dt indicates the derivative evaluated with respect to O') is the rate of change of $\vec{r}$ as observed from O', which is the change $\vec{\delta r}$ in δt as observed from O plus an additional vector change due to the rotation of the reference frame in space at the rate $\vec{\omega}$:

$$\frac{d'\vec{r}}{dt} = \frac{\vec{dr}}{dt} + \vec{\omega} \times \vec{r} \tag{6-69}$$

from which a generalized procedure is obtained for determining the relationship between time derivatives of a vector quantity in the two coordinate systems:

$$\frac{d'(\ )}{dt} = \frac{d(\ )}{dt} + \vec{\omega} \times (\ ) \tag{6-70}$$

[28] As in Section 6.14, velocities in the relative (moving) frame of reference will be denoted as w and absolute velocities by v.

With (6-70), the absolute acceleration of the element at P can be obtained from the time derivative of the velocity,

$$\vec{a}_{abs} = \frac{d'\vec{V}}{dt} = \frac{d'}{dt}(\vec{W} + \vec{V}_f + \vec{\omega} \times \vec{r})$$

$$= \frac{d\vec{W}}{dt} + \vec{\omega} \times \vec{W} + \frac{d'\vec{V}_f}{dt} + \vec{\omega} \times \frac{d\vec{r}}{dt} + \frac{d\vec{\omega}}{dt} \times \vec{r} + \vec{\omega} \times (\vec{\omega} \times \vec{r})$$

or

$$\vec{a}_{abs} = \frac{d\vec{W}}{dt} + 2\vec{\omega} \times \vec{W} + \frac{d'\vec{V}_f}{dt} + \frac{d\vec{\omega}}{dt} \times \vec{r} + \vec{\omega} \times (\vec{\omega} \times \vec{r})$$

since $d\vec{r}/dt = \vec{W}$. With $d\vec{W}/dt = \vec{a}$ and $d\vec{\omega}/dt = \vec{\alpha}$,

$$\vec{a}_{abs} = \vec{a} + \underbrace{\vec{a}_f + 2\vec{\omega} \times \vec{W} + \vec{\alpha} \times \vec{r} + \vec{\omega} \times (\vec{\omega} \times \vec{r})} \tag{6-71}$$

ACCELERATION DUE TO RELATIVE MOTION OF C.V. COORDINATE SYSTEM WITH RESPECT TO SPACE-FIXED COORDINATE SYSTEM

ACCELERATION OF ELEMENT AT POINT P WITH RESPECT TO RELATIVE COORDINATE SYSTEM

The term $2\vec{\omega} \times \vec{W}$ is the Coriolis acceleration, and $\vec{\omega} \times \vec{\omega} \times \vec{r}$ is the centripetal acceleration.

Expressing Newton's second law for the fluid element,

$$\sum \overrightarrow{\Delta F} \propto \text{TIME RATE OF CHANGE OF MOMENTUM (WITH RESPECT TO SPACE-FIXED COORDINATES OR AN INERTIAL COORDINATE SYSTEM)}$$

where $\sum \overrightarrow{\Delta F}$ implies all the forces acting on the mass element Δm. The rate of change of momentum can be expressed as the rate of change with respect to the relative coordinate system O, as previously developed for (6-45), plus the added effect due to the motion of the control volume (and relative coordinate system) through space. Newton's second law for the fluid within the control volume can then be written

$$\sum \vec{F} = \underbrace{\frac{1}{g_c} \oint\!\!\oint \vec{w}(\rho \vec{w} \cdot d\vec{A}) + \frac{1}{g_c} \oint\!\!\oint\!\!\oint \frac{\partial \rho \vec{w}}{\partial t}\, dV}_{\text{RATE OF CHANGE OF MOMENTUM WITH RESPECT TO RELATIVE COORDINATE SYSTEM}}$$

$$+ \underbrace{\frac{1}{g_c} \oint\!\!\oint\!\!\oint [\vec{a}_f + 2\vec{\omega} \times \vec{w} + \vec{\alpha} \times \vec{r} + \vec{\omega} \times (\vec{\omega} \times \vec{r})]\rho\, dV}_{\text{MOMENTUM TERM DUE TO NONUNIFORM MOTION OF CONTROL VOLUME}} \tag{6-72}$$

The inertial effects of the relative motion of volume (and its coordinate system) with respect to the space-fixed coordinate system involves only the mass of the fluid in the control volume at any instant. The rate of change of momentum with respect to the relative coordinate system involves both the momentum transport through the surface of the control volume and the time rate of change of momentum within the control volume as before.

The momentum term due to the nonuniform motion of the control volume may be considered separately and (6-72) rewritten in the form

$$\sum \vec{F} = \frac{1}{g_c} \oint\!\!\oint\!\!\oint \frac{\partial \rho \vec{w}}{\partial t}\, dV + \frac{1}{g_c} \oint\!\!\oint \vec{w}(\rho \vec{w} \cdot d\vec{A}) + \vec{H}' \tag{6-73}$$

where

$$\vec{H}' = \frac{1}{g_c} \oint\!\!\oint\!\!\oint [\vec{a}_f + 2\vec{\omega} \times \vec{w} + \vec{\alpha} \times \vec{r} + \vec{\omega} \times (\vec{\omega} \times \vec{r})]\rho\, dV \tag{6-74}$$

is the rate of change of momentum of the control volume. Since $\vec{a}_f$ is the instantaneous acceleration of the coordinate system origin O, and $\vec{\omega}$ is the instantaneous rotational velocity of the control volume with respect to O, these quantities do not vary over the control volume. $\vec{H}'$ may then be written[29]

$$\vec{H}' \equiv \frac{1}{g_c} m\vec{a}_f + \oint\!\!\oint\!\!\oint [2\vec{\omega} \times \vec{w} + \vec{\alpha} \times \vec{r} + \vec{\omega} \times (\vec{\omega} \times \vec{r})]\rho\, dV \tag{6-75}$$

[29] It is not practical to remove a vector-product factor from an integral. This removal is best carried out after the products are expressed in specific coordinate components.

where $m = \iiint \rho \, dV$ is the mass of fluid within the control volume at any instant.

★6.17 ILLUSTRATIVE EXAMPLE 6-5: Noninertial Effects on Rocket Propulsion

It is desired to determine the effect of flight maneuvers (flight path, velocity, and acceleration) of a rocket on the thrust mechanics of the rocket engine. The rocket at an arbitrary instant in its propelled flight is shown in Fig. 6-24. Because of transient pitch and yaw motions of the missile and possible application of control maneuvers, its axis will not necessarily be collinear with the flight path (trajectory) direction. In a previous example (Example 6-2), the static firing of a rocket was considered with the objective of determining the thrust for the engine. In the present case, it is desired to determine the effect of the rocket motion on the acceleration of the engine exhaust gases and the consequent effect on the thrust.

As a first approximation, the flow in the control volume will be considered essentially steady. Although the inertial forces will affect this assumption, the quantitative effects of this approximation will be considered later. The coordinate-system origin in the moving frame (the rocket) is best taken at the center of mass of the rocket. Also assume that the relative flow in the control volume is quasi-one-dimensional and nearly uniform at any cross section perpendicular to the flow direction. Only terms with components in the engine-axis or x direction will be considered. The angular velocity at any point will be the resultant of the three possible modes of rocket rotation in space: (1) roll about the x axis, (2) pitch, and (3) yaw of the missile axes with respect to the flight path. The direction of $\vec{\omega}_{\text{roll}}$ will be nearly parallel to $\vec{\text{v}}_f$. The angular acceleration modes caused by the maneuvering and aerodynamic forces applied to the rocket would be similar to the angular velocity modes. Each angular acceleration component is then determined from the relation $T = (I/g_c)\alpha$, where I is the appropriate moment of inertia[30] measured with respect to the relative origin O. Since the roll moment of inertia is much smaller than the other two (small radius of gyration), only the pitch and yaw accelerations will initially be considered significant. Also $\vec{\omega}_{\text{roll}} \times \vec{\text{w}} = 0$. Using these conditions, (6-72) can be written

$$\vec{F} = \frac{1}{g_c} \oiint \vec{\text{w}}(\rho\vec{\text{w}} \cdot d\vec{A}) + \frac{m\vec{a}_f}{g_c} + \frac{2}{g_c} \oiiint \vec{\omega} \times \vec{\text{w}}\rho \, dV$$

$$+ \frac{1}{g_c} \oiiint [\vec{\alpha} \times \vec{r} + \vec{\omega} \times (\vec{\omega} \times \vec{r})]\rho \, dV$$

[30] The moment of inertia is a tensor, so $T = I\alpha/g_c$ is actually $\vec{T} = I \cdot \vec{\alpha}/g_c$.

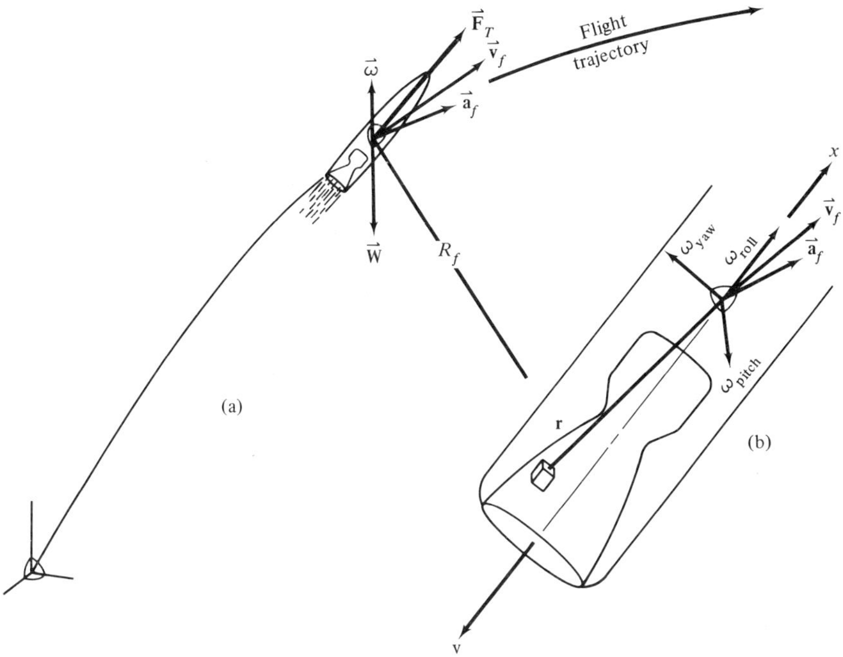

Fig. 6-24. Representation of forces, velocities, and accelerations on a rocket in flight. $\vec{F}_T$ is the engine thrust, $\vec{W}$ the gravity force, R_f the flight path radius of curvature, $\vec{v}_f$ and $\vec{a}_f$ the flight velocity and acceleration, and $\vec{\omega}$ the angular velocity (made up of roll, pitch, and yaw components.)

The x component of $\vec{\alpha} \times \vec{r}$ is negligible and $\vec{\omega} \times (\vec{\omega} \times \vec{r})$ is approximately $\bar{x}\bar{\omega}^2$, where $\bar{x}$ is the distance from O to the centroid of the engine and $\bar{\omega}$ is the magnitude of $\vec{\omega}$. The x-component equation then becomes

$$F_x = \frac{\dot{m}}{g_c}\,\mathrm{w}_e + \frac{ma_{f,x}}{g_c} + \frac{m\bar{\omega}^2\bar{x}}{g_c}$$

The first term is the same as that obtained for the static thrust. The order of magnitude of the second term with respect to the first can be obtained by comparing the average relative acceleration of a gas element inside the engine with the maximum possible flight acceleration a_f. The ratio of the second term to the first is

$$\frac{ma_f}{\dot{m}\mathrm{V}_e} = \frac{ma_f}{\rho_e A_e \mathrm{w}_e^2}$$

where m is the mass of gas within the engine control volume ($m = \iiint \rho\, dV$). It is difficult to determine this ratio directly because of problems involved in calculating m. However, the average acceleration of the gas as it passes through the engine control volume can be approximated by the inertial acceleration $\dot{m}w_e$. The gas within the control volume is accelerated from zero relative velocity to w_e in the engine. For a representative engine 4 ft long and a w_e approximately 5000 ft/sec,[31] the average acceleration would be $a_{\text{gas}} \approx w_e^2/2l$, where l is the engine length. For these representative values

$$a_{\text{gas}} = \frac{(5000)^2}{8} \approx 3 \times 10^6 \text{ ft/sec}^2 \qquad \text{or} \qquad \frac{a_{\text{gas}}}{g} \approx 10^5$$

The maximum flight acceleration developed by a rocket this size would be about $30g$. The ratio of a_f/a_g is then about 3×10^{-4}. The maximum pitch, roll, or yaw angular velocity would be on the order of 10 rad/sec for the combined vector magnitude. A representative value for $\bar{x}$ would be 10 ft. Note that large $\bar{\omega}$ (small rocket size and moment of inertia) and small x values go together, and conversely. The value of $\bar{\omega}^2\bar{x}/g$ is then on the order of 30, which is negligible with respect to 10^5.

It is then concluded that maneuvering accelerations of a rocket would have only a very small effect on the effective thrust as obtained from static determinations.

★6.18 ENERGY EQUATION FOR A NONINERTIAL CONTROL VOLUME

The principle of energy conservation formulated as a rate statement for a flow field within a control volume is given by (5-3) or (5-6). From these equations, a work–energy relation can be obtained as

$$\frac{\partial}{\partial t} \oiiint \rho e \, dV + \oiint e\rho \vec{w} \cdot d\vec{A} = \dot{W}_p + \dot{W}_{\text{etc.}} + \dot{Q}_H$$

where e is again the sum of all mass-associated forms of energy (in units of energy per unit mass). $\dot{W}_p$ is the pressure work rate and $\dot{W}_{\text{etc.}}$ is all the other forms of power to be considered (mechanical, shear where applicable, and so on). The first term on the left side is the instantaneous rate of increase of internal energy within the control volume. The second term gives the transport rate of energy across the control surface. Since they are determined only by the thermodynamic state of the fluid, such energy

[31] For similar (shorter) rocket engines the pressure and temperature are usually higher than for larger sizes, so w_e essentially increases as the size decreases; therefore, the acceleration of the gas within the engine increases rapidly as the size decreases. The above velocities are representative of average gas acceleration for a medium-sized rocket.

components as the thermodynamic internal energy $\tilde{u}$ are not influenced by the state of motion of the control volume. However, the kinetic energy is defined as $m\mathrm{v}^2/2g_c$ and is therefore a property of the motion of the mass and must be determined by the *absolute* velocity. The specific mass-associated energy e is then

$$e = \tilde{u} + \frac{\mathrm{v}^2}{2g_c} + e_{\text{chem}} + e_{\text{etc.}}$$

All other components of e are independent of the reference system with which they are measured. The velocity appearing in the transport or propagation rate $\vec{\mathrm{w}} \cdot \overrightarrow{dA}$ is the velocity by which a property (mass, energy, momentum, and so on) is transported through the control surface and is therefore the velocity relative to the control volume or moving control surface.

The pressure work $\dot{W}_p$ on the cross-sectional area of a streamtube where it leaves the control volume was defined as

$$\Delta W_p = \overrightarrow{\Delta F_p} \cdot \overrightarrow{\delta s} = -p\, \overrightarrow{\Delta A} \cdot \overrightarrow{\delta s}$$

Although the choice of control volume and relative coordinate system is arbitrary, there can be no arbitrariness in determining the pressure work. Therefore the distance through which the pressure force acts to produce the pressure work must be the *absolute* distance. The pressure work rate is then

$$\Delta \dot{W}_p = -p\vec{\mathrm{v}} \cdot \overrightarrow{\Delta A}$$

With $\vec{\mathrm{v}}$ from (6-68), the energy equation written for velocities measured with respect to the relative coordinate system is

$$\oiiint \frac{\partial}{\partial t}\left(\tilde{u} + \frac{\mathrm{v}^2}{2g_c} + e_{\text{etc.}}\right) dV + \oiint \left(\tilde{u} + \frac{\mathrm{v}^2}{2g_c} + e_{\text{etc.}}\right) \rho\vec{\mathrm{w}} \cdot \overrightarrow{dA}$$
$$= -\oiint p\vec{\mathrm{v}} \cdot \overrightarrow{dA} + \dot{W}_{\text{etc.}} + \dot{Q}_H \quad (6\text{-}76)$$

Writing the $\dot{W}_p$ term with the use of (6-68) gives

$$-\dot{W}_p = \oiint p(\vec{\mathrm{w}} + \vec{\omega} \times \vec{r} + \vec{\mathrm{v}}_f) \cdot \overrightarrow{dA}$$
$$= \oiint \frac{p}{\rho} \rho\vec{\mathrm{w}} \cdot \overrightarrow{dA} + \oiint p(\vec{\omega} \times \vec{r} + \vec{\mathrm{v}}_f) \cdot \overrightarrow{dA}$$

Combining the first part of this expansion with (6-76) gives

$$\oiiint \frac{\partial \rho e}{\partial t}\, dV + \oiint \left(\tilde{h} + \frac{\mathrm{v}^2}{2g_c} + e_{\text{etc.}}\right)\rho \vec{\mathrm{w}}\cdot d\vec{A} + \oiint p(\vec{\omega} \times \vec{\imath} + \vec{\mathrm{v}}_f)\cdot d\vec{A} = \dot{W}_{\text{etc.}} + \dot{Q}_H \quad \text{(6-77)}$$

or

$$\oiiint \frac{\partial \rho e}{\partial t}\, dV + \oiint \left[\left(\tilde{h} + \frac{\mathrm{v}^2}{2g_c} + e_{\text{etc.}}\right)\vec{\mathrm{w}} + \frac{p}{\rho}(\vec{\omega} \times \vec{\imath} + \vec{\mathrm{v}}_f)\right]\cdot \rho\, d\vec{A} = \dot{W}_{\text{etc.}} + \dot{Q}_H \quad \text{(6-77a)}$$

With $\vec{\mathrm{v}}$ in expanded form (6-68), (6-77) is a modified energy equation in terms of quantities measured with respect to the moving control volume and quantities determined by the trajectory of the control volume coordinate system origin (O). An example of the use of this energy equation for a noninertial system will be presented following Section 6.19.

★6.19 TORQUE AND ANGULAR MOMENTUM FOR A NONINERTIAL SYSTEM

The torques on and angular momentum of a moving control volume are determined in the same way that they were for the inertial system, that is, by taking the cross product of the relative position vector $\vec{\imath}$ with each elemental force and momentum of each elemental mass. Performing this vector multiplication for each element and summing over the control volume and control surface [or taking the cross product of $\vec{\imath}$ with each term of the integrand of (6-72)] gives

$$\sum \vec{T} = \frac{1}{g_c}\oiiint \frac{\partial(\rho \vec{\imath} \times \vec{\mathrm{w}})}{\partial t}\, dV + \frac{1}{g_c}\oiint (\vec{\imath} \times \vec{\mathrm{w}})(\rho \vec{\mathrm{w}}\cdot d\vec{A})$$
$$+ \frac{1}{g_c}\oiiint [\vec{\imath} \times \vec{\mathrm{a}}_f + 2\vec{\imath} \times (\vec{\omega} \times \vec{\mathrm{w}}) + \vec{\imath} \times \vec{\alpha} \times \vec{\imath} + \vec{\imath} \times \vec{\omega} \times (\vec{\omega} \times \vec{\imath})]\rho\, dV \quad \text{(6-78)}$$

which can also be rewritten

$$\sum \vec{T} = \frac{1}{g_c}\oiiint \frac{\partial(\rho \vec{\imath} \times \vec{\mathrm{w}})}{\partial t}\, dV + \frac{1}{g_c}\oiint (\vec{\imath} \times \vec{\mathrm{w}})(\rho \vec{\mathrm{w}}\cdot d\vec{A}) + \vec{M}' \quad \text{(6-79)}$$

where

$$\vec{M}' \equiv \frac{1}{g_c}\oiiint [\vec{\imath} \times \vec{\mathrm{a}}_f + 2\vec{\imath} \times (\vec{\omega} \times \vec{\mathrm{w}}) + \vec{\imath} \times (\vec{\alpha} \times \vec{\imath}) + \vec{\imath} \times \vec{\omega} \times (\vec{\omega} \times \vec{\imath})]\rho\, dV \quad \text{(6-79a)}$$

is the rate of change of angular momentum of the control volume. Again note that $\vec{a}_f$, $\vec{\omega}$, and $\vec{\alpha}$ are constants over the limits of integration and could therefore be taken outside the integral. However, this can be done formally only after the equation is decomposed into its coordinate component equations prior to actual numerical evaluation.

In some problems it is necessary to determine the power of a rotating system. The power is a scalar product of the torque and angular velocity $P = \vec{T}\cdot\vec{\omega}$, or

$$
\begin{aligned}
P &= \sum \vec{T}\cdot\vec{\omega} \\
&= \frac{1}{g_c}\oiiint \vec{\omega}\cdot\frac{\partial\rho\vec{r}\times\vec{w}}{\partial t}\,dV + \frac{1}{g_c}\oiint(\vec{\omega}\cdot\vec{r}\times\vec{w})(\rho\vec{w}\cdot\vec{dA}) \\
&\quad + \frac{1}{g_c}\oiiint[\vec{\omega}\cdot\vec{r}\times\vec{a}_f + 2\vec{\omega}\cdot\vec{r}\times(\vec{\omega}\times\vec{w}) + \vec{\omega}\cdot\vec{r}\times(\vec{\alpha}\times\vec{r})]\rho\,dV
\end{aligned} \tag{6-80}
$$

In this expression for the power input to the fluid in a control volume by the action of the external forces (or torques produced), the product of $\vec{\omega}$ with the last term of (6-78) or (6-79a) has been omitted. It can be shown that the direction of $\vec{r}\times[\vec{\omega}\times(\vec{\omega}\times\vec{r})]$ is tangential (perpendicular to the rotation $\vec{\omega}$), so that the dot product of $\vec{\omega}$ with this resultant vector is identically zero (see Problems 6-55 and 6-56). The product $\vec{\omega}\times\vec{r}$ is the tangential velocity denoted by $\vec{u}$. Since the dot and cross product are interchangeable in the triple scalar product,

$$
\vec{\omega}\cdot\vec{r}\times(\quad) \equiv \vec{\omega}\times\vec{r}\cdot(\quad) \equiv \vec{u}\cdot(\quad)
$$

Making this rearrangement and substitution in (6-80) gives

$$
\begin{aligned}
P_{\substack{\text{APPLIED TO}\\ \text{FLUID IN C.V.}}} &= \frac{1}{g_c}\oiiint\frac{\partial\rho\vec{u}\cdot\vec{w}}{\partial t}\,dV + \frac{1}{g_c}\oiint(\vec{u}\cdot\vec{w})(\rho\vec{w}\cdot\vec{dA}) \\
&\quad + \frac{1}{g_c}\oiiint[\vec{u}\cdot\vec{a}_f + 2\vec{u}\cdot(\vec{\omega}\times\vec{w}) + \vec{u}\cdot(\vec{\alpha}\times\vec{r})]\rho\,dV
\end{aligned} \tag{6-81}
$$

★6.20 ILLUSTRATIVE EXAMPLE 6-6: Radial-Flow Pump

Another example of a noninertial control volume is the flow field within a radial-flow turbomachine (radial-flow pump, compressor, or turbine).

The design or operating relationships among the torque and flow field quantities (pressure, velocity, and so on) will be investigated.

It is first assumed that trajectory velocity ($\vec{v}_f$) and acceleration ($\vec{a}_f$) effects are negligible compared to the inertial forces within the control volume, so the relative control volume is assumed to be rotating with a constant velocity $\vec{\omega}$ with the rotation axis fixed in space. The inertial motion of the control volume is then one of rotation only, as illustrated in Fig. 6-25. Assume also that a steady-state operating condition has been attained. The torque–angular momentum equation (6-79) gives

$$\sum \vec{T} = \frac{1}{g_c} \oint\!\!\oint (\vec{r} \times \vec{w})(\rho \vec{w} \cdot \vec{dA}) + \vec{M}'$$

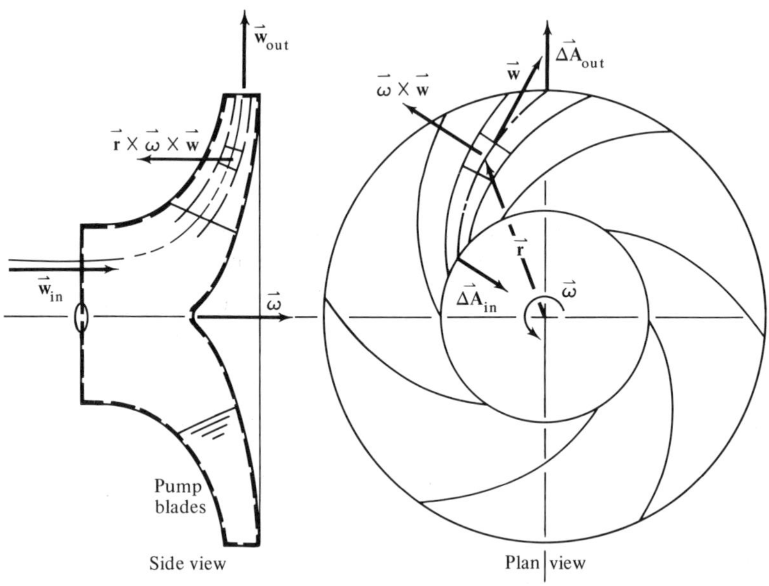

Fig. 6-25. Radial-flow pump showing a streamtube and relationships among vectors. The control volume boundary is indicated by the heavy dashed line.

Assuming quasi-one-dimensional flow, this equation reduces to

$$\sum \vec{T} = \frac{\dot{m}}{g_c} [(\vec{\beta r} \times \vec{w})_{out} - (\vec{\beta r} \times \vec{w})_{in}] + \vec{M}'$$

$\vec{M}'$ is obtained from (6-79a) as

$$\vec{M}' \equiv \frac{1}{g_c} \oint\!\!\oint\!\!\oint [2\vec{r} \times (\vec{\omega} \times \vec{w}) + \vec{r} \times \vec{\omega} \times (\vec{\omega} \times \vec{r})]\rho \, dV$$

If the control volume is chosen as indicated in Fig. 6-25, $\vec{\omega} \times \vec{w}$ will have a maximum value at the outlet, with a direction as shown in the figure. Near the outlet the direction of $\vec{r} \times (\vec{\omega} \times \vec{w})$ will be axial as indicated in the side view.

The torque rate or power is usually the quantity of primary interest in evaluating turbomachine performance. This power is obtained from the relation

$$P = \vec{T} \cdot \vec{\omega} \qquad \text{or} \qquad P = \frac{\dot{m}}{g_c}[(\beta \vec{\omega} \times \vec{r} \cdot \vec{w})_{out} - (\beta \vec{\omega} \times \vec{r} \cdot \vec{w})_{in}] + P'$$

where $P' = \vec{\omega} \cdot \vec{M}'$ and the dot and cross products have been interchanged as before. The power P' due to the nonuniform control volume motion is

$$P' = \vec{\omega} \cdot \vec{M}' = \frac{1}{g_c} \oiiint 2\vec{\omega} \cdot [\vec{r} \times (\vec{\omega} \times \vec{w})] \rho \, dV$$

$$= \frac{2}{g_c} \oiiint \vec{u} \cdot (\vec{\omega} \times \vec{w}) \rho \, dV$$

where $\vec{u} = \vec{\omega} \times \vec{r}$ is the tangential velocity.[32] The integrand in the inertial power (P') term varies from zero at the C.V. inlet to a maximum at the outlet, so the effect of the Coriolis acceleration ($\omega \times w$) is not negligible. The power applied to change the angular momentum of a fluid (and therefore "pump it") is

$$P_{\text{FROM BLADES AND C.V. BOUNDARY TO FLUID}} = \frac{\dot{m}}{g_c}[(\beta \vec{u} \cdot \vec{w})_{out} - (\beta \vec{u} \cdot \vec{w})_{in}] + \frac{2}{g_c} \oiiint \vec{u} \cdot (\vec{\omega} \times \vec{w}) \rho \, dV \tag{6-82}$$

which is also the power input by the drive shaft. The force is actually applied to the boundaries of the fluid control volume at the blade–fluid interfaces and appears as a pressure difference across the blades acting on the blade area. Not all this input power appears as useful or available energy and pressure work difference across the pump. Some of it is used to overcome the friction forces between the fluid and blades. Equation (6-82) can also be written

$$P = \frac{\dot{m}}{g_c}[(\beta u w_u)_{out} - (\beta u w_u)_{in}] + \frac{2}{g_c} \oiiint \vec{u} \cdot (\vec{\omega} \times \vec{w}) \rho \, dV \tag{6-82a}$$

[32] Note that $\vec{\omega} \times \vec{r} = \vec{\omega} \times \vec{r}$, where $\vec{r}$ is the position vector from an arbitrarily chosen C.S.O., as in Fig. 6-25, and $\vec{r} = \hat{r}r$ is the radius measured from the axis of rotation for a cylindrical C.S.

where w_u is the u or tangential component of the relative velocity $\vec{w}$ and $\vec{u} = \vec{\omega} \times \vec{r}$ is the tangential "wheel" speed.

A determination of the Coriolis term is seen to be difficult and depends on the relation between the velocity field and the fluid passage geometry between the blades. The overall evaluation of this term will be made below.

Rather than considering the control volume fixed with the rotor as was done above, take the control volume completely surrounding the fluid, and consider the absolute flow velocity as it enters and leaves the impeller. A contradiction of the conditions of steady state and that the control volume must be chosen entirely of fluid elements now occurs. If the control volume is around only the fluid within the impeller, the blades and impeller shrouds (walls) form the boundaries. But the blades are rotating, so if attention is focused on a point in space within the control volume, the flow is unsteady due to interruption by the solid blades. This unsteady flow can be reduced to steady flow by considering the impeller as an actuator disc (as introduced in Section 5.2), which produces the net overall change of angular momentum observed from inlet to outlet. The torque input to the fluid by the rotor will then change the fluid momentum as determined by the mass flow rate at the inlet and outlet velocities according to the relation

$$\sum_{\substack{\text{ON FLUID}\\\text{BY ROTOR}}} \vec{T} = \frac{\dot{m}}{g_c} [(\beta \vec{\imath} \times \vec{v})_{out} - (\beta \vec{\imath} \times \vec{v})_{in}] \tag{6-83}$$

The power input is

$$P_{\substack{\text{INTO}\\\text{FLUID}}} = \frac{\dot{m}}{g_c} [(\beta \vec{u} \cdot \vec{v})_{out} - (\beta \vec{u} \cdot \vec{v})_{in}] \tag{6-84}$$

or

$$P = \frac{\dot{m}}{g_c} [(\beta u v_u)_{out} - (\beta u v_u)_{in}] \tag{6-84a}$$

It is now of interest to examine the details of the flow as it enters and leaves the impeller control volume. Consider the correspondence among the velocity vectors as the flow leaves the impeller as illustrated in Fig. 6-26. The universal vector relationship among the absolute, relative, and translational velocities at a point is

$$\vec{v} = \vec{w} + \vec{u} \tag{6-85}$$

The tangential component of this vector equation is

$$v_u = w_u + u \tag{6-85a}$$

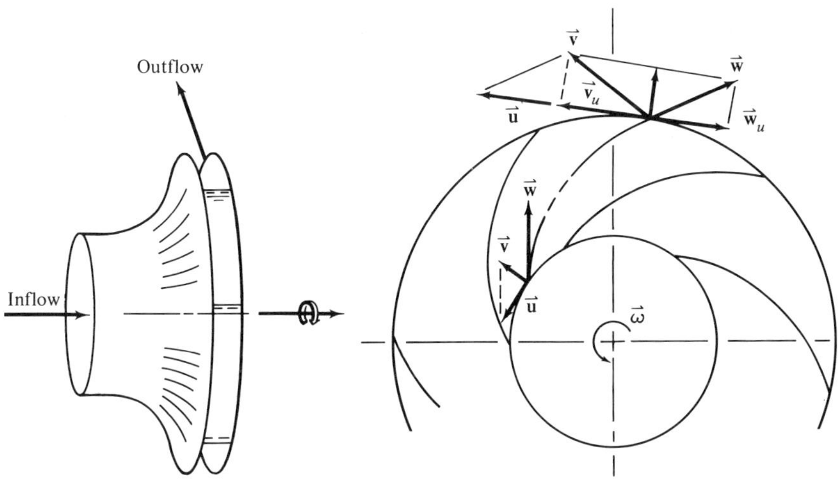

Fig. 6-26. Relation among absolute and relative velocities and tangential wheel velocity $(\vec{u} = \vec{\omega} \times \vec{r})$ at blade inlet and outlet.

If this equation is substituted into (6-84a), the power is obtained in terms of the relative velocity components and the wheel speed as

$$P = \frac{\dot{m}}{g_c}\,[(\beta u w_u)_{out} - (\beta u w_u)_{in}] + (\beta u^2)_{out} - (\beta u^2)_{in} \tag{6-86}$$

Comparing the result with (6-82a) indicates that the term $(\dot{m}/g_c)[(\beta u^2)_{out} - (\beta u^2)_{in}]$ is the net equivalent of the Coriolis torques and the power input therefrom; that is,

$$\dot{m}[(\beta u^2)_{out} - (\beta u^2)_{in}] \equiv 2 \oiiint \vec{u}\cdot(\vec{\omega} \times \vec{w})\rho\, dV$$

In many works on turbomachinery the $\Delta[\beta u^2]$ term is referred to as a centrifugal-force effect. But as was previously pointed out, the centrifugal forces are radial and cannot possibly produce a torque about the rotational axis. This term is actually seen to be the effect of the Coriolis forces.

Turbines, turbopumps, and compressors are often used in aircraft and missiles. In these cases, the trajectory motion and its effect on the fluid flow through the control volume should be considered to properly determine the operation of the turbomachine unit. However, as in the example of the rocket engine, a detailed investigation will indicate that the inertia forces associated with the angular velocity component and linear acceleration of the flight trajectory are usually negligible compared to the forces and angular accelerations produced within the control volume.

For axial-flow turbomachinery, where the flow is through the blade passages on streamlines of constant radius, this condition $r =$ const. implies that u $=$ const., and therefore the Δu^2 term of (6-86) drops. It also follows that the Coriolis force term must then be zero. It is left as an exercise to show that if the flow has no radial component within the control volume then

$$\oiiint \vec{u}\cdot(\vec{\omega} \times \vec{w})\rho \, dV = 0$$

It is also usually necessary to use information obtained from the energy equation when investigating problems in the application or design of turbomachinery. The flow rate through a turbine, pump, or compressor is usually so rapid that the energy transferred by the process of heat transfer through the solid boundaries of the machine is negligible with respect to the net energy conversion to work, so the control volume can be considered to be adiabatic ($\dot{Q}_H = 0$). Assuming steady-state operation, $v_f = 0$, and neglecting $e_{\text{etc.}}$, the energy equation (6-77a) gives

$$\dot{m}\left[\tilde{h} + \frac{\alpha V^2}{2g_c}\right]_{\text{out}} - m\left[\tilde{h} + \frac{\alpha V^2}{2g_c}\right]_{\text{in}} + \oiint p(\vec{\omega} \times \vec{r})\cdot d\vec{A} = \langle \dot{W}_m \rangle$$

By considering the correspondence and direction of the vectors $\vec{\omega}$, $\vec{r}$, and $d\vec{A}$ over the inlet and outlet areas, $(\vec{\omega} \times \vec{r})\cdot d\vec{A}$ is zero over all these areas. On the shroud surfaces, $\vec{\omega} \times \vec{r}$ must be a fluid velocity which cannot be parallel to the surface direction $d\vec{A}$ (the velocity cannot be *through* the solid surface), so $\vec{\omega} \times \vec{r}\cdot d\vec{A}$ must be zero everywhere. Investigating the product $\vec{\omega} \times \vec{r}$ will show that it is tangential and therefore perpendicular to $d\vec{A}$ everywhere. For an axial-flow machine, the product $(\vec{\omega} \times \vec{r})\cdot d\vec{A}$ is also identically zero for all elements of the control surface. The energy equation for a turbomachine under the above conditions and assumptions then reduces to

$$\left(\tilde{h} + \frac{\alpha V^2}{2g_c}\right)_{\text{out}} - \left(\tilde{h} + \frac{\alpha V^2}{2g_c}\right)_{\text{in}} = \langle w_m \rangle \tag{6-87}$$

For the particular case of *incompressible* flow, (6-87) takes the form

$$\left(\tilde{u} + \frac{p}{\rho} + \frac{\alpha V^2}{2g_c}\right)_{\text{out}} - \left(\tilde{u} + \frac{p}{\rho} + \frac{\alpha V^2}{2g_c}\right)_{\text{in}} = \langle w_m \rangle \tag{6-88}$$

For adiabatic flow, the relations developed in (6-23b) and (6-24) allow (6-88) to be written

$$\left(\frac{p}{\rho} + \frac{\alpha V^2}{2g_c}\right)_{\text{out}} - \left(\frac{p}{\rho} + \frac{\alpha V^2}{2g_c}\right)_{\text{in}} + \frac{\Delta p_{\text{loss}}}{\rho} = \langle w_m \rangle \tag{6-89}$$

The initial or one-dimensional design of a pump or turbine (or of a stage of a multistage machine) usually starts with (6-83) or (6-84) and (6-87) and the relations afforded by the velocity triangles corresponding to the vector relations $\vec{v} = \vec{w} + \vec{u}$ at the inlet and outlet blade stations. The details of determining optimum blade shapes and angles, three-dimensional effects, and losses comprise the study of turbomachinery.

6.21 SUMMARY

1. Newton's second law must be applied as stated: The rate of change of the momentum *of the fluid* in a designated control volume is proportional to the sum of all forces acting *on the fluid* in the C.V. Accordingly, the C.V. must be designated to contain only fluid, although its boundaries may be adjacent to solid duct or machine boundaries. Since this law involves vector quantities, coordinate directions must be specified and the positive direction conventions maintained during all derivations and calculations.
2. Use of Bernoulli's equation in the form

$$\frac{p}{\rho} + \frac{v^2}{2g_c} + \frac{gz}{g_c} = \text{const.}$$

   is subject to the restrictions listed in Section 6.6.
3. Loss of availability (unfortunately often referred to as "energy loss" or "pressure loss") is a consequence of the conversion from available energy and work potential (p/ρ) into unavailable internal energy by irreversible work (friction) and heat-transfer processes. In an incompressible flow (or in the flow of a hypothetical incompressible fluid) all shear stress work and all energy transferred to the fluid as heat is in the form of unavailable energy.
4. The process of producing work on a fluid in a C.V. implies a boundary displacement, and work rate implies a boundary velocity and an unsteady flow field. No "machine-type" work is then possible for a fixed C.V. or a steady-state flow. Pump-and-turbine-type work is therefore not possible for a fixed C.V. Such work can only exist for unsteady flow in a C.V., of which at least a part is not fixed. This work can be applied only at a C.V. boundary where the fluid is in contact with the actuating, machine-C.V. interface. In machines where the work producing actuation is periodic, the "quasi-steady" time-varying C.V. may be replaced by a time-mean effect. This actuator is sometimes considered to be replaced by an *actuator disc* in a hypothetical overall fixed C.V. The same steady work input rate and the same mean force–momentum rate exchange are produced by this actuator disc.

5. A uniform normal positive stress (negative pressure) produces a volume increase or dilatation. A nonuniform normal stress system produces a dilatation and normal strain rates. In a nonprinciple coordinate axis system, shear stress produces a shear (angular) deformation or strain. In an incompressible flow, the dilatation is zero.
6. Mathematical statements of the energy, momentum, and so on, principles are most accurately and interpretatively written (or derived) as integral equations using the most appropriate mathematical stators, such as vectors, tensors, and so on. Differential equations can be deduced from the integral equations where necessary.
7. The fluid mechanics of turbomachinery involves noninertial C.V.'s and must be treated accordingly to obtain a correct interpretation of the actions occurring.

REFERENCES

1. L. M. Milne-Thompson, *Theoretical Hydrodynamics*, The Macmillan Company, New York, 4th ed., 1960.
   A rigorous treatise in fluid dynamics by an applied mathematician.
2. S. Goldstein (ed.), *Modern Developments in Fluid Dynamics*, vols. I and II, Oxford University Press, New York, 1938.
   Contains much historical development, expository material, and experimental comparison with theoretical results.
3. L. Prandtl and O. Tietjens, *Fundamentals of Hydro- and Aero-Mechanics*, McGraw-Hill, Inc., New York, 1934 (also Dover Publications, Inc., New York).
   A transcription of Prandtl's lectures by Tietjens.
4. H. Rouse, *Advanced Mechanics of Fluids*, John Wiley & Sons, New York, 1959.
5. H. Schlichting, *Boundary Layer Theory*, McGraw-Hill, Inc., New York, 6th ed., 1968.
6. Sir Horace Lamb, *Hydrodynamics*, Dover Publications, Inc., New York.
   A very comprehensive treatise on many phases of hydrodynamics. (First ed., 1879.) Somewhat difficult to read.
7. R. Aris, *Vectors, Tensors, and the Basic Equations of Fluid Mechanics*, Prentice-Hall, Inc., Englewood Cliffs, N.J., 1962.

PROBLEMS

6-1 A simple diffuser is mounted between two flexible couplings as shown. The inlet air velocity is 200 ft/sec at 14 psia and 75°F. The inlet area is 4 ft^2 and the area ratio is 5. The diffuser loss coefficient K is 0.2, where

$$\frac{p_{02} - p_{01}}{\rho} \equiv K \frac{(\mathrm{v}_1 - \mathrm{v}_2)^2}{2g_c}$$

Find the force required to hold the diffuser in place.

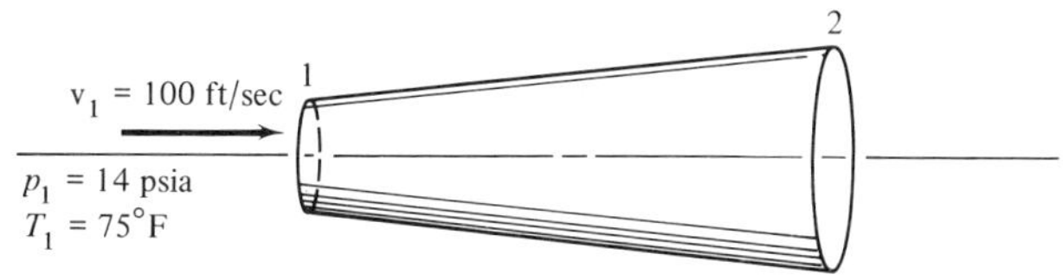

Fig. P6-1.

6-2 Determine the magnitude and direction of the force exerted by the water jet on the wall. The flow rate is 2.5 ft³/sec, and the nozzle diameter is 2 in.

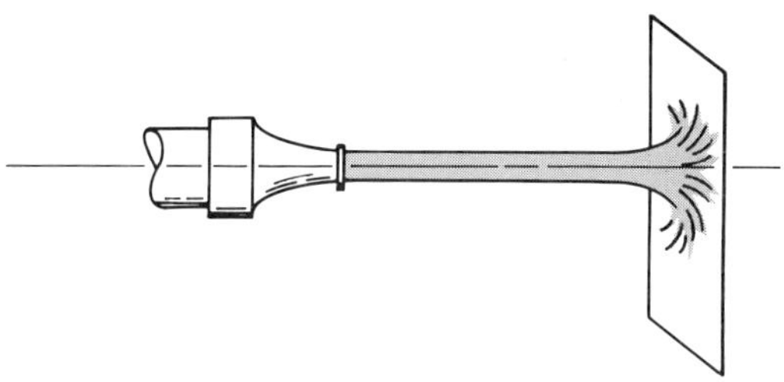

Fig. P6-2.

6-3 A two-dimensional jet of width b_1 strikes a plate making an angle θ with the initial jet direction. Show that the jet is deflected into two streams by the plate so that $b_2 = (b_1/2)(1 + \cos\theta)$ and $b_3 = (b_1/2)(1 - \cos\theta)$. Neglect friction.

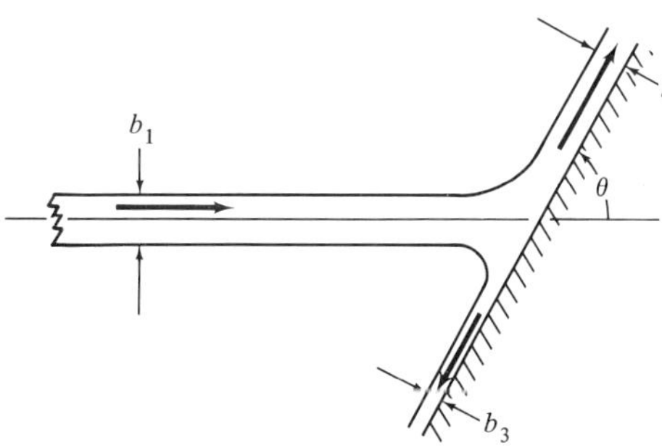

Fig. P6-3.

6-4 (a) If a water jet leaves a nozzle at a velocity of 100 ft/sec as shown and strikes a plane wall dead on, what is the force exerted on the wall?

(b) If the same jet is lofted and strikes the wall as shown, what is the force exerted on the wall?
(c) Find the force for this case also.

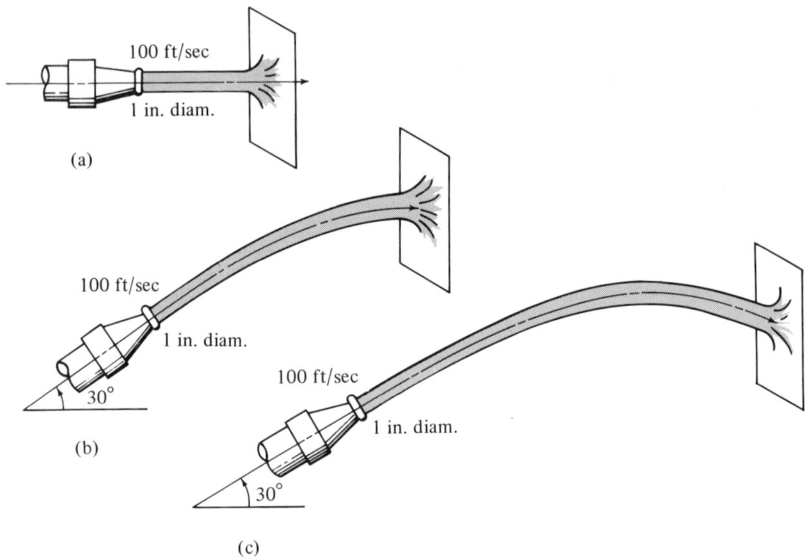

Fig. P6-4.

6-5 An Indian fakir starts his rope trick with a coiled rope having a mass of 0.4 lb_m/ft. The rope uncoils at a constant velocity as it is pulled up into the air by a mystic vertical force. If the velocity of the end of the rope during the uncoiling is 2 ft/sec, what is the magnitude of the mystic sky force on the rising end of the rope (a) just as it begins to rise, and (b) when the end is 20 ft up? (c) Find a general expression for the force as a function of the altitude of the end of the rope, z.

6-6 Two jet boats were constructed as shown. The pumps produced a total head of 12 in. at the discharge ports. Find the thrust obtained by both boats. All pipe lines are 5 in. in diameter.

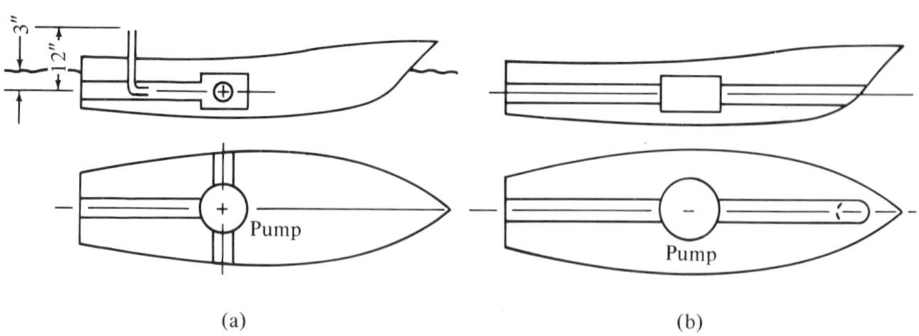

Fig. P6-6.

6-7 Two possible configurations are being considered for a jet-propelled torpedo (see the sketch). In the first, water is drawn in through the pump and is ejected through an exhaust jet pipe (same diameter as inlet pipe). In the second, the inlets are placed on the side as shown. You are asked to evaluate the two designs and to determine the one that would be the most effective with respect to the ducting systems indicated. As a consultant, specify which of the two designs is superior and substantiate your decision.

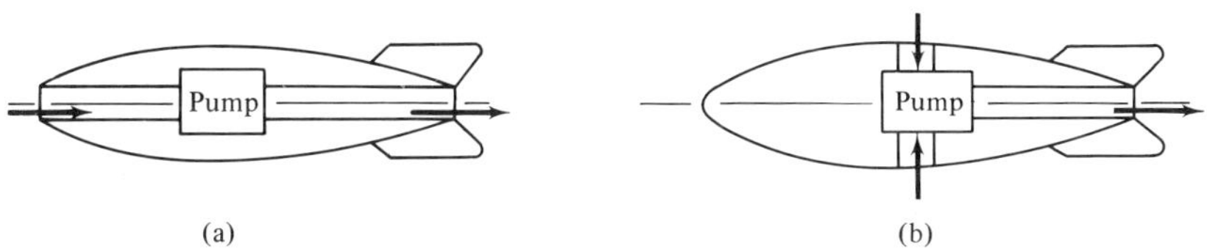

Fig. P6-7.

6-8 A ramjet is sometimes referred to as a flying stovepipe. Such a stovepipe would take in air into which fuel would be injected and burned as shown in (a). The hot, expanded exhaust gases would issue from the rear. If the inlet velocity is 600 ft/sec (at 0°F and 10 psia), the exhaust velocity is 750 ft/sec, and the fuel/air mass flow ratio is 1/35, what is the thrust developed by the engine? Explain and justify your answer.

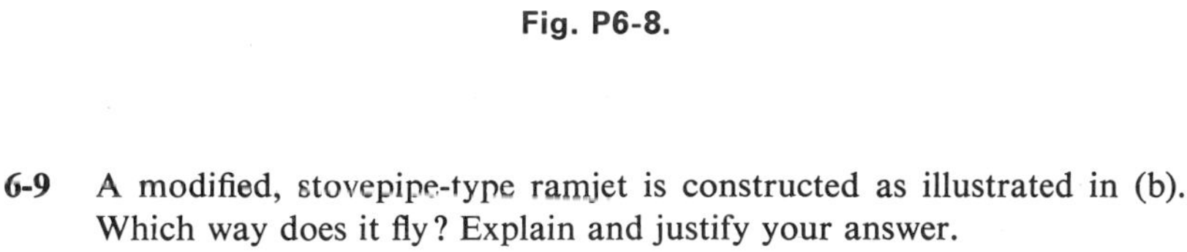

Fig. P6-8.

6-9 A modified, stovepipe-type ramjet is constructed as illustrated in (b). Which way does it fly? Explain and justify your answer.

6-10 A rocket sled weighing 3000 lb_m is slowed by a water-brake arrangement consisting of two scoops that are dropped into water troughs at the end of the sled run. If the sled is traveling at a speed of 400 ft/sec at the end of the run when the scoop brakes are dropped, what is the initial braking force on the sled? (b) Find the initial deceleration of the sled. (c) Find the time required to slow the sled from 400 to 100 ft/sec.

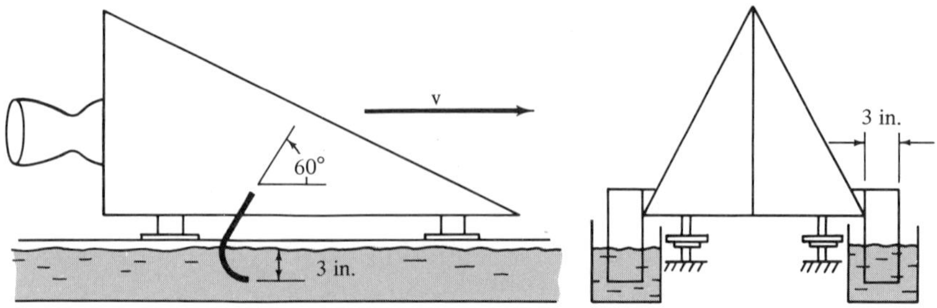

Fig. P6-10.

6-11 A train picks up water from a trough alongside the track by means of an open scoop as shown. (a) For a train speed of 60 ft/sec, find the drag that the scoop exerts on the train. (b) Find the velocity (relative to the moving train) of the water as it enters the tank. (c) What horsepower is required to take on the water? (d) Would the results differ if the scoop were a pipe? Why?

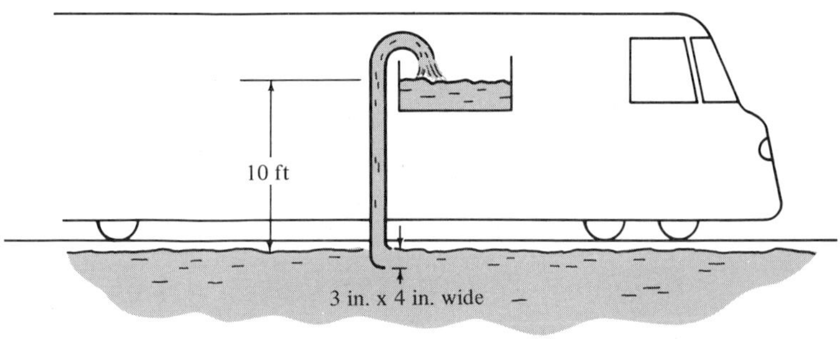

Fig. P6-11.

6-12 A train traveling at 40 mph lowers a 3-in.-diameter pipe scoop into a water trough to pick up water. (See the figure for Problem 6-11.) Find the flow rate into the tank. What is the pressure just inside the scoop inlet?

6-13 For the train in Problem 6-12, find the velocity with which the water enters the tank if the train is traveling at 15 mph.

6-14 For the apparatus shown, find the tension force in the hose in the axial direction. Neglect friction effects.

6-15 The pressure at a fireplug is 100 psi as measured on a gage as shown. (a) Calculate the velocity of the discharge jet. If it is necessary to make any assumptions, state these explicitly. Specify what you believe to be the accuracy of your result. (b) Calculate the force on the coupling between the nozzle and the hose coupling and specify its magnitude and direction (that is, tension or compression).

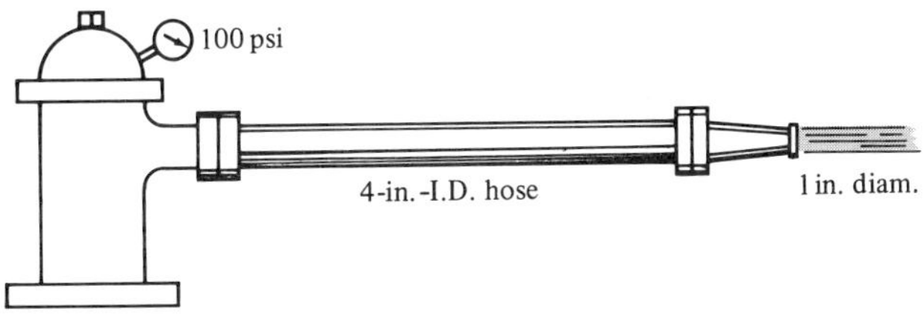

Fig. P6-14, 15.

6-16 Find the force (direction and magnitude) that a fireman must apply to hold a nozzle in the position shown. The jet velocity is 100 ft/sec through a 1-in.^{2} area and the hose diameter is 3 in. List the effects that must be considered to solve this problem.

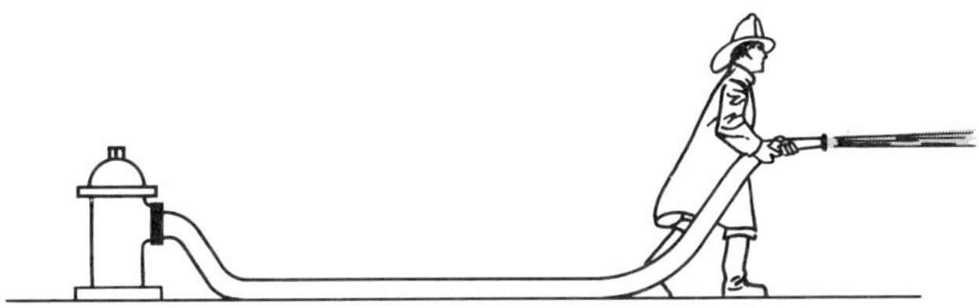

Fig. P6-16.

6-17 The V-2 rocket engine was designed for an exhaust jet velocity of 6540 ft/sec. The rocket carried 12,900 lb_m of fuel, which was burned in 65 sec. Assuming the exhaust pressure to be equal to atmospheric pressure, find the thrust produced by this engine.

6-18 In Example 6-2 it was stated that the ratio of the exhaust velocity to the inlet velocity (which is also the ratio of the momentum transport rates) was of the order of 10^{-5}. Assume that (1) the ratio $A_e/(A_{ox} + A_f)$ is 20:1, (2) the average molecular weight of the exhaust products is 30, (3) the fuel ($\rho_f = 50\ lb_m/ft^3$) to liquid oxygen ($\rho_{ox} = 71.3\ lb_m/ft^3$) mass flow ratio is 1/3.5, and (4) the exhaust temperature T_e is 1000°R. Determine the approximate ratio of exhaust velocity to liquid fuel–oxygen inlet velocity for these conditions.

6-19 (a) In Example 6-2 consider that the fuel and oxidant are contained in pressurized tanks such that no pumps are required. Neglecting friction in the supply piping, consider a control volume surrounding all the fluids within the system. Apply the principle of force and momentum and show that the thrust is still the value given in (6-50). (b) Now consider that the friction forces in the piping system must be accounted for and show that the thrust is still given by (6-50).

6-20 In Example 6-2 a pumping system is to be used. Since fuel and oxygen vapor exist in the space above the tanks, the pressure will be relatively low. Neglecting friction in the system, consider a control volume of fluid from the tanks through the pumps and show that the engine thrust can still be given by (6-50). For simplicity, assume that the pump inlet and outlet pipes are the same diameter.

6-21 In most large liquid-propellant rockets, propellant feed pumps are used to pressurize the fuel feeds into the combustion chamber. The fuel and oxygen are therefore tanked at low pressure in lightweight tanks. If the tank pressure, pump feed pressure, and chamber pressure are all different, consider the forces and momentum involved and show that the thrust for this system is given by (6-50).

6-22 Propeller blades exert thrust on the fluid as it is accelerated through the plane of the propeller. Although the flow field in the vicinity of the propeller must be unsteady, its action can be replaced by an accelerating "actuator disc" to reduce the problem to an equivalent steady one. The flow field is then approximately as shown with $v_1 < v_2$ and $p_1 = p_2$. For this idealized theory, show that $\bar{v}$ is the average velocity $(v_1 + v_2)/2$.

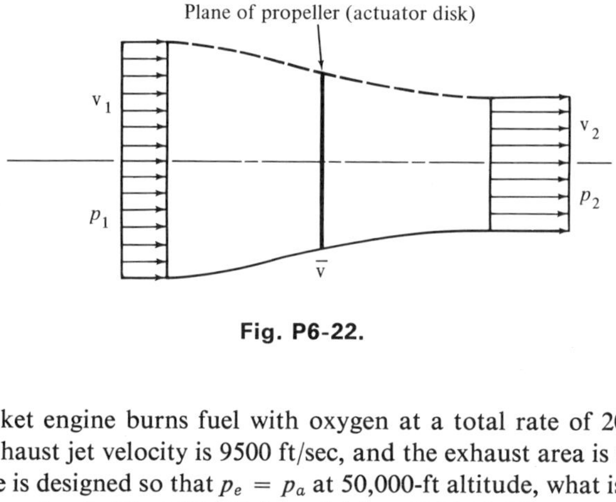

Fig. P6-22.

6-23 A rocket engine burns fuel with oxygen at a total rate of 200 lb_m/sec; the exhaust jet velocity is 9500 ft/sec, and the exhaust area is 6 ft^2. If the engine is designed so that $p_e = p_a$ at 50,000-ft altitude, what is the thrust (a) at sea level, (b) at 50,000 ft, and (c) at 200,000 ft?

6-24 If the rocket in the above problem carries enough propellant for 200 sec of operation and its dry weight is 5000 lb_m, what maximum altitude will it reach on a vertical flight? What is the acceleration at liftoff and at burnout?

6-25 If the flight velocity of a rocket is v_f, find an expression for the propulsive efficiency [η_p = output/(output + residual K.E.)] as a function of v_f/v_e, where v_e is the jet exhaust velocity relative to the rocket. Discuss the physical significance of an efficiency defined in this manner.

6-26 Expand the product $\vec{v} \times \nabla \times \vec{v}$ for a two-dimensional plane cartesian (x, y) coordinate system.

6-27 Expand the product $\vec{v} \times \nabla \times \vec{v}$ for a two-dimensional plane polar (r, θ) coordinate system.

6-28 Write Euler's equations in coordinate form by determining the expanded vector form of (6-36c) for a three-dimensional cartesian (x, y, z) coordinate system. Use the form of $\vec{a}$ from (6-35c).

6-29 Expand the vector form of Euler's equation, (6-36c), for a cylindrical (r, θ, z) coordinate system. Use the form of $\vec{a}$ from (6-35c).

6-30 (a) For a fluid rotating as a solid body about a fixed axis, find $\vec{\omega}$ anywhere in the fluid (that is, find the $\vec{\omega}$ field) by determining ω_1 and ω_2 at any arbitrary point. (b) Find the strain rate $\dot{\gamma}$.

6-31 For a fluid rotating about a fixed axis according to the relationship $vr = C$, calculate the rotation and strain rates as functions of r and C (C = constant).

6-32 For a fluid rotating about a fixed axis so that the tangential velocity is everywhere constant, find the rotation and strain rate fields, that is, find ω and $\dot{\gamma}$ as functions of r.

6-33 For "couette" flow (laminar, two-dimensional flow between two surfaces, one moving with respect to the other as shown) find the rotation $(\vec{\omega})$ and strain rate $(\dot{\gamma})$ as functions of x and y.

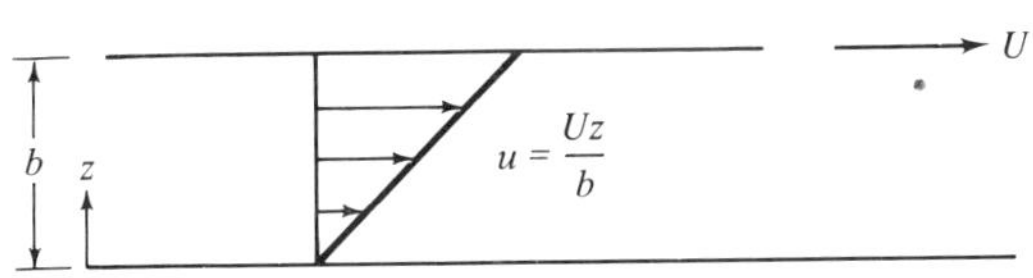

Fig. P6-33.

6-34 For laminar flow between two stationary plates as shown, determine the rotation $(\vec{\omega})$ and strain rate $(\dot{\gamma})$ as functions of x and y.

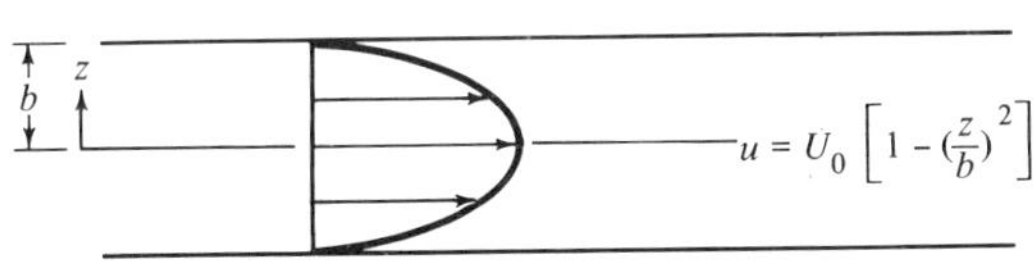

Fig. P6-34.

6-35 For laminar flow in a round pipe, find the rotation $(\vec{\omega})$ and strain rate $(\dot{\gamma})$ as functions of the cylindrical coordinates r, θ, z. The velocity distribution is given by $u = U_0[1 - (r/r_0)^2]$.

6-36 A free vortex is one that is defined by an irrotational flow in concentric circles about a fixed axis. Find (a) the velocity distribution as a function of radius, and (b) the pressure distribution for an incompressible flow as a function of radius assuming that at some radius r_1 the pressure is p_1.

6-37 Assuming that the flow halfway around a 90° bend behaves as an irrotational vortex flow, find an expression giving the mass flow rate as a function of pipe radius, mean bend radius, density, and pressure difference measured between the inside and outside of the bend.

6-38 A storm carrying a twister was traveling at a speed of 30 mph. Velocity readings at two points separated by a distance of 200 yards on a line perpendicular to the direction of travel of the storm indicated 40 and 15 mph in the direction of the travel as the eye of the twister passed somewhere between these two points. The local barometer reading was 28.9 in. Hg and the temperature was 78°F. (a) Locate the center of the "eye." (b) Assuming that the core of this twister was about 50 ft in diameter and that the flow outside this core is essentially irrotational, find the velocity and pressure at the edge of the core. Assume incompressible flow, but comment on this assumption.

6-39 A tornado transported by a storm moving at a ground speed of 40 ft/sec due west was determined to be nearly irrotational down to a 90-ft-diameter "core." (The core is a region of nearly constant angular velocity.) Air-speed meters recorded winds of 67.4 and 51.7 ft/sec at distances of 150 and 350 yards from the center of the core on a line running south from the path. Plot the pressure distribution (psf) across the path to a distance of 500 yards on each side of the center. (This approximation to a real vortex by an irrotational plus a "solid-body" vortex is referred to as a "combined vortex.")

6-40 In Example 6-3, it was stated that the time-mean velocity v in the wake region behind the orifice plate was zero. Consider the flow geometry in the jet and wake regions and describe any flow in the wake that could possibly occur from pressure gradients in the fluid downstream of the orifice.

6-41 It is assumed that the pressure measure at wall tap 2 (Fig. 6-14) is the same as the pressure in the jet at this station. State reasons why this assumption is either valid or false.

6-42 Explain how the average pressure at the vena contracta or at section 2 could be determined (refer to Fig. 6-14).

6-43 An orifice is to be sized such that the pressure drop is to be just equal to the upstream dynamic pressure ($\rho v_1^2/2g_c$). Find the necessary ratio of orifice diameter to upstream pipe diameter (d_0/d_1) to obtain this condition. The orifice Reynolds number will always be greater than 10^5.

6-44 A 1.000-in.-diameter orifice is used in a 1½-in. pipe (I.D. = 1.63 in.) to determine water flow. The pressure differential is to be determined from a mercury manometer. If the manometer reads 3.13 in., what is the flow rate in gpm if the water temperature is 68°F?

6-45 For the setup in Problem 6-44, make up a calibration curve of flow rate in gpm versus the manometer reading in in. Hg.

6-46 An orifice meter is to be used to determine air flows in a 3-in.-diameter duct, supplied by a blower. The blower discharge conditions are 450 standard ft^3/min at 18 psia and 108°F. A 6-in. micromanometer using isopropyl alcohol ($S = 0.802$) to measure the pressure across an orifice is to be used to measure the exact flow rate. What is the minimum-sized orifice that can be used?

6-47 (a) For the "impulse turbine" blading section shown, construct the inlet and outlet velocity triangles. Show and label all components and angles. Indicate the through-flow or Q/A component on both triangles. In laying out the triangles, take u, the blade velocity, as 1 in. in length and make the triangles as near to scale as possible. (b) Find the flow rate, Q, for $n = 1200$ rpm and the dimensions given on the sketch. (c) Find the horsepower if the fluid is water.

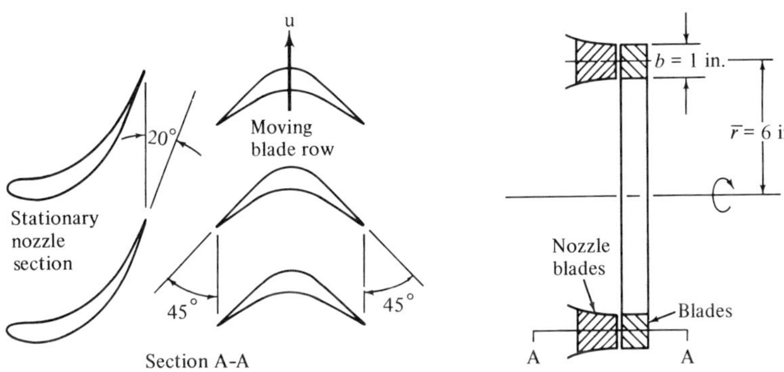

Fig. P6-47.

6-48 A Pelton water turbine is constructed as shown. For a jet velocity of 240 ft/sec and a mass flow rate of 600 lb/sec, (a) find the horsepower for a turbine speed of $n = 800$ rpm. (b) What is the "runaway" (no-load) speed (rpm) of this turbine? (c) Is this turbine operating at maximum efficiency? Explain.

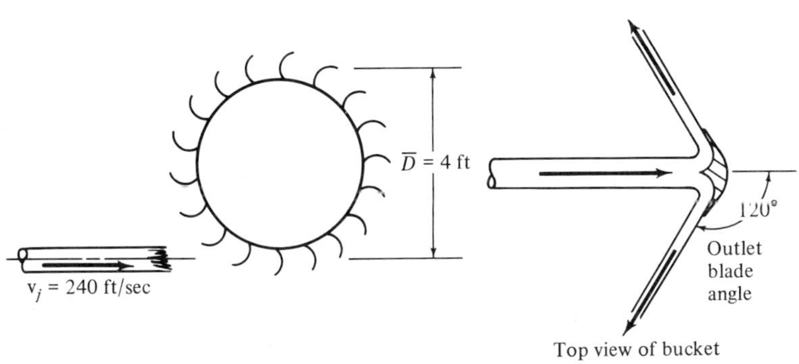

Fig. P6-48.

6-49 An axial-flow fan is mounted in a straight duct section between two flexible couplings as shown. The pressure rise across the fan corresponds to 6 in. of water. The drive motor is mounted in a set of flow-straightening vanes so that the flow is axial at 2 as well as at 1. Assuming that the air velocity is 180 ft/sec and the density at 1 is 0.75 lb_m/ft^3, find the thrust force on the flexible couplings. What is the torque on the couplings?

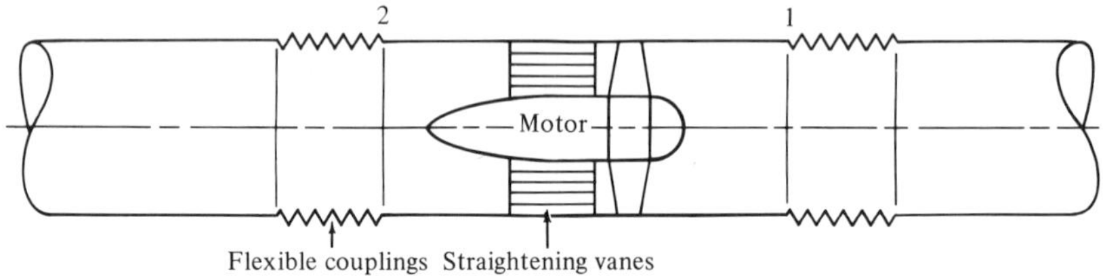

Fig. P6-49.

6-50 Determine an expression for the velocity coefficient C_v for a Venturi (or flow nozzle) meter as a function of the upstream and throat kinetic-energy correction factors, the area ratio, the pressure differential, and the pressure loss; that is, find $C_v(\alpha_1, \alpha_2, A_2/A_1, \Delta p, \Delta p_{\text{loss}})$.

6-51 At time $t = 0$, the cork is pulled out of the opening of the tank shown. The discharge coefficient for this opening is 1.00. Find the height (z) and the force on the scale (F) as functions of time. The diameter of the opening is d and diameter of the tank is D.

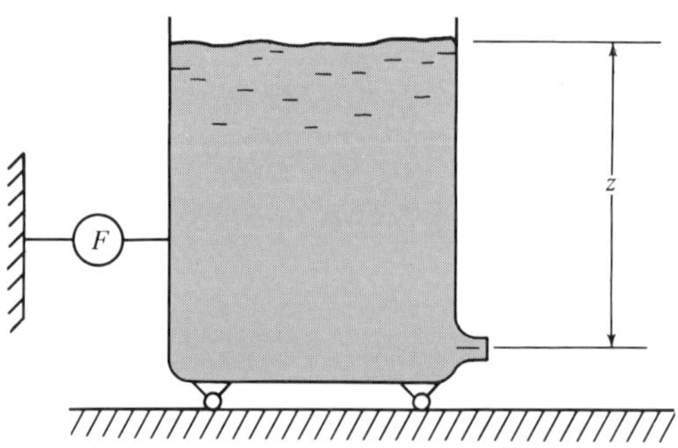

Fig. P6-51.

6-52 A very small pump is constructed by drilling two holes in a disc as shown. The disc is rotated at a velocity ω. Assuming an efficiency

$$\eta = \frac{\text{hydraulic output}}{\text{shaft input}} = \frac{\dot{m}\,\Delta p_0/\rho}{T} = \frac{Q\,\Delta p_0}{T} = 30\%$$

find the pressure rise across the pump.

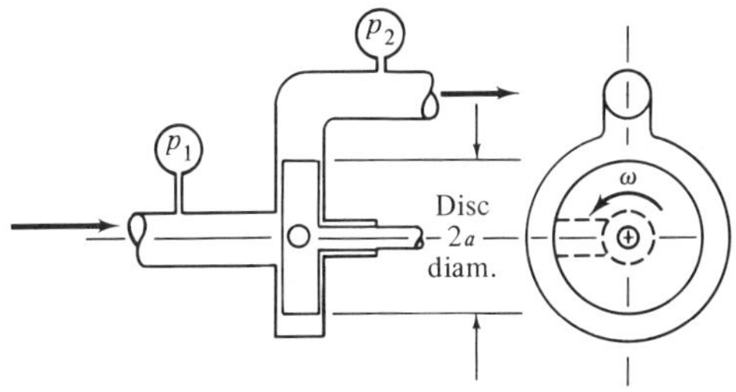

Fig. P6-52.

6-53 The two-dimensional pump impeller illustrated develops a differential head of 286 ft of water at a flow rate of $Q = 6$ ft³/sec. Assume radial inlet flow and that the relative flow enters and leaves tangent to the blades: (a) Draw the entrance velocity triangle relating absolute, blade ($u = \omega r$), and relative velocities. (b) Draw the exit triangle. (c) Circle all the quantities on these two triangles that are given or can be computed from the given quantities. (d) Find the rotational speed n. (e) Find the angle β_2'.

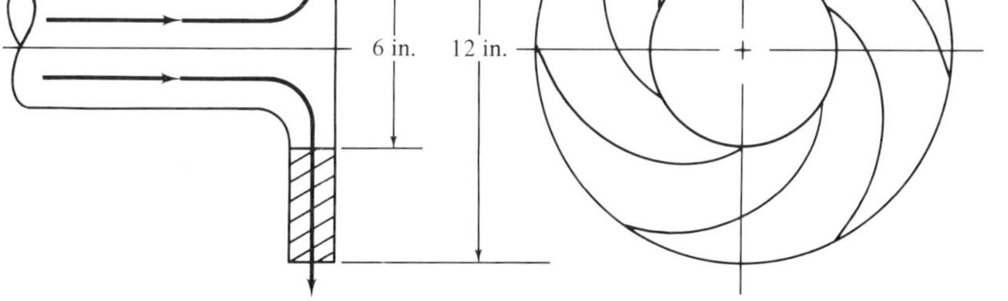

Fig. P6-53.

6-54 A radial-flow pump with the geometry shown is pumping 15 ft³/sec of water. What is the pressure rise ($p_2 - p_1$) across the pump? Assume that the water enters the rotor with zero angular velocity and leaves with a direction relative to the rotor determined by the exit blade geometry.

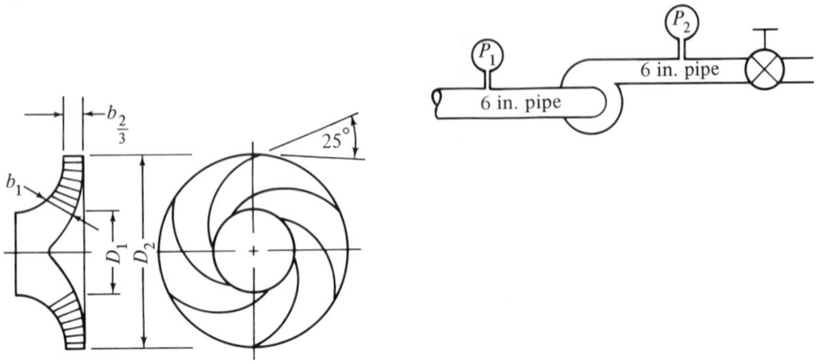

Fig. 6-54.

6-55 Use a 1,2,3 right-hand coordinate system (x_1, x_2, x_3); take $\vec{\omega} = \hat{x}_1\omega_1$ and $\vec{r} = \hat{x}_1 x_1 + \hat{x}_2 x_2$ and find the coordinate direction components of (a) $\vec{\omega} \times \vec{r}$, (b) $\vec{\omega} \times (\vec{\omega} \times \vec{r})$, and (c) $\vec{r} \times [\vec{\omega} \times (\vec{\omega} \times \vec{r})]$. (d) If x_1, x_2, and x_3 are, respectively, the radial (r), tangential (θ), and axial (z) directions (a cylindrical coordinate system), what are the directions of the above vector products? Comment on this result.

6-56 Using an (r, θ, z) cylindrical coordinate system and the geometry for a radial-flow pump (Figs. 6-24 and 6-25), indicate the direction choice giving $\vec{\omega} = \hat{z}\omega$ and $\vec{\imath} = \hat{r}r + \hat{z}z$. The velocity $\vec{w}$ will have all three components $\vec{w} = \hat{r}u + \hat{\theta}v + \hat{z}w$. Determine the direction of the vector products $\vec{\imath} \times (\vec{\omega} \times \vec{w})$ and $\vec{\imath} \times \vec{\omega} \times (\vec{\omega} \times \vec{\imath})$ appearing in $\vec{M}'$ in Section 6.20. Determine $P' = \vec{\omega} \cdot \vec{M}'$ and comment on the result with respect to (6-82a).

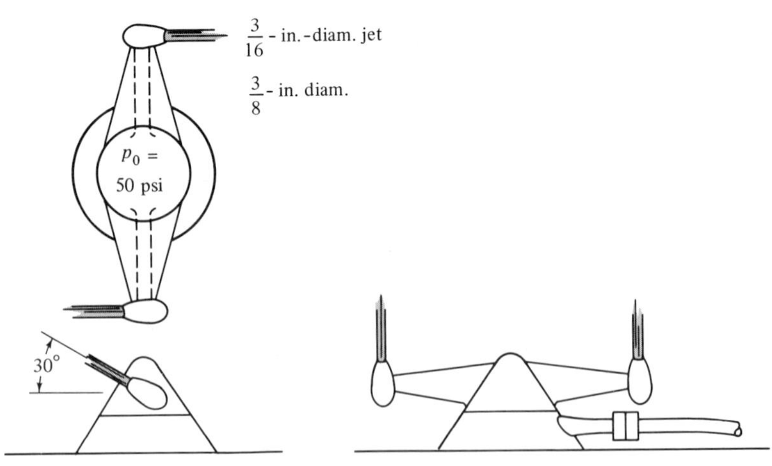

Fig. P6-57.

6-57 A lawn sprinkler is constructed as illustrated. The radial supply pipes are ⅜ in. in I.D. and the jet openings are 3/16 in. in I.D. The radius is 4.5 in. If the available supply pressure is 50 psi (measured at a point of negligible velocity), what is the maximum rpm if the bearing and air friction effects are neglected? The loss of availability between the supply (at 50 psi) and the jet exit is $2.5(\rho v_s^2/2g_c)$, where v_s is the velocity in the ⅜-in.-diameter supply line. What is the flow rate at the equilibrium rpm and when the sprinkler head is held stationary? What is the torque required to hold the sprinkler head stationary? Note that the jets are tilted at an angle of 30° with the horizontal plane.

6-58 A small ramjet helicopter is constructed with ramjet engines at the two blade tips. The blades are 7.5 ft long (radius) and the operating tip speed is 250 ft/sec. The fuel is picked up from a tank with a transfer pump that maintains a 5-psi pressure at the transfer line inlet at the hub. The fuel-supply rate is 0.2 lb_m/sec to each rotor (fuel density is 42 lb_m/ft^3) through ¼-in.-I.D. lines. The fuel is sprayed into the combustion chamber in the relative direction of the air stream through the engine. The nozzle velocity coefficient for the spray apparatus is $C_v = 0.8$. (a) Find the pressure in the fuel line where it enters the spray head. (b) Find the spray velocity. (c) The effective momentum factor to account for the spray direction distribution is 0.8 referred to the spray jet velocity. Find the net torque and horsepower due to the fuel supply of the system. Does this torque add to or subtract from the torque developed by the engine?

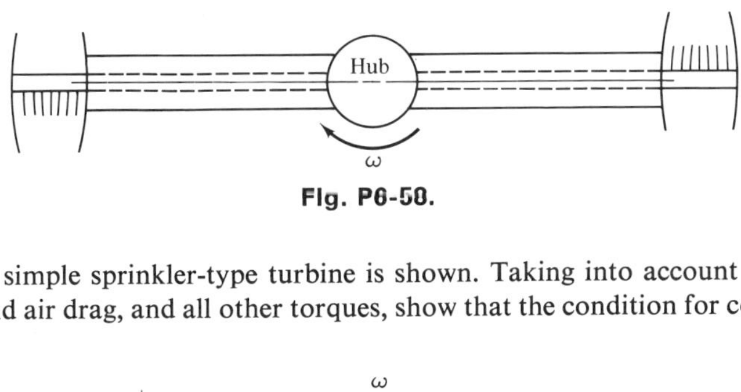

Fig. P6-58.

6-59 A simple sprinkler-type turbine is shown. Taking into account bearing and air drag, and all other torques, show that the condition for constant-

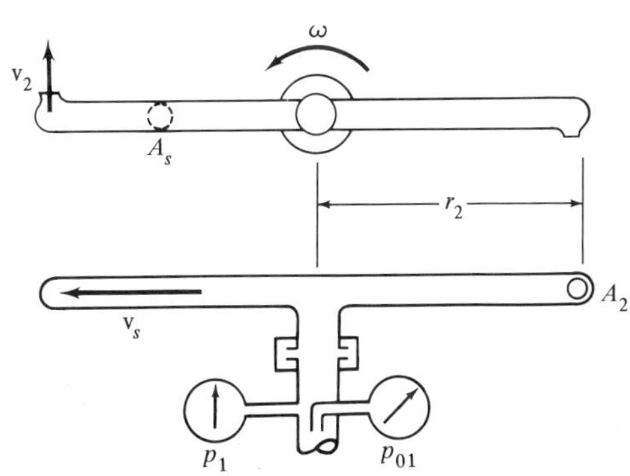

Fig. P6-59.

speed operation is $(mr_2/g_c)(\mathrm{v}_2 - \mathrm{u}_2) - C_T\omega^{1.2} = 0$, where $\mathrm{u}_2 = \omega r_2$, v_2 is the jet velocity relative to the arm at 2, and $C_T\omega^{1.2}$ is the air plus bearing drag torque. (C_T is an overall drag torque coefficient for the apparatus and it has been determined experimentally that the drag torque is proportional to $\omega^{1.2}$.)

6-60 For the apparatus shown in the figure, if $A_s = 4A_2$ ($A_2 = 0.2$ in.^{2}), $r_2 = 6$ in., and p_{01} is maintained constant at 20 psi, find the angular speed, neglecting friction effects.

6-61 For the conditions of Problem 6-60, what is the running speed ω if p_1 rather than p_{01} is maintained constant? Interpret your results.

6-62 For the sprinkler apparatus shown in Problem 6-59, $A_2 = 0.02$ in.^{2}, $A_s = 4A_e$, $p_{01} = 20$ psi, the loss coefficient for the flow from 1 to the nozzle exit is 2.0 [that is, $p_{02} - p_{01} = 2(\rho \mathrm{v}_s^2/2g_c)$]. Under these conditions, the steady running speed was found to be 620 rpm. What is the value of the drag torque coefficient C_T defined in Problem 5-59, assuming that the drag torque is $C_T\omega^{1.2}$? What are the dimensions of C_T?

6-63 A model of Hero's turbine (ca. 600 B.C.) was constructed as shown. The sphere diameter is 6 in. and the jet outlets are on a 4.5-in. radius and are 3/16 in. in I.D. If the boiler pressure is maintained at 1 psi, what would be the speed of rotation if the friction torque is ½ of the torque measured when the rotor is held stationary under conditions of the same boiler pressure (1 psi)?

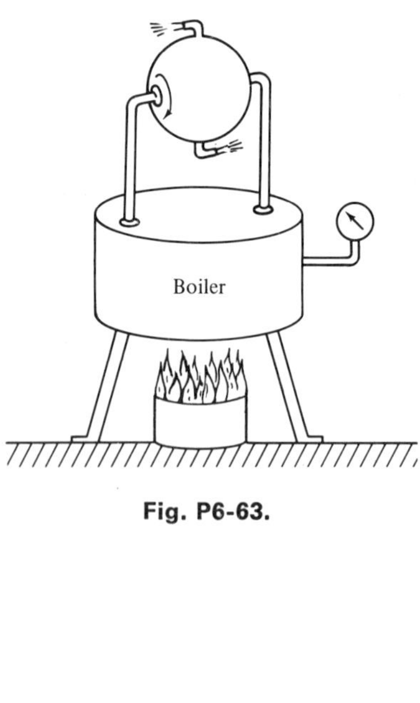

Fig. P6-63.

CHAPTER 7

CONTENTS

DIMENSIONAL ANALYSIS

When problems arise that cannot be solved by analytical methods and when physical system behavior is too complicated to be formulated mathematically into a set of equations, the methods of dimensional analysis allow the variables to be arranged into physically significant dimensionless groups. These groups can be used to simplify the mathematical equations and their solutions, or to simplify and give direction to experimental investigations and the formulation of semiempirical relationships. The complexity of problems in fluid mechanics has led to the use of dimensional analysis in the formulation of important relationships where analytical formulations or solutions are not possible. The most notable example is the formulation of approximate relationships for the behavior of turbulent flow (considered in Chapter 10).

7.1 DIMENSIONAL HOMOGENEITY

The dimensions of variables in an equation written to determine a desired resultant quantity must combine to yield the dimensions of the quantity to be determined. For example, the area of a surface is a function of a combination of lengths. These lengths must combine to give the dimensions of $(\text{length})^2$, the dimensions of an area. Consider the area of the composite shape shown in Fig. 7-1. The surface area is given by the relation

$$A = L_1L_2 + \frac{\pi}{8}L_1^2 + \frac{1}{2}L_1L_3 + \frac{2}{3}L_2L_4 \tag{7-1}$$

Each group on the right is made up of length times length, or $(\text{length})^2$, which is the dimension of area. The fact that each and every separate group of the equation has the same dimension (or resultant combination of dimensions) is an expression of dimensional homogeneity. To have any physical meaning, an equation must be dimensionally homogeneous. Also, the type of dimensions used must be homogeneous; all dimensions

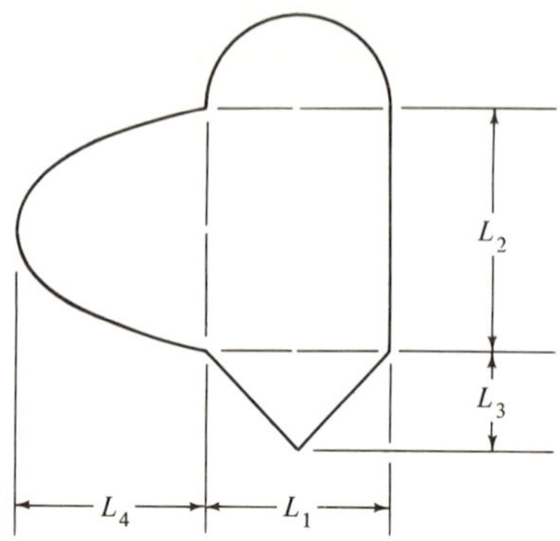

Fig. 7-1. Composite surface area.

must be compared to the same standard of length measurement, or a conversion factor must be used such that all lengths (L_1, L_2, and so on) are referred to the same standard. An equation must then be completely dimensionally homogeneous in both the basic types of dimensions and in the units of measurement of the dimensions.

The dimensions of quantities used in fluid mechanics can be expressed in terms of combinations of five standard dimensions: mass, force, length, time, and temperature.[1] Heat units are sometimes used, but this addition is unnecessary. "Heat" is defined as transient energy and therefore has the dimensions of energy rate. A relation between temperature and energy is also available by way of Boltzmann's constant. Although it is easy to represent the dimensions of temperature in energy units (for example, joules or ft-lb_f), the current use of temperature as a fundamental quantity makes this substitution undesirable.

7.2 DIMENSIONAL ANALYSIS

The mathematical formulation of a physical phenomenon is a functional relationship among the physical quantities involved. These quantities are measured in terms of arbitrary but standardized reference units. It is an absolute necessity that the standard or basis of measurement for these unit measurements be consistent throughout the functional relationship (mathematical equation) in order for the equation to possess any sense or meaning.

The above example for area as a function of various lengths is quite simple, since all quantities are lengths and area is in terms of $(\text{length})^2$.

[1] If electromagnetic effects are to be considered, dimensions of charge and magnetism must be included.

However, consider another example: By noticing the wind resistance on a person's hand held out of a moving vehicle, an observer may conclude that wind force or pressure is a function of velocity. Further consideration would indicate that the density of the fluid would be relevant (that is, at the same velocity in water, a larger pressure would be expected). A general functional relation,

$$f(p, \rho, \mathrm{v}) = 0 \tag{7-2}$$

among these variables could be investigated. Equation (7-2) gives an implicit relation for the pressure p. The zero on the right side indicates that this relation has zero dimensions. An explicit relation for p can be written

$$p = p(\rho, \mathrm{v}) \tag{7-3}$$

Since the left side of (7-3) has the dimensions of pressure, the functional relation on the right side must form a combination of the physical variables ρ and v, which also has the dimensions of pressure (that is, force/area, or F/L^2 in fundamental units). In a mechanical system, the fundamental units would be chosen as mass (M), force (F), length (L), and time (T). The following notation implies that pressure has the units of F/L^2, and so on:

$$[p] = \frac{F}{L^2} \qquad [\mathrm{v}] = \frac{L}{T} \qquad [\rho] = \frac{M}{L^3}$$

A mathematical function such as (7-2) can be expanded into a power series of terms containing each of the variables in a manner that leaves the powers and constant coefficients as unknowns to be determined by whatever methods are available. For example,

$$f(p, \rho, \mathrm{v}) = C_1 p^{a_1} \rho^{b_1} \mathrm{v}^{c_1} + C_2 p^{a_2} \rho^{b_2} \mathrm{v}^{c_2} + \cdots + C_n p^{a_n} \rho^{b_n} \mathrm{v}^{c_n} = 0 \tag{7-4}$$

which states that the variables in each term must form a group that has dimensions consistent with the rest of the equation. These dimensions must be the same as those on the right side, or zero. The C's, a's, b's, and c's are dimensionless constants that could be determined if an analytic solution were possible. The requirement of physical homogeneity of an equation allows some (if not all) of the unknown exponents to be determined.

The dimensions of p, v, and ρ expressed in fundamental units are indicated above. Noting that M, F, and T each occur only once, no relation with zero dimensions can be found. Reference to Newton's second law in the form

$$F = \frac{1}{g_c} \frac{d}{dt}(m\mathrm{v})$$

indicates the necessity of including the dimensional factor g_c in dynamic analyses where F and m occur. In general, g_c enters in dynamic problems involving force and mass (or density). Equation (7-2) can then be rewritten

$$f(p, \rho, \text{v}, g_c) = 0 \tag{7-5}$$

so (7-4) becomes

$$f(p, \rho, \text{v}, g_c) = C_1 p^{a_1} \rho^{b_1} \text{v}^{c_1} g_c^{d_1} + \cdots = 0 \tag{7-6}$$

Each term of (7-6) has identically the same form and each term must have zero dimensions, as dictated by the zero on the right side. This statement can be translated into the mathematical requirement

$$[p^a][\rho^b][\text{v}^c][g_c^d] = M^0 F^0 L^0 T^0 \tag{7-7}$$

which states that the quantities p, ρ, v, and g_c must combine to form a product (or group of products) that must have zero dimensions. This requirement may be restated as an equation among the order of the dimensions:

$$\left(\frac{F}{L^2}\right)^a \left(\frac{M}{L^3}\right)^b \left(\frac{L}{T}\right)^c \left(\frac{ML}{FT^2}\right)^d = M^0 F^0 L^0 T^0 \tag{7-8a}$$

or

$$M^{b+d} F^{a-d} L^{-2a-3b+c+d} T^{-c-2d} = M^0 F^0 L^0 T^0 \tag{7-8b}$$

This relation is valid only if the exponents of each of the dimensions are identical. The equations relating the exponents a, b, c, and d are then

$$\begin{aligned} b + d &= 0 \\ a - d &= 0 \\ -2a - 3b + c + d &= 0 \\ -c - 2d &= 0 \end{aligned} \tag{7-9}$$

These relations form a set of four equations in four unknowns. However, the set is not linearly independent; that is, one equation can be obtained as a linear combination of the others. This condition is also indicated by considering (7-8a). The fourth group can be obtained as a combination of the first three; that is,

$$\frac{ML}{FT^2} = \frac{(M/L^3)(L^2/T)}{F/L^2}$$

Consequently equations (7-9) cannot be solved, since the set is redundant by one equation; no unique value for the variables can be determined.

However, it is possible to determine any three of the unknowns in terms of a fourth.

The solution to (7-9) is easily carried out by substitution. Systematic solution of more complicated sets of equations is possible using matrix methods as presented in the appendix to this chapter, where the solution to (7-9) is used as an example.

Determining a, b, and c in terms of d gives $a = d$, $b = -d$, and $c = -2d$. The desired result indicated by (7-6) is then

$$f(p, \rho, \mathrm{v}, g_c) = C_1\left(\frac{g_c p}{\rho \mathrm{v}^2}\right)^{d_1} + C_2\left(\frac{g_c p}{\rho \mathrm{v}^2}\right)^{d_2} + \cdots = 0 \qquad (7\text{-}10)$$

Dividing through by the first group gives

$$C_1 + C_2\left(\frac{g_c p}{\rho \mathrm{v}^2}\right)^{\delta_1} + \cdots = 0$$

where $\delta_1 = d_2 - d_1$, and so on. If the series is terminated with two terms, $g_c p/\rho \mathrm{v}^2 = K$ or $p = K(\rho \mathrm{v}^2/g_c)$ is the final result. In either case $g_c p/\rho \mathrm{v}^2$ is shown to be the important dimensionless group. An experiment to determine K would give a value of about ½.

The above procedure is known as the *Buckingham pi method.* The dimensionless groups obtained are termed π groups; that is, $\pi \equiv g_c p/\rho \mathrm{v}^2$.

The same result would be obtained if the observer moved his hand through water at a velocity v. However, if the experiment were tried with a "very viscous" liquid such as corn syrup, the results may be different, and the viscosity of the fluid would be a significant quantity.

As a more general case, consider a body of significant dimensions A, B, and C moving in a fluid of density ρ and viscosity μ at a velocity v and angle of attack α measured between some defined axis and the direction v. What will be the primary factors affecting the pressure measured at any point on the body? The general functional relation can be written

$$f(p, \rho, \mathrm{v}, \mu, A, B, C, \alpha, g_c) = 0 \qquad (7\text{-}11)$$

The number of quantities is $n = 9$ and the number of dimensions is $m = 4$. The number of dimensionless groups characterizing the flow behavior would then be 5 at most. Before proceeding, some obvious shortcuts may be used to good advantage. The quantity α is an angle and therefore already dimensionless. A, B, and C are of identical dimensions and dimensionless ratios B/A and C/A can immediately be formed by inspection. The problem can then be reduced to the relation

$$\left[(p^a \rho^b \mathrm{v}^c \nu^d A^e g_c^f)(\alpha)^g \left(\frac{B}{A}\right)^h \left(\frac{C}{A}\right)^i\right] = M^0 P^0 L^0 T^0 \qquad (7\text{-}12)$$

It is necessary to deal with only the first term in the brackets, since the last three groups are already dimensionless. Another simplification has been included by using ν instead of μ for the viscosity quantity. By definition $\nu \equiv g_c(\mu/\rho)$; ρ and g_c are also in the set, so either μ or ν is a satisfactory choice. Since the dimensions of ν are more direct, simpler, fewer, and less ambiguous, it is easier to use in cases where ρ is a necessarily included quantity.

The functional relationship to be investigated is $\varphi(p^a \rho^b \mathrm{v}^c \nu^d A^e g_c^f) = 0$. The equations for the exponents of the dimensions are

$$\begin{aligned} b + f &= 0 \\ a - f &= 0 \\ -2a - 3b + c + 2d + e + f &= 0 \\ c + d + 2f &= 0 \end{aligned} \tag{7-13}$$

Since there are only four equations, only four of the exponents can be determined in terms of two others. The easiest solution is to determine *a* through *d* in terms of *e* and *f*, indicating that there will be two dimensionless groups. Solving (7-13) gives

$$\begin{aligned} a &= f \\ b &= -f \\ c &= d + f \\ d &= -e \end{aligned}$$

The indicated solution is then

$$f\left(\frac{g_c p}{\rho \mathrm{v}^2}, \frac{\mathrm{v}A}{\nu}, \alpha, \frac{B}{A}, \frac{C}{A}\right) = 0$$

or

$$p = K\frac{\rho \mathrm{v}^2}{2g_c}\varphi\left(N_R, \alpha, \frac{B}{A}, \frac{C}{A}\right)$$

In this result, the pressure p is given explicitly as the reference dynamic pressure $(\rho \mathrm{v}^2/2g_c)$ times some function of the geometry $(\alpha, B/A, C/A)$ and the Reynolds number (N_R). The function of the Reynolds number would still have to be determined. Instead of dealing with the dimensional quantity p it is advantageous to use the dimensionless group $p/(\rho \mathrm{v}^2/2g_c)$ defined as a *pressure coefficient* c_p, or $c_p = \varphi(\alpha, B/A, C/A, N_R)$.

In the above example, the dimensionless groups were immediately recognized as significant parameters (pressure coefficient and Reynolds number). In many cases the results will not produce recognizable groups. The arbitrariness comes from the choice of the order in which the variables are written down in (7-11) or (7-12). It follows that the π groups are not unique in arrangement. However, since any two or more dimensionless groups can be combined (by multiplying, dividing, squaring, and so on) to form another group, the set of π groups can be manipulated to obtain another set of π's which have physically significant interpretations.

7.3 NONDIMENSIONALIZATION OF MATHEMATICAL EQUATIONS

The nondimensionalization of the mathematical equations governing the physical behavior of a system gives an alternative method of obtaining significant dimensionless groups. This method yields both a simplification of the mathematical analysis by nondimensionalizing the equations and allows the important parameters in the solution to be determined.

Consider the flow around a two-dimensional airfoil of chord length c at an angle of attack measured with respect to the direction of the relative approach velocity U, as indicated in Fig. 7-2. It is desired to determine

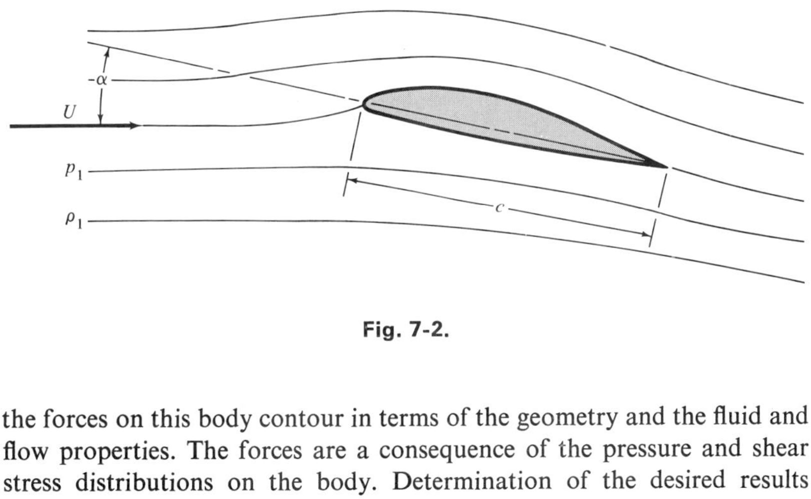

Fig. 7-2.

the forces on this body contour in terms of the geometry and the fluid and flow properties. The forces are a consequence of the pressure and shear stress distributions on the body. Determination of the desired results would require simultaneous solution of the coordinate momentum equations, the energy equation, and the continuity equation under the restrictions of the physical boundary conditions. This set of equations is nonlinear to the extent that a solution for the pressure and shear distributions at a given angle of attack is not possible. Therefore, the momentum equations will be nondimensional to see what information can be obtained from this procedure. Consider first the momentum equation in the streamline direction as given in Chapter 6:

$$\frac{1}{g_c}\left(\frac{\partial \mathrm{v}}{\partial t} + \mathrm{v}\frac{\partial \mathrm{v}}{\partial s}\right) + \frac{1}{g_c}\frac{\partial gz}{\partial s} + \frac{1}{\rho}\frac{\partial p}{\partial s} = \frac{1}{\rho}\frac{\partial \tau}{\partial s} = \frac{1}{\rho}\frac{\partial}{\partial s}(2\mu\dot{\gamma}_{ns}) \qquad (7\text{-}14)$$

where

$$\dot{\gamma}_{ns} = \frac{1}{2}\left(\frac{\partial \mathrm{v}}{\partial n} - \frac{\mathrm{v}}{R_s}\right)$$

Nondimensional variables can be defined with respect to *constant reference quantities* that have a particular significance to the problem, for example, p_1, ρ_1, U_1, and c. The nondimensional variables will be indicated by the prime superscript:

$$p' \equiv \frac{p}{p_1} \quad \mathrm{v}' \equiv \frac{\mathrm{v}}{U_1} \quad \rho' \equiv \frac{\rho}{\rho_1} \quad s' \equiv \frac{s}{c} \quad z' \equiv \frac{z}{c} \quad t' \equiv \frac{U}{c}t \quad g' \equiv \frac{g}{g_0}$$

Making this substitution gives

$$\left(\frac{U_1^2}{g_c c}\right)\frac{\partial \mathrm{v}'}{\partial t'} + \left(\frac{U_1^2}{g_c c}\right)\mathrm{v}'\frac{\partial \mathrm{v}'}{\partial s'} + \left(\frac{g_0}{g_c}\right)\frac{\partial g'z'}{\partial s'} + \left(\frac{p_1}{\rho_1 c}\right)\frac{1}{\rho'}\frac{\partial p'}{\partial s'} = \left(\frac{\mu}{\rho_1}\frac{U_1}{c^2}\right)\frac{1}{\rho'}\frac{\partial \dot{\gamma}'}{\partial s'}$$

Multiplying through by $g_c c/U^2$ will make the resulting equation dimensionless:

$$\frac{\partial \mathrm{v}'}{\partial t'} + \mathrm{v}'\frac{\partial \mathrm{v}'}{\partial s'} + \left(\frac{g_0 c}{U^2}\right)\frac{\partial g'z'}{\partial s'} + \left(\frac{g_c p_1}{\rho U_1^2}\right)\frac{1}{\rho'}\frac{\partial p'}{\partial s'} = \left(\frac{\mu g_c}{\rho_1 U_1 c}\right)\frac{1}{\rho'}\frac{\partial \dot{\gamma}'}{\partial s'} \quad (7\text{-}15)$$

Since the primed quantities are all dimensionless, the groups formed by the combinations of the physical dimensional quantities (the groups in parentheses) must be dimensionless. They are.

The group on the right side of (7-15) is the reciprocal of the Reynolds number. In this application, it is the Reynolds number based on approach values of density and velocity with the airfoil chord as the reference dimension. As explained in Chapter 6, (7-15) applies only for the particular case of incompressible flow, so that $\rho = \text{const.} = \rho_1$; consequently, $\rho' = 1$. The quantity $g_0 c/U^2$ is the reciprocal of the square of the Froude number ($N_F \equiv U/\sqrt{g_0 c}$). The Froude number can be interpreted as the ratio of the inertia forces to the square root of the gravitational forces. The quantity $g_c p_1/\rho_1 U^2$ is a dimensionless pressure coefficient sometimes referred to as the Euler number.

The solutions to a nondimensionless equation such as (7-15) also depend on the boundary conditions, which are the given reference quantities p_1, ρ_1, U, μ, d and the geometry as specified by the size c and shape of the physical boundaries and α. The only parameters in (7-15) that depend specifically on these quantities are the dimensionless groups in the coefficients and α. The results expressed as dimensionless quantities (pressure, shear stress, lift, and drag coefficients) will be identical independent of the specific size of the geometry and properties of the fluid. The significance of these coefficients and the coefficient parameters of (7-15) will be discussed further below.

Another arbitrary choice is possible. Instead of the reference pressure p_1, an equivalent quantity with the dimension of pressure could be used; that is, the choice could have been made that $p' \equiv g_c p/\rho_1 U^2$, thereby eliminating p_1 as a reference quantity and indicating that the solution is independent of p_1. This result is not surprising considering that a uniform pressure field would have no effect on the dependent variables, the pressure p', and velocity v′, which are the quantities obtained from the solution. The ambient field pressure p_1 is only significant in connection with the buoyant force. In this case the buoyant force and ambient pressure variation are in equilibrium (as treated in Chapter 3). Therefore, the coefficient of the dimensionless pressure term is unity, and this coefficient loses its significance as a dimensionless group. However, the dimensionless pressure $p' \equiv g_c p/\rho_1 U^2$ is of greater significance.

If the body forces in this particular problem are only of gravitational type (as indicated), they are balanced by the ambient static-pressure field. The pressure p would then be interpreted as the sum of the dynamic and hydrostatic effects as previously considered. The hydrostatic variation would be that due to the buoyant effects, and these effects could be removed from the dynamic equation and calculated separately, leaving

$$\frac{\partial \mathrm{v}'}{\partial t'} + \mathrm{v}' \frac{\partial \mathrm{v}'}{\partial s'} + \frac{\partial p'}{\partial s'} = \frac{1}{N_R} \frac{\partial \dot{\gamma}'}{\partial s'} \tag{7-15a}$$

for the s-component equation. Note that the Froude number also disappears, indicating that the gravitational force is no longer of significance in the solution of this problem. The n-component equation could also be nondimensionalized, but it is obvious that no new information would be gained.

The solution of (7-15a) will now depend only on the magnitudes of the *reference Reynolds number* as long as geometric similarity is maintained. A solution to this equation in combination with the other equations would yield the dimensionless pressure distribution on the body. As long as the Reynolds numbers are the same and no forces other than those considered (for example, electromagnetic forces) are significant, this dimensionless pressure distribution solution (from which the pressure forces can be obtained) is then the same for any size of geometrically similar airfoil at the same angle of attack α in any fluid at any flight speed U.

The above procedure has produced two valuable pieces of information. First, the nondimensional equation is more general, yields a general solution, and is simpler in form than the dimensional equation. Second, the important parameter of the solution is the Reynolds number. The second consequence is of importance with respect to experimental investigations. An experiment can be designed to model test airfoil performance (lift and drag) as functions only of Reynolds number, angle of attack, and geometry rather than each of the dimensional quantities that make up the Reynolds number.

A similar problem can be considered where a hydrofoil is running at a reference depth d as indicated in Fig. 7-3. If the z coordinate is chosen as the vertical (altitude) coordinate and z' is now defined with respect to d, $z' \equiv z/d$ the coefficient of the gravitational force term will be $g_0 d/U^2$ (the reciprocal of the square of the Froude number based on running depth). The fact that the free surface is relatively close to the hydrofoil ($z \approx d \approx c$) indicates that this parameter may be important. The conclusion drawn in the case of the airfoil (that is, that the hydrostatic forces were independent of the gravitational forces) is no longer valid because the shape of the free

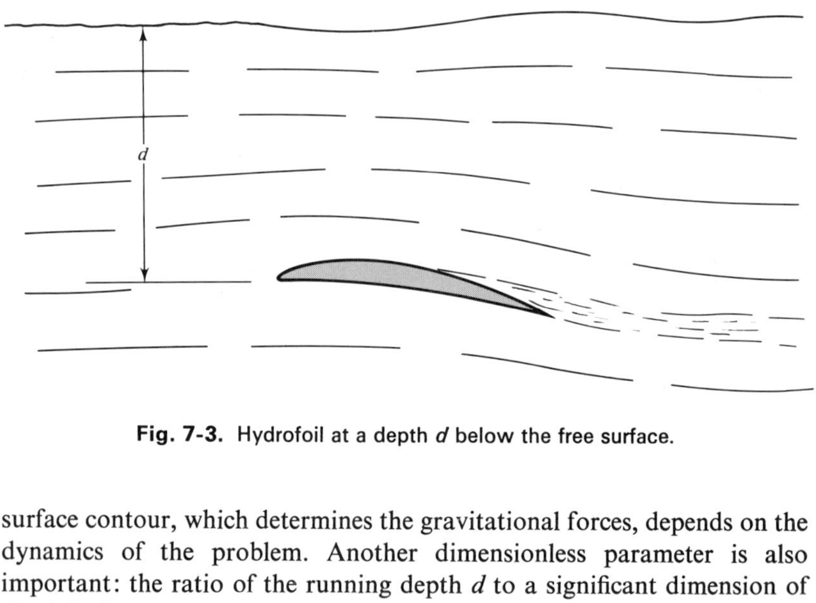

Fig. 7-3. Hydrofoil at a depth *d* below the free surface.

surface contour, which determines the gravitational forces, depends on the dynamics of the problem. Another dimensionless parameter is also important: the ratio of the running depth d to a significant dimension of the foil, for example, c; this parameter is then d/c.

It is not possible to draw any immediate conclusions with respect to the magnitude of the dimensionless groups. For example, if the Reynolds number is "quite large" (so that its reciprocal is very small with respect to unity), it cannot be deduced that viscous terms can be neglected. A direct consequence of a large magnitude of Reynolds number is that the shear strain rate (velocity gradient at the body surface) is quite large, so the product of the two terms is still significant.

Another development and interpretation is possible if the time variations that affect the first term are periodic. If the frequency is ω, nondimensionalization of the first term with $t' \equiv t\omega$ (instead of that used above) gives

$(\omega U/g_c)(\partial v'/\partial t')$ which, when multiplied by $g_c c/U^2$, leaves the parameter $\omega c/U$, the ratio of two characteristic speed quantities. If it is rewritten $\omega/(U/c)$, it can be interpreted as the ratio of two characteristic frequencies. In this second form it is referred to as the Strouhal number, N_{Str}.[2]

Still another interpretation of the pressure term is possible. In (7-20) the $(1/\rho)(\partial p/\partial s)$ term can be rewritten $(\partial p/\partial \rho)(1/\rho)(\partial \rho/\partial s)$. The density and density gradient are then nondimensionalized to give

$$\left(\frac{1}{c}\frac{\partial p}{\partial \rho}\right)\frac{1}{\rho'}\frac{\partial \rho'}{\partial s'}$$

When this term is multiplied by $g_0 c/U^2$ (as in the above procedure to nondimensionalize the entire equation), the result is

$$\frac{g_c(\partial p/\partial \rho)}{U^2}\frac{1}{\rho'}\frac{\partial \rho'}{\partial s'}$$

The quantity $g_c(\partial p/\partial \rho)$ is the square of the velocity of sound a, so this coefficient (which must be dimensionless) is $a^2/U^2 \equiv 1/M^2$. The quantity M^2 (or M, the *Mach number*) is seen to be significant when the variations of density $\Delta\rho/\rho$ are significant. It will be shown in Chapter 12 that Mach-number variations over a range from zero to 0.3 in air flows (air speeds of 0 to 350 ft/sec) produce a relative compressibility $\Delta\rho/\rho$ of only 2 percent. Air (and gas) *flows* are then approximately "incompressible" within this range of Mach numbers.

7.4 SIMILARITY

For two flow patterns to be similar, the first requirement is that the boundary geometries be similar. The flow around a model airfoil is geometrically similar to the full-scale prototype only if the model is an exact scale model of the prototype and is tested in a similar geometric surrounding. Geometric similarity also requires that the angle of attack of the model be identical to that of the prototype for a given flight condition. Under these conditions, the model airfoil of Fig. 7-4b would be geometrically similar to that of the prototype (Fig. 7-4a). The flow field patterns would be geometrically similar if photographs of the streamlines were taken and these also satisfied the required condition of geometrical similarity. If the flow patterns of Fig. 7-4 are similar, dynamic similarity exists among the flow properties: pressure, velocity, and so on. This

[2] A third interpretation is possible if the ratio is written as $(c/U)/(1/\omega)$ or $(c/U)/T$, where T is the period of the frequency ($T = 2\pi/\omega$). The ratio then relates two significant periods.

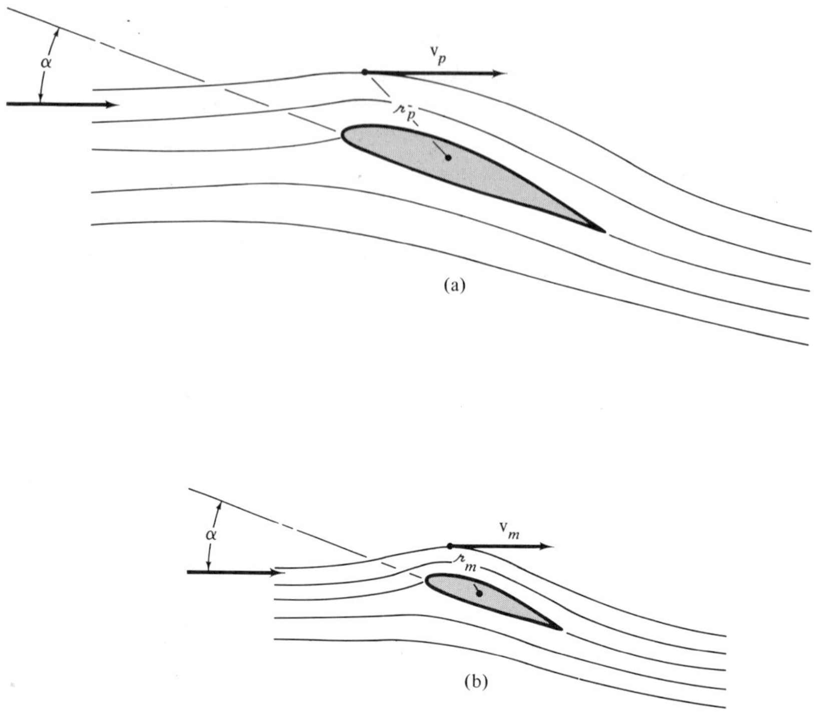

Fig. 7-4. Prototype airfoil (a) and geometrically similar model (b).

dynamic similarity depends upon the principles (continuity, momentum, and so on), which govern the flow behavior. These principles governing the dynamics of the flow must be considered to determine the requirements for dynamic similarity.

There are three methods for obtaining the parameters, which must be identical if the flow fields are to be similar: (1) an intuitive consideration of the actions that determine the flow field, (2) nondimensionalization of the equations derived from the applicable physical principles, and (3) the formal procedure of dimensional analysis of the variables that enter into the problem. An example of the intuitive approach was given in Chapter 4. When the viscous and dynamic (inertia) forces are of primary importance, equivalence of the dimensionless parameter Reynolds number in two flow cases of identical geometry indicates that the flow fields are dynamically similar.

7.5 DYNAMIC SIMILARITY

Dynamic similarity will be discussed in connection with the example of determining the power required to propel a ship. The quantities of

TABLE 7-1

Quantity	*Symbol*	*Dimensions*
Power	P	FL/T
Speed	U	L/T
Length	l	L
Beam	b	L
Draft	d	L
Propeller speed	ω	$1/T$
Fluid density	ρ	M/L^3
Surface tension	σ	F/L
Viscosity	μ (or ν)	FT/L^2 (or L^2/T)
Fluid modulus	E	F/L^2
Propeller pitch	β	0
Propeller diameter	D	L
Gravitation acceleration	g	L/T^2
Dynamic constant	g_c	ML/FT^2

importance are given in Table 7-1. The mass (or weight) of the ship is not included in the table as an independent quantity, because it is a direct function of ρ, b, d, and l (the weight is equal to the weight of the displaced fluid in the "displacement volume"). Some immediately recognizable dimensionless groups are β, l/b, d/b, and D/b. The quantities β, d, l, and D will then be omitted from further consideration in the analysis, while the dimension b is retained. There are then six (or less) π groups expected from the remaining ten variables, since these variables are all expressible in four units. Using the above method of dimensional analysis these groups are:

1. The power coefficient or power number $C_P \equiv g_c P/\rho U^3 b^2$.
2. The propeller tip speed to ship speed ratio:

$$\frac{\omega b}{U}\frac{D}{b} = \frac{\omega D}{U} = N_{\text{Str}}$$

3. The Reynolds number $N_R = Ub/\nu$.
4. The Mach number $M = U\sqrt{g_c E/\rho}$.
5. The Froude number $N_F = U/\sqrt{gb}$.
6. The Weber number $N_W = g_c\sigma/\rho U^2 b$.

These parameters are all "dynamic" dimensionless numbers in that they all contain the velocity U. Indeed, they are all ratios of the forces considered to the reference inertia force. The only group not determined from the previous nondimensionalization of the force–momentum equation is the Weber number. In that equation there was no way of considering surface tension forces in a simple manner in the formulation of the momentum equation. The implicit reduced functional relationship is then

$$f\left(C_P, N_{\mathrm{Str}}, N_R, M, N_F, N_W, \beta, \frac{l}{b}, \frac{d}{b}, \frac{D}{b}\right) = 0 \tag{7-16a}$$

or, explicitly,

$$C_P = \varphi\left(N_{\mathrm{Str}}, N_R, M, N_F, N_W, \beta, \frac{l}{b}, \frac{d}{b}, \frac{D}{b}\right) = 0 \tag{7-16b}$$

The second relation states that the dimensionless power number will be the same (same functional relationship) for all geometrically identical ships (same β, l/b, d/b, D/b) under identical dynamic running conditions as determined by identical values of N_{Str}, N_R, M, N_F, and N_W. An analytical study would then attempt a solution in terms of these dimensionless parameters. An analytical study that would give the power coefficient in terms of these parameters would be either too complicated or too approximate or both. In any case, a model study is usually made as a final check. Equation (7-16) then gives the important variables to be correlated. The power coefficient for a model study would be determined as a function of the five dynamic similarity parameters. The functional relationships obtained would be graphical results of power coefficient as a function of each of the five parameters.

Some further thought and experience can be used to simplify the order of (7-16) by reducing the number of significant parameters. Any combination of the dimensionless groups is a dimensionless group. Therefore, the relative importance of one effect (force) with respect to the others may be determined by forming ratios different from those above. For example, the Reynolds number represents the ratio of inertia to viscous forces, while the Weber number represents the ratio of inertia to surface tension forces. The Weber number divided by the Reynolds number then indicates the relative importance of viscous forces with respect to surface tension forces. The surface tension is important in two ways, by the actual surface tension force at the water line and by the production of so-called surface tension or capillary waves. In general, these forces are relatively small with respect to both the viscous shear forces, or the gravitational forces resulting from the production of large gravity waves. Experience indicates that surface tension forces are not significant with respect to gravity or shear forces for a model of usual size (on the order of 2 ft or greater waterline length). Also water is relatively incompressible, so the effects contained with Mach number can be neglected. The relationship to be determined is then reduced to $C_P(N_R, N_F, \omega d/U, \text{geometry})$.

It is quite difficult to test a scale model with both Reynolds number and Froude number the same as for the prototype (see Problem 7-13). However, previous experience with model testing indicates that for bodies at large Reynolds numbers, the dependence of the dimensionless drag force (the force that must be equaled by the propulsion system) on Reynolds number varies only at relatively low values of Reynolds number and becomes constant (and therefore independent of Reynolds number) above a certain *critical Reynolds number*. The full-scale prototype is so large that it can be expected to be operating above this critical Reynolds number. If insufficient experience on ranges of critical Reynolds numbers for ship geometries is available, the specific model can be tested and its critical Reynolds number obtained. If all significant data are then obtained at a Reynolds number greater than this critical value, Reynolds number will no longer be a significant parameter for the model study. This does not say that the friction forces are not important, only that they are a fixed function of the inertia forces or the dynamic motion. The relation is now reduced to $C_P(N_F, \omega D/U$, geometry). The geometry must, of course, be identical. As stated above, the propulsion system supplies only sufficient power to overcome the drag force. If the model is tow-tested to determine this power, the test is independent of the power-producing system and therefore the group $\omega D/U$ is eliminated. However, if it is desired to determine the drive characteristics with consideration of the interdependence of the propulsion system (propeller), the model would have to be self-propelled and the Froude and Strouhal numbers would have to be the same (see Problem 7-14). This situation is nearly impossible to achieve except in the case of a full-scale model, so that tow-tank tests are usually made to determine the desired data.

Dimensional analysis and consideration of the relative importance of dimensionless groups has reduced the problem to determining the dependent quantity C_P as a function of only one independent parameter. The significance of this particular independent parameter is that it indicates the relative importance of gravity forces and inertia forces. The gravity forces in this problem produce the large wave system accompanying the motion of a body on the surface of a liquid. The geometry of the wave pattern is a function of the speed of the body U. The significant dimensionless parameter is then the Froude number.

7.6 SUMMARY

1. A general functional relationship among the physical quantities relevant to a physical phenomena can be written in the form $f(q_1, q_2, q_3, \ldots, q_n) = 0$. By dimensional analysis a simplified expression relating the significant dimensionless π groups can be obtained as $\varphi(\pi_1, \pi_2, \pi_3, \ldots, \pi_{n-r}) = 0$. There will be $n - r$ groups resulting from the dimensional analysis, where r is the rank of the matrix formed by the equations

expressing the fact that the dimensions of the first equation above are zero. This rank will be equal to or less than the number of fundamental dimensions (n) in which the quantities (q's) are measured.

2. Determination of the quantities to be included in the dimensional analysis depends on the investigator's understanding of the physical phenomena being studied.
3. The dimensionless groups greatly simplify the design of an experimental or semiempirical solution to an investigation. Consideration of the physical significance of the dimensionless groups adds further significance to, and a possible further simplification of, the investigation.
4. In some cases *similarity variables* (combinations of the independent variables with dependent variables and dimensionless parameters) can be developed that simplify the mathematical analyses. These methods are best illustrated by way of examples found in the solution of the boundary-layer equations, and so on, in the following chapters.

7.7 APPENDIX: Matrix Solutions of Simultaneous Equations

The methods of matrix algebra and the use of determinants simplify the solution of sets of equations and provide more logical means of handling such sets. The development of the set (7-9) from the dimensional relations (7-7) or (7-8) can be done in the following manner. The variables are written across the top of a tabular array and the dimensions down the left side:

	p^a	ρ^b	V^c	g_c^d
M	0	1	0	1
F	1	0	0	-1
L	-2	-3	1	1
T	0	0	-1	-2

(7-17)

The matrix array is filled in with the numerical coefficients of the exponents. For example, the first *column* indicates that pressure has the dimensions of FL^{-2}. The top *row* is seen to give the coefficients of the equation for the condition that the dimensions of the mass terms of (7-8) or (7-9) are zero. Equations (7-9) can be determined directly from this matrix array. The array has m rows and n columns. An abbreviation notation for the elements of this matrix is α_{ij}, where $i = 1, 2, \ldots, m$ and $j = 1, 2, \ldots, n$. For example, $\alpha_{12} = 1$ (row 1, column 2), $\alpha_{44} = -2$, and so on. The equation for the requirement of dimensional homogeneity of mass units in the above dimensional equations is

$$\alpha_{11}a + \alpha_{12}b + \alpha_{13}c + \alpha_{14}d = 0$$

For the set of $m = 4$ of these equations to have a solution (for a, b, c, and d), the determinant of the matrix must not vanish. The determinant (Δ) of this matrix is a number,

$$\Delta \equiv |\alpha_{ij}| = \begin{vmatrix} \alpha_{11} & \alpha_{12} & \alpha_{13} & \alpha_{14} \\ \alpha_{21} & \alpha_{22} & \alpha_{23} & \alpha_{24} \\ \alpha_{31} & \alpha_{32} & \alpha_{33} & \alpha_{34} \\ \alpha_{41} & \alpha_{42} & \alpha_{43} & \alpha_{44} \end{vmatrix} = \begin{vmatrix} 0 & 1 & 0 & 1 \\ 1 & 0 & 0 & -1 \\ -2 & -3 & 1 & 1 \\ 0 & 0 & -1 & -2 \end{vmatrix} \tag{7-18}$$

This number is evaluated most easily by factoring the determinant. The order of the determinant ($n = m = 4$) is reduced by factoring rows or columns until a 2×2 (or 3×3) determinant is obtained, which is easily calculated. For example, each element of the top row of the above determinant can be factored out and multiplied by a quantity from the remainder of the determinant as follows:

$$\Delta = \alpha_{11}C_{11} + \alpha_{12}C_{12} + \alpha_{13}C_{13} + \alpha_{14}C_{14} \tag{7-18a}$$

The C's are the *cofactors* of the corresponding *element*. The cofactors are defined as $(-1)^{i+j}M_{ij}$ of the element α_{ij}, where the *minor* M_{ij} is the determinant (or array) obtained by striking out the *row* and *column* in which the term α_{ij} occurs. Equation (7-12) can then be rewritten

$$\alpha_{11}(-1)^{1+1}M_{11} + \alpha_{12}(-1)^{1+2}M_{12} + \alpha_{13}(-1)^{1+3}M_{13} + \alpha_{14}(-1)^{1+4}M_{14}$$

This procedure gives

$$\Delta = \alpha_{11}(-1)^2 \begin{vmatrix} \alpha_{22} & \alpha_{23} & \alpha_{24} \\ \alpha_{32} & \alpha_{33} & \alpha_{34} \\ \alpha_{42} & \alpha_{43} & \alpha_{44} \end{vmatrix} + \alpha_{12}(-1)^3 \begin{vmatrix} \alpha_{21} & \alpha_{23} & \alpha_{24} \\ \alpha_{31} & \alpha_{33} & \alpha_{34} \\ \alpha_{41} & \alpha_{43} & \alpha_{44} \end{vmatrix}$$
$$+ \alpha_{13}(-1)^4 \begin{vmatrix} \alpha_{21} & \alpha_{23} & \alpha_{24} \\ \alpha_{31} & \alpha_{33} & \alpha_{34} \\ \alpha_{41} & \alpha_{43} & \alpha_{44} \end{vmatrix} + \alpha_{14}(-1)^5 \begin{vmatrix} \alpha_{21} & \alpha_{22} & \alpha_{23} \\ \alpha_{31} & \alpha_{32} & \alpha_{33} \\ \alpha_{41} & \alpha_{42} & \alpha_{43} \end{vmatrix} \tag{7-19}$$

The M_{ij} minor determinants are further factored by the same means. For example, M_{11} can be evaluated (by factoring the first *column*) as

$$M_{11} = \alpha_{22} \begin{vmatrix} \alpha_{33} & \alpha_{34} \\ \alpha_{43} & \alpha_{44} \end{vmatrix} - \alpha_{32} \begin{vmatrix} \alpha_{23} & \alpha_{24} \\ \alpha_{43} & \alpha_{44} \end{vmatrix} + \alpha_{42} \begin{vmatrix} \alpha_{23} & \alpha_{24} \\ \alpha_{33} & \alpha_{34} \end{vmatrix} \tag{7-20}$$

The value of a 2 × 2 determinant is defined as

$$\begin{vmatrix} \beta_{11} & \beta_{12} \\ \beta_{21} & \beta_{22} \end{vmatrix} = \beta_{11}\beta_{12} - \beta_{12}\beta_{21} \tag{7-21}$$

It is obvious that the most direct way to factor a determinant is first to factor the column or row with the most zeros. Factoring out the top row of the above numerical determinant (7-18) gives

$$\Delta = -1\begin{vmatrix} 1 & 0 & -1 \\ -2 & 1 & 1 \\ 0 & -1 & -2 \end{vmatrix} + (-1)\begin{vmatrix} 1 & 0 & 0 \\ -2 & -3 & 1 \\ 0 & 0 & -1 \end{vmatrix} = 0$$

which is an interesting result. Since the first two and fourth equations of (7-9) can be combined to give the third, one entire row is reduced to zeros. When factored, this row then yields zero for the value of the determinant.

The unknowns in a set of equations such as (7-9) are obtained by (1) replacing the *column* in the matrix (7-18) in which the desired unknown appears by the constant term column from the right side of the equation, and (2) dividing the determinant resulting from this substitution by the determinant (Δ) of the coefficients of the unknowns. For example, for the more general set

$$\begin{aligned} \alpha_{11}x_1 + \alpha_{12}x_2 + \alpha_{13}x_3 + \alpha_{14}x_4 &= k_1 \\ \alpha_{21}x_1 + \alpha_{22}x_2 + \alpha_{23}x_3 + \alpha_{24}x_4 &= k_2 \\ \alpha_{31}x_1 + \alpha_{32}x_2 + \alpha_{33}x_3 + \alpha_{34}x_4 &= k_3 \\ \alpha_{41}x_1 + \alpha_{42}x_2 + \alpha_{43}x_3 + \alpha_{44}x_4 &= k_4 \end{aligned} \tag{7-22}$$

the unknowns are determined as

$$\begin{aligned} x_1 &= \frac{(-1)^{1+1}}{\Delta}\begin{vmatrix} k_1 & \alpha_{12} & \alpha_{13} & \alpha_{14} \\ k_2 & \alpha_{22} & \alpha_{23} & \alpha_{24} \\ k_3 & \alpha_{32} & \alpha_{33} & \alpha_{34} \\ k_4 & \alpha_{42} & \alpha_{43} & \alpha_{44} \end{vmatrix} \\ x_3 &= \frac{(-1)^{1+3}}{\Delta}\begin{vmatrix} \alpha_{11} & \alpha_{12} & k_1 & \alpha_{14} \\ \alpha_{21} & \alpha_{22} & k_2 & \alpha_{24} \\ \alpha_{31} & \alpha_{32} & k_3 & \alpha_{34} \\ \alpha_{41} & \alpha_{42} & k_4 & \alpha_{44} \end{vmatrix} \end{aligned} \tag{7-23}$$

In the above example, since all the k's are zero and $\Delta = |\alpha_{ij}| = 0$, all the unknowns ($x_1 = a$, $x_2 = b$, and so on) are indeterminant. This result is, of course, a direct consequence of the fact that the set (7-9) is not independent. Because this set of four equations is not linearly independent, there are three (or less) independent equations. Consequently, any three unknowns may be determined in terms of the other by solving the *reduced set* of *any three* equations (provided that these three equations are linearly independent).

The number of linearly independent equations in a set such as (7-15) is determined by the *rank* (r) of the matrix of the set. The rank of a matrix is defined as *the order of* the *largest of any nonzero determinant that can be formed from that matrix.* By inspection of (7-18) it can be seen that all 3×3 determinants within the matrix are nonzero; therefore, $r = 3$. From these observations of the behavior of the set of dimensional equations, the following theorem was derived: *The number of dimensionless groups which can be formed by a set of n variables having m dimensions is $n - r$, where r is the rank of the matrix formed by the dimensional coefficients.* If there are no redundant (linearly dependent) relations, $r = m$ or n, whichever is smaller. In general, $r \leq m, n$.

In the above example, there are $n = 4$ variables and the rank of the matrix is $r = 3$; therefore, one dimensionless group is expected. The easiest choice for the three independent equations is

$$\begin{aligned} b + d &= 0 \\ a - d &= 0 \\ -c - 2d &= 0 \end{aligned}$$

which is a set of three equations and four unknowns. The most simple solution is to determine a, b, and c in terms of d. However, the following set is selected to better illustrate the calculation procedure and should give the same results:

$$\begin{aligned} b + d &= 0 \\ -2a - 3b + c + d &= 0 \\ -c - 2d &= 0 \end{aligned}$$

Assume that it is desired to solve for a, b, and c in terms of d. The variable d is now assumed to be a "known" or nonvariable quantity and, with the appropriate coefficients, forms the k's of (7-15) and (7-16). This set can then be written

$$\begin{aligned} b &= -d \\ -2a - 3b + c &= -d \\ -c &= 2d \end{aligned}$$

with a matrix given by

a	b	c	k
0	1	0	$-d$
-2	-3	1	$-d$
0	0	-1	$2d$

From the form of (7-16), a is obtained by replacing the first row by the "k row" (the last row) and dividing by the determinant found from the coefficients of a, b, and c:

$$a = \frac{1}{\Delta}\begin{vmatrix} -d & 1 & 0 \\ -d & -3 & 1 \\ 2d & 0 & -1 \end{vmatrix} = \frac{d}{\Delta}\begin{vmatrix} -1 & 1 & 0 \\ -1 & -3 & 1 \\ 2 & 0 & -1 \end{vmatrix}$$

where

$$\Delta = \begin{vmatrix} 0 & 1 & 0 \\ -2 & -3 & 1 \\ 0 & 0 & -1 \end{vmatrix} = -2$$

(The factor d is common to an entire column and therefore can be factored as indicated.) The factoring can be continued to give

$$a = \frac{d}{\Delta} -1 \left\{ \begin{vmatrix} -3 & 1 \\ 0 & -1 \end{vmatrix} - 1 \begin{vmatrix} -1 & 1 \\ 2 & -1 \end{vmatrix} \right\} = \frac{d}{2}(-3 \quad +1) = d$$

The unknown b is obtained by replacing the b column by the k column and dividing by Δ:

$$b = \frac{1}{\Delta}\begin{vmatrix} 0 & -d & 0 \\ -2 & -d & 1 \\ 0 & 2d & -1 \end{vmatrix} = \frac{d}{2}\begin{vmatrix} 0 & 1 & 0 \\ -2 & 1 & 1 \\ 0 & -2 & -1 \end{vmatrix} = -d$$

Similarly, c is determined as

$$c = \frac{1}{\Delta}\begin{vmatrix} 0 & -1 & -d \\ -2 & -3 & -d \\ 0 & 0 & 2d \end{vmatrix} = \frac{d}{2}\begin{vmatrix} 0 & 1 & 1 \\ -2 & -3 & 1 \\ 0 & 0 & -2 \end{vmatrix} = -2d$$

The matrix for the dimensional relation (7-12) is

	p^a	ρ^b	v^c	ν^d	A^e	g_c^f
M	0	1	0	0	0	1
F	1	0	0	0	0	−1
L	−2	−3	1	2	1	1
T	0	0	−1	−1	0	−2

Since at least one fourth-order determinant can be found that is not zero, the resulting 4×6 matrix has a rank of four. The number of dimensionless groups will then be two, or the coefficients can be determined in terms of any two. These two are then assumed "known" and are the k's of (7-22) and (7-23). To obtain the form (7-22) in order to write the solutions in the form of (7-23), the "constants" must be transposed to the right side of the equation. For this example, a, b, c, and d will be the unknowns to be found in terms of e and f. The quantities e and f are then inserted into the matrix (as indicated below) with opposite signs according to the requirement that they are transferred to the right side of the equations to agree with the form (7-22):

$$\begin{array}{rrrr|rr} 0 & 1 & 0 & 0 & 0 & -f \\ 1 & 0 & 0 & 0 & 0 & f \\ -2 & -3 & 1 & 2 & -e & -f \\ 0 & 0 & -1 & -1 & 0 & 2f \end{array} \qquad \begin{aligned} k_1 &= -f \\ k_2 &= f \\ k_3 &= -e - f \\ k_4 &= 2f \end{aligned}$$

Note that the number of rows is determined by the number of dimensions and, in most cases, will be less than the number of unknowns (or columns). The constant (k) terms will then be made up of combinations of possibly several coefficients. In these cases it will be easier to use k_1, k_2, k_3, and so on, as indicated and then substitute the k values after the solution for the unknown has been made. For this example the solution will be expected to be in the form $p^a \rho^b v^c \nu^d L^e g_c^f = (\pi_1)^e (\pi_2)^f$, where π_1 and π_2 are the two resulting dimensionless groups. The value of the denominator determinant is easily determined as $\Delta = -1$. The exponent a is then determined as

$$a = -1 \begin{vmatrix} k_1 & 1 & 0 & 0 \\ k_2 & 0 & 0 & 0 \\ k_3 & -3 & 1 & 2 \\ k_4 & 0 & -1 & -1 \end{vmatrix} = k_2 = f$$

The remaining exponents are determined in a similar manner.

PROBLEMS

7-1 The pressure on a moving body was determined as a function of ρ and U (and g_c). Form a dimensional analysis to determine the force as a function of the density, velocity, g_c, and a reference area A (for example, the frontal projected area) of the object. From this result show that a force coefficient can be determined as $C_F = F/[(\rho U^2/2g_c)A]$.

7-2 Show that each term of (7-20) is dimensionless.

7-3 The pressure loss, Δp_{loss} (loss of availability), in a length L of pipe is a function of the diameter D, the density ρ, and the velocity v. By dimensional analysis show that these variables can be grouped into the functional relation

$$\varphi\left(\frac{\Delta p_{\text{loss}}}{\rho \text{v}^2/2g_c}, \frac{L}{D}\right) = 0$$

The simplest functional relation is a constant, for example,

$$\frac{\Delta p_{\text{loss}}}{\rho \text{v}^2/2g_c} = f\frac{L}{D}$$

where f is termed the "pipe friction factor."

7-4 The friction characteristics for flow in a pipe would be expected to be a function of ρ, v, D, μ (or ν), and roughness ϵ. If the dimensionless friction coefficient is defined as f, make a dimensional analysis of the variables involved and show that f is a function of Reynolds number and the relative roughness ϵ/D.

7-5 Consider the flow of water through an orifice of diameter d in a pipe D. List the other quantities that you expect to be of importance and perform a dimensional analysis to determine the effect of these quantities on the pressure drop across the orifice.

7-6 The force required to propel a body through a fluid at a velocity v depends upon the density of the fluid and the geometry of the body as specified by three characteristic lengths, L_1, L_2, and L_3. Perform a dimensional analysis on these factors to determine the groups of primary importance.

7-7 The lift force F_L that an airplane wing attains at a speed v in a fluid of density ρ and viscosity μ also depends on the plane form area A and the span s of the wing as well as the angle of attack α. Perform a dimensional analysis on these variables and show that

$$\varphi\left(\frac{F_L}{(\rho \text{v}^2/2g_c)A}, \frac{A}{s^2}, \frac{\text{v}s}{\nu}, \alpha\right) = 0$$

7-8 In analyzing turbulent flow velocity profiles to determine U as a function of y, the skin friction τ_0 and the fluid properties (density and viscosity) are important. Show that these variables can be grouped into the dimensionless quantities $U/\sqrt{g_c\tau_0/\rho}$ and $y\sqrt{g_c\tau_0/\rho}/(g_c\mu/\rho)$. Note that the quantity $g_c\tau_0/\rho$ has the dimensions of velocity squared and is given the designation u_τ. The above relation can then be written $\varphi(u_+, y_+) = 0$, where u_+ and y_+ are used as abbreviations to designate the quantities indicated.

7-9 Perform a general nondimensionalization using the π method to determine the importance of nondimensional parameters for a hydrofoil, where p, ρ, U, ν, σ, g, g_c, c, and d are significant quantities (c and d are the chord and running depth, respectively).

7-10 A force coefficient is defined as $C_F = F/[(\rho v^2/2g_c)A]$. Without making a dimensional analysis, show that a power coefficient can be obtained as $C_P = P/[(\rho v^3/2g_c)A]$.

7-11 Derive the π groups as given in the example of the model-ship propulsion test.

7-12 Discuss the parameters of importance for determining stability criteria for a jet airplane on landing.

7-13 Investigate the scale factors for model testing a ship where it is necessary to consider both Froude number N_F and Reynolds number N_R.

7-14 Discuss the model testing of a self-propelled model where it is necessary to maintain N_F and the N_{Str}.

7-15 In the case of the hydrofoil illustrated in Fig. 8-3, nondimensionalization of the momentum equation indicated that N_R and N_F are both significant. Determine and discuss the requirements for a scale-model experiment.

7-16 For a 1:100 scale model of a ship with a 400-ft length at water line and a 60-ft beam, what is the model speed to model the performance of the prototype cruising at 20 knots? (1 knot = 1.689 ft/sec; 1 nautical mile = 6080 ft). What are the beam Reynolds numbers of model and prototype?

7-17 In Section 7-5 it was shown that N_F and N_{Str} would have to be the same for modeling a self-propelled ship model. What are the velocity ratio v_m/v_p and the propeller speed ratio ω_m/ω_p in terms of the scale factor l_p/l_m? Would you expect the propeller performance to be independent of Reynolds number from prototype to model? Explain.

7-18 A 1:100 scale ship model is to be self-propelled. If the prototype is to cruise at 15 knots with a propeller speed of 70 rpm, what are the cruising speed and propeller speed of the model?

7-19 For a model test where M and N_R must be the same for prototype and model, show that the thickness of the wind-tunnel walls is independent of the scale. (*Tip*: Consider the relation for the wall stress in a cylinder.)

7-20 A prototype transport is to be designed to fly at $M = 3$ at 80,000 ft. At what Mach number must a 1:50 scale model be tested to provide dynamic similarity? If the model fluid is air at the same conditions as those for the prototype, what is the ratio of the power required to fly the prototype to that required by the tunnel power system to overcome the drag of the model?

7-21 "High-density" wind tunnels have been used to model test airplane components in transonic and supersonic flow. What is the relation between the geometric scale and density ratio for model testing, where both N_R and M are significant? Assume that in the temperature range of both prototype and model the relation $\mu/\mu_0 = \sqrt{T/T_0}$ is valid, where μ_0 is a constant reference viscosity at T_0.

7-22 Investigate the possibility of model testing a ship hull with both N_R and N_F the same. If the prototype is to be used in water, is it possible to find a liquid in which to test a quarter-scale model? (Refer to Table 2-1.)

7-23 For a V-notch weir as shown in the figure, make a dimensional analysis to determine the functional relationship among the important variables and show that your results can be put in the form

$$\frac{Q}{\sqrt{gH^5}} = \varphi\left(\frac{\sqrt{gH^3}}{\nu}, \frac{gH^2}{g_c\sigma/\rho}, \theta, \frac{H}{D}\right)$$

Consider and discuss this relation to imply that $Q = KH^{5/2}$ could be assumed to give fairly accurate results for a specified weir installation and state the conditions for which the assumptions are valid.

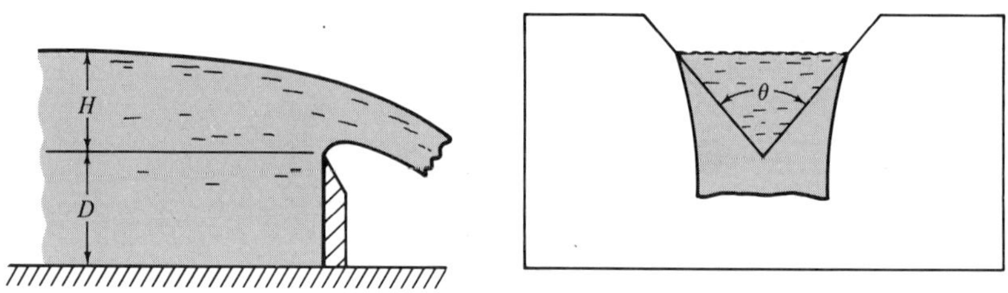

Fig. P7-23.

7-24 The power developed by a turbine can be expressed as $P = KQH$, where K is a dimensional constant, Q is the flow rate (ft^3/sec), and H is the total head or energy drop across the turbine. (a) Show that K is a function of ρ, g, and g_c. (b) Find the numerical value of K for water so that P will be determined in horsepower.

7-25 Pump power P is a function of ρ, Q, D, μ (or ν), and g_c. Using dimensional analysis find the power as a function of groups of the remaining variables. From this result derive a dimensionless power coefficient.

7-26 Compressor performance data can be expressed in terms of dimensionless groups to simplify experimental testing and to make the results more general and easier to express. If the important variables are mass flow rate $\dot{m}$, rotational speed N, pressure rise across the compressor Δp, and inlet temperature T, find the dimensionless groups that would best allow "mapping" of compressor performance. Density ρ and reference quantities p_a and ρ_a are also important in determining the performance parameters. However, if the performance parameters are referred to standard atmospheric conditions p_a and T_a (and therefore ρ_a), show that the compressor performance map for a given size and geometry compressor (that is, for a given D) can be plotted with $\Delta p/n^2$ versus $\dot{m}\sqrt{T}/n$ with $n/\sqrt{T}$ as a curve parameter.

7-27 The speed of an elastic wave through a medium would be expected to be a function of the temperature (molecular agitation) and density (molecular inertia) of the medium. Find the expected functional relationship between a, T, ρ and g_c. Discuss the results.

7-28 For operation at subsonic flow ($M < 0.3$), tests of aircraft and missile components may be made in a water tunnel as well as a wind tunnel. Consider the testing of a 4-in.-chord airfoil model in both water and air. (a) If the wind tunnel test data is to cover a range of speeds of 50 to 300 ft/sec (atmospheric pressure and temperature test section), what is the range of water-tunnel speeds for the Reynolds number range to be the same? (b) Discuss the results. (c) What is the ratio of the forces measured in the wind tunnel to the forces measured in the water tunnel? (d) At zero angle of attack, the drag coefficient is $C_D =$ drag/$[(\rho v^2/2g_c)A]$, which is in the neighborhood of 0.01 for the airfoil being tested. The reference area is the chord times span area. The span is 12 in. Find the drag-force range for the wind-tunnel and water-tunnel tests and comment on the results.

7-29 Cavitation tests are to be made on a supercavitating hydrofoil in a free-surface water tunnel. Specify the relevant variables and perform a dimensional analysis to determine the important dimensionless groups. (The cavity pressure, or the fluid vapor pressure p_v, will be one of the relevant quantities.) Discuss the relative importance of the dimensionless groups.

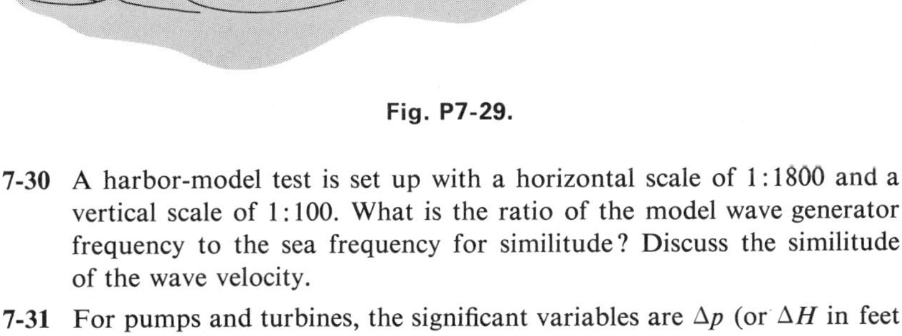

Fig. P7-29.

7-30 A harbor-model test is set up with a horizontal scale of 1:1800 and a vertical scale of 1:100. What is the ratio of the model wave generator frequency to the sea frequency for similitude? Discuss the similitude of the wave velocity.

7-31 For pumps and turbines, the significant variables are Δp (or ΔH in feet of head), flow rate Q, power P, rpm N, impeller diameter D, impeller width at exit b, blade angle θ, torque T, and fluid properties ρ and μ

(or ν) and g_c. Find the dimensionless groups of significance. Show that these can be obtained as

$$\pi_1 = \frac{g_c \, \Delta p}{\rho N^2 D^2} \quad \left(\text{or } \frac{gH}{N^2 D^2}\right) \qquad \text{pressure or head coefficient}$$

$$\pi_2 = \frac{Q}{ND^3} \qquad \text{flow coefficient}$$

$$\pi_3 = \frac{g_c P}{\rho N^3 D^5} \qquad \text{power coefficient}$$

$$\pi_4 = \frac{ND^2}{\nu} \qquad \text{Reynolds number}$$

$$\pi_5 = \frac{g_c T}{\rho N^2 D^5} \qquad \text{torque coefficient}$$

and others. Note that

$$\frac{N\sqrt{Q}}{H^{3/4}} = \frac{\sqrt{\pi_2}}{\pi^{3/4}} = N_{S,P}$$

the pump *specific speed*, which is a pump geometry factor. For example, a low-flow-rate high-head performance (as obtained with a "centrifugal" or radial-flow pump) is indicated by a low specific speed. An axial-flow or propeller-type pump delivers a large flow at a relatively small head and is therefore a high-specific-speed pump.

Show that

$$N_{S,T} = \frac{N}{(gH)^{3/4}} \sqrt{\frac{g_c P}{\rho}}$$

which is defined as the turbine specific speed. The g and g_c are sometimes dropped to yield relevant, but not dimensionless, operating coefficients.

7-32 A 1:10 scale model test of an airfoil is to be conducted in a wind tunnel (atmospheric test-section conditions of T and p). If the prototype sea-level velocity is to be 250 mph, what must be the test velocity in the wind tunnel? Discuss your results.

7-33 For a high-speed wind-tunnel test, consider modeling at supersonic speeds with a 10:1 scale and discuss practical methods to obtain model test data. Discuss the possibility of using various gases for modeling purposes where the prototype is to fly in air (air average molecular mass number $\mathscr{M} = 28.9$). What would be the maximum obtainable scale factor using mercury vapor in a supersonic wind tunnel operating at atmospheric pressure and temperature ($\mathscr{M} = 200$)?

PART III REAL FLOW TOPICS

The simplified analyses made possible by the assumptions defining one-dimensional and quasi-one-dimensional flow allow simplified solutions to problems that are quite difficult in many complete (three-dimensional) formulations. The results are applicable to a wide range of fluid flow systems. In Chapter 8, example flow fields are analyzed using the ideas, definitions, principles, and equations developed in the previous chapters.

Pipes, ducts, nozzles, and diffusers are examples of common flow geometries that can be treated as quasi-one-dimensional and analyzed to yield useful relations among flow rate or average velocity, pressure, and friction force. In most cases, exact analytical solutions are obtainable for *laminar* flow only. Solutions of *turbulent* flows usually require a semi-empirical approach using information obtained from experimental observations.

This class of flows in pipes, channels, ducts, and so on, is termed internal flow. These basic internal flow geometries can often be combined to make up the overall flow fields in flow systems or machines such as heat exchangers, pumps, turbines, and rocket nozzles. In the more complicated geometries, the quasi-one-dimensional results can be corrected with efficiency (or loss) factors. These corrections may be experimental, empirical, semiempirical, or analytically derived from two- and three-dimensional analyses. For example, the blades in a turbine or pump change both the direction and magnitude of the velocity passing between them. A quasi-one-dimensional analysis assumes that the flow is turned through an angle determined by the curvature of the blade passage. The availability loss might be approximated by the equivalent channel loss or boundary-layer friction loss over an equivalent area of a flat plate, which is easily calculated. More exact, analytical two- and three-dimensional analyses give correct results for each specific case. Results of these solutions are used as correction data to the one-dimensional flow. These corrections for flow through pump or turbine rotors can be determined as a function of the inlet and outlet flow angles and the ratio of blade spacing to cord length. In many cases, the simplified (or approximate) quasi-one-dimensional analyses are sufficient to yield information on the initial design suitability of a piece of machinery.

The class of flows about completely immersed bodies is termed *external* flow. The flow about basic geometries such as flat plates, cylinders, spheres, and airfoils falls in this class. General analytical solutions to the fluid mechanical equations for even these simple geometries are not possible.[1] Solutions can be obtained for certain "truncated" statements of the problem, where some of the effects, or terms in the equations, are cut off or dropped by assuming that these effects are secondary. The solutions obtained are good only in cases where the assumptions are valid. The flow field about a sphere can be obtained analytically for very slow or viscous (low-Reynolds-number) flows, where the nonlinear terms can be dropped or linearized to yield an equation that can be solved. For a wide range of Reynolds numbers, the drag on a sphere or cylinder is obtained experimentally. For airfoils, the lift as a function of angle of attack can be calculated from velocity and pressure distributions if friction effects are neglected. The friction or viscous effect is confined to a fluid layer so much thinner than the body that its direct effect on determining the pressure distribution is neglected. The solution is then a hypothetical one, the assumptions of which must be tested if there is not sufficient work or data on similar geometries to indicate that the solution is valid within the range of the assumptions. The insight gained into the basic behavior of flow fields, both internal and external, by considering certain hypothetical classes of mathematical formulations and solutions is valuable. In many cases, different solutions can be pieced together to yield an overall solution that provides valuable design information. The airfoil lift is obtained by solving the simplified set of equations for the pressure distribution and integrating the pressure distribution over the body surface. The drag is determined from an approximate solution to the equations for the boundary layer by integration of the shear force. The elements of these types of solutions will be treated in the following chapters.

[1] Computer solutions are possible using methods of numerical analysis.

CHAPTER 8

CONTENTS

LAMINAR, VISCOUS, INCOMPRESSIBLE FLOW: Quasi-One-Dimensional Internal Flows with Friction

Laminar, viscous flow is necessarily two- or three-dimensional because of the required condition that the relative velocity of the fluid adjacent to a solid boundary must be zero. The assumption of quasi-one-dimensional flow yields a number of direct and exact solutions of the general (Navier–Stokes) equations for laminar flow by simplifying both the equations and the boundary conditions. Equations in only one independent variable are obtained.

The following cases can be termed "classical" examples because they are some of the first problems to be solved exactly and appear in nearly every text on fluid mechanics. They are repeated here to illustrate methods by which such problems are handled. Also, the results (and modifications thereof) have many applications in flow systems.

8.1 FLOWS BETWEEN PARALLEL PLATES AND CONCENTRIC CYLINDERS

The simple example of laminar flow between two plates, one of which is in motion relative to the other, was considered in Chapter 2 to describe the action of viscous shear. In Fig. 8-1, the plate length and width are assumed to be significantly greater than the uniform separation distance b, allowing end conditions at the plate boundaries to be neglected. A general velocity distribution is indicated. Its actual shape will be determined from the information available. With the simple geometry indicated, this problem will be developed first from the basic principles of continuity, energy, and force and momentum. It is sometimes advantageous to formulate the problem from the beginning in this manner, because of the additional insight into the way the principles and boundary conditions are applied. The alternative procedure of starting with the general equations expressing the physical principles available and dropping the terms that are not applicable will be considered later.

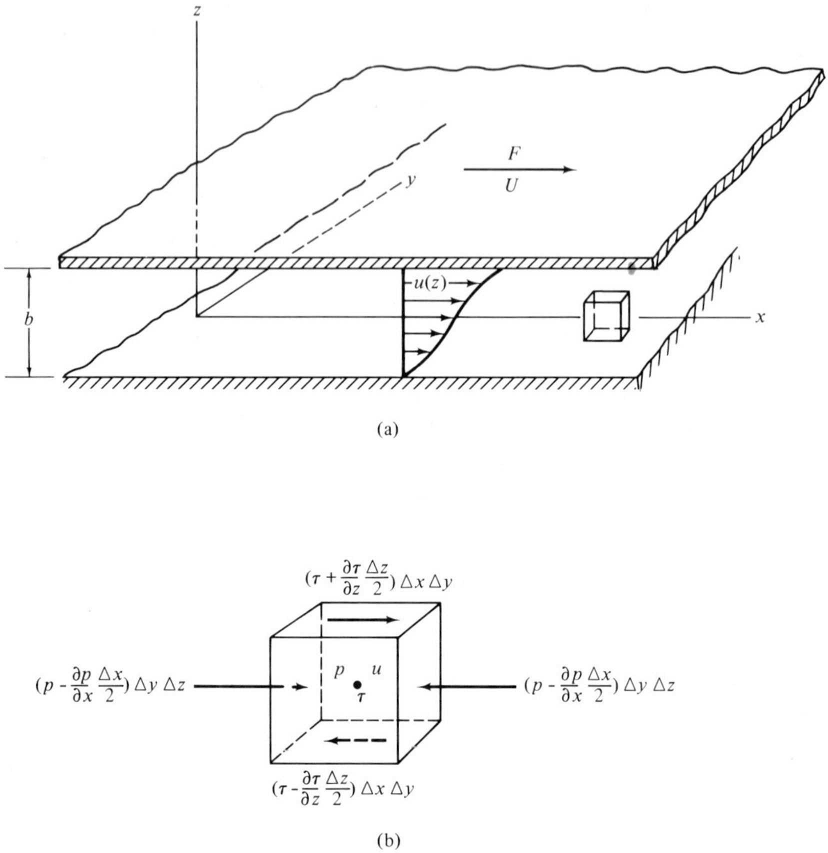

Fig. 8-1. (a) Flow field produced between two parallel flat plates; the top plate is dragged by the force F at the velocity U. The force on the fluid by the upper plate is directed toward the right. The viscous drag on the plate is to the left. The external force necessary to maintain the plate motion is to the right. (b) A fluid element showing the forces acting in the x direction.

First the coordinate system that best suits the boundaries is chosen as indicated in Fig. 8-1a. The flow field is assumed to be (1) steady, and (2) incompressible. Additional conditions to be imposed on the field are: (3) the fluid must have zero velocity relative to an adjacent solid surface, (4) the motion of the upper plate is in the x direction, (5) there are no external force fields acting on the fluid other than (6) gravity acting in the negative z direction, and (7) there are no variations of any fluid properties in the y direction. Stated mathematically, these imposed conditions are:

1. $\partial(\ \ )/\partial t = 0$.
2. $\rho = \text{const.}$

3. At $z = 0$, $u = 0$.
4. At $z = b$, $u = U$.
5. $\partial p/\partial x = 0$.
6. $f_{b,z} = -gz/g_c$.
7. $\partial(\ )/\partial y = 0$.

The basic principles will be applied with these conditions to determine the unknown information:

1. The velocity (u, v, w) field.
2. The pressure (p) field.
3. The shear stress (τ) field.
4. The force acting on the top and bottom plates.

Considering a small element of fluid within the flow field, the continuity equation derived for this element under the above conditions (steady, incompressible flow) yields

$$\frac{\partial u}{\partial x} = 0$$

Since the velocity components in the y and z directions are initially assumed to be zero, this equation can be integrated to give

$$u = f(z \text{ only})$$

Applying the force–momentum relationship in each of the three component directions, the y-direction equation yields nothing, since there are no forces and no motion in this direction. The z-component equation reduces to a simple hydrostatic balance, since there is no motion in this direction. The differential equation in the x direction is obtained by considering the forces on, and motion of, the differential control volume in Fig. 7-1b. Since the flow is steady and since u cannot vary with x as determined above, the rate of change of momentum in the x direction is zero. The x- and z-momentum equations then reduce to the force balances

$$-\frac{\partial p}{\partial x} + \frac{\partial \tau}{\partial z} = 0$$

$$-\frac{\partial p}{\partial z} - \frac{\rho g}{g_c} = 0$$

In the first of these equations, the x-pressure gradient is assumed to be zero by condition 5. The second of these equations is the hydrostatic balance in the z direction. The only usable significant result is

$$\frac{\partial \tau}{\partial z} = 0 \qquad \text{or} \qquad \tau = \text{const.}$$

(τ cannot be a function of y or z, since there is no motion in these directions.)

Assuming Newtonian flow behavior,

$$\tau = \mu \frac{du}{dz} = \text{const.} = C_1$$

or

$$u = \frac{C_1 z}{\mu} + C_2$$

Applying boundary condition 3 gives $C_2 = 0$; boundary condition 4 gives $C_1 = \mu U/b$, so the velocity distribution is linear; therefore

$$u(z) = \frac{U}{b} z$$

or expressed as a nondimensional equation, $u/U = z/b$. The shear stress is constant and is

$$\tau = \mu \frac{U}{b}$$

The velocity, pressure, and shear stress fields have been determined. The magnitude of the force on an area A of either the top or bottom plates is τA. The work rate or power necessary to maintain the plate motion at the velocity U is FU or τAU. Since there is no acceleration of the fluid anywhere within the field, conservation of energy yields the result that this input power is converted into internal energy and appears as a temperature increase of the fluid (heat transfer is neglected).

Alternatively, the characteristics of this flow can be determined by considering the relations developed in the preceding chapters. The procedure would be as follows. The applicable mathematical relations obtained in Chapters 4 through 6 are used, plus the physical boundary conditions for the flow field. The equation of continuity for an incompressible flow field is

$$\text{div}\,\vec{\text{v}} = 0 \qquad \text{or} \qquad \frac{\partial u}{\partial x} + \frac{\partial v}{\partial y} + \frac{\partial w}{\partial z} = 0$$

This equation reduces to

$$\frac{\partial u}{\partial x} = 0 \qquad \text{or} \qquad u = u(z \text{ only})$$

as previously obtained. The natural coordinate force–momentum equation is used with x as the streamline direction s; z is equivalent to the normal coordinate n. In the coordinates of Fig. 8-1, the component equations (6-28) become

$$\frac{u}{g_c}\frac{\partial u}{\partial x} = -\frac{1}{\rho}\frac{\partial p}{\partial x} + \frac{1}{\rho}\frac{\partial \tau}{\partial z} + f_{b,x}$$

$$\frac{1}{g_c}\frac{u^2}{R_s} = -\frac{1}{\rho}\frac{\partial p}{\partial z} + f_{b,z}$$

The x-velocity component u is a function of z only, so

$$\frac{\partial u}{\partial x} = 0$$

From the second equation $R_s = \infty$ and u^2 is finite within the entire flow field, so

$$\frac{1}{\rho}\frac{\partial p}{\partial z} - f_{b,z} = 0 \qquad \text{or} \qquad \frac{p}{\rho} + \frac{gz}{g_c} = \text{const.}$$

These equations reduce to the condition that $\partial\tau/\partial z = 0$ or $\tau = \text{const.}$

Note that the shear stress (and therefore the shear force) is the same for any layer of fluid parallel to the plates (x-y plane), because the layers of fluid are in equilibrium under the action of the shear force only (x direction) and hydrostatic body forces (z direction), and no fluid element experiences any net acceleration.

The type of flow considered above (and illustrated in Figs. 2-3 and 8-1) is not usually encountered in any real apparatus. However, it does closely approximate the flow between two concentric cylinders, with one cylinder rotating with respect to the other for small b/r_1 and small b/l, as illustrated in Fig. 8-2.[1] The radius of curvature is then very large with respect to the gap or flow field width b.

When using this apparatus for experimental determination of viscosity, the outer cylinder is usually rotated and the center cylinder is supported in a torque-measuring mount, as illustrated in Fig. 8-2. If the center cylinder is rotated, the inside layers of fluid have a larger centrifugal force than the outside ones, setting up a centrifugal force field that produces an unstable equilibrium condition. The flow is then unstable between the centrifugal and viscous force fields and breaks down, first into a system of alternate annular vortices, and then into a random, turbulent

[1] This type of flow between concentric cylinders as indicated in Fig. 8-2 (and approximated by that in Fig. 8-1) was first investigated by M. Couette and is therefore referred to as Couette flow.

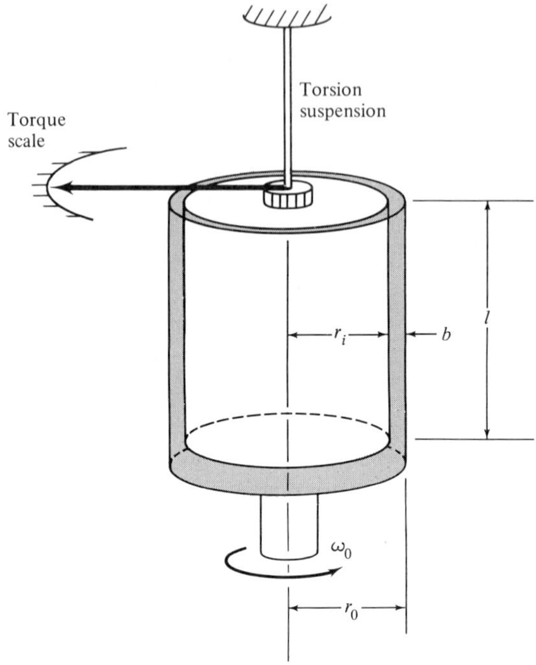

Fig. 8-2. Schematic representation of a rotating cylinder viscometer. The outer cylinder is rotated with an angular speed ω_0. The viscous torque on the inner cylinder is measured by means of the torsion suspension. The calibration relationship for a given viscometer can be calculated or determined experimentally from measurements made with a fluid of known viscosity.

motion. The limit of stability for laminar motion is determined by the condition[2]

$$N_{R,i}\sqrt{\frac{b}{r_i}} > 41.3 \tag{8-1}$$

where $N_{R,i}$ is the Reynolds number based on the inside cylinder velocity $r_i\omega_0$ and gap width b ($N_{R,i} = r_i\omega_0 b/\nu$). This type of flow instability is termed Taylor instability, after G. I. Taylor, who discovered it and explained its occurrence.[3] If the outside cylinder is rotated and the inside cylinder is held stationary, the concentric cylinder flow field is stable under all conditions (neglecting end conditions).

[2] See, for example, chap. 17 in H. Schlichting, *Boundary Layer Theory*, McGraw-Hill, Inc., New York, 6th ed., 1968.

[3] G. I. Taylor, Stability of a Viscous Liquid Contained between Two Rotating Cylinders, *Phil. Trans. Roy. Soc. London*, **A223**, 289–293 (1923).

If the torque scale of Fig. 8-2 is calibrated to read torque directly, the relation between the torque on the cylinder (T_c), the geometry, and the viscosity is (see Problem 8-2)

$$T_c = \frac{2\pi r_i^3 \omega_0 l}{b} \mu. \tag{8-2}$$

If the inside cylinder is solid, there is an additional viscous torque on the bottom of the cylinder. The relations for determining this torque between two circular plates (one rotating with respect to the other) can be set up and the torque on the bottom obtained as (see Problem 8-10)

$$T_b = \frac{\pi r_i^4 \omega_0}{t} \mu \tag{8-3}$$

where t is the separation distance between the plates. The total viscous torque is the sum of these two torques. The viscosity is then determined from the relationship

$$\mu = \frac{T}{\pi r_i^3 \omega_0 [(2l/b) + (r_i/t)]} \tag{8-4}$$

An alternative method of construction is to form the inside cylinder from an inverted cylindrical "cup." The trapped airspace will exclude the fluid from the inside of the cup except for a small rise due to the air compression by the fluid static head (equal to the height of the cup, l). Alternatively, air (or a gas) can be injected into the cup until the fluid level is even with the bottom of the inside cylinder. The thin layer of fluid between the rotating cylinder and the air space rotates with about the angular speed of the cylinder because the shear stress at the liquid–air interface is negligible.

The velocity distribution for the fluid within the gap is obtained by considering the equations of motion of the fluid. A force–momentum analysis for a fluid element within the annular gap yields the result that $\tau_{r\theta} r^2 = \text{const.}$ (see Problems 8-4 and 8-5). The shear stress–strain rate relation gives $\tau_{r\theta} = 2\mu\dot{\gamma}_{r\theta}$. The similarity of the cylindrical coordinate system with the natural coordinate system can now be used to obtain the strain rate $\dot{\gamma}_{r\theta} = -\dot{\gamma}_{sn}$. The minus sign occurs because the n direction corresponds to the negative r direction (see Problems 8-8 and 8-9). The shear stress is[4]

$$\tau_{r\theta} = \mu\left(\frac{\partial v}{\partial r} - \frac{v}{r}\right) = \mu r \frac{\partial}{\partial r}\left(\frac{v}{r}\right) \tag{8-5}$$

[4] For motion with a radial velocity component u as well as the tangential velocity component v,

$$2\dot{\gamma}_{r\theta} = \frac{\partial u}{r\,\partial\theta} + r\frac{\partial}{\partial r}\left(\frac{v}{r}\right)$$

The equation of motion for this flow is then

$$\tau r^2 = \left[\mu r \frac{d}{dr}\left(\frac{v}{r}\right)\right] r^2 = C_1$$

which can be integrated to give (with $C_1' = C_1/\mu$)

$$v = \frac{2C_1'}{r} + C_2 r \tag{8-6}$$

If the inner cylinder of radius r_i is stationary and the outer cylinder of radius r_o rotates with a speed $v_0 = r_o\omega_0$,

$$\frac{v}{v_0} = \frac{r_o}{r}\frac{(r/r_i)^2 - 1}{(r_o/r_i)^2 - 1} \tag{8-7}$$

If y is specified as the "gap coordinate" measured outward from the inner cylinder ($y = r - r_i$) and b is the gap ($b = r_o - r_i$), the velocity distribution can be written

$$\frac{v}{v_0} = \frac{r_o}{y + r_i}\frac{y(y + 2)}{b(b + 2)} \tag{8-8}$$

This result can be compared with the approximate linear velocity distribution,

$$\frac{v}{v_0} = \frac{r - r_i}{r_o - r_i} = \frac{y}{b}$$

It is of interest to compare the error involved by the approximation (see Problems 8-13 and 8-14).

The radial pressure distribution is determined from the condition for radial equilibrium,

$$\frac{\partial p}{\partial r} = \frac{\rho v^2}{g_c r}$$

which gives

$$p - p_0 = \frac{\rho v^2}{2g_c} \tag{8-9}$$

where p_0 is the pressure at $r = r_o$.

8.2 LAMINAR FLOW BETWEEN PLATES WITH AN APPLIED PRESSURE GRADIENT

The flow field for the geometry illustrated in Fig. 8-3 will be analyzed to determine relationships among the velocity field, flow rate, shear stress,

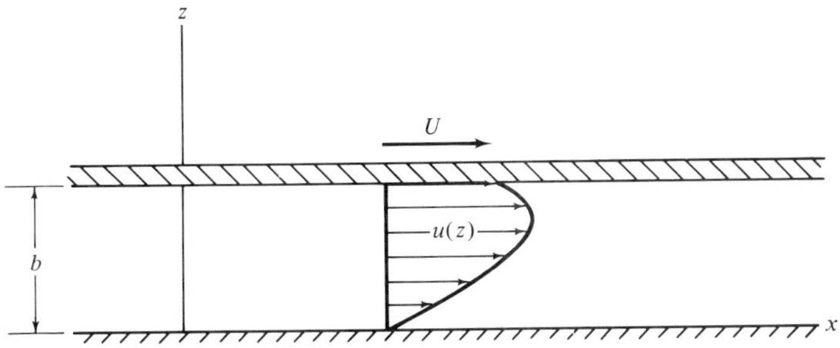

Fig. 8-3. Flow between two parallel plates with an applied pressure gradient. For the velocity distribution shown, the applied pressure gradient direction is to the right. The viscous drag on the plate is to the right (compare with the direction of the fluid drag on the plate in Fig. 8-1).

pressure gradient, and geometry (length l and spacing b). The motion of one plate with respect to the other produces a flow rate $Q/c = Ub/2$, where c is the width of the plate and Q/c is the volume flow rate per unit width. A pressure gradient $\partial p/\partial x$ applied to the flow will produce an additional flow rate. The net flow indicated in Fig. 8-3 is obtained with a negative value of $\partial p/\partial x$ in addition to the upper plate velocity U. This type of flow geometry has application in hydrostatic lubrication and in "viscosity pumps and turbines" (see Problem 8-18).

It is assumed that the flow is (1) steady, (2) fully developed, and (3) laminar, and (4) that end effects will be neglected. The velocity relative to the solid walls or surfaces is zero, giving the boundary conditions (5) $u = 0$ at $z = 0$, and (6) $u = U$ at $z = b$. Assumption (4) is a boundary condition that simplifies the problem to quasi-one-dimensional flow and holds mathematically only if the plates are infinite in extent. This assumption is approximately valid if the plate length and width are each much greater than the separation b.

In order to determine the velocity in terms of the applied pressure gradient, the plate velocity U, and the fluid properties, the motion of the differential fluid element control shown in Fig. 8-1b will be considered. There is no rate of change of momentum, since the flow is fully developed and steady. The equation of motion therefore reduces to $\sum F_x = 0$, or

$$-\frac{\partial p}{\partial x}\Delta x\,\Delta z + \frac{\partial \tau}{\partial z}\Delta x\,\Delta z = 0$$

or

$$\frac{d\tau}{dz} = \frac{\partial p}{\partial x} \tag{8-10}$$

The shear stress gradient is expressed as a total derivative, since there is no motion in either the y or z direction; consequently, there is no shear force or variation of τ in these directions. The pressure varies in the x and z directions because of the applied pressure gradient and the hydrostatic pressure gradient (assumed to be produced by a gravity force field parallel to the z direction). The hydrostatic pressure gradient produces no motion. For this type of flow the shear stress is given by

$$\tau = \mu \frac{du}{dz}$$

Equation (8-10) is then

$$\frac{d}{dz}\left(\mu \frac{du}{dz}\right) = \frac{\partial p}{\partial x} \tag{8-11a}$$

Equation (8-10) or (8-11a) can be integrated to give

$$\tau = \frac{\partial p}{\partial x} z + C_1 \tag{8-11b}$$

or

$$\frac{du}{dz} = \frac{1}{\mu}\frac{\partial p}{\partial z} z + C_1' z + C_2 \tag{8-11c}$$

The shear stress varies linearly with z. Assuming that the viscosity is constant, (8-11c) can be integrated to give the velocity distribution

$$u = \frac{1}{\mu}\frac{\partial p}{\partial x}\frac{z^2}{2} + C_1' z + C_2 \tag{8-12a}$$

Applying the boundary conditions $u = 0$ at $z = 0$ and $u = U$ at $z = b$ to determine the constants of integration gives

$$u(z) = \frac{Uz}{b} - \frac{1}{2\mu}\frac{\partial p}{\partial x} z(b - z)$$

or, as a nondimensional equation,

$$\frac{u}{U} = \frac{z}{b} - \frac{b^2}{2\mu U}\frac{\partial p}{\partial x}\frac{z}{b}\left(1 - \frac{z}{b}\right) \tag{8-12b}$$

If the pressure gradient is zero, the velocity distribution is linear, as previously indicated. If the plate velocity U is zero, the velocity profile is a parabola, symmetric about the center of the gap. Since $\partial p/\partial x$ can be either positive or negative, the net mass flow can be either in the positive or negative x direction. The flow rate between the plates is obtained by integrating the velocity distribution over the passage width b, that is,

$Q = \int uc\,dz$, where c is the plate width perpendicular to the flow (x) direction. The flow per unit plate width is

$$\frac{Q}{c} = \frac{Ub}{2}\left(1 - \frac{b^2}{6\mu U}\frac{\partial p}{\partial x}\right) \tag{8-13}$$

which may be integrated, giving the pressure as a linear function of x,

$$p - p_1 = \frac{12\mu}{b^3}\left(\frac{Ub}{2} - \frac{Q}{c}\right)(x - x_1) \tag{8-14}$$

where $p = p_1$ at $x = x_1$.

8.3 LAMINAR FLOW IN DUCTS

The flow field for pipe, tube, or duct flow will be analyzed to obtain relations among the flow rate, shear stress, pressure gradient, and geometry (length and diameter). The results obtained will apply only to duct flows where the flow is laminar as determined by the Reynolds number, which must be less than the critical value for transition from laminar to turbulent flow (approximately 2000) for this geometry.

The flow is again assumed to be (1) fully developed, (2) steady, (3) laminar, and (4) incompressible. If the duct is round (pipes, tubes, and so on) and is long enough to attain fully developed flow, the velocity distribution will be symmetric about the centerline, or *axisymmetric*. If the flow is incompressible, there is no velocity in the radial direction, so all flow field properties will be constant on concentric cylindrical surfaces.

The x-component equation of motion for a control volume of length Δx and thickness Δr (Fig. 8-4) gives

$$\begin{aligned}\sum F_x = {} & \left(p - \frac{\partial p}{\partial x}\frac{\Delta x}{2}\right)2\pi r\,\Delta r - \left(p + \frac{\partial p}{\partial x}\frac{\Delta x}{2}\right)(2\pi r\,\Delta r) \\ & - \left(\tau - \frac{\partial \tau}{\partial r}\frac{\Delta r}{2}\right)2\pi\left(r - \frac{\Delta r}{2}\right)\Delta x \\ & + \left(\tau + \frac{\partial \tau}{\partial r}\frac{\Delta r}{2}\right)2\pi\left(r + \frac{\Delta r}{2}\right)\Delta x = 0\end{aligned}$$

or

$$\frac{1}{r}\frac{\partial}{\partial r}(\tau r) - \frac{\partial p}{\partial x} = 0 \tag{8-15}$$

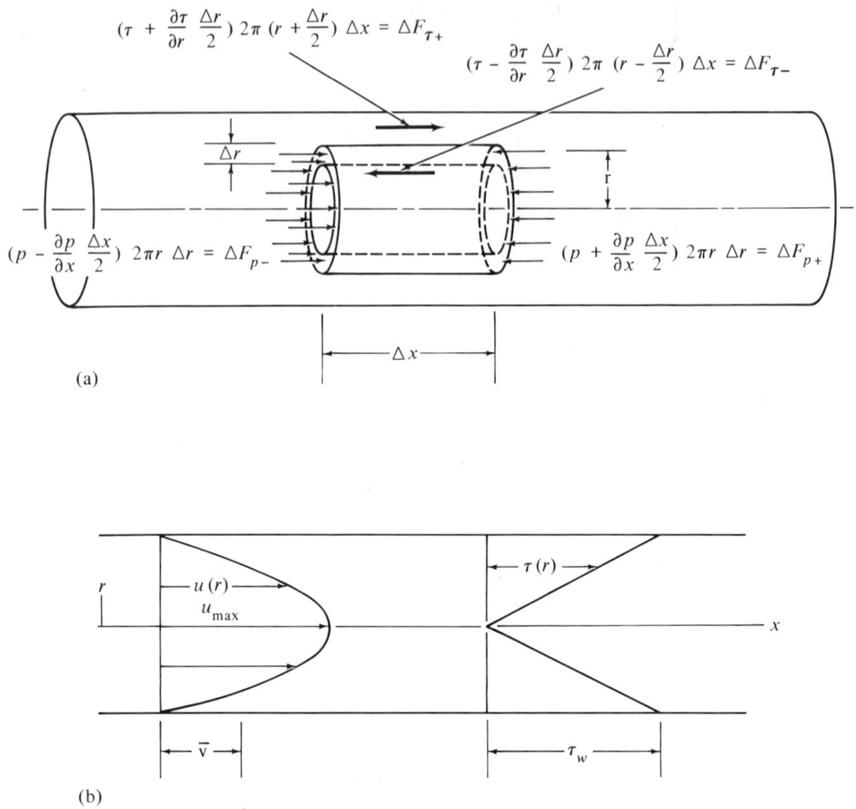

Fig. 8-4. (a) Representation of an annular fluid element in the flow in a round tube with flow symmetry about the axis. The pressure and shear stress are p and τ at a given x (at the center of the element) and r. Only x-direction forces are shown. (b) Velocity and shear stress distributions for fully developed flow in a round tube.

Since the fully developed flow is axisymmetric, the only change of pressure is in the x direction (except for hydrostatic variations), so (8-15) can be integrated to give

$$\tau r = \frac{r^2}{2}\frac{\partial p}{\partial x} + C_1 \tag{8-15a}$$

or

$$\tau = \frac{r}{2}\frac{\partial p}{\partial x} + \frac{C_1}{r} \tag{8-15b}$$

PIPE OR TUBE FLOW

For a pipe or tube, the constant C_1 in (8-15a) can be determined from the conditions at $r = 0$, giving $C_1 = 0$. The stress–strain rate relation for the flow field of Fig. 8-4 is

$$\tau = \tau_{xr} = 2\mu\dot{\gamma}_{xr}$$

The shear strain rate $\dot{\gamma}_{xr}$ is

$$\dot{\gamma}_{xr} = \frac{1}{2}\left(\frac{\partial u}{\partial r} + \frac{\partial v}{\partial x}\right)$$

where u and v are the velocity components in the x and r directions, respectively. For this flow field, $v = 0$ and $u = u(r)$ only. Since the flow is incompressible, the same volume flow rate exists at each section (at each x station) of the tube. Since the flow is fully developed, the velocity profile shape is the same at any x. These two conditions of identical shape and flow rate (integral of the velocity distribution over the area) imply that $u =$ const. at a given r (that is, along a given streamline). The shear stress can then be written simply as

$$\tau = \mu \frac{du}{dr}$$

With this value in (8-15a),

$$\mu \frac{du}{dr} = \frac{r}{2}\frac{\partial p}{\partial x}$$

If the streamlines are straight and parallel, p and $\partial p/\partial x$ are not functions of r. This equation can then be integrated directly to give

$$u = \frac{r^2}{4\mu}\frac{\partial p}{\partial x} + C_2 \tag{8-16}$$

C_2 can be evaluated with either of the two conditions

$$u = 0 \quad \text{at} \quad r = r_0 \qquad \text{or} \qquad u = u_0 \quad \text{at} \quad r = 0$$

The first gives $C_2 = -(r_0^2/4\mu)(\partial p/\partial x)$, so (8-16) becomes

$$u = -\frac{1}{4\mu}\frac{\partial p}{\partial x}(r_0^2 - r^2) \tag{8-16a}$$

The second gives $C_2 = u_0$, or

$$u_0 = -\frac{r_0^2}{4\mu}\frac{\partial p}{\partial x} \tag{8-16b}$$

Eliminating the quantity $(1/4\mu)(\partial p/\partial x)$ then gives

$$\frac{u}{u_0} = 1 - \left(\frac{r}{r_0}\right)^2 \tag{8-17}$$

which indicates that the velocity distribution for laminar flow in a round duct is parabolic, as indicated in Fig. 8-4b.

Integrating the velocity distribution over the area ($A_0 = \pi r_0^2$) gives the mass flow rate as

$$\dot{m} = \frac{\rho A_0 u_0}{2} \qquad \text{and} \qquad Q = \frac{A_0 u_0}{2}$$

Since the axial pressure gradient $\partial p/\partial x$ is constant, (8-16b) gives

$$-\frac{\partial p}{\partial x} = \frac{p_1 - p_2}{L} \equiv \frac{\Delta p}{L} = \frac{4\mu u_0}{r_0^2} \tag{8-18a}$$

or

$$\frac{\Delta p}{L} = \frac{8\mu Q}{\pi r_0^4} = \frac{8\nu \dot{m}}{\pi g_c r_0^4} \tag{8-18b}$$

The pressure gradient necessary to produce a given flow rate is proportional to the viscosity and inversely proportional to the fourth power of the tube diameter. The average velocity ($\bar{\text{v}} = Q/A_0$) is

$$\bar{\text{v}} = \frac{r_0^2}{8\mu}\frac{\Delta p}{L} \tag{8-19}$$

The work-and-energy-rate relation for steady flow can be applied to the elemental control volume of Fig. 8-4. The rate at which the pressure forces do work on the control volume is pu. The rate at which the shear forces do work is τu. The quantities p, τ, and u can be considered as those at the center reference plane, as in the previous developments. The work-and-energy-rate principle can then be considered in the form: *The rate of increase of energy in Δm is due to the rate of energy addition by heat transfer ($\Delta \dot{Q}_H$) plus the rate at which work is being done on the fluid in the control volume:*

$$\dot{m}\,\Delta\tilde{u} = \Delta\dot{Q}_H + \left(-\frac{\partial pu}{\partial x}\,\Delta x\right)2\pi r\,\Delta r - \left[\frac{1}{r}\frac{\partial}{\partial r}(\tau u r)\,\Delta r\right]2\pi r\,\Delta x$$

Dividing this equation by $\Delta m = \rho 2\pi r\,\Delta r\,\Delta x$ gives the rate at which "availability" of energy (and work potential) is converted into unavailable internal energy, or the *irreversibility* rate, $\dot{\text{Ir}}$:

$$\Delta\tilde{u} - \dot{q} \equiv \dot{\text{Ir}} = -\frac{1}{\rho}\frac{\partial pu}{\partial x} + \frac{1}{r}\frac{\partial \tau u r}{\partial r}$$

Since $\partial u/\partial x = 0$ for this flow field geometry,

$$\dot{\text{Ir}} = -\frac{1}{\rho}\left(u\frac{\partial p}{\partial x} + \frac{1}{r}\frac{d\tau ur}{dr}\right) = -\frac{1}{\rho}\left[u\left(\frac{dp}{dx} + \frac{1}{r}\frac{d\tau r}{dr}\right) + \tau\frac{du}{dr}\right]$$

which, with (8-15), gives[5]

$$\dot{\text{Ir}} = -\frac{\tau}{\rho}\frac{du}{dr} = \frac{\mu}{\rho}\left(\frac{du}{dr}\right)^2 = \frac{\nu}{g_c}\left(\frac{du}{dr}\right)^2 \tag{8-20}$$

The general units are energy rate per unit mass (or power per unit mass). Generally, in a quasi-one-dimensional flow field,

$$\dot{\text{Ir}} = \frac{\nu}{g_c}\left(\frac{d\text{v}}{dy}\right)^2 \tag{8-20a}$$

where y is the coordinate normal to the stream or v direction. For the case of laminar flow in a round duct,

$$\dot{\text{Ir}} = \frac{\nu}{g_c}\left(\frac{du}{dr}\right)^2 = \frac{4\nu u_0^2 r^2}{g_c r_0^4} = Cr^2 \tag{8-21}$$

so that the irreversibility rate, or rate of conversion of available work and energy into unavailable (internal) energy ("dissipation" of available work and energy), is proportional to r^2.

From (8-15b) a relation can be obtained for the pressure gradient and the shear stress at the wall by considering a control volume of length L and radius r_0,

$$\frac{\Delta p}{L} = \frac{2\tau_w}{r_0} \tag{8-22}$$

From (6-23) the energy dissipation rate per unit mass flow rate (irreversibility rate) is $\Delta p/\rho_0$, or

$$\frac{\Delta p}{\rho} = \frac{4\tau_w}{\rho D_H} = \dot{\text{Ir}}$$

The energy-dissipation rate for a mass flow rate $\dot{m}$ is then

$$\frac{\dot{m}\,\Delta p}{\rho} = \rho A\text{v}\,\frac{\Delta p}{\rho} = \dot{m}\,\frac{4\tau_w}{\rho d_0} = \frac{Q\,\Delta p}{L} \tag{8-23}$$

[5] Note the similarity of procedure for obtaining (8-19) with that leading up to (6-23) in Chapter 6: The irreversibility is obtained originally from the energy equation; then the momentum equation is combined with this equation to determine the expression for the irreversibility.

The energy-dissipation rate per unit length of pipe is then[6]

$$\frac{\Delta\tilde{\dot{U}} - \dot{Q}_H}{L} = Q\,\Delta p$$

If the ducting system is insulated, $\dot{Q}_H = 0$ and the temperature rise in the length of duct in which Δp occurs is

$$\Delta T = \frac{1}{c_v}\frac{\Delta p}{\rho} \tag{8-24}$$

where c_v is the specific heat at constant volume (therefore, constant density) for the fluid.

FLOW IN AN ANNULUS

If the ducting is an annular space between two pipes or tubes (a geometry sometimes used for simple heat exchangers and chemical reactors), the same development illustrated in connection with Fig. 8-4 is valid, except that the boundary conditions are changed. $u = 0$ at $r = r_i$ (the inside radius), and, as before, $u = 0$ at $r = r_0$. The first boundary condition gives a nonzero value for C_1, which cannot be determined until the expression for $u(r)$ is obtained.

With C_1 not equal to zero, (8-15b) is written

$$\tau = -\mu\frac{du}{dr} = -\frac{\partial p}{\partial x}\frac{r}{2} + \frac{C_1}{r}$$

or

$$\frac{du}{dr} = \frac{1}{2\mu}\frac{\partial p}{\partial x}r - \frac{C_1}{\mu r}$$

which can be integrated to give

$$u = \frac{1}{4\mu}\frac{\partial p}{\partial x}r^2 - \frac{C_1}{\mu}\ln r + C_2 \tag{8-25a}$$

The two boundary conditions allow evaluation of C_1 and C_2, giving the velocity distribution as

$$u(r) = -\frac{1}{4\mu}\frac{\partial p}{\partial x}\left(r_o^2 - r^2 + \frac{r_o^2 - r_i^2}{\ln r_o/r_i}\ln\frac{r_o}{r}\right) \tag{8-25b}$$

Although an equation for u/u_0 as a function of r/r_o can be determined (see Problem 8-40), there is no simple expression for u as was obtained for pipe flow.

[6] $\dot{Q}_H$ is the heat-transfer rate and $Q \equiv A\bar{v}$ the volume flow rate.

8.4 TRANSITION TO FULLY DEVELOPED FLOW

The above developments yield results that agree with experimental observations for values of Reynolds number less than the critical Reynolds number above which the flow becomes turbulent. For the flow between fixed plates or in a round tube, the lower value of Reynolds number below which all turbulent disturbances are damped out is approximately 2000, based on plate spacing or tube diameter. It is possible to obtain fully developed laminar flow in tubes for higher Reynolds numbers by maintaining ideal conditions: no mechanical vibrations and feeding the tube through a well-shaped nozzle from a reservoir where the fluid is perfectly still. Since conditions are usually not met except in a laboratory, laminar flow is usually assumed to occur for Reynolds numbers below approximately 2000 for the geometries mentioned.

For a flow entering a pipe where the Reynolds number based on flow rate and diameter is less than 2000, a certain length of pipe is required for the viscous forces to slow the fluid layers and produce the parabolic profile derived above for fully developed flow. Because this boundary-layer growth to fill the tube is an exponential (growth-rate) type of process, the length for its completion can never be determined exactly. However, several investigators[7] indicate that this "transition length" required for a laminar velocity profile to become 99 percent fully developed from an initially uniform profile is given by the expression

$$\frac{L}{d_0} \approx 0.06 N_R \qquad (N_R \leq 2000) \tag{8-26}$$

where the Reynolds number is based on average velocity (equal to uniform inlet velocity) and tube diameter. The pressure drop Δp obtained from experimental results is

$$\Delta p = \frac{32\mu\bar{v}L}{d_0^2} + 2.3\,\frac{\rho\bar{v}^2}{2g_c} \tag{8-27}$$

where $d_0 = 2r_0$ is the tube diameter and $\bar{v}$ is the average velocity ($Q/A = \bar{v}$). This result corresponds fairly well with that calculated from an analytical development that gives approximately 2.2 in place of the 2.3 above. The added correction factor is due to the redistribution of kinetic energy from the flat profile to the fully developed parabolic profile plus an additional shear stress increment (see Problem 5-16).

Experimental results for turbulent flow at a pipe inlet indicate that the axial pressure gradient does not vary much from the value calculated

[7] See, for example, W. D. Campbell and J. C. Slattery, Flow in the Entrance in a Tube, *Trans. ASME*, **D85** (1), 41–46 (1963).

assuming fully developed pipe flow, although more than 50 pipe diameters are required for the transition from a uniform inlet flow to a 99 percent fully developed turbulent profile.[8] Because the fully developed turbulent profile does not vary significantly from a uniform profile and because the actual pressure gradient does not vary significantly, the 50-pipe-diameter requirement for full development of the turbulent profile is not of great significance in engineering design calculations or corrections for inlet conditions to metering devices (Venturies, orifices, and so on).

The above developments and results also apply to ducts of rectangular, elliptical, and other cross-sectional shapes if the ratio of $\sqrt{A}/D_H$ is not large (that is, less than 2), and to curved channels, where $R_s/D_s > 10$ (D_s is the channel width in the direction of R_s).

8.5 SUMMARY

1. For steady, laminar, viscous, fully developed flows, the equations of motion (Newton's laws) reduce to a simple force balance on a fluid element, or overall control volume. Application of this force–equilibrium condition yields the relation between shear stress and pressure gradient. The stress–strain rate relationship, $\tau = 2\mu\dot{\gamma}$, is then used to obtain an equation for the velocity field. The geometric boundary conditions provide conditions for evaluating the constants of integration.
2. The upper limit for laminar flow between parallel plates or in ducts, tubes, and pipes is determined from "lower critical Reynolds number," below which turbulent flow cannot be obtained. The value of this lower critical Reynolds number for these geometries is approximately 2000, based on average velocity and spacing or diameter. For rectangular, triangular, and so on, duct cross sections, the correlating "diameter" is the hydraulic diameter defined as $4A/Per$.
3. The flow field between concentric cylinders where the gap/cylinder radius is much less than unity ("Couette flow") can be approximated by the flow between parallel plates or by a linear velocity profile. This geometry then provides an excellent means for determining viscosity (see Problems 8-6 and 8-7). Because of centrifugal-force field instabilities produced in the rotating fluid when the inner cylinder is rotated, this type of apparatus is usually constructed with the outer cylinder rotating as illustrated in Fig. 8-2.
4. Fully developed, laminar, viscous flow in round ducts produces a parabolic velocity profile. The streamwise pressure gradient is proportional to the mass flow rate and inversely proportional to the fourth power of the diameter.

[8] A. R. Barbin and J. B. Jones, Turbulent Flow at the Inlet Region of a Smooth Tube, *Trans. ASME*, **D85** (1), 29–34 (1963).

5. The viscous flow between plates and cylinders approaches the conditions obtained in hydrodynamic lubrication in "slipper" and journal bearings. For a specific development of hydrodynamic bearing lubrication, see, for example, chap. 12 in J. C. Hunsaker and B. G. Rightmire, *Engineering Applications of Fluid Mechanics*, McGraw-Hill, Inc., New York, 1947.

PROBLEMS

8-1 What is the "one dimension" in each of the one-dimensional flows illustrated in Figs. 8-1 through 8-4?

8-2 Derive (8-2) assuming the approximate relationships for the shear stress and strain rate for the flow between rotating concentric cylinders where $b \ll r_i$.

8-3 Find the vector rotation field $\vec{\omega}$ as a function of the field coordinates; that is, find the components ω_x, ω_y, and ω_z of $\vec{\omega}(x_1 y_1 z) = \hat{x}\omega_x + \hat{y}\omega_y + \hat{z}\omega_z$ for the one-dimensional flow of Fig. 8-1.

8-4 From Newton's laws for torque and angular momentum, consider the flow field between two concentric cylinders with one rotating at a constant speed with respect to the other and show that the shear stress on the inner and outer cylinders must satisfy the relation $\tau_o r_o^2 = \tau_i r_i^2$ and further deduce that $\tau r^2 =$ const. gives the shear stress distribution for the flow field between the cylinders. Is this result limited to laminar flow?

8-5 By considering the forces and motion in a fluid element in a cylindrical coordinate system, derive the steady-state condition $\tau_{r\theta} r^2 =$ const. for the fluid within the gap between two concentric cylinders with one cylinder rotating with respect to the other.

8-6 A circular cylinder viscometer (Fig. 8-2) is to be used for determining the viscosity of water for the temperature range 0 to 100°C. The inner cylinder is to be rotated. The inner cylinder is 50 mm long and 50 mm in diameter, and the gap is 2 mm. What is the maximum speed with which the cylinder can be rotated? What is the maximum torque that will be obtained?

8-7 If the torque scale of the viscometer illustrated in Fig. 8-2 is in degrees, find the "gain" or calibration factor K of the instrument as a function of the dimensions; that is, find K in the relation $\mu = K\theta$, where θ is the scale reading in degrees.

8-8 For the motion of a fluid in concentric circles in plane two-dimensional flow, obtain the relationship for the shear strain rate

$$\dot{\gamma}_{r\theta} = \frac{1}{2}\left(\frac{\partial v}{\partial r} - \frac{v}{r}\right)$$

Set up a fluid element and determine the angular deformations of the elemental lengths Δr and $r\,\Delta\theta$ in a plane polar coordinate system similar to the manner in which the strain rate $\dot{\gamma}$ was determined for the natural coordinate system.

8-9 The shear strain rate component for a shearing motion in the r-θ plane in cylindrical coordinates is

$$\dot{\gamma}_{r\theta} = \frac{1}{2}\left(\frac{1}{r}\frac{\partial u_r}{\partial \theta} + \frac{\partial v_\theta}{\partial r} - \frac{v_\theta}{r}\right)$$

where u_r is the radial velocity component and v_θ is the tangential velocity component. For two concentric cylinders with the outer cylinder rotating with an angular speed ω_0 and the inner cylinder stationary, find the velocity distribution for the flow within the cylindrical gap; that is, find v_θ as a function of r assuming $u_r = 0$. Find the shear stress distribution. Comment on the results.

8-10 Set up a fluid element for the fluid field between the bottom of the two concentric cylinders and determine the relationship for the torque on the circular disc (bottom of the cylinder) given by (8-4). Consider the force on an element of width dr, determine the corresponding torque, and integrate it to get the result.

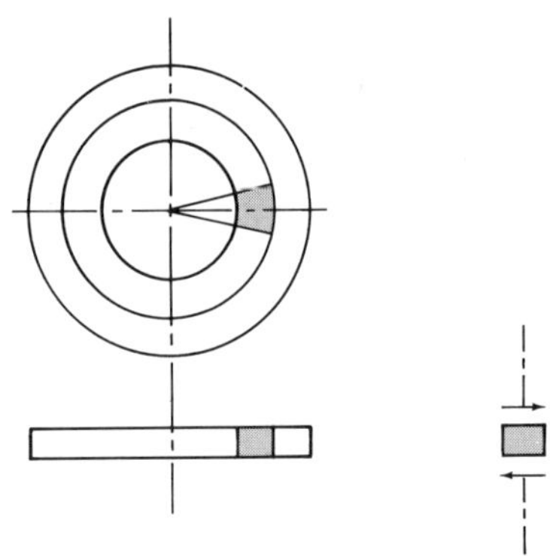

Fig. P8-10.

8-11 For an inverted cup viscometer, find the relative rise $(\Delta l/l)$ of liquid inside the cup due to the static liquid head in the annular gap between the cylinders as a function of the liquid density and the geometry of the viscometer.

8-12 Perform the integration necessary to obtain (8-6) and evaluate the constants to obtain (8-7).

8-13 For a concentric cylinder apparatus with the outer cylinder rotating and with a radius ratio $r_o/r_i = 1.1$ (gap to inside radius ratio 1:10), compare the velocity ratio (v/v_0) at the center of the gap as determined from the exact and approximate results. What is the percentage error in this center-radius velocity introduced by assuming the velocity distribution to be linear? What is the error for radius ratios of 1.2 and 1.5?

8-14 Equation (8-8) can be nondimensionalized further with respect to the inner cylinder radius r_i ($y' \equiv y/r_i$, $r_o' \equiv r_o/r_i$, $b' \equiv b/r_i$, and so on). With this substitution, obtain (8-8) in the form (with $v' \equiv v/v_0$)

$$v' = \left(\frac{1 + b'}{1 + y'}\right) \frac{y'(y' + 2)}{b'(b' + 2)}$$

For a dimensionless gap with $v' = 0.2$ (that is, $r_o/r' = 1.2$), plot the actual velocity distribution v' versus y' from this equation and compare with the linear distribution $v' = y'$. What is the maximum percentage error in v' obtained by assuming a linear distribution?

8-15 Derive the radial pressure distribution given by (8-9) for the flow between concentric cylinders with the outer cylinder rotating (assume that the gap is small with respect to the radius r_i so that the velocity distribution is approximately linear). It is stated explicitly that this is the "radial" pressure distribution. What is the total expression for the pressure distribution?

8-16 Derive the velocity and pressure distributions for the flow rate between two concentric cylinders for the case in which the outer cylinder is stationary and the inner cylinder rotating with the velocity ω_i. Discuss the stability of the flow of one case as compared with the other. Assume that the gap to inside radius ratio is much less than 1.

8-17 Find the rotation distribution $\omega(r)$ for the fluid between concentric cylinders where one cylinder is rotating relative to the other.

8-18 A viscosity pump for pumping viscous liquids is constructed as shown. The viscous shear force on the inner cylinder "drags" the fluid and therefore pumps it from the inlet to the outlet. Find the equation relating pressure rise ($p_2 - p_1$) to flow rate in terms of the pump geometry and fluid properties. The pump width is c. Determine a relationship for the efficiency. What is the maximum efficiency?

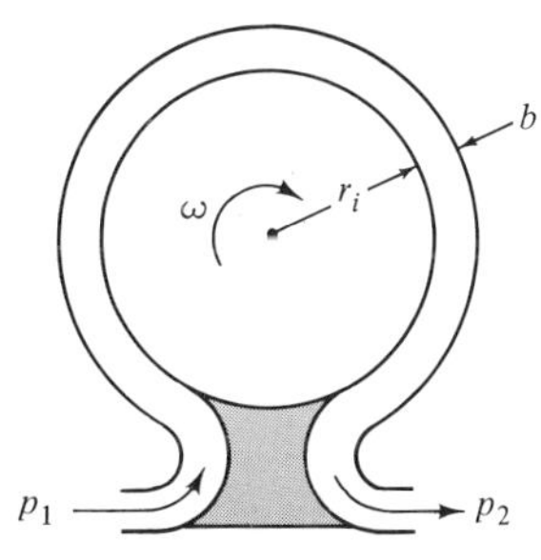

Fig. P8-18.

8-19 A roller feeder for a polymer-sheet-casting process is used to "pump" the liquid polymer by means of a belt as illustrated. Determine the equation relating the flow rate to belt speed, fluid properties, and the geometry shown.

8-20 Determine the constants of integration in (8-12a) to obtain (8-12b).

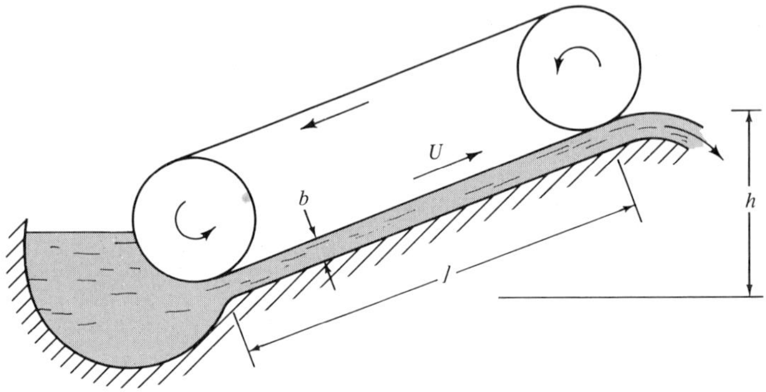

Fig. P8-19, 20.

8-21 (a) Integrate the velocity distribution for the flow between two plates with one in motion relative to the other and with a flowwise pressure gradient as given in (8-12b) to obtain the flow rate per unit width Q/c as given in (8-13).
(b) Plot a nondimensionalized velocity distribution ($u' = u/U$, $z' = z/b$) for the case of zero flow rate.

8-22 (a) For the flow between parallel plates with one plate moving with an applied pressure gradient as illustrated in Fig. 8-3, plot $z' = z/b$ versus $u' = u/U$ for the values of

$$P \equiv \frac{b^2}{2\mu U}\frac{\partial p}{\partial x} = 0, \pm 1, \pm 2, \pm 3$$

(b) What is true of the shear stress on the moving plate when $P = +1$ and on the stationary plate when $P = -1$?

8-23 A hydrostatic journal bearing of diameter d is supplied with high-pressure oil at p_1 as shown. The ambient pressure is p_a. For no rotation find the relationship between p_1, p_a, b, d, W, and the oil properties for which the journal is floated at a constant gap width b. If p_1 is maintained constant, is this position one of stable equilibrium?

8-24 For the conditions of Problem 8-23, except that the journal is rotating with an angular speed ω, what is the maximum possible ω for lubrication to be maintained?

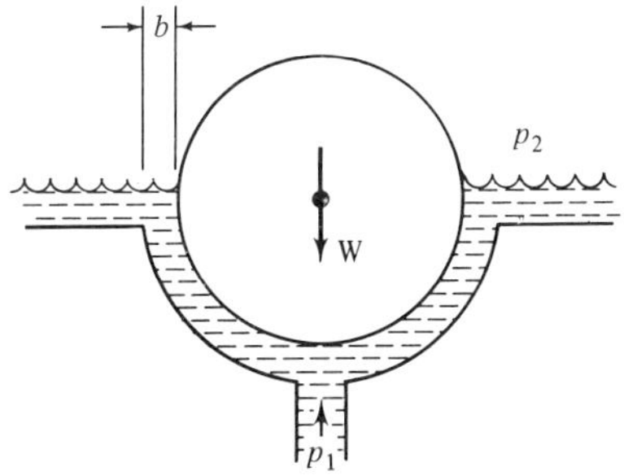

Fig. P8-23, 24.

8-25 Starting with the equation immediately above (8-15), derive (8-15) and (8-15a).

8-26 Consider a two-dimensional fluid element in the r-x plane for the flow field in Fig. 8-4 and derive the relationship that

$$\dot{\gamma}_{xr} = \frac{1}{2}\left(\frac{\partial u}{\partial r} + \frac{\partial v}{\partial x}\right)$$

(For this flow field, $\partial v/\partial x = 0$.)

8-27 Obtain the radial shear stress relation for laminar flow in a pipe. What is the value of the shear stress at the wall and at the centerline in terms of u_0 and in terms of $\partial p/\partial x$?

8-28 Integrate the flow distribution in the round pipe to obtain the given value of $\dot{m} = \rho A_0 u_0/2$ and $Q = A_0 u_0/2$. This result indicates that the average velocity $\bar{v} = u_0/2$ (that is, the centroid of the paraboloid of revolution is at one half of its axial length).

8-29 Lubricating oil ($T = 60°F$, $S = 0.95$, $\mu = 10^{-2}$ lb$_f$-sec/ft^2) is pumped through a ½-in. pipe by a transfer pump from a refinery sampling tap to a testing station 200 ft from the tap. The maximum discharge pressure available at the pump is 100 psi. If the testing station is 20 ft below the tap location, what is the maximum flow rate possible? (The I.D. of ½-in. pipe is 0.628 in.) What is the Reynolds number of the flow in the pipe?

8-30 As a first approximation, assume that μ = const. and that the pipe is insulated and calculate the discharge temperature of the oil (specific "heat" $c = 0.50$) for the setup of Problem 8-29.

8-31 For the capillary type of viscometer shown, calculate the time for the measured volume of water at 60°F to flow through the capillary. What is the Reynolds number for the flow in the capillary tube?

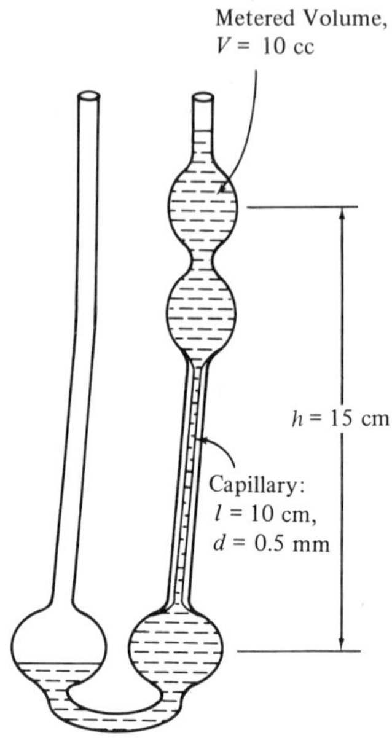

Fig. P8-31, 32, 33.

8-32 The V, h, L, and d are all geometry constants for the viscometer shown in Problem 8-31. Based on h as an average head, and the other dimensions, the relationship for the volume V of flow through the capillary tube can be reduced to an expression of the form $\nu = Ct$ if end effects, head variation, variations due to thermal expansion, and so on, are neglected. For these assumptions, calculate the viscometer calibration constant C in terms of V, h, L, d, and so on.

8-33 A capillary-tube type of viscometer as illustrated for Problem 8-31 yields ν directly as a consequence of the flow whereas the concentric-cylinder type (Fig. 8-2) yields μ. Describe the reasons for these results.

8-34 For the flow in a round duct, plot the velocity, shear stress, and irreversibility rate as functions of radius.

8-35 For the flow between two stationary parallel plates, plot the velocity, shear stress, and irreversibility rate as a function of z.

8-36 For the flow between two parallel plates with one plate moving with a velocity U and with an applied pressure gradient, plot the velocity, shear stress, and irreversibility rate as a function of z.

8-37 What are the units of c_v in (8-24)?

8-38 Substitute the boundary conditions into (8-25a) and obtain (8-25b).

8-39 Calculate and plot (sketch) the velocity and shear stress distributions for laminar, annular flow represented by (8-25b). Find the radius at which the velocity is a maximum.

8-40 Determine the relationship between flow rate Q and the pressure gradient $\Delta p/L$ and μ, r_i, and r_o) for flow in an annulus.

8-41 Find the equation giving u/u_{max} as a function of r/r_o and r_o/r_i for flow in an annulus.

8-42 Calculate the kinetic-energy rate and the kinetic-energy correction factor for flow in an annulus as given by (8-25b).

8-43 For the viscometer tube of Problem 8-31, find the transition-length fraction of the total capillary length and the percentage correction that must be added to the laminar friction for this entrance transition length.

CHAPTER 9

CONTENTS

THE TWO-DIMENSIONAL LAMINAR BOUNDARY LAYER

9.1 BOUNDARY-LAYER CHARACTERISTICS

A major advance in the understanding of fluid flow phenomena was made when Prandtl "discovered" and introduced the concept of the boundary layer.[1] In the case of flow around a body at "large" Reynolds numbers, he showed that the region of significant viscous effects was restricted to only a thin "boundary layer" next to the solid boundary. In many cases, the thickness of the boundary-layer region is much smaller than any significant dimension of the body. The flow field can then be considered as consisting of two regions, one wherein the viscous effects are not important and the other where they are. The flow analysis is made in two separate parts: The "external" flow outside the boundary layer can be analyzed as an ideal or potential flow around the body neglecting the boundary-layer thickness, and the thin boundary layer is treated as a thin, viscous flow field. The fact that the viscous flow is restricted to a very thin region adjacent to the body allows a simplification in the equations and their solution.

The existence of the boundary layer can be visualized by considering a thin, sharp-edged flat plate propelled through a fluid at a constant speed U, as illustrated in Fig. 9-1a, or the equivalent relative situation of an otherwise undisturbed flow at speed U past the plate at rest, Fig. 9-1b. The stationary flow illustrated in Fig. 9-1b is more easily considered. It corresponds to the "wind-tunnel" point of view as compared with the "free-flight" viewpoint of Fig. 9-1a. It is a physical requirement for the flow of any *real* fluid past a body that the "ordered continuum" velocity of the fluid particles immediately adjacent to the body must have the velocity of the body.[2] If the body is at rest, the ordered velocity of these particles must be zero. If the fluid is a gas (for example, air) it can be regarded as a very numerically dense population of particles in random

[1] L. Prandtl, Uber Flussigkeitsbewegung, bei zehr kleiner Reibung, *Proc. 3rd Intern. Math. Congr., Heidelberg, 1904*. See also *NACA TM 452*, 1929.

[2] The random molecular motion does not, of course, cease at the boundary.

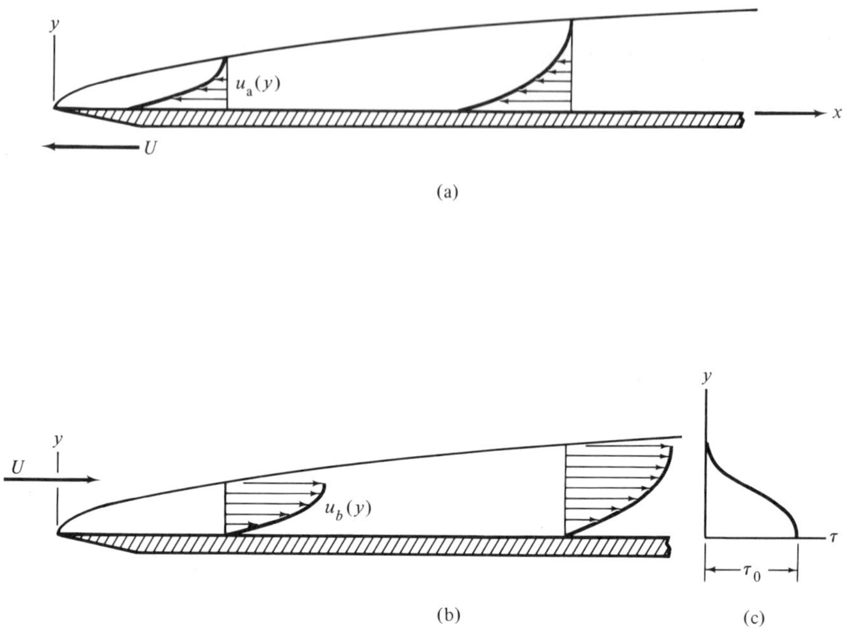

Fig. 9-1. (a) Laminar boundary layer on a plate moving with velocity U into a fluid at rest. (b) Laminar boundary layer on a plate at rest in a fluid stream with velocity U. Note that $u_a + u_b = U$; u_a can be considered as a velocity defect due to the viscous retarding action. (c) Shear stress distribution in the boundary layer. Note that $\partial\tau/\partial y = dp/dx = 0$ at the plate surface.

motion on which is superimposed the ordered fluid (continuum) velocity. The condition of zero-ordered velocity at the wall implies that the statistical average of the velocity (relative to the solid wall boundary) is zero (see Appendix C).

The action of viscosity as described in Chapter 2 produces a varying velocity in a direction perpendicular to the plate. Referring to Fig. 9-1b, as the particles in a plane parallel with the plate just pass its leading edge, those immediately adjacent to the plate will have their momentum reduced to zero. Since they still retain their thermal agitation momentum (random motion component), some will have a vertical velocity component. Some of the particles with velocity U which approach immediately above the plate will "collide" with these zero x-directed momentum particles that have collided with the plate and rebounded into the flow. An exchange of momentum will take place whereby the particles from the plate are accelerated and the others slowed. At the leading edge, this momentum exchange layer will be of the order of only one mean free path in thickness. As the flow proceeds back along the plate, the momentum exchange from one "layer" to the next continues and is propagated perpendicularly outward from the plate.

This perpendicular momentum exchange is the action, which is the shear stress. The perpendicular momentum exchange depends primarily on random molecular motion (thermal agitation state of the gas). This momentum exchange rate was shown to be characterized by the viscosity (a molecular momentum exchange coefficient) as described in Chapters 2 and 4 and Appendix C). The rate at which the shear layer grows as the fluid passes back along the plate depends on the relation of this perpendicular momentum transport to the directed momentum of the flow field. The random momentum transport velocity is the average random molecular speed and is therefore approximately constant within the fluid and depends primarily on the temperature. If the directed approach flow velocity U is increased, the particles above the plate will travel a correspondingly further distance in the x direction before colliding with the zero-directed velocity particles moving up from the plate. As a result, the retarding effect will take place farther back along the plate. Consequently, the shear layer will grow more slowly (will be thinner at a given x) as the free stream velocity U is increased. The ratio of this directed momentum transport to the random momentum transport is again indicated by the Reynolds number. If the Reynolds number is large, the random momentum transport is less than the directed, so the shear layer thickness grows more slowly than for the case of a small Reynolds number.[3]

Within this shear zone, the velocity must vary from zero at the plate to the "external field velocity" U as indicated in Fig. 9-1b. The shear stress must decrease to zero at the outer limit of the shear layer, as indicated in Fig. 9-1c. The velocity profile must then have a vertical tangent at the outer edge of the boundary layer. The boundary-layer thickness δ is a function of x, as described above. Although a finite boundary-layer thickness has been indicated in Fig. 9-1, the shear stress must fall off exponentially, since the formation or growth of the boundary layer is a diffusion process. The absolute limit of the shear effect is then at $y = \infty$. However, a finite boundary-layer thickness can be defined nominally as the distance required to attain 99 percent (or 99.5 percent) of the external or free stream velocity U. Another characteristic of the velocity profile in the boundary layer is that the tangent $(\partial u/\partial y)$ at the boundary must be finite (nonzero), as indicated by a finite angle of the velocity distribution at $y = 0$; otherwise the shear stress at the plate (at $y = 0$) would be infinite, which is not a physically allowable condition.

The above discussion considered a gas because the mechanism of molecular momentum exchange (viscosity) can be idealized for a perfect

[3] For this particular case as illustrated in Fig. 9-1, there is no characteristic length of the "body" on which to base the Reynolds number. The significant Reynolds number is then based on the distance from the leading edge: $N_R = Ux/\nu$.

gas. For a liquid, the mechanism of viscosity is more complicated (and not fully understood), but the action of viscous momentum transport is essentially independent of the directed momentum transport (as above), so the results are the same with respect to the formation of the boundary layer. From the above discussion, it is seen that both the directed momentum transport and the shear stress are required to produce a boundary layer. The shear varies from a maximum to zero from the wall outward through the boundary layer while the directed momentum (or inertia) varies from zero to a maximum, so the net effect of each is significant. Consider the force–momentum principle applied to the control volume formed by a section of boundary layer as indicated in Fig. 9-2.

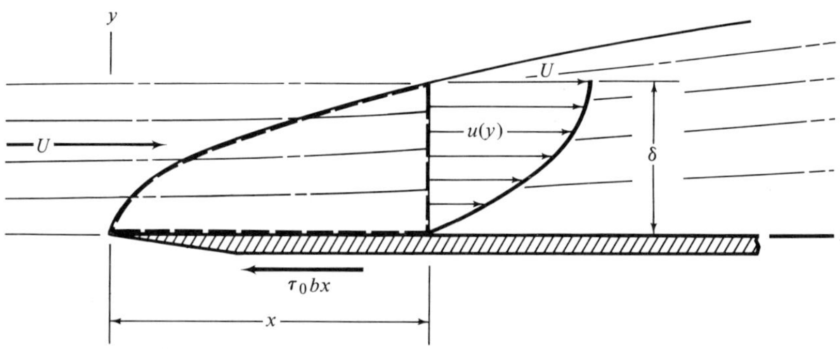

Fig. 9-2. Laminar boundary layer on a flat plate. Note that the streamlines diverge slightly in the boundary-layer region as a result of flow deceleration near the wall and that the boundary layer intersects the leading edge; consequently, the edge of the boundary layer cannot be a streamline. The heavy dashed line is the control volume boundary; the dot-dashed lines are streamlines.

There is no shear stress on the control volume boundary adjacent to the external flow. The only significant force and momentum transports are in the x direction. For steady flow, an approximate momentum equation can be considered

$$\begin{bmatrix} x\text{-MOMENTUM} \\ \text{TRANSPORT OUT} \end{bmatrix} - \begin{bmatrix} x\text{-MOMENTUM} \\ \text{TRANSPORT IN} \end{bmatrix} = \sum F_x$$

$$\frac{1}{g_c}\iint_{A_{\text{out}}} u\rho\vec{V}\cdot d\vec{A} - \frac{1}{g_c}\iint_{A_{\text{in}}} U\rho\vec{V}\cdot d\vec{A} = -\tau A_{\text{surface}} \tag{9-1}$$

The in-momentum rate over the boundary is approximately equal to (but less than) $\rho U^2 b\delta$, where δ is the boundary-layer thickness at x and b is the plate width. The out-momentum rate is $\int_0^\delta \rho b u^2\,dy$, which is somewhat less than the inlet rate, since $u_{\max} = U$ at $y = \delta$. Equation (9-1) then states that the shear force is equal to the momentum defect in the flow. Equation (9-1) can be written in approximate form as

$$\frac{\rho U^2 b\delta}{g_c} - \frac{k\rho U^2 b\delta}{g_c} = \tau_0 bx \tag{9-2}$$

where τ_0 denotes the shear at $y = 0$ and k is a number less than unity. For laminar flow of a Newtonian fluid, $\tau_0 = \mu(du/dy)_{y=0}$. The velocity gradient at the plate is not constant but varies with x, so that τ_0 is also a function of x. The value of τ_0 in (9-2) is therefore an average value. The velocity gradient at the wall $(du/dy)_{y=0}$ is $\sim U/\delta$, say, $(du/dy)_{y=0} = K(U/\delta)$, where $K > 1$; (9-2) then gives

$$\frac{1}{g_c}(1 - k)\rho U = K\mu \frac{U}{\delta} x$$

Solving for δ,

$$\delta = \sqrt{\frac{K}{1-k}}\sqrt{\frac{\nu x}{U}}$$

or

$$\frac{\delta}{x} = \frac{C}{\sqrt{Ux/\nu}} = \frac{C}{\sqrt{N_{R,x}}} \tag{9-3}$$

indicating that the boundary-layer thickness δ increases with decreasing or with increasing viscosity. Experimentation or an exact analysis gives

$$\frac{\delta}{x} = \frac{5}{\sqrt{N_{R,x}}} \tag{9-3a}$$

for δ measured to the point where $u/U = 0.99$. This result indicates that the boundary layer varies parabolically with x, with the vertex of the parabola at the leading edge, as illustrated in Figs. 9-1 and 9-2.

The characteristics and behavior of the boundary layer will be discussed further in connection with the flow fields about submerged bodies in Chapter 11.

9.2 BOUNDARY-LAYER EQUATIONS

With the assumption that the viscous shear region exists only in a very narrow layer of fluid adjacent to a solid boundary, the equations of motion can be simplified so that an exact (series) solution is possible for the flow over a body surface. However, the method of solution is not simple. For curved surfaces, exact solutions are quite difficult and limited to certain simple classes of surfaces or bodies. The reasons for these difficulties will be discussed later.

For a flat or gently curved surface as illustrated in Fig. 9-3, a two-dimensional coordinate system is chosen such that the body streamline coordinate s_0 or the x coordinate is measured along the surface and the n or y axis is perpendicular to the surface. This coordinate system is selected because of the initial assumption that the entire flow region considered will be in a layer of thickness much less than any significant dimension of the body, for example, its local radius of curvature. The s-n or x-y coordinate system may be considered to be a quasi-cartesian coordinate system, since the curvature and δ are small.

The fundamental assumptions necessary for evaluation of this flow are:

1. The Reynolds number is large enough that the boundary layer will be thin, that is, $\delta \ll c$ (where c is the body length as illustrated in

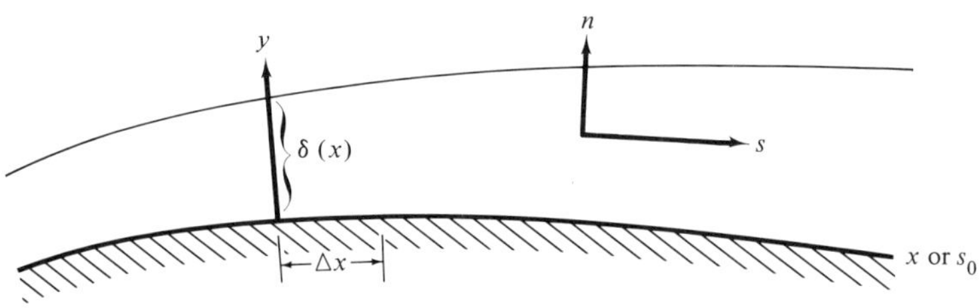

(a)

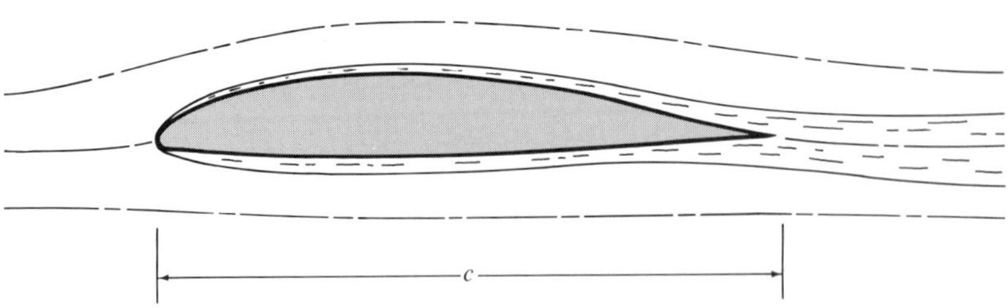

(b)

Fig. 9-3. (a) Representation of boundary-layer coordinate system: x or s_0 is the wall-boundary abscissa and y or n is the ordinate. (b) Boundary layer on a body with reference dimension c, the chord length.

Fig. 9-3b, or any significant dimension affecting the flow field characteristics, for example, radius of curvature of the body).
2. The flow is incompressible (ρ = const.) and steady.
3. The fluid viscosity μ is constant. Therefore with ρ = const., ν = const.

The first is the basic boundary-layer assumption. The objective will be to obtain the velocity distribution, shear stress, and shear force.

NONDIMENSIONALIZATION OF EQUATIONS: ORDER OF MAGNITUDE, COMPARISON OF TERMS

Assuming two-dimensional, steady, incompressible flow in a constant gravitational field, equations (6-28) are

$$\frac{\mathrm{v}}{g_c}\frac{\partial \mathrm{v}}{\partial s} = -\frac{1}{\rho}\frac{\partial p}{\partial s} - \frac{\partial gz}{\partial s} + \frac{1}{\rho}\frac{\partial \tau}{\partial n}$$
$$\frac{\mathrm{v}^2}{g_c R} = \frac{1}{\rho}\frac{\partial p}{\partial n} - \frac{\partial gz}{\partial n} \tag{9-4}$$

With the assumption that ρ = const., an equivalent pressure p_t can be defined as $p_t = p + (\rho gz/g_c)$. Dropping the t subscript for convenience, and using the relation between stress and strain rate for a Newtonian fluid [with the strain rate given by (6-33a)], (9-4) becomes

$$\frac{\mathrm{v}}{g_c}\frac{\partial \mathrm{v}}{\partial s} = -\frac{1}{\rho}\frac{\partial p}{\partial s} + \frac{\mu}{\rho}\frac{\partial}{\partial n}\left(\frac{\partial \mathrm{v}}{\partial n} - \frac{\mathrm{v}}{R_s}\right)$$
$$\frac{\mathrm{v}^2}{g_c R_s} = \frac{1}{\rho}\frac{\partial p}{\partial n} \tag{9-5}$$

These equations are nondimensionalized with the nondimensional variables defined as $\mathrm{v}' \equiv \mathrm{v}/U$, $p' \equiv p/(\rho U^2/g_c)$, $s' \equiv s/c$, $R_s' \equiv R_s/c$, $\rho' \equiv \rho/\rho_{\text{ref}} = 1$. For a curved body, $U_e(x)$ is the external velocity and c is a characteristic body dimension (see Fig. 9-3). Comparing order of magnitude terms (as in Chapter 7), (9-5) becomes

$$\mathrm{v}'\frac{\partial \mathrm{v}'}{\partial s'} = -\frac{\partial p'}{\partial s'} + \frac{1}{N_R}\frac{\partial}{\partial n'}\left(\frac{\partial \mathrm{v}'}{\partial n'} - \frac{\mathrm{v}'}{R_s'}\right)$$
$$1 \sim \frac{\partial p'}{\partial s'} + \frac{1}{N_R}\frac{1}{\delta}\left(\frac{1}{\delta} + \frac{1}{1}\right) \tag{9-6}$$
$$\frac{\mathrm{v}'^2}{R_s'} = \frac{\partial p'}{\partial n'}$$
$$1 \sim \frac{\partial p'}{\partial n'}$$

The dimensionless velocity v′ varies from zero to an order of 1 in the flow field (boundary layer) under consideration and can be taken to be ~ 1. The change in the s coordinate $\Delta s'$ would also be ~ 1; also $\Delta \mathrm{v}' \sim 1$, so that $\partial \mathrm{v}'/\partial s' \sim 1$. The maximum order of variation of $\delta n'$ is $\sim \delta'$, where $\delta' \equiv \delta/c$ and $\delta' \ll 1$ is assumed. The radius of curvature is assumed to be of first order, $R_s' \sim 1$. The order of magnitude of the terms of (9-6) is indicated below each corresponding term of the equation. The first comparative equation indicates that $\partial \mathrm{v}'/\partial n'$ is an order of magnitude greater than v'/R_s', so that v'/R_s' can be neglected in further calculations. For the last term to be of equal order with the inertia and pressure gradient terms, it is necessary that $N_R \sim 1/\delta^2$. This result was previously obtained. The comparison then indicates that $\partial p'/\partial s' \sim 1$ or $(\Delta p')_s \sim \Delta s' \sim 1$. The second relation indicates $(\Delta p')_n \sim \Delta n' \sim \delta'$, or the variation of pressure through the boundary is much less than the flow- or streamwise pressure gradient ($\partial p/\partial n < \partial p/\partial s$) and can be neglected. The original set of equations, (9-5), is then reduced to

$$\mathrm{v}\frac{\partial \mathrm{v}}{\partial s} = -\frac{g_c}{\rho}\frac{\partial p}{\partial s} + \nu\frac{\partial^2 \mathrm{v}}{\partial n^2}$$

$$\frac{\partial p}{\partial n} \sim 0 \tag{9-7}$$

The significance of $\partial p/\partial n \sim 0$ is that the pressure variation through the boundary layer (from the wall to the external flow) is negligible with respect to the other actions being considered (that is, the shear stress variation, the streamwise pressure gradient, and the inertia force or momentum). This result is a direct consequence of the fact that only flows where $\delta \ll R_s$ (or δ is much less than any significant dimension of the body) will be considered as boundary-layer flows, which in turn is a necessary consequence of the condition that $N_R \gg 1$. If (9-7) is broken up into mutually perpendicular (and therefore independent) component equations in the x and y directions, the system of equations equivalent to the Navier–Stokes equations (6-59) in an x-y coordinate system is obtained,[4]

$$u\frac{\partial u}{\partial x} + v\frac{\partial u}{\partial y} = -\frac{g_c}{\rho}\frac{\partial p}{\partial x} + \nu\left(\frac{\partial^2 u}{\partial x^2} + \frac{\partial^2 u}{\partial y^2}\right)$$

$$u\frac{\partial v}{\partial x} + v\frac{\partial v}{\partial y} = -\frac{g_c}{\rho}\frac{\partial p}{\partial y} + \nu\left(\frac{\partial^2 v}{\partial x^2} + \frac{\partial^2 v}{\partial y^2}\right)$$

[4] It is also necessary to use the defining condition for the streamline $\vec{\mathrm{v}} \times \vec{ds} = \hat{z}(u\,dx - v\,dy) = 0$ or $u\,dx = v\,dy$.

If these equations are nondimensionalized, the x- and y-momentum equations reduce to

$$u\frac{\partial u}{\partial x} + v\frac{\partial u}{\partial y} = -\frac{g_c}{\rho}\frac{\partial p}{\partial x} + \nu\frac{\partial^2 u}{\partial y^2}$$

$$\frac{\partial p}{\partial y} \sim 0 \tag{9-8}$$

and the continuity equation remains as

$$\frac{\partial u}{\partial x} + \frac{\partial v}{\partial y} = 0 \tag{9-9}$$

(See, for example, Problem 9-2.)

Since the transverse pressure gradient component is negligible in either coordinate system, $\partial p/\partial s = dp/ds$ or $\partial p/\partial x = dp/dx$. There are two further developments as a consequence of this result. Support is given for the coordinate system chosen with the s or x coordinate along the (gently curved) body surface and the n or y coordinate everywhere perpendicular to it; the pressure within the boundary layer is the same as the pressure on the external edge, where the shear flow "joins" the flow region, where the viscous effect is negligible. In the external region the pressure distribution can be determined from Bernoulli's equation

$$\frac{dp}{dx} = -\frac{\rho}{g_c}U\frac{dU}{dx}$$

where $U(x)$ is determined from the analysis of the ideal flow around the body. The boundary layer is so much thinner than the body that it can be neglected (as a first approximation) in determining the external velocity $U(x)$ and the pressure $p(x)$ distribution over the body. The flow about a "streamlined" body is determined by first calculating the ideal flow field to obtain the external velocity field and pressure field, and then determining the solution to the boundary-layer problem to obtain the shear stress distribution. In the case of attached (nonseparated) flow about the airfoil illustrated in Fig. 9-3b, the integral of the pressure and shear stress force components perpendicular and parallel to the approach flow (U) direction would give the lift and drag on the body.

The three boundary conditions required for the solution to the set of equations (9-8 and 9-9) are

(1) At $y = 0$, $u = 0$
(2) At $y = 0$, $v = 0$ (9-10)
(3) At $y = \infty$, $u = U$

Three boundary conditions are necessary because the equations for the solution to (9-8) and (9-9) are of third order. The first derivative term of (9-8) is of second order, but there are two equations from which one of the independent variables u or v must be eliminated to obtain a solution. This elimination procedure results in one third-order equation (in place of one first-order and one second-order equation). An auxiliary boundary condition is (4) at $y = \infty$, $\tau = 0 = \partial u/\partial y$. Another condition that is a consequence of applying (1) and (2) of (9-10) to (9-8) is (5) at $y = 0$,

$$\frac{\partial^2 u}{\partial y^2} = \frac{1}{\mu}\frac{dp}{dx} = \frac{U}{\nu}\frac{dU}{dx}$$

It can also be required that the shear stress at $y = \infty$ is not discontinuous or that (6) at $y = \infty$, $\partial\tau/\partial y = 0 = \partial^2 u/\partial y^2$.

The simplest case that can be visualized is the determination of the velocity distribution and shear stress distribution in the boundary layer on a flat plate at zero incidence angle (the angle between the flow direction and the plane of the plate). Although this problem is simplified by the condition that $dp/dx = (\rho U/g_c)(dU/dx) = 0$, the solution of the set of equations (9-8) and (9-9) is still moderately difficult. The solution was first worked out by Blasius[5] and is well documented.[6] His results for the boundary-layer velocity distribution as shown in Fig. 9-4 agree quite well experimental results.

The set of boundary-layer equations (9-8) and (9-9) can be combined into a single equation. The continuity condition (9-9) implies the existence of a stream function Ψ such that $u = \partial\Psi/\partial y$ and $v = -\partial\Psi/\partial x$, which allows the two independent quantities u and v to be replaced by the single stream function $\Psi(x, y)$.[7] Using Ψ, (9-8) becomes

$$\Psi_y\Psi_{xy} - \Psi_x\Psi_{yy} = \frac{-g_c}{\rho}\frac{dp}{dx} + \nu\Psi_{yyy} \qquad (9\text{-}11)$$

The subscripts imply partial differentiation with respect to the indicated variable. For the flat plate at zero incidence, $dp/dx = 0$. Equation (9-10) is a third-order, nonlinear partial differential equation. It is possible to obtain a total differential equation if a suitable transformation can be found.

If a dimensional analysis is made on the fundamental variables of the boundary-layer problem Ψ, ν, x, y, U_∞, the two dimensionless groups

[5] P. R. H. Blasius (1883–), student of Prandtl at Göttingen. His thesis on Prandtl's boundary layer was the first analytical development: Grenzschichten in Flussigkeiten mit kleiner Reibung, *Z. Math. Phys.*, **56**, 1 (1908). Translated as *NACA TM 1256*.

[6] See, for example, H. Schlichting, *Boundary Layer Theory*, McGraw-Hill, Inc., New York, 6th ed., 1968, chap. VII.

[7] The stream function will be treated in Chapter 14.

$\Psi/\sqrt{\nu U_\infty x}$ and $y\sqrt{U_\infty/\nu x}$ can be obtained (see Problem 9-7). The first contains the dependent variable Ψ, and the second, only the independent variables (and constants). The general relation can then be written $\Psi/\sqrt{\nu U_\infty x} = f(y\sqrt{U_\infty/\nu x})$. Defining the independent variable as $\eta \equiv y\sqrt{U_\infty/\nu x}$, the dependent variable is $f(\eta) \equiv \Psi/\sqrt{\nu U_\infty x}$. By definition $u = \partial\Psi/\partial y$; it can easily be determined that $u/U_\infty = df/d\eta = f'(\eta)$. This reduction of the partial differential equation to a total differential equation in nondimensionalized variables then reduces all flat-plate boundary-layer problems to a single problem. The results will likewise all be *similar*, or the same when expressed in nondimensional or coefficient form. The transformation from the $\Psi(x, y)$ equation to $f(\eta)$ is then a *similarity transformation*. By making the substitution of variables, the derivatives can be determined as $\partial\Psi/\partial x = (\partial\Psi/\partial\eta)(\partial\eta/\partial x)$, and so on, and (9-11) reduces to (see Problem 9-8)

$$2f''' + ff'' = 0 \tag{9-12}$$

This nonlinear differential equation can be solved numerically: A series solution is assumed and the number of terms carried to any desired degree of accuracy. The function and its first and second derivatives are indicated in Fig. 9-4 and Table 9-1. As given above, $f' = u/U_\infty$. The significance of

TABLE 9–1

TABULATED SOLUTIONS TO THE PROBLEM OF THE BOUNDARY LAYER ON A FLAT PLATE AT ZERO INCIDENCE

η	$f(\eta)$	$f'(\eta)$	$f''(\eta)$
0	0	0	0.3321
0.2	0.0066	0.0664	0.3320
0.4	0.0266	0.1328	0.3315
0.6	0.0597	0.1989	0.3301
0.8	0.1061	0.2647	0.3274
1.0	0.1656	0.3298	0.3230
1.5	0.3716	0.4867	0.3023
2.0	0.6500	0.6298	0.2668
2.5	0.9953	0.7518	0.2173
3.0	1.3968	0.8460	0.1614
3.5	1.8282	0.9128	0.1080
4.0	2.3058	0.9555	0.0642
4.5	2.7852	0.9794	0.0343
5.0	3.2833	0.9916	0.0159
5.5	3.7806	0.9969	0.0068
6.0	4.2796	0.9990	0.0024
6.5	4.7793	0.9997	0.0008
7.0	5.2793	0.9999	0.0002

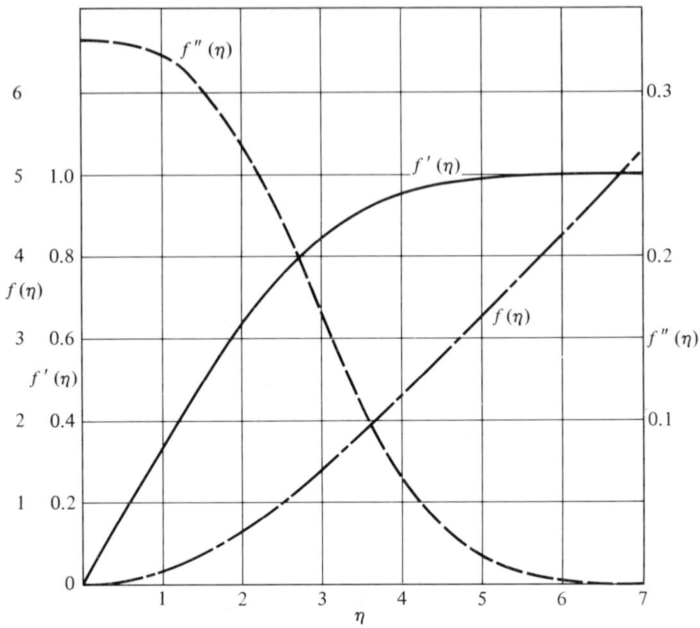

Fig. 9-4. Boundary-layer functions: Dimensionless stream function $f(\eta)$, dimensionless velocity $f'(\eta) = u/U$, and dimensionless wall friction coefficient $f''(\eta)$.

f'' is that

$$\frac{\tau(x)}{\rho U_\infty^2/2g_c} \equiv c_f = \frac{2f''}{U_\infty x/\nu}$$

or the wall friction coefficient

$$c_f = \frac{\tau_0(x)}{\rho U_\infty^2/2g_c} = \frac{2f''(0)}{\sqrt{U_\infty x/\nu}} = \frac{0.664}{\sqrt{N_{R,x}}}$$

The 99.5 percent value of u/U_∞ is attained at $\eta = 5.3 = \delta/x\sqrt{N_{R,x}}$ (or $\eta_{0.99} = 5.0$) or the dimensionless boundary-layer thickness can be taken as $\delta/x \approx 5/\sqrt{N_{R,x}}$. The drag coefficient C_D for a flat plate of length l can be obtained from the defining relations

$$C_D \equiv \frac{D}{(\rho U_\infty^2/2g_c)A} \qquad \text{and} \qquad D = \int \tau_0 \, dA = \int_0^l \tau_0 b \, dx$$

where b is the width of the plate. For the drag D on one side of area $A = bl$

(or equivalently, the drag on both sides with respect to the wetted area $2bl$),

$$C_D = 2c_{f,l} = \frac{1.328}{\sqrt{N_{R,l}}}$$

where $N_{R,l} = U_\infty l/\nu$.

9.3 CHARACTERISTIC BOUNDARY-LAYER PROPERTIES

WALL SHEAR STRESS

The physical boundary-layer thickness δ is not of much direct significance with respect to the important consequences of the existence of the boundary layer. The wall shear stress is a result of primary importance. If the velocity distribution can be determined, the wall velocity gradient $(\partial u/\partial y)_{y=0}$ can be calculated, and the shear stress is then determined from $\tau_0 = \mu(\partial u/\partial y)_{y=0}$.

BOUNDARY-LAYER DISPLACEMENT THICKNESS δ^*

The boundary-layer velocity *defect* $U - u$ has the effect of displacing the mass flow rate slightly outward from the solid boundary as compared with the ideal or potential flow that would exist in the absence of this shear layer. A boundary-layer displacement thickness δ^* can then be defined as an additional "wall thickness" that would have to be added to compensate for this defect. This displacement thickness can be determined by considering the equivalence of the two flows illustrated in Fig. 9-5.

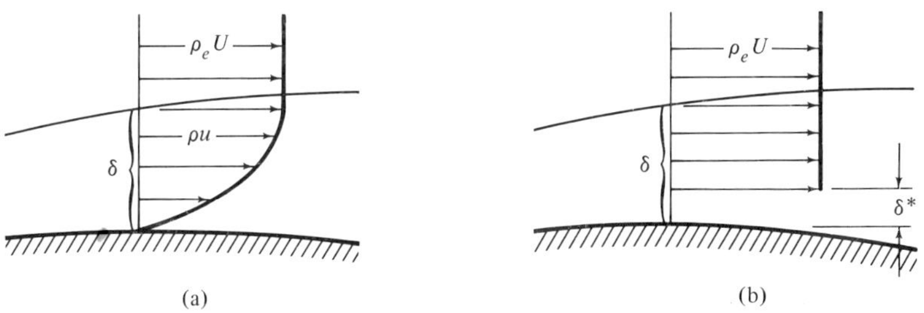

Fig. 9-5. (a) Boundary-layer mass flow rate per unit area distribution; (b) equivalent mass flow rate.

The mass flow rate through the area $b\delta$ (b is the width of flow perpendicular to the plane of the velocity distribution) is

$$\dot{m}_a = \int_0^\delta \rho u b\, dy$$

The equivalent flow rate of Fig. 9-5b is $\dot{m}_b = \rho_e bU(\delta - \delta^*)$, where ρ_e is the density in the local external flow. Setting the two mass flow rates equal and writing $\delta \equiv \int_0^\delta dy$,

$$\delta^* = \int_0^\delta \left(1 - \frac{\rho u}{\rho_e U}\right) dy \tag{9-13a}$$

If the density is constant,

$$\delta^* = \int_0^\delta \left(1 - \frac{u}{U}\right) dy \tag{9-13b}$$

MOMENTUM THICKNESS θ

The momentum is important in connection with pressure and shear forces. A momentum defect thickness can be defined in a manner similar to the mass flow rate displacement thickness. Figure 9-6a indicates the

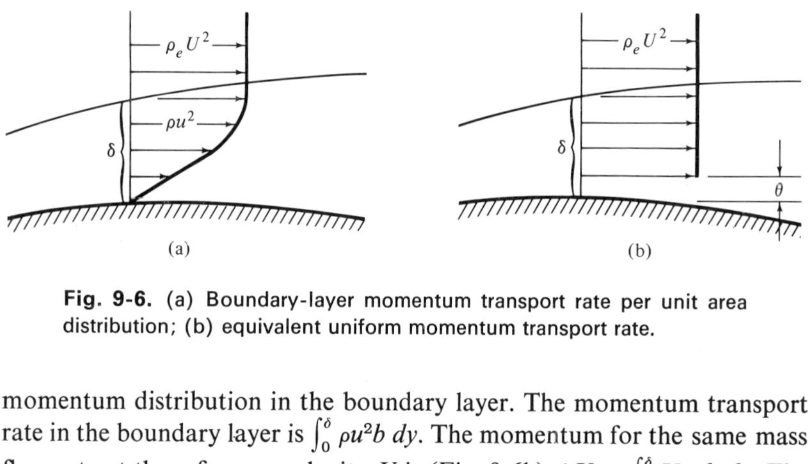

Fig. 9-6. (a) Boundary-layer momentum transport rate per unit area distribution; (b) equivalent uniform momentum transport rate.

momentum distribution in the boundary layer. The momentum transport rate in the boundary layer is $\int_0^\delta \rho u^2 b\, dy$. The momentum for the same mass flow rate at the reference velocity U is (Fig. 9-6b) $\dot{m}U = \int_0^\delta U\rho u b\, dy$. The momentum defect thickness referred to the reference external quantities is $\rho_e U^2 \theta$. Equating these momentum rates gives

$$\rho_e U^2 b\theta + U \int_0^\delta \rho u b\, dy = \int_0^\delta \rho u^2 b\, dy$$

or

$$\theta = \int_0^\delta \frac{\rho u}{\rho_e U}\left(1 - \frac{u}{U}\right) dy \tag{9-14a}$$

For incompressible flow, $\rho = \rho_e =$ const.,

$$\theta = \int_0^\delta \frac{u}{U}\left(1 - \frac{u}{U}\right) dy \tag{9-14b}$$

Because the shear boundary layer is an exponential growth diffusion process, the upper limit of integration should be $\delta \to \infty$ for the exact mathematical formulation and determination of these quantities from an analytical result.

If analytical functions are available for $u(x, y)$, the upper limit is ∞. For complicated geometries where analytic velocity distribution $u(x, y)$ cannot be formulated so that approximate relations or numerical techniques are used, δ must be taken as a finite quantity. If δ is not known, it is varied until τ varies from zero by only a specified negligible amount.

If $u(x, y)$ can be expressed as an analytic function and the upper limit of integration is ∞, the quantities of primary interest τ, δ^*, and θ are functions of x only. That is, δ^* and θ are integral functions and τ will be shown to be a function of these and the pressure gradient dp/dx or $U(x)$. These integral quantities depend only on the *shape* of the boundary layer as

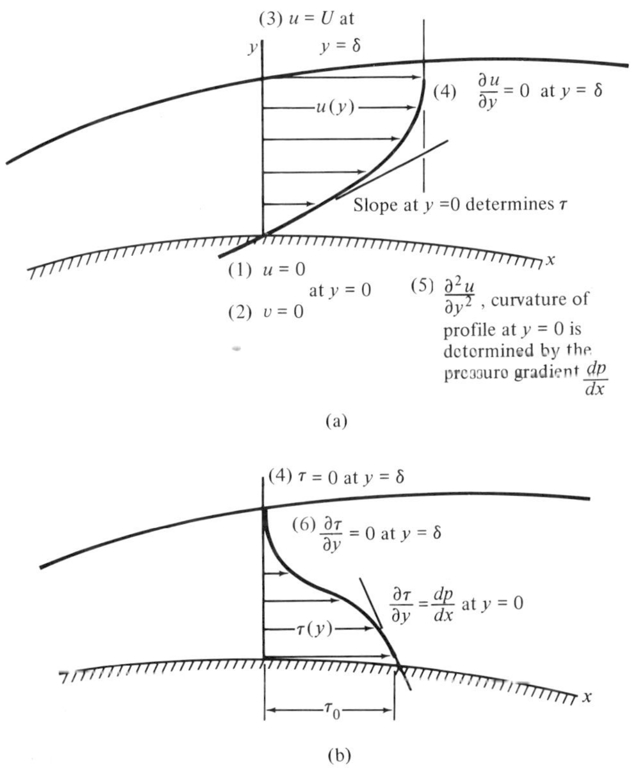

Fig. 9-7. (a) Velocity profile and its boundary conditions. Note that since $\partial\tau/\partial y = dp/dx$ at $y = 0$, $\partial^3 u/\partial x^3 = 0$ there. (b) The shear stress profile and its determining conditions. Since τ is zero for $y > \delta$, all the y derivatives of τ must be zero for $y > \delta$.

determined by the shape of the velocity profile distribution $u(y)$ (or on $u(y)/U$). The shape is subject to the conditions 1 through 5 in and following (9-10), as illustrated in Fig. 9-7.

9.4 APPROXIMATE METHODS: The Karman Integral Method

Because of the difficulty of solving even the apparently simple set of equations and boundary conditions (9-8 through 9-10) for an arbitrary two-dimensional body, a simpler and more direct method is required. Since the integral quantities are of primary interest, a momentum relation in the form of an integral equation might be used rather than the differential equation (9-8). This method was first used by Theodore von Karman[8] to derive the boundary-layer integral relationship that bears his name. The momentum equation is integrated with respect to y over the boundary-layer thickness. An alternative method starting with the momentum principle and deriving the result for a control volume boundary-layer element will be carried out here as another example of the application of the momentum principle. The control volume is chosen as an arbitrary boundary-layer element of length Δx as shown in Fig. 9-8. At the center of the element, the reference quantities are p, δ, the mass flow rate $\dot{m}$, and the momentum transport rate through the area $b\delta$ (where b is the width of the boundary layer perpendicular to the plane of the figure). For the dynamic equilibrium between the x component of the forces and the rate of change of momentum,

$$\sum F_x = \frac{1}{g_c}(\text{MOMENTUM RATE OUT} - \text{MOMENTUM RATE IN}) \tag{9-15}$$

The x-component forces on the control volume boundaries are the shear force at the wall, the pressure forces on the front and back, and an x component of the pressure force on the top, as shown in Fig. 9-8b,

$$\sum F_x \equiv \left(p - \frac{dp}{dx}\frac{\Delta x}{2}\right)\left(\delta - \frac{d\delta}{dx}\frac{\Delta x}{2}\right)b - \left(p + \frac{dp}{dx}\frac{\Delta x}{2}\right)\left(\delta + \frac{d\delta}{dx}\frac{\Delta x}{2}\right)b$$

$$+ pb\,\Delta x\left(\frac{\Delta\delta}{\Delta x}\right) - \tau_0 b\,\Delta x \tag{9-16}$$

[8] Theodore von Karman (1881–1963), Hungarian. Active and prolific in all fields of mechanics. He had a great facility for understanding the physical phenomena he observed and formulating these into the most direct, accurate, and simple mathematical relations. He was an influential teacher and leader in the fields of mechanics, fluid mechanics, aerodynamics, and astronautics. The von Karman integral method was first published as Uber laminare und turbulente Reibung, *Z. Angew. Math. Mech.*, **1**, 233–252 (1921), and was translated as *NACA TM 1092*, 1946.

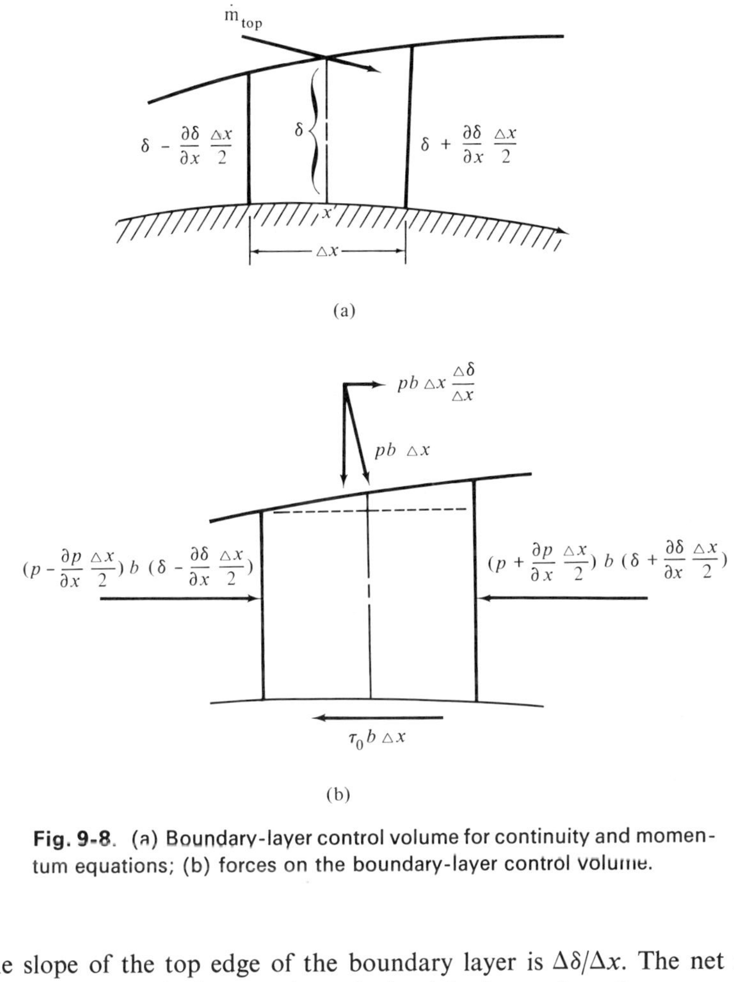

Fig. 9-8. (a) Boundary-layer control volume for continuity and momentum equations; (b) forces on the boundary-layer control volume.

The slope of the top edge of the boundary layer is $\Delta\delta/\Delta x$. The net momentum rate out is that out through the right face, minus the momentum in through the left and top faces. Momentum is transported in through the top face by the mass flow, which has a velocity U, because the top edge of the boundary layer is not a streamline (as previously discussed). The momentum rate at the station x is $\int_0^\delta \rho u^2 b\, dy$. The momentum rate out through the right face minus the momentum in through the left face is just the rate of change with respect to x multiplied by Δx, or

$$\frac{d}{dx}\left[\int_0^\delta \rho u^2 b\, dy\right]\Delta x$$

The momentum transport rate in through the top is $\dot{m}_{top}U$. The mass flow in through the top is determined from the continuity condition

$\sum \dot{m}_{\text{in}} = \sum \dot{m}_{\text{out}}$, so that the variation is $d/dx[\int_0^\delta \rho ub\, dy]\,\Delta x$, which is identical to $\dot{m}_{\text{R.H.}} - \dot{m}_{\text{L.H.}}$. The net momentum transport rate out is then

$$\frac{d}{dx}\left(\int_0^\delta \rho u^2 b\, dy\right)\Delta x - U\frac{d}{dx}\left(\int_0^\delta \rho ub\, dy\right)\Delta x$$

and the momentum equation is

$$-\frac{dp}{dx}b\delta\,\Delta x - \tau_0 b\,\Delta x = \frac{1}{g_c}\frac{d}{dx}\left(\int_0^\delta \rho u^2 b\, dy\right)\Delta x - \frac{U}{g_c}\frac{d}{dx}\left(\int_0^\delta \rho ub\, dy\right)\Delta x \tag{9-17}$$

Several modifications can be made: (1) From the fact that the shear is negligible in the external flow, $dp/dx = -(\rho U/g_c)(dU/dx)$; (2) δ can be written as $\delta \equiv \int_0^\delta dy$; (3) the last term on the right side can be expressed as

$$U\frac{d\int}{dx} = \frac{d(U\int)}{dx} - \frac{dU}{dx}\int$$

and (4) the equation can be divided by $(\rho b\,\Delta x)/g_c$ (ρ = const.). These modifications applied to (9-17) give

$$\frac{-g_c\tau_0}{\rho} = -U\frac{dU}{dx}\int_0^\delta dy + \frac{d}{dx}\int_0^\delta u^2\, dy - \frac{d}{dx}\left(U\int_0^\delta u\, dy\right) + \frac{dU}{dx}\int_0^\delta u\, dy$$

Regrouping and noting that U is a function of x only and can be placed inside or outside the integral,

$$\frac{g_c\tau_0}{\rho} = U\frac{dU}{dx}\int_0^\delta\left(1 - \frac{u}{U}\right)dy + \frac{d}{dx}\left[U^2\int_0^\delta \frac{u}{U}\left(1 - \frac{u}{U}\right)dy\right] \tag{9-18}$$

Reference to (9-11b) and (9-12b) shows that (9-18) can be expressed in the abbreviated form

$$\frac{g_c\tau_0}{\rho} = \delta^* U\frac{dU}{dx} + \frac{d}{dx}(U^2\theta) \tag{9-19}$$

which is the desired resultant form of the *Karman integral equation.* All the quantities τ_0, U, δ^*, and θ are functions of x only. The y dependence has been integrated out. Equation (9-19) is easily nondimensionalized by dividing by $U^2/2$ to give

$$c_\tau = 2\frac{\delta^*}{U}\frac{dU}{dx} + \frac{2}{U^2}\frac{d}{dx}(U^2\theta) \tag{9-20}$$

where a dimensionless wall shear stress coefficient has been defined as

$$c_\tau \equiv \frac{\tau_0}{\rho U^2/2g_c} \tag{9-21}$$

Equation (9-19) or (9-20) gives the shear stress coefficient as a function of the external velocity distribution U and the displacement and momentum thicknesses δ^* and θ, all of which are functions of x only. $U(x)$ "along the body" must be known from other sources of information or obtained from a separate analysis of the ideal velocity field around the body. This final equation is explicitly independent of y and the "physical" boundary-layer thickness δ. The quantities δ^* and θ are, however, functions of δ and the velocity distribution $u(y)$ must be known to evaluate the integrals for δ^* and θ.

In the development of the above equation, no approximations were made other than those that define the boundary layer. An approximate velocity is now introduced. Five conditions similar to the boundary conditions listed above give five requirements on the "shape" of the velocity distribution function. The form of the velocity distribution function suggested most readily is then a fourth-order polynomial

$$u(y) = a_0 + a_1 y + a_2 y^2 + a_3 y^3 + a_4 y^4 \tag{9-22a}$$

or, as a dimensionless expression,

$$u' = a_0 + a_1\eta + a_2\eta^2 + a_3\eta^3 + a_4\eta^4 \tag{9-22b}$$

where $u' \equiv u/U$, $\eta \equiv y/\delta$. The five coefficients are determined by conditions given below. This polynomial method was used by Pohlhausen to obtain approximate solutions to boundary-layer problems for various geometries.[9] Note that once the approximate boundary-layer profile is assumed, another expression is available for the shear stress in the laminar boundary-layer case in addition to (9-19),

$$\tau_0 = \mu\frac{du}{dy} = \frac{\mu U}{\delta}\frac{du'}{d\eta}$$

It is necessary to use both expressions for τ_0 to eliminate δ. The procedure is illustrated in the example below.

9.5 LAMINAR BOUNDARY LAYER ON A FLAT PLATE

The five conditions that can be applied to determine an approximate boundary-layer profile shape such as given in (9-22) are the following two

[9] See, for example, H. Schlichting, *Boundary Layer Theory*, McGraw-Hill, Inc., New York, 6th ed., 1968, chap. XII.

boundary conditions:

(1) At $y = 0, u = 0$ or at $\eta = 0, u' = 0$

(2) At $y = \delta, u = U$ or at $\eta = 1, u' = 1$

and three auxiliary conditions:

$$\text{(3) At } y = \delta, \frac{du}{dy} = 0 \quad \text{or} \quad \eta = 1, \frac{du'}{d\eta} = 0 \tag{9-23}$$

$$\text{(4) At } y = 0, \frac{d^2u}{dy^2} = 0 \quad \text{or} \quad \eta = 0, \frac{d^2u'}{d\eta^2} = 0$$

$$\text{(5) At } y = \delta, \frac{d^2u}{dy^2} = 0 \quad \text{or} \quad \eta = 1, \frac{d^2u'}{d\eta^2} = 0$$

Note that boundary condition (2) of (9-10) does not appear, since the y dependence and therefore the v component of velocity have been integrated out of the momentum relation.

The second boundary condition gives the result that $dp/dx = 0$. This boundary condition with the auxiliary condition $(d^2u/dy^2)_{y=0} = (1/\mu)(dp/dx)$ indicates that the curvature of the velocity distribution at the boundary is zero, or there is an inflection point there. This result is shown above as condition (4).

For an approximate analysis, only the two boundary conditions are absolutely necessary. Condition (3) is the next most important. Several possible functions can be chosen to fit these three conditions [for example, parabolic and trigonometric (sine)]. The parabola is simply the second-order polynomial $u' = a_0 + a_1\eta + a_2\eta^2$ [where the variables $u' = u/U$, $\eta = y/\delta$ are chosen to give a dimensionless function which is most easily used to evaluate δ^* and θ from (9-13) and (9-14)]. Third- and fourth-order polynomials can also be assumed, since there are a maximum of five conditions from which a maximum of five coefficients can be determined.

The first boundary condition will always give $a_0 = 0$ for any of the polynomial forms assumed. As a numerical example, consider the parabolic distribution given by $u' = a\eta + b\eta^2$. Conditions (2) and (3) of (9-23) determine a and b to give $u' = 2\eta - \eta^2$. Equation (9-19) gives

$$\frac{g_c \tau_0}{\rho U^2} = \frac{d\theta}{dx}$$

The wall shear stress τ_0 can also be evaluated from the condition that the boundary layer is laminar, $\tau_0 = \mu(du/dy)_{y=0}$. Evaluating these quantities and δ^* and θ from (9-13b) and (9-14b) gives

$$\frac{\delta^*}{\delta} = \int_0^1 (1 - u')\, d\eta = \int_0^1 (1 - 2\eta + \eta^2)\, d\eta = \frac{1}{3}$$

$$\frac{\theta}{\delta} = \int_0^1 u'(1 - u')\, d\eta = \int_0^1 (2\eta - \eta^2)(1 - 2\eta + \eta^2)\, d\eta = \frac{2}{15}$$

or $$\frac{\delta^*}{\theta} = 2.5$$

$$\tau_0 = \mu\left(\frac{du}{dy}\right)_{y=0} = \left(\frac{\mu U}{\delta}\frac{du'}{d\eta}\right)_{\eta=0} = \frac{2\mu U}{\delta}$$

and

$$\tau_0 = \frac{\rho U^2}{g_c}\frac{d\theta}{dx} = \frac{\rho U^2}{g_c}\frac{d\theta}{d\delta}\frac{d\delta}{dx} = \frac{2}{15}\frac{\rho U^2}{g_c}\frac{d\delta}{dx}$$

Note the direct relation here between τ and $d\theta/dx$. With c_τ as defined in (9-21) [or from (9-20)], $c_\tau = 2(d\theta/dx)$. Each of the above quantities is a function of δ. The two relations for τ_0 give an expression from which $\delta(x)$ can be determined:

$$\frac{2}{15}\frac{\rho U^2}{g_c}\frac{d\delta}{dx} = \frac{2\mu U}{\delta}$$

which can be written as the differential equation

$$\delta\, d\delta = 15\frac{g_c\mu}{\rho U}\, dx$$

which gives

$$\frac{\delta}{x} = \sqrt{\frac{30\nu}{Ux}} = \frac{5.47}{\sqrt{N_{R,x}}}$$

This result again shows that $\delta(x)$ is a parabola as indicated in Figs. 9-1 and 9-2. The quantities δ^* and θ expressed as dimensionless ratios referred to the boundary-layer length x become

$$\frac{\delta^*}{x} = \frac{1.82}{\sqrt{N_{R,x}}} \qquad \frac{\theta}{x} = \frac{0.730}{\sqrt{N_{R,x}}} \qquad H \equiv \frac{\delta^*}{\theta} = 2.5$$

The ratio $H \equiv \delta^*/\theta$ appears in determinations of overall boundary-layer effects and is one type of boundary-layer *shape parameter*. The shear stress is

$$\tau_0 = \frac{2\mu U}{\delta} = 0.365\frac{\mu U}{x}\sqrt{\frac{Ux}{\nu}} \qquad \text{or} \qquad c_\tau = \frac{0.730}{\sqrt{N_{R,x}}}$$

The drag force on a plate of length l and width b is determined by integrating the shear stress over the area. The drag force for one side of a plate is

$$D = \int_0^l \tau_0 b\, dx = 0.365\mu U\sqrt{\frac{U}{\nu}}\, b \int_0^l x^{-1/2}\, dx = 2(0.365)\mu U b\sqrt{\frac{Ul}{\nu}}$$

A force coefficient is defined as the ratio of the force to a reference dynamic pressure force:

$$C_F \equiv \frac{F}{(\rho U^2/2g_c)A}$$

The drag coefficient defined with respect to the area $A = bl$ is

$$C_D \equiv \frac{D}{(\rho U^2/2g_c)A} = \frac{1.46}{\sqrt{N_{R,l}}}$$

The drag coefficient C_D for a plate length l is just $2c_{\tau,l}$ (the shear stress coefficient at $x = l$). An alternative method of obtaining C_D is to form the coefficient before performing the integration,

$$C_D = \frac{D}{(\rho U^2/2g_c)bl} = \frac{2g_c}{\rho U^2 bl}\int_0^l \tau_0 b\, dx = \int_0^1 c_\tau\, d\left(\frac{x}{l}\right)$$

which is easier to handle for the integration procedure. Note that C_D for one side is identical to C_D for both sides, since both D and A are doubled.

The results for a number of assumed profiles are presented in Table 9-2 with the results from the exact analysis of Blasius for comparison. Note that for the flat plate $\theta/x\sqrt{N_{R,x}} = c_\tau\sqrt{N_{R,x}}$. The largest error in c_τ for any of the assumed profiles is of the order of 10 percent.

TABLE 9-2

COMPARISON OF FLAT-PLATE BOUNDARY-LAYER PARAMETERS FOR VARIOUS ASSUMED VELOCITY PROFILES

Velocity profile	$\frac{\delta^*}{\delta}$	$\frac{\theta}{\delta}$	$\left(\frac{du}{d\eta}\right)_{\eta=0}$	$\frac{\delta^*}{x}\sqrt{N_{R,x}}$	$\frac{\theta}{x}\sqrt{N_{R,x}}$	$\frac{\delta^*}{\theta}$
1. Exact (Blasius)	(~0.34)	(~1.3)	(~1.7)	1.729	0.664	2.61
2. $u' = 2\eta - \eta^2$	1/3	0.133	2.0	1.82	0.730	2.5
3. $u' = \frac{3}{2}\eta - \frac{1}{2}\eta^3$	3/8	0.139	1.50	1.740	0.646	2.70
4. $u' = 2\eta - 2\eta^3 + \eta^4$ [a]	3/10	0.118	2.0	1.752	0.686	2.55
5. $u' = \sin \pi\eta$	0.364	0.136	1.57	1.741	0.654	2.66

[a] Pohlhausen polynomial profile.

The above results are for a laminar boundary layer on a flat plate at zero angle of incidence relative to the approach flow. For streamline bodies with some curvature, the shear stress and drag can be approximated by assuming an equivalent area of flat plate and using the above results.

A further advantage of the approximate method (which arises from the fact that the velocity profile is assumed) is that it can also be used for turbulent boundary layers by a suitable choice of velocity profile function. The determination of the $u(y)$ function depends on the characteristics of turbulent flow considered in Chapter 10.

A similar procedure is necessary for curved surfaces. The calculation is more complicated because the relations for eliminating δ are also functions of $U(x)$ and dU/dx.

9.6 CHARACTERISTICS OF THE BOUNDARY LAYER: Boundary-Layer Separation

Temporarily neglecting the effect of the boundary layer on the flow over a body, various representative pressure and velocity distributions near the surface are shown in Fig. 9-9. From the principle of continuity, the average velocity between two streamlines is inversely proportional to the distance between them. On the surfaces of the profiles of Fig. 9-9, the velocity is seen to decrease below the approach stream velocity at the nose, increase along the body to its midpoint, and then decrease again. At the nose the velocity is zero (a stagnation point). Therefore, the pressure at the nose is a maximum and decreases with x over the forward part of the body. For ideal flow the pressure would again increase to the stagnation value at the aft end. The pressure distributions for these profile shapes are also given in Fig. 9-9 by the pressure coefficient c_p, where

$$c_p \equiv \frac{p - p_\infty}{\rho U_\infty^2 / 2g_c}$$

where p_∞ is the pressure of the "free-stream" approach flow with velocity U_∞. The value of 1.0 corresponds to the stagnation pressure, and a negative value implies a pressure less than the free stream pressure.

The external flow is accelerated extremely rapidly from zero velocity to a maximum over a short distance near the nose of the body, as indicated by the pressure gradient and reference to the momentum equation in the form

$$\frac{\vec{a}}{g_c} = -\frac{\nabla p}{\rho}$$

where $\vec{a}$ is the vector acceleration. Here the pressure gradient is strongly negative. The flow is accelerated to a velocity greater than the free stream

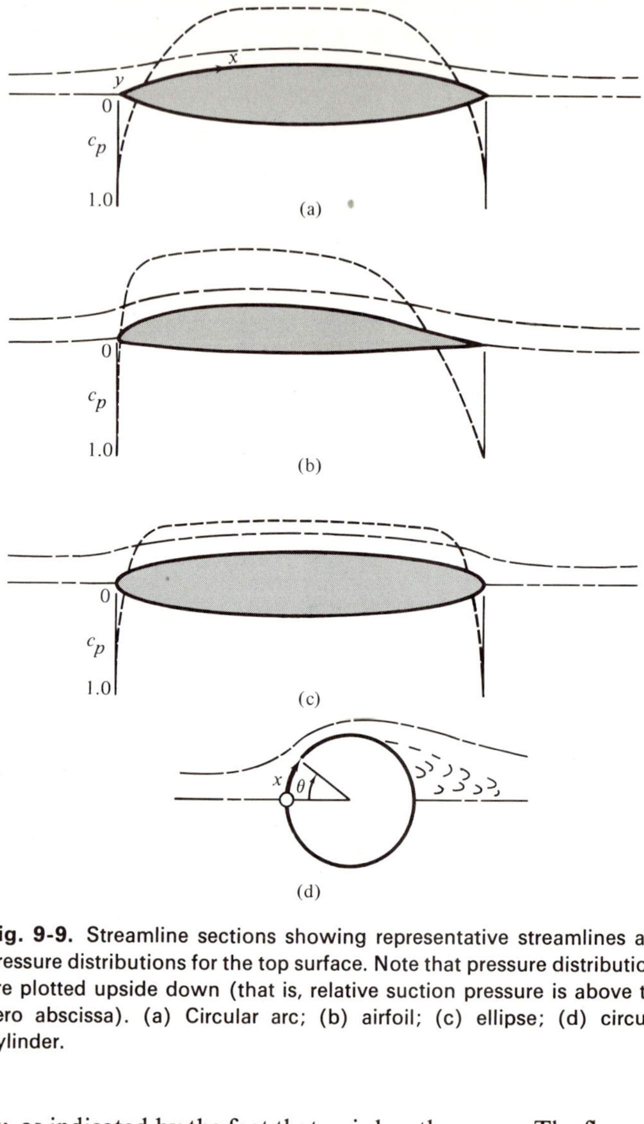

Fig. 9-9. Streamline sections showing representative streamlines and pressure distributions for the top surface. Note that pressure distributions are plotted upside down (that is, relative suction pressure is above the zero abscissa). (a) Circular arc; (b) airfoil; (c) ellipse; (d) circular cylinder.

velocity, as indicated by the fact that c_p is less than zero. The flow acceleration from zero to a velocity greater than the free stream velocity produces a thinner boundary layer than obtained on a flat plate for the same x, as can be seen from consideration of the quantitative effects of viscous momentum transport and directed momentum transport as discussed at the beginning of the chapter. Over the portion of the body where the pressure is about constant, the boundary-layer behavior is similar to that on a flat plate and the thickness (δ, δ^*, and θ) increases nearly as the square root of the x distance. The boundary-layer region then appears as in

Fig. 9-10. When the flow is decelerated, the directed momentum decreases so that the viscous layer increases more rapidly in thickness.

Another important characteristic can be determined with reference to the boundary-layer equation (9-8), or condition (5) of (9-10). The curvature of the velocity distribution at the wall is

$$\left(\frac{\partial^2 u}{\partial y^2}\right)_{y=0} = \frac{1}{\mu}\frac{dp}{dx}$$

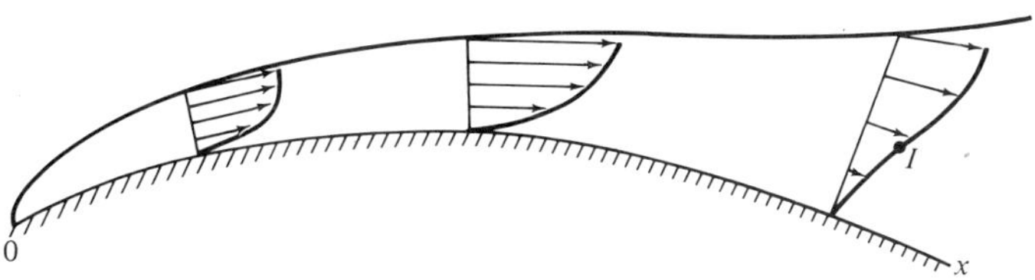

Fig. 9-10. Velocity profiles on a curved wall. In the deceleration (adverse pressure gradient) region, an inflection point *I* will occur.

In the external flow that determines dp/dx,

$$\frac{dp}{dx} = -\frac{\rho U}{g_c}\frac{dU}{dx} \qquad \nabla p = -\frac{\rho \vec{a}}{g_c}$$

If the flow is accelerated (dU/dx positive), then dp/dx is negative; consequently, the curvature of the velocity profile $u(y)$ at the wall is negative. Over the region where p is about constant ($\vec{a} \sim 0$), the curvature at the wall is zero; therefore, an inflection point exists there. In the region where the flow is *decelerated*, the velocity profile curvature at the wall will be positive, as indicated in the third profile of Fig. 9-10. Since the profile must then curve into the external flow with zero slope, there must be an intermediate inflection point in the boundary layer, as indicated.

If the region of positive pressure gradient is extensive in length and the positive gradient is large in magnitude (for example, over the rear half of the cylinder of Fig. 9-9d), the boundary layer will continue to grow rapidly, and the curvature at the wall will become more positive, as shown in the series of profiles of Fig. 9-11 (which indicate the sequence of profiles obtained for the laminar boundary layer on a circular cylinder). As the flow continues to decelerate, the curvature region becomes larger; consequently, the slope of the profile decreases until the slope is zero at the boundary (at point x_s, Fig. 9-11). If the deceleration continues (pressure gradient increases) beyond this point, a reverse flow will be obtained near the wall as shown. The flow then *separates* from the wall at the point x_s where the velocity profile has zero slope. Because the positive pressure

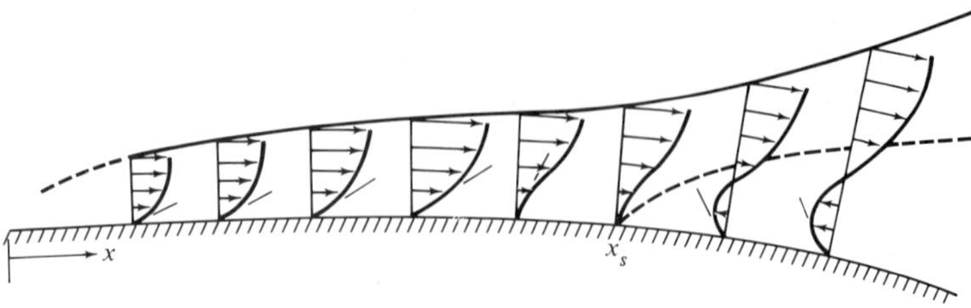

Fig. 9-11. Velocity profiles on a strongly curved surface such as a circular cylinder. Note decreasing profile slopes at the wall as separation at x_s is approached. Downstream of the separation point, a recirculating backflow is produced.

gradient region tends to produce separation, it is often termed a region of *adverse* pressure gradient. The backward or reverse flow shown downstream from the separation point indicates the formation of a *separation eddy*.

The series of events described above have been shown to occur experimentally. A very interesting set of photographs of flows around various objects can be found in the back of L. Prandtl and O. Tietjens, *Applied Hydro- and Aero-Mechanics* (Dover Publications, Inc., New York).

The flow with separation is then completely different from the ideal, attached flow pattern originally predicted. Consequently, the ideal solution for external flow is no longer valid, nor is there any way to predict or calculate flow around bodies where separation occurs. Moreover, the order-of-magnitude approximations from which the boundary-layer equations (9-7) through (9-9) were obtained are no longer valid in the region of separation. The quantities Δy, v, $\partial v/\partial x$, and so on, are no longer of second order within the boundary-layer flow region in the neighborhood of separation. For similar reasons, the boundary-layer equations do not give valid solutions near a leading edge or stagnation point where $x \sim 0$ (since it was assumed that Δx, x, and so on, were of the order of l). Separated flows and their correlation with drag and Reynolds number will be considered further in Chapter 11.

A physical argument for the boundary-layer separation can be made as follows. The boundary layer is a viscous dissipation region. The fact that the boundary-layer region thickens along a plate indicates that energy is continuously converted to unavailable internal energy in the boundary layer by the viscous action.[10] In regions of accelerating flow (negative pressure gradient), the available pressure force is used to do work in accelerating the fluid. Some available kinetic energy is converted to internal

[10] A kinetic-energy thickness (or kinetic-energy defect) can be defined and determined in a manner similar to the momentum thickness. It, too, would increase proportional to $\sqrt{x}$ for the flat-plate laminar boundary layer.

energy by the shear-force work. In regions of decelerating flow (positive or adverse pressure gradient) a reverse conversion from kinetic energy into available work potential in the form of pressure is necessary. However, the kinetic energy approaches zero as the boundary is approached (as $y \rightarrow 0$). Therefore, there is little or no energy available to provide the pressure gradient required to deviate the flow in the direction of the body curvature near the trailing edge. The flow must then take a curvature that is different from (smaller than) that of the body, or the flow separates from the body. The velocity variation from zero at the wall to a maximum velocity at the external boundary-layer edge can be considered to be the result of an equilibrium momentum transport process between the layers at the wall which produce the retarding effect, and those at the external edge which supply the directed momentum. In the region of positive pressure gradient and flow deceleration, the velocity distribution does not obtain zero slope at the wall at the point where the pressure gradient becomes positive. There is still a directed velocity component (U) at the edge of the boundary layer, so some transverse momentum is transported toward the wall. In laminar flow, the transverse (molecular viscous) momentum transport rate is very slow, so the boundary layer separates easily.

9.7 BOUNDARY LAYER ON A CIRCULAR CYLINDER

The pressure distribution $p(x)$ is easily determined for the ideal or potential flow around a circular cylinder (Chapter 14). It is possible to obtain either an exact solution of (9-8) and (9-9) or an approximate solution using the Karman integral equation.[11] Both solutions yield separation at an angle of 110° measured from the direction of the approach flow. Experimental measurements of the location of the separation point given an angle of about 82°. The large discrepancy is due to the fact that the pressure distribution produced by the flow field with separation is much different from the assumed ideal pressure distribution. It is interesting that if the experimentally obtained pressure distribution is used in the solution of the equations, the experimentally obtained separation angle of 82° is predicted. The interesting conclusion of this observation is that it is necessary to know the answer before the problem can be solved. A second conclusion is that the assumptions made in obtaining the boundary-layer equations (9-8) are satisfactory, so the method will yield valid results where the adverse pressure gradient is not too severe.

[11] These solutions are reviewed in H. Schlichting, *Boundary Layer Theory*, McGraw-Hill, Inc., New York, 6th ed., 1968, chaps. IX and XI. The exact method is quite tedious, requiring the solution of several simultaneous equations to determine series expansion coefficients. The approximate solution is also somewhat tedious.

9.8 TRANSITION TO TURBULENT BOUNDARY LAYER

As the boundary layer increases in length, it tends to become less stable to small disturbances. For a smooth flat plate, the boundary layer becomes turbulent at a distance from the leading edge determined by a transition Reynolds number value between 3×10^5 and 10^6. This transition Reynolds number depends on the degree of smoothness of the plate and on external disturbances such as approach flow turbulence or mechanical vibrations of the plate. The subject of transition and the origin of turbulence will be considered further in Chapter 10.

9.9 SUMMARY

1. The boundary layer is a consequence of a balance between the directed inertia forces (or momentum) and the viscous forces (transverse momentum) with the condition that the relative velocity at a solid boundary is zero.
2. For a boundary-layer type of flow to exist, the Reynolds number (ratio of directed to transverse random momentum transport) must be large.
3. Approximate or dimensional considerations of the above effects (1) indicate that the laminar boundary-layer thickness varies with $\sqrt{x}$.
4. From the condition that the boundary-layer region is thin with respect to any other significant dimension of the body, a "boundary-layer" set of equations can be formulated which permit more tractible methods of solution than the complete equations of motion (Navier–Stokes equations). The assumptions required for the determination of these equations have been shown to be valid by comparing the solutions (for a given physical geometry) with experimental measurements.
5. The Karman integral method can be used to give more readily determinable solutions for shear stress (θ, δ^*, and so on) than the exact method. Because the velocity distribution is not determined nor explicitly specified, the Karman integral method can also be applied to turbulent boundary layers. This method indicates that separation occurs too soon on gently curved (streamlined) surfaces and too late on bodies with large curvature.
6. Boundary-layer separation is a consequence of a positive (adverse) pressure gradient.

PROBLEMS

9-1 Assuming δ to be given by (9-3a), calculate the boundary-layer thickness 1 ft back from the leading edge of a flat plate for the following cases:

(a) the flow of water at 10 ft/sec, (b) the flow of air at 10 ft/sec, (c) the flow of air at 100 ft/sec.

9-2 Assuming the boundary-layer thickness to be given by (9-3a), what is the boundary-layer thickness on a flat plate 6 in. back from the edge for an air velocity of 200 ft/sec ($p = 1$ atm, $T = 30°$F, and 200°F)? For the same plate and a water velocity of 40 ft/sec, what is the boundary-layer thickness at 6 in. ($T = 60°$F)?

9-3 If the Reynolds number ($N_{R,x}$) at which the boundary layer becomes turbulent on a flat plate is $\sim 5 \times 10^5$, what is the maximum laminar boundary-layer length that can exist for (a) an air speed of 100 ft/sec, and (b) for a water speed of 40 ft/sec?

9-4 It is desired to make laminar boundary-layer measurements on a 2-ft-long plate in a low-turbulence wind tunnel. What is the maximum wind-tunnel speed if the transition Reynolds number ($N_{R,x}$) is 10^6 ($T = 80°$F)? What is the dynamic pressure variation (as indicated in inches of water) through the boundary layer?

9-5 A belt of sheet metal is brought through a sealing slit at A and out through another slit at B, as shown. Sketch the boundary-layer growth on this plate in the region A-B. Sketch in three or four representative streamlines of the boundary layer and a representative velocity midway along the plate. Where would such geometries be found in practice?

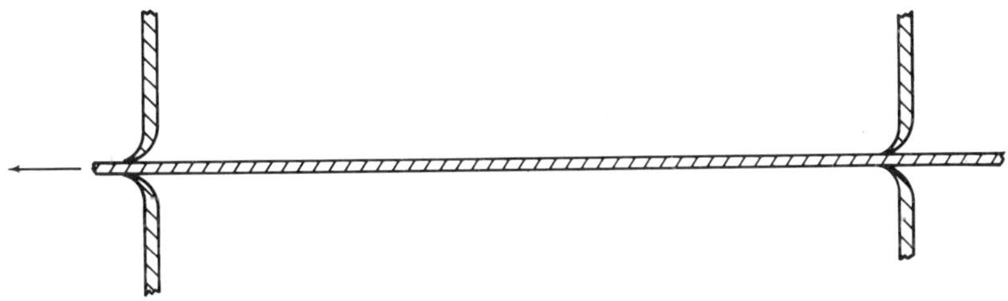

Fig. P9-5.

9-6 Starting with the Navier–Stokes equations (6-59), specialize these to a two-dimensional flow and nondimensionalize the resulting set of equations. Make the order of magnitude approximations for a boundary-layer flow field and obtain equations (9-8). Do the same for the continuity equation for steady incompressible flow to get (9-9).

9-7 Perform a dimensional analysis on the quantities Ψ, x, y, U, and ν letting the variable quantities have a different arbitrary dimension, say, $\Psi \sim A$, $x \sim B$, $y \sim C$, to obtain the dimensionless groups $\eta = y\sqrt{U/\nu x}$, $f = \Psi/\sqrt{\nu U x}$. It will be necessary to use (9-10) with $dp/dx = 0$ to obtain the dimension of ν in terms of A, B, and C; also the dimension of U is determined from the relation $u = \partial\Psi/\partial y$. Since there are five quantities and three dimensions, two dimensionless groups are expected.

9-8 Perform the necessary transformation of variables from $\Psi(x, y)$ to $f(\eta)$ to obtain (9-12) from (9-11).

9-9 Show that $C_D = \int_0^1 c_\tau d(x/l)$ and obtain the result that $C_D = 1.328/\sqrt{N_{R,l}}$.

9-10 Show that $c_\tau = \tau_0/(\rho U^2/2g_c)$ can be determined as $c_\tau = 2f''(0)/\sqrt{U_\infty x/\nu}$.

9-11 For a flat plate in air (standard sea-level conditions) at 50 ft/sec, find the drag per foot of width for the first foot of length, the second foot, the fifth foot, and the tenth foot.

9-12 If $u' = u/U$ and $\eta = y/\delta$, show that (9-13b) and (9-14b) can be written

$$\frac{\delta^*}{\delta} = \int_0^1 (1 - u')\, d\eta \qquad \frac{\theta}{\delta} = \int_0^1 u'(1 - u')\, d\eta$$

9-13 Make the modifications listed between (9-17) and (9-18) and show that the results (9-18) and (9-19) are valid.

9-14 For the laminar boundary-layer flow past a flat plate at zero incidence, show that $c_\tau = 2(d\theta/dx)$ and $C_D = \int_0^l c_\tau d(x/l) = 2\theta/l$.

9-15 Find the expression giving the drag coefficient per unit width for an arbitrary geometry by integrating (9-19) over the length of the body.

9-16 What conditions are used to obtain the cubic velocity distribution $u' = (\eta/2)(3 - \eta^2)$ given in line 3 of Table 9-2?

9-17 Apply the available conditions to the fifth-order polynomial velocity distribution of (9-22b) to obtain the coefficients. The result is given as line 4 of Table 9-2.

9-18 Determine δ^*/x, θ/x, and c_τ as functions of $N_{R,x} \equiv Ux/\nu$ for the velocity distribution of line 3, Table 9-2.

9-19 Same as Problem 9-18 for line 4 of Table 9-2.

9-20 Same as Problem 9-18 for line 5 of Table 9-2.

9-21 Determine the air drag on a rectangular flat plate 1 ft long and 2 ft wide in an air stream of 100 ft/sec under standard atmospheric conditions (assume that the boundary layer is laminar over the entire length). What is the length Reynolds number?

9-22 Determine the laminar air drag on a square plate of dimension a in air with velocity, density, and viscosity U, ρ, and μ.

9-23 Assuming that the boundary layer at each station along the surface perpendicular to the flow behaves independently of the adjacent boundary layers on a triangular plate and calculate the drag on the plate as a function of a, U, ρ, and μ.

9-24 Assuming two-dimensional laminar boundary-layer flow, calculate the drag on a thin circular disc of radius a as a function of a, U, ρ, and μ.

9-25 The function $u' = 1 - e^{-5\eta}$ satisfies some of the conditions (9-23) and can be used as an approximate boundary-layer velocity distribution. Which conditions does it satisfy exactly and which ones approximately? How approximately?

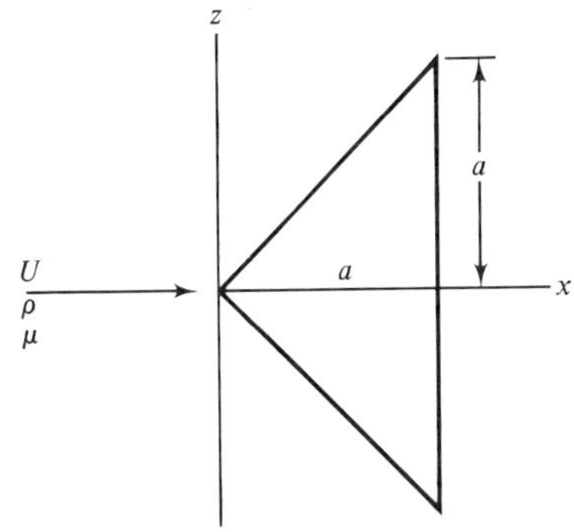

Fig. P9-23.

9-26 Using the velocity distribution given in Problem 9-25, calculate δ^*, θ, and c_τ as functions of x and $N_{R,x}$. Compare the results with those given in Table 9-2.

9-27 Calculate approximately the maximum length of flat plate over which a boundary layer will remain laminar for an air speed of 100 and 200 ft/sec (air at standard sea-level conditions).

9-28 For an air speed of 50 ft/sec over a flat plate, what is the boundary-layer thickness 1 ft back from the leading edge? What is the Reynolds number there? Is $\delta/x \ll 1$?

9-29 At what distance x measured from the leading edge is $\delta/x = 0.1$ at a speed of 50 ft/sec? What are δ, x, and $N_{R,x}$ at this point?

9-30 Find δ, δ^*, $N_{R,l}$, C_D, and the drag on a 1-ft-long flat plate in air and in water at standard conditions for a speed of 20 ft/sec. Compare the results.

9-31 Find δ, δ^*, N_R, C_D, and the drag on a 2-in.-long flat plate in water and in mercury for a liquid speed of 5 ft/sec. Comment on the results.

9-32 (a) Calculate the approximate drag for an airplane wing with a 6-ft chord and 40-ft span flying at 200 mph. (b) What is the horsepower required for the wing? (c) The minimum, ideal, two-dimensional drag or drag coefficient is about two times this ideal flat-plate value, and the drag or drag coefficient for an entire aircraft with the wing is about ten times this value because of three-dimensional and interference effects and drag of other components. Calculate the drag and horsepower for the above airfoil mounted on an aircraft.

9-33 A submarine can be idealized for drag calculations as a cylindrical surface with a length and diameter equal to those of the submarine. Calculate the drag and power required for a submarine 100 ft long and 30 ft maximum diameter if (a) the boundary layer is completely laminar, and (b) the boundary layer is completely turbulent. Compare and discuss the results.

CHAPTER 10

CONTENTS

TURBULENCE

Turbulent flow is characterized by a random eddy motion of small masses of fluid within a flowing stream or flow field. Experimental measurements show that the velocity at a point can be considered to be made up of a mean component and a randomly varying component (see Section 4.4 and Figs. 4-6 and 4-7). The eddies are continuously formed, dissipated by viscous effects, and new eddies formed. It is interesting to note that these eddies are damped, dissipated, and "lose their identity" by viscous action, and that the transport of the eddies cannot take place through a solid boundary. Turbulence is consequently decreased in the region of a wall. On the other hand, turbulence in a flow cannot originate unless a solid boundary is present[1]; the origin of turbulence requires a shear flow. When the shear disturbance is of sufficient magnitude, it is propagated into the flow and a turbulent field is produced.

The strong, rapid, relatively high-frequency turbulence in a flow (for example, the flow in a pipe) produces a very effective fluid mixing process. Fluid mass, momentum, and energy are rapidly mixed by this turbulence. This mixing is many times more rapid or intense than for laminar flow, where the mixing is produced only by the random motion of individual molecules[2] (Appendix C). Observations of the laminar mixing of two parallel streams of fluid in a straight pipe at relatively large subcritical Reynolds numbers (500 to 1500) have shown that mixing was incomplete even after a flow length of 2000 diameters. On the other hand, turbulent mixing of these two parallel streams is nearly complete in 5 to 10 diameters. The random transport of molecular momentum transverse to the mean directed flow was shown to produce the laminar parabolic profile in a

[1] Turbulence may also originate at shear layers between two flows, for example, at the boundaries of a jet.

[2] The mean random molecular speed is quite large (much larger than the mean directed flow speed in liquids or the incompressible flow of a gas), but only one molecule is effective in the momentum (mass or energy) exchange process in laminar flow. In turbulent flow, the mean eddy speed may be only about 1 percent of the directed stream speed, but the eddying mass will contain more than 10^8 molecules, so the property transport rate is much greater than in laminar flow.

pipe. The random transport of turbulent eddies is much more effective in the "open" regions of the fluid, for example, in the center or core region of a tube or channel flow. Consequently, a more uniform velocity or momentum profile is produced. Figure 10-1 shows a comparison of a fully developed laminar profile with a turbulent profile for the same mass flow rate.

Another property of turbulent flow is now evident from an inspection of Fig. 10-1. The shear stress at the wall is determined by the slope of the

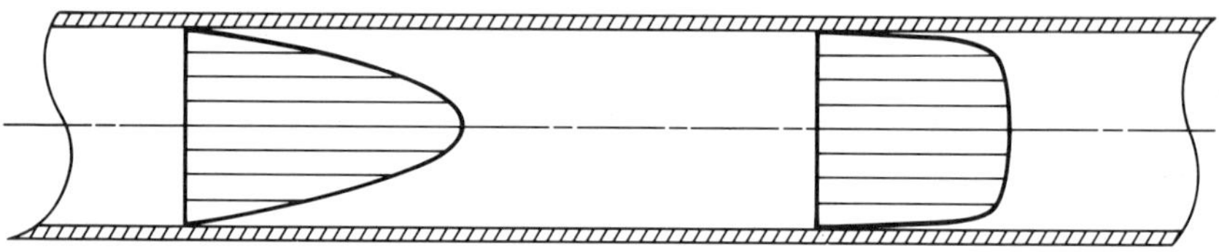

Fig. 10-1. Laminar (parabolic) and turbulent velocity distributions for equal flow rates in a pipe.

velocity profile there $(du/dy)_{y=0}$. This slope is much greater for turbulent flow, which indicates that the shear stress and skin friction drag in turbulent flow are much larger than in laminar flow (at the same mass flow rates in tubes, or for the same external velocities in the case of a boundary-layer flow).

10.1 CHARACTERIZATION OF TURBULENCE

For the continuing production of the eddies in a directed flow, an energy supply is necessary. This energy is available from the kinetic energy of the mean directed flow, which in turn is supplied by the work done by the pressure forces in the case of an internal (for example, tube, channel, or pipe) flow and by the propelling force necessary to overcome the drag for bodies being propelled in an otherwise undisturbed fluid. An analogy can be made with a simple damped mechanical oscillator as illustrated in Fig. 10-2. The damping is assumed to be directly proportional to the velocity $(A_2\omega)$ of the motion, and the spring force is proportional to the displacement from the neutral position. For this system, there is a critical or resonant frequency (or range of frequencies) with which the mass m' will oscillate at an amplitude A_2 much larger than the disturbing amplitude A_1. At frequencies much lower than this critical value, the mass oscillates with nearly the same amplitude as the disturbing amplitude A_1. At frequencies much larger than the critical, the damping force predominates (since it is

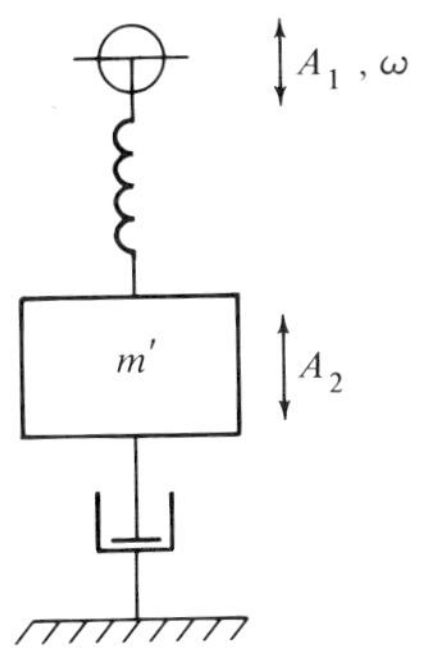

Fig. 10-2. Analog representation for excitation of eddy fluid mass. A_1 and ω are the disturbing amplitude and frequency. A_2 and m' are the eddy "oscillation" and mass.

proportional to velocity) and the responding motion of the mass is negligible.

In a flow, and even in a "quiet" fluid, there is always a spectrum of small-disturbance wavelengths and frequencies.[3] In a particular type of initially laminar flow (usually in a boundary layer), a range of natural or characteristic frequencies will exist which, if excited by an external force of sufficient strength or energy, will "resonate"; that is, they will be amplified at some point and will propagate into the entire flow field. The energy source in a flow is the kinetic energy. This available energy must be sufficiently larger than the work done by viscous damping at the characteristic frequencies before the system will resonate. This condition is indicated directly by the magnitude of the Reynolds number. When the Reynolds number is great enough and the characteristic disturbance frequencies and wavelengths are present, the disturbances at the characteristic frequencies are amplified, the fluid motion becomes unstable, and the instability is amplified and propagated into the other regions of the flow. Once the disturbance is of sufficient amplitude to be amplified into a major disturbance or eddy, it becomes a source of disturbance from which other disturbances and eddys are created. The turbulence then propagates into the flow. The unstable wave motion interacts with the mean flow to produce the eddying turbulent motion. If a stationary position is observed in a passing flow, the flow appears "wavy" at this point, as illustrated in Fig. 10-3. If the observer moves with the mean directed velocity of the fluid, a circulating eddying motion is observed.

[3] For example, the amplified playback of a recording made by placing a sensitive microphone next to a violin in a still room will sound much like a bagpipe band as fundamental (critical frequency) tones and harmonics of the strings are excited by disturbances of undeterminable origin. Nature is dynamic; there is no perfect peace and quiet in nature.

Fig. 10-3. Appearance of turbulent flow from fixed reference and one moving with the flow.

It has been shown that a shear flow is the easiest type of flow to perturb into a turbulent condition. Since shear flows occur primarily in boundary layers (or regions where there is a viscous action) the action responsible for damping the motion in a fluid is also responsible for the origination of turbulence. In this connection it is important to note that a velocity profile in which a point of inflection occurs is inherently unstable. Laminar boundary-layer profiles in positive streamwise ("adverse") pressure gradients are therefore inherently unstable (Fig. 9-11). The boundary layer in a zero pressure gradient (for example, on a flat plate at zero incidence) has a zero curvature point or inflection point at the wall, and instability and transition are then easily induced. The range of disturbance frequency and wavelength and necessary forcing energy to produce instability and transition have been the subject of much analytical and experimental study.[4] A good correlation exists between predicted analytical results and experimental data.

NOTATION

A large variety of symbols pertaining to velocities and velocity components is necessary for describing turbulent flow. The notation for this chapter must then vary from the other chapters. In Figs. 4-6 and 4-7, it was indicated that the velocity $\vec{v}$ at a point could be considered to be made up of a time-mean component $\langle\vec{v}\rangle$ plus a time-random component $\vec{v}'$. In this chapter the local time-mean velocity and its components will be denoted by capital letters, $\langle\vec{v}\rangle \equiv \vec{V}$ with components $\langle u\rangle \equiv U$, and so on. The *instantaneous* fluctuation velocity will remain as $\vec{v}'$ with components u', and so on. The local instantaneous velocity will then be

$$\vec{v} = \vec{V} + \vec{v}'(t)$$

[4] See, for example, chaps. XVI and XVII of reference 1 at the end of the chapter for a discussion of this work.

By definition from Section 4.3,

$$\langle \vec{V} \rangle \equiv \langle \vec{v} \rangle \equiv \frac{1}{\Delta t} \int_{\Delta t} \text{v}\, dt$$

Because $\vec{v}'$ is defined as the *random* component,

$$\langle \vec{v}' \rangle \equiv \frac{1}{\Delta t} \int_{\Delta t} \vec{v}'(t)\, dt \equiv 0$$

The root mean square of the random velocity and its components are not zero and are important in turbulence. For ease of notation, the rms velocity and its components will be denoted by lowercase letters,

$$\vec{v}^2 \equiv \langle \vec{v}'^2 \rangle \equiv \frac{1}{\Delta t} \int_{\Delta t} \vec{v}'^2\, dt, \qquad u^2 \equiv \langle u'^2 \rangle \equiv \frac{1}{\Delta t} \int_{\Delta t} u'^2\, dt, \qquad \cdots$$

Relative local intensity components are then denoted by u/U, and so on.

10.2 SEMI-EMPIRICAL RELATIONS FOR TURBULENT FLOWS

The most direct attack on the problem of formulating analytical relations for turbulent flows in and around given geometries would be by means of statistical analyses, since turbulent variations are primarily random. Much effort has been expended in this direction. The most satisfactory formulation made by G. I. Taylor[5] for the particular case of *isotropic* turbulence where the turbulence fluctuation velocity components are the same for (or independent of) the orientation of the coordinate system chosen (that is, the direction from which the flow is viewed). It follows that the mean values of these fluctuation components must then be equal. This restriction admits a solution for only a few very simple flows, where the assumption of isotropy is valid. One example is the turbulent flow produced downstream of a wire screen or grid.

Certain semi-empirical approaches have produced results that correlate surprisingly well over a broad range of flow conditions. The first of these proposed by Boussinesq[6] in 1877 was a simple linear model that stated

[5] G. I. Taylor, The Statistical Theory of Turbulence, *Proc. Roy. Soc. (London)*, **A151**, 421–478 (1935).

[6] Joseph Boussinesq (1842–1929), Research Professor at the Sorbonne. Worked in soil mechanics, thermodynamics, hydrodynamics, and hydraulics. Student of philosophy and metaphysics. His works on hydraulics are in a single volume of 700 pages, *Essay on the Theory of Water Flows*.

that the total shear stress could be represented as the sum of the laminar and turbulent contributions

$$\tau = \tau_l + \tau_t = \mu \frac{dU}{dy} + A_t \frac{dU}{dy} \tag{10-1}$$

A_t is termed the turbulent "eddy viscosity," which is not a fluid property, as is μ, but also depends on the flow and the geometry (therefore, on the Reynolds number). Because it is a dependent property, it is not easily used in the formulation of fluid-flow problems. Except very near a wall or solid boundary, the order of magnitude of the turbulent shear stress $A_t(dU/dy)$ is much greater than the laminar part $\mu(du/dy)$ or $A_t \gg \mu$. A quantitative comparison between A_t and μ will be made later after a method has been developed by which A_t (or τ_t) can be determined.

PRANDTL'S MIXING-LENGTH THEORY

The hypothesis that has provided the most satisfactory correlation with experimental results, was proposed by Prandtl in 1925.[7] He proposed a mechanism of momentum exchange between turbulent "eddies" of fluid over a momentum exchange distance l' that is similar to the mechanism of molecular momentum exchange over the mean free path distance between molecular collisions (as described in Chapter 2 and Appendix C).

In the following development of Prandtl's theory, only the turbulent shear stress (turbulent momentum exchange) is to be considered. If the turbulent stress τ_t becomes very small (with respect to the laminar shear) as the wall is approached, the results of this development apply only to the region where turbulence is significant. The actual wall shear stress and the velocity and momentum transport immediately adjacent to the wall (at $y = 0$) must be determined from the consideration that the shear layer is predominantly laminar there. However, the value or magnitude of this laminar shear is directly dependent on the turbulent mechanism in the overall shear layer. The problem of determining the shear stress in a fluid near and at a surface is then one of determining the mechanism by which the shear stress is produced, and then combining this turbulent shear stress action with the predominantly laminar shear layer immediately adjacent to the wall.

A set of general equations of motion for turbulent flow similar to the Navier–Stokes equations (6-59) for laminar flow can be formulated by postulating that the instantaneous fluid motion at any point has a time-mean velocity $\vec{V}(\vec{r})$ plus a time-varying or turbulent component $\vec{v}'(\vec{r}, t)$ (see Section 4.4 and Figs. 4-6 and 4-7). The random turbulent fluctuations

[7] Ludwig Prandtl, Uber die ausgebildete Turbulenz, *Z. Angew. Math. Mech.*, **5**, 136 (1925). Translated as *NACA TM 1231*, 1949.

give rise to a set of dynamic or inertial shear stresses that are combinations of the coordinate components of $\vec{v}'$. Prandtl's first simple analysis considered only a two-dimensional flow and the turbulent or eddy momentum exchange between "layers" of fluid. A two-dimensional velocity distribution in a turbulent flow near a solid boundary (or a wall) is represented in Fig. 10-4. It is desired to determine the shear stress components on the control volume of fluid indicated. In addition to the x-directed mean mass

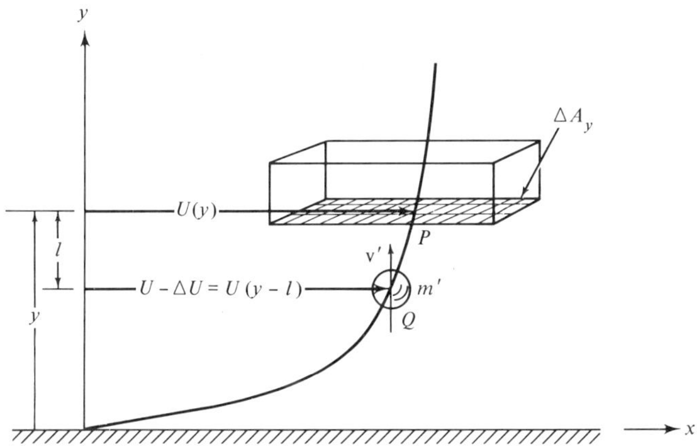

Fig. 10-4. Velocity profile near a wall. An eddy fluid mass m' at Q is transported to P by the turbulent velocity component v′.

momentum and energy transport carried by the time-mean mass flow in the x direction, there will be a random turbulent transport of mass, momentum, and energy in the x, y, and z directions due to the random eddy motion of the small eddy masses of fluid. The turbulent shear stress will be evaluated as a turbulent momentum exchange at the bottom surface (ΔA) of this control volume due to the random y-transported mass entering this surface and exchanging x-momentum there. This momentum exchange is a consequence of the difference in the x-velocity component (ΔU) of the small eddy mass originally at a location $y - \Delta y$ as it is transported by the random component v' up to ΔA at y, where the momentum exchange is assumed to take place. The time-average component of velocity at P is U with time-varying turbulent components u' and v'. A small mass of fluid at $y - \Delta y$ with average velocity $U - \Delta U$ can be transported by the random component v' up to P. It will "mix" with the fluid in the region of P. Since the fluid at Q initially had an x-velocity component $U - \Delta U$, less than the velocity U at P, a momentum exchange will take place due to accelerating this mass of fluid to the velocity U. A force then acts on the eddy mass in the $+x$ direction; an equal and opposite retarding force is produced on the fluid control volume as the eddy enters across ΔA. The distance Δy required for the mixing to take place is then termed the turbulent mixing length l. The turbulent shear force results from the momentum exchange rate at ΔA. For the element approaching $\Delta A = \Delta A_y$ from below, the force on this fluid is[8]

$$F_x = \tau_{yx}\,\Delta A_y = \frac{\dot{m}}{g_c}[U - (U - \Delta U)] = \frac{\rho\,\Delta A v'}{g_c}\Delta U$$

The velocity variation Δv over the distance l is proportional to the random component u', the momentum velocity-change component produced by the turbulent mixing. This random component can be determined in terms of the mixing length and the mean velocity profile slope. From Fig. 10-4, $\Delta U/l \approx \Delta U/\Delta y \approx dU/dy$, since l is small ($l \ll y$). Also when v' is up, $\Delta U \approx -u'$. The shear stress on the area ΔA at P is equal and opposite to that on the element,

$$\tau_{yx} = -\frac{\rho u'v'}{g_c} = \frac{\rho v'}{g_c}\,l\frac{dU}{dy} \tag{10-2}$$

Since the turbulent eddies are formed and dissipated in a mutual interaction between themselves and the mean flow, a correlation can be expected between u' and v' (although they need not necessarily be equal). This statement can be formulated as $|u'| = k|v'|$, where k is some constant $0 < k < 1$. This assumption is shown to be nearly valid from experimental measurements of turbulence for several different cases. As indicated in Fig. 10-5, the shapes of the curves for the mean (rms) turbulence components for pipe flow are about the same. Using this relation in (10-2) gives

$$\tau_{yx} = \frac{\rho k l^2}{g_c}\left(\frac{dU}{dy}\right)^2 = \frac{\rho l'^2}{g_c}\left(\frac{dU}{dy}\right)^2 \tag{10-3}$$

where the proportionality constant k has been included with the mixing length to yield an effective or "correlated" mixing length l'.[9] Consider

[8] The subscript notation on τ denotes first the "face" or surface area on which the force acts (as determined by the normal to this surface) and, second, the direction in which it acts. The orientation of an area is uniquely described by its normal; the y subscript then denotes that ΔA_y is in the x-z plane. When two identical subscripts appear in a single term, their effect cancels; the force F_τ is then in the remaining, or x, direction.

[9] Since l or l' must be determined by a semi-empirical correlation with experimental results, the inclusion of a constant is of no consequence. This expression for τ is more correctly written

$$\tau_{yx} = \frac{\rho l'^2}{g_c}\left|\frac{dU}{dy}\right|\frac{dU}{dy}$$

If dU/dy is negative, τ is also negative. An example of a negative velocity gradient and τ is the backflow part of the profile just downstream of the separation streamline of Fig. 9-11.

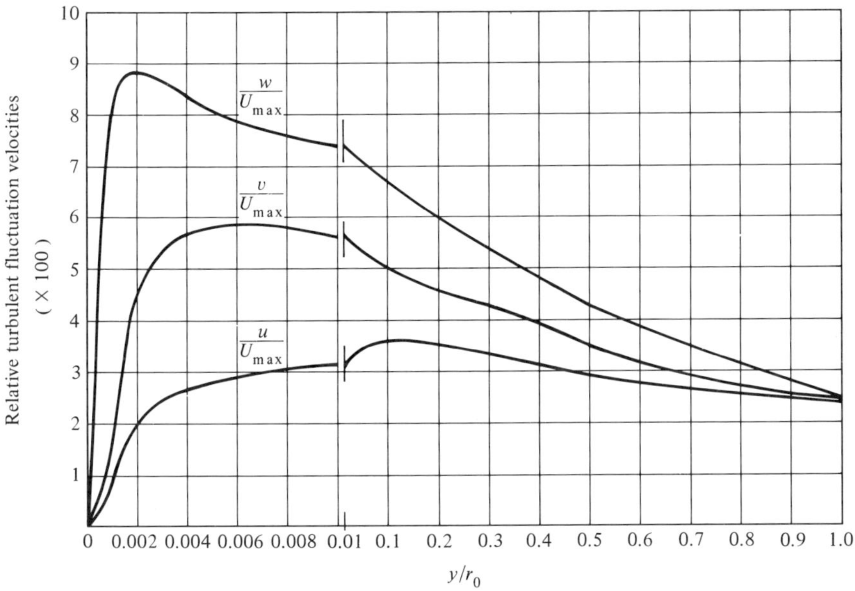

Fig. 10-5. Distribution of turbulent velocity components in pipe flow. u is the rms value of the radial turbulent component, v is the rms tangential component, and w is the rms axial component. Note change of scale at $y/r_0 = 0.1$.

(10-2) written

$$\tau_{yx} = -\frac{\rho\langle u'v'\rangle}{g_c}$$

where $\langle u'v'\rangle$ implies the time average of the $u'v'$ product. The minus sign is necessary for the following reason: If the instantaneous v' is positive at P, Fig. 10-4, the u' component of the fluid eddy will be negative (opposite to the direction shown in the figure); if v' is negative, u' will be positive, so that $\langle u'v'\rangle$ is always negative for positive shear.

If the turbulent shear stress as given in (10-3) is compared with the turbulent component of the shear stress from (10-1), the turbulent eddy viscosity A_t is obtained as

$$A_t = \frac{\rho l'^2}{g_c}\frac{dU}{dy} \tag{10-4}$$

or a kinematic viscosity can be defined as

$$\varepsilon = \frac{g_c A_t}{\rho} = l'^2\frac{dU}{dy} \tag{10-5}$$

This eddy viscosity is therefore a function of the turbulent mixing length and the mean velocity profile; both of these would be expected to be functions of position (y coordinate) and a characteristic Reynolds number of the flow.

10.3 TURBULENT FLOW NEAR A WALL

If τ and l' can be expressed as functions of y, (10-3) can be integrated to yield a velocity profile for turbulent flow. Without prior knowledge of the shear stress distribution, the simplest assumption (made by Prandtl) is to take $\tau = \text{const.} = \tau_{y=0} \equiv \tau_0$ (the yx subscript will be dropped) for a small region in the immediate neighborhood of the wall. The mixing length l' must approach zero at the wall ($y = 0$), because the v' fluctuations cannot penetrate the wall. A simple assumption for l' is then a linear relation (κ is a constant)

$$l' = \kappa y \tag{10-6}$$

With these assumptions, (10-3) can be written

$$\tau_0 = \frac{\rho}{g_c} \kappa^2 y^2 \left(\frac{dU}{dy}\right)^2 \tag{10-7}$$

The quantity $g_c \tau_0/\rho$ (like $g_c p/\rho$) has the dimensions of velocity squared. As an abbreviation it is then called the shear or friction velocity $\mathrm{u}_\tau \equiv \sqrt{g_c \tau_0/\rho}$. With this substitution, (10-7) becomes

$$\frac{1}{\mathrm{u}_\tau} \frac{dU}{dy} = \frac{1}{\kappa y} \tag{10-8}$$

This equation can be integrated to give

$$\frac{U}{\mathrm{u}_\tau} = \frac{1}{\kappa} \ln y + C \tag{10-9}$$

The constant can be evaluated at some position where the velocity is known. For example, if a maximum velocity U_{max} exists at $y = h$,

$$\frac{U_{\text{max}}}{\mathrm{u}_\tau} = \frac{1}{\kappa} \ln h + C \tag{10-9a}$$

substituting the value of C determined from (10-9a) into (10-9) gives

$$\frac{U_{\text{max}} - U}{\mathrm{u}_\tau} = \frac{1}{\kappa} \ln \frac{h}{y} \tag{10-10}$$

The quantity $(U_{max} - U)/u_\tau$ is seen to be the *velocity defect*, which appears as the independent variable or ordinate of Fig. 10-6a. This solution does not give the entire velocity distribution directly, because the primary turbulence parameter l' disappears as the wall is approached [according to the assumption (10-4)]. The above development gives only the effect of turbulence on the velocity distribution. If a combined mechanism such as expressed by (10-1) is assumed to exist, the shear stress at the wall cannot approach zero as l' approaches zero but should be a maximum there. The

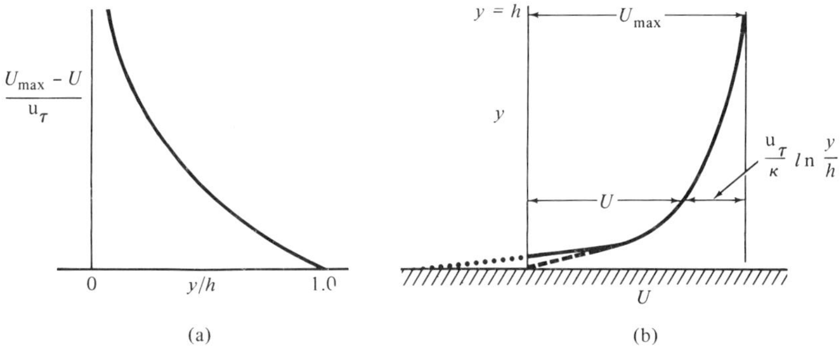

Fig. 10-6. (a) Velocity defect distribution, and (b) velocity distribution according to (10-10).

only answer to this apparent paradox is that the shear stress mechanism at the wall must be predominantly laminar. This conclusion has led to the assumption that a "laminar shear layer" exists at the wall. Although the shear stress at the wall is predominantly laminar, it is difficult to see how a completely laminar layer could exist adjacent to a turbulent flow, particularly when the actual turbulent mechanism must be three-dimensional (even though the mean flow may be only two-dimensional). Experimental results indicating the degree of laminar flow and turbulence will be presented below for some specific flow geometries.

If the velocity distribution of (10-10) and Fig. 10-6a is represented as the velocity U (abscissa) near a surface, Fig. 10-6b is obtained. If only the turbulent mechanism is considered, the velocity approaches minus infinity (dashed line) and the shear stress approaches infinity at $y = 0$. This apparent paradox was expected earlier (as stated in Section 10.2) as a consequence of the hypothesis required to determine the turbulent behavior. It is necessary to "fit" a primarily laminar shear limit profile at $y = 0$ to determine the behavior (dU/dy and τ) at the wall. Such an approximate fit is indicated by the dashed line in Fig. 10-6b.

It is emphasized here that the velocity distribution (10-10) was obtained under assumptions that make it valid only for the *turbulent* region *near* the wall. For this reason, (10-10) is sometimes referred to as "the law of the wall." Since no specific flow geometry has been assumed, the application of (10-10) is *universal.* The test of the validity of this universal distribution law is its correlation with experimental results. Because the constant κ must also be evaluated from experimental results, the final form for the velocity distribution is a semi-empirical relation.

For a further analysis of turbulent flow near a body or a flat plate, it is necessary to consider what further information can be obtained from considerations of the flow and boundary geometry. In the above relationships the thickness length h appears. As in the case of the laminar boundary layer, h is a function of a significant x coordinate. In fully developed pipe or channel flow, the distance h, where $U = U_{max}$, is just the distance to the centerline and therefore is a constant and does not depend on x. This simpler flow geometry may be considered to determine which results might apply to more general flow cases.

10.4 TURBULENT FLOW IN PIPES

Equation (10-9) was derived assuming a mixing-length hypothesis, and further that the mixing length near the wall is directly proportional to y ($l' = \kappa y$). If it is assumed that the fully developed turbulent profile in a smooth round pipe possesses a unique shape for all axial stations in the pipe and therefore that the turbulent profiles are *similar* (or identical) then (10-9) will satisfactorily represent this profile *near* (but not *at*) the wall.[10] With this similarity assumption, τ_0 is a "constant" of the flow; therefore, so is u_τ. Equation (10-7) may then be written

$$\frac{U}{u_\tau} = \frac{1}{\kappa} \ln \frac{u_\tau y}{\nu} + C_1 \tag{10-11}$$

where $C_1 = C + (1/\kappa) \ln (u_\tau/\nu)$. The dimensionless groups in (10-11) may be defined as a dimensionless velocity $u_+ \equiv U/u_\tau$ and a dimensionless y coordinate $y_+ \equiv u_\tau y/\nu$, which has the form of a Reynolds number. For fully developed turbulent flow in a pipe, the dimensionless velocity distribution u_+ as a function of the dimensionless coordinate y_+ (measured from the wall) depends only on the constants of the flow and fluid, κ, u_τ, and ν.

The above development has assumed that the mixing length is given by the relation $l' = \kappa y$ in the turbulent flow region under consideration

[10] For example, all fully developed pipe-flow laminar profiles are parabolic: therefore, they are "similar." In a broader sense, similarity implies that a mathematical scale function $S(x)$ exists by which all profiles may be reduced to one.

("near the wall") and that τ is a constant (τ_0). It will be shown later that τ (the total shear stress) is directly proportional to the radial coordinate, $r = r_0 - y$:

$$\tau = \tau_0\left(\frac{r}{r_0}\right) = \tau_0\left(1 - \frac{y}{r_0}\right) \tag{10-12}$$

since $y = r_0 - r$. The shear stress is shown plotted in Fig. 10-7. The mixing length l' (and therefore l) can be determined from experimental measurements of the velocity distribution $u(y)$ and the relations (10-3) and (10-12). This calculation was made by Nikuradse (a student of Prandtl), who also made extensive experimental measurements on turbulent flows. The results of this calculation give

$$\frac{l}{r_0} = 0.4\frac{y}{r_0}\left[1 - 1.1\frac{y}{r_0} + 0.6\left(\frac{y}{r_0}\right)^2 - 0.15\left(\frac{y}{r_0}\right)^3\right] \tag{10-13}$$

for Reynolds numbers greater than $\sim 10^5$ and closely approximate the experimental data for Reynolds numbers from 10^5 down to the critical value of 2×10^3. The variation of mixing length with radius is also plotted in Fig. 10-7. Since it has been assumed that $l' = \kappa y$, the experimental value of $\kappa = 0.4$ is obtained as the first term approximation for the slope of the l' versus y distribution at $y = 0$.

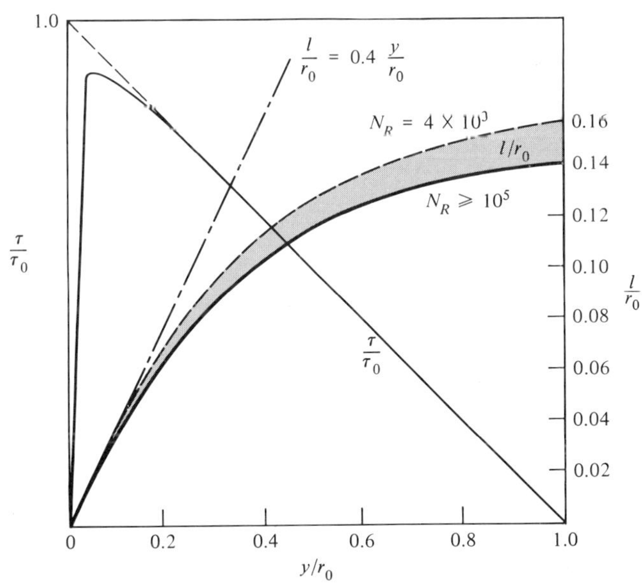

Fig. 10-7. Shear stress distribution and mixing-length distribution in pipes. Solid mixing-length curve corresponds to (10-13).

For $l' = \kappa y$ and $\kappa = 0.04$, the universal velocity distribution (10-11) may be written

$$\frac{U}{u_\tau} = 2.5 \ln \frac{u_\tau y}{\nu} + C_1 \tag{10-14a}$$

or using the dimensionless variables u_+ and y_+,

$$u_+ = 2.5 \ln y_+ + C_1 \tag{10-14b}$$

This relation is usually plotted in terms of base-ten logarithms or

$$u_+ = 5.75 \log y_+ + C_1 \tag{10-14c}$$

The constant C_1 must also be evaluated from experimental data because of the implicit form of (10-14) for u_τ as a dependent parameter [a function of the $u(y)$ distribution]. The velocity distributions in a pipe in terms of the explicit physical quantities U_{max} (centerline velocity) and r_0 are shown in Fig. 10-8, where two values of Reynolds number with the profile for laminar flow shown for comparison. The effects of the greater transverse momentum transport on "flattening" the profile and on increasing the wall shear stress are quite apparent. These same data are plotted as

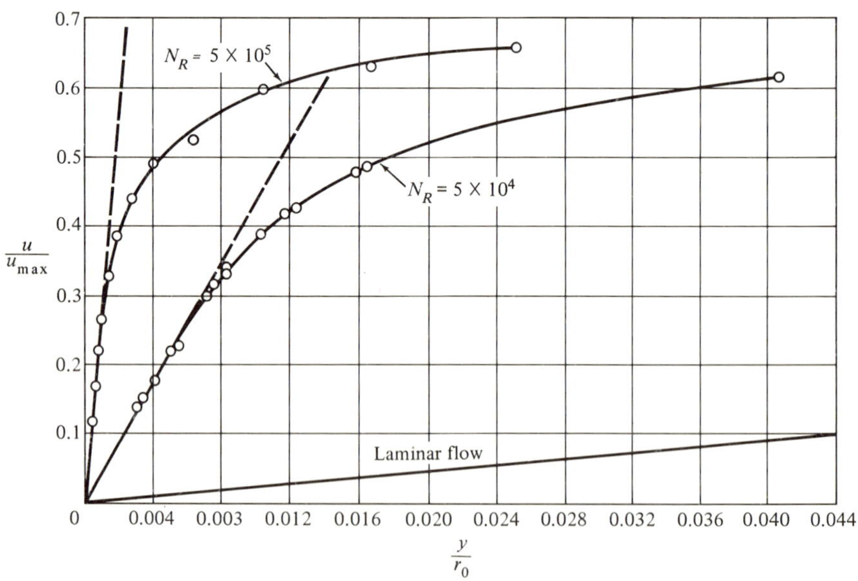

Fig. 10-8. Velocity distribution in a round duct as determined by Laufer (reference 4).

dimensionless linear coordinates in Fig. 10-9. Equation (10-14) indicates that a semilog representation of the data should plot as a straight line. A plot of the experimental data for u_+ versus y_+ is shown in Fig. 10-10. All the experimental data to date[11] lie on (and would completely obliterate) the curve shown in this figure for $y_+ > 70$.

C_1 in (10-14) can be determined from the slope of the straight-line portion of Fig. 10-10 to give

$$u_+ = 5.75 \log y_+ + 5.5 \tag{10-15}$$

These experimental results also indicate that the profile is essentially linear, and consequently that the shear stress is predominantly laminar for $0 < y_+ < 5$ or $0 < u_+ < 5$. In the region $5 < y_+ < 70$, both the molecular and turbulent shear (momentum exchange) are significant.

The significance of the *universal* velocity distribution of (10-15) is that correlations with experimental data verify that such a universal "law" does indeed exist. An actual velocity distribution (U as a function of y or r) cannot be determined or constructed from this equation alone. It is necessary to know τ_0 from some other source. On the other hand, the quantity of primary interest for application is the shear stress. The above determined relation can be used to determine τ_0 or a dimensionless shear stress coefficient c_τ (or friction factor f) as a function of the Reynolds number for the flow in a pipe or tube.

The shear stress coefficient c_τ is defined as

$$c_\tau \equiv \frac{\tau_0}{\rho \overline{\mathrm{V}}^2/2g_c} \tag{10-16}$$

where $\overline{\mathrm{V}}$ is the mass-averaged flow velocity

$$\overline{\mathrm{V}} = \frac{\dot{m}}{\rho A} = \frac{Q}{A} = \frac{1}{A}\int \mathrm{V}\, dA$$

The Darcy–Weissbach friction factor f is defined in terms of the pressure drop or pressure gradient $\Delta p/L$ for a length L of straight pipe from the defining relation,

$$\frac{\Delta p}{L} \equiv f \frac{L}{D}\frac{\rho \overline{\mathrm{V}}^2}{2g_c} \tag{10-17}$$

where $D = 2r_0$. The relation between Δp and τ_0 for steady flow in a straight length of pipe is easily determined by a force balance on a control volume of length Δx and radius r_0 within the pipe, as shown in Fig. 10-11.

[11] J. Nikuradse, *VDI-Forschungstr.*, **356** (1932). T. E. Stanton and J. R. Pannel, *Phil. Trans. Roy. Soc. London*, **A214**, 199 (1914). J. Laufer, *NACA Report 1174*, 1954.

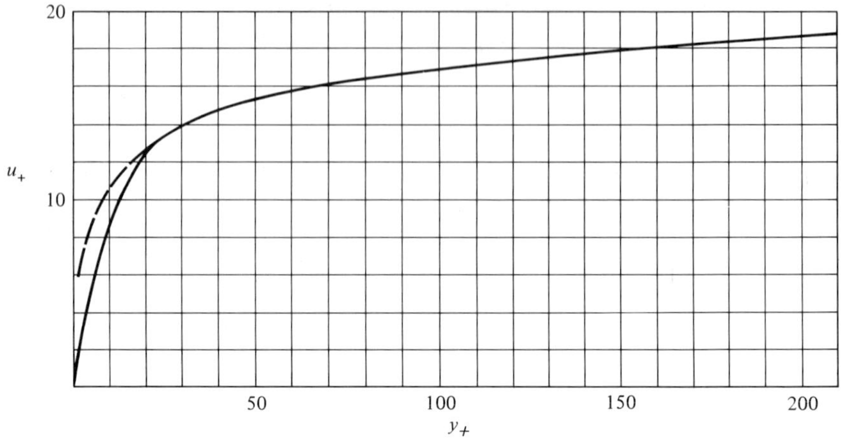

Fig. 10-9. Plot of "law of the wall" universal velocity distribution for smooth pipes (linear scale). Dashed line is an extension of (10-15). Solid line is experimental data and (1,0-15).

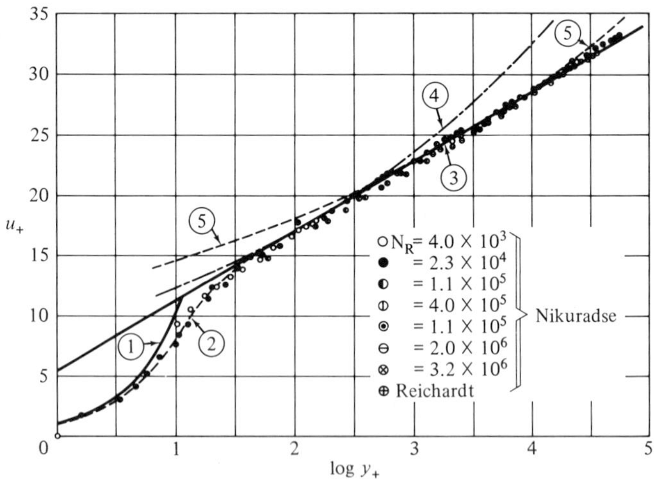

Fig. 10-10. Universal "law of the wall" velocity distribution for turbulent flow in smooth pipes plotted on semilog coordinates. Curve 1 is for a straight-line laminar profile, $u_+ = y_+$, curve 2 is (10-15), curve 3 is (10-26), and curve 4 is for a 1/10-power-law profile (see Fig. 10-14). The dashed curve is experimental data.

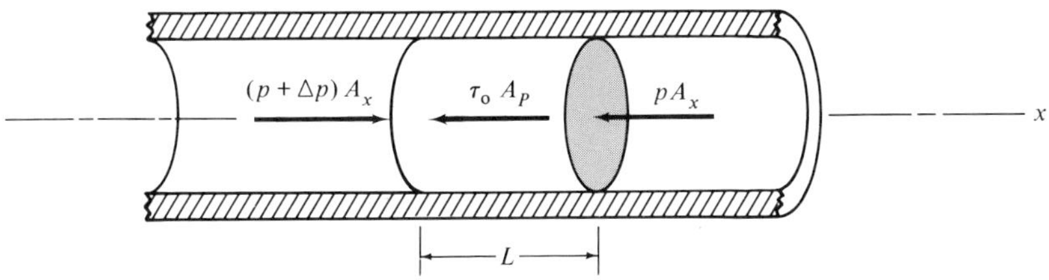

Fig. 10-11. Force balance on fluid control volume in a pipe. A_x is cross-sectional area; A_p is peripheral area.

This force balance gives $|\Delta p| r_0 = \tau_0 2L$ or $|\Delta p| = (2L/r_0)\tau_0$. From these relations

$$c_\tau = \frac{f}{4} \tag{10-18}$$

From the defining relation for u_τ,

$$u_\tau^2 = \frac{g_c \tau_0}{\rho} = \frac{\overline{V}^2}{2} \frac{\tau_0}{\rho \overline{V}^2/2g_c} = \frac{\overline{V}^2}{2} c_\tau = \frac{\overline{V}}{8} f \tag{10-18a}$$

the relation for the friction factor in terms of $\overline{V}$ and u_τ is obtained,

$$\frac{\overline{V}}{u_\tau} = \sqrt{\frac{8}{f}} \tag{10-19}$$

A relation between $\overline{V}$ and u_τ can be obtained from the velocity distribution (10-15) and the above definition of $\overline{V}$ by integrating the velocity distribution over the pipe cross-sectional area. Equation (10-15b) is easier to use and can be expanded in the form (with $D = 2r_0$)

$$\frac{\overline{V}}{u_\tau} = 2.5 \ln \frac{u_\tau D}{2\nu} + 2.5 \ln \frac{y}{r_0} + 5.5$$

Now with $N_R = VD/\nu$ and (10-19),

$$\frac{U}{u_\tau} = 2.5 \ln \frac{1}{2}\sqrt{\frac{f}{8}}\, N_R + 2.5 \ln \frac{y}{r_0} + 5.5$$

Also $\overline{V}/u_\tau = 1/A \int (v/u_\tau)\, dA$; performing the integration yields

$$\frac{\overline{V}}{u_\tau} = 2.5 \ln N_R\sqrt{f} - 2.58 \tag{10-20}$$

From (10-15) U_{max}/u_τ can be evaluated and used with (10-19) and (10-20) to give

$$\frac{U_{\max}}{\overline{V}} = 1 + 1.33\sqrt{f} \tag{10-21}$$

This result is of interest in that it gives an approximate relationship for the flatness of the velocity profile. For example, a representative value of $f = 0.02$ gives $U_{\max}/\overline{V} = 1.19$ (as compared with 2.0 for a parabolic profile).

Using (10-19), (10-20) can be expressed

$$\frac{1}{\sqrt{f}} = 0.883 \ln N_R\sqrt{f} - 0.913 \tag{10-22a}$$

The velocity distribution assumed for this determination is illustrated in Fig. 10-6b. Since the conditions in the immediate neighborhood of the wall are not truly represented by this velocity distribution, (10-22a) will not represent the correct f versus N_R relation. If the constants are determined from experimental data, they are found to be relatively close to those obtained by the approximate evaluation above giving (with conversion to base-ten logs)

$$\frac{1}{\sqrt{f}} = 2.0 \log N_R\sqrt{f} - 0.8 \tag{10-22b}$$

Equation (10-22b) is also a "universal" result for friction factor as a function of Reynolds number for turbulent flow in *smooth* pipes and tubes. A plot of $1/\sqrt{f}$ versus $\log N_R\sqrt{f}$ gives a straight line as shown in Fig. 10-12. Experimental data by numerous investigators also fall on and

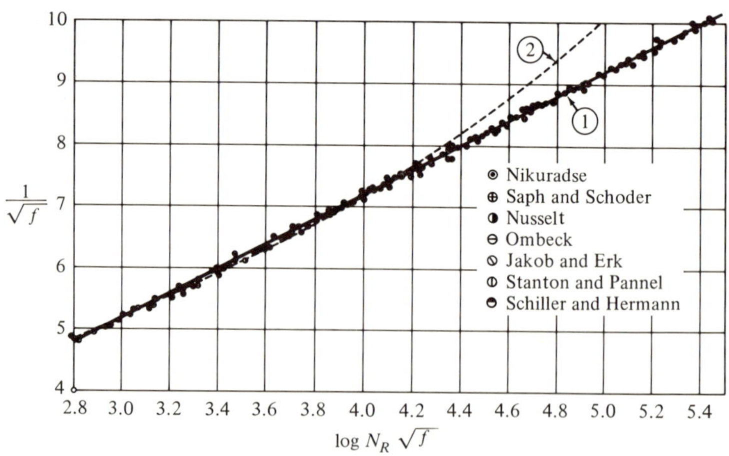

Fig. 10-12. Wall friction relation for "the law of the wall" for smooth pipes. Curve 1 is (10-22b), curve 2 is (10-25).

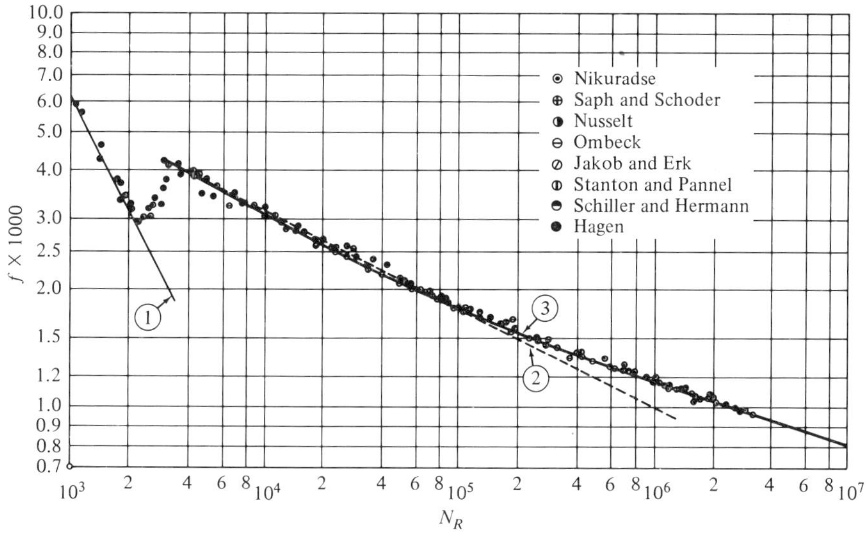

Fig. 10-13. Friction factor versus Reynolds number for smooth pipes as determined from Fig. 10-12. Curve 1 is for laminar flow, curve 2 is (10-25), and curve 3 is (10-22b).

obliterate this line, as plotted from (10-22b). Equation (10-22b) can be solved explicitly for N_R as a function of f and represented as f versus N_R, or f as a function of N_R can be determined from the plot of Fig. 10-13.

10.5 ILLUSTRATIVE EXAMPLE 10-1: Relative Magnitude of μ and A_t (or ν and ε) for Flow in a Smooth Pipe

Starting with the definition of shear stress from (10-1),

$$\tau = (\mu + A_t)\frac{dU}{dy}$$

and the relation for τ in a round pipe, (10-12), $\tau = \tau_0(1 - y/r_0)$, and the velocity distribution for turbulent flow (10-14) for $y_+ > 50$

$$u_+ = 2.5 \ln y_+ + 5.5$$

The ratio of eddy viscosity to molecular viscosity can be derived (see Problem 10.14) as

$$\frac{\varepsilon}{\nu} = \frac{A_t}{\mu} = \left(1 - \frac{y}{r_0}\right)(0.4y_+) - 1 \tag{10-23}$$

With the definition $y_+ = yu_\tau/\nu$, (10-23) may be written

$$\frac{\varepsilon}{\nu} = 0.4\left(1 - \frac{y}{r_0}\right)\left(\frac{y}{r_0}\right)\left(\frac{r_0 u_\tau}{\nu}\right) - 1$$

Using (10-19) and with $N_R = \overline{V}D/\nu$, this equation becomes

$$\frac{\varepsilon}{\nu} = 0.4\left(1 - \frac{y}{r_0}\right)\left(\frac{y}{r_0}\right)N_R\sqrt{\frac{f}{32}} - 1 \qquad (10\text{-}24)$$

(The factor 1 may be neglected since it is quite small with respect to the first term in the region where turbulence is significant, that is, for $y_+ > 100$.) The function of y/r_0 is seen to have a maximum at $y/r_0 = \frac{1}{2}$. Using this value for y/r_0 and arbitrarily taking a representative value of $f = 0.02$ and $N_R = 10^5$ (Fig. 10-12) gives $\varepsilon/\nu = 249$. If y/r_0 is arbitrarily taken as $\frac{1}{10}$, then $\varepsilon/\nu = 89$. (See Problems 10-15 and 10-16 for further examples.)

10.6 POWER-LAW RELATIONS

The friction factor–Reynolds number data were fitted with the purely empirical relation

$$f = 0.3164N_R^{-1/4} \qquad (10\text{-}25)$$

by H. Blausius in 1913.[12] This result fits moderately well only between the limits $4000 < N_R < 10^5$ but is simple to use. For example, with (10-19) and (10-25), the relation

$$\frac{\overline{V}}{u_\tau} = 6.99\left(\frac{r_0 u_\tau}{\nu}\right)^{1/7}$$

can be derived (see Problem 10-21). This relation suggests a velocity profile of the form $U \propto y^{1/n}$ ($y = r_0 - r$ and $0 < y < r_0$). For a velocity profile of the type $U \propto y^{1/n}$, the result that

$$\frac{\overline{V}}{U_{max}} = \frac{2n^2}{(n + 1)(2n + 1)}$$

is directly obtained by integration of the velocity distribution. For $n = 7$, the equation gives $\overline{V}/U_{max} = 0.817 \approx 0.8$. If this quantity is used with the

[12] H. Blausius, Das Ahnlichkeit Gesetz bei Reibungs vorgangen in Flussigkeiten, *Forsch.-arb. Ingenieuw.-Wes. No. 131* (1913).

above expression for $\overline{V}/u_\tau$,

$$\frac{U_{max}}{u_\tau} = 8.74\left(\frac{r_0 u_\tau}{\nu}\right)^{1/7}$$

from which the general distribution

$$\frac{U}{u_\tau} = 8.74\left(\frac{y u_\tau}{\nu}\right) \qquad \text{or} \qquad u_+ = 8.74 y_+^{1/7} \tag{10-26}$$

can be deduced. This result is valid for $N_R \approx 10^5$ and $30 < y_+ < 1000$. The turbulence and therefore the value of n in the exponent is a function of N_R, as shown in Fig. 10-14.

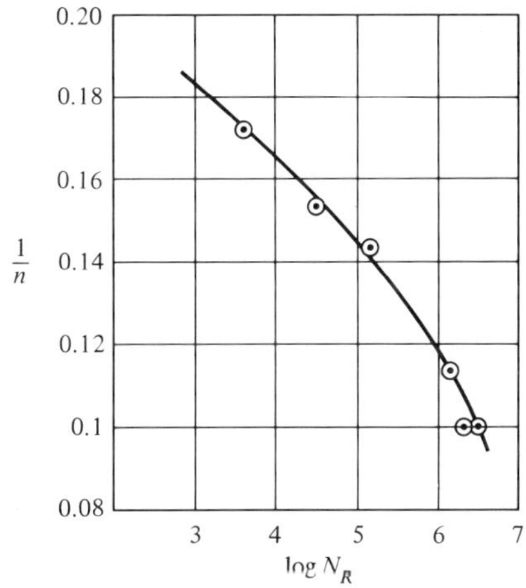

Fig. 10-14. Experimentally determined values of the exponent n as a function of Reynolds number for exponential representation of the velocity distribution near a smooth wall; $U/U_{max} = (y/r_0)^{1/n}$.

10.7 TURBULENT FLOW IN ROUGH PIPES

Since most manufactured pipes tubes and ducts are not "smooth," the relations for flow in rough pipes must be investigated. First it is necessary to determine a relative measure of roughness. From Figs. 10-6 through 10-10 it is seen that the laminar viscous effect is predominant in a region immediately adjacent to the wall $0 < y_+ < 5$. For $5 < y_+ < 70$, both effects are significant and for $y_+ > 70$ the turbulent viscous effect is predominant. Note that both the distance from the wall y and the characteristic turbulence velocity u_τ are required to determine the parameter indicating the extent of the laminar and viscous regions of influence.

For example, as the Reynolds number increases, the wall shear stress τ_0 becomes larger, so that u_τ becomes larger. The extent $y = y_+\nu/u_\tau$ of the predominantly laminar layer would then become smaller. Roughness elements with mean dimension ϵ within the range $0 < \epsilon_+ < 5$ (where $\epsilon_+ = u_\tau\epsilon/\nu$) would then lie within the region where the turbulence is negligible. The maximum roughness element Reynolds number at $y_+ = 5$ can be determined from $u_+ = 5$. The product of these characteristic parameters is a Reynolds number based on maximum y and maximum U for this layer of small turbulence and is

$$u_+ y_+ = \frac{U}{u_\tau}\frac{yu_\tau}{\nu} = \frac{yU}{\nu} = 25$$

If the characteristic size of a roughness element $\epsilon_+ < 5$, the Reynolds number based on the velocity at the top and the height of the roughness element for the flow past this roughness element is <25. This roughness or projection Reynolds number for a body immersed in a flow will have more significance with respect to the material to be covered in Chapter 12. However, the flow about a single roughness element would be essentially laminar for roughness projection Reynolds numbers less than 25. The flow conditions described above are illustrated in Fig. 10-15.

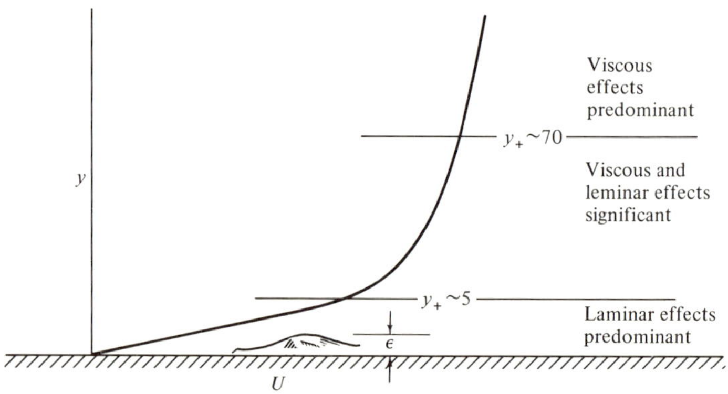

Fig. 10-15. Representation of turbulent boundary layer on a relatively smooth surface. The roughness element of height ϵ lies within the predominantly laminar region, so the wall is effectively smooth.

The roughness on a commercially manufactured pipe surface is not uniform in size, shape, or distribution. It is therefore necessary to characterize such roughness by comparison with actual tests where the mean or average roughness is measured or inferred. Because of the difficulty of measuring and specifying roughness parameters for arbitrary roughness, the inferential method yields the most useful results. This method will be described in connection with experimental tests on artificially roughened pipe and the results compared with tests on representative samples of commercial pipe.

If the roughness size is large enough that the characteristic Reynolds number is quite large (of the order of 100), the flow at this roughness boundary will also be turbulent. In this case, the friction parameter c_τ or f will be proportional to the square of the velocity, so that f (or c_τ) will be independent of Reynolds number. The state of the mean pipe flow depends on the pipe Reynolds number $\overline{V}D/\nu$, while the roughness flow depends on the roughness Reynolds number $\overline{V}\epsilon/\nu$. The state of the flow for rough pipes can then be determined in terms of these two dimensionless parameters, or in terms of an equivalent set of dimensionless groups, the pipe Reynolds number, and the relative roughness ϵ/D (which is just the ratio of the two Reynolds numbers). This second set of parameters is the easiest to use in actual calculations.

EXPERIMENTAL RESULTS

The first significant experimental work on rough pipes was performed by Nikuradse[13] on pipes artificially roughened by gluing sand grains of a uniform size on the surface of a smooth pipe. The results obtained are shown in Fig. 10-16. The flat portions of each curve at high Reynolds

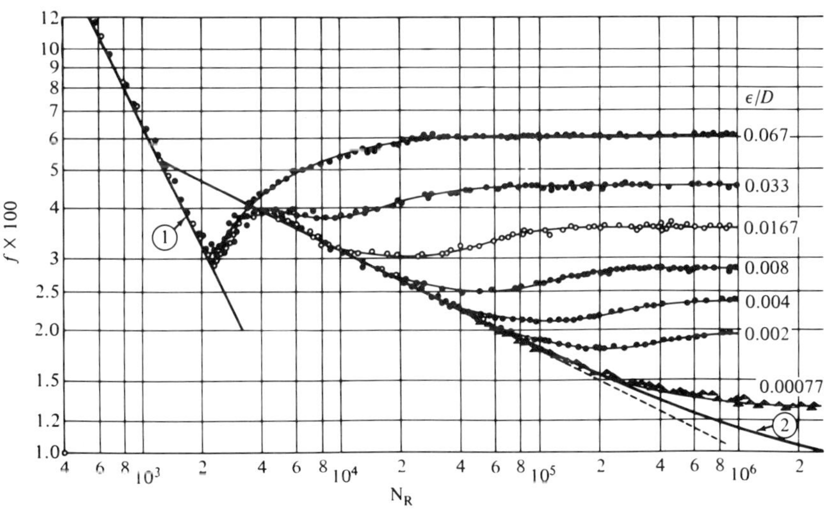

Fig. 10-16. Friction factor as a function of Reynolds number and relative roughness for artificially sand-roughened pipes as determined by Nikuradse. Curve 1 is for laminar flow; curve 2 is for smooth pipes from (10-22b).

[13] J. Nikuradse, Strömungsgesetze in rauben Rohren, *Forsch.-arb. Ingenieuw-Wes. No. 361* (1933).

numbers indicate that the flow is predominantly turbulent everywhere (there is no predominantly laminar layer). The sand grains in this region of Reynolds numbers are significantly larger than the limiting value determined by $y_+ = y\mathrm{u}_\tau/\epsilon = 5$; that is, $\epsilon\mathrm{u}_\tau/\nu \gg 5$ or nominally >30.[14] At relatively low Reynolds numbers note that the roughened pipe curves "fair into" the relation determined for a smooth pipe. In this region of coincidence, the sand grain size and the flow conditions are such that $\epsilon\mathrm{u}_\tau/\nu < 5$, so the roughness is submerged in the laminar layer and does not cause a significant disturbance. A transition regime occurs in the region $5 < \epsilon\mathrm{u}_\tau < 70$. For $\epsilon_+ > 100$, the roughness elements act primarily as bodies submerged in a flow, so that the drag on these bodies accounts for a large part of the "skin friction."

UNIVERSAL VELOCITY DISTRIBUTION FOR ROUGH PIPES

If it is assumed that the turbulent flow near the wall is still determined by the mixing-length hypothesis, then from (10-8) the flow in this turbulent zone is still determined by (see Fig. 10-6b)

$$\frac{U}{\mathrm{u}_\tau} = \frac{1}{\kappa}\int_{y_1}^{y}\frac{dy}{y} = \frac{1}{\kappa}\ln y - \frac{1}{\kappa}\ln y_1 \tag{10-27a}$$

or adding and subtracting the constant $(1/\kappa)\ln(\mathrm{u}_\tau/\nu)$,

$$\frac{U}{\mathrm{u}_\tau} = \frac{1}{\kappa}\ln\frac{y\mathrm{u}_\tau}{\nu} - \frac{1}{\kappa}\ln\frac{y_1\mathrm{u}_\tau}{\nu} \tag{10-27b}$$

or

$$\mathrm{u}_+ = \frac{1}{\kappa}\ln y_+ - \frac{1}{\kappa}\ln y_{1+} \tag{10-27c}$$

If y_1 is greater than ϵ, the mixing-length hypothesis still gives the same value of κ. It is also expected that the lower limit for which y_1 is determined would be a function of the roughness ϵ; that is,

$$y_1 = f(\epsilon) \tag{10-28}$$

Equation (10-27c) can then be written in the form

$$u_+ = 2.51\ln y_+ - g(\epsilon) \tag{10-29}$$

For large roughness ($\epsilon_+ \gg 5$) (Fig. 10-15), it may be assumed that a simple relation

$$y_{1+} = C\epsilon_+ \tag{10-30}$$

[14] For $y_+ = 30$ in Fig. 10-9, the deviation from fully developed turbulent flow is less than 10 percent.

exists. Equation (10-27c) can then be written

$$u_+ = 2.5(\ln y_+ - \ln \epsilon_+) - \ln C \tag{10-31a}$$

or with $B = -\ln C$,

$$u_+ = 2.5 \ln \frac{y}{\epsilon} + B \tag{10-31b}$$

This relation correlates well with Nikuradse's data for a value of $B = 8.5$ in the region where $\epsilon_+ > 70$, that is, where the flow is completely rough. For $\epsilon_+ < 70$, B would be expected to be a function of ϵ_+. For a smooth pipe flow ($\epsilon_+ < 5$), a comparison of (10-31) with (10-15) gives

$$B = 5.5 + 2.5 \ln \epsilon_+ \tag{10-32}$$

This relation for B for a smooth pipe is only useful as a verification of the validity of the development of the velocity distribution, (10-31b). A plot of the relation for B as a function of ϵ_+ is shown in Fig. 10-17. Note that the

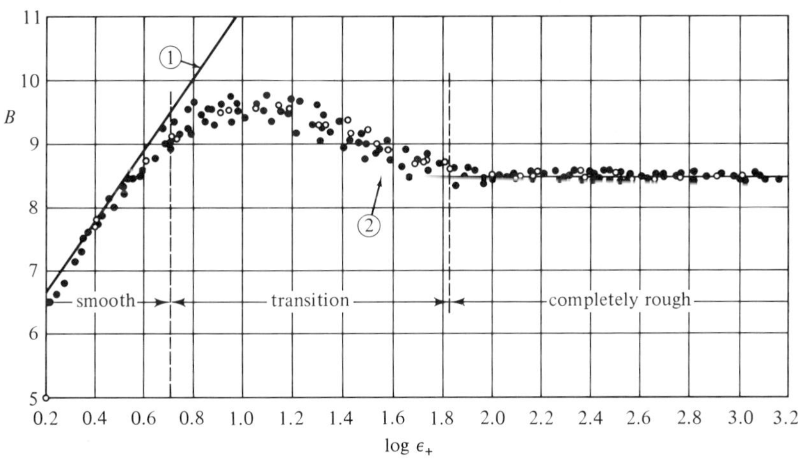

Fig. 10-17. Plot of roughness function B as a function of ϵ_+. Curve 1, effectively "smooth," (10-32); curve 2, completely rough, $B = 8.5$.

limit of the "smooth" correlation corresponds to $y_+ = \epsilon_+ = 5$ ($\log \epsilon_+ = 0.7$), giving $B = 9.5$ and, for the lower limit of the completely rough zone, $y_+ = \epsilon_+ = 70$ ($\log \epsilon_+ = 1.84$). Also note that the roughness has no effect on determining the velocity distribution for $\epsilon_+ < 5$.

The dimensionless velocity distribution for "completely rough" pipe is then given by

$$u_+ = 2.5 \ln y_+ + 8.5 \tag{10-33a}$$

or

$$u_+ = 5.75 \log y_+ + 8.5 \tag{10-33b}$$

FRICTION COEFFICIENT FOR ROUGH PIPES

It is of primary interest to calculate the friction factor for the completely rough flow region, where the velocity distribution is given by (10-33). The asymptotic limit of f for fixed sand grain roughness values as shown on Fig. 10-17 can then be correlated with the asymptotic values for various commercial grades of pipe to yield an equivalent sand grain roughness characterizing the effective roughness of these pipes.

With $h = r_0$ in (10-10),

$$\frac{U_{\max} - U}{U} = 2.5 \ln \frac{y}{r_0} \tag{10-34}$$

is valid for $y > y_1$ or $y > \epsilon_1$ as determined from $\ln(\epsilon_1 \mathrm{u}_\tau/\nu) = 1.84$. By integration of the velocity distribution, the average flow velocity $\overline{\mathrm{V}} = Q/A$ is determined as

$$\overline{\mathrm{V}} = U_{\max} - 3.75\mathrm{u}_\tau \tag{10-35}$$

Substituting the value of $U_{\max}$ from (10-34) into (10-31b) gives

$$\frac{\overline{\mathrm{V}}}{\mathrm{u}_\tau} = 2.5 \ln \frac{r_0}{\epsilon} + 8.5 \tag{10-36}$$

Using the definitions $\mathrm{u}_\tau^2 = g_c\tau_0/\rho$ and $f = 8g_c\tau_0/\rho\overline{\mathrm{V}}^2$ from (10-18a),

$$\sqrt{\frac{8}{f}} = 2.5 \ln \frac{r_0}{\epsilon} + 4.75$$

$$f = \left(0.885 \ln \frac{r_0}{\epsilon} + 1.68\right)^{-2} \tag{10-37}$$

$$f = \left(2.04 \log \frac{r_0}{\epsilon} + 1.68\right)^{-2}$$

PIPE FRICTION FACTORS

L. G. Moody accumulated a great deal of data on friction factors as a function of Reynolds number for many types of commercial pipes. By cross plotting the results in the completely rough zone, a value of ϵ was obtained from a correlation of these experiments with (10-37). A representative single roughness factor was then assigned for each of several types of pipe tested. A complete plot of these data as obtained by Moody is presented in Fig. 10-18.[15]

Only a single curve is needed for Reynolds numbers below the lower critical value of 2000. In this region the flow is fully laminar with a parabolic velocity distribution and $f = 64/N_R$. Although it is possible to obtain laminar flow at higher Reynolds numbers as discussed in Chapter 4, most commercial pipes are not smooth, and there are geometry transitions and vibrational disturbances that produce early transition to turbulent flow. In the instability region for "smoother" pipes at Reynolds numbers above 2000, the degree of roughness, the system geometry, and the state of vibration in the system all affect the state of the flow. In the transition region, the friction is a function of both the Reynolds number and the roughness. In the fully developed region the flow is "fully rough" and a function of relative roughness only (see also Fig. 10-17). The lower limiting "smooth" curve in the supercritical Reynolds number region corresponds to (10-22b).

For pipes and ducts of noncircular cross section the results for f as presented above are correlated very well if the pipe diameter is replaced with the "hydraulic diameter," defined as $D_H = 4A/Per$ (where Per is the "wetted perimeter"), as long as the aspect ratio[16] of the duct is not too large (nominally less than 7).

10.8 TRANSITION LOSSES

The preceding material has considered some of the phenomenological or semi-empirical approaches to the study of some specific turbulent flows. The results for incompressible pipe flow are best presented in the graphical representation of the Moody diagram, Fig. 10-18.

In piping and ducting systems, turbulent friction losses of availability also occur in various standard types of fittings such as elbows, tees, valves, and reducers. Most of these losses are of the type discussed in connection with orifice losses (Example 6-3). In most of these fittings the flow is too complicated and the minor variations in geometry are too extensive to warrant analytic attack. Consequently, the losses are expressed in terms of

[15] L. F. Moody, Friction Factors for Pipe Flow, *Trans. ASME*, **66**, 671–684 (1944).

[16] The aspect ratio can be defined as the square of the major dimension of the body divided by the area. For an ellipse with major diameter a and minor diameter b, the aspect ratio is $4a/\pi b$. For an isosceles triangle with altitude h and base b, the aspect ratio is $2h/b$.

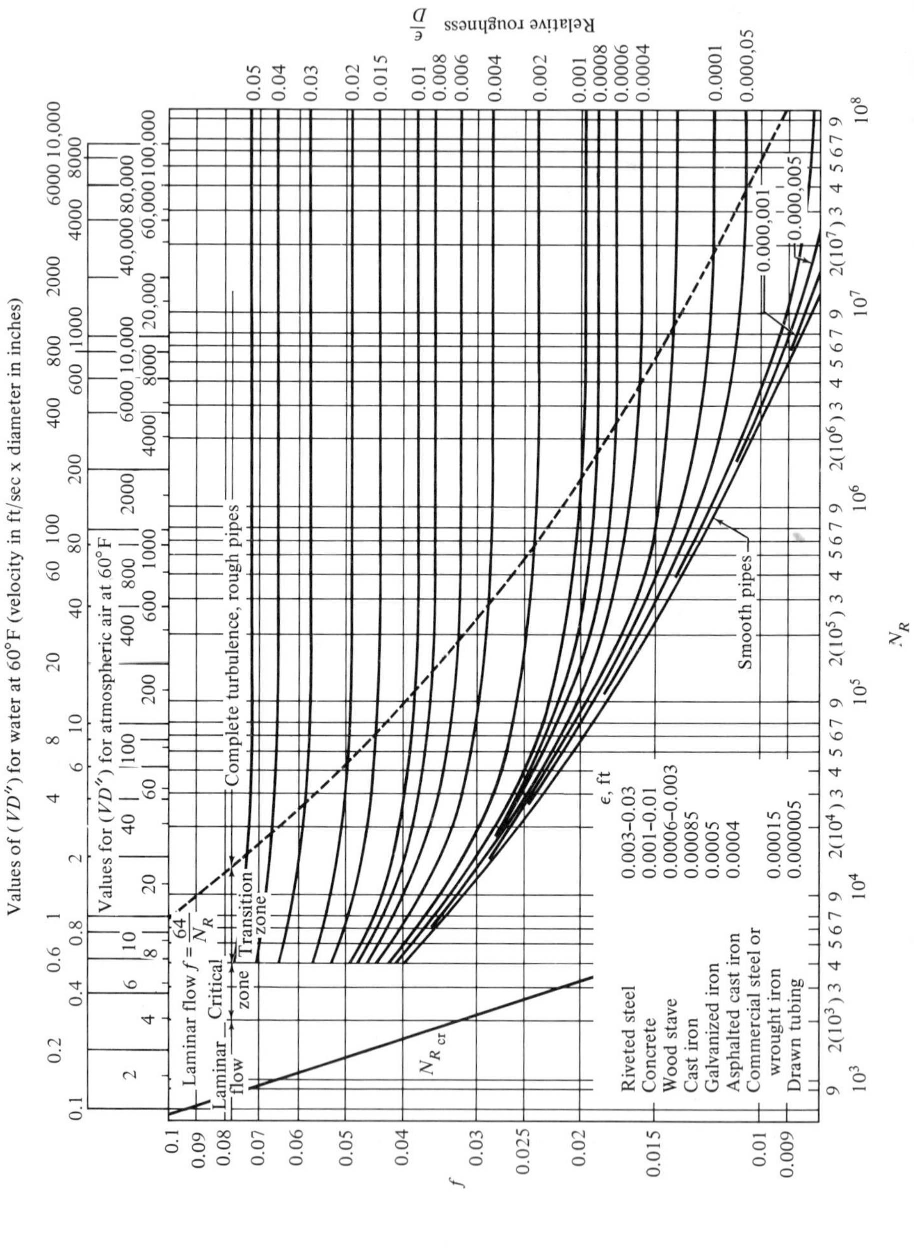

Fig. 10-18. Moody diagram.

loss factors defined as

$$\Delta p_{\text{loss}} = K\frac{\rho\overline{V}^2}{2g_c} \qquad \text{or} \qquad \Delta h_{\text{loss}} = K\frac{\overline{V}^2}{2g} \tag{10-38}$$

TABLE 10-1

LOSS FACTORS FOR SOME REPRESENTATIVE PIPE FITTINGS

Fitting	*Loss factor K*
Globe valve fully open	8–10
Angle valve fully open	4–6
Swing-check valve	~2.5
Gate valve fully open	0.2
Standard 90° elbow	0.8–1.0
Medium-sweep 90° el	0.75
Long-sweep 90° el	0.6
45° el	0.4–0.5
Standard tee	1.8–2
Close return bend	2.2–2.5

The variation in loss factors indicated in Table 10-1 is primarily due to variations in methods of installation. Such loss factors are usually determined in a test setup with a long run (50 diameters or more) of pipe upstream of the unit being tested. If one fitting is close to another fitting, the loss factors may be larger than those indicated in Table 10-1. The losses in elbows, return bends, tees, and so on, are due to separation and secondary flows. It is interesting that the loss for the close return bend is greater than that for two elbows. The additional loss is that associated with a larger magnitude of secondary flow.

SUDDEN ENLARGEMENT

Some separated flow geometries can be analyzed if suitable (experimentally verifiable) assumptions are made. One such example is the flow at an abrupt increase in cross section as illustrated in Fig. 10-19. It is possible to analyze the flow in the control volume indicated by the heavy dashed line. It is assumed that the flow separates at the shoulder with a zero radius of curvature here. Consequently, the Reynolds number must be large enough (the inertia forces relatively large) that this condition and assumption is valid. Applying (6-25) to this control volume gives

$$\frac{p_1}{\rho} + \frac{V_1^2}{2g_c} - \frac{\Delta p_{\text{loss}}}{\rho} = \frac{p_2}{\rho} + \frac{V_2^2}{2g_c} \tag{10-39}$$

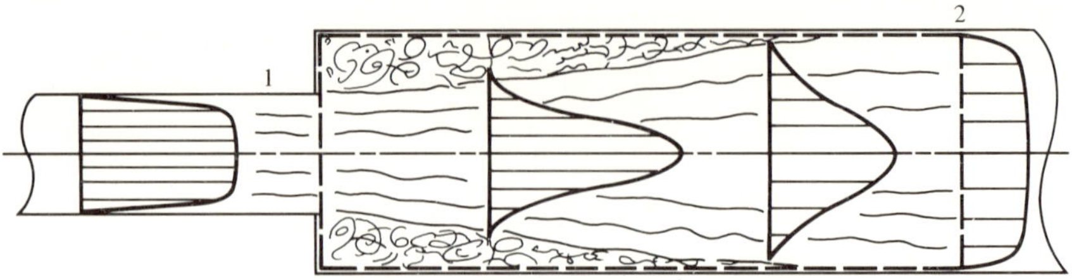

Fig. 10-19. Separation of flow at a sudden enlargement of flow area.

The assumption that N_R is large is also seen to be necessary, so $\alpha_1 \approx \beta_1 \approx 1.0$. It is also assumed that section 2 is far enough downstream that $\alpha_2 \approx \beta_2 \approx 1.0$.

The momentum equation for the control volume is

$$p_1A_2 - p_2A_2 = \frac{\dot{m}}{g_c}(\mathrm{V}_2 - \mathrm{V}_1) \tag{10-40}$$

The pressure p_1 is uniform over the annular area at 1 since the flow is assumed to separate in a straight jet here. The wake region adjacent to the annular area $(A_2 - A_1)$ is a region of negligible mean velocity such that the pressure here is determined by and is therefore equal to the pressure in the let p_1.

The continuity equation gives $Q_1 = Q_2$ or $A_1\overline{\mathrm{V}}_1 = A_2\overline{\mathrm{V}}_2$. Combining these equations gives

$$\frac{\Delta p_{\text{loss}}}{\rho} = \frac{(\overline{\mathrm{V}}_1 - \overline{\mathrm{V}}_2)^2}{2g_c} = \frac{\overline{\mathrm{V}}_1^2}{2g_c}\left(1 - \frac{A_1}{A_2}\right)^2 \tag{10-41}$$

From (10-36) the loss factor is then

$$\left(1 - \frac{A_1}{A_2}\right)^2 = K_{\text{S.E.}} \tag{10-42}$$

The value calculated from this simple analysis agrees very well with experimental results. A detailed examination of the flow conditions in a sudden expansion also indicates that the assumptions are valid. The primary loss is dynamic as described previously. In the wake region there is very little directed flow velocity, so the skin friction on the cylindrical control volume boundary is negligible. In an experimental setup station 2 must be downstream just far enough that the velocity distribution is approximately uniform but not so far that wall friction becomes significant.

SUDDEN CONTRACTION

A sudden contraction is actually a contraction and expansion, as seen in the flow pattern sketched in Fig. 10-20. The loss is primarily a consequence of the expansion from the vena contracta area A_j to A_2. This loss is

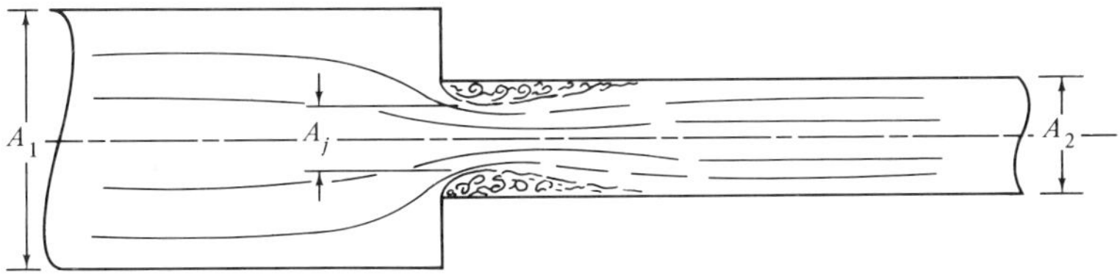

Fig. 10-20. Sudden contraction loss indicated as a separation and subsequent enlargement loss.

usually defined in terms of V_2, since it is possible for A_1 to become infinitely larger than A_2 for flow entering a pipe from a reservoir. Assuming that the loss from the pipe entrance to the vena contracta is negligible, the pressure loss is given by

$$\frac{\Delta p_{\text{loss}}}{\rho} = \frac{(V_1 - V_2)^2}{2g_c} = \frac{V_2^2}{2g_c}\left(\frac{A_2}{A_j} - 1\right)^2 = \frac{V_2^2}{2g_c}\left(\frac{1}{C_c} - 1\right)^2$$

where C_c is defined as the contraction coefficient. The loss factor is then

$$K_{\text{S.C.}} = \left(\frac{1}{C_c} - 1\right)^2 \tag{10-43}$$

The mathematical analysis involved in determining the contraction ratio or coefficient $C_c = A_j/A_0$ as a function of A_2/A_1 is quite complicated. The analytical result for $A_1/A_2 \to \infty$ (that is, flow from a large tank or reservoir into a pipe with a sharp entrance) gives $C_c = 0.6$ or $K = 4/9$ (usually taken as 0.5). Table 10-2 gives some representative values for C_c

TABLE 10-2

A_2/A_1	0	0.2	0.4	0.6	0.8	1.0
C_c	0.6	0.632	0.659	0.712	0.813	0

for this geometry, from which the loss factor can be calculated. For a smoothly rounded entrance from a tank or reservoir with a radius of 1 pipe radius or diameter, the entrance loss is quite small; $K = 0.03$ to 0.05.

DIFFUSERS

In many ducting and piping systems it is desirable to decrease the velocity and increase the pressure by diffusing the flow. Some examples are the diffusers downstream of a high-speed test section in a wind tunnel, the inlet diffusers on jet engines, and diffusers on pump discharges. Simple types of diffusers are two-dimensional with only one side diverging, and a conical diffuser used in piping or round ducting systems. Although these geometries are quite simple, the conditions that govern the flow patterns are difficult to formulate and solve mathematically. Consequently, their performance is usually presented graphically. One method for expressing the loss is to give it as a fraction of the loss for an equivalent sudden enlargement,

$$\frac{\Delta p_{\text{loss}}}{\rho} = K_D \frac{(V_1 - V_2)^2}{2g_c} \tag{10-44}$$

where K_D is the diffuser loss factor, which depends upon the diffuser angle, the diameter ratio, and the Reynolds number. If the Reynolds number is relatively large (for example, the inlet Reynolds number is greater than 4000), the loss factor is nearly independent of Reynolds number. The diameter ratio effect is minimized by the definition of K_D, which is then primarily a function of the included angle θ. It is expected that for $\theta = 180°$ (sudden enlargement as considered above) $K_D = 1.0$. Experimental values of K_D as a function of diffuser angle are presented in Fig. 10-21. A minimum loss is obtained for an included angle of around 7 or 8°. For smaller angles the diffuser is quite long and a high friction loss is obtained. For $\theta > 8°$, the loss is predominantly a consequence of separation (discussed in Section 6.12), as illustrated in Fig. 10-22a. The 8° angle then corresponds to an optimum between these two effects. Some widely quoted previous data taken over 50 years ago with no reference to the instrumentation and little description of the experimental setup indicate a value greater than 1.0 in the region of $\theta = 60°$. There is little reason to expect such behavior, and a rerun of conical and two-dimensional diffusers indicates that no such maximum exists. The lower curve of Fig. 10-21 was obtained by rounding the entrance break with a radius of 1 upstream pipe diameter. For all angles greater than 20° the effect is seen to allow some attachment to be maintained (Fig. 10-22b) in the rounded portion, yielding some pressure recovery and an efficiency improvement of about 35 percent for large angles.

Good diffusion and better flow stability can be maintained with a vaned diffuser, which is essentially a parallel system of small angle diffusers, as illustrated in Fig. 10-23. The inner diffusion angle can be made somewhat larger than the minimum 7 or 8° indicated in Fig. 10-21, since there is no approach boundary layer through this section. The same reasoning indicates that θ_2 must be the smallest angle ($\sim$6 or 7°). Also, the angle that the flow must turn through at the leading edge of each diffuser increases for each successive vane from the center. The study of vane

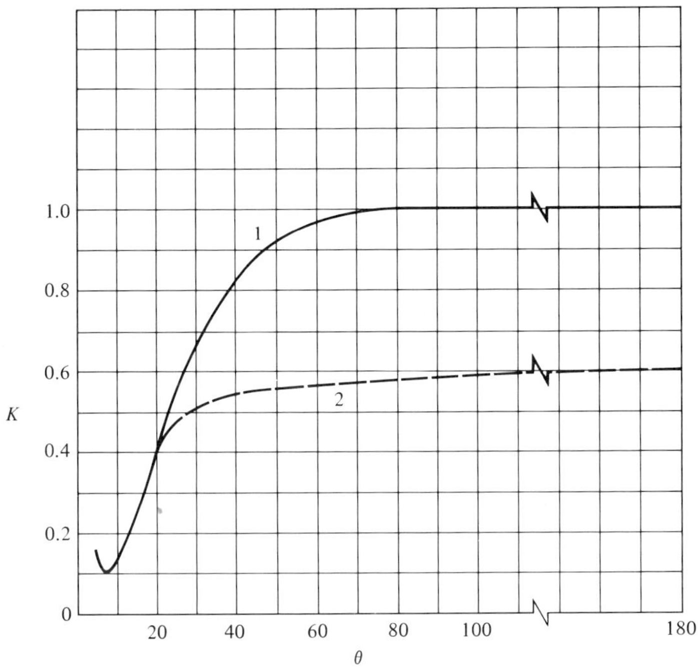

Fig. 10-21. Loss factor K for conical diffusers as a function of diffuser included angle θ. Curve 1, straight-sided diffuser; curve 2, diffuser inlet rounded with radius equal to inlet pipe diameter.

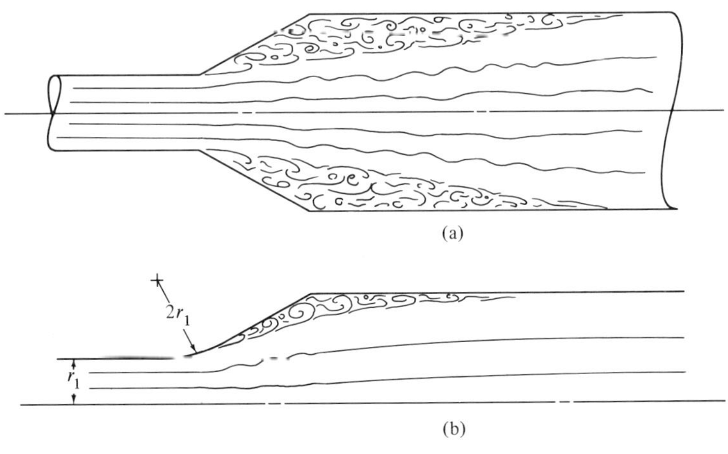

Fig. 10-22. Patterns for flow in a wide-angle diffuser: (a) the flow separates at the discontinuity corner and diffuses freely with a large turbulent separation loss; (b) some attachment is maintained by rounding the sharp discontinuity, thereby reducing the separation loss.

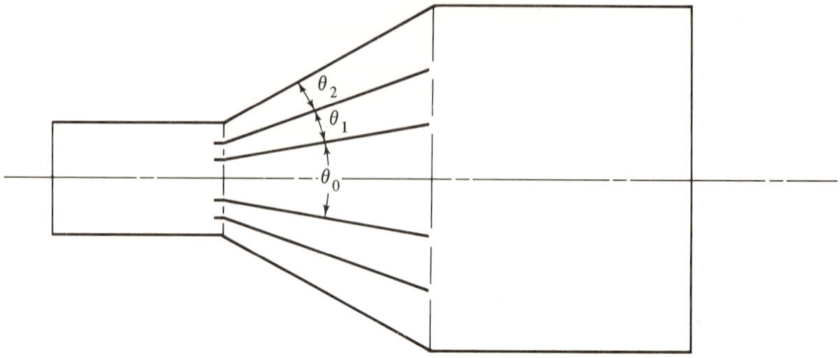

Fig. 10-23. Vaned diffuser.

diffusers was an outgrowth of the study of separation losses in compressor cascades. The literature on the subject of cascade losses and on vaned diffusers is voluminous.

10.9 FLOW IN PIPING AND DUCTING SYSTEMS

The majority of internal flow problems involve turbulent flow. Laminar "viscous" flows are to be found in chemical process systems, for example, process flows of monomers and polymers. Only a few such substances can be handled with the methods developed in Chapter 7, since many are non-Newtonian and present complicated problems with respect to determining their flow mechanics. Other problems, such as the flow of blood in the vascular system are extremely complicated. Not only is the flow unsteady and the fluid non-Newtonian, but the walls are elastic and actively alive, owing to a layer of controllable muscle tissue.

The flow of air, water, and other relatively low kinematic viscosity fluids in piping and ducting systems can be determined by applying the principles previously developed and using empirical and semi-empirical data as presented in Tables 10-1 and 10-2 and in Figs. 10-18 and 10-21. Simplifying assumptions such as one-dimensional, fully developed flow, and so on, are usually made.

Piping problems can be classified into three cases:

Case I

Find the pressure loss Δp_{loss} for a given flow rate Q in a given piping system D and L. This case is the easiest, since the flow rate Q, piping lengths and area, and so on, are given, allowing the velocity and Reynolds number to be determined immediately. The fitting losses are essentially independent of Reynolds number, but the piping friction loss depends on Reynolds number, which is known.

CASE II

Find the flow rate Q for a given pipe system D and L and a given available Δp. In this case the velocity (and therefore the Reynolds number) is not known, so a friction factor f for the pipe cannot be determined directly. However, it is possible to estimate a value of f. As a first estimate, the most logical value is one in the fully rough region. Since the piping is specified, the relative roughness is known. The value of f for fully turbulent flow is then obtained directly from Fig. 10-18. The velocity and Q can then be calculated from the relations available. If the value of N_R determined for this velocity is less than that for fully turbulent flow, a value of f can be determined for the known ϵ/D and a Reynolds number slightly less than the first value. Since f increases as N_R decreases, the error in the initial estimate of f will give a velocity larger than the correct one. This trial-and-error procedure is repeated until the values of N_R agree. The convergence on the correct answer is quite rapid because of the flatness of the f versus the N_R curves. (Note that on the log–log Moody diagram, an f-scale cycle is three times the N_R scale.)

CASE III

Find the piping size D for a given available Δp and flow rate Q. This case is the most difficult because no entry information is available. Again a value of f must be assumed. Even if the pipe material is specified (as is usually the case), the size is not known. A good first estimate for this type of problem is to take $f = 0.02$ or 0.03. The procedure is then similar to that given in case II.

10.10 ILLUSTRATIVE EXAMPLE 10-2: Piping System Discharge

It is necessary to find the maximum flow available at the discharge of a sillcock (hose-connection globe valve) for the distribution system shown in Fig. 10-24. The reservoir surface is 300 ft above the valve discharge. The distribution manifold is large enough that the available total pressure head at b is $0.9H$ under normal demand periods. The piping system from b to the discharge is essentially horizontal. With a known pressure at 2, the equations available can be applied between this point and the discharge. Applying the energy equation from (6-25) and assuming one-dimensional flow,

$$\frac{p_2}{\rho} + \frac{V_2^2}{2g_c} - \frac{\Delta p_{\text{loss}}}{\rho} = \frac{p_4}{\rho} + \frac{V_4^2}{2g_c}$$

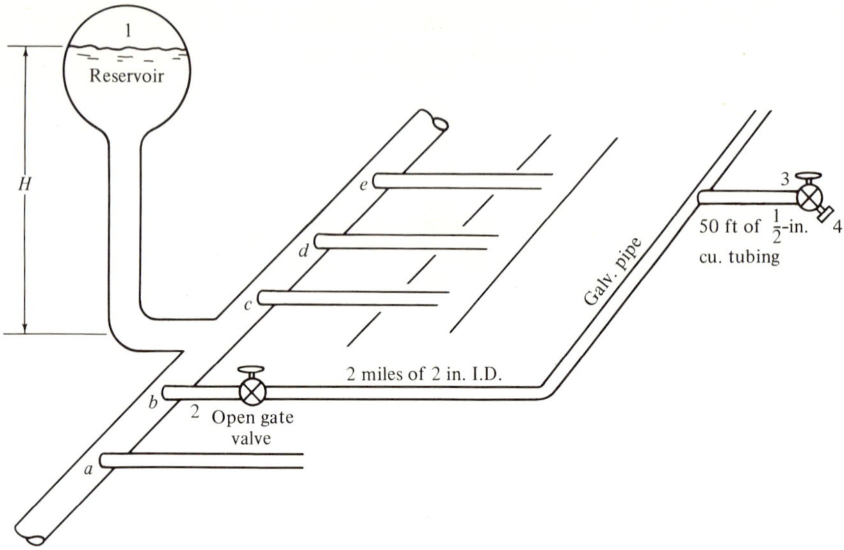

Fig. 10-24. Piping distribution system.

where

$$gH_2 = \frac{p_2}{\rho} + \frac{V_2^2}{2g_c} = 0.9(h_{01} - h_{04})$$

is given. The availability loss is the sum of all the losses between 2 and the discharge at 4:

$$\frac{\Delta p_{\text{loss}}}{\rho} = K_{\text{gate}}\frac{V_2^2}{2g_c} + f_2\frac{L_2}{D_2}\frac{V_2^2}{2g_c} + f_3\frac{L_3}{D_3}\frac{V_3^2}{2g_c} + K_{\text{globe}}\frac{V_3^2}{2g_c}$$

$$= \frac{V_2^2}{2g_c}\left[K_{\text{gate}} + f_2\frac{L_2}{D_2} + f_3\frac{L_3}{D_3}\left(\frac{D_2}{D_3}\right)^4 + K_{\text{globe}}\left(\frac{D_2}{D_3}\right)^4\right]$$

where the continuity equation has been used: $A_2V_2 = A_3V_3$. For the 2-in. pipe, $\epsilon/D = 0.003$, which gives $f_2 \approx 0.026$ in the fully turbulent region. For the ½-in. tubing, $\epsilon/D = 0.0012; f_3 \approx 0.02$. The discharge is to the atmosphere, so $p_4 = 0$. The above relations give (assuming $V_4 \approx V_3$)

$$\frac{0.9g(h_{01} - h_{04})}{g_c} = \frac{0.9gH}{g_c} = \frac{V_4^2}{2g_c} + \frac{\Delta p_{\text{loss}}}{\rho} = \frac{V_2^2}{2g_c}\left[\left(\frac{D_2}{D_3}\right)^4 + \sum K_i\right]$$

which can be solved for V_2 and the discharge $Q = A_2V_2$.

The magnitudes of the various loss terms K_i in the bracketed term are interesting to investigate. (The lower limit for the globe-valve loss factor is assumed, since sillcocks have slightly larger passages than standard globe valves.) Solving the above equation for V_2 gives

$$V_2^2 = \frac{0.9gH}{K_{\text{gate}} + [f(L/D)]_2 + (D_2/D_3)^4\{1 + [f(L/D)]_3 + K_{\text{globe}}\}}$$

$$= \frac{(0.9)(32.2)(300)}{0.2 + 0.026[(2)(5280)(12)/2] + 4^4\{1 + 0.02[(50)(12)/2] + K_{\text{globe}}\}}$$

The major loss factors are (when related to V_2 as $KV_2^2/2g_c$)

Unit	*K*
2-in. pipe	1647
½-in. pipe	6144
Globe valve	2048
Exit K.E.	256

The ½-in. pipe and globe valve are seen to account for more than 80 percent of the pressure drop. The velocity V_2 in the 2-in. line is 0.923 ft/sec and $V_3 = V_4 = (A_2/A_3)V_2 = 16V_2 = 14.8$ ft/sec. The discharge rate is $Q = A_2V_2 = 0.0218$ ft³/sec or about 3 gpm (1 ft³/sec is equivalent to about 448 gpm).

The smaller Reynolds number is the more important in determining the effect on the f's. For a given flow rate in a system with different areas, the Reynolds number is inversely proportional to the diameter, so N_R will be smaller in the larger pipe ($N_{R2} \approx 10^4$ for water at 60°F). For $(\epsilon/D)_2 = 0.003$, this combination would indicate $f_2 \approx 0.035$ instead of the 0.026 initially assumed. The value of $N_{R3} \approx 4 \times 10^4$ indicates that $f_3 \approx 0.024$ instead of 0.02. The 30 percent increase in f_2 and 5 percent increase in f_3 would decrease V_2 to about 0.9 ft/sec.

More complicated problems involving series-parallel networks of pipes can be solved by usual network techniques. These involve the added complications of writing flow equations for each junction with the possibilities that the flow directions at the junctions are not known. For example, one of the manifold junction points such as a in the above example may be a tie-in from another reservoir. If the tie-in connections were open, the flow could be in either direction, depending on the system characteristics. Such problems then involve the solution of a set of equations.

10.11 INTENSITY AND SCALE OF TURBULENCE

In turbulent-flow measurements, it is advantageous to take the primary coordinate direction (1 or x direction) in the direction of the local mean flow: $\vec{v} = \vec{U} + \vec{v}'$. The coordinate velocity fluctuation components are then, u_1', u_2', and u_3', and the mean velocity has only the one component, $\langle \vec{v} \rangle = \vec{U}$.

Relative component intensities (for example, Fig. 10-5) are usually expressed in terms of the rms fluctuations as

$$\frac{u_1}{U}, \quad \frac{u_2}{U}, \quad \frac{u_3}{U}$$

For example, these relative intensities for pipe flow are u/U, v/U, and w/U, where u, v, and w are the rms fluctuation components in the axial, radial, and tangential directions, respectively. The total relative intensity is

$$\frac{\sqrt{u_1^2 + u_2^2 + u_3^2}}{U}$$

The fluctuating components are measured with a hot-wire anemometer. A very fine wire 0.0002 to 0.001 in. in diameter and about 1 to 5 mm long welded (or soldered) to a two-needle tipped probe is heated to a temperature slightly greater than the gas-flow temperature. (Hot-film probes are used for liquid fluctuation measurements.) The wire forms one leg of a resistance bridge. By suitable electronic circuitry, either temperature, voltage, or current is held constant. For example, in a constant-current hot-wire circuit, the current is held constant and the voltage is varied to compensate for the variation in wire resistance produced as the air flow fluctuates across the wire, causing a fluctuating cooling effect. Instantaneous measurements can be observed with an oscilloscope or recorded on an oscillograph. A squaring and averaging circuit can be used to give $\langle u'^2 \rangle \equiv u^2$. The square root of this value can then be taken electronically to yield u.

Visualizing the structure of a turbulent flow as a stream filled with eddies, as an eddy passes a fixed reference point there will be an instantaneous correlation of fluctuation velocity within a small region about the reference point. As the distance from the reference point is increased, the correlation will decrease. This correlation can be defined and measured as a relative quantity

$$f_1(x) = \frac{\langle u'(x_1)u'(x_1 + x) \rangle}{\langle u'^2 \rangle}$$

where $f_1(x)$ is the correlation function of the x component at reference point 1 with coordinate x_1. Some obvious characteristics of $f_1(x)$ are: (1) at $x = 0$, $f_1(0) = 1$; (2) as x gets very large ($x \to \infty$), there will be no correlation, or $f(\infty) = 0$; and (3) from a physical consideration of the eddy as it passes point 1 the correlation should be the same as it approaches and when it passes this point or $f(-x) = f(+x)$, and therefore (4) at $x = 0$ the slope $f'(x)$ should be zero. A function with these properties is illustrated in Fig. 10-25. Such a mathematically derived curve would require $x = \infty$ for $f(x) = 0$; therefore, no characteristic mean eddy size or *turbulent scale* is immediately evident. For a time-growth process [such as that for the charging (or discharging) of a capacitor] a characteristic time is defined with respect to the voltage-time slope at $t = 0$. This procedure is not possible for $f(x)$ because of the zero slope at $x = 0$. However, the second derivative can be used to obtain a characteristic length by fitting a parabola that is tangent to the curve at $x = 0$. The x intercept then gives a characteristic scale of turbulence termed the

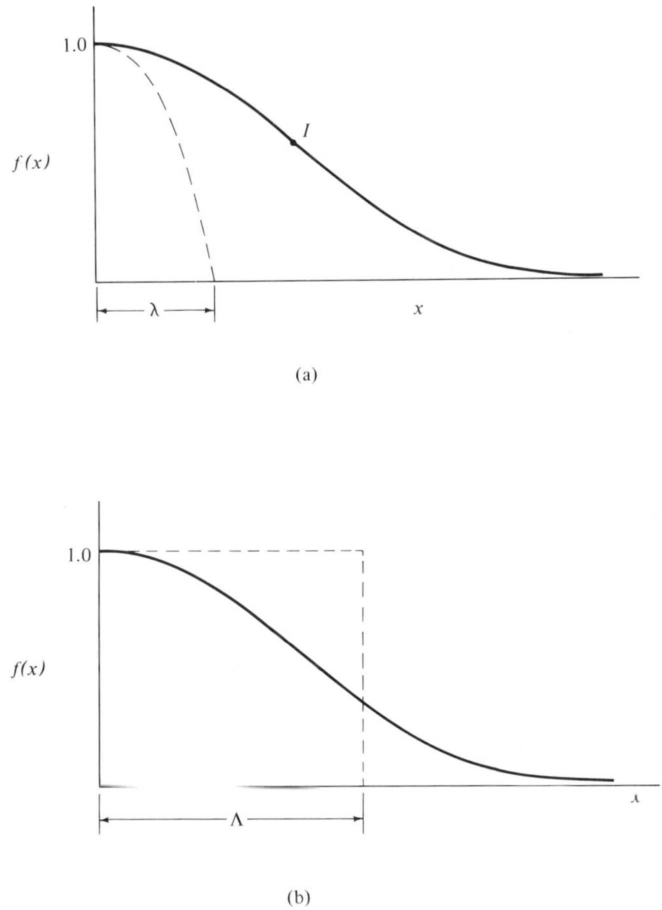

Fig. 10-25. Graphical representation of the microscale λ(a) and the macroscale Λ(b) of turbulence.

microscale λ, as shown in Fig. 10-25a.[17] The integral or macroscale Λ is another characteristic scale of turbulence, which can be defined with respect to the integral property of the curve:

$$\Lambda \equiv \int_0^\infty f_1(x)\,dx$$

As illustrated in Fig. 10-25b, this quantity is the mean abscissa of the curve.

The intensity and scale are important in rapid, turbulent mixing processes, where reactants or suspensions are mixed in transport piping rather than in agitating mixing vessels.

10.12 TURBULENT FLOW ON A SMOOTH FLAT PLATE

The wall region governs the turbulence conditions, as previously described and illustrated in connection with Fig. 10-4. The relations (10-2) through (10-10) are consequently valid for this case. With the same assumptions as those for pipe flow, the value of κ remains the same ($\kappa = 0.4$). Equation (10-10) can be applied to the turbulent boundary layer flow past a flat plate where $h = \delta$ (the boundary-layer thickness) and U_{max} is the external velocity U_e. The analysis of the turbulent flow past a plate is more complicated for the following reasons:

1. Both δ and U are functions of x, the distance along the plate from the leading edge.
2. An initial length of laminar boundary layer will exist which is determined by the degree of turbulence of the free stream and the state of quietness of the system; therefore, δ and U_e in the turbulent boundary-layer region back along the plate will depend on this initial length of laminar boundary layer x_{lam} and the succeeding length of turbulent boundary layer x_{turb}, which comprise the total x distance.
3. The value of τ_0 and therefore u_τ also depend on x.

The universal velocity distribution law of the wall is still valid. If the analysis is made to determine f, a form similar to that for a pipe may be expected, $1/\sqrt{f} = C \ln N_R\sqrt{f}$, when the boundary layer is completely

[17] Mathematically, λ is given by the relation

$$\frac{1}{\lambda^2} = \frac{1}{2}\left(\frac{\partial^2 f}{\partial x^2}\right)_{x=0} = \frac{1}{2u^2}\left\langle\left(\frac{\partial u}{\partial x}\right)^2_{x=0}\right\rangle$$

See, for example, reference 2 at the end of the chapter.

turbulent for the entire plate length. A semiempirical expression for f was obtained by Schultz–Grunow by matching the experimental velocity profile for the flat-plate boundary layer:

$$f = 0.427(\log N_R - 0.407)^{-2.64} \tag{10-45}$$

where N_R is the length Reynolds number based on the plate length and f is the integrated value for the entire plate. Prandtl and Schlichting proposed a simpler relation,

$$f = 0.455(\log N_R)^{-2.58} - \frac{A}{N_R} \tag{10-46}$$

where the value of A depends on the length of the laminar boundary layer. Unfortunately this length varies with many conditions (disturbances in the mechanical system and fluid flow field, plate surface condition, and so on). Figure 10-26 indicates the results of experimental measurements compared with the above relations. An example curve for transition at $N_R = 5 \times 10^5$ is also given which fits both experimental data and (10-46) for this condition. The laminar curve is for the Blasius result. The graphical representation is seen to be the most useful for presenting the friction factor results in this case of the smooth flat plate.

The rough-plate problem is complicated by the fact that the relative roughness ϵ/δ varies along the plate and also affects the transition point

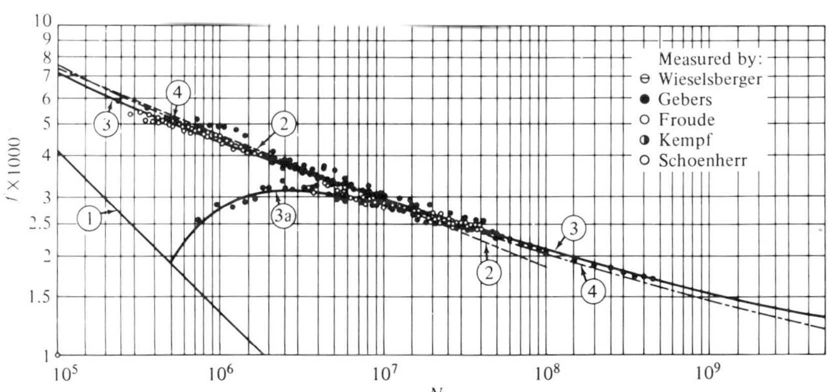

Fig. 10-26. Drag coefficient for a smooth flat plate obtained by integrating the friction coefficient over a length L of the plate. $N_R = U_\infty L/\nu$. Curve 1 is for laminar flow; curve 2, $f = 0.074N_R^{1/5}$ for a $\frac{1}{7}$-power velocity distribution law; curve 3, $f = 0.455(\log N_R)^{-2.58}$; curve 3a is (10-46); curve 4 is (10-45).

(or Reynolds number). The problem of correlating experimental data for naturally rough plates with that from controlled sand grain roughness is consequently more difficult. For a further discussion of flow past flat plates refer to Chapter 21 of reference 1 at the end of the chapter.

10.13 MATHEMATICAL TREATMENT OF TURBULENT FLOWS

In order to develop general mathematical analyses of turbulent flow, the general (Navier–Stokes) equations of motion are written with the velocity components represented in the form $u = U + u'$, and so on, and the time mean of these equations is then taken. The turbulent components then yield terms of the form $\rho\langle u'^2\rangle$, $\rho\langle u'v'\rangle$, and so on, as developed and used previously. The possible permutations of terms of this type yield nine "turbulent shear stress" components. These "Reynolds stresses," or the Reynolds stress tensor, can then be combined with the laminar shear stress tensor (or used in place of it, since the turbulent shear has been shown to be many times greater than the laminar shear stress). (See, for example, references 1 and 2 at the end of the chapter.)

The mixing-length derivations used in this chapter yield the law of the wall for the mean profiles and friction coefficients by considering the behavior of the mean profile with assumed deductions concerning the wall shear stress and mixing-length behavior. An alternative procedure, considered to be superior in concept by some investigators, is to develop the relations from the premise that the turbulence and velocity distribution are both manifestations of the *vorticity* in the flow. Consequently, the problem should be approached using the vorticity rather than the velocity as the primary independent variable. So far, the vorticity developments have not yielded correlations as good as those obtained from Prandtl's mixing-length hypothesis and development.

10.14 SUMMARY

1. The random fluctuations called turbulence originate as a consequence of instabilities at boundaries or in shear layers.
2. Turbulent eddy transport is orders of magnitude more effective in mass, momentum, and energy transport than laminar molecular transport. Consequently, a greater wall shear stress and greater energy dissipation is obtained. Turbulence, therefore, has advantages or disadvantages, depending on the desired result.
3. Since the turbulence treated in this chapter is considered to have its origin in the shear layer near a wall (Fig. 10-4), it is termed *wall*

turbulence. The relations derived, (10-15) and (10-22), are therefore referred to as "the law of the wall." These relations hold in the predominantly turbulent region, $50 < y_+ < 1000$. For (1) y_+ very large (nominally greater than 1000), (2) a flow with zero pressure gradient, or (3) "free" or for "wake" turbulence, it is necessary to hypothesize and derive additional relationships.[18]

4. Because the semi-empirical relations are implicit [for example, (10-22)], the combined analytical and experimental results are best expressed graphically (Figs. 10-10, 10-13, and 10-18).
5. It is surprising (amazing) that Prandtl's simplifying hypotheses that $\tau = \tau_0 =$ const. and $l = \kappa y$ produced relations (10-15) and (10-22), which agree so closely with experimental results (Figs. 10-10 and 10-13) when it is seen that in reality $\tau = \tau_0[1 - (y/r_0)]$ and l' varies as indicated in (10-13) and Fig. 10-7. It is of interest to consider this paradox. The velocity distribution is derived from (10-3),

$$\frac{g_c \tau}{\rho} = l'^2 \left(\frac{dU}{dy}\right)^2$$

If l' is taken as given in (10-13) in the form

$$\frac{l'}{r_0} = \kappa \frac{y}{r_0} f\left(\frac{y}{r_0}\right)$$

and $\tau = \tau_0[1 - (y/r_0)]$, then the above differential equation can be written in the integral form (with $\kappa = 0.04$ and $u_\tau^2 = g_c \tau_0/\rho$) as

$$\frac{U - U_{\max}}{u_\tau} = \frac{1}{\kappa} \int_{y/r_0 = 0+}^{1} \left[\frac{\sqrt{1 - y/r_0}}{f(y/r_0)}\right] \frac{d(y/r_0)}{y/r_0}$$

If the term in the brackets,

$$g\left(\frac{y}{r_0}\right) = \frac{\sqrt{1 - (y/r_0)}}{f(y/r_0)}$$

is constant with the value unity for the entire integration (that is, $0 < y/r_0 \leq 1$), then the inaccuracies of the two assumptions will cancel. Indeed, the value of $g(y/r_0)$ varies only a few percent from 1.00, which explains the surprising fit of the derived equations with the experimental data.

[18] Donald Coles, The Law of the Wake in the Turbulent Boundary Layer, *J. Fluid Mech.*, **1** (pt. 2), 191–226 (1956).

REFERENCES

1. H. Schlichting, *Boundary Layer Theory*, McGraw-Hill, Inc., New York, 6th ed., 1968.
2. J. O. Hinze, *Turbulence*, McGraw-Hill, Inc., New York, 1959.
   This text gives a comprehensive mathematical treatment of turbulence with generous examples of experimental correlation material. (It does not treat the subject of stability and the origin of turbulence.)
3. S. Goldstein (ed.), *Modern Developments in Fluid Dynamics*, vols. I and II, Oxford University Press, New York, 1938.
   These volumes contain presentations of various mixing-length theories and vorticity theories of turbulence with experimental correlating data.
4. J. Laufer, The Structure of Turbulence in Fully Developed Pipe Flow, *NACA Technical Report 1154*, 1954.
5. A. A. Townsend, *The Structure of Turbulent Shear Flow*, Cambridge University Press, New York, 1956.
6. G. K. Batchelor, *The Theory of Homogeneous Turbulence*, Cambridge University Press, New York, 1956.
   The last two references are primarily mathematical treatises dealing with isotropic turbulence.

PROBLEMS

10-1 For fully developed turbulent flow in a pipe the velocity distribution can be approximated by $U(r) = U_0[1 - (r/r_0)]^{1/7}$, where U_0 is the centerline velocity at $r = 0$ and r_0 is the pipe radius. Find the value of $U_0/\overline{V}$, where $\overline{V}$ is the average velocity $= Q/A$.

10-2 Considering a control volume of length L and radius r_0 in a pipe, (a) show that a force momentum balance for steady flow gives

$$\tau_0 = \frac{-r_0}{2}\frac{\partial p}{\partial z} \equiv \frac{-r_0}{2}\frac{\Delta p}{L}$$

for a length L of pipe. (b) Is this result valid for compressible as well as incompressible flow?

10-3 Defining $c_\tau = \tau_0(\rho\overline{V}^2/2g_c)$ and f from the defining relationship

$$\Delta p \equiv f\frac{L}{D}\frac{\rho\overline{V}^2}{2g_c}$$

show that (a) $f = 4c_\tau$, and (b) for the special case of laminar flow, $f = 64/\sqrt{N_R}$.

10-4 Work out the details for the shear stress relation for a fluid "eddy element" entering the control volume of Fig. 10-4 from above and show that the results (10-2) are obtained.

10-5 Form a dimensional analysis on the variables U, τ_0, y, ρ, and ν (and g_c) to obtain the dimensionless groups

$$\frac{U}{\sqrt{g_c\tau_0/\rho}} \equiv u_+ \quad \text{and} \quad \frac{y\sqrt{g_c\tau_0/\rho}}{\nu} \equiv y_+$$

10-6 Consider the action of a pure shear stress (as an effect independent of rotation) on the element shown and show that $\tau_{yx} = \tau_{xy}$.

10-7 At the wall ($y \to 0$) the flow is predominantly laminar. The shear stress is then obtained from the relation $\tau_0 = \mu(dU/dy)_{y=0}$. Calculate τ_0 from (10-8) and comment on the result.

10-8 Perform the integration necessary to obtain (10-20).

10-9 For flow in a pipe, $N_R = 50{,}000$ and $D = 1$ in., calculate (a) the predominantly laminar thickness y_1 for $y_+ = 5$, and (b) y for the transition thickness at $y_+ = 70$. What are $\overline{V}$ values for air and water at these conditions?

10-10 For a flow of 200 gpm in a smooth 2-in. pipe, determine (a) N_R, (b) f, (c) y for $y_+ = 5$ and 70, and (d) $U_{max}/\overline{V}$. (e) What is the N_R based on the actual y at $y_+ = 5$? Comment on this result.

10-11 Show that the relation $u_+ = y_+$ yields $f = 64/N_R$ and discuss this result with respect to the state of the flow implied.

10-12 Discuss the applicability of using the pipe flow results obtained for velocity distribution and pressure loss (with respect to friction factor) as design information for the test section ducting in a "low-speed" (up to 200 ft/sec) wind tunnel or in a water tunnel.

10-13 Perform the necessary integration to arrive at the result given in (10-21) and then derive (10-22).

10-14 Derive the expression given in Example 10-1 for the ratio of eddy viscosity to molecular viscosity as given in (10-23).

10-15 For nominal values of $f = 0.02$ and $N_R = 10^5$, plot ε/ν for $0 \leq y/r_0 \leq 1$ and comment on the resultant curve. What is the significance of the negative values of ε obtained? Refer to Figs. 10-5 and 10-7 for your discussion.

10-16 Find the ratio ε/ν for flow in a smooth duct for $N_R = 10^6$ at $y/r_0 = 0.1$ and 1.0.

10-17 Referring to Example 10-1, show that $y_+ = (y/r_0)N_R\sqrt{f/32}$, where $N_R = \overline{V}D/\nu$, and derive (10-23).

10-18 The relationships of Example 10-1 can be used to obtain the relation

$$\frac{\varepsilon}{r_0 u_\tau} = 0.4\left(1 - \frac{y}{r_0}\right)\left(\frac{y}{r_0}\right) - \frac{1}{N_R}\sqrt{\frac{32}{f}}$$

for $N_R > 2000$ ($f < 0.05$). The last term is less than 0.013, so ε is nearly independent of N_R and this term may be neglected. Plot the above equation and compare with the results of Nikuradse as given in Fig. 20-8 of reference 1.

10-19 The relations given in Example 10-1 can be used to obtain ε/ν in terms of y_+. Show that the approximate expression $\varepsilon/\nu = 0.4y_+$ can be obtained.

10-20 Find the ratio ε/ν for flow in a smooth duct for $N_R = 5 \times 10^6$ at $y_+ = 5$ and $y_+ = 500$. Comment on the result for $y_+ = 5$.

10-21 Using (10-19) and (10-25), derive the result that $\overline{\mathrm{V}}/\mathrm{u}_\tau = 6.99(r_0\mathrm{u}_\tau/\nu)^{1/7}$.

10-22 For an average value of 20 ft/sec in a 1-in.-diameter pipe, find the allowable roughness (sand equivalent) for the pipe to be considered smooth for water and for air flow.

10-23 Integrate the velocity for rough pipes to obtain (10-23).

10-24 Perform the necessary indicated steps to obtain (10-35).

10-25 In Example 10-2 find the discharge if the sillcock is removed.

10-26 Hose manufacturers rate hoses by discharge. Assuming 80 psi available ahead of a sillcock, find the open-valve discharge rates for 50-ft lengths of 7/16-in, ½-in., and ⅝-in. hose. For this problem assume that the sillcock loss factor is 5.

10-27 (a) Find the pressure p_2 in Example 10-2. (b) Find the pressure p_3 just upstream of the sillcock. (c) If the sillcock were closed, what would be the pressure p_3? Comment on this last result.

10-28 Why does f increase as the velocity decreases for laminar flow in a pipe?

10-29 The velocity profile at a certain point in a ducting system is determined as elliptical, as given by $(U/U_{\max})^2 + (r/r_0)^2 = 1$. Sketch this profile and calculate the wall shear stress. Discuss the results.

10-30 For laminar flow in a pipe or tube, derive an expression for the drag per unit length as a function of viscosity, flow rate, and pipe radius.

10-31 (a) Since you have taken a course in fluid mechanics, your neighbor asks for some consulting help on the design of a pumping system. The setup is as shown. The tank header pressure is to be maintained at 10 psi. The maximum flow requirement is to be about 22.5 gpm (1/20 cfs). A pump with a maximum efficiency of 65 percent is obtainable. What size of motor will be required for the pump drive?
(b) The local hardware dealer has an oversupply of ½-in. pipe on hand and says he will give you the pipe if you buy the pump from him. If his prices on pumps are as good as can be obtained, is this offer worth taking up?

10-32 Calculate the pump horsepower required for a flow rate of 1800 gpm (4 ft^3/sec) in the system shown.

10-33 Show that the entrance loss factor for flow from a reservoir through a square-edged entrance into a round pipe is $\approx$½.

10-34 Show that the exit loss factor from a pipe into a reservoir is 1.0. What is the equivalent length of pipe ($f = 0.02$) that would give this same loss?

10-35 Find the equivalent length of pipe (expressed as number of diameters) that would give a loss factor equivalent to that for a fully opened globe valve.

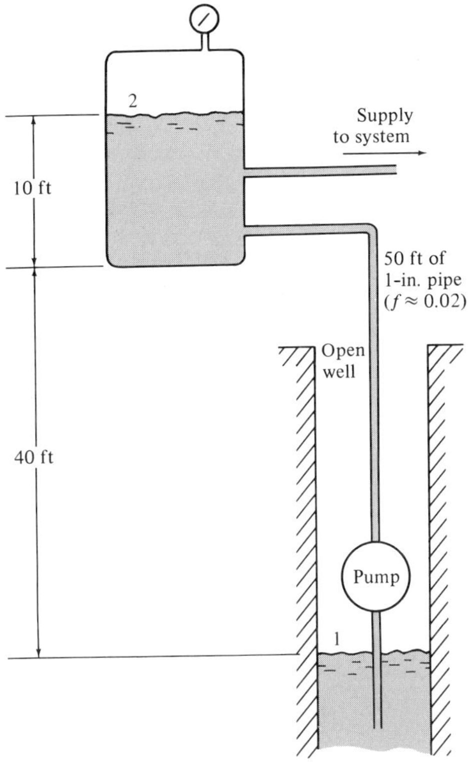

Fig. P10-31.

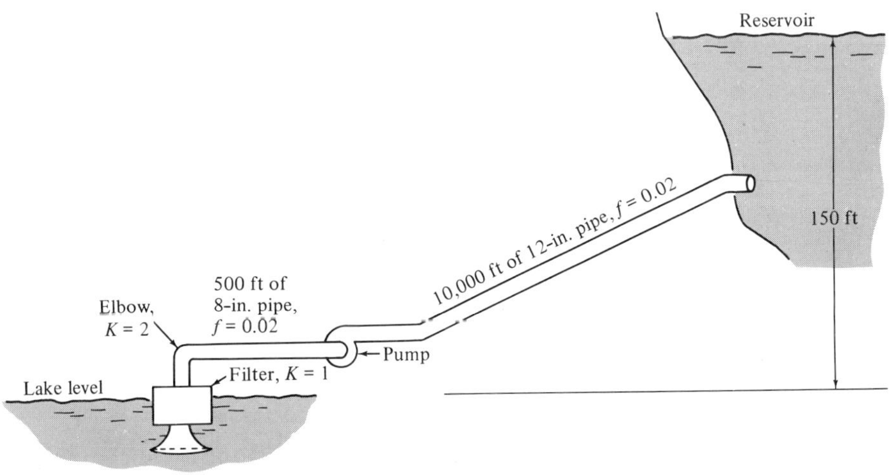

Fig. P10-32.

10-36 (a) The cross section through a standard pipe reducer bushing is illustrated. If the flow is from D_1 to D_2, specify a loss factor and justify the value. (b) If the flow is from D_2 to D_1, specify a loss factor and justify the value.

Sketch flow patterns for both cases. Assume that N_R is large (that is, supercritical) for the flow in D_2.

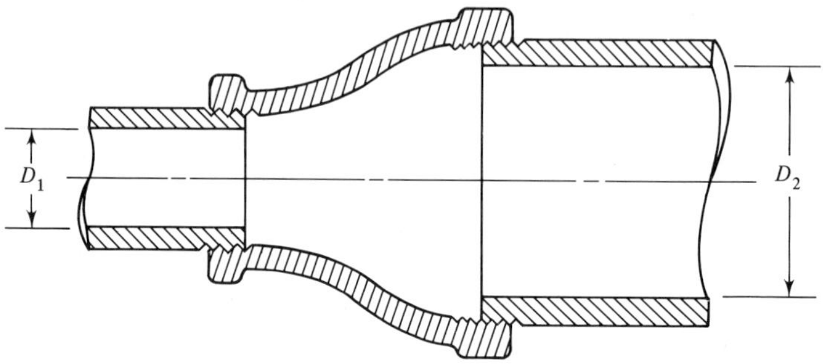

Fig. P10-36.

CHAPTER 11

CONTENTS

EXTERNAL FLOW: LIFT, FRICTION, AND DRAG

11.1 INTRODUCTION

The force that acts on a body in flight, or in a wind tunnel type flow past it, is a consequence of the pressure and shear stress distributions on the body. The attached streamline flow about an airfoil is illustrated in Fig. 11-1. The flow velocity is greater over the top of the airfoil than over the

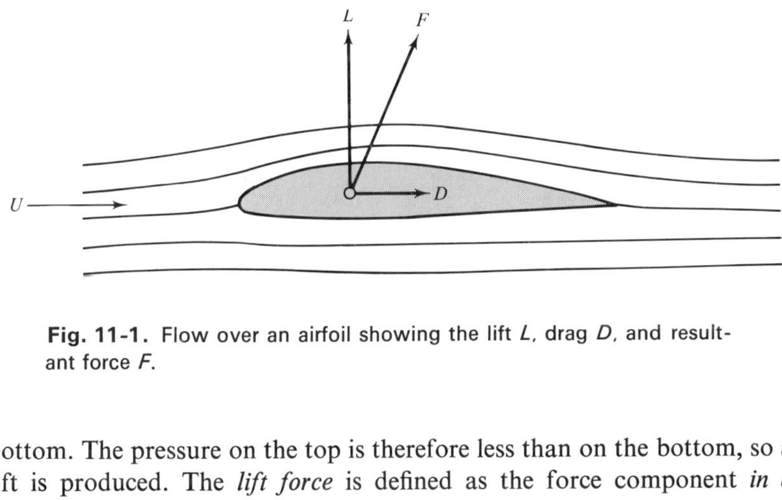

Fig. 11-1. Flow over an airfoil showing the lift L, drag D, and resultant force F.

bottom. The pressure on the top is therefore less than on the bottom, so a lift is produced. The *lift force* is defined as the force component *in a direction perpendicular to the flight* direction or relative approach velocity U. *Drag* is the force *in the direction of* U. If the airfoil were ideally designed and at a small angle of attack, the flow would be attached as indicated and the only drag force would be that due to viscous friction. The friction drag force is the same as that considered previously under laminar and turbulent friction on a flat plate. Calculation of the local shear stress or friction coefficient is complicated by the curvature of the boundary. The fact that $\partial p/\partial s \neq 0$ makes any analysis much more complicated.

In the streaming flow past relatively blunt bodies (cylinders, spheres, and so on) or airfoil shapes at moderate angles of attack where the flow is required to decelerate on the downstream side of the body, the consequences of the boundary shear layer cause the flow to separate, as illustrated in Fig. 11-2. A turbulent wake region behind the body is produced. The flow separates at points S on the body where the external velocity at the edge of the boundary layer is approximately equal to or greater than the free stream velocity. The pressure at separation (and therefore the pressure in the wake) is less than free stream pressure. The maximum pressure on the forepart of the body is the stagnation pressure.

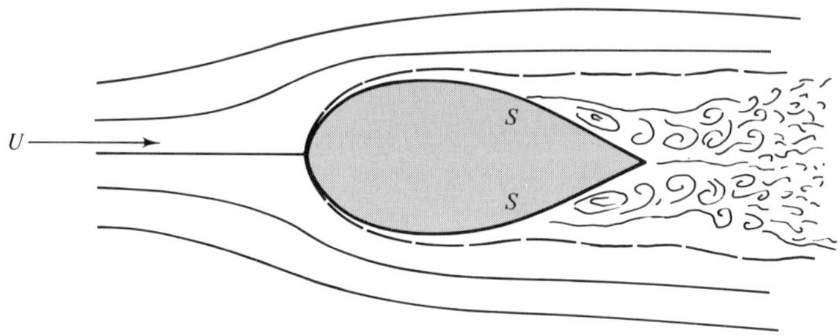

Fig. 11-2. Flow past a relatively blunt body. Separation occurs at S. The boundary layer is shown as the dashed line. The separated laminar shear flow between the separation streamline from S and the outer boundary-layer boundary rapidly becomes unstable and undergoes transition to a turbulent wake.

If the flow did not separate, the pressure at the trailing edge would be nearly stagnation pressure. However, since the flow separates at a relatively low pressure point and the mean motion in the wake is negligible, the pressure recovery to stagnation pressure is not achieved: a pressure drag that is termed a separation drag results. The total drag is then the sum of the viscous shear stress distribution and this separation or pressure drag.

11.2 FLOW ABOUT A CIRCULAR CYLINDER

The pressure drag is produced by the indirect action of viscous shear in causing the flow to separate in regions of increasing pressure. As the flow decelerates, $\partial p/\partial s$ becomes positive as $\partial v/\partial s$ becomes negative. Consider, for example, the flow around a circular cylinder. Streamlines for ideal, nonviscous flow about a cylinder are illustrated in Fig. 11-3. The local velocity is inversely proportional to the distance between the streamlines as a consequence of the continuity condition. The flow adjacent to the cylinder boundary first stagnates at a, accelerates to a maximum velocity

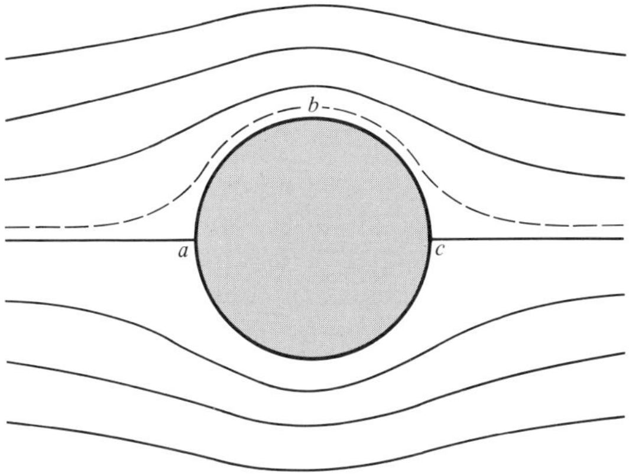

Fig. 11-3. Streamlines for ideal (nonviscous) flow around a cylinder. The stagnation points are at *a* and *c*.

at b, and decelerates to a second stagnation at c. A particle or fluid element traveling along the dashed path would slow down as it approaches a, accelerate to a maximum velocity at the top, and decelerate as it approaches c. Subsequently, the flow accelerates back to a velocity equal to the approach velocity several diameters downstream from the cylinder. As stated in Chapter 9 the boundary layer, and therefore the shear stress behavior, depends on the pressure gradient, which is a direct consequence of the flow curvature. For the simple case of the flat plate, no pressure gradient exists and the laminar boundary-layer thickness is proportional to the distance along the plate and the distance Reynolds number, $\delta/x \propto \sqrt{N_{R,x}}$. A rough approximation of the boundary-layer growth can be obtained from this expression. The length $x = a\theta$ (a is the cylinder radius); the velocity can be approximated by that determined from potential flow,

$$\text{v} = 2U_\infty \sin\theta$$

(as will be obtained in Chapter 14). The boundary-layer thickness (or the displacement or momentum thickness) can be expressed as

$$\frac{\delta}{a}\sqrt{\frac{U_\infty D}{\nu}} \propto \frac{\theta}{\sin\theta}$$

where $D = 2a$ is the cylinder diameter. This expression indicates that thickness increases with θ, or that the boundary layer continuously thickens even though the flow is accelerated for the first 90°. The above approximation and more exact solutions, and experimental measurements indicate that the laminar boundary-layer growth over the forward quarter of the cylinder is as indicated in Fig. 11-4. The normal or perpendicular distance scale for the boundary-layer distributions is greatly exaggerated with respect to the cylinder size, so $\delta \ll a$, in accordance with the necessary boundary-layer-theory approximation for the existence of a boundary-layer type of flow. As the flow starts over the back side of the cylinder (or around the back of any body), it decelerates with a corresponding pressure increase (positive pressure gradient). Immediately adjacent to the boundary, however, the velocity in the boundary layer is very small, so only a small positive pressure gradient is possible. When the transformation from streamwise directed momentum to pressure force utilizes

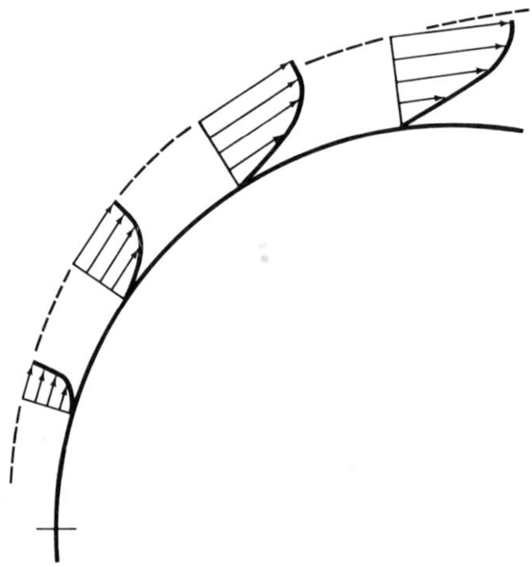

Fig. 11-4. Development of the boundary layer over the forward upper quarter of a circular cylinder or sphere.

all the available momentum, no further pressure increase is possible and the flow separates. This procedure is also affected by the momentum exchange in the transverse (normal) direction. This normal momentum transport is the same as that which produces the shear layer, but when the boundary flow slows rapidly due to the positive pressure gradient, the net transverse momentum transport is toward the wall, tending to increase the available momentum and energy of the flow in the boundary layer adjacent to the wall. For a laminar boundary layer, this momentum transport is very slow and of little effect. The streamwise momentum equation illustrates some interesting behavior. For steady flow, (6-12) gives

$$\mathrm{V}\frac{\partial \mathrm{V}}{\partial s} = -\frac{g_c}{\rho}\frac{\partial p}{\partial s} + \frac{g_c}{\rho}\frac{\partial \tau}{\partial n}$$

with the boundary-layer assumptions

$$\frac{\partial}{\partial s}\left(\frac{\mathrm{v}^2}{2} + \frac{g_c p}{\rho}\right) = \nu \frac{\partial^2 \mathrm{v}}{\partial n^2}$$

The velocity v in the boundary layer and immediately adjacent to the wall is very small, so the flow behavior can be evaluated by considering

$$\frac{g_c}{\rho}\frac{\partial p}{\partial s} \approx \nu \frac{\partial^2 \mathrm{v}}{\partial n^2}$$

In the boundary layer

$$\frac{\partial p}{\partial n} \ll \frac{\partial p}{\partial s}$$

so

$$\frac{g_c}{\rho}\frac{\partial p}{\partial s} = -U_e \frac{\partial U_e}{\partial s}$$

where U_e is the external velocity at the edge of the boundary layer. Combining the above relations gives

$$U_e \frac{\partial U_e}{\partial s} \approx -\nu \frac{\partial^2 \mathrm{v}}{\partial n^2}$$

for the streamline immediately adjacent to the wall. As the external flow U_e accelerates near the top of the cylinder, becomes maximum, and decelerates beyond this point; $\partial U_e/\partial s$ is positive, becomes zero, and then is negative. The *curvature* of the velocity profile is correspondingly negative, zero, and positive, while v is always zero at $n = 0$. These velocity distributions are illustrated in Fig. 11-5. As the flow decelerates, producing the

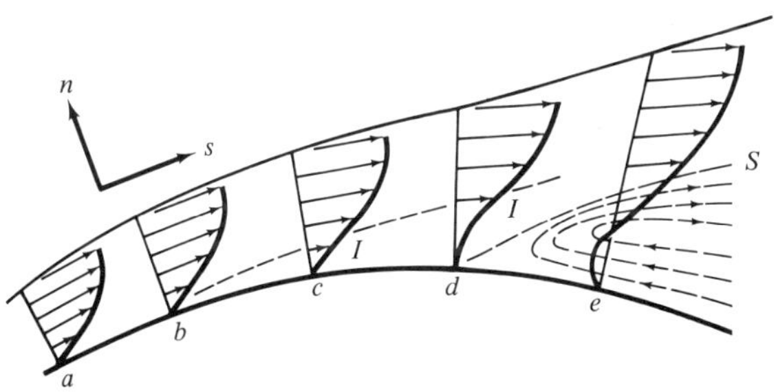

Fig. 11-5. Effect of acceleration and deceleration on boundary-layer behavior. The flow begins to decelerate at *b*, producing an inflection point *I* in the profile. The dot-dashed line is the locus of the profile inflection points. At *d* the slope $\partial \mathrm{v}/\partial n$ is zero at the wall, indicating incipient separation. In the region of continued main stream deceleration *d-e*, a backflow exists inside the separation streamline *S*. The recirculating backflow streamlines are shown dashed. The light line indicating the edge of the boundary layer is not a streamline.

inflexed distribution indicated at and beyond *c* (Fig. 11-5), the transverse (*n*-directed) molecular momentum transport (viscous action) now tends to accelerate the flow next to the wall and delay the above process so that the point of zero curvature *d* is delayed beyond the maximum external velocity point *c*. The continued deceleration of U_e (negative $\partial U_e/\partial s$) produces a backward curved profile, as indicated at *e*. The point of zero velocity and zero curvature is then defined as the *separation point* of the streamwise or *s*-directed flow. An eddying backflow is obtained beyond this point. If the development of this flow field is considered as the body accelerates from rest (or the relative approach U_∞ is increased from zero), the above description is accurate for the latter stages of flow development. Because the viscous diffusion of shear is also a time-dependent process, the separation and circulating eddy flow are initiated near the rear stagnation point. The separation point *d* then moves forward or upstream as time increases.

In the above discussion it was assumed that the boundary-layer approximations are valid; that is, the boundary layer is much thinner than the body curvature or the distance from the boundary-layer origin. The boundary-layer thickness was shown to increase with *x* or *s* and decrease with increasing U_e. The increasing or *adverse pressure gradient* also increases the displacement boundary-layer thickness as a consequence of the effect of the decreasing curvature of the velocity distribution at the wall.[1] The rapid boundary-layer thickening and the reverse flow in the separated flow region cause the separation point to move forward. For example, laminar boundary-layer separation on a circular cylinder is established at an angle of 82° from the forward stagnation point.

11.3 EFFECT OF TURBULENCE ON TRANSITION, SEPARATION, AND DRAG

A method exists whereby the zero-velocity-gradient separation condition can be delayed and the drag decreased. It was stated that as the boundary-layer velocity distribution starts to inflect, the viscous transverse momentum transfer tends to accelerate the flow below the inflection points. (See discussions of viscous shear effect in Chapters 7 and 9 and Appendix D, the discussion on the turbulent shear layer in Chapter 10, and Problem 11-2). In the turbulent boundary layer, the random momentum transport is much more effective than for laminar flow. Therefore, if the boundary layer is turbulent, the increased transverse momentum transport yields a much steeper velocity gradient at the wall. Since this gradient is initially steeper, a larger distance is required for the zero-curvature condition to develop. Also the indicated transverse momentum transport will tend to

[1] Both δ^* and θ will increase as the profile becomes more triangular.

delay the zero-curvature occurrence for a longer period as compared with laminar flow. As an example, for a turbulent boundary layer on a cylinder, the separation point will be as far back as 120° from the forward stagnation point. The corresponding decrease in drag is considerable. The details of the flow around cylinders, spheres, and other bodies are presented in Section 11.5.

11.4 FORCE COEFFICIENTS

Mathematical calculations of the combined external and boundary-layer flow field problem are quite detailed, tedious, and in most cases difficult. For example, the solution to the laminar boundary-layer flow around a cylinder has not yet been obtained. Solution to the boundary-layer problem requires the external pressure (or velocity U_e) distribution to be known or predictable. The external flow problem is a nonviscous or potential flow problem that is not difficult in itself, but the external flow (and pressure distribution) around the body plus separated region is different from that which exists for flow around the body without separation. The two problems are therefore coupled; one cannot be solved without the solution to the other, yet neither one can be obtained by itself. Using experimentally determined pressure distributions as inputs for the boundary-layer equations, the solution of these equations then predicts velocity distributions and separation point locations that agree quite well with measured values. The success achieved using this procedure therefore indicates that the approximations made in "truncating" the complete equations to obtain the boundary-layer equations are valid for more complicated shapes than the flat plate. In cases where the curvature is not as severe as that of a cylinder or sphere (for example, elliptic cylinders, ellipsoids, airfoils, and other more "streamlined" shapes) the boundary-layer analysis gives results that are in fairly good agreement with experimental results when the external potential flow is determined for the body without consideration of the effect of separation. If the body is designed so that separation will not occur,[2] the agreement between analytically and experimentally obtained characteristics is excellent.

The force (lift or drag) on a body submerged in an infinite surrounding of fluid will be a function of (1) geometry of the body, (2) its attitude or angle of attack α, and (3) the fluid properties density ρ and viscosity μ (or ν). If the geometry is characterized by a representative area (for example, the planform area of an airfoil or the frontal projected area of a

[2] For example, well-designed airfoils. A method of designing airfoil sections and comparisons of analytic characteristics with measured results are presented in I. H. Abbott, A. E. von Doenhoff, and L. S. Stivers, Jr., Summary of Airfoil Data, *NACA Report 824* (1945).

missile or automobile) and a representative length (for example, the span of an airfoil or diameter of a missile), the variables to be considered are F, U, A, l, ρ, ν, α, and g_c. A dimensional analysis on these variables allows grouping as

$$\frac{F}{(\rho U^2/2g_c)A} = C_F = \varphi\left(\frac{Ul}{\nu}, \frac{l^2}{A}, \alpha\right) \tag{11-1}$$

The first group is a dimensionless force coefficient, for example, the lift coefficient C_L or drag coefficient C_D. The dimensional analysis indicates that the force coefficient is then a function of Reynolds number, aspect ratio, and angle of attack. The aspect ratio is defined as l^2/A. For example, a two-dimensional wing of span l and chord c has an aspect ratio (AR)

$$\text{AR} \equiv \frac{l^2}{lc} = \frac{l}{c} \qquad \text{or} \qquad \frac{l}{\bar{c}}$$

($\bar{c}$ is the mean chord $= A/l$ for a three-dimensional airfoil.) For two-dimensional or cylindrical bodies (such as a cylinder or airfoil of constant diameter or chord tested in a wind tunnel with the body spanning a test section), the aspect ratio is not significant and the coefficient is expressed as

$$C_F = \frac{F/l}{(\rho U^2/2g_c)c}$$

and the results are significant with respect to the force per unit length or span of body.

11.5 FLOW AROUND SPHERES AND CYLINDERS

Because of the excessive difficulty in obtaining analytical solutions for geometries with separated flows, it has been necessary to determine the force coefficients experimentally. The cylinder and sphere are again excellent examples of basic geometries where much experimental measurement work has been done to obtain accurate results. The drag coefficient versus Reynolds number data for flow about a cylinder and sphere are presented in Fig. 11-6.

On the initial portions of the log–log plots, C_D varies inversely with Reynolds number, or the drag is directly proportional to velocity for Reynolds numbers less than 1.0. An analysis by G. G. Stokes for attached viscous flow around a sphere gives the solution

$$D = 6\pi\mu r_0 U \qquad \text{or} \qquad C_D = \frac{24}{N_R} \tag{11-2}$$

For $0 < N_R < 1$, this flow is completely attached and is not a boundary-layer flow; the viscous shear region occupies the entire flow field. As the Reynolds number increases, the inertia forces become relatively large and separation occurs. The mechanism by which separation occurs at low Reynolds numbers is similar to that described above in connection with boundary-layer separation; however, the shear layer is not thin and the

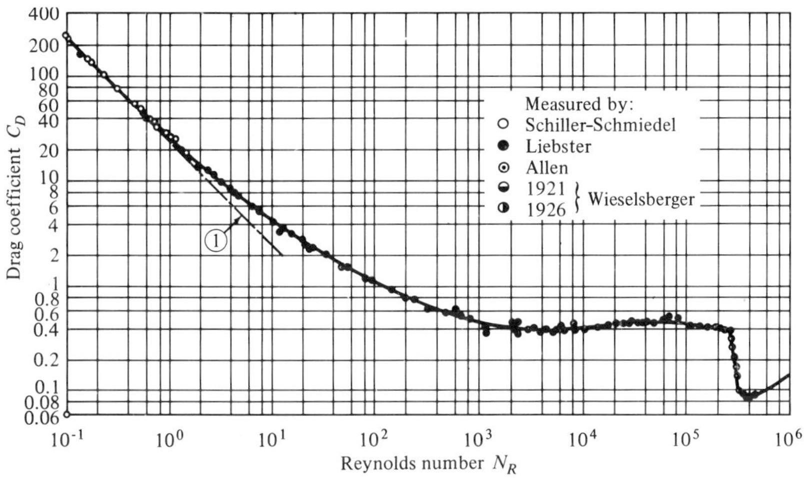

Fig. 11-6 (a). Drag coefficient as a function of Reynolds number for spheres. Curve 1 is (11-2). (Reproduced from reference 3.)

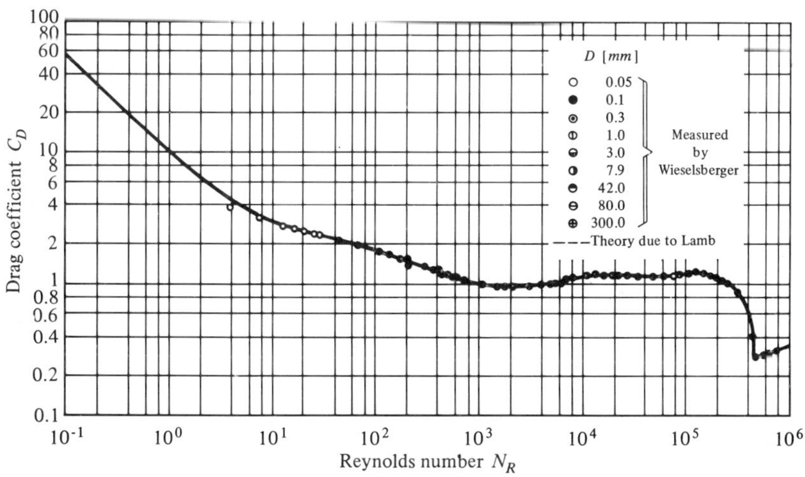

Fig. 11-6 (b). Drag coefficient as a function of Reynolds number for circular cylinders. (Reproduced from reference 3.)

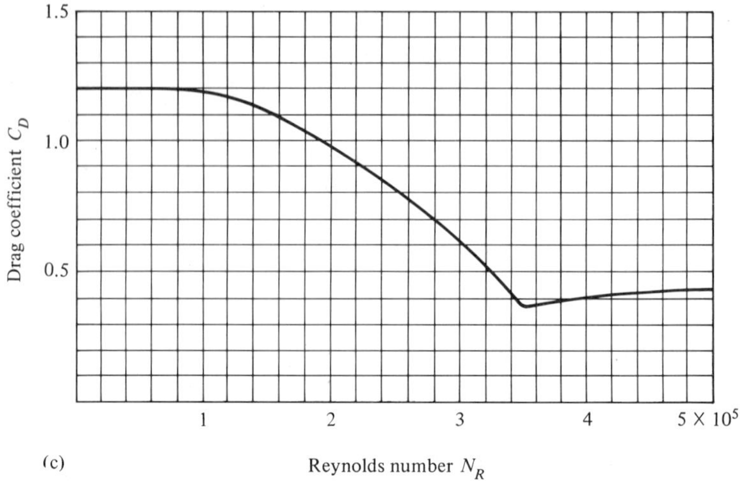

Fig. 11-6 (c). Drag coefficient as a function of Reynolds number for circular cylinders; linear scales.

flow cannot be classed as a boundary-layer flow. At values of Reynolds numbers between 1 and 60, the separated flow forms an attached vortex pair immediately behind the cylinder. This flow pattern is illustrated in Fig. 11-7. The vortex pair wake grows as the Reynolds number is increased within this range. Above $N_R \approx 50$ the separated shear flow and stagnation point in the wake region give rise to a flow instability first noticed as a waviness of the centerline downstream of the vortex pair. The wave increases in amplitude until it becomes unstable and breaks, forming the alternating, periodic vortex wake of Fig. 11-8. This wake is termed a Karman vortex street after Theodore von Karman, who first analyzed its motion.

This street phenomenon persists with nearly constant wavelength (or Strouhal number) up to Reynolds numbers of about 150. Within this Reynolds number range, each vortex of the pair breaks off alternately

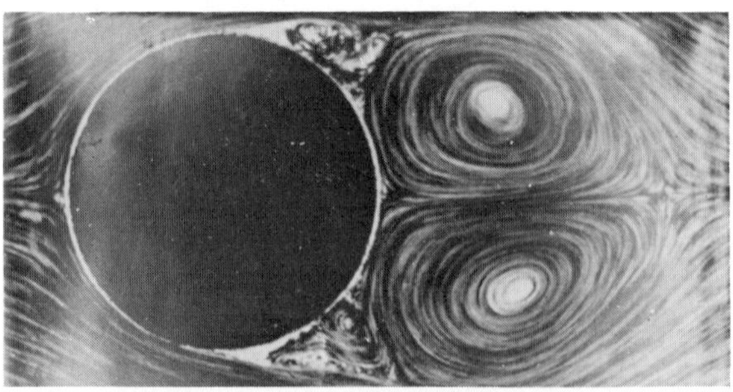

Fig. 11-7. Vortex pair behind a cylinder.

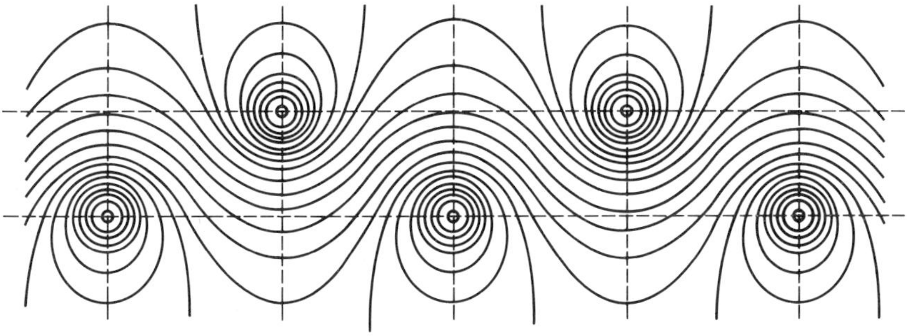

(a)

Fig. 11-8 (a). Streamlines in a vortex street. The view is relative to the moving vortices.

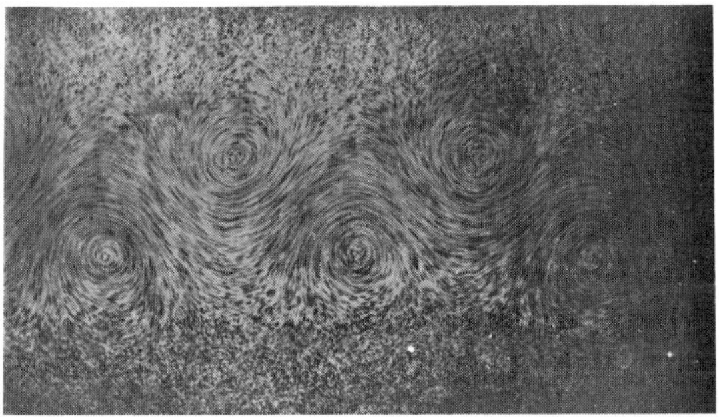

Fig. 11-8 (b). Photograph of a vortex street. The camera and viewer are moving with the vortices. (From J. G. Richards, *British Aircirculation Research Council Reports and Memoranda* 1590, 1934.)

from the cylinder to form the street. The vortex spacing in the wake is constant. The shedding frequency is then directly proportional to the stream velocity. Roshko[3] measured the characteristics of wakes behind cylinders and gives the shedding frequency in the Reynolds-number range 50 to 150 as

$$F = 0.212N_R - 4.5$$

where $F = nd^2/\nu$ and n = number of cycles or vortex pairs per second. Since the accuracy of this formula is within 1 percent, it provides a relation that allows measurement of very low velocities. A small microphone, hot wire, or other frequency-sensitive pickup can be placed in the wake of a small cylinder to determine free stream velocity as a function of shedding frequency. (See Problems 11-5 and 11-7.)

[3] A. Roshko, On the Development of Turbulent Wakes from Vortex Streets, *NACA Report 1191* (1954).

In the Reynolds-number range 150 to 300, a secondary instability develops on the street vortices at some distance downstream from the body and the frequency becomes irregular. A regular shedding frequency again exists for Reynolds numbers between 300 and about 2000, which correlates within 1 percent with the relation

$$F = 0.212N_R - 2.7$$

up to Reynolds numbers of 2000, and within 4 percent up to Reynolds numbers of 10^4. In the range $300 < N_R < 10^4$, the oscillatory wake flow results as the boundary layers separate, become unstable, and roll up into alternating vortices. For Reynolds numbers less than 300, the flow cannot be categorized as a boundary-layer flow because the shear layer is not thin with respect to the cylinder diameter. For Reynolds numbers greater than 10^4, the separated boundary layer immediately behind the cylinder shows some periodic characteristics. The increasing kinetic energy as the Reynolds number increases makes any instabilities that exist in the separated flows more unstable, so there is a rapid transition to a random, turbulent motion that produces a turbulent wake. The shear layers are primarily unstable as they separate, since an inflection point in a velocity profile is a sufficient condition for the flow to be unstable. The vortices themselves are unstable in the surrounding flow above a certain critical Reynolds number[4]; secondary separations off the vortices occur, and the flow rapidly develops into a turbulent state.

The flow around a cylinder at Reynolds numbers on the order of 10^4 can be characterized as a boundary-layer flow with a laminar boundary layer separating at ~82° from the front stagnation point as described in Section 11.2. The characteristics of the boundary layer on and separating from the cylinder do not change much in the range $10^4 < N_R < 10^5$. For $N_R > 2 \times 10^5$ the boundary layer on the cylinder begins to become unstable. This instability is evidenced by a "waviness" at the edge of the layer, as indicated in Fig. 11-9; the amplitude decreases as the body surface is approached. An increase in the Reynolds number (increase in the flow kinetic energy) will cause this instability to break down into turbulence, as discussed in Chapter 10. The point of initiation of instability point I, Fig. 11-9, and the transition point X, Fig. 11-9, depend on the Reynolds number. As the Reynolds number increases, the transition range decreases in length and more of the boundary layer becomes turbulent. The greater transverse momentum transfer increases in the turbulent length of boundary layer allows a longer length of adverse pressure gradient flow along the body. Consequently, the boundary layer is said to *reattach* and the separation point S moves back along the cylinder. As the separation point moves

[4] A critical Reynolds number is generally defined as the Reynolds number at which a change in flow behavior occurs.

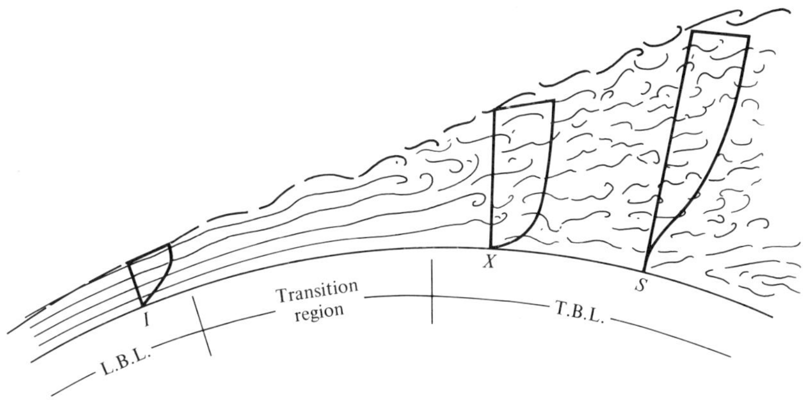

Fig. 11-9. Development of a boundary-layer instability through the transition region into turbulence on a body with a positive pressure gradient. The laminar boundary layer (L.B.L.) has an inflection point at *I*; however, before a laminar separation (Fig. 11-5) can be produced, the boundary layer becomes unstable and subsequently turbulent. The turbulent boundary layer (T.B.L.) finally separates at *S*.

aft of the meridian, a pressure recovery over the back side is obtained and the drag coefficient decreases. As the Reynolds number increases in the range 10^5 to $\sim 5 \times 10^5$ the transition region and transition point move forward, producing a longer distance of turbulent boundary layer, which further delays separation, giving a decreased C_D. This mechanism of drag reduction for the cylinder is so effective that, not only does the drag coefficient decrease, but the *drag force* decreases in this critical range (see Problem 11-8).

At $N_R \sim 5 \times 10^5$ the boundary layer is fully developed; that is, the transition range has moved forward to a stable position so that a further increase in Reynolds number has little effect on the lengths of transition and turbulent boundary layer or on the flow field configuration. A slowly increasing drag coefficient with increasing N_R for Reynolds numbers greater than this upper critical value of 5×10^5 is due primarily to the increase of shear stress as the wall velocity gradient (and therefore τ_0) increases.

The location of the instability, transition, and separation points are not fixed but oscillate both tangentially and axially along a cylindrical body. In the upper Reynolds number, vortex street separation, this oscillation is periodic and traverses a relatively large arc on the surface. This separation is asymmetric (like the vortex street separation), so the corresponding drag force variation is not too large. On the other hand, a relatively large periodic lift force is produced. If the Reynolds number is such that the vortex shedding frequency is at the natural frequency of the cylinder (or a near harmonic), as often occurs in wires and structures (for example, smokestacks), an enforced critical oscillation can be set up. An alternating vortex pattern can be shed from many relatively blunt geometries. Such a resonant condition occurred with the Tacoma Narrows Bridge in the State of Washington. The bridge bucked and twisted until it collapsed.

The above discussion of the flow around a cylinder is also applicable to most blunt bodies where boundary-layer separation exists. It is also applicable to more streamlined shapes such as struts and airfoils. On the more streamlined shapes the oscillating street-type separation does not occur on the body because of the interference of the body extension behind the separation point. However, oscillating street wakes have been observed downstream of airfoils and thin flat plates in certain Reynolds-number

TABLE 11-1

DRAG COEFFICIENTS FOR VARIOUS BODY SHAPES FOR $N_R > 10^4$

Body	C_D
Disc perpendicular to flow	1.12
Tandem discs	
Spacing/diameter = 1	0.93
2	1.04
3	1.54
Plate perpendicular to flow	
Length/width = 1	1.16
5	1.20
20	1.50
∞	1.95
Circular cylinder	
Axis parallel to flow l/d = 1	0.91
2	0.85
4	0.87
7	0.99
Cube	1.0
Hollow hemisphere	
	1.33
	0.34
Cylindrical shapes (all $l/b \sim \infty$)	
	2.0
b	1.72
(60°)	2.2
	1.39
(30°)	1.8
	1.0
	2.3
	1.12
	2.0
	1.6

ranges. If the body shape has a natural transition point, such as a sharp break or corner, the drag coefficient will be essentially independent of Reynolds number over a large range above the critical Reynolds number, where the flow becomes detached at the corner. Examples of drag coefficient values for a number of shapes are presented in Table 11-1.

11.6 EFFECTS OF FREE STREAM TURBULENCE AND SURFACE CONDITION

If there is some turbulence in the approach flow (as is the case in most wind tunnels), instability and transition are excited or tripped sooner or at a lower Reynolds number. A rough surface will also tend to trip the instability and transition. It is also possible to produce transition to a turbulent boundary layer with a single trip device such as a wire along the surface of a cylinder or around a sphere ahead of the maximum diameter. These effects are indicated in Fig. 11-10. Because of the indicated effect of stream turbulence on the C_D of spheres, the N_R for which $C_D = 0.30$ on a smooth sphere has been used as an indication of wind-tunnel turbulence. A turbulence factor was defined as the ratio of the N_R at which $C_D = 0.30$ was obtained in the tunnel to the maximum N_R obtainable in free flight at which $C_D = 0.30$, nominally $N_R = 385{,}000$.

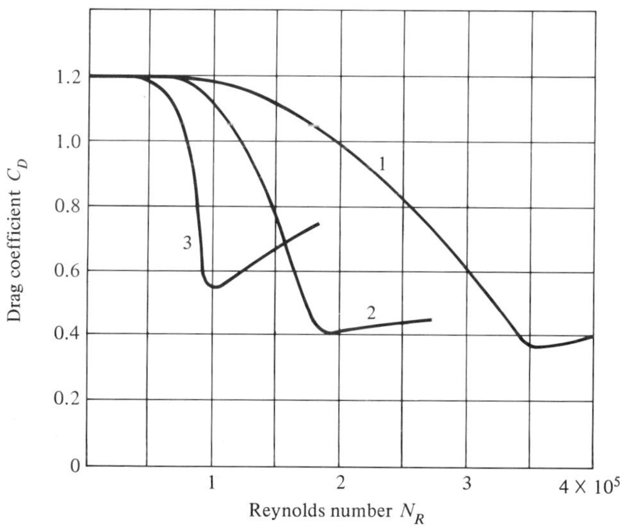

Fig. 11-10. Drag coefficient as a function of Reynolds number and roughness for circular cylinders. Curve 1, smooth cylinder; curves 2 and 3, increasing roughness. Note that C_D for cylinder 3 at $N_R = 10^5$ is less than half the C_D for the smooth cylinder.

11.7 LIFT AND DRAG ON STREAMLINE SHAPES

If a body is excessively blunt (for example, a circular cylinder or sphere), flow separation will produce a large form or profile drag. If it is excessively slender, the skin friction drag will be large. Figure 11-11 shows the drag coefficient based on thickness as a function of slenderness ratio (t/c) for a strut profile. This comparison of drag for struts of the same thickness indicates a minimum drag at $t/c = 0.27$ or $c/t = 3.7$. If the data are plotted as C_D based on chord ($C_{D,c}$) the dashed curve is obtained. It is interesting that this curve fairs into the value for a flat plate at the Reynolds number for which the strut data were taken, $N_R = 4 \times 10^5$. The magnitudes of C_D, the shape of the curve, and the t/c for minimum C_D depend on body shape, N_R, and surface condition, as discussed above.

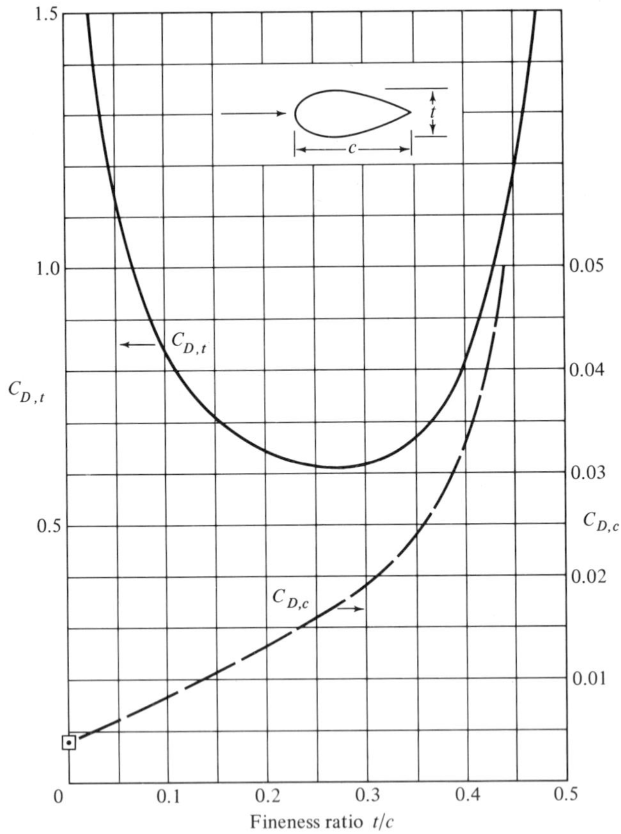

Fig. 11-11. Experimental data for airfoil strut shapes at a Reynolds number $Uc/\nu = 4 \times 10^5$. The optimum fineness ratio is 0.27, or the optimum c/t ratio is 3.7. The data approach the drag coefficient for laminar flow over a flat plate as $t/c \to 0$.

For the same profile shape, the C_D based on frontal projected area for bodies of revolution is less than that for the two-dimensional (cylindrical) body based on the same area in the Reynolds-number region where separation occurs. The primary reason for this behavior is that the separated area is relatively smaller for a body of revolution. The separation condition is not as severe, since the pressure on the three-dimensional body is less than that on the two-dimensional body in the critical or separation region. For example, the *ideal* minimum pressure coefficient on a cylinder is -3, corresponding to a maximum velocity at the top and bottom of $u = 2U_\infty$. On a sphere $c_{p,\min} = -1.25$ at $u = \frac{3}{2}U_\infty$. The pressure gradients are therefore less severe on a body of revolution.

The pressure distributions on a cylinder and a sphere are presented in Fig. 11-12. The relation between C_D and c_p as functions of N_R and boundary-layer behavior can be deduced from Figs. 11-6 and 11-11. At $N_R = 4.25 \times 10^5$, separation of the turbulent boundary layer from the sphere is at $\theta = 140$ to $145°$.

On more streamlined bodies, separation may not occur. The objective of good airfoil design is to obtain a flow and pressure distribution that does not produce separation. The only drag then is skin friction. Since it has been shown that the wall shear stress for laminar flow is much less than that for turbulent flow, it is desirable to maintain laminar flow over as much of the surface as possible. Laminar flow will be maintained as long as the pressure gradient is "favorable" (decreasing, or with an accelerating external velocity), which provides a continuing stabilizing type of condition in the boundary layer. Since the airfoil or body shape must be closed

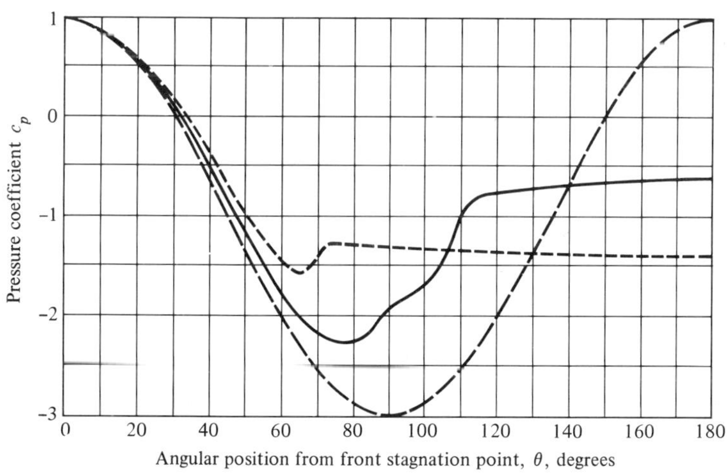

Fig. 11-12 (a). Pressure distribution for a circular cylinder. Solid curve, $N_R = 2.1 \times 10^5$; dashed curve, $N_R = 7 \times 10^4$; dotted curve, potential flow.

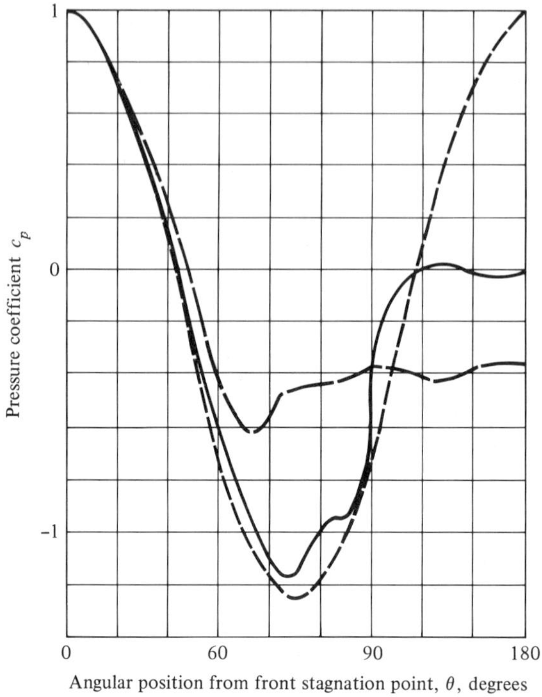

Fig. 11-12 (b). Pressure distribution for a sphere. Solid curve, N_R = 3×10^5; dashed curve, $N_R = 1.6 \times 10^5$; dotted curve, potential flow.

at the aft end, a length of "adverse" pressure gradient flow is unavoidable. It is therefore desirable to maintain as long a length of laminar boundary layer as possible followed by transition to a short turbulent boundary layer which will aid in delaying separation.

It is interesting to examine the flow about an elliptical cylinder with these objectives in mind. The velocity and pressure distributions for 20 percent elliptical sections are illustrated in Fig. 11-13. (The v/U and c_p data for an ellipse are easily calculated using the methods of Chapter 14.) The points marked S are the calculated laminar boundary-layer separation points for Reynolds numbers of the order of 10^4. For sufficiently large Reynolds numbers, the boundary layer would undergo transition to a turbulent boundary layer at a position aft of the maximum thickness point as determined by the Reynolds number. (For a discussion of transition as affected by shape and N_R, see chaps. XVI and XVII of reference 3 at the end of the chapter.) In Fig. 11-13c a representative low-drag airfoil shape has been constructed using a 20 percent ellipse for the top forward half of the airfoil and a 5 percent ellipse for the lower forward half. The afterpart has been faired off to produce transition to a uniformly increasing pressure gradient,

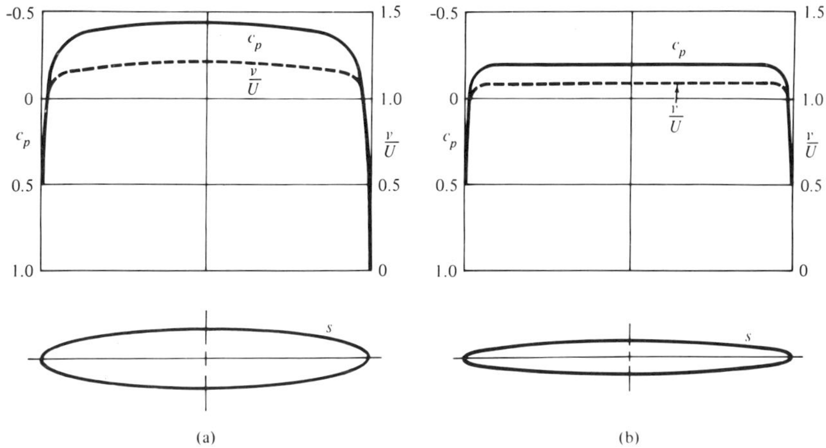

Fig. 11-13 (a) and (b). Ideal flow velocity and pressure distributions for 20 percent thick (a) and 10 percent thick (b) elliptical cylinders. *S* indicates the calculated laminar boundary-layer separation points.

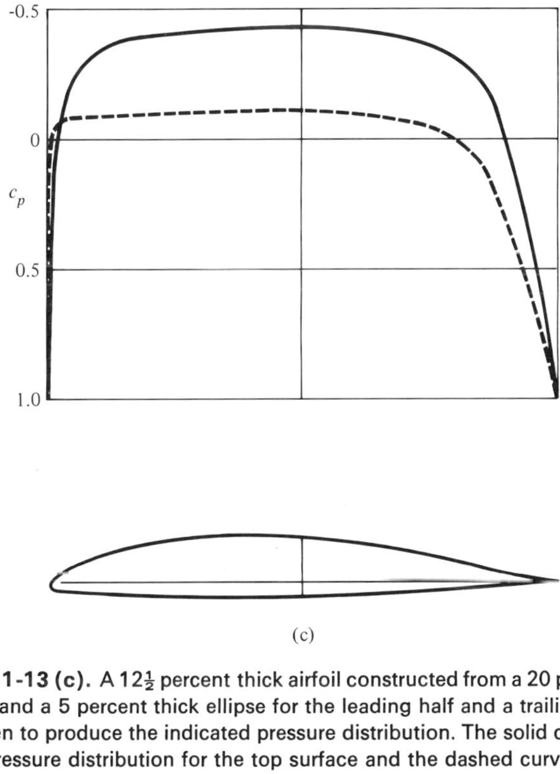

Fig. 11-13 (c). A $12\frac{1}{2}$ percent thick airfoil constructed from a 20 percent thick and a 5 percent thick ellipse for the leading half and a trailing half chosen to produce the indicated pressure distribution. The solid curve is the pressure distribution for the top surface and the dashed curve is for the lower surface.

as shown on the c_p curve. The resulting airfoil is 12½ thick with a design $C_L = 0.47$ at minimum drag.

The flat-plate friction coefficient data presented in Fig. 10-26 can be used to approximate the C_D for a slender body such as an airfoil in nonseparating flow. For an airfoil section such as illustrated in Fig. 11-13c, it can be assumed that a laminar boundary layer exists on the front half (or portion up to the point of maximum thickness). Half of the C_D can then be estimated taking the C_D for an equivalent surface of flat plate at the flight Reynolds number based on flight speed and airfoil chord. At large Reynolds numbers the boundary layer on the rear half will be mostly turbulent; the C_D for a turbulent plate at the flight Reynolds number can be used to estimate the other half of the C_D. The overall C_D is then nominally the sum of the laminar and turbulent flat plate C_D's at the flight N_R. Obviously, this procedure is only an approximation; for example, the laminar boundary layer τ_0 on the airfoil will be greater than that on a flat plate because the velocity is greater than the free stream velocity. A small amount of separation will also have a large effect on the results.

Drag coefficient data for a number of airfoils are given in Fig. 11-14. The above approximate procedure for estimating C_D is seen to be qualitatively satisfactory for most of the "laminar airfoils." Note that the airfoil C_D is based on chord length, while τ_0 of the airfoil depends on the actual surface area. The surface area of a 20 percent thick air foil is about 25

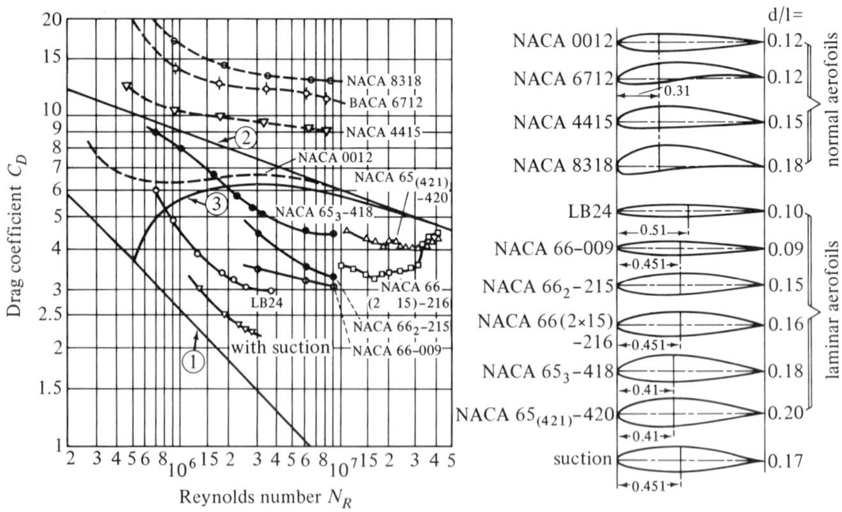

Fig. 11-14. Drag coefficients for a number of airfoils. C_D curves 1, 2, and 3 are for a flat plate at zero incidence in laminar flow (1), fully developed turbulent flow (2), and a transition condition (3). (Reproduced from reference 3.)

percent greater than the chord-plane area. Low-turbulence wind-tunnel tests of laminar airfoil model wing sections give $L/D\,(\equiv C_L/C_D)$ coefficients of ~ 150 at design Reynolds numbers of $\sim 10^7$. The lift of a three-dimensional wing is less than that of a two-dimensional section because of end effects. The bottom to top pressure difference causes a spanwise flow that produces trailing vortices from the wing tips and decreases the lift coefficient. This resulting flow pattern also adds an "induced" drag effect. Consequently, the maximum L/D obtainable on a production aircraft (including fuselage, empenage, and engine interference effects) is ~ 20. For a discussion of airfoil characteristics, see reference 5 at the end of the chapter.

EFFECT OF ROUGHNESS

The boundary-layer flow is dependent on Reynolds number with respect to determining effects of transition and separation. In connection with the discussion of flow about a cylinder, it was stated that roughness and small protuberances are also important on geometries where transition is possible. Roughness can produce either favorable or adverse conditions, depending on the Reynolds number. If the flow on a laminar-flow airfoil is at a Reynolds number much less than the design value, transition may not take place and the laminar boundary layer will separate yielding a C_D much larger than the design value. In this case, a certain amount of roughness may produce transition and reattachment of the boundary layer with a noticeable decrease in C_D. On the other hand, at the design N_R, the same roughness would produce an increase in C_D (that is, an increase in τ_0) over that for the smooth surface. If the boundary layer is thin enough, the roughness acts as additional bodies about which there is an additional drag component. A small step such as a plate overlap may itself separate the boundary layer. One example is a knuckle ball or spitball. The flow separates off the baseball seams as it travels with only a slight rotation. The seam pattern then produces separation such that the ball actually "dances" in its flight.

11.8 BOUNDARY-LAYER CONTROL

The use of roughness is one method of controlling boundary-layer behavior. Another effective method is boundary-layer suction. If the low-momentum boundary layer can be removed in regions of positive pressure gradient, the flow will remain attached and laminar. Continuous suction can be applied by the use of porous surfaces, or continual suction is possible with a series of transverse slots. The data for a laminar boundary-layer profile with suction slots are shown in Fig. 11-14. Aircraft using suction slots have shown a 30 percent net increase in performance (range or payload) where suction power and weight of equipment is included in the overall evaluation. The equipment and slot construction involve significant increases in initial cost, however. Other methods, such as moving skin surfaces and utilization of the Magnus effect, have been tried but are impractical because of mechanical complications.

11.9 SUMMARY

1. Because of the complexity of boundary-layer flows around nonsimple body shapes, force-coefficient data as a function of N_R, angle of attack, and geometry are still semi-empirical or empirical. This statement is particularly true for separating flows and composite body shapes where interference and secondary flows occur.
2. When giving and using force coefficient data, it is necessary to be particularly careful of the reference dimension or area on which the coefficient is based.

REFERENCES

1. S. Goldstein (ed.), *Modern Developments in Fluid Dynamics*, vols. I and II, Oxford University Press, New York, 1938.
   A very comprehensive collection of interestingly presented theoretical and experimental material with correlations, discussions, and evaluations of results.
2. L. Prandtl and O. Tietjens, *Fundamentals of Hydro- and Aeromechanics, and Applied Hydro- and Aeromechanics.*, McGraw-Hill, Inc., New York, 1934 (also Dover Publications, Inc., New York).
   A transcription of Prandtl's lectures with much supplemental material.
3. H. Schlichting, *Boundary Layer Theory*, McGraw-Hill, Inc., New York, 6th ed., 1968.
   A very broad coverage of boundary-layer theory correlated with experimental results.
4. A. Roshko, On the Development of Turbulent Wakes from Vortex Streets, *NACA Report 1191*, 1954.
5. A. M. Kuethe and J. D. Schetzer, *Foundations of Aerodynamics*, John Wiley & Sons, Inc., New York, 2nd ed., 1959.

PROBLEMS

11-1 Plot the dimensionless boundary-layer thickness given by the relation

$$\frac{\delta}{a}\sqrt{N_R} = \frac{\theta}{\sin\theta}$$

as developed in Section 11.2.

11-2 Consider the inflected velocity distributions c and d in Fig. 11-5 and discuss the transverse laminar and turbulent momentum transport and its effects on each side of the inflection point. Show that below I the transverse momentum transport tends to maintain a positive velocity while the streamwise force and momentum action are decelerating the flow.

11-3 Perform the indicated dimensional analysis to obtain the dimensionless groups in (11-1).

11-4 Stokes flow. Calculate the largest size of sphere that will fall under its own weight with completely viscous (or Stokes) flow in a fluid of viscosity ν. If $\nu = 0.02$ ft^2/sec, calculate the radius of the sphere. What is the difference if the sphere material is steel or plastic?

11-5 A cylinder of what diameter must be used to make air-speed measurements in the speed range of 0.5 to 1 ft/sec to be determined from Karman vortex street shedding frequencies?

11-6 (a) Calculate the manometer height h for a Pitot-tube measurement of a free stream velocity of 1 in./sec and comment on the result. (b) Determine the cylinder diameter from which this velocity can be calculated from measurements of shedding frequency in the Karman vortex street range.

11-7 It is desired to measure free-stream air speeds (at atmospheric pressure and temperature) on the order of 1 cm/sec. (a) Determine the manometer Δh on an alcohol (sp. gr. $= 0.8$) manometer from a Pitot tube and comment on the result. (b) Determine the cylinder diameter from which this velocity can be calculated from measurements of shedding frequency in the Karman vortex street region.

11-8 From the data given in Fig. 11-6c, plot drag force versus velocity for a cylinder of 2 in. radius in water for the Reynolds number range 10^4 to 3.5×10^5.

11-9 Although the boundary layer is indicated to thicken rapidly in the turbulent range (Fig. 11-9), the flow is not displaced outward as rapidly in the turbulent boundary layer as in the laminar boundary layer. To illustrate this fact, calculate the ratio $\delta^*_{\text{turb}}/\delta^*_{\text{lam}}$ for a $\frac{1}{7}$ power-law turbulent profile.

11-10 What is the critical Reynolds number for flow about a cylinder or sphere? Is there more than one critical Reynolds number? If there is, describe the "critical condition" for each.

11-11 Discuss the possibility of the formation of a vortex street behind a sphere.

11-12 Considering the geometry of a current automobile, refer to Table 11-1 and estimate an applicable drag coefficient based on frontal area. Using this C_D, calculate the drag horsepower required at speeds of 30, 50, 70, and 90 mph.

11-13 Show that the drag coefficient of a flat plate at $N_R = 4 \times 10^5$ does give the limiting point at $t = 0$ for the strut data of Fig. 11-11.

11-14 Discuss the drag coefficient to be expected at low Reynolds numbers (less than 10^5) for the airfoils of Figs. 11-13 and 11-14. Note that the coefficient increases rapidly with decreasing Reynolds numbers for the airfoils tested at Reynolds numbers less than 10^6. What is responsible for this behavior?

11-15 In knocking golf balls around the rocky moors, the Scots found that after a new smooth ball was quite nicked up, it could be driven almost twice as far as when it was new. Thereafter, golf balls were made with dimples. Calculate the N_R for a 1½-in.-diameter sphere in atmospheric air at speeds between 100 and 200 ft/sec and discuss the effect of roughing or dimpling the ball.

11-16 Estimate the C_D for the elliptical foil of Fig. 11-13 at $N_R = 5 \times 10^6$. The design coefficient for this airfoil was specified as $C_L = 0.47$. What is the design L/D ratio for this airfoil? What would be the cruise horsepower required for a 100,000-lb airplane if the wing loading is 40 lb_f/ft^2?

11-17 Discuss the possibility of polishing the surface of a sailboat hull to reduce the skin friction. It has been shown that rubbing a polished sailboat hull with colloidal graphite will give a slightly faster boat. How can this fact be explained?

11-18 The relation for c_p on a sphere in ideal, nonviscous or "potential" flow is $c_p = 1 - \frac{9}{4} \sin^2 \varphi$, where φ is the polar angle. Assume that the flow about a sphere corresponds to the potential flow for $0 \le \varphi \le 145°$ and that the pressure in the separated wake corresponds to the pressure obtained from the above expression for $\varphi = 145°$. Integrate this pressure distribution over the sphere to obtain an approximation to the drag coefficient. Check this result with the drag coefficient data of Fig. 11-6.

11-19 The c_p for a cylinder in ideal, nonviscous *potential* flow is $c_p = 1 - 4 \sin^2 \theta$. Assume that the flow about a circular cylinder corresponds to potential flow up to the separation point at 120° (turbulent boundary-layer separation at minimum C_D) and that the wake pressure is constant and the same as that at the separation point, and calculate the C_D for the cylinder, and compare it with the data of Fig. 11-6. Evaluate, compare, and discuss this approximation with respect to the actual flow around the cylinder.

11-20 Formula 1 and Indianapolis 500 racing cars have a frontal body area of ~6.8 ft². In addition, the frontal area of the front tires is ~160 in.² and the rear wheels ~350 in.² each. Estimate a C_D for these cars and calculate the maximum speed for horsepowers of 200, 400, 600, and 800. Discuss the lift and drag effects of the aerodynamics of the wheels and body.

11-21 A fast baseball leaves the pitcher's hand at ~120 ft/sec. The distance between the pitcher's rubber and home plate is ~62 ft. Calculate the approximate speed with which the ball crosses the plate. (Neglect gravity effects.)

11-22 Discuss the aerodynamics of a curve-pitched baseball, a knuckle ball, and a spitball.

11-23 For an airfoil of finite length, the C_L is given by the relation $C_L = C_{L\infty}(1 - 1/\pi AR)$, where $C_{L\infty}$ is the section (or infinite AR C_L). For low-speed ($M < 0.3$) flight, $C_{L\infty} = 2\pi\alpha$, where α is the angle of attack. Estimate the gross weight and the maximum and cruising speed of a small, light airplane with a 35-ft span, $AR = 6.5$ wing, and a 75-hp engine.

11-24 For Mach numbers from 0 to 0.3, the compressibility effects are negligible. For $0.3 < M < \sim 0.6$, the C_L and C_D can be approximated from incompressible values by $C_F = C_{F0}/\sqrt{1 - M_\infty^2}$, where C_{F0} is the force coefficient determined for incompressible flow (at $M \approx 0$) and M_∞ is the flight Mach number. Calculate the angle of attack α for takeoff and climb at 130 mph at sea level and the cruise α at 600 mph at 20,000-ft altitude for an airplane with a wing loading of 40 lb_f/ft^2 and $AR = 6$. (Refer also to Problem 11-23.)

11-25 If the cruise condition on a four-jet engine transport aircraft corresponds to a thrust of 8000 lb_f/per engine, estimate the maximum cruise load.

PART IV SPECIAL TOPICS

The flow equations in complete form are not generally solvable for internal and external flow fields. Approximations must be made by excluding some effects (dropping terms in the equations) and/or by approximate methods of solutions. Some of the specific approximation flow classes are presented in this part.

Usable information is obtained by considering particular types of flow. These particular solutions may be used for partial flow field solutions, or their use may be extended with the aid of small perturbation analyses, and so on. The flow field around a body may be assumed to be frictionless and irrotational to yield an idealized, potential flow solution from which a pressure distribution and lift force may be obtained. This potential flow can then be used in calculating the boundary-layer field yielding the skin friction and drag force on the body. The incompressible potential flow solutions can also be extended to low-Mach-number compressible flows around certain restricted ("slender") geometries.

Open-channel flow is important in water-transportation systems and general hydraulics. The primary resistance of ships is produced by the wave system generated. Analysis of these wave systems is quite complicated.

Some problems encountered in unsteady flows are treated in Chapter 15. Methods for the design of pulsation dampers are presented.

Chapter 16 introduces the predominate electromagnetic fluid-dynamic effects in order to present an understanding of the fundamentals behind electromagnetic flowmeters, generators, and thrustors.

CHAPTER 12

CONTENTS

COMPRESSIBLE FLOW

It was previously mentioned that flow fields for fluids of relatively large compressibility may be considered *incompressible* if the dynamic or inertia forces are not large. As an example, for air flow in a field where the speed does not exceed 300 ft/sec, the maximum density variation will not exceed $3\frac{1}{2}$ percent (at 100 ft/sec, the maximum $\Delta\rho/\rho$ is about 0.4). Consequently, the error in force and other determinations caused by neglecting compressibility effects is small. The important parameter in determining whether compressibility effects are significant is the ratio of the dynamic or inertia force (or stress) to the compressibility force (or stress). The dynamic force is a direct consequence of the accelerating force (pressure) that produces it. Since the compressibility effect is also a consequence of the normal stress (pressure), the two effects are directly connected. Considering only dynamic effects the momentum relation can be written

$$(pA)_2 - (pA)_1 = \frac{\dot{m}}{g_c}(\mathrm{v}_2 - \mathrm{v}_1) = \frac{1}{g_c}(\rho A\mathrm{v}^2)_2 - (\rho A\mathrm{v}^2)_1$$

or

$$\Delta(pA) = \Delta\left(\frac{\rho A\mathrm{v}^2}{g_c}\right) \equiv \Delta F_{\mathrm{dyn}}$$

The bulk modulus is a parameter relating the volume (or density) change to the applied pressure change:

$$E_B = -\frac{\Delta p}{\Delta V/V} = \frac{\Delta p}{\Delta\rho/\rho}$$

The ratio of dynamic to elastic ΔF is then

$$\frac{\rho A\mathrm{v}^2/g_c}{\rho\,\Delta p/\Delta\rho A} = \frac{\mathrm{v}^2}{g_c(\Delta p/\Delta\rho)} = \frac{\mathrm{v}^2}{g_c(\partial p/\partial\rho)}$$

The *elastic* deformations are produced by normal (conservative) stresses, so the elastic compression process is isentropic or this significant ratio is defined as the square of the Mach number,

$$M^2 = \frac{\text{v}^2}{g_c(\partial p/\partial \rho)_s} \tag{12-1}$$

The quantity $g_c(\partial p/\partial \rho)_s$ has the dimension of velocity (or speed) squared. It will be shown that this characteristic speed is the speed of a small intensity (isentropic) wave propagating through the medium: $a^2 = g_c(\partial p/\partial \rho)_s$, so (12-1) can be written $M = \text{v}/a$. Since the pressure (or Δp) involved with the dynamic force is the same that produces the elastic deformations, the Mach number has a quantitative significance.[1] The *critical* Mach number is then expected to occur when the two effects are equal, or $M_{\text{cr}} = 1$ when the local velocity equals the local elastic wave speed.

Analysis of compressible flow involves consideration of an additional variable, the density ρ, and is therefore more complicated than the incompressible flow cases considered previously. In addition, when the flow speed exceeds the wave speed (supersonic flow), the flow characteristics become entirely different. Wave-propagation phenomena are then predominant and analysis is by means of wave mechanics. The treatment of nonviscous, compressible flow phenomena (and in particular supersonic flow) is therefore a subject requiring a completely different type of mathematical analysis and evaluation. The addition of another variable makes an intuitive evaluation more difficult. For incompressible flow, the continuity equation implies that the velocity decreases as the area increases. Bernoulli's equation then yields the information that the velocity will increase with decreasing pressure. However, when the density becomes an additional variable, the overall behavior is not obvious, since ρ is now involved in both the momentum (or energy) equation and the continuity equation, so at least two simultaneous, dependent principles govern the flow behavior (the two equations and their implications are *coupled* with two variables).

Because of additional mathematical complications caused by considering viscous effects, only nonviscous flows will be considered in this chapter. In the light of Chapter 11, it does not appear that any valuable results could be obtained with this assumption. The high-speed flow about a sphere produces separation and a flow field much different from the ideal flow pattern. On the other hand, high-speed nonseparating flow with a thin boundary layer can be closely approximated by ideal flow, from which the pressure field can be calculated. The results of nonviscous (isentropic) flow analyses are then valuable in making approximate evaluations of and understanding compressible flow.

[1] This behavior is unlike that of the Reynolds number, the absolute significance of which depends on an arbitrary geometric dimension.

12.1 COMPRESSIBILITY EFFECTS

It was implied above that the compressibility effect can be interpreted as a consequence of the relation between the wave speed and the local flow speed. A body traveling through a fluid produces a disturbance in the fluid (for example, water surface waves). A disturbance is essentially a pressure signal and so is propagated into the fluid with the wave speed a. If the flow past a body at rest is considered (for example, the ideal flow past a sphere similar to the ideal cylinder flow illustrated in Fig. 11-3), the velocity disturbance varies as $1/r^2$ from the center of the cylinder and therefore it becomes zero at $r = \infty$ and also propagates into the entire flow field. If the maximum speed attained on the body is much less than the sonic speed a, this signal will travel to all parts of the field before the body moves very far. If an incompressible fluid is hypothesized ($\Delta\rho = 0$), this propagation speed is infinite. The disturbance produced by the sphere is then felt instantaneously throughout the entire flow field.

If the body is traveling at a speed about equal to but less than the disturbance propagation (wave) speed, the disturbance travels into the fluid at a speed depending on the wave speed and the body velocity. Figure 12-1a shows a small body traveling at a velocity $U = a/2$. The pattern represents the signal-wave propagation at time t_6 when the body has traveled from position 0 to 6. The circles represent the positions of disturbance wave signals that occurred at each of the marked points. The disturbance that occurred at $t = t_0$ (when the body was at O) would have radiated by the process of sound-wave propagation out to the radius of $6a\ \delta t$, since a time $6\ \delta t$ has elapsed. The signal from position 5 propagated to a distance $a\ \delta t$. If the waves represent a sound signal of frequency f being radiated by the body, an observer at A would pick up twice this frequency and an observer at B would pick up one half of this frequency. This effect of the relative motion on received frequency is termed the *Doppler effect*. The speed of the body can be calculated by recording the frequency as the body approaches and passes. (See Problems 12-1 and 12-2).

Figure 12-1b illustrates the pattern for a body traveling at two times the speed of sound. The radiated pattern is seen to be a cone with a half-angle μ equal to

$$\mu - \sin^{-1}\frac{a}{U} = \sin^{-1}\frac{1}{M}$$

The angle μ is then termed the *Mach angle*. The above discussion applies only to weak disturbances ($\Delta p/p_{\text{amb}} \ll 1$) which are considered to be propagated without much dissipation in the propagating process. Large "blast" or shock waves propagate at larger speeds, as will be determined later. The wave pattern from an airplane traveling at supersonic speed is

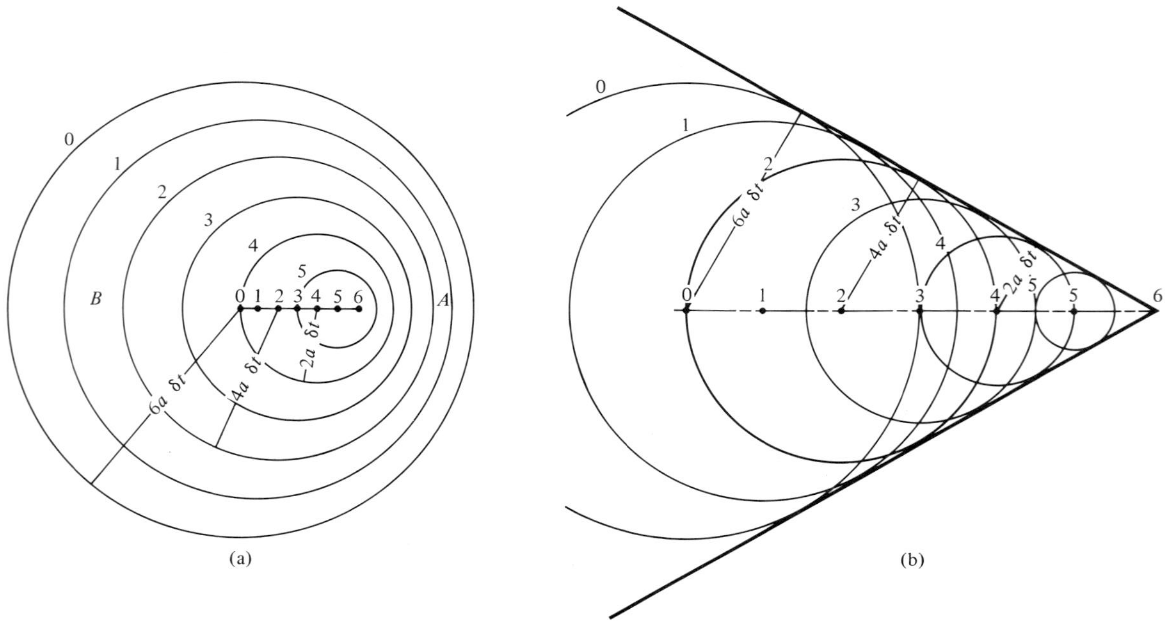

Fig. 12-1. (a) Signal pattern for a subsonic disturbance. The numbered waves indicated can be considered as "beeps" emitted when the source was at the correspondingly numbered position. The radius of each circle is $na\ \delta t$, where n is the number of the point at which the signal was emitted. The entire pattern is shown for the source at point 6. The first signal for the pattern was emitted at point 0. For the pattern shown, $U = a/2$ or $M = 0.5$. (b) Signal pattern for a supersonic disturbance where $U = 2a$ or $M = 2$. The entire signal pattern is confined within a cone of semivertex angle $\mu = \sin^{-1}(1/M)$.

also not isentropic in the neighborhood of the plane; however, at a few plane lengths away, the disturbance intensity has decreased to a level where it propagates at very nearly the speed of sound.

The same pattern would be obtained if the body were considered to be stationary in a uniform flow field from right to left at the speed U. The signals would then be simultaneously propagated radially outward with the speed a and washed downstream by the velocity field $-U$.

In the following developments, the thermodynamic energy quantities (enthalpy, internal energy, and so on) will be understood to be in mechanical units (for example, ft-lb_f/lb_m in the English system and joules in the mks system); therefore, it will not be necessary to carry in the equations the constant for converting heat units to mechanical units.

12.2 WAVE-PROPAGATION SPEED

A sound disturbance propagating in a medium is a pressure disturbance. Consider a small pressure disturbance traveling at a speed v_w in a medium as indicated in Fig. 12-2a. Since it is a disturbance, changes in pressure, density, and velocity will occur across the wave. In previous chapters a fixed control volume has been considered with steady flow therein. If a control volume is sketched in the fluid, the flow therein will be unsteady. However, a steady flow frame of reference can be obtained by superimposing a constant velocity field $+v_w$ on Fig. 12-2a. The top and bottom boundaries can be considered as stream-tube boundaries, or the control volume taken as short enough that friction effects can be neglected. Since pressure and density changes exist, a temperature difference is also expected; however, the time required for the wave to pass a point is much smaller than the time required for any significant molecular-energy diffusion (heat transfer), so the flow is very nearly adiabatic.

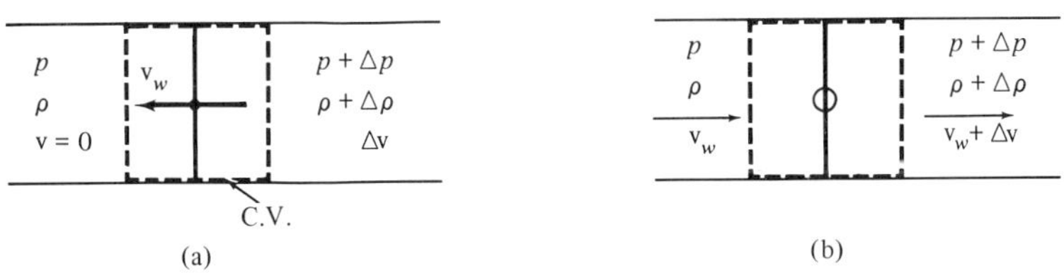

Fig. 12-2. (a) Wave propagating with speed v_w into a fluid at rest in a duct or streamtube; (b) the same flow transformed to a steady flow by addition of a uniform flow velocity v_w to the right. The wave is stationary in the control volume with an approach flow v_w.

Applying the continuity and momentum equations to the control volume of Fig. 12-2b,

$$\rho\,\Delta\mathrm{v} + \mathrm{v}_w\,\Delta\rho = 0$$

$$p + \frac{\rho \mathrm{v}_w^2}{g_c} = (p + \Delta p)\,\frac{\rho \mathrm{v}_w}{g_c}\,(\mathrm{v}_w + \Delta\mathrm{v})$$

Eliminating $\rho\,\Delta\mathrm{v}$ between these equations gives

$$\mathrm{v}_w^2 = g_c\left(\frac{\partial p}{\partial \rho}\right)_s \tag{12-2}$$

The subscript s implies isentropic, since the conditions of frictionless, adiabatic (or isentropic) flow have been considered. This particular speed is also seen to be a characteristic thermodynamic property of the substance, depending primarily on its equation of state. Since equations of state for most substances are quite complicated, the characteristic wave speed can be determined from experimental values of the compressibility (see Chapter 2). In dealing with most gases near standard atmospheric conditions (more specifically, for gases at conditions significantly removed from the critical point) perfect gas behavior can be assumed, for which the equation of state is

$$\frac{p}{\rho} = \frac{R_0}{\mathscr{M}}\,T$$

or

$$\frac{p}{\rho} = RT \tag{12-3}$$

where R_0 is the universal gas constant and $\mathscr{M}$ is the molecular mass number. For a given gas, a specific gas constant $R = R_0/\mathscr{M}$ is more convenient to use. The isentropic *process relation* for a perfect gas is[2]

$$\frac{p}{\rho^\gamma} = \text{const.} \tag{12-3a}$$

Evaluating $(\partial p/\partial \rho)_s$ for the conditions of (12-3) and (12-4) gives

$$a^2 = g_c\,\frac{\gamma p}{\rho} = \gamma g_c RT \tag{12-4}$$

or, for a perfect gas, the characteristic wave speed is a simple function of temperature only.

[2] $\gamma = c_p/c_\mathrm{v}$ = const. for a perfect gas. For perfect diatomic gases (for example, air), $\gamma = 1.4$; for monatomic gases $\gamma = 1.67$.

Experimental measurements of the actual speed of sound propagation in a perfect gas give results within a fraction of a percent of values calculated using (12-4). The speed of sound (propagation of disturbances such that $\Delta p/p \ll 1$) is then essentially identical to this idealized (isentropic) characteristic wave speed.[3]

12.3 STEADY, ONE-DIMENSIONAL DUCT FLOW

In the process of accelerating a flow from rest (or from some small initial speed) in a nozzle as represented in Fig. 12-3, the nozzle flow is stabilized with respect to the production of turbulence, or existing turbulence intensity is increased and therefore damps out rapidly. At high

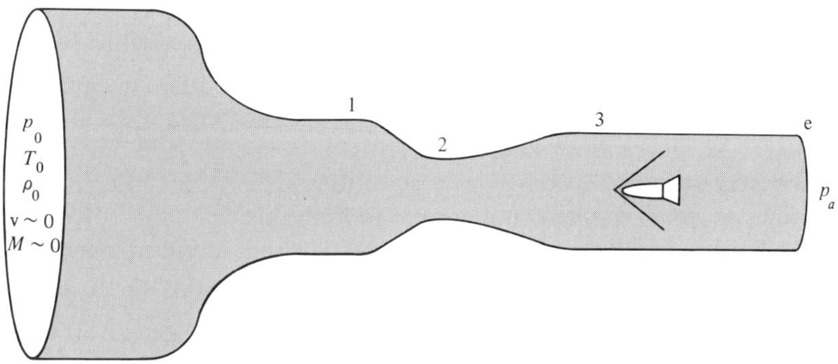

Fig. 12-3. Example duct system for isentropic flow determinations. A minimum area is indicated at 2 and a body at 3.

speeds and relatively low densities, the kinematic viscosity is low enough that the Reynolds numbers are large and the boundary layers are relatively thin. Therefore, most flow processes outside the boundary layers in ducts with relatively small L/D may be considered as essentially reversible. If the area and flow direction changes are relatively gentle, the flow can also be approximated as one dimensional (see also Section 4.9). If the heat-transfer rate across the boundaries is negligible compared with the kinetic-energy rate at any point in the ducting system, the flow may also be considered to be adiabatic. With these assumptions and assuming no

[3] Until experimental techniques were available by which the speed of sound in a gas could be measured accurately, it was not known whether sound propagated adiabatically or isothermally.

external work on the fluid, the steady-flow energy equation reduces to[4]

$$\tilde{h} + \frac{v^2}{2g_c} = \text{const.} = \tilde{h}_0 \tag{12-5}$$

and the Euler momentum equation is

$$\int \frac{dp}{\rho} + \frac{v^2}{2g_c} = \text{const.} \tag{12-6}$$

Since the process is isentropic, $\tilde{h} = \int dp/\rho$ and the momentum and energy principles are identical. The applicable equation of continuity is

$$\dot{m} = \rho A v = \text{const.} \tag{12-7}$$

For a perfect gas, $\tilde{h}$ is a function of T only. The above two independent equations then involve the independent variables T, v, ρ, and A as a function of the stream position or dimension x. One of the equations (12-5) or (12-6) expresses a dynamic condition and (12-7) expresses the conservation-of-mass requirement. The *process* by which the state variables change gives another condition: $\tilde{s}$ = const. is a third condition or equation. This relation combined with the first law of thermodynamics to give the form (12-3a) in terms of state variables is more useful. Addition of the equation of state (12-3) then gives a set of four independent equations for the four variables, so the set is complete and solvable.

It has been stated that the Mach number is also a useful parameter. If it is included as a variable, $M = v/a$ and two more variables, M and a, are added to the set. The sound speed is easily determined from $a^2 = \gamma g_c RT$. A set of six equations and six variables is now obtained. The above variables are all dependent variables; the one independent variable is the distance x along the flow. These equations are tabulated here for convenience.[5]

	Integral form	*Differential form*
(1) Energy or momentum	$\tilde{h} + \frac{v^2}{2g_c} = \text{const.}$	$d\tilde{h} + \frac{v\,dv}{g_c} = 0$ or $\frac{dp}{\rho} + \frac{v\,dv}{g_c} = 0$
(2) Continuity	$\dot{m} = \rho A v = \text{const.}$	$\frac{d\rho}{\rho} + \frac{dA}{A} + \frac{dv}{v} = 0$
(3) State	$\frac{p}{\rho} = RT$	$\frac{dp}{p} - \frac{d\rho}{\rho} = \frac{dT}{T}$

(12-8)

[4] The "one dimension" will be assumed to be in the direction of the mean flow.

[5] Determination of derivative or differential equations where products and powers are involved is simplified by first taking the logarithm of the equation and then taking the derivative. This process is known as taking the log-derivative. For example, p/ρ^γ = const.; $\ln p - \gamma \ln \rho = \ln C$, and then $(dp/p) - \gamma(d\rho/\rho) = 0$.

(4) Process $\dfrac{p}{\rho^\gamma} = \text{const.}$ $\qquad \dfrac{dp}{p} - \gamma\dfrac{d\rho}{\rho} = 0$

(5) Mach number $M = \dfrac{\text{v}}{a}$ $\qquad \dfrac{dM}{M} = \dfrac{d\text{v}}{\text{v}} - \dfrac{da}{a}$

(6) Sonic speed $a^2 = \gamma g_c RT$ $\qquad 2\dfrac{da}{a} = \dfrac{dT}{T}$

These are the fundamental equations from which steady, one-dimensional isentropic flow behavior can be determined.

Consider first the determination of the Mach number–velocity behavior in a variable area duct. The only relation in which A occurs is the continuity equation. The differential form of (12-6) can be used with the isentropic relation to give

$$\frac{dp}{d\rho}\frac{d\rho}{\rho} + \frac{\text{v}\, d\text{v}}{g_c} = 0 \qquad \text{or} \qquad a^2\frac{d\rho}{\rho} + \text{v}^2\frac{d\text{v}}{\text{v}} = 0$$

Substituting $d\rho/\rho$ from the continuity equation gives

$$\left(\frac{\text{v}^2}{a^2} - 1\right)\frac{d\text{v}}{\text{v}} = \frac{dA}{A} \qquad \frac{d\text{v}}{\text{v}} = \frac{1}{M^2 - 1}\frac{dA}{A} \qquad \frac{d\text{v}}{\text{v}} = \frac{-1}{1 - M^2}\frac{dA}{A} \tag{12-9}$$

The implications of (12-9) are shown here graphically. Consideration of the

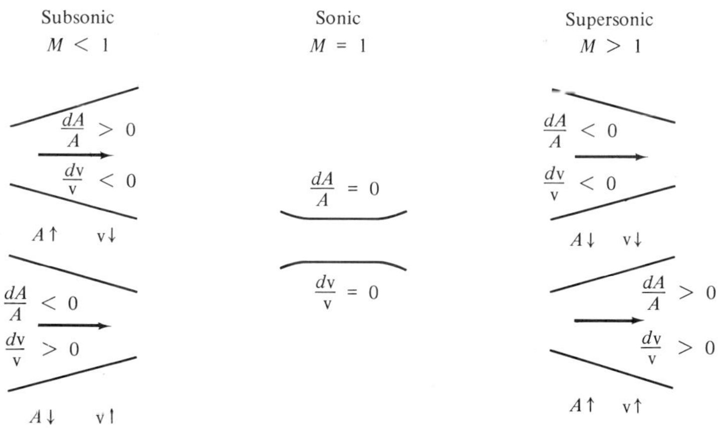

simple continuity relation for incompressible flow indicates that an increase in area results in a decrease in velocity. The Bernoulli equation indicates that the velocity increases as the pressure decreases, which is also indicated by the momentum equation for compressible flow. For compressible flow, the fact that the velocity increases in an increasing area supercritical (supersonic) flow implies that the density must be decreasing faster than the velocity is increasing, that is, for diverging ($dA/A > 0$) supersonic ($M > 1$) flow, $(d\text{v}/\text{v})/(d\rho/\rho) > 1$ (see Problem 12-9).

It would be possible to eliminate either M or v (preferably v, or $d\text{v}/\text{v}$) in terms of the other variable to obtain an equation giving the area as a function of M only. This procedure is possible, but a more direct approach will be developed later (see Problem 12-6).

To accelerate a flow from rest to supersonic speeds then requires (1) a converging subsonic passage, (2) a "sonic throat," (3) a diverging supersonic section, and (4) a force large enough to accelerate the gas to the required speed. This force is supplied by the pressure difference across the accelerating section of the duct.

12.4 ISENTROPIC FLOW RELATIONS

Consider the isentropic (reversible) flow of a perfect gas from a reservoir through a directing system such as illustrated in Fig. 12-3. Since the velocity (therefore momentum and kinetic energy) is negligible in the reservoir, the reservoir values of pressure, temperature, and density are the reference stagnation values p_0, T_0, and ρ_0. T_0 and p_0 are easily measured by conventional means, as is the local static pressure in the duct. The local stagnation pressure can be determined using a Pitot tube.

The flow is accelerated from the reservoir to the velocity in the duct by the pressure difference $p_0 - p_a$. The local p, ρ, v, and T may be determined in terms of the stagnation p_0 and T_0 and the Mach number M, or M may be determined in terms of the ratios of p_0/p, and so on.

For a perfect gas (constant c_p, c_v, and γ) the energy equation (12-5) can be written

$$c_pT_0 = c_pT + \frac{\text{v}^2}{2g_c}$$

Dividing by c_pT will yield the temperature ratio directly,

$$\frac{T_0}{T} = 1 + \frac{\text{v}^2}{2g_cc_pT}$$

This equation is easily written in terms of M, since T is directly proportional to a^2 ($T = a^2/\gamma g_cR$),

$$\frac{T_0}{T} = 1 + \frac{\gamma g_cR\text{v}^2}{2g_cc_pa^2}$$

For a perfect gas, $c_p - c_v = R$ and $\gamma = c_p/c_v$ giving

$$\frac{T_0}{T} = 1 + \frac{\gamma - 1}{2}M^2 \qquad (12\text{-}10)$$

The equation of state and isentropic process relation for a perfect gas give

$$\frac{T_0}{T} = \left(\frac{p_0}{p}\right)^{(\gamma-1)/\gamma} = \left(\frac{\rho_0}{\rho}\right)^{\gamma-1} \tag{12-11}$$

so

$$\frac{p_0}{p} = \left(1 + \frac{\gamma - 1}{2} M^2\right)^{\gamma/(\gamma-1)} \tag{12-12}$$

$$\frac{\rho_0}{\rho} = \left(1 + \frac{\gamma - 1}{2} M^2\right)^{1/(\gamma-1)} \tag{12-13}$$

These relations are shown graphically in Fig. 12-4.

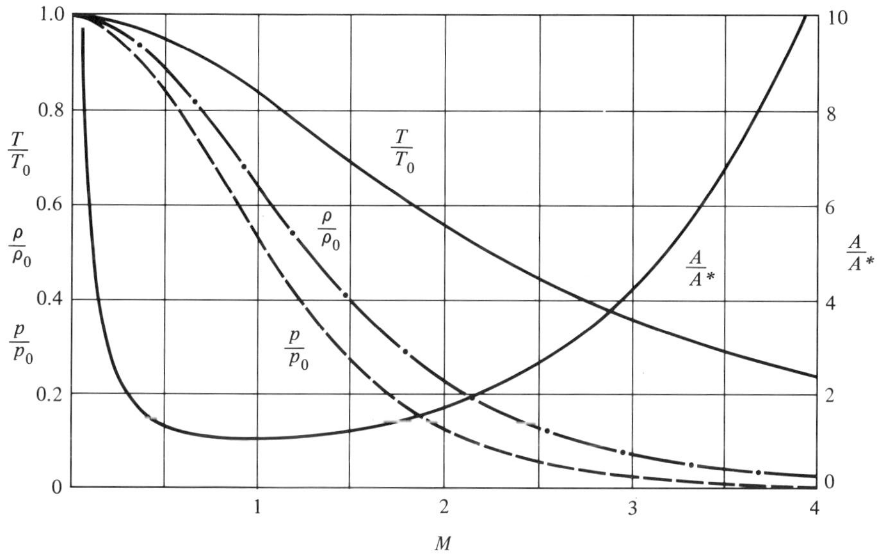

Fig. 12-4. Isentropic relationships plotted as a function of Mach number M.

The series-expansion forms of (12-12) and (12-13) are sometimes useful for values of $M^2 < 1$. The binomial expansion of (12-12) gives

$$\frac{p_0}{p} = 1 + \frac{\gamma M^2}{2} + \frac{\gamma M^4}{8} + \frac{\gamma(\gamma - 2)}{48} M^6 + \cdots \tag{12-12a}$$

By definition the critical flow condition is reached when $M = 1$. The above ratios are then functions of γ only at this critical condition. This unique local condition is then denoted by an asterisk (but usually referred to as "star," since "star" is more expedient in conversation than "asterisk," for example, "p-star"). The critical pressure ratio

$$\frac{p_0}{p^*} = \left(\frac{\gamma + 1}{2}\right)^{\gamma/(\gamma - 1)}$$

is then the necessary pressure ratio required to accelerate the flow from rest to sonic speed. For a diatomic perfect gas ($\gamma = 1.4$), the critical ratios are

$$\frac{\tilde{h}_0}{\tilde{h}^*} = \frac{T_0}{T^*} = \left(\frac{a_0}{a^*}\right)^2 = \frac{\gamma + 1}{2} = 1.20 \qquad \text{(12-14a)}$$

$$\frac{p_0}{p^*} = (1.2)^{3.5} = 1.892 \qquad \text{(12-14b)}$$

Another necessary condition for obtaining sonic speed is the existence of a "sonic throat" or constriction. Sonic or critical speed may be obtained without a well-shaped nozzle. For example, if the critical pressure is applied across an orifice, a contraction will occur (as described in Section 6.12). Critical speed will then be obtained through an area $A_{\text{v.c.}} = A^*$, which is less than the orifice area. In addition, the contraction coefficient C_c now depends on Mach number (as well as Reynolds number).

The isentropic stagnation pressure p_0 will be obtained whenever the flow is brought to rest isentropically anywhere within the ducting system. For example, if a Pitot tube or body is placed in the duct, the stagnation pressure on the nose will be p_0 as long as the local Mach number is less than 1. If $M > 1$, a shock wave will form in front of the body. The "shock process" is not isentropic, so the stagnation pressure p_0' behind the shock will be less than p_0. Shock waves will be considered in Section 12.8.

Since no heat transfer is assumed, the stagnation enthalpy $\tilde{h}_0$ is a constant for the flow. T_0, $\tilde{h}_0$, and a_0 are then constants of the flow whether it is reversible (isentropic) or not as long as the heat transfer to the flow is negligible and no external work is applied. On the other hand, p_0 and ρ_0 are constant only if the flow is reversible (isentropic).

The quantities $\tilde{h}^*$, T^*, $a^* = \text{v}^*$ are also constant *reference* quantities. The ratios T/T^*, p/p^*, and ρ/ρ^* may then be obtained merely as ratios of the ratios of (12-10) through (12-13) to those of (12-14), for example,

$$\frac{p}{p^*} = \frac{p}{p_0} \cdot \frac{p_0}{p^*} = \left\{\frac{(\gamma + 1)/2}{1 + [(\gamma - 1)/2]M^2}\right\}^{\gamma/(\gamma - 1)}$$

12.5 MASS FLOW RATE

The mass flow rate through any section of area A may be obtained in terms of the local Mach number and the reference conditions:

$$\dot{m} = \rho A \mathrm{v} = \rho A a M = \frac{p}{RT} AM\sqrt{\gamma g_c RT} = pAM\sqrt{\frac{\gamma g_c}{RT}}$$

$$\dot{m} = p_0 AM\sqrt{\frac{\gamma g_c}{RT_0}\frac{p}{p_0}}\sqrt{\frac{T}{T_0}} \tag{12-15}$$

or

$$\frac{\dot{m}}{p_0 A}\sqrt{\frac{RT_0}{\gamma g_c}} = M\frac{p}{p_0}\sqrt{\frac{T_0}{T}}$$

There are difficulties connected with taking measurements of static temperature T in a high-speed stream. However, for flow with negligible curvature, the static pressure may be measured by a wall tap and pressure indicator (gage, manometer, transducer, and so on). The expression giving the mass flow rate $\dot{m}$ as a function of directly measurable quantities is then [using (12-11)]

$$\frac{\dot{m}}{p_0 A}\sqrt{\frac{RT_0}{\gamma g_c}} = M\left(\frac{p}{p_0}\right)^{(\gamma+1)/2\gamma}$$

Next, it is desirable to replace M in terms of the pressure ratio [using (12-12)]

$$\frac{\dot{m}}{p_0 A}\sqrt{\frac{RT}{\gamma g_c}} = \sqrt{\frac{2}{\gamma - 1}\left[\left(\frac{p}{p_0}\right)^{2/\gamma} - \left(\frac{p}{p_0}\right)^{(\gamma+1)/\gamma}\right]} \tag{12-16}$$

This relation gives $\dot{m}$ directly as a function of the reference reservoir quantities and the local pressure and area. If tabulated values of p/p_0 and T/T_0 (or the inverse ratios) are available, the dimensionless $\dot{m}$ may easily be obtained. Using (12-10) and (12-11) or (12-12) the mass flow parameter $\dot{m}'$ may be written as an explicit function of M as

$$\dot{m}' \equiv \frac{\dot{m}}{p_0 A}\frac{RT_0}{\gamma g_c} = \frac{M}{\{1 + [(\gamma - 1)/2]M^2\}^{(\gamma+1)/2(\gamma-1)}} \tag{12-17}$$

This relation is plotted in Fig. 12-5.

For given reservoir conditions and flow through a given area A, $\dot{m}$ has a maximum at $M = 1$. Since the pressure is the directly controllable variable on a ducting system (for example, Fig. 12-3), it is of interest to investigate variations of the flow rate as a function of the pressure across the system (p_0 to p_a), or the pressure ratio p_0/p_a. This evaluation is most direct if p_0 is assumed to be constant and the surrounding pressure is decreased.

First consider the flow through a simple converging nozzle (Fig. 12-3 cut off at section 1). From (12-17) and Fig. 12-4, a maximum is seen to occur at $M = 1$ corresponding to

$$\frac{p_0}{p} = \left(\frac{\gamma + 1}{2}\right)^{\gamma/(\gamma-1)}$$

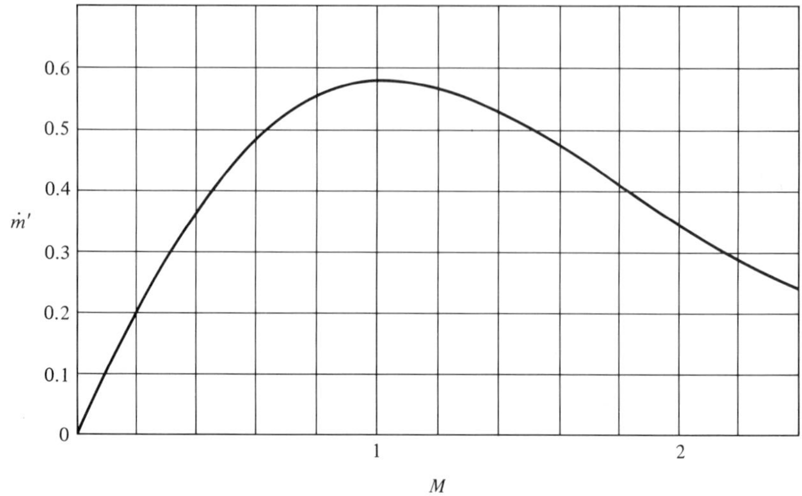

Fig. 12-5. Dimensionless mass flow rate as a function of Mach number from (12-17).

or for air (or a diatomic gas) $p/p_0 = 0.528$. A maximum $\dot{m}'$ will then occur at $p/p_0 = 0.528$; the value of $\dot{m}$-maximum is 0.578 [from Fig. 12-4 or (12-17) at $M = 1$]. The quantity $\dot{m}'/\dot{m}'_{\text{max}}$ can then be plotted as a function of p/p_0 for values of p/p_0 decreasing from 1.0. If (12-16) is plotted as $\dot{m}'/\dot{m}'_{\text{max}}$, the right-hand dashed and left-hand solid curves of Fig. 12-6 are obtained. However, experimental flow measurements would give the solid curve with $\dot{m}'/\dot{m}'_{\text{max}} = 1.0$ for $p_a/p_0 = 0.528$. The flow rate would increase with decreasing exhaust pressure until the maximum Mach number reaches 1 (the exit flow is sonic). Further decrease of the external pressure would have no further effect on the flow up to the nozzle exit. If the ambient pressure is dropped below $0.528p_0$, there is no way that this decrease can affect the upstream flow. One simple intuitive expanation is that since the flow at the nozzle exit is sonic, the signal that the external conditions have been changed cannot be propagated back into the upstream flow. The flow is then fixed or "choked" at the critical condition. If the ambient pressure is decreased below the critical p^*, the gas surely must continue to expand, but it does so by a supersonic wave expansion process outside the nozzle while the pressure at the nozzle exit plane remains critical at p^*.

The dimensionless mass flow parameter $\dot{m}'_{\text{max}}$ evaluated at $M = 1$ is calculated as 0.578. The flow at the area is then critical, or $A = A^*$, and (12-15) can be written

$$\frac{\dot{m}}{p_0 A^*}\sqrt{\frac{RT_0}{\gamma g_c}} = 0.578 \qquad \text{(12-18a)}$$

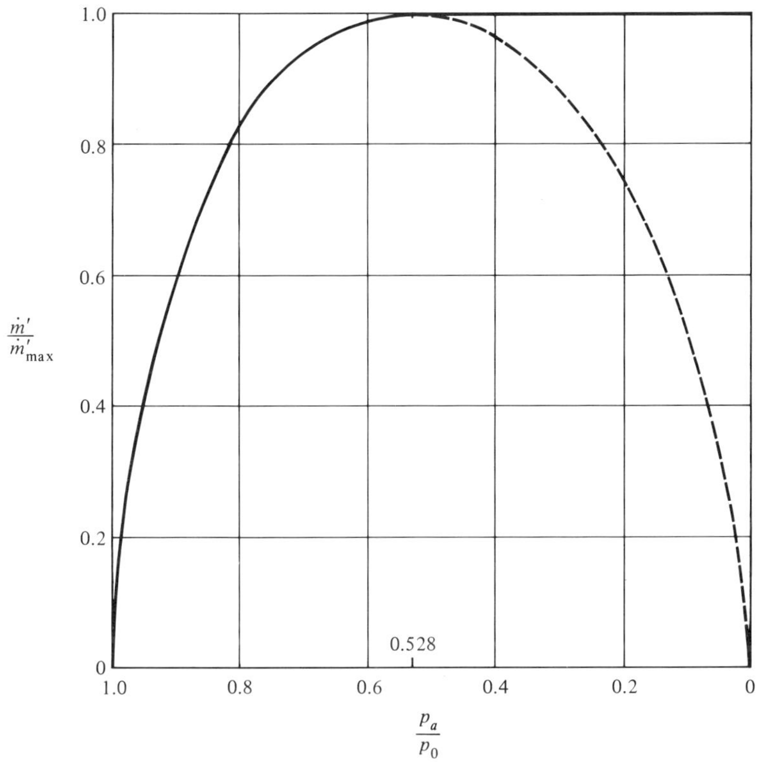

Fig. 12-6. Relative mass flow rate as a function of pressure ratio.

For the particular case of air as the gas, all the constants can be combined to give a relation for the critical mass flow rate for air as

$$\frac{\dot{m}\sqrt{T_0}}{p_0 A^*} = 0.532 \tag{12-18b}$$

For steady ($\dot{m}$ = const.) one-dimensional isentropic flow of a perfect gas in a duct, the relation between A and M is now obtained from (12-17) and (12-18) as

$$\frac{A}{A^*} = \frac{1}{M}\left[\frac{2}{\gamma + 1}\left(1 + \frac{\gamma - 1}{2} M^2\right)\right]^{(\gamma+1)/2(\gamma-1)} \tag{12-19}$$

This relation is plotted in Fig. 12-4. The importance of (12-19) is that for compressible flow the Mach number is determined as a function of the area ratio and γ only.

12.6 ILLUSTRATIVE EXAMPLE 12-1: Converging–Diverging Nozzle

Next consider the flow from the reservoir through the throat section to and into section 3 as the ambient pressure p_a is decreased (Fig. 12-7). The flow through this system will be similar to that through a Venturi tube as long as $p_2 > p^*$. For the purpose of this example, assume that

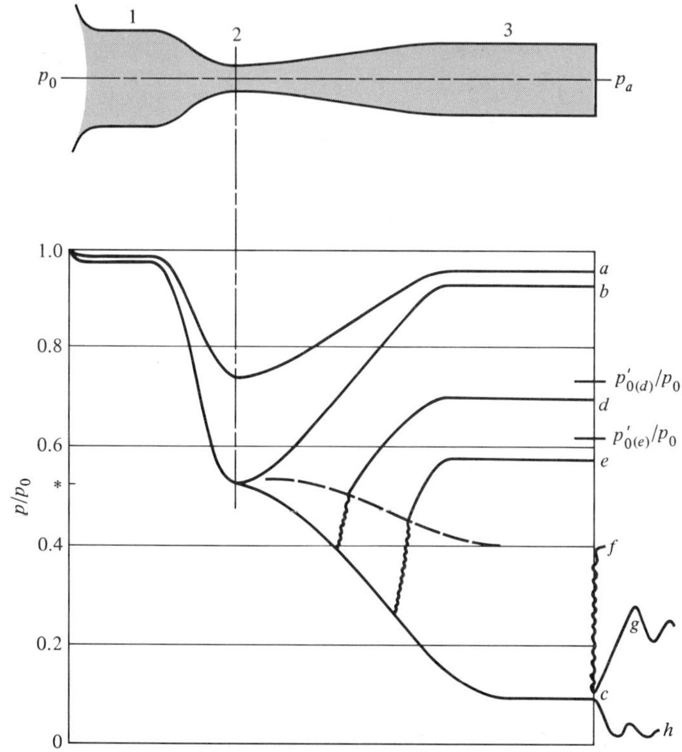

Fig. 12-7. Pressure as a function of axial distance for flow in a converging–diverging duct for a fixed supply pressure p_0 and various back pressures p_a.

$A_1/A_2 = 4$ and $A_3/A_2 = 2$. As a starting condition, assume that $p_a/p_0 = p_e/p_0 = 0.95$, for which the exit Mach number $M_e = 0.27$ from (12-12) (or Fig. 12-4). The Mach number at 2 can be determined from the ratio

$$\frac{A_3}{A_2} = \frac{A_3/A^*}{A_2/A^*} \qquad \text{or} \qquad \frac{A_2}{A^*} = \frac{A_3}{A^*}\frac{A_2}{A_3} = \frac{2.24}{2} = 1.12$$

The corresponding Mach number at 2, $M_2 = 0.67$, and $p_2/p_0 = 0.74$, for which $M_1 = 0.075$ and $p_1/p_0 = 0.997$ (curve a). When the flow at the throat just becomes critical, $p_2/p_0 = p^*/p_0 = 0.528$; $M_2 = 1 = M^*$

(curve b). M_1 and p_1/p_0 are then determined from $A_1/A^* = 4$; $M_1 = 0.146$, $p_1/p_0 = 0.985$. Similarly, for $A_3/A_2 = A_3/A^* = 2$, $M_3 = 0.305$, and $p_3/p_0 = p_a/p_0 = 0.937$ (curve b). As p_a is further decreased, the flow between the reservoir and the throat will remain "choked."

Next the pressure will be determined for supersonic flow in 3. For the area ratio $A_3/A^* = 2$, $M_3 = 2.20$; the required pressure ratio is $p_a/p_0 = p_3/p_0 = 0.0935$ (curve c). This case can be described as the supersonic design condition.

If the exhaust pressure is somewhere between (b) and (c), for example, at (d) or (e), the pressure difference is sufficient to produce supersonic flow beyond the throat but not enough to achieve the Mach number corresponding to the existing area ratio. The flow then accelerates beyond the throat to a supersonic Mach number at which a shock compression occurs that will just allow continuing subsonic flow out the exhaust at the applied p_a. A shock compression is a nonisentropic process. The decrease is indicated as a decrease in p_0, as indicated by $p'_{0(\mathrm{d})}$. If the pressure ratio is less than that at (c), the flow conditions are still those of curve c; the additional expansion occurs outside the nozzle (curve h).

12.7 ILLUSTRATIVE EXAMPLE 12-2: Rocket Engine

In a rocket engine, as illustrated in Fig. 12-8, an energy-conversion process first generates gases with a high available energy, as indicated by the value of $\tilde{h}_0$. The energy conversion can be the combustion of liquid or solid propellants, or an energy addition by means of direct nuclear bombardment (heat transfer) from a nuclear "heater." The available force, as indicated by the pressure difference, $p_0 - p_a$, is used to accelerate the gases through the nozzle. A reaction thrust is produced on the engine

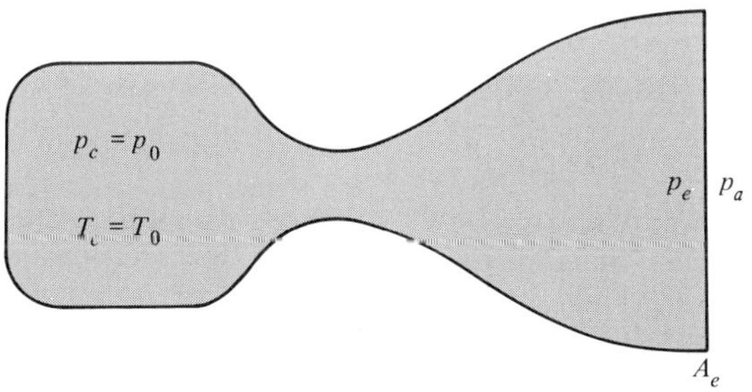

Fig. 12-8. Representation of a rocket engine. The exit plane pressure in the flow p_e is not necessarily the same as the surrounding ambient pressure p_a.

as described in Section 6.11. Equation (6-50) for the thrust can be written

$$F_T = \frac{\dot{m}}{g_c} M_e a_e + p_a\left(\frac{p_e}{p_a} - 1\right)A_e$$

Because the chamber velocity v_c is much less than the exit velocity v_e, $p_c \approx p_0$, $T_c \approx T_0$, and

$$F_T = \frac{\dot{m}}{g_c} M_e\left(\frac{a_e}{a_0}\right)a_0 + p_a\left(\frac{p_e}{p_a} - 1\right)A_e \tag{12-20}$$

The ratio $a_e/a_0 = \sqrt{T_e/T_0}$ is a function of M_e. If the chamber pressure p_0 and the exhaust plane pressure p_e are specified, M_e is determined. Note that both quantities p_0 and T_0 must be known to specify the state of the gas in the chamber.

The exit velocity v_e may be determined from the adiabatic energy equation (assuming one-dimensional, adiabatic flow)

$$v_e = \sqrt{2g_c c_p(T_0 - T_e)} = \sqrt{2g_c c_p T_0\left(1 - \frac{T_0}{T_e}\right)}$$

$$= \sqrt{2g_c c_p T_0\left[1 - \left(\frac{p_e}{p_0}\right)^{(\gamma-1)/\gamma}\right]} \tag{12-21}$$

where the last form is valid only for isentropic flow. Assuming perfect gas behavior, c_p can be written in terms of γ and R as

$$c_p = \frac{R}{\gamma - 1} = \frac{R_0}{\mathcal{M}(\gamma - 1)}$$

The rocket nozzle walls must be cooled by some process, such as "regenerative" or transpiration cooling by liquid propellants or ablative "cooling" in the case of solid propellants. However, the energy-transfer rate of the cooling process is much smaller than the energy rate of the nozzle flow, so the assumption of adiabatic flow is satisfactory for a first approximation.

It is interesting to examine (12-20). Assume that $p_e \approx p_a$ or that only the first term is to be considered. A thrust parameter referred to as the "specific impulse" is defined as

$$I_{sp} = \frac{F_T}{\dot{m}} \frac{\text{lb}_f\text{-sec}}{\text{lb}_m}$$

This parameter is useful as an indication of the overall thrust capability of the engine and fuel combination, particularly when evaluated as vacuum specific impulse, where it is assumed that the nozzle would expand the gases ideally to $p_e = p_a = 0$. M_e for this condition is obviously infinity, so that (12-20) can be used to give (see also Problem 12-28)

$$I_{sp,vac} = \frac{V}{g_c} = \frac{1}{g_c}\sqrt{2g_c c_p T_0} = \sqrt{\frac{2\gamma R T_0}{(\gamma - 1)g_c \mathcal{M}}} \tag{12-22}$$

T_0 is usually limited by nozzle cooling and available materials. For a maximum attainable T_0, the best performance then primarily depends on $\mathcal{M}$ (little variation in γ is possible). Equation (12-22) then indicates that the gas should have as small an $\mathcal{M}$ as possible. On the other hand, a small $\mathcal{M}$ implies a low density, so that a large tanking volume (and tank mass) is necessary even if the propellant is stored in liquid form. For a monopropellant (for example, in a nuclear rocket) hydrogen would have the highest available I_{sp}, but the density of liquid hydrogen is only about 4.4 lb_m/ft^3. For a chemical combustion rocket, oxygen and hydrogen would give maximum performance; the explosion hazard and excessive T_0 make use of this combination extremely difficult. Choice of propellants, type of propellant, and type of engine are determined by optimizing I_{sp} with overall size, mass, complication, and cost for a given mission. Nominal values of I_{sp} for liquid propellants (chemical combustion) are $I_{sp} \sim 400$, and for solid propellants $I_{sp} \sim 300$.

It is also interesting to examine (12-20) in the form (neglecting the over- or underpressure term)

$$F_T = \frac{\dot{m}}{g_c} \frac{M_e a_0}{\sqrt{1 + [(\gamma - 1)/2]M_e^2}}$$

F_T is seen to have an asymptotic value as M_e is increased (for a given T_0). For example, as $M_e \to \infty$,

$$F_T = \frac{\dot{m}}{g_c}\sqrt{\frac{2}{\gamma - 1}}\,a_0 \qquad \text{or} \qquad I_{sp} \to \sqrt{\frac{2}{\gamma - 1}}\frac{a_0}{g_c} = I_{sp,vac}$$

which is just the vacuum specific impulse. Therefore, based on considerations of increasing engine size and mass with increasing A_e, there is an optimum M_e to which the gas should be expanded.

12.8 NORMAL SHOCK WAVES

Although the adiabatic compression process by means of a normal shock (perpendicular to the flow or flight direction) is nonisentropic, analysis of the fluid mechanics is only complicated with respect to the extent of the algebra required. The normal shock is essentially a discontinuity in the flow quantities. Since an entropy increase is involved, the reversible process or normal expansion discontinuity would imply an entropy decrease, which is impossible for an adiabatic process. The analysis including the evaluation of the entropy change is simplified by the fact that the only irreversibility is within the shock discontinuity and need not be considered explicitly in the evaluation.

Consider the control volume for flow through a normal shock in a duct such as illustrated in Fig. 12-9. Assuming perfect gas behavior, the equations applicable to this case are

(*a*) Continuity: $\rho_1 \mathrm{v}_1 = \rho_2 \mathrm{v}_2$

(*b*) Momentum: $p_1 + \dfrac{\rho_1 \mathrm{v}_1^2}{g_c} = p_2 + \dfrac{\rho_2 \mathrm{v}_2^2}{g_c}$

(*c*) Energy: $\tilde{h}_0 = \text{const.}: \; c_p T_1 + \dfrac{\mathrm{v}_1^2}{2g_c} = c_p T_2 + \dfrac{\mathrm{v}_2^2}{2g_c}$

$$\frac{a_1^2}{\gamma - 1} + \frac{\mathrm{v}_1^2}{2}\frac{a_2^2}{\gamma - 1} + \frac{\mathrm{v}_2^2}{2} = \frac{\gamma + 1}{2(\gamma - 1)} a^{*2} = \frac{a_0^2}{\gamma - 1} \tag{12-23}$$

(*d*) State: $\dfrac{p}{\rho} = RT$

(*e*) Sonic speed: $a^2 = \dfrac{g_c \gamma p}{\rho} = \gamma g_c RT$

(*f*) Mach number: $M = \dfrac{\mathrm{v}}{a}$

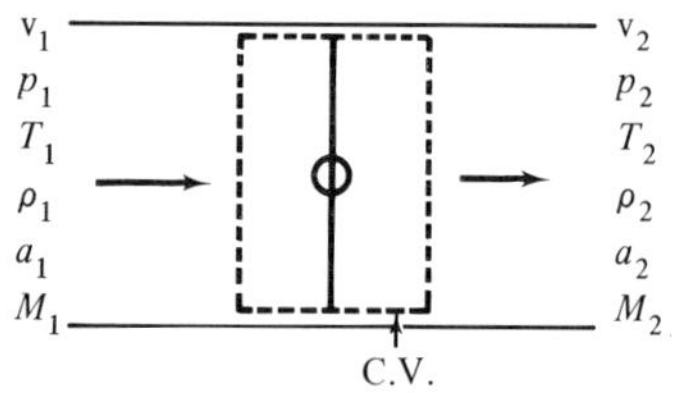

Fig. 12-9. Stationary shock wave in a duct or streamtube. The control volume is quite short in the flow direction, so that friction irreversibilities at the control volume surface may be neglected with respect to the shock irreversibilities.

Equations (*a*) through (*d*) form an independent set for determining p, ρ, T, and v at station 2 downstream of the shock in terms of the approach conditions at station 1. In place of T, the sonic speed may be introduced from (*e*), giving the second equation of (*c*). The Mach number M may also be substituted for the velocity with the use of (*f*). The distinction between this problem and that of determining the characteristic, *isentropic* wave behavior is that isentropic flow is not assumed; consequently, the momentum and energy equations are independent. The control volume is chosen quite short, so it is not necessary to consider friction on the sides of the duct. The critical speed $\mathrm{v}^* = a^*$ has been chosen as the reference quantity rather than a_0, as indicated in (12-23). Since $a_0 = \text{const.}$ ($\tilde{h}_0 = \text{const.}$), $a^* = \text{const.}$ As stated previously, it is usually most expedient to work with, or solve for, the desired variables (or ratios) in terms of Mach numbers. Dividing (12-23a) into (*b*) and using (*f*) gives

$$\frac{a_1^2}{\gamma \mathrm{v}_1} + \mathrm{v}_1 = \frac{a_2^2}{\gamma \mathrm{v}_2} + \mathrm{v}_2 \qquad \text{or} \qquad \frac{\mathrm{v}_1}{\gamma}\left(\frac{1}{M_1^2} + \gamma\right) = \frac{\mathrm{v}_2}{\gamma}\left(\frac{1}{M_2^2} + \gamma\right)$$

It is advantageous to use $M^* = \mathrm{v}/a^*$ and write (*c*) in the form

$$\frac{1}{M^2} = \frac{\gamma + 1}{2M^{*2}} - \frac{\gamma - 1}{2} \tag{12-24}$$

which is used in the above result to eliminate M_1 to obtain an equation in M^* only,

$$\frac{1}{\gamma M_1^*}\left(\frac{\gamma + 1}{2M_1^{*2}} + \gamma - \frac{\gamma - 1}{2}\right) = \frac{1}{2M_2^*}\left(\frac{\gamma + 1}{2M_2^{*2}} + \gamma - \frac{\gamma - 1}{2}\right)$$

where both sides have also been multiplied by a^*. This equation is easily reduced to

$$\frac{1}{M^*}(M_1^{*2} + 1) = \frac{1}{M_2^*}(M_2^{*2} + 1)$$

or

$$M_1^* M_2^*(M_2^* - M_1^*) = (M_2^* - M_1^*) \tag{12-25}$$

One trivial solution is $M_1^* = M_2^*$ or $\mathrm{v}_1 = \mathrm{v}_2$; the flow is continuous with no change across the control volume. The other significant solution is the simple Prandtl–Meyer relationship,

$$M_1^* M_2^* = 1 \qquad \text{or} \qquad M_1^{*2} M_2^{*2} = 1 \tag{12-26}$$

The relation between M^2 and M^{*2} is unique and monotonic.[6] The desired result giving the downstream Mach number M_2 in terms of the upstream Mach number M_1 can be obtained by solving (12-24) in terms of M^{*2} (or the reciprocal thereof) and substituting in (12-26):

$$M_2^2 = \frac{1 + [(\gamma - 1)/2]M_1^2}{\gamma M_1^2 - [(\gamma - 1)/2]} \tag{12-27}$$

[6] As M increases, M^* increases; note also that when $M = 1$, $M^* = 1$.

A very useful form of (12-23b) is

$$p_1(1 + \gamma M_1^2) = p_2(1 + \gamma M_2^2) \tag{12-28}$$

from which the pressure ratio across a normal shock can be obtained,

$$\frac{p_2}{p_1} = \frac{1 + \gamma M_1^2}{1 + \gamma M_2^2}$$

or, with (12-27),

$$\frac{p_2}{p_1} = \frac{2\gamma M_1^2}{\gamma + 1} - \frac{\gamma - 1}{\gamma + 1} \tag{12-29}$$

(a direct parabolic relation).

The temperature ratio is most directly obtained using the adiabatic condition for a perfect gas $T_{02} = T_{01}$, which gives

$$\frac{T_2}{T_1} = \frac{T_2/T_{02}}{T_1/T_{01}} = \frac{1 + [(\gamma - 1)/2]M_1^2}{1 + [(\gamma - 1)/2]M_2^2} \tag{12-30}$$

Using (12-27), (12-30) becomes

$$\frac{T_2}{T_1} = \frac{2(\gamma - 1)}{(\gamma + 1)^2 M_1^2}\left(1 + \frac{\gamma - 1}{2} M_1^2\right)\left(\frac{2\gamma}{\gamma - 1} M_1^2 - 1\right) \tag{12-31}$$

The density ratio and velocity ratio are then obtained from (12-29), (12-30), and the equation of state,

$$\frac{\rho_2}{\rho_1} = \frac{\mathrm{V}_1}{\mathrm{V}_2} = \frac{p_2/p_1}{T_2/T_1} \tag{12-32}$$

The first law of thermodynamics with the definition of entropy gives

$$d\tilde{s} = \frac{d\tilde{h}}{T} - \frac{dp}{\rho T}$$

(There are no friction forces, so pressure is the only stress on the control volume.) Assuming a perfect gas with $\tilde{h} = c_p T$ and $p/\rho = RT$,

$$\tilde{s}_2 - \tilde{s}_1 = c_p \ln \frac{T_2}{T_1} - R \ln \frac{p_2}{p_1}$$

or

$$\frac{\tilde{s}_2 - \tilde{s}_1}{R} = \frac{\gamma}{\gamma - 1} \ln \frac{T_2}{T_1} - \ln \frac{p_2}{p_1}$$

or

$$e^{\Delta \tilde{s}_0/R} = \left(\frac{T_2}{T_1}\right)^{\gamma/\gamma - 1} \frac{p_1}{p_2} \tag{12-33}$$

The quantities p_0, T_0, and so on, are defined as the stagnation quantities obtained by stagnating the gas isentropically; therefore, it follows immediately that $\Delta\tilde{s} = \Delta\tilde{s}_0$. Equation (12-33) can then be written

$$e^{\Delta\tilde{s}_0/R} = \left(\frac{T_{02}}{T_{01}}\right)^{\gamma/\gamma-1} \frac{p_{01}}{p_{02}}$$

and for adiabatic flow $T_{02} = T_{01}$, so

$$e^{\Delta\tilde{s}/R} = e^{\Delta\tilde{s}_0/R} = \frac{p_{01}}{p_{02}} \tag{12-34}$$

For a real physical adiabatic process, the second law of thermodynamics requires that $\Delta\tilde{s} \geq 0$; consequently, $p_{02} \leq p_{01}$ or the stagnation pressure can only decrease in the direction of the flow (or remains constant for isentropic flow). Since the stagnation pressure is defined as

$$p_0 = p\left(1 + \frac{\gamma - 1}{2} M^2\right)^{\gamma/\gamma-1}$$

the stagnation pressure ratio can be determined in terms of M from (12-27) and (12-29) as

$$\frac{p_{02}}{p_{01}} = \left\{\frac{[(\gamma + 1)/2]M_1^2}{1 + [(\gamma - 1)/2]M_1^2}\right\}^{\gamma/\gamma-1} \bigg/ \left(\frac{2\gamma M_1^2}{\gamma + 1} - \frac{\gamma - 1}{\gamma + 1}\right)^{1/\gamma-1} \tag{12-35}$$

from which the entropy change can be determined with (12-34). Some of the above results are plotted versus M in Fig. 12-10.

It is interesting to expand $\Delta\tilde{s}/R = \ln p_{01}/p_{02}$ into a series expansion in terms of the small quantity $\epsilon \equiv M_1^2 - 1$, which is then valid for M_1 slightly greater than unity. The first nonzero term of this series expansion is

$$\frac{\Delta\tilde{s}}{R} = \frac{2\gamma(M_1^2 - 1)^3}{3(\gamma + 1)^2} \sim \epsilon^3 \tag{12-36}$$

which is quite small. A *weak shock* ($M_1 < 1.3$) is then nearly isentropic. This result also indicates that M_1 must be greater than unity; no discontinuity is possible in subsonic flow. It is also of interest that no continuous solutions can exist in supersonic flow. It is then necessary to obtain solutions to supersonic flow fields by numerical, stepwise procedures. These solutions are by means of the "method of characteristics," which is described in more advanced texts, for example, reference 1 at the end of the chapter.

The relations developed in this chapter are valid only if the gas process and state indicate perfect gas behavior. At low temperatures and high

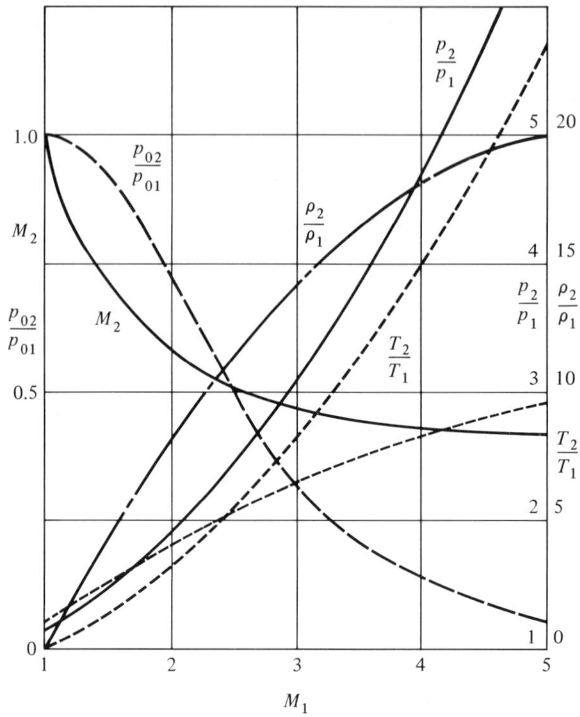

Fig. 12-10. Normal shock relations for $\gamma = 1.4$. The very light dashed line is p_2/p_1 for isentropic compression corresponding to the ρ_2/ρ_1 ratio.

pressures, as the critical point is approached, the intermolecular forces will begin to become significant and the fluid will no longer behave as a perfect gas. At very high temperatures and reduced pressures (low density, high temperature) the vibrational modes of polyatomic gas molecules will become excited, resulting in an increase in the specific heat and a decrease in γ. At still higher temperatures, the molecules may start to come apart or "dissociate." In the plasma range of very high temperatures and/or low densities, the electrons are stripped off the atoms, giving a mixture of ion and electron gases. The physics and mechanics of dissociated gases and plasmas are interesting advanced fields of study within the realm of fluid mechanics.

12.9 ILLUSTRATIVE EXAMPLE 12-3: Supersonic Wind Tunnel

As seen in Sections 12.4 and 12.7, a converging–diverging geometry is required to obtain supersonic velocities, and the Mach number in a supersonic section is a function of A/A^* only. Two types of supersonic wind tunnels are used, closed circuit and single pass. The closed-circuit tunnel uses a compressor as an energy source. The "blow-down" single-pass tunnel uses a high-pressure tank at the inlet and/or an evacuated tank at the outlet. In either type it is necessary to have as efficient a flow channel as possible. A representative tunnel geometry is illustrated in Fig. 12-11. The flow is accelerated to a supersonic Mach number M_2 in the test section. M_2 is a function of $A_2/A_{1t} = A_2/A_1^*$ (and γ). Downstream of the test section, the flow is decelerated through a converging supersonic diffuser

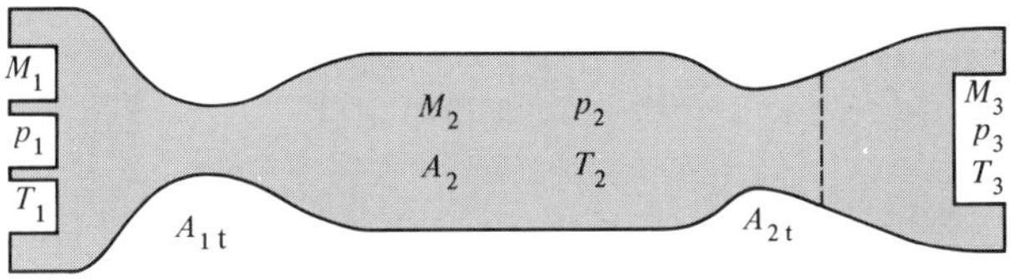

Fig. 12-11. Two-throat wind-tunnel configuration with a test section of uniform Mach-number flow at 2. The flow exhausts at 3 to a vacuum tank or compressor inlet.

and a diverging subsonic diffuser and returned to a low-speed condition at 3. If the flow were not diffused to $M_3 < 1$ for the blow-down case, a larger and lower pressure exhausting tank would be necessary. In the closed-circuit case, the flow must be subsonic for proper functioning of the heat-exchange apparatus and the compressor inlet.[7]

The transient starting procedure is interesting and somewhat complicated, involving the transit of a shock wave down the first diverging section, which is then "swallowed" through the second throat. During this phase of starting, the second throat must be larger than the first because of the irreversibility across the shock. After the shock has been swallowed, the second throat can be decreased in size to give higher efficiency, steady-state operating conditions. For a discussion of the transient conditions and power requirements and efficiency at start and operation and for various operating conditions, see reference 1 at the end of the chapter.

Once maximum-efficiency, steady-state operating conditions are established, the second throat would be about the same size as the first. It must be somewhat larger to account for (1) "duct loss" in the throat to throat section, (2) irreversibilities from model drag, and (3) to provide stable flow conditions. For $M_{2t} = 1.0$ and subsonic flow downstream therefrom, the flow will be unstable and the shock can be "regurgitated"

[7] The compressors used for supersonic wind tunnels are radial-flow or multistaged axial-flow types.

into the accelerating nozzle. A stable condition can be obtained by having the flow sonic in the second throat, accelerating to $M \approx 1.2$ and dropping to $M \approx 0.8$ across a shock in the second diffuser as indicated by the dashed line in Fig. 12-11 and discussed in Section 12.7. The flow is then stable to small transient disturbances in the system.

For nonideal or nonisentropic adiabatic (T_0 = const.) operating conditions, (12-17) and (12-18) yield the condition that for the steady flow of a perfect gas,

$$p_0 A^* = \text{const.}$$

The throat ratio A_2^*/A_1^* is then determined by[8]

$$\frac{A_2^*}{A_1^*} = \frac{p_{01}}{p_{02}} \tag{12-37}$$

The difference in throat-stagnation pressures, $p_{01} - p_{02}$, is found from the irreversibilities occurring between the two throats. These irreversibilities and the Δp_0 are found from data on the particular geometry and from the model drag coefficient and interaction effect between the model and the wind tunnel. Note that the requirement $p_{02} < p_{01}$ indicates that $A_2^* > A_1^*$.

The above discussion implies that an efficiently operating wind tunnel requires an adjustable second or diffuser throat. In addition, since M is a function of A/A^* only, the nozzle throat must also be variable if variable-Mach-number operation is desired. Mechanical complications indicate that variable-throat wind tunnels must necessarily be two-dimensional or plane-walled.

Since accelerating flow occurs in the nozzle, the boundary layer is stable and laminar, so "losses" are relatively small there. Diffusion, with the accompanying adverse pressure gradients and possible separation, is less efficient. The sudden pressure rise across the shock in the diffuser also causes shock–boundary layer interaction detachment, so efficient design of the subsonic portion of the diffuser is quite difficult.

Consider design and operation of a wind tunnel for testing at $M_2 = 3$ (Fig. 12-11). The area ratio is 4.23 from (12-19) or tabulated data. The worst case is during transient start-up, giving the maximum Δp_0 when a normal shock occurs in the test section at M_2. The second throat can then be determined from (12-37). The stagnation pressure ratio across the $M = 4.23$ shock is $p_{02}/p_{01} = 0.328 = A_1^*/A_2^*$. It is interesting that $A_2/A_1^* = 4$, but $A_2/A_2^* = 1.314$ ($M_{2t} = 1 = M_2^*$); the second throat is then only slightly smaller than A_2 during start-up.

If the flow were isentropic through the test section of Fig. 12-11, the second throat could be decreased to A_1^*, after the shock passes through it.

[8] Note that the p_0/p condition in (12-17) is either for isentropic flow, or defines the local isentropic stagnation pressure in terms of the local pressure and Mach number.

However, if the model drag and duct friction losses would give $p_{02}/p_{01} = 0.8$, for example, then $A_2^*/A_1^* = 1.25$. For stability of operation, A_{2t} will usually be greater than A_2^*, so a supersonic flow will still exist in the second throat; for example, if $M_{2t} = 1.1$, then $A_{2t}/A_2^* = 1.008$. A_{2t} is then the throat opening and A_2^* is the hypothetical throat size for $M_2^* = 1.0$. If the flow is further expanded to $M = 1.2$ downstream of the second throat, then goes through a normal shock before subsonic diffusion, the Mach number downstream of the shock is 0.842 and $p_{03}/p_{02} = 0.993$. It is then desirable to subsonically diffuse the flow to $M \approx 0.1$.

12.10 DIFFUSER EFFICIENCY

An overall adiabatic diffuser efficiency can be defined for diffusion processes with reference to the $\tilde{h}$-$\tilde{s}$ diagram for the process, Fig. 12-12. If the diffusion were isentropic, the kinetic energy $v^2/2g_c$ could be converted to pressure-work availability at p_2' and h_{02}. However, the actual "pressure

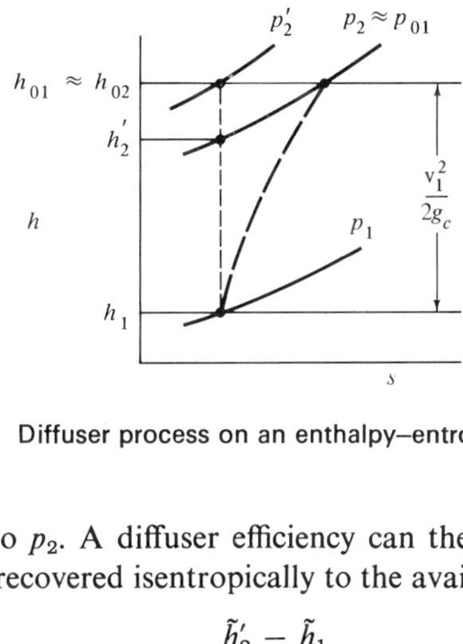

Fig. 12-12. Diffuser process on an enthalpy–entropy diagram.

recovery" is only to p_2. A diffuser efficiency can then be defined as the ratio of the energy recovered isentropically to the available kinetic energy,

$$\eta_D = \frac{\tilde{h}_2' - \tilde{h}_1}{v_1^2/2g_c} \tag{12-38}$$

If the kinetic energy at 2 is small, $M_2 \approx 0.1$, $\tilde{h}_2 \approx \tilde{h}_{02}$, and $p_2 \approx p_{02}$. Noting that $v_1^2/2g_c = \tilde{h}_{02} - \tilde{h}_1$, (12-38) can be written

$$\eta_D = \frac{(\tilde{h}_2'/\tilde{h}_1) - 1}{v_1^2/2g_c\tilde{h}_1}$$

Since $\tilde{h}_2'$ is on the isentrope from 1 at p_2,

$$\eta_D = \frac{(p_2/p_1)^{(\gamma-1)/\gamma} - 1}{[(\gamma - 1)/2]M_1^2} \tag{12-39}$$

Then noting that $\tilde{h}'_2 \approx \tilde{h}'_{02}$,

$$\eta_D = \frac{(\tilde{h}'_{02}/\tilde{h}_{01})(\tilde{h}_{01}/\tilde{h}_1) - 1}{[(\gamma - 1)/2]M_1^2} = \frac{(p_{02}/p_{01})^{(\gamma-1)/\gamma}\{1 + [(\gamma - 1)/2]M_1^2 - 1\}}{[(\gamma - 1)/2]M_1^2} \tag{12-39a}$$

This expression is of value in that it gives the stagnation pressure in terms of the diffuser efficiency and upstream Mach number. For example, for $\eta_D = 0.8$ and $M_1 = 3$, $p_{02}/p_{01} = 0.872$. These diffuser relations are valid for diffusion from either a subsonic or supersonic Mach number M_1.

12.11 MOVING SHOCK WAVE

It is also of interest to determine the relations for a wave moving into a medium at rest. The results of the above development can be used by superimposing a uniform velocity v_1 (and M_1) from right to left in Fig. 12-9. The flow conditions are then illustrated in Fig. 12-13, where the

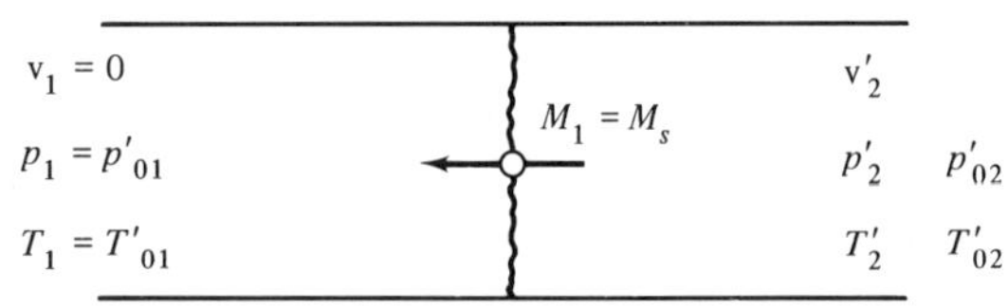

Fig. 12-13. Representation for determining conditions behind a shock moving into a gas at rest. Quantities in this system are primed.

primed quantities refer to conditions in this coordinate representation. T_1 and p_1 are considered to be the same static conditions as in Section 12.8 and are also the stagnation conditions. The "shock Mach number" M_s is identical to M_1, as indicated. The shock moves into the stationary gas at v_1 and $M_s = v_1/a_1$. M_s is then the primary flow parameter. The downstream velocity is $v'_2 = v_2 - v_1$. Since the static quantities (pressure and temperature) are "processed" by the same shock in either system, $p'_2 = p_2$ and $T'_2 = T_2$ ($a'_2 = a_2$).[9] The Mach number behind the shock is then

$$M'_2 = \frac{v_2 - v_1}{a_2} = \frac{v_1 - v_2}{a_2} = \frac{a_1}{a_2} M_1 - M_2 \tag{12-40}$$

[9] Addition of the uniform field motion v_1 will not affect static conditions measured relative to the shock wave.

The stagnation pressure and temperature behind the shock are

$$
\begin{aligned}
p'_{02} &= p_2\left(1 + \frac{\gamma - 1}{2} M_2'^2\right)^{\gamma/\gamma - 1} \\
T'_{02} &= T_2\left(1 + \frac{\gamma - 1}{2} M_2'^2\right)
\end{aligned}
\tag{12-41}
$$

The above relationships are only valid for one-dimensional, plane shock waves. For a spherical shock wave, the flow area (or radius) is another parameter, which makes the analysis more complicated. However, the above relations are still valid across the wave at any point. The additional relation from the analysis of an expanding spherical wave would give the Mach number as a function of radius.

12.12 SUMMARY

1. An upper limit exists for the Mach number and mass flow rate that can be obtained through a ducting system. This limit is attained when the fluid speed at any section in the duct first becomes sonic. The Mach-number distribution throughout the system will then remain fixed. The mass flow rate can only increase if the reservoir or supply density is increased.
2. In isentropic supersonic flow, the Mach number and the thermodynamic flow quantity ratios are a function of the area only.
3. The Mach number can only decrease through a normal shock. A shock can only occur if the upstream Mach number is greater than unity.
4. Weak shock waves ($M_1 < 1.3$) are nearly isentropic. Therefore, if a specified compression ratio p_1/p_2 could be obtained by a series of weak shocks rather than a single strong shock, it would be possible to obtain efficient shock compression. This process is utilized in supersonic-engine inlet diffusers, where a series of oblique shocks followed by a weak perpendicular shock are used to obtain efficient supersonic diffusion. Unfortunately, it is not physically possible to design a compression system where a series of weak perpendicular shocks can be employed. Further information on shocks and shock compressions is given in reference 1.
5. The characteristics of supersonic flow are completely different from, and in some ways opposite to, those of subsonic or incompressible flow. The equations governing supersonic flow are also nonlinear in form. The controlling parameter is $M^2 - 1$, which is plus or minus, depending on whether M is greater or less than unity. The supersonic equations of motion for ideal (nonviscous) flow are hyperbolic in character. The

"solutions" are by characteristics; a continuous analytical solution does not exist, but all flow properties are constant along characteristic lines or surfaces which are inclined at the local Mach angle to the flow direction. The method of characteristics provides solutions for supersonic flow fields by knowing the boundary conditions on the physical boundary (or other given curve). The flow conditions of velocity direction and magnitude, pressures, and so on, are then constant along Mach lines drawn from this boundary curve. Since these characteristic waves are reflected or canceled by surfaces and are reflected and refracted by other waves, the detailed design of supersonic nozzles requires that all such waves be canceled at the nozzle exit. This and other problems of compressible and supersonic flow are treated in more detail in advanced texts.

REFERENCES

1. A. H. Shapiro, *The Dynamics and Thermodynamics of Compressible Fluid Flow*, The Ronald Press Company, New York, 1953.
   An interestingly written engineering approach to compressible flow with many illustrations and applications to compressor and supersonic flow geometries (jet-engine systems, wind tunnels, and so on).
2. Ames Research Staff, Equations, Tables and Charts for Compressible Flow, *NACA Report 1135*, 1953.
   Isentropic, normal-shock, and oblique-shock equations and tabulated values and graphical results are presented. Effects of nonperfect gas behavior are also given.

PROBLEMS

12-1 (a) As the sound-radiating body illustrated in Fig. 12-1a passes the observer at point A, describe the variation of sound with respect to time. (b) Describe the variation of the tone heard at C as the sound source approaches from far off to the left. (c) Write an equation for the frequency at C as a function of sending frequency f_n and angle between C and the normal to the travel line of the source.

12-2 In a manner similar to that used for the construction of Fig. 12-1, sketch the wave pattern for the critical case where $u = a$ and comment on the result. If the object were sending out a tone at a fixed frequency, explain what an observer at A would hear.

12-3 Derive a method for determining the speed of an object by recording the frequency as it approaches and passes a fixed recorder.

12-4 A train approaches an observer who hears the fundamental of the tone as the note A (440 cycles/sec). After it passes, the pitch changes to G. An octave implies a frequency change by a factor of 2 and there are

12 notes in an octave; from G to A is actually two notes (two half-steps). Each note is then a variation of $2^{1/12}$ in frequency. If the sound speed is 1100 ft/sec, what is the train speed?

12-5 An airplane is observed directly overhead. It is first heard 6 sec later and observed at an angle of 30° with respect to the vertical. For an average speed of sound of 1100 ft/sec, calculate the Mach number, altitude, and speed of the plane.

12-6 Fill in the necessary steps to obtain (12-9).

12-7 Using the equations available in the set (12-8), reduce (12-9) to an equation in A and M only. Integrate the resulting equation between any arbitrary section and a critical section A^* (real or hypothetical) where $M = 1$.

12-8 From (12-8), determine the differential relation between M and v.

$$\frac{d\text{v}}{\text{v}} = \frac{2}{\gamma - 1} \frac{dM}{M\{1 + [(\gamma - 1)/2]M^2\}}$$

12-9 Using the result of Problem 12-7 and the continuity equation in the form

$$\frac{d\text{v}}{\text{v}} - \frac{d\rho}{\rho} = -\frac{dA}{A}$$

show that, for a diverging duct ($dA > 0$),

$$\frac{d\text{v}}{\text{v}} - \frac{d\rho}{\rho} = \frac{(1 - M^2)\,dM}{M\{1 + [(\gamma - 1)/2]M^2\}} = -\frac{dA}{A}$$

Discuss the cases of subsonic, sonic, and supersonic flow in this duct section.

12-10 Develop the series expansion for ρ_0/ρ in terms of M.

12-11 Determine a series expansion for p/p_0 as a function of M.

12-12 Determine a series expansion for either a p_0/p or p/p_0 as function of M for $M^2 > 1$.

12-13 Solve (12-11) for M as a function of p_0/p. Discuss the flow conditions and ducting geometry for maintaining isentropic flow in the limit as p approaches zero.

12-14 Are the pressures and temperatures in (12-10) through (12-13) absolute or gage quantities? Explain.

12-15 Obtain (12-12) from (12-6) and (12-4).

12-16 Air is accelerated from a large tank where the temperature is $T_0 = 60$°F and the $p = p_0$. The exhaust is to the atmosphere at $p_a = 14.7$ psia. What is the tank pressure for an exhaust Mach number of $M = 0.90$? What is the exit temperature? If you were to place your hand in the discharging stream, what temperature would you sense? Explain.

12-17 The p, ρ, and T ratios can also be determined in terms of a Mach number, $M_0 \equiv v/a_0$ (referred to the stagnation or reference reservoir sonic speed rather than the local value). Determine p_0/p, ρ_0/ρ, and T_0/T as functions of M_0 and γ.

12-18 Discuss the variation in mass flow rate $\dot{m}$ for a fixed ambient pressure (for example, 1 atm) through a diverging nozzle as the reservoir pressure is increased.

12-19 An air compressor is rated at 500 standard ft^3/min (that is, 500 ft^3/min at 1 atm and 68°F). What is the maximum critical nozzle area that can be used to obtain sonic flow if the compressor reservoir tank is to be maintained at 100 psi and the discharge from the tank is to the atmosphere?

12-20 For inlet sections to the compressor of Problem 12-19 from the atmosphere at standard conditions, what is the critical choking area A^*?

12-21 What are the units of 0.532 in (12-17b)?

12-22 For a nozzle similar to that in Fig. 12-7, except with $A_1/A_2 = A_3/A_2 = 2$, it is desired to maintain a uniform pressure gradient through the throat region. Draw a straight-line pressure gradient on graph paper connecting the two points from p/p_0 for subsonic $A/A^* = 2$ to p/p_0 for supersonic $A/A^* = 2$ through the point $p^*/p_0 = 0.528$. Carefully plot the area ratio A/A^* versus x for this pressure distribution. Be particularly accurate in the critical throat region with enough points to well define the area and the rate of change of area there. Comment on the resulting nozzle shape.

12-23 The design Mach number for impulse turbine and nozzles is about 1.7. For an air turbine, determine the overall pressure ratio for two stages. The pressure is constant through the rotor of an impulse stage; the pressure staging is then only in the nozzles. What is the nozzle-area ratio? What is the necessary inlet temperature for an exhaust temperature of 0°F?

12-24 Work the above problem for superheated steam with design inlet conditions of 600°F. What is the required inlet pressure if the exhaust is to be at saturated conditions? Assume the steam to be a perfect gas ($\gamma = 1.28$).

12-25 For the conditions of the above problem but with mercury vapor initially at 1200°F and 100 psi, what are the exit pressure and temperature? What is the nozzle exit velocity?

12-26 A supersonic wind tunnel is to be designed for testing at $M = 4$. If the test-section pressure must be maintained above 0.1 psia and the temperature above 0°F, then (a) what are the requirements on p_0 and T_0? (b) What is the test section to throat area ratio? (c) What is the Reynolds number per foot (the Reynolds number for a characteristic length of 1 ft)? (d) What is the horsepower represented by the test-section flow if the test-section area is 1 ft^2? (e) Sketch the geometry. (f) What is the approach velocity in an area equal to the test-section area?

12-27 A nuclear rocket using hydrogen is to have a thrust of 10^6 $\mathrm{lb_f}$. The design exhaust pressure is 2 psia and the chamber pressure is 1000 psia. Determine the throat and exit plane areas.

12-28 Determine the thrust force F_T, the I_{sp}, and $I_{sp,vac}$ in terms of $M_{0e} \equiv v_e/a_0$. Note that for expansion of the gas into a vacuum, $M_e \to \infty$, but $M_0 \to$ const. $= M_{0,max}$. Evaluate $M_{0,max}$ for a monatomic, a diatomic, and a triatomic gas.

12-29 For a representative $T_c = 6000°\mathrm{R}$, determine the $I_{sp,vac}$ for (a) air, (b) chemical combustion products that are essentially triatomic with a mean $\mathscr{M} = 28$, (c) solid propellant products with $\gamma = 1.22$ and $\mathscr{M} = 60$, and (d) hydrogen.

12-30 Discuss the I_{sp} of an ion rocket. What is the I_{sp} of an electron gas?

12-31 Determine M as a function of M_0, and the inverse relation, M_0, as a function of M.

12-32 Starting with the energy equation, (12-22a), determine M as a function of M^*.

12-33 Verify (12-26) and (12-28).

12-34 From Fig. 12-9 it appears that M_2 approaches a nonzero asymptotic value as M_1 approaches infinity. Find this value if it exists.

12-35 Determine the ratio M_2/M_1 across a normal shock by noting that it also can be determined using the continuity equation and the condition $M_2/M_1 = (v_2/v_1)(\sqrt{T_1/T_2})$.

12-36 In determining the compressor efficiency, it was stated that $p_2 \approx p_{02}$ and $\tilde{h}_2 \approx \tilde{h}_{02}$ since $M_2 < 1$. Find the error in η_D produced by neglecting M_2 for $M_2 = 0.1$ and 0.2 if $M_1 = 2.0$.

12-37 Sketch nonideal nozzle flow (from $M_1^2 \ll 1$ to M_2) on an $\tilde{h}$-$\tilde{s}$ diagram. Indicate $\tilde{h}_{01} = \tilde{h}_{02} \approx \tilde{h}_1$, $p_1 \approx p_{01}$, isentropic, and real process lines. Determine the nozzle efficiency, η_N, in terms of the stagnation pressure ratio and nozzle Mach number M_2 in a manner similar to that by which η_D was determined in (12-39).

12-38 Assuming a wind-tunnel configuration such as that of Fig. 12-11, determine T_2 and p_2 for $M_2 = 3.0$ and standard atmospheric inlet conditions at 1 ($A_1 \gg A_{1t}$).

12-39 For a test-section Mach number $M_2 = 3$, $p_{03}/p_{01} = 0.92$ determined as a result of model and duct "losses" and $\eta_D = 0.8$ for the diffuser section 3–4: Find (a) M_3, (b) p_{04}/p_{01}, and (c) the mass flow rate $\dot{m}$. (d) It will be helpful to sketch the processes 1 through 4 on an $\tilde{h}$-$\tilde{s}$ diagram. (e) The power required for this test-section flow if $A_3 = 1\ \mathrm{ft}^2$ and standard atmospheric conditions exist at 1.

12-40 For an explosion wave traveling into atmospheric air, the propagation Mach number is measured as 4.0 at a certain radius. What are p_0' and T_0' behind this wave?

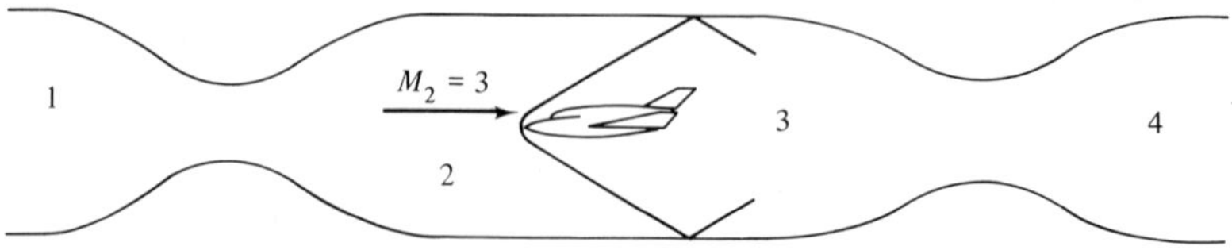

Fig. P12-39.

CHAPTER 13

CONTENTS

OPEN-CHANNEL FLOW

Flow in aqueducts, flumes, and rivers is the oldest aspect of the study of hydraulics. The remaining portions of Roman aqueducts are evidence of the ability of the Roman hydraulic and structural engineers. In the early and middle eighteenth century, the need again arose to increase the water supply to population centers such as Paris. A number of relations for the flow velocity in an open channel were proposed. The first recorded development of a satisfactory relation for the flow in an open channel is credited to Anton Chezy,[1] who reasoned that the friction would be about proportional to $v^2 P$ ($P \equiv$ wetted perimeter) and the moving force due to gravity would be proportional to the flow area A and the slope S, so that for steady, fully developed flow

$$v \propto \sqrt{\frac{AS}{P}} \qquad \text{or} \qquad v \propto \sqrt{RS} \tag{13-1a}$$

where the "hydraulic radius" $R \equiv A/P$ (for a circle of radius r_0, $R = r_0/2$). This relationship was written as an equation with a factor of proportionality (which depends on the channel conditions) and is termed the *Chezy equation*:

$$v = C\sqrt{RS} \tag{13-1b}$$

A modification of this relation in the form

$$v = \frac{1.486}{n} R^{2/3} S^{1/2} \tag{13-2}$$

is known as *Manning's formula* (or the Chezy–Manning formula) for steady water flow in open channels, although Manning did not present it in this form. The factor n is a roughness correlation factor.

[1] Anton Chezy (1718–1789), civil engineer of Paris. Developed the above relation while studying the best way to increase the water supply to Paris from the Yvette River.

Because of its inherent simplicity, (13-2) has been widely used in hydraulics work for closed-channel flows of other fluids as well as open-channel water flow. Volumes of modification factors have been developed for other fluids and conditions for which the formula was not initially intended. The advantages of the apparent simplicity of (13-2) are then exceeded by the disadvantages of the supplemental material required because of the lack of more exact and rigorous developments.

In the flow through closed conduits (ducts, pipes, and so on) the geometry is regular and specified, or is to be determined as part of the design problem (for example, the diameter or hydraulic radius of the flow channel). For open-channel flow, the area is only partially specified by the geometry, and the shape and slope of the free surface are not known. The running depth as well as the velocity will in general be functions of the slope and surface conditions of the channel.

Surface wave motion is another class of free-surface flows. These wave phenomena are similar to those encountered in supercritical compressible flows discussed in Chapter 12.

13.1 ANALYSIS OF STEADY FLOW IN OPEN CHANNELS

Consider the flow in an open channel such as a river or stream bed with an irregular bottom as represented in Fig. 13-1. The possibility of any significant density variation for a liquid flowing with a free surface is negligible. The continuity equation can then be used in the simple form

$$Q = \iint \vec{\text{v}} \cdot \vec{dA} \equiv A\bar{\text{v}} \tag{13-3}$$

The significant component of the momentum equation is in the x direction along the stream bed, which is at an average slope angle θ with the horizon or constant gravity surface. The x-component forces acting on the control volume defined in Fig. 13-1 are

$$\sum F_x = F_{p,x} + F_{g,x} + F_\tau$$

The pressure force on the control volume is

$$F_{px} = -\iint_{A_x} p \, dA$$

where the plane of A_x is taken perpendicular to the flow direction or nominally perpendicular to x. The control volume is taken with a small incremental length Δx. The channel wall curvature in the flow direction is usually quite small for open-channel flow; consequently, the y, or normal,

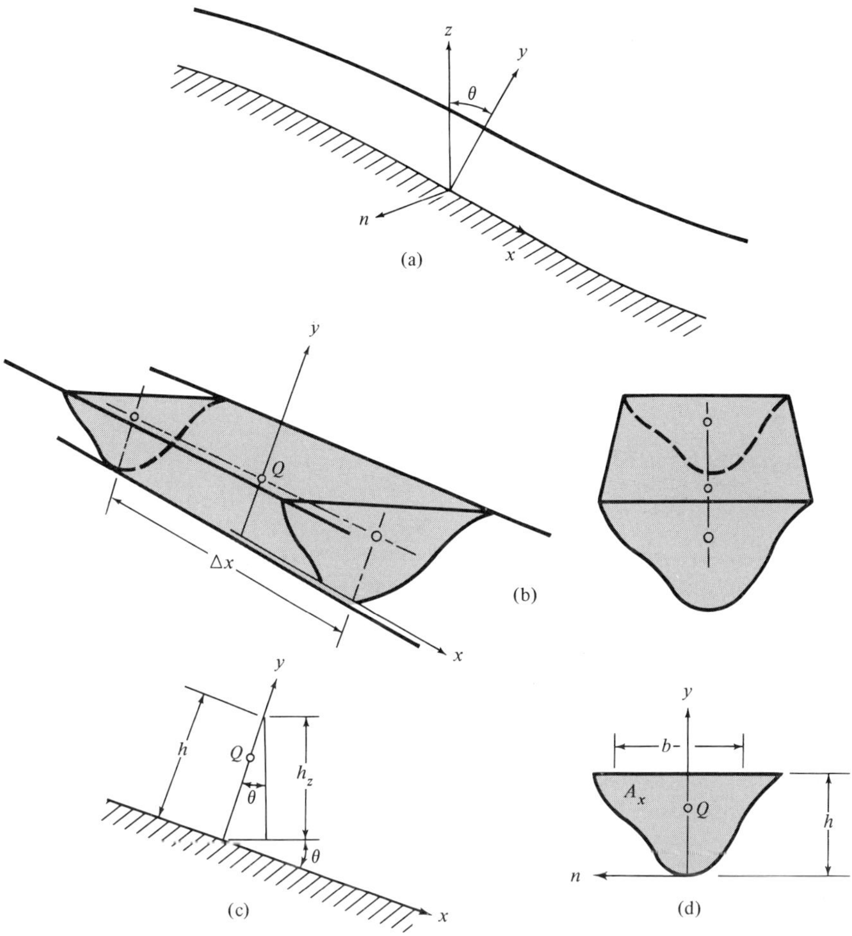

Fig. 13-1. (a) Coordinate representation for flow in an open channel. (b) Two views of the control volume for determination of fluid mechanical equations for open-channel flow. At point Q, at the center of the element, the variables have the values p, ρ, g, h, A, and P. (c) Coordinate direction representation. (d) Area section perpendicular to the flow at Q. See also Fig. 13-6.

components of the terms of the force momentum equation in these directions will be negligible with respect to the x terms. The pressure-force term has one contribution from the forces on the end (A_x) surfaces and one from the pressure force on the side or peripheral (A_P) surface. The pressure force on the A_x surface at the center reference point is just $\bar{p}A_x \equiv \iint_{A_x} p\, dA$. The pressure forces on the right and left faces are, respectively,

$$pA_x + \frac{d(pA_x)}{dx}\frac{\Delta x}{2} \qquad \text{and} \qquad pA_x + \frac{d(pA_x)}{dx}\left(-\frac{\Delta x}{2}\right)$$

Total derivatives are indicated, since the other two component effects are assumed to be negligible. The net pressure force on these faces is then

$$-\frac{d(pA_x)}{dx}\Delta x$$

The x-pressure force on the wetted peripheral face is just the pressure at the reference center times the x projection of the peripheral area which is

$$\frac{dA_x}{dx}\Delta x.$$

The gravity force is

$$F_{g_x} = \frac{\rho g}{g_c} A\,\Delta x \sin\theta$$

and it is noted that $\Delta z = -\Delta x \sin\theta$. The shear on the peripheral area (neglecting the "free-surface" shear) is

$$F_\tau = \tau_0 A_P$$

where A_P = peripheral area, $A_P \equiv P\,\Delta x$.

The momentum flux through the center plane is

$$\iint_{A_x} \rho \mathrm{v}^2\, dA = \beta\rho\bar{\mathrm{v}}^2 A_x = \beta\dot{m}\bar{\mathrm{v}}$$

(The bar will be dropped from the average velocity and henceforth v will be understood to denote average velocity, $\mathrm{v} \equiv Q/A$.) The net momentum flux out of the control volume is

$$\frac{d}{dx}(\beta\dot{m}\mathrm{v})\,\Delta x = \beta\dot{m}\frac{d\mathrm{v}}{dx}\Delta x$$

assuming that the rate of change of channel shape is small so that $\beta \approx$ const. Open-channel flows are usually of sufficiently large Reynolds number that $\beta \approx 1.0$. The x component of the force–momentum equation for the indicated control volume can now be written

$$-\frac{d}{dx}(pA_x)\,\Delta x + p\frac{dA}{dx}\Delta x + \frac{\rho g}{g_c}A_x\,\Delta x \sin\theta - \tau_0 A_P = \frac{\beta\dot{m}}{g_c}\frac{d\mathrm{v}}{dx}\Delta x$$

Inserting the above-defined quantities and simplifying gives

$$-A_x \frac{dp}{dx} \Delta x - \frac{\rho g}{g_c} A_x \frac{dz}{dx} \Delta x - \tau_0 P \, \Delta x = \frac{\dot{m}}{g_c} \frac{d\mathrm{v}}{dx} \Delta x \qquad \text{(13-4a)}$$

The pressure is only the hydrostatic pressure and can be expressed as

$$p = \frac{\rho g}{g_c} (h - y) \cos \theta$$

For open-channel flows, ρ and g can be assumed to be constant; (13-4a) can then be divided by the *weight* of the element $(\rho g/g_c)A_x \, \Delta x$ to give

$$\frac{dh_z}{dx} + \frac{dz}{dx} + \frac{g_c \tau_0}{\rho g R} + \frac{\mathrm{v}}{g} \frac{d\mathrm{v}}{dx} = 0 \qquad \text{(13-4b)}$$

where $h_z = h \cos \theta$ is the vertical projection of h and $R \equiv A_x/P$. Note that the hydraulic radius defined as the cross-sectional area divided by the wetted perimeter is one quarter of the hydraulic diameter, $R \equiv D_H/4$. The continuity equation can be used to eliminate $d\mathrm{v}$, giving

$$\frac{dh_z}{dx} + \frac{dz}{dx} + \frac{g_c \tau_0}{\rho g R} - \frac{\mathrm{v}^2}{g} \frac{dA}{A \, dx} = 0 \qquad \text{(13-5)}$$

The area may be defined arbitrarily as $A \equiv \bar{b} h$ (Fig. 13-1b). With this substitution for A, (13-5) gives

$$\frac{dh_z}{dx} - \frac{\mathrm{v}^2}{gh} \frac{dh}{dx} + \frac{dz}{dx} + \frac{g_c \tau_0}{\rho g R} - \frac{\mathrm{v}^2}{g\bar{b}} \frac{d\bar{b}}{dx} = 0 \qquad \text{(13-6)}$$

For relatively small slope, $h_z \approx h$ ($\cos \theta \approx 1$) and

$$\frac{dh}{dx}(1 - N_F^2) + \frac{dz}{dx} + \frac{g_c \tau_0}{\rho g R} - \frac{\mathrm{v}^2}{g\bar{b}} \frac{d\bar{b}}{dx} = 0 \qquad \text{(13-7)}$$

where N_F is the local *Froude number* $N_F \equiv \mathrm{v}/\sqrt{gh}$ based on local values of v and h.

CRITICAL FLOW BEHAVIOR

Consider frictionless, "one-dimensional" (b = const.) rectangular channel flow for which (13-7) gives

$$\frac{dh}{dx} = \frac{dz/dx}{N_F^2 - 1} \qquad \text{(13-8)}$$

and the continuity equation is $dv/v = -dh/h$, or dv/dx behaves in the opposite sense to dh/dx. The behavior of dh/dx (and dv/dx) for "downhill" ($dz/dx < 0$) and "uphill" ($dz/dx > 0$) flows is indicated in Fig. 13-2. "Uphill" flow can only exist for relatively short distances following an accelerating downhill flow that has produced a supercritical flow

The manner in which N_F indicates different flow behaviors is similar to the behavior of compressible flow, as indicated by the Mach number. As in compressible flow, at least three predominant actions simultaneously

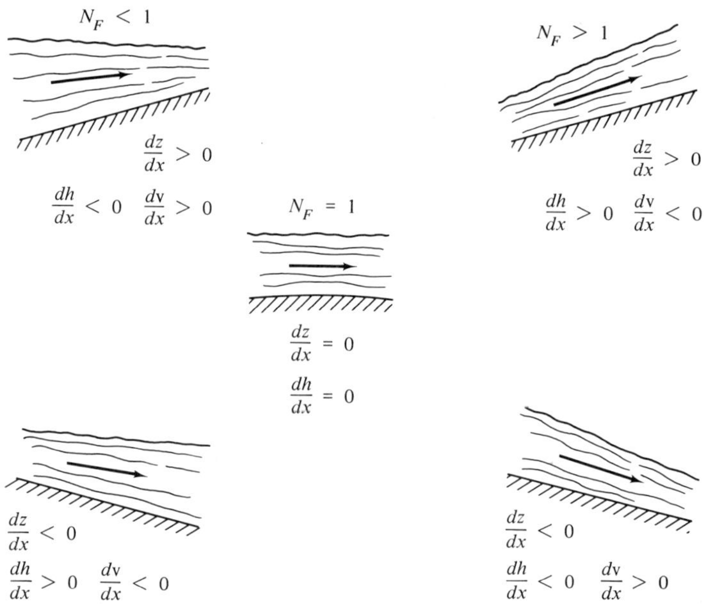

Fig. 13-2. Graphical representation of flow-depth relationships: channel slope–flow depth–Froude number behavior.

affect open-channel flow: (1) the momentum, (2) the gravitational force due to the slope of the bottom, and (3) the hydrostatic pressure force (which is a consequence of the gravitational field). These three effects are coupled, so that it is again difficult to draw any "intuitive" conclusions regarding the flow behavior.

Another critical behavior exists for constant depth ($dz/dx = 0$), variable width, and channel flow. Again, for the idealized case of frictionless flow, these conditions in (13-6) give

$$\frac{dh}{h} = \frac{N_F^2}{N_F^2 - 1} \frac{db}{b} \tag{13-9}$$

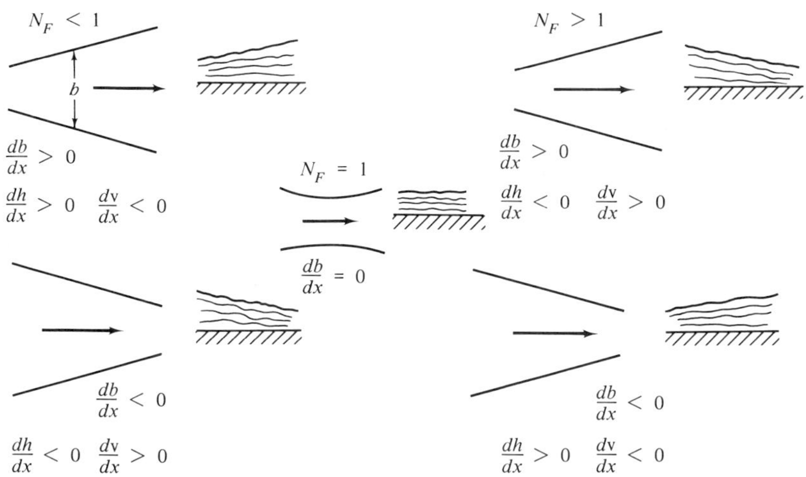

Fig. 13-3. Flow behavior as a function of Froude number in a channel of varying width.

which indicates the flow conditions illustrated in Fig. 13-3. This behavior is directly analogous to that of one-dimensional compressible gas flow, indicating the possibility of an open-channel analog to compressible flow. This analogy has been used to great advantage in initial studies of supersonic geometries. Such a *water table* is more expedient and economical to construct than a supersonic wind tunnel. One limitation of the method is that the $\gamma \equiv c_p/c_v$ ratio for the surface-table analog is $\gamma = 2.0$, which does not correspond to any real gas.

SMALL-DISTURBANCE WAVES

Since the wave-propagation speed of a small disturbance in a compressible medium was determined by the condition that the wave Mach number $= 1$ ($M_w = v_w/a_w$), it may be anticipated that the propagation speed of a small surface wave would be given by the condition $N_F = 1$ or $v_w = \sqrt{gh}$.

Consider a wave ($\Delta h/h \ll 1$) propagated along a channel with a speed v_w into a liquid at rest as indicated in Fig. 13-4a (or the stationary flow relative to the wave front illustrated in Fig. 13-4b). The continuity and momentum equations can be applied to the control volume to give the result (see Problem 13-8)

$$v_w = \sqrt{gh} \tag{13-10}$$

Note that dh/dx can be either positive or negative, depending on the direction of v_w. If dh/dx is positive, dv/dx is negative and conversely.

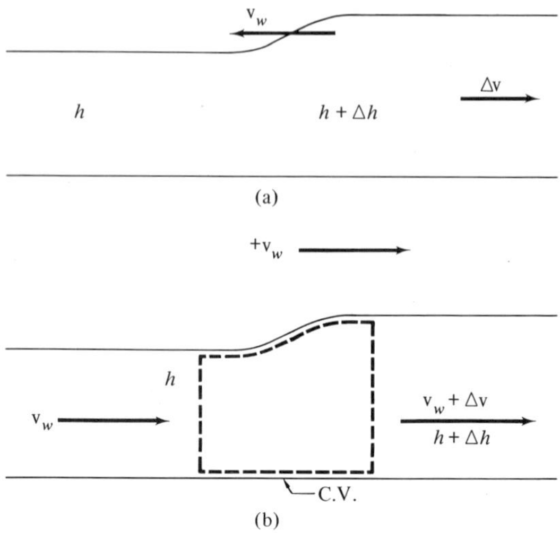

Fig. 13-4. Control volume for determining the wave-propagation speed of a small surface wave: (a) wave moving into a fluid at rest; (b) the wave motion transformed for a steady-state control volume by superposition of a uniform field v_w to the right on the flow field of (a).

This derivation is for frictionless flow ($\tau = 0$) in a horizontal, constant-width channel ($dz/dx = 0$, $db/dx = 0$), leaving

$$\frac{dh}{dx}(1 - N_F^2) = 0$$

which has the trivial solution $dh/dx = 0$, or the solution $N_F = 1$. Note that the only restriction on this simple derivation for determining v_w is that $\Delta h/h \ll 1$. Therefore, the reflection, refraction, and interference of waves (including light and sound) in various media can be studied in a shallow ripple tank.[2] Although the above characteristic wave relation was derived for a one-dimensional model, it is also applicable to a circular wave on an infinite fluid surface, and to wave systems (see Problem 13-12). So far there is no indication in the relations used that $\Delta h/h \ll 1$ is a limitation pertaining to characteristic wave propagation. As in the case of compressible flow, it is necessary to consider the energy and process relations to quantitatively indicate the limiting behavior for characteristic or Froude waves.

[2] One of the best constructions is a shallow tank with a transparent bottom, floodlit from the bottom with an overhead projection screen. The light diffraction through the water will project the wave patterns as shadow patterns on the screen. A water depth of 1 in. is sufficient with ripples on the order of 0.1 in.

It is interesting to note that the relation for a "small" wave is entirely relative. It is equally valid for a 0.1-in. wave on a 1-in.-deep ripple tank, or a 30-ft tidal wave on a 3000-ft-deep ocean.

In general, flow conditions are termed *tranquil* for $N_F < 1$, *rapid* for $N_F > 1$, and *critical* for $N_F = 1$.

13.2 UNIFORM FLOW: Fully Developed Flow in a Channel of Uniform Shape

For the flow in a channel with constant slope and cross-sectional area far from any transitions in the geometry, the flow and its velocity distribution can be considered to be constant along the channel. These assumptions define the following conditions:

1. Fully developed or uniform flow: $dh/dx = 0$, $h_x = \text{const.}$
2. Uniform channel shape: $db/dx = 0$, $b = \text{const.}$
3. Uniform channel slope: $dz/dx = \text{const.} = -S_u$ (uniform slope).
4. $\text{v} = \text{const.}$

Equation (13-6) then gives

$$\tau_0 = \frac{\rho g}{g_c} R S_u \qquad \text{or} \qquad S_u = \frac{g_c \tau_0}{\rho g R} \tag{13-11}$$

Under the assumed conditions, the shear stress can also be defined in terms of a friction coefficient,

$$c_f \equiv \frac{\tau_0}{\rho \text{v}^2 / 2g_c} \tag{13-12}$$

In general, c_f is a function of Reynolds number, roughness, shape (S, R, and so on). However, most open channels are relatively rough and sufficiently large that the flow is essentially independent of Reynolds number. Some empirical work from the nineteenth century led to the formula for expressing the average flow velocity in an open channel by Manning's formula (13-2),

$$\text{v} = \frac{1.49}{n} R^{2/3} S_u^{1/2}$$

The roughness parameter n is obtained experimentally. Some representative values of n are tabulated in Table 13-1. The above three expressions

TABLE 13.1

ROUGHNESS PARAMETER n

Channel Material	n
Planed wood	.009–.011
Unplaned timber	.012–.013
Finished plaster or concrete	.010–.012
Unfinished concrete	.012–.014
Brick and sewer pipe	.013–.016
Cast iron	.012–.015
Stonework, rough brick	.017–.022
Earth (river bottom)	.025
Earth with stones and weeds	.03–.04
Gravel, crushed rock	.027–.03

can be combined to give an equation for c_f in terms of the roughness parameter n and the hydraulic radius

$$c_f = \frac{2gn^2}{(1.49)^2 R^{1/3}} = \frac{29.1n^2}{R^{1/3}} \tag{13-13}$$

13.3 HYDRAULIC JUMP

The flow conditions will be considered for a constant-width channel (b = const.). The combination of the energy equation with the momentum and continuity equations will be examined for possible interesting flow behavior predictions. As in the case of compressible flow, the existence of a large-scale, irreversible discontinuity may be expected, since an analogous critical flow behavior ($N_F = 1$, $M = 1$) has been noted. Because of the complexity in more complicated geometries, the flow will be considered to be steady, one-dimensional, and fully developed (or initially uniform). These restrictions necessarily imply $dz/dx = 0$ when applied to open-channel flow. The surface friction will be neglected; any irreversibility that occurs will initially be assumed to be predominantly "internal" (again analogous to the "shock-wave" irreversibility). The flow control volume is indicated in Fig. 13-5. The momentum equation gives

$$\frac{\rho Q}{g_c}(\mathrm{v}_2 - \mathrm{v}_1) = \frac{\rho g b}{2g_c}(h_1^2 - h_2^2)$$

which, with the continuity equation,

$$Q = bh_2\mathrm{v}_2 = bh_1\mathrm{v}_1 = \text{const.}$$

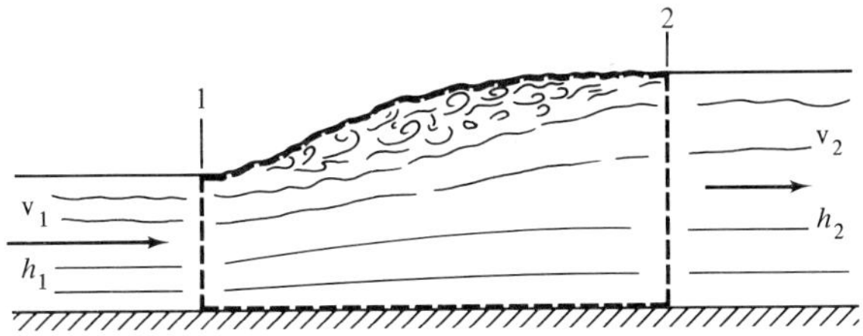

Fig. 13-5. Control volume representation for a finite discontinuity or hydraulic jump.

can be written in the form

$$(h_1 - h_2)\left(h_1 + h_2 - \frac{2Q^2}{gb^2h_1h_2}\right) = 0 \tag{13-14}$$

One trivial solution is $h_1 = h_2$. The second solution for h_2, or the ratio h_2/h_1, is obtained by arranging (13-14) in the form

$$\left(\frac{h_2}{h_1}\right)^2 + \frac{h_2}{h_1} - 2N_{F1}^2 = 0$$

which gives

$$\frac{h_2}{h_1} = \tfrac{1}{2}(\sqrt{1 + 8N_{F1}^2} - 1) \tag{13-15a}$$

The depths h_1 and h_2 before and after the jump are termed the *conjugate depths*. The continuity equation and the definition of N_F can be combined to give

$$\frac{N_{F2}}{N_{F1}} = \left(\frac{h_1}{h_2}\right)^{3/2}$$

allowing (13-15) to be solved for N_{F2} as a function of N_{F1},

$$N_{F2} = N_{F1}[\tfrac{1}{2}(\sqrt{1 + 8N_{F1}^2} - 1)]^{-3/2}$$

or

$$\left(\frac{N_{F2}}{N_{F1}}\right)^2 = \frac{8}{(\sqrt{1 + 8N_{F1}^2} - 1)^3} \tag{13-15b}$$

Note that for $N_{F1} = 1$, $h_2 = h_1$, which corresponds to the solution for a small disturbance or "characteristic" wave (similar to the situation for a characteristic Mach wave in a compressible flow).

Physically, this discontinuity occurs when an initial supercritical flow exists but where there is not enough channel slope to maintain the supercritical condition. Figure 13-10 and Section 13.7 illustrate this situation.

13.4 ENERGY EQUATION FOR OPEN-CHANNEL FLOW

The energy equation as developed in Chapters 5 and 6 can be applied to the control volume of Fig. 13-6. Assuming adiabatic flow (negligible heat transfer) the work–energy relation is

$$\oiint e\rho \vec{v} \cdot \vec{dA} = \sum \text{WORK RATE ON C.V. FLUID} \qquad \text{(13-16a)}$$

The applicable terms of the internal energy per unit mass e are the kinetic energy and the internal thermal energy $\tilde{u}$. Work is done on the control volume by pressure and gravity forces. The channel surface friction work is zero, since the velocity is zero on the surface. With consideration of these terms, (13-16a) becomes

$$\oiint \left(\frac{v^2}{2g_c} - \Omega + \tilde{u}\right)\rho \vec{v} \cdot \vec{dA} = -\oiint p\vec{v} \cdot \vec{dA}$$

where the body-force potential Ω has been introduced and transposed to the left side of the equation. Taking $\vec{v}$ perpendicular to the area plane (Fig. 13-6), with $dA = b\,dz$, and expanding the integrals gives

$$\int_0^{h_2} \frac{v^3}{2g_c}\rho b\,dz - \int_0^{h_1} \frac{v^3}{2g_c}\rho b\,dz + \int_0^{h_2} \frac{g}{g_c}(z_2 + y\cos\theta_2)\rho v b\,dy$$
$$- \int_0^{h_1} \frac{g}{g_c}(z_1 + y\cos\theta_1)\rho v b\,dy + \int_0^{h_2} \tilde{u}\rho v b\,dz - \int_0^{h_1} \tilde{u}\rho v b\,dz$$
$$= -\int_0^{h_2} \frac{\rho g}{g_c}(h - y)\cos\theta_2\, vb\,dz + \int_0^{h_1} \frac{\rho g}{g_c}(h - y)\cos\theta_1\, vb\,dz \qquad \text{(13-16b)}$$

The zero subscript is used to indicate conditions at the channel bottom. As in the evaluation of the momentum equation, the velocity profiles at 1 and 2 are assumed to be sufficiently uniform that the kinetic-energy correction factor $\alpha \approx 1.0$. Integrating (13-16b), combining terms, and using the continuity equation $\dot{m} = \text{const.}$ gives

$$\frac{\dot{m}}{2g_c}(v_2^2 - v_1^2) + \frac{\dot{m}g}{g_c}[(z + h\cos\theta)_2 - (z + h\cos\theta)_1] + \dot{m}(\tilde{u}_2 - \tilde{u}_1) = 0 \qquad \text{(13-16c)}$$

As described in Chapter 6, the change in the internal energy $\tilde{u}$ is a consequence of the irreversible work done within the control volume and can be

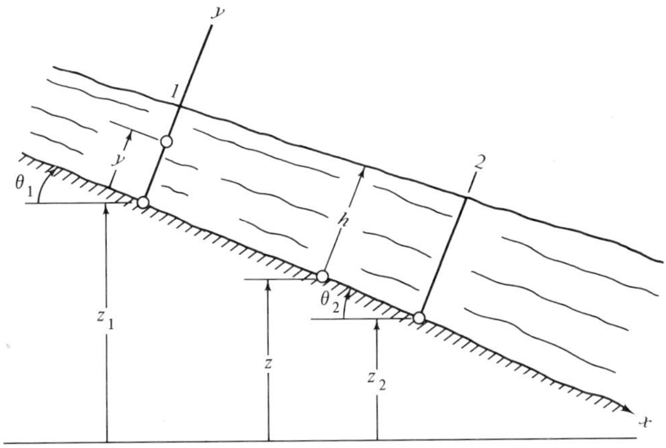

Fig. 13-6. Control volume and coordinate representation for determination of the energy equation.

considered as a loss of availability and defined as $\Delta\tilde{u} \equiv g\,\Delta h_{\text{loss}}/g_c$. The energy equation then becomes (see Problem 13-18)

$$\frac{v_1^2}{2g} + z_1 + h_1 \cos\theta_1 - \Delta h_{\text{loss}} = \frac{v_2^2}{2g} + z_2 + h_2 \cos\theta_2 \qquad \text{(13-16d)}$$

For most cases where channel flows correspond to the initial assumptions of negligible streamwise curvature and nearly uniform velocity profiles at any section, the bed slope is relatively small, so $\cos\theta \approx 1$ the *elevation* of the water surface. The energy equation for these assumed conditions can then be written

$$\frac{v_1^2}{2g} + z_1 + h_1 - \Delta h_{\text{loss}} = \frac{v_2^2}{2g} + z_2 + h_2 \qquad \text{(13-16e)}$$

The substitution $z = z + h$ has not been used in this equation because both z and h are functions of x and therefore are dependent variables. Note that for channels with nearly uniform slope, $\Delta z/\Delta x = \Delta z_0/L = S_0$, the channel-bed slope. Defining the quantity h_0 as the total available energy and work,

$$h_0 \equiv \frac{v^2}{2g} + h\cos\theta + z_0$$

then (13-16d) or (13-16e) can be interpreted as[3]

$$h_{01} - h_{\text{loss}} = h_{02} \qquad \text{(13-16f)}$$

[3] This h_0 is then seen to be analogous to the stagnation enthalpy.

In the hydraulic jump, the nature of the irreversibility is similar to that in the orifice as described in Section 6.12. A turbulence wave (stationary "breaker") accompanies the surface jump. The ratio $\Delta h_{loss}/h_1$ can be obtained from (13-16e) as

$$\frac{\Delta h_{loss}}{h_1} = 1 - \frac{h_2}{h_1} + \frac{N_F^2}{2}\left[1 - \left(\frac{h_1}{h_2}\right)^2\right] \tag{13-17}$$

where the continuity equation $v_1 h_1 = v_2 h_2$ has been used and the elevation change (Δz) for the length of the jump has been neglected. This result may be determined as a function of N_{F1} only by substituting h_2/h_1 from (13-15). A consequence of this combined solution is that Δh_{loss} is less than zero if $N_F < 1$. This result is impossible, which indicates that a hydraulic jump can occur only if the approach flow is initially supercritical (see Problems 13-16 and 13-17). Also if $N_F = 1$, $\Delta h_{loss} = 0$.

13.5 CRITICAL FLOW

As stated in the introductory section, critical flow can be obtained by accelerating a subcritical flow, for example, from a reservoir as indicated in Fig. 13-7, over an adverse gradient from 1 to 3, and then over a favorable gradient from 3 to 4. If friction is neglected between the reservoir and the

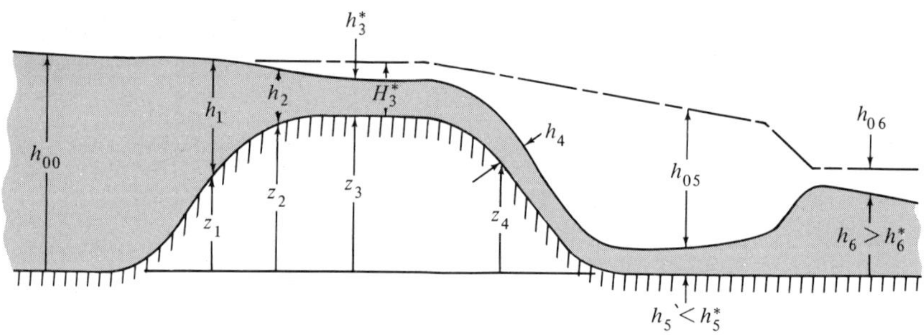

Fig. 13-7. Open-channel profile with various stages of flow.

critical flow point, the total head h_0 is constant and the energy equation may be written

$$h_0 = \frac{v_3^2}{2g} + z_3 + h_3 = \frac{v^{*2}}{2g} + z_3 + h_3^* \tag{13-18}$$

where the asterisk denotes critical conditions at which location $N_F = 1$, or $v^{*2} = gh^*$. This substitution gives

$$h_0 - z_3 \equiv H_3^* = 1.5h_3^* \tag{13-19}$$

where the quantity $H \equiv h_0 - z$ is sometimes termed the "specific head." This result states that the *local critical depth* is two thirds the *reference* reservoir depth measured from the local channel bed as illustrated in Fig. 13-7. In general, the variation of the *local* H with h can be investigated by means of the definition of the *local* stagnation head h_0 and the local specific head H. With the definition of H and Q, (13-18) can be written

$$H = \frac{Q^2}{2gb^2h^2} + h \tag{13-20a}$$

This equation is a cubic in h as a function of Q and H. For steady flow (Q = const.) it has only two real roots. Equation (13-20a) can be put into dimensionless form by dividing by h^*, giving

$$\frac{H}{h^*} = \frac{Q^2}{2gbh^2h^*} + \frac{h}{h^*} \tag{13-20b}$$

For constant Q, the reference value $Q^2 = b^2h^{*2}v^{*2}$ can be substituted. Then since $v^{*2} = gh^*$, (13-20b) reduces to

$$\frac{H}{h^*} = \frac{1}{2}\frac{h^*}{h} + \frac{h}{h^*} \tag{13-20c}$$

which is plotted in Fig. 13-8. The vertex of the curve occurs at the point where $dH/dh = 0$ (or $h/h^* = 1.0$); this condition gives

$$h = \left(\frac{Q^2}{gb^2}\right)^{1/3} = \left(\frac{v^2h^2}{g}\right)^{1/3} = h^*$$

or $v^* = \sqrt{gh^*}$, so the conditions at the vertex are seen to be critical, as indicated by the asterisk. It is again indicated that $h^* = \frac{2}{3}H^*$ and that this vertex is the critical point. For a given H (total available energy with respect to the local channel bed), two values of h will exist [the two real roots of (13-20a)]. Similarly, for a given h_0 (or H), a critical value of the flow rate can be defined as

$$Q^* = A^*v^* = bh^*v^* = \sqrt{g}\, bh^{*3/2} = \sqrt{g}\, b(\tfrac{2}{3}H^*)^{3/2} \tag{13-21}$$

The critical flow rate is also the maximum that can occur for a given H (or h_0 and z). A dimensionless flow rate can be obtained by referring the equation for Q from (13-20a) to $Q^* = Q_{\max}$,

$$\left(\frac{Q}{Q^*}\right)^2 = \frac{2gb^2h^2(H-h)}{gb^2h^{*3}}$$

or

$$\frac{Q}{Q^*} = \sqrt{\frac{h}{h^*}} \tag{13-22}$$

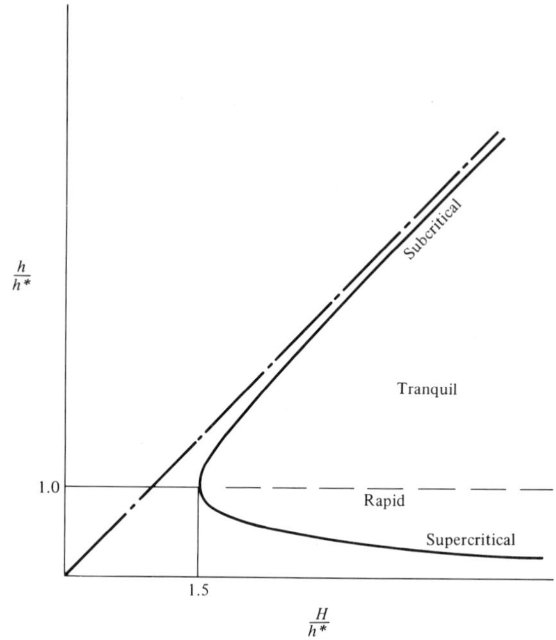

Fig. 13-8. Local depth h as a function of specific head H represented dimensionlessly as h/h^* versus H/h^*.

13.6 CLASSIFICATION OF SURFACE PROFILES

The surface profiles of the water level with respect to the channel bed $h(x)$ are termed the *backwater curves.* The bed profile or bed slope is specified as $z_0(x)$ or $S_0 = -\Delta z_0/\Delta x$. The variation of the water surface with respect to the channel-bed surface dh/dx can be obtained from (13-7). Considering channels with approximately uniform width (b = const.),

$$\frac{dh}{dx} = -\frac{(dz/dx) + (g_c\tau_0/\rho g R)}{1 - (\mathrm{v}^2/gh)} = S_0 \frac{1 - (g_c\tau_0/\rho g R)/S_0}{1 - (\mathrm{v}^2/gh)} \tag{13-23}$$

Using the continuity equation, and since $\mathrm{v}^{*2}/gh^* = 1$, the square of the Froude number $N_F^2 = \mathrm{v}^2/gh$ can be written as the ratio

$$\frac{\mathrm{v}^{*2}}{gh^*}\frac{\mathrm{v}^2}{\mathrm{v}^{*2}}\frac{h^*}{h} = \left(\frac{h^*}{h}\right)^3 \tag{13-24}$$

The shear term can be evaluated if the Manning relation is assumed to be valid; (13-12) and (13-13) can be used to give

$$\frac{g_c\tau_0}{\rho g R} = \frac{g_c}{\rho g R} c_f \frac{\rho \mathrm{v}^2}{2g_c} = \frac{\mathrm{v}^2}{2gR}\frac{2gn^2}{(1.49)^2 R^{1/3}} = \frac{n^2\mathrm{v}^2}{(1.49)^2 R^{4/3}} \tag{13-25}$$

Uniform flow conditions at a reference depth h_u can be assumed as a reference quantity allowing (13-11) to be used as

$$S_0 = S_u = \left(\frac{g_c \tau_0}{\rho g R}\right)_u \tag{13-26}$$

The shear ratio term can then be written

$$\frac{\tau_0}{\tau_{0u}} = \frac{n^2 b^2 h_u^2 R_u^{4/3} Q^2}{n_u^2 b^2 h^2 R^{4/3} Q_u^2}$$

The roughness factor will be assumed to be constant for the given channel. Also $Q = Q_u$, since h_u and v_u have been assumed to be reference quantities for the flow. For the purposes of this investigation assume that the channel width b is significantly greater than the channel depth h; then $R = R_u \approx h$. With these substitutions (13-23) becomes

$$\frac{dh}{dx} = S_0 \frac{1 - (h_u/h)^{10/3}}{1 - (h^*/h)^3} \tag{13-27}$$

which is important in indicating the rate of change of depth as a function of the three quantities S_0, h/h_u, and h/h^*. The bed conditions are defined with respect to the behavior of S_0:

Horizontal (H):	$S_0 = 0$
Mild (M):	$S_0 > 0$ and $h_u/h^* > 1$ (the uniform flow is subcritical)
Critical (C):	$S_0 > 0$ and $h_u/h^* = 1$ (the uniform flow is critical)
Steep (S):	$S_0 > 0$ and $h_u/h^* < 1$ (the uniform flow is supercritical)
Adverse (A):	$S_0 < 0$

In general, three subcategories can be defined with respect to the regimes:

1. Flow is initially subcritical and slower than uniform $h > h_u \geq h^*$.
2. Flow is initially subcritical and faster than uniform $h < h_u$, $h > h^*$.
3. Flow is initially supercritical and faster than uniform $h < h_u$, $h < h^*$.

As an example, consider flow over a horizontal bed ($S_0 = 0$); (13-23) gives (see Problem 13-20)

$$\frac{dh}{dx} \sim \frac{1}{1 - (h^*/h)^3} \tag{13-28}$$

If the flow is initially subcritical, $h^*/h < 1$ (type H2 flow) the slope is negative, and it is positive if the flow is initially supercritical (type H3). As h approaches h^* the slope approaches minus or plus infinity, a hypothetical limiting case is approached as $x \to \infty$. These characteristic slopes are indicated in Fig. 13-9a. In the case of a mild slope ($h_u > h^*$), all three

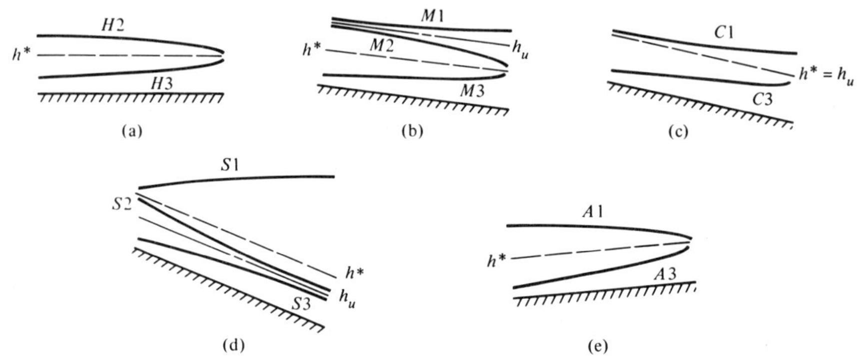

Fig. 13-9. Surface-profile curves for different bed-slope conditions.

subcategories are possible. For example, if the flow is initially subcritical and $h > h_u$, $dh/dx > S_0$. The two conditions $h > h_u$ and $dh/dx > S_0$ indicate that the tranquil flow is approaching a reservoir. For the M2 case, where the flow is initially faster than uniform, (13-27) and the M2 conditions indicate that $dh/dx < S_0$ and again that $dh/dx \to 0$ as $h \to h^*$. The M3 conditions for initially supercritical flow indicate $dh/dx > S_0$ with the same limiting conditions. This family of M curves is sketched in Fig. 13-9b. The other flow categories are also shown in Fig. 13-9.

13.7 ILLUSTRATIVE EXAMPLE 13-1: Sluice-Gate Discharge

Water flow from a reservoir into a horizontal flume is controlled by a sluice gate, as shown in Fig. 13-10. The flume length L is 1000 ft, after which the water drops abruptly into a river bed. It is desired to find the flow rate and the flume flow conditions as functions of the gate rise h_g and the reservoir level above the flume bed h_{01}, for $h_{01} = 9$ ft, $b = 10$ ft, $h_g = 2$ ft, and a coefficient of contraction C_c for the gate discharge of 0.6. The concrete flume surface has a Manning roughness factor $n = 0.012$.

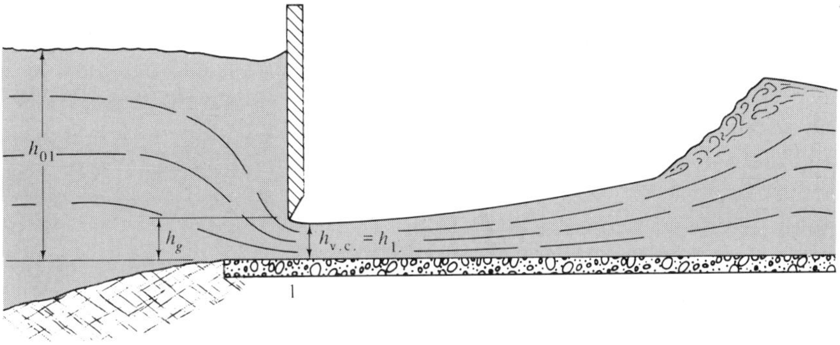

Fig. 13-10. Discharge under a sluice gate with a hydraulic jump downstream of the gate.

The distance measured downstream from the gate is x. Since the distance from the gate to the vena contracta is small (about 1 ft), the vena contracta is considered to be located at the gate with respect to significant distances along the flume. The point 1 where $x = 0$ will then be at the vena contracta.

The flow rate can be determined from the flow conditions at the vena contracta,

$$Q = (A\text{v})_{\text{v.c.}} = bC_c h_g \text{v}_1$$

The velocity can be obtained from the energy equation. Neglecting losses from the reservoir to the vena contracta, and assuming that the profile is about uniform at 1,

$$h_{01} = \frac{\text{v}_1^2}{2g} + h_1 \qquad \text{or} \qquad \text{v}_1 = \sqrt{2g(h_{01} - h_1)}$$

With $h_1 = C_c h_g$, $\text{v}_1 = 8.02\sqrt{7.8} = 22.4$ ft/sec and $Q = (10)(1.2)(22.4) =$ 269 ft^3/sec.

The Froude number is $N_{F1} = \text{v}_1/\sqrt{gh_1}$. However, it has been noted from the developed equations that N_F^2 (like M^2) is the significant quantity. It is also interesting to note for this case that

$$N_{F1}^2 = \frac{\text{v}_1^2}{gh_1} = \frac{2g(h_{01} - h_1)}{gh_1} = 2\left(\frac{h_{01}}{h_1} - 1\right) = 13$$

In general this relation can be applied to any flow under any conditions where h_0 is the *local* stagnation head, $h_0 = (\text{v}^2/2g) + h + z$. Thus the

Froude number (as determined directly from the given conditions of h_{01} and h_1) is supercritical, indicating the possibility that a hydraulic jump may occur somewhere in the flume. It is then necessary to calculate the $h(x)$ profile down the flume to determine whether the distance to the jump, x_j, is greater than the flume length (L). If $x_j > L$, the flow will discharge in a supercritical state. If $x_j < L$, a jump will occur somewhere in the flume, and the water will leave the end with the condition $N_{F,L} = 1.0$, since the abrupt dropoff can be assumed to produce acceleration to critical flow in its immediate vicinity. The end conditions at $x = 0$ and $x = L$ are then determined, and the jump location (x_j) depends on simultaneous solution of the two flows, one downstream from the gate and the other "back-calculated" upstream from the end, $x = L$.

FLOW DOWNSTREAM FROM THE GATE

It is first desired to determine h as a function of x for the given parametric values of n and Q, noting that the flow is one of constant b and zero bed slope, $dz/dx = S_0 = 0$. Either the momentum equation (13-7) or the energy equation (13-16d) can be used with the local friction term assumed to be represented by the Manning friction relation (13-25),

$$\frac{g_c \tau_0}{\rho g R} = \frac{n^2 \mathrm{v}^2}{2.22 R^{4/3}}$$

For this illustration, the momentum equation (13-7) reduces to

$$\frac{dh}{dx}(1 - N_F^2) + \frac{g_c \tau_0}{\rho g R} = 0 \tag{13-29}$$

or

$$\frac{dh}{dx}(1 - N_F^2) + \frac{n^2 \mathrm{v}^2}{2.22 R^{4/3}} = 0$$

The hydraulic radius is

$$R = \frac{bh}{b + 2h} = \frac{h}{1 + 2h/b}$$

and will vary along the flume. Equation (13-29) contains the dependent variables h, N_F, and v (R has been determined in terms of h). With Q and b constant, the continuity equation gives a direct relation between v and h (Q and b constant),

$$Q = bh\mathrm{v}$$

Using N_F as the primary variable, the continuity equation and the definition of N_F then give

$$N_F^2 = \frac{\text{v}^2}{gh} = \frac{Q^2}{gb^2h^3} \qquad \text{or} \qquad N_F^2 \propto \frac{1}{h^3} \tag{13-30a}$$

which gives the differential relation

$$\frac{dN_F^2}{N_F^2} = -3\frac{dh}{h} \tag{13-30b}$$

Equation (13-29) can then be put in the form

$$-\frac{1}{h}\frac{dh}{dx}(1 - N_F^2) = \frac{gn^2}{2.22R^{4/3}}N_F^2$$

and, using (13-30b),

$$\frac{1}{3}\frac{dN_F^2}{N_F^4}(1 - N_F^2) = \frac{gn^2}{2.22R^{4/3}}dx$$

which can be written as an explicit relation in x giving x as a function of N_F or of h. With (13-30a),

$$dx = \frac{2.22R^{4/3}}{3gn^2}\frac{1 - N_F^2}{N_F^4}dN_F^2 \tag{13-31a}$$

The hydraulic radius R is still an unspecified function of N_F or of x. The depth h can be determined in terms of N_F from (13-30a), giving $R(N_F)$ as

$$R = \frac{h}{1 + (2h/b)} = \frac{(Q^2/gb^2N_F^2)^{1/3}}{1 + (2/b)(Q^2/gb^2N_F^2)^{1/3}}$$

so x as an explicit function of N_F is

$$dx = \frac{0.74R^{4/3}}{gn^2}\left(\frac{Q^2}{gb^2}\right)^{4/9}\left(1 + \frac{2}{b}\frac{Q^2}{gb^2}N_F^{-2}\right)^{4/9}\frac{1 - N_F^2}{N_F^4}d(N_F^2) \tag{13-31b}$$

NUMERICAL SOLUTION

Equation (13-31b) is difficult to integrate and even complicated to evaluate numerically because of the fractional powers of N_F^2. A numerical procedure is indicated using a finite-difference method. A shortcut procedure for numerical integration of (13-31a) by hand or machine can be set up which requires less computer time (larger increments of N_F^2 or h).

Reference to (13-31a) indicates an easy analytical solution for the $R^{4/3}$ term. However, h and therefore R vary slowly with x. R varies essentially as h while N_F^2 varies as h^{-3}; therefore, the solution or integration procedure is much less dependent on the R behavior than on the N_F^2 behavior. Integrating (13-31a) over a small Δx with $\overline{R^{4/3}}$ taken as the average value over the increment $x_2 - x_1$ gives

$$x_2 - x_1 = \frac{0.74\overline{R^{4/3}}}{gn^2}\left[-\left(\frac{1}{N_{F2}^2} - \frac{1}{N_{F1}^2}\right) - \ln\left(\frac{N_{F2}}{N_{F1}}\right)^2\right]$$

or

$$x_2 - x_1 = \frac{0.74\overline{R^{4/3}}}{gn^2}\left\{\ln\left(\frac{N_{F1}}{N_{F2}}\right)^2 - \frac{1}{N_{F1}^2}\left[\left(\frac{N_{F1}}{N_{F2}}\right)^2 - 1\right]\right\} \tag{13-32}$$

The terms in the brackets have been arranged in this form since the supercritical H3 curve downstream of the gate will be one of increasing h or decreasing N_F ($N_{F1} > N_{F2}$); the terms then have a positive value.

For an arbitrarily small incremental step, (13-32) can be written

$$\Delta x_i = \frac{0.74\overline{R_i^{4/3}}}{gn^2}\left\{\ln\left(\frac{N_{Fi}}{N_{F(i+1)}}\right)^2 - \frac{1}{N_{Fi}^2}\left[\left(\frac{N_{Fi}}{N_{F(i+1)}}\right)^2 - 1\right]\right\} \tag{13-33}$$

Since both R and N_F are given directly as functions of h,

$$R_i = \frac{h_i}{1 + (2h_i/b)} \qquad \text{and} \qquad \left(\frac{N_{Fi}}{N_{F(i+1)}}\right)^2 = \left(\frac{h_{i+1}}{h_i}\right)^3$$

incremental steps of h_i can be specified and Δx_i calculated directly. The calculation procedure is greatly simplified if equal increments (Δh_i) of h_i are taken; that is,

$$h_{i+1} = h_i + \Delta h_i \qquad \text{or} \qquad \frac{h_{i+1}}{h_i} = 1 + \frac{\Delta h_i}{h_i} \equiv 1 + \delta$$

where δ is a constant smaller than unity. Since the bracketed term of (13-33) will give exact variations, the degree of approximation of the numerical procedure is only in the $\overline{R_i^{4/3}}$ evaluation, so δ need not be "very small"; for example, $\delta = 0.1$ is a satisfactory increment. With these substitutions, (13-33) becomes

$$\Delta x_i = \frac{0.74\overline{R_i^{4/3}}}{gn^2}\left\{3\ln\frac{h_{i+1}}{h_i} - \frac{1}{N_{Fi}^2}\left[\left(\frac{h_{i+1}}{h_i}\right)^3 - 1\right]\right\} \tag{13-34a}$$

where

$$\overline{R_i^{4/3}} \equiv \frac{R_i^{4/3} + R_{i+1}^{4/3}}{2} \qquad \text{and} \qquad N_{Fi}^2 = N_{F1}^2\left(\frac{h_1}{h_i}\right)^3$$

With $h_{i+1}/h_i = 1 + \delta$, which also gives $h_i/h_1 = (1 + \delta)^{i-1}$,

$$\Delta x_i = \frac{0.74\overline{R_i^{4/3}}}{gn^2}\left\{3 \ln (1 + \delta) - \frac{(1 + \delta)^{3(i-1)}}{N_{F1}^2}[(1 + \delta)^3 - 1]\right\} \quad (13\text{-}34b)$$

With $\delta = 0.1$, (13-34a) becomes (with $g = 32.2$ ft/sec², $n = 0.012$, and $N_{F1}^2 = 13$)

$$\Delta x_i = 159.7\overline{R_i^{4/3}}[0.286 - 0.02546(1.331)^{i-1}] \quad (13\text{-}35)$$

Because the value of $\overline{R_i^{4/3}}$ depends on the average of two successive values of R_i, it is left as a separate calculation. The computation procedure is then set up as indicated in Table 13-2. Note that the determination of h_i can be hand-multiplied as a successive pyramid faster than it can be programmed and put onto a machine. The backwater curve of h_i versus x_i can now be plotted downstream from the gate.

TABLE 13-2

CALCULATION FOR SLUICE-GATE DISCHARGE OF FIGS. 13-10 AND 13-11

i	h_i	R_i	$R_i^{4/3}$	$\overline{R_i^{4/3}}$	Δx_i	$x_i = \Delta x_{i-1}$	N_{Fi}^2	h_j	N_{Fj}^2
1	1.2	0.968	0.9576	1.014	44.5	0	13	5.54	0.133
2	1.32	1.052	1.070	1.121	45.2	44.5	9.76	5.207	0.159
3	1.452	1.127	1.172	1.227	47.2	89.7	7.35	4.89	0.193
4	1.597	1.212	1.282	1.351	48.8	136.9	5.52	4.57	0.236
5	1.757	1.301	1.420	1.490	49.0	185.7	4.14	4.27	0.288
6	1.933	1.396	1.560	1.633	46.8	234.7	3.12	3.96	0.371
7	2.126	1.493	1.706	1.789	41.1	281.5	2.34	3.66	0.459
8	2.338	1.594	1.861	1.963	30.5	322.6	1.76	3.37	0.586
9	2.572	1.698	2.064	2.128	11.4	354.1	1.30	3.10	0.611
10	2.830	1.806	2.210	—	—	[365.5]	[0.983]	[2.78]	[1]

The conjugate depth h_j can be obtained from (13-15),

$$h_j = \frac{h_i}{2}[\sqrt{1 + 8N_{Fi}^2} - 1]$$

in the form

$$h_j = \frac{h_1}{2}(1 + \delta)^{i-1}[\sqrt{1 + 8N_{F1}^2/(1 + \delta)^{3(i-1)}} - 1]$$

The conjugate depths are also tabulated in Table 13-2, allowing the conjugate depth curve to be plotted.

The H2 backwater curve for the discharge over the end of the flume can be determined by working upstream from the end condition where $N_{FL} = 1$. In the backward direction, h_{i+1}/h_i will again be greater than 1 and δ can be taken as 0.1. The numerical equation for Δx_i is

$$\Delta x_i = 159.7\overline{R_i^{4/3}}[0.286 - 0.331(1.331)^{i-1}] \tag{13-36}$$

It is seen that Δx_i is negative, since the coordinates have not been changed from the original equation, (13-32) or (13-34a). The results for this tail-off backwater curve are given in Table 13-3.

TABLE 13-3

CALCULATIONS FOR BACKWATER CURVE FROM FLUME DISCHARGE

i	h_i	R_i	$R_i^{4/3}$	$\overline{R_i^{4/3}}$	$-\Delta x_i$	$x_i = 1000 + \Delta x_{i-1}$	N_{Fi}^2
1	2.880	1.828	2.235	2.325	16.7	1000	1
2	3.170	1.94	2.420	2.518	62.3	983.3	0.750
3	3.483	2.055	2.615	2.712	130.5	921	0.564
4	3.835	2.170	2.810	2.912	231	790.5	0.423
5	4.22	2.290	3.015	3.117	375	559.5	0.318
6	4.64	2.405	3.220	3.33	583	185	0.239
7	5.10	2.524	3.440	3.56	884	−398	0.180
8	5.61	2.650	3.670	—	—	—	0.135

N_{Fi}^2 is tabulated in Table 13-2 to indicate the limit of the calculation. For the supercritical discharge (H3) curve, N_{Fi} has no significance beyond the limiting value of 1.0. In the calculations in Table 13-3, no such limit exists. Hypothetically the curve can approach a horizontal asymptote as N_{Fi} approaches zero and z_i and Δx_i approach minus infinity. These results for Δx_i are then transferable; that is, they can be used for the tail-off for any length of flume with the same Q, n, and b.

The distance to the jump is obtained by plotting the results (or by machine comparison routine) to determine the location for which the H2 curve intersects the $H3_j$ (conjugate H3) curve, as illustrated in Fig. 13-11. The jump is seen to occur in the vicinity of 115 ft from the gate. If the flume were shorter, the distance to the jump would be longer. If L were more than 1000 ft, the jump would be closer to the gate. The H2 curve can be transferred to the right until it just intersects $H3_j$ at $x = 0$. This

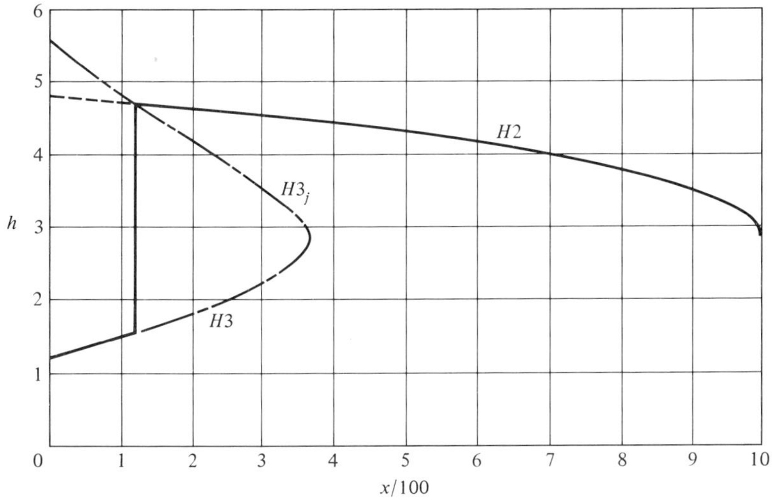

Fig. 13-11. Sluice discharge profile curves for flow into the horizontal bed of Fig. 13-10.

limiting case would occur for $L \approx 1800$ ft. If L is greater than 1800 ft, the discharge is "submerged."

TRANSIENT CASE

As the gate is first opened, $N_{F1}^2 \gg 1$ and R is small (in the limit as $h/b \ll 1$, $R \to h$). Reference to Tables 13-2 and 13-3 indicates that R and h are of the same order of magnitude, but the variation of N_F^2 is more significant. Also, h_L is determined primarily by h_1 for $h_{01}/h_1 \gg 1$ (see Problem 13-27). From this behavior it can be deduced that the jump will occur farther downstream at lower flow rates (smaller h_{gate}, same h_{01}).

LOSS OF AVAILABILITY

The loss of availability in a hydraulic jump can be quite large. Since all the downstream or conjugate quantities (h_j, N_{Fj}^2, and so on) depend only on h_1 and N_{F1} immediately upstream of the jump, the ratio h_{loss}/h_1 can be determined as a function of N_{F1} only. In general,

$$h_{\text{loss}} = h_{01} - h_{02} = \left(\frac{V_1^2}{2g} + h_1 + z_1\right) - \left(\frac{V_2^2}{2g} + h_2 + z_2\right)$$

which can be written

$$h_{\text{loss}} = h_1\left(\frac{N_{F1}^2}{2} + 1\right) - h_2\left(\frac{N_{F2}^2}{2} + 1\right)$$

The difference in the z term can usually be neglected, since the jump occurs over a relatively short distance (on the order of $6h_2$). Using the preceding relationships, the loss can be obtained as (see Problem 13-31)

$$\frac{h_{\text{loss}}}{h_1} = \frac{N_{F1}^2}{2} + 1 - \frac{N_{F1}^2}{2}\left(\frac{h_1}{h_2}\right)^2 - \frac{h_2}{h_1} = \frac{N_{F1}^2}{2}\left[1 - \left(\frac{h_1}{h_2}\right)^2\right] + \left(1 - \frac{h_2}{h_1}\right) \tag{13-37}$$

13.8 ILLUSTRATIVE EXAMPLE 13.2: Water Table for Hydraulic Analogy

It is desired to build a water table to model supercritical wave phenomena corresponding to $N_{F2} = 3$ at a depth of $h_2 = 1$ in. with a channel width $b_2 = 10$ in. It is necessary to find the throat conditions h^* and b^*, the reservoir condition h_0, and the flow rate Q. The channel plane and elevation views are sketched in Fig. 13-12. Friction will be neglected. The friction effects of the boundary layer may be considered as another problem wherein compensations are made to account for boundary-layer displacement and friction resistance.

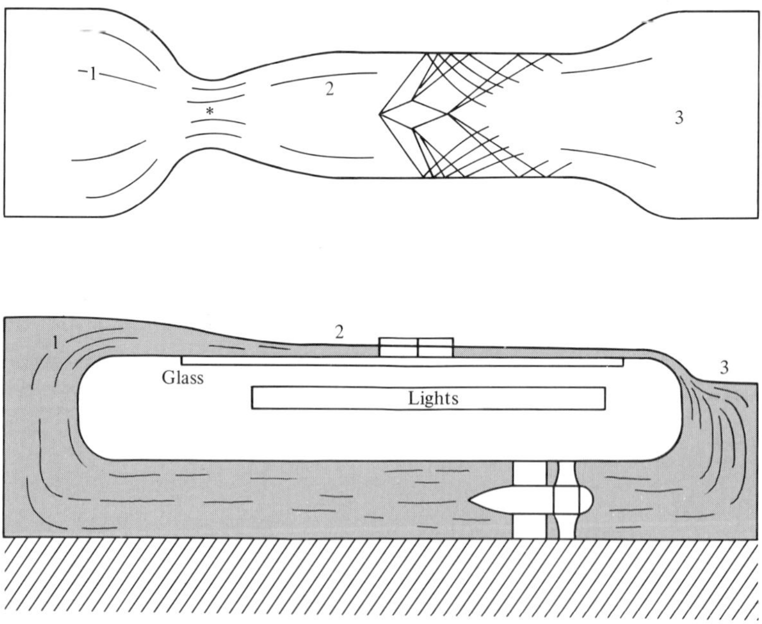

Fig. 13-12. Plan (top) and elevation (side) views of a water table used for supersonic flow analogy.

Since the flow in the test section is to be uniform at $N_{F2} = 3$, the bottom will be horizontal. The momentum equation for the above condition is (13-9)

$$dh(1 - N_F^2) - N_F^2 \frac{db}{b} = 0$$

which is a functional relation among b, h, and N_F, or b, h, and v, discussed in connection with Fig. 13-4. One of the variables can be eliminated by use of the continuity equation $Q = bh\text{v}$. If db/b is eliminated, the resulting equation can be integrated to give

$$h + \frac{\text{v}^2}{2g} = h_0 = \text{const.}$$

This equation can be divided by h^* (the local critical h at $N_F = 1$) to give

$$\frac{b}{h^*}\left(1 + \frac{N_F^2}{2}\right) = \frac{3}{2} \tag{13-38}$$

With the continuity equation and the definition of N_F,

$$\frac{h}{h^*}\left[1 + \frac{1}{2}\left(\frac{b^*}{b}\right)^2\left(\frac{h^*}{h}\right)^3\right] = \frac{3}{2} \tag{13-39}$$

which is an equation between b and h in terms of the reference critical parameters b^* and h^*.

From (13-38), the depth h^* is

$$h^* = \frac{2}{3}h\left(1 + \frac{N_{F2}^2}{2}\right) = \frac{2}{3}\left(1 + \frac{9}{2}\right) = \frac{11}{3} = 3.67 \text{ in.}$$

The reservoir depth is then $h_0 = 3h^*/2 = 5.5$ in. The throat width is determined from (13-39),

$$\left(\frac{b^*}{b}\right)^2 = \left(3\frac{h^*}{h} - 1\right)\left(\frac{h}{h^*}\right)^3 = 0.2026 \qquad \text{or} \qquad b^* = 4.50 \text{ in.}$$

The flow rate Q is most easily determined at the test section, since b_2, h_2, and N_{F2} are specified:

$$Q = b_2h_2\text{v}_2 = b_2h_2N_{F2}\sqrt{gh_2} = \sqrt{g}\, b_2N_{F2}h_2^{3/2} = 0.34 \text{ ft}^3/\text{sec}$$

or since it is also of interest to determine v_2, $\text{v}_2 = \sqrt{gb_2}\, N_{F2} = 4.9$ ft/sec. Then $Q = b_2h_2\text{v}_2 = 0.34$ ft^3/sec (153 gpm).

The table discharges supercritically over the end ramp into the outlet reservoir at 3, as indicated in Fig. 13-12. The primary head and power loss arises from the free discharging of this supercritical flow into a reservoir where the velocity is essentially zero. The pump speed is adjusted to maintain the water level at 3 a few inches below the table level.

13.9 ILLUSTRATIVE EXAMPLES 13-3: Weirs

A weir is an open-channel device for measuring flow rate or discharge rate from a flume or canal.

BROAD-CRESTED WEIR

The broad-crested weir is usually used at the end of a flume with a width equal to that of the upstream flume. Since the discharge is unchecked (as illustrated in Fig. 13-13), critical conditions will exist at a point on the crest. It is only necessary to know either the upstream depth h_0 or the critical depth h^*. The depth h is measured from the weir surface. Since

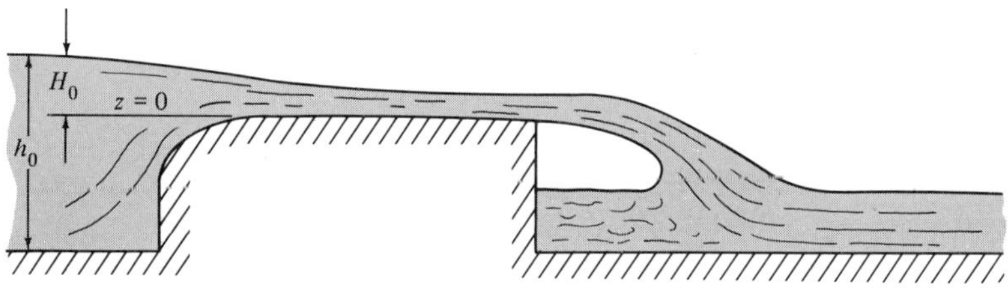

Fig. 13-13. Cross section through the flow over a broad-crested weir.

the upstream velocity is smaller, it is easier to measure h_0 with a hook, float, or manometer gage. It is assumed that the upstream kinetic energy is much less than the critical kinetic energy. The discharge is then

$$Q = C_d A \text{v} = C_d A^* \text{v}^* = C_d b h^* \text{v}^* = C_d b h^* \sqrt{2gh^*}$$

or

$$Q = C_d \sqrt{\tfrac{2}{3}g}\, \tfrac{2}{3} b H_0^{3/2} = 3.09 C_d b H_0^{3/2} \qquad (13\text{-}40)$$

where C_d is the coefficient of discharge, which depends on the velocity distribution and is therefore a function of the Reynolds number based on hydraulic radius. The discharge coefficient for large broad-crested weirs is approximately 0.98.

SHARP-CRESTED WEIR

The above method of calculation also applies to each of the weirs illustrated in Fig. 13-14. For example, consider the V-notched weir with the

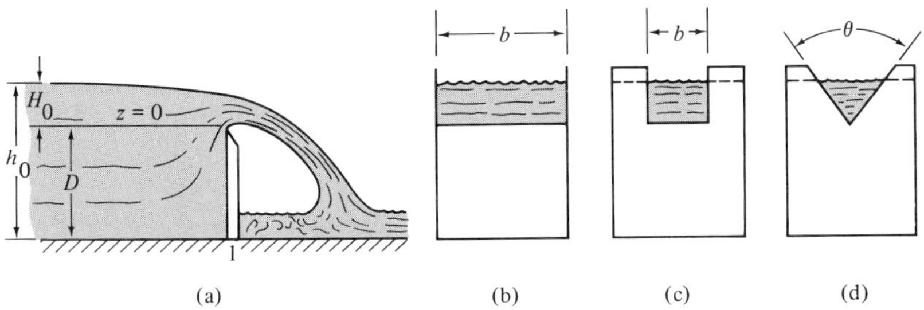

Fig. 13-14. Representation of the flow over sharp-crested weirs: (a) section view; (b) full width uncontracted; (c) contracted; (d) V notch.

condition that the discharging area is much smaller than the flume or approach area ($v^{*2} \gg v_0^2$), so the approach kinetic energy can be neglected. Neglecting friction effects, the energy equation gives

$$v^* = C_v\sqrt{gh^*} = C_v\sqrt{\tfrac{2}{3}gH_0}$$

where C_v is the velocity coefficient which accounts for the assumptions of negligible friction and uniform velocity profile. The flow rate is $Q = A^*v^*$. Since the critical condition will not occur at the weir plate (owing to stream contraction), the area A^* is not known exactly but can be taken as $A^* = C_cA_V$, where A_V is the flow area through the V notch. The area $A_V = 2h_V^2 \tan \theta/2$. The depth h_V will be proportional to $h^* = 2H_0/3$. All the constants for a given weir can be lumped into a discharge coefficient C_d to give, finally,

$$Q = C_dH_0^{5/2} \tag{13-41}$$

For example, for a 90° V-notched weir, $C_d = 2.5$.

For sharp-crested weirs, the weir plates should be smooth and sharp (metal), as indicated in Fig. 13-14. Atmospheric pressure should exist above and below the discharging nappe and the approach flow should be smooth and uniform. If these conditions are not attained, it will not be possible to calculate the discharge from the formulas given for a specified weir.

13.10 SUMMARY

1. The Froude number is the parameter by which open-channel flows are characterized.
2. Because of the critical nature and the corresponding wave characteristics of an open-channel flow at $N_F = 1$, a similar or analogous behavior

exists with compressible gas flows, where M is the characteristic parameter. Use of this similarity is made in the hydraulic analogy.

3. Open-channel flow behavior is more complicated than that of one-dimensional flows considered previously because the depth h enters as an additional dependent variable.
4. Open-channel flow analysis is also complicated by the exponential form of the Manning relation used for determining friction forces and availability loss. Also for analyses where it is necessary to use the Manning relation, the results are strongly dependent on the Manning roughness factor n, which usually can only be specified to ± 5 or ± 10 percent.
5. Most long "reach" open-channel flows are uniform (constant h) or have relatively slowly varying depths, so accelerations are negligible with respect to effects on velocity profile. Average or "uniform flow" values of friction and other characteristics are then applicable to such "gradually varied" flows. In the case of rapidly varied flows (for example, through a hydraulic jump or over a spillway), the wall friction effects are either usually negligible or the flow is too complicated to yield analytical results.
6. Because of the complicated nature of open-channel (or free-surface) flows, it is necessary in many cases to rely on a semi-empirical analysis where general behavior is indicated as the result of analytical procedures. Deviations from the assumptions (idealizations) necessary to obtain an analytical solution must be obtained by experiment and specific coefficients must be obtained by semi-empirical correlation with experimental results.

REFERENCES

1. F. M. Henderson, *Open Channel Flow*, The Macmillan Company, New York, 1966.
2. V. T. Chow, *Open Channel Hydraulics*, McGraw-Hill, Inc., New York, 1959.
3. Hydraulic Design of Stilling Basins and Basins for Rocket Energy Dissipators, *Engineering Monograph 25*, Hydraulics Lab. Branch U.S. Bureau of Reclamation, Denver, 1958.

PROBLEMS

13-1 What are the dimensions of C in (13-1a)?

13-2 Consider dynamic equilibrium of steady flow in a gradually sloping channel as sketched. Draw a control volume and determine the Chezy relation. Find C in terms of a friction coefficient $c_f = \tau_0(\rho\bar{V}^2/2g_c)$.

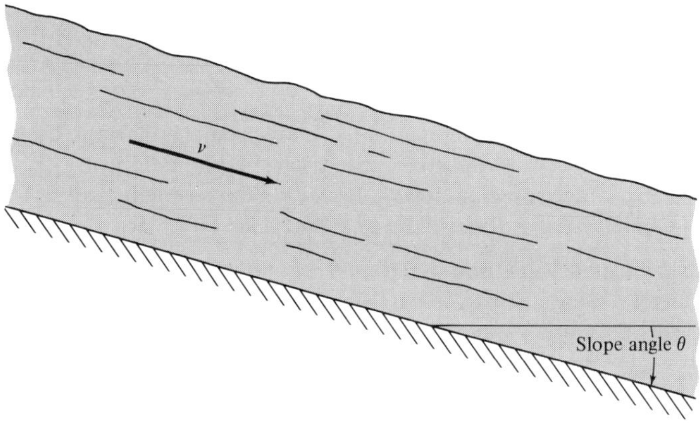

Fig. P13-2.

13-3 Consider the force–momentum equation applied to a rectangular control volume of length Δx, width b, and depth h. Take all quantities referred to those at the center of the control volume and derive the differential equation (13-5) directly. Note that b and h are functions of x; that is, the control volume is tapered in the x direction.

13-4 Given a 10- by 10-ft flume cross section with velocity profile map as shown. The numbers give the fraction of maximum velocity contours. For this distribution find $\overline{\mathrm{v}}/\mathrm{v}_{max}$ and α.

Answer: $\overline{\mathrm{v}}/\mathrm{v}_{max} \approx 0.75$, $\beta \approx 1.02$, $\alpha \approx 1.05$

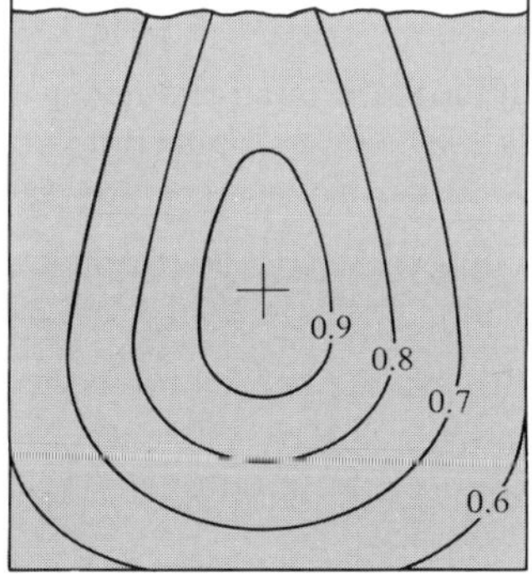

Fig. P13-4.

13-5 What are the dimensions of 1.49 in Manning's formula (13-2)?

13-6 For fully developed flow in a channel, the velocity is expected to be a function of the gravity force, which depends on g, the slope S, and the hydraulic radius R. By a dimensional analysis, derive the Chezy form for flow in open channels $v = C\sqrt{RS}$, where C is the Chezy coefficient.

13-7 For uniform, fully developed flow in an open channel, the flow rate or average velocity $\bar{v}$ would be expected to be a function of ρ, g, μ, R, S, and an average roughness size ϵ. Perform a dimensional analysis to obtain an explicit relation between v and the other variables.

13-8 Using momentum and continuity equations for the conditions and control volume illustrated in Fig. 13-2, derive the wave-propagation speed for a small wave $v_w = \sqrt{gh}$.

13-9 For flow along a constant-width horizontal channel with friction, show that the effect of friction on the flow depth (and velocity) is opposite for subcritical and supercritical flow and that the profile shown in Fig. 13-7 is correct.

13-10 For the case of steady, frictionless channel flow over a horizontal bed, obtain the relationships among b, N_F, and v,

$$(1 - N_F^2)\frac{dv}{v} + \frac{db}{b} = 0$$

and interpret this relation.

13-11 For the same case as in Problem 13-10, eliminate dh/b. Integrate the resulting equation and interpret the results.

13-12 Consider a small cylindrical surface wave ($\Delta h/h \ll 1$), use the momentum and continuity conditions for a sector of angle $\Delta\theta$, and show that the speed of a small characteristic wave is $v_w = \sqrt{gh}$.

13-13 (a) Calculate the speed of a 0.1-in. wave on a 1-in.-deep tank. (b) Calculate the speed of a 30-ft wave over a 3000-ft-deep ocean.

13-14 It is desired to pitch the slope of a rectangular, wood flume for uniform flow. If the flume is 3 by 3 ft and is to run two thirds full, what is the slope for flow at $N_F = \frac{1}{2}$? What is the flow rate?

13-15 Plot h_2/h_1 as a function of N_F for the hydraulic jump (13-15a). What is the behavior of the curve for large N_{F1}? What is the slope for large N_{F1}?

13-16 To illustrate that a hydraulic jump cannot occur for $N_{F1} < 1$, show that $\Delta h_{loss} < 0$ for $N_{F1} = 0.8$.

13-17 Substitute h_2/h_1 from (13-15) into (13-17) to obtain the relation for $\Delta h_{loss}/h_1$ as a function of N_{F1}.

13-18 Consider the momentum equation in the form (13-4b), integrate, and show that for steady, quasi-one-dimensional flow with $\alpha \approx 1.0 \approx \beta$, $\Delta\tilde{u} = \int_L \tau_0 \, dx/\rho R$ ($L = \Delta x$). Deduce that $\Delta h_{loss} = \int_L (g_c\tau_0/\rho g R)\, dx$.

13-19 Find N_F in terms of local H_0 and local h.

13-20 Starting with (13-23) and taking $S_0 = dz/dx = 0$ for a horizontal channel, show that (13-28) can be obtained in the form

$$\frac{dh}{dx} = \frac{-C}{1-(h/h^*)^3} = \frac{-C}{1-N_F^2} = \frac{-C}{1-(Q^2/gb^2h^3)}$$

and that $C = n^2/(1.49)^2b^2$ for constant channel breadth b and for $b \gg h$.

13-21 Complete the calculations for the shapes of the depth or "backwater profiles" as indicated in Fig. 13-9c, d, and e.

13-22 Discuss the similarity between Mach number M and Froude number N_F.

13-23 Integrate the momentum equation (13-7) using the continuity equation and Manning substitution (13-25) to obtain an implicit relation for $h(x)$, or determine Δx explicitly as a function of h. Obtain Δx as $f(Q, g, b, n, h)$. Assume $b \gg h$.

13-24 The relation between the *local* isentropic stagnation enthalpy and the local static $\tilde{h}$ for compressible flow of a perfect gas was $\tilde{h}_0/\tilde{h} = 1 + [(\gamma - 1)/2]M^2$. Find the equivalent expression for open-channel flow over a horizontal bed. Compare the two expressions and find the equivalent γ for the open-channel flow. Discuss the results.

13-25 The relation between the local stagnation enthalpy $\tilde{h}_0$ and the local critical value for the compressible flow of a perfect gas was $\tilde{h}_0/\tilde{h}^* = (\gamma + 1)/2$ from Chapter 12. By comparing the local value of h_0/h^* for open-channel flow over a horizontal bed, find the equivalent of γ for the analogy between compressible perfect gas flow and this open-channel flow. Discuss the results.

13-26 Find the gate height in Example 13-1 that will just give critical flow at the vena contracta. As h_{01}/h_1 increases, the C_c will increase and approach unity as $h_{01}/h_1 \to 1$. Take the contraction coefficient as $C_c = 0.75$ for critical flow. Comment on the result.

13-27 Find h_L as a function of h_1 and h_{01} only for the case where $N_{FL} = 1$. Show that h_L varies nearly as $h_1^{5/3}$ for $h_{01}/h_1 \gg 1$.

13-28 (a) Plot $h_0 = h + (v^2/2g)$ as a function of x for Example 13-1 and discuss the shape of the curve. Note that the curve of h^* can also be determined easily, since $h^* = \frac{2}{3}h_0$.

13-29 In Table 13-2, the values of N_{Fj}^2 are also given. With the rest of the information in the table already calculated, what is the most direct way to calculate N_{Fj}^2.

13-30 Discuss the "backwater curve" combination in the flume of Example 13-1 as L is made greater than 1000 ft and as L is made very large (limiting case as $L \to \infty$).

13-31 The jump in Example 13-1 occurred at $x = 115$ ft. At this point $N_F^2 = 6.44$ and $N_{Fj}^2 = 0.215$. Derive an expression for the loss of available energy per unit mass (or loss rate/mass flow rate) across a hydraulic jump and calculate this specific energy loss and the total loss in horsepower for this jump. Show that the head loss across the jump can be determined in terms of N_{F1} and h_1/h_j as given in (13-37). The ratio h_1/h_j can be determined as a function of N_{F1} from (13-15) so that h/h_1 can be determined as a function of N_{F1}^2.

13-32 (a) In the case of a shallow channel $h/b \ll 1$, show that $R \approx h$. (b) In this case the momentum equation for constant Q and constant b can be integrated directly. Show that for such a case

$$\Delta x = \frac{2.22h_1^{4/3}N_{F1}^{8/9}}{3gn^2}\,\Delta\left[3N_{F1}^{7/3}\left(\frac{N_{F1}^2}{16} - \frac{2}{7}\right)\right]$$

13-33 In the development of Example 13-1, it was stated that the energy equation could be used to obtain the same results as given in (13-31). Start with the energy equation in the form

$$\frac{V_1^2}{2g} + h_1 - \Delta h = \frac{V_2^2}{2g} + h_2$$

and show that it can be put into a finite-difference form equivalent to that of (13-31a).

13-34 Find the horsepower dissipated in the hydraulic jump of Example 13-1.

13-35 With reference to Fig. 13-12, find b^*, h^*, and h_0 for $N_{F2} = 2$. What is the horsepower for this case if $h_3 = -4$ in.?

13-36 Find the pump horsepower for operation of the water table under the conditions of Example 13-2 with $h_3 = -4$ in.

13-37 Discuss and sketch depth profiles $h(x)$ for (a) subcritical flow over a bump in a channel bed, and (b) for flow over a small depression.

13-38 A V-notch weir is to be used to meter water from an irrigation canal into a farm irrigation ditch. If the water is to be metered at 0.1 acre-ft/day, at what depth H should the weir-notch apex be set?

13-39 Develop the relations for the discharge Q for the weirs shown in Fig. 13-34a and b. Show that Q for these weirs is of the form $Q = C_d b h^{3/2}$ and discuss and estimate the value of C_d to be expected for each.

13-40 In the formulas developed for weir discharge rates, it was always assumed that the flow attained critical conditions. Is it possible to have a weir discharge such that the overflow does not become critical? Explain.

CHAPTER 14

CONTENTS

POTENTIAL FLOW

So far, only methods of analysis of one-dimensional flows have been treated (with Chapter 9 excepted). Where pressure or velocity fields are required for most external and some internal flows, two- or three-dimensional problems must be solved. The external pressure or velocity fields must be known to obtain solutions to boundary-layer problems.

The nonlinear nature of the general motion and energy equations renders general solutions with given boundary conditions impossible. Some problems specify conditions by which the equations of motion may be simplified such that exact solutions of the approximation problem can be obtained. For example, if the flow is "extremely viscous" ($N_R \lesssim 1$), so that the viscous forces are much greater than the inertia forces, the nonlinear inertia terms can be neglected with respect to the viscous terms. The Prandtl boundary-layer hypothesis allows "exact" solutions to the "truncated" problem obtained by order of magnitude considerations whereby certain terms are "cut off" from the complete equations.

Another hypothesis by which exact solutions can be obtained to an approximate formulation of the equations is obtained by neglecting the viscous action. This class of flow field solutions for a hypothetical *inviscid* or *ideal* flow may be used to approximate conditions of a real flow only where the approximation conditions are valid.

In the derivation of the equations of motion, the viscous force term was seen to be the most difficult to handle. The stress–strain rate relations are difficult to formulate and include in flow solutions even for a "Newtonian" fluid where a simple linear relation exists between stress and strain rate. If the friction or viscous action could be neglected, the equations would be greatly simplified to allow analytical solutions (and classes of solutions) to be obtained. Great care must be taken in interpreting these solutions and applying them to *real* flow cases. The primary effect of viscosity in producing a thin boundary layer at large Reynolds numbers allows the flow external to the boundary layer to be considered as frictionless. However, for "blunt" bodies (for example, cylinders and spheres) the secondary effect of friction in producing separation yields a flow field much different from that predicted by ideal or potential flow.

For most external and some internal flows, Reynolds numbers are sufficiently large that the Prandtl boundary-layer hypothesis is valid: The region of significant shear stress (or viscous action) is only a very thin boundary layer on the body. In the flow external to this boundary layer, the shear stress is negligible so the assumptions of frictionless, ideal, irrotational, or potential flow are valid. If there is no friction, no dissipation mechanism exists and the force fields are conservative and can be derived from a force potential. The relation between a potential and a conservative field is: The change in the force potential = the work done by a conservative force along a given path,

$$W = \int_a^b \vec{F} \cdot \vec{ds} = \Phi_{F,a} - \Phi_{F,b} \tag{14-1}$$

where Φ_F is the force potential. If the forces are conservative, there is no dissipation, so the work around a closed path must be zero,

$$W = \oint \vec{F} \cdot \vec{ds} = \lim_{a \to b} (\Phi_{F,a} - \Phi_{F,b}) = 0 \tag{14-2}$$

The derivative relation then exists,[1]

$$\vec{F} = -\text{grad } \Phi_F \tag{14-3}$$

which is proved by insertion into (14-2),

$$\oint \vec{F} \cdot \vec{ds} = -\oint \vec{ds} \cdot \nabla \Phi_F = -\oint d\Phi_F = 0$$

[1] The minus sign is the usual convention with a force potential, so $W = \Phi_{F,a} - \Phi_{F,b}$, or the work is positive for a "downhill-running" potential (an unnecessary but predominant convention).

This is the mathematical condition for the force field to be conservative and for the existence of the force field potential or potential-energy function Φ.

Friction forces are dissipative or nonconservative and are the only mechanical forces by which a fluid element can be rotated (in the sense of rotation as treated in Chapter 6). If the viscous or shear forces are then neglected, there can be no change in the state of rotation of a fluid element. If the fluid is initially irrotational, it must then remain irrotational. These conditions imply the possible existence of a velocity potential Φ,[2]

$$\vec{v} = \text{grad}\ \Phi \tag{14-4}$$

The existence of such a potential is verified by the condition that the flow field is irrotational, that is, $2\vec{\omega} = \text{curl}\ \vec{v} = 0$, for the curl of the gradient of any function is identically zero,

$$2\vec{\omega} = \nabla \times \vec{v} = \nabla \times \nabla\Phi_v \equiv 0 \tag{14-5}$$

This restriction can be relaxed somewhat by allowing a singular point (or singular region) to exist where $\vec{\omega} \neq 0$, but then by direct consideration of Newton's laws, only that fluid for which it is specified that $\vec{\omega} \neq 0$ can ever have any rotation. If there is no possibility of a shear stress existing anywhere else within the flow field (that is, no viscous action is possible), there can be no diffusion of vorticity into that part of the field considered devoid of viscous action (or irrotational). The singular irrotational point therefore remains fixed with that "fluid mass point" or is isolated with this point. The conditions of irrotationality and a potential field are then satisfied throughout the flow field except at this singular point. On the other hand, a shear strain rate can exist, but the velocity field must be one where $\vec{\omega} = 0$, or the sum of the rotations of two perpendicular axes in any plane of the fluid element must be zero (see Section 6.8).

A potential function is introduced for the primary reason of reducing the number of dependent variables to just one—the potential function. The three velocity (or force) components are determined in terms of the scalar potential function.

All potential flows are similar for all geometrically similar bodies: The pattern formed by lines of constant potential, the velocity distribution derived from the potential field, the streamlines, and the pressure distribution obtained from the velocity distribution by use of Bernoulli's

[2] The convention for the sign of the velocity-potential relation is not as uniform as that for the force potential. The English have adopted the negative sign, while Germany and the United States have usually used a positive sign, as indicated here. It is just as easy to consider the velocity increasing in the direction of increasing Φ as it is to have the velocity "run downhill"; also the necessity of carrying the minus sign is eliminated.

equation are all similar. For a given body, it is necessary only to apply suitable scale factors in geometry and a reference (for example, free-stream) velocity to obtain the flow about any body of the same shape and orientation. This similarity property will be illustrated below in examining potential flow fields.

CLASSES OF POTENTIAL FLOW PROBLEMS

Three types of boundary problem involving the potential equation arise:

1. *Direct.* A potential Φ is chosen or given with the necessary boundary conditions, and $\vec{v}$ and p are determined directly by calculating the velocity from $\vec{v} = \nabla\Phi$, and p from Bernoulli's equation.

2. *Indirect 1.* The physical or geometric boundary conditions and a reference $\vec{v}$ or p distribution are given at one flow field (control volume) boundary; Φ, $\vec{v}$, and p are to be determined for the rest of the field.

3. *Indirect 2.* The p or v distribution is specified for an unknown contour or surface and a reference boundary condition is given for p or $\vec{v}$ as in 2; the contour or surface is to be determined.

1. If the potential function Φ and the boundary geometry are given, the flow boundary geometry is specified along with the function from which the p and $\vec{v}$ fields are determined directly. For example, if Φ is given for a cylinder in a uniform (or otherwise undisturbed) free stream, the $\vec{v}$ and p fields are determined by direct calculations.
2. A given boundary shape (for example, a cylinder in a free stream) represents one mathematical boundary condition. Since there can be no flow across this physical boundary, one mathematical condition on the velocity field is that the velocity component normal to this boundary must be zero, $v_{n,b} = 0$, or the condition on Φ is $(\partial\Phi/\partial n)_b = 0$ (the subscript b implies on the body or boundary). The other boundary condition is that far away from the body (approaching infinity) the velocity is the reference velocity, say, U_∞ or $(\partial\Phi/\partial s)_\infty = U_\infty$ (where s is the "streaming" direction). Solution of this problem involves assuming a general form for the solution of $\nabla^2\Phi = 0$, for example, a series solution for which the coefficients are determined by the boundary conditions. An internal flow example would be the flow field in a wind-tunnel nozzle where the approach velocity distribution and the nozzle contour are given, and the p and v fields are to be determined in and on the boundaries of the nozzle. There are certain very simple problems of this type in which the entire flow field is represented by a family of similar streamlines or potential lines, for example, circles, as in the case of potential source and vortex flows, to be developed below.

3. If a specific p or $\vec{v}$ distribution is desired for an unknown body contour or surface in a flow field with some reference velocity or pressure at another location (boundary condition), the problem of determining the shape is quite complicated. The NACA No. 6 series airfoils were determined by this method. The desired pressure distribution for an airfoil in a uniform free stream was specified. The airfoil shape was then calculated.

The optimum conditions for converging nozzles for wind tunnels, and so on, are specified in terms of the pressure gradient on the nozzle boundary and the desired velocity distribution across the flow at the exit for a given approach velocity distribution (usually taken as uniform). This inverse problem of determining the contour is solved to yield the nozzle shape.

The indirect problems involve complications beyond the scope of an introductory text. So only the direct problem will be treated here, with the exception of a simple example of the first type of indirect problem.

14.1 INCOMPRESSIBLE POTENTIAL FLOW

The mechanical motion condition on the field is established by the assumption that the rotation is zero. The flow field must also be continuous in the region under consideration. The "process" condition for cases to be considered in this chapter is ρ = const. (that is, incompressible flow). The pertinent equations are then:

(1) Motion (irrotational): $\vec{\omega} = 0$

(2) Continuity (with ρ = const.): $\nabla \cdot \vec{v} = 0$ (14-6)

(3) Process: $\rho = \text{const.}$

The first condition is used to indicate the existence of Φ and also gives $\vec{v} = \nabla\Phi$. The continuity equation then becomes

$$\nabla \cdot \nabla\Phi \equiv \nabla^2\Phi = 0 \tag{14-7}$$

which is Laplace's equation for Φ. When solved with the boundary conditions, (14-7) will yield the potential function $\Phi(\vec{r}, t)$, from which the velocity field can be obtained. The symbol ∇^2 ("del-squared") to denote $\nabla \cdot \nabla$ (div grad) is the Laplace operator or the Laplacian.

Equation (14-7) is a second-order, *linear* partial differential equation. (The coefficients of the highest-order derivatives are functions of the independent variables or are constants.) Its form is more evident if written out in a coordinate system. For example, it can be written in cartesian coordinates by determining ∇^2 directly for a cartesian coordinate system, or by development from (14-6b),

$$\nabla \cdot \vec{v} = \nabla \cdot (\nabla \Phi) = \left(\hat{x}\frac{\partial}{\partial x} + \hat{y}\frac{\partial}{\partial y} + \hat{z}\frac{\partial}{\partial z}\right) \cdot \left(\hat{x}\frac{\partial}{\partial x} + \hat{y}\frac{\partial}{\partial y} + \hat{z}\frac{\partial}{\partial z}\right)\Phi(x, y, z) = 0$$

or

$$\left(\frac{\partial^2}{\partial x^2} + \frac{\partial^2}{\partial y^2} + \frac{\partial^2}{\partial z^2}\right)\Phi \equiv \frac{\partial^2 \Phi}{\partial x^2} + \frac{\partial^2 \Phi}{\partial y^2} + \frac{\partial^2 \Phi}{\partial z^2} = 0 \quad (14\text{-}8)$$

From these forms, the Laplacian is seen to be a linear, second-order scalar operator.

An important characteristic of the solutions of linear differential equations is that additional solutions can be constructed using linear combinations of existing solutions. For example, if Φ_1, Φ_2, and Φ_3 are solutions to (14-7), then $\Phi = A\Phi_1 + B\Phi_2 + C\Phi_3$ (where A, B, and C are constants) is also a solution. In general, there are no specific solutions to a partial differential equation such as (14-7), or, stated another way, (14-7) is not a complete statement of the problem; it is also necessary to specify two boundary conditions or limits of integration, since the equation is second order. However, certain *classes* of general solutions exist that depend on the coordinates in which (14-7) is expressed. These solutions are determined by the method of separation of variables to be considered later.

Laplace's equation is referred to as a harmonic equation. Its solutions are harmonic functions: sine, cosine, and exponential functions.[3]

In cylindrical coordinates, (14-7) is

$$\nabla^2 \Phi(r, \theta, z) = \frac{\partial^2 \Phi}{\partial r^2} + \frac{1}{r}\frac{\partial \Phi}{\partial r} + \frac{\partial^2 \Phi}{r^2\, \partial \theta^2} + \frac{\partial^2 \Phi}{\partial z^2} = 0 \quad (14\text{-}9)$$

The general class of solutions to this equation involves exponential, simple harmonic (sine and cosine) and cylindrical harmonic (Bessel) functions.

In spherical polar coordinates, Laplace's equation is

$$\nabla^2 \Phi(r, \varphi, \theta) = \frac{1}{r^2}\frac{\partial}{\partial r}\left(r^2 \frac{\partial \Phi}{\partial r}\right) + \frac{1}{r \sin \varphi}\frac{\partial}{\partial \varphi}\left(\sin \varphi \frac{\partial \Phi}{r\, \partial \varphi}\right) + \frac{1}{\sin^2 \theta}\frac{\partial^2 \Phi}{r^2\, \partial \theta^2} = 0 \quad (14\text{-}10)$$

The solutions to this equation are power series, simple harmonic, and spherical harmonic (Legendre polynomial) functions. Equations (14-9)

[3] See, for example, chap. IV of J. Artley, *Fields and Configurations*, Holt, Rinehart and Winston, New York, 1965.

and (14-10) can be derived by first evaluating $\nabla \cdot \vec{v}$ by the methods presented in Chapter 4 and Appendix B and then substituting for the components of $\vec{v}$, which are directly determined from the relation $\vec{v} = \nabla\Phi$,

$$u_i = \frac{\partial\Phi}{\partial x_i} \tag{14-11}$$

where Δx_i are the coordinate components of $\vec{\Delta\imath}$, a general elemental line segment. For example, in cylindrical polar coordinates,

$$u_r = \frac{\partial\Phi}{\partial r} \qquad v_\theta = \frac{\partial\Phi}{r\,\partial\theta} \qquad w_z = \frac{\partial\Phi}{\partial z} \tag{14-12}$$

and in spherical polar coordinates,

$$u_r = \frac{\partial\Phi}{\partial r} \qquad v_\varphi = \frac{\partial\Phi}{r\,\partial\varphi} \qquad v_\theta = \frac{\partial\Phi}{r \sin\varphi\,\partial\theta} \tag{14-13}$$

14.2 USE OF POTENTIAL FUNCTIONS AND SOLUTIONS OF LAPLACE'S EQUATION

There are four methods for determining potential functions that can be used to represent possible flow fields:

1. A general class of solutions to the Laplace equation may be obtained, for example, by the method of separation of variables, or assumption of certain types of functions (for example, exponential).
2. Hypothetical flow fields may be postulated and velocity or other field or boundary conditions may be integrated to obtain Φ.
3. Arbitrary functions of the coordinates may be examined to see whether they satisfy $\nabla^2\Phi = 0$.
4. Electric and magnetic fields (and all other potential fields) also satisfy Laplace's equation. The voltage equipotential lines in an electric field are equivalent lines of constant velocity potential. Electric, magnetic, fluid, and steady-state heat-conduction fields are all analogous (are all represented mathematically by Laplace's equation). Since the electric field is the easiest to set up, it is used extensively as a two- and three-dimensional analog for potential field analyses.

14.3 EXAMPLE OF A THREE-DIMENSIONAL POTENTIAL: The Potential Source

As an example of determining a potential function from the second method listed above, consider a source (or sink) flow hypothesized as a singular point source of spontaneous mass creation (or annihilation). Every other point in the flow field must satisfy the requirements of continuity and irrotationality. No exact physical counterpart exists to this type of flow, but it can be linearly combined with other flow fields to obtain more complete fields for use in approximating physical or real fields. If no other field or action is present, the "fluid" will emanate radially from the source point; the velocity is everywhere radial and of equal magnitude on spherical surfaces. A spherical polar coordinate system then most directly represents the flow. The rate of mass creation ($\dot{m} = \rho Q$) is assumed constant, or, for an incompressible flow, $Q =$ const. The flow rate through a sphere of radius r is given by

$$Q = Au_r \tag{14-14}$$

where $u_r = \partial\Phi/\partial r$ is the radial component of the velocity (meridianal v_φ and azimuthal w_θ components are zero) and $A = 4\pi r^2$ is the surface area perpendicular to the velocity u_r. Equation (14-14) then becomes

$$Q = 4\pi r^2 \frac{\partial\Phi}{\partial r}$$

or

$$\Phi = -\frac{Q}{4\pi r} \tag{14-15}$$

Since the velocity components are of primary importance, the constant of integration is omitted. Continuity of mass flow (4-19a) is satisfied for any surface not enclosing the origin point $r = 0$. If the source point is included in the flow field, the continuity equation including the source in the C.V. boundary is

$$\oiiint \frac{\partial\rho}{\partial t}\, dV + \oiint \rho\vec{v}\cdot\vec{dA} = \dot{m}_s \tag{14-16a}$$

where $\dot{m}_s$ is the rate of mass creation within the C.V. For an incompressible flow, (14-16a) reduces to

$$\operatorname{div}\vec{v} = \frac{Q}{V} \tag{14-16b}$$

Surfaces of constant potential are perpendicular to the flow direction or streamlines, as seen from the defining relation for Φ, $\vec{v} = \operatorname{grad}\Phi$. For example, from (14-14), the surfaces of constant Φ are given by the equation[4]

$$r = \frac{-Q}{4\pi\Phi}$$

[4] The minus sign is of little consequence in this case, since a negative radius implies a negative polar angle, and this expression is independent of the angle.

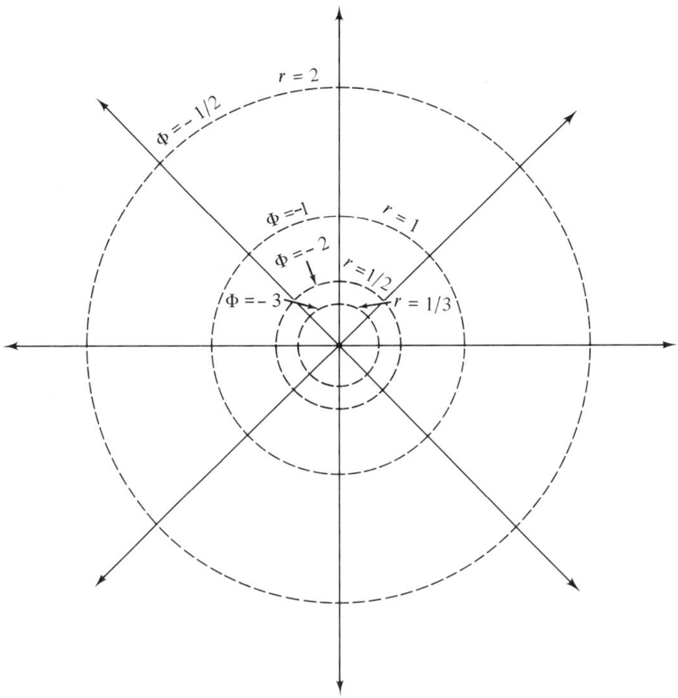

Fig. 14-1. Streamlines and potential lines for a plane section through a three-dimensional source flow.

Figure 14-1 illustrates such a source flow; $\Phi = 0$ corresponds to a sphere of $r = \infty$, and $\Phi = \infty$ corresponds to $r = 0$ ($Q/4\pi$ has arbitrarily been taken equal to unity).

14.4 TWO-DIMENSIONAL OR CYLINDRICAL PROBLEMS:

Two-Dimensional or Line Source

If the flow boundaries are essentially two-dimensional or cylindrical, a single plane perpendicular to the cylindrical axis will represent the entire flow field. The determination of airfoil section characteristics is an example of such a flow. A model of the airfoil is constructed with constant chord and thickness and is installed in a wind tunnel such that it spans the width of a rectangular wind-tunnel test section. The flow across all spanwise stations is identical, so the results are independent of the axis station or z coordinate.[5] A two-dimensional or plane representation is then sufficient

[5] The flow is two-dimensional except for a region near the walls that is in, or is affected by, the wind-tunnel-wall boundary layer.

for formulating the mathematical representation.

The flow field for a steady incompressible two-dimensional line (or point) source is again determined by the condition that the velocity is radial only and the source strength Q is constant.[6] For a line length b,

$$Q = Au_r = 2\pi rbu_r = 2\pi rb\,\frac{\partial \Phi}{\partial r}$$

or

$$\frac{Q}{b} = 2\pi r\,\frac{\partial \Phi}{\partial r}$$

where u_r is the radial velocity component (the tangential or θ component v_θ and the axial or z component w_z are zero). The potential field is then given by

$$\Phi = \frac{Q/b}{2\pi}\ln r \tag{14-17}$$

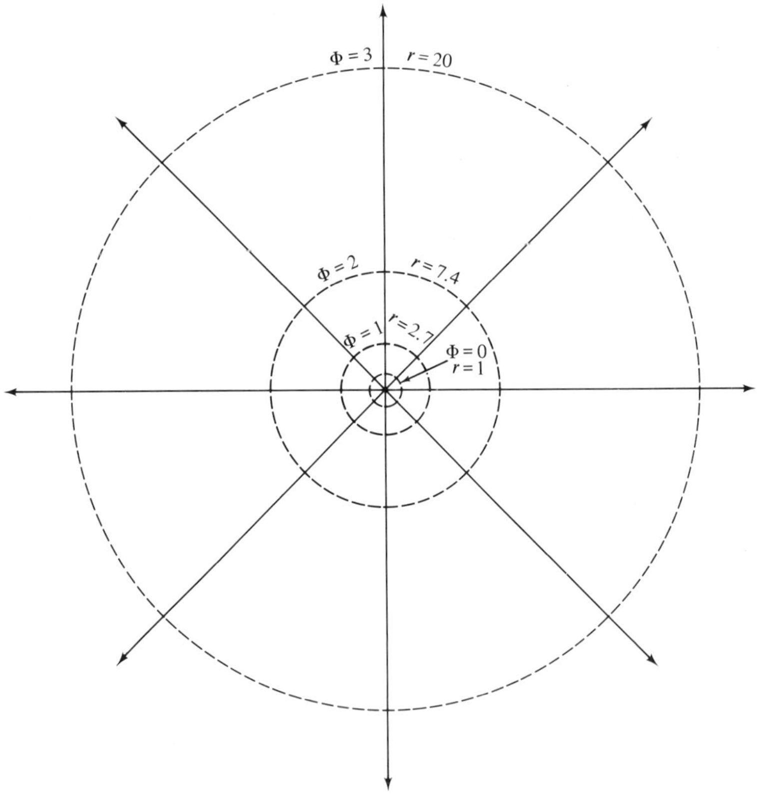

Fig. 14-2. Streamlines and potential lines for a plane section perpendicular to a line or two-dimensional source flow.

[6] A line source is a source uniformly distributed along a straight line, Fig. 14-2.

Lines of constant Φ are seen to be circles, logarithmically distributed with respect to radius. Figure 14-2 represents the constant Φ lines and streamlines for this flow field ($Q/2\pi b$ has again arbitrarily been taken as unity). Values of $0 < r < 1$ imply negative values of Φ (and negative values of r imply imaginary Φ).

14.5 THE STREAM FUNCTION Ψ

The potential function is derived from the condition that the flow is irrotational; which, with the continuity equation, gave the field equation $\nabla^2\Phi = 0$ for an incompressible steady flow (or it gives the instantaneous field for an incompressible unsteady flow in which case Φ is also a function of time).

Since streamlines are defined as lines of constant flow direction, they are continuous if the flow is continuous. It therefore appears possible to determine a *stream function* representing this continuity condition. For example, for two-dimensional flow, the mass flow rate per unit width b between any pair of streamlines must be constant. Consider a streamline through the point P and two neighboring streamlines as shown in Fig. 14-3. Since the streamlines are lines in space, they can be represented as functions of a chosen coordinate system, for example $\Psi(x, y)$, where Ψ is the defined stream function. Since the flow is everywhere tangent to the streamlines, there is no flow across a streamline; consequently, the mass flow rate between any pair of streamlines, for example, $\Psi + \Delta\Psi/2$ and $\Psi - \Delta\Psi/2$, will be constant for any point $P(x, y)$ along $\Psi(x, y)$. This

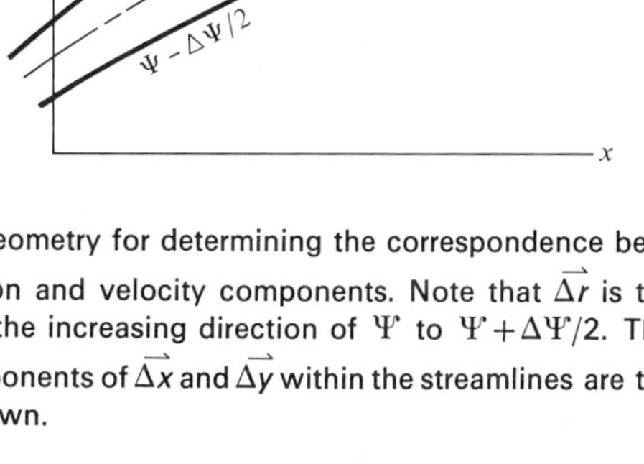

Fig. 14-3. Geometry for determining the correspondence between the stream function and velocity components. Note that $\vec{\Delta r}$ is taken from $\Psi - \Delta\Psi/2$ in the increasing direction of Ψ to $\Psi + \Delta\Psi/2$. The corresponding components of $\vec{\Delta x}$ and $\vec{\Delta y}$ within the streamlines are then in the directions shown.

continuity condition can be expressed in terms of the scalar quantities with due consideration of their directions as

$$\Delta \dot{m} = \rho \vec{\mathrm{v}} \cdot \overrightarrow{\Delta A} \propto \Delta \Psi$$

or

$$\rho_0 \, \Delta \Psi = \rho \vec{\mathrm{v}} \cdot \overrightarrow{\Delta A} \tag{14-18}$$

where ρ_0 is a reference density somewhere in the flow field. From the geometry of Fig. 14-3,

$$\rho_0 \, \Delta \Psi \equiv \rho_0 \frac{\partial \Psi}{\partial x} \Delta x + \rho_0 \frac{\partial \Psi}{\partial y} \Delta y = \rho u b \, \Delta y - \rho v b \, \Delta x$$

so the velocity components are determinable in terms of the stream function as

$$u = \frac{\rho_0}{\rho} \frac{\partial \Psi}{\partial y} \qquad \text{and} \qquad v = -\frac{\rho_0}{\rho} \frac{\partial \Psi}{\partial x} \tag{14-19a}$$

(where the width b has been arbitrarily incorporated into Ψ according to usual convention). For incompressible flow, ρ = const. and

$$u = \frac{\partial \Psi}{\partial y} \qquad v = -\frac{\partial \Psi}{\partial x} \tag{14-19b}$$

The stream function may also be a function of time, in which case a plot of the streamlines could be made for a given instant. This graphical representation would then change from one instant to another. For steady flow, the pattern remains fixed or *stationary* in space.

By the physical definition that the streamlines (lines of constant Ψ) are parallel to the direction of the velocity vector v and that the potential function is defined by the relation $\vec{\mathrm{v}} = \nabla \Phi$, it follows that lines of constant Φ are perpendicular to the flow of $\vec{\mathrm{v}}$ direction, and that the stream function and velocity-potential function are consequently orthogonal functions. This fact is also easily shown using the mathematical descriptions of these functions (see Problem 14-8).

For a cartesian coordinate system, (14-10) and (14-19b) can be compared to show that

$$\frac{\partial \Phi}{\partial x} = \frac{\partial \Psi}{\partial y} \qquad \text{and} \qquad \frac{\partial \Phi}{\partial y} = -\frac{\partial \Psi}{\partial x} \tag{14-20a}$$

Representation of the flow field in a plane polar coordinate system is indicated in Fig. 14-4. By a development similar to that for the cartesian coordinate system (Problem 14-10),

$$u_r = \frac{\partial \Psi}{r \, \partial \theta} = \frac{\partial \Phi}{\partial r} \qquad v_\theta = -\frac{\partial \Psi}{\partial r} = \frac{\partial \Phi}{r \, \partial \theta} \tag{14-20b}$$

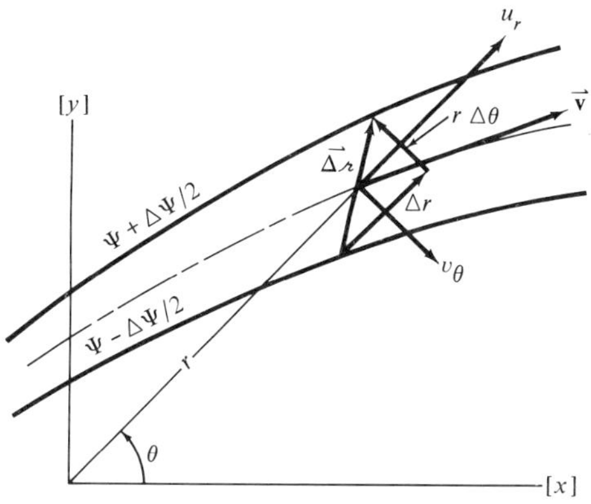

Fig. 14-4. Geometry for determining the correspondence between the stream function and velocity components for a plane polar coordinate system.

These mathematical conditions on the first derivatives of Φ and Ψ are the necessary and sufficient conditions that $\Phi(x, y)$ and $\Psi(x, y)$ exist, and represent satisfactory potential and stream functions for incompressible flow. For fields to be represented by Φ and Ψ, all the physical conditions stated with respect to their original definitions must be satisfied. Expressed mathematically: Φ and Ψ and their first derivatives must exist and be continuous everywhere within the field or region under consideration. These relations (14-20) are known as the Cauchy–Reimann conditions on Φ and Ψ (or on the equivalent complex potential; see Section 14.17) with respect to existence and uniqueness. The relations (14-20) indicate that the stream function must satisfy the conditions by which the potential function is defined, that is, irrotationality. If a two-dimensional flow is irrotational, this condition of irrotationality can be expressed in terms of Ψ as

$$2\omega_z = \frac{\partial v}{\partial x} - \frac{\partial u}{\partial y} = -\frac{\partial^2 \Psi}{\partial y^2} - \frac{\partial^2 \Psi}{\partial x^2} = -\nabla^2 \Psi = 0 \qquad (14\text{-}21)$$

Continuous irrotational flow is then represented by Laplace's equation with Φ and/or Ψ as the dependent variable.

LINE SOURCE

The stream function for the flow field in the neighborhood of a line source can be determined from (14-21) either from the definition of source flow or by calculation from (14-17) if $\Phi(r, \theta)$ is given. From the defined physical considerations

$$Q = A\text{v} = 2\pi r u_r \qquad \text{or} \qquad \frac{\partial \Psi}{r\,\partial \theta} = \frac{Q}{2\pi r} \qquad \text{giving } \Psi = \frac{Q}{2\pi}\theta \quad (14\text{-}22)$$

The streamlines, defined as lines of constant Ψ, implying θ = const., are radial lines, as illustrated in Fig. 14-2 for sequential values of θ at 45° intervals. The streamlines and equipotential lines are seen to be everywhere orthogonal.

In general, the streamlines for a given stream function $\Psi = \Psi(x, y)$ may be plotted by assigning sequential values (constants) to Ψ and solving for, or plotting, the function $\Psi(x, y)$ or $\Psi(r, \theta)$. For example, in (14-22),

$$\theta = \frac{2\pi\Psi}{Q} \text{ (rad)} \qquad \text{or} \qquad \theta° = 360\frac{\Psi}{Q}$$

Since Q is a constant (at any instant), intervals of Ψ/Q can be assigned to different streamlines. Note that the intervals must be fractions with the total interval in $\Psi/Q < 2\pi$ or the function will no longer be single-valued, and an ambiguity would result.

AXISYMMETRIC FLOW

A stream function can be defined for a pseudo-three-dimensional or *axisymmetric* flow as illustrated in Fig. 14-5. The velocity components are then taken as u and v in the axial (x) and radial (η) directions, respectively. In this case, the mass flow is constant between stream surfaces which are surfaces of revolution about the x axis. The condition of constant mass flow between adjacent surfaces is then

$$\begin{aligned}\Delta\dot{m} &= \rho\vec{\mathrm{v}}\cdot\overrightarrow{\Delta A} = \rho u 2\pi\eta\,\Delta\eta + \rho v 2\pi\eta(-\Delta x)\\ &= 2\pi\rho_0\,\Delta\Psi = 2\pi\rho_0\left(\frac{\partial\Psi}{\partial x}\Delta x + \frac{\partial\Psi}{\partial y}\Delta y\right)\end{aligned}$$

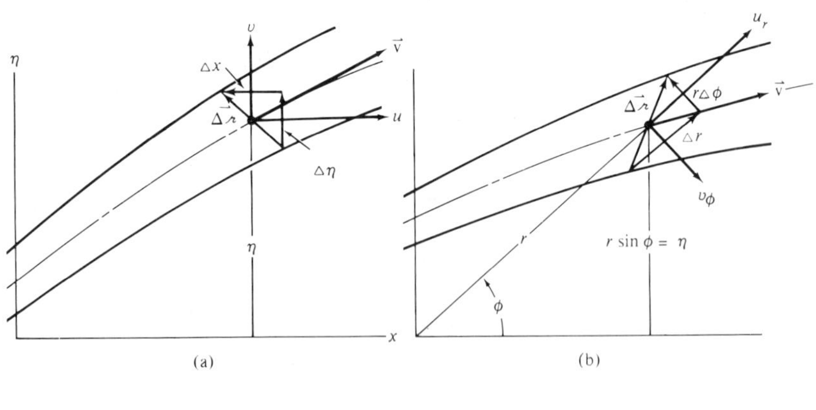

Fig. 14-5. Geometry for determining the correspondence between the stream function and velocity components for axisymmetric flow: (a) cartesian components; (b) polar components.

giving

$$u = \frac{\rho_0}{\rho\eta}\frac{\partial\Psi}{\partial x} \qquad v = -\frac{\rho_0}{\rho\eta}\frac{\partial\Psi}{\partial\eta} \tag{14-23a}$$

or, for incompressible flow,

$$u = \frac{\partial\Psi}{\eta\,\partial x} \qquad v = -\frac{\partial\Psi}{\eta\,\partial\eta} \tag{14-23b}$$

Two simple axisymmetric flows are the three-dimensional potential source flow considered above and conical flows. Although the source flow was stated to be three-dimensional, and the potential function determined for spherical polar coordinates, any straight line through the origin is also an axis of symmetry. It is not advantageous to try to represent spherical source flow in terms of $\Psi(x, \eta)$ as in (14-23), since a similar development can be made directly in terms of spherical coordinates with the z axis as the symmetry axis. Then from Fig. 14-5b,

$$2\pi r \sin\varphi r\, d\varphi u_r = 2\pi\, d\Psi$$

or, for a potential source in terms of $\Psi(r, \varphi)$,

$$u_r(r, \varphi) = \frac{1}{r^2 \sin\varphi}\frac{\partial\Psi}{\partial\varphi} \tag{14-24a}$$

With $Q = 4\pi r^2 u_r$, the stream function is

$$\Psi = -\frac{Q}{4\pi}\cos\varphi \tag{14-25}$$

For a *cylindrically* symmetric flow where v_φ also exists, it can be determined that

$$v_\varphi(r, \varphi) = -\frac{1}{r\sin\varphi}\frac{\partial\Psi}{\partial r} \tag{14-24b}$$

For this type of flow, there must be no tangential flow component, $v_\theta = 0$. This flow component would represent a tangential flow about the cylindrical axis of symmetry that could not have a mass flow component and therefore could not be represented by means of a stream function.

14.6 EXAMPLES OF OTHER SIMPLE FLOWS

POTENTIAL AND STREAM FUNCTION FOR A UNIFORM FLOW

For uniform flow with a velocity U_∞ parallel to the x axis and approaching from $+\infty$, the velocity is everywhere constant with only the component

$u = -U_\infty$, giving

$$\Phi(x, y, z) = -U_\infty x \qquad \Psi(x, y) = -U_\infty(y) \tag{14-26}$$

in cartesian coordinates. In plane polar coordinates, integration of (14-12) and (14-21) or a coordinate transformation from (14-26) gives

$$\Phi(r, \theta, z) = -U_\infty r \cos\theta \qquad \Psi(r, \theta) = -U_\infty r \sin\theta \tag{14-27}$$

In spherical polar coordinates, the free stream is most easily taken along the polar (z) axis; $u_r = -U_\infty \cos\varphi$, from which the potential function can be determined from (14-13) as

$$\Phi(r, \varphi, \theta) = -U_\infty r \cos\varphi \tag{14-28}$$

POTENTIAL VORTEX

The source was seen to be a hypothetical flow field satisfying continuity everywhere except at the origin. The streamlines are radial and the equipotential lines are circles for two-dimensional flow, or the equipotential surfaces are spherical for three-dimensional flow. The equipotential lines and streamlines are orthogonal, as indicated by (14-20). Another basic and simple flow field can be obtained by inverting the potential and stream surfaces (or lines) for a two-dimensional flow. This procedure is possible because the potential and stream functions will remain orthogonal. This procedure gives

$$\Phi = C_1\theta \qquad \Psi = C_2 \ln r \tag{14-29}$$

This flow field no longer represents a source flow, so Q is no longer a field constant. The streamlines (and therefore the flow) are circular, and the equipotential lines are radial. The velocity has only one tangential component (in a plane polar coordinate system) given by

$$v_\theta = \frac{\partial\Phi}{r\,\partial\theta} = -\frac{\partial\Psi}{\partial r} = -\frac{C_2}{r} \tag{14-30}$$

This flow is seen to be irrotational by determining the rotation,

$$2\vec{\omega} = \nabla \times \vec{v}(r, \theta) = \hat{z}\left(\frac{\partial v_\theta}{\partial r} + \frac{v_\theta}{r} - \frac{\partial u_r}{r\,\partial\theta}\right) \quad \text{or} \quad 2\omega_z = \frac{1}{r}\left(\frac{\partial r v_\theta}{\partial r} - \frac{\partial u_r}{r\,\partial\theta}\right)$$

which is zero (the radial component u_r is identically zero). The streamline or tangential velocity v_θ is a hyperbolic function of r.

At the beginning of the chapter, the conservative force field and irrotational velocity field were compared. The work done by a conservative force around a closed path is zero. Similarly, the line integral of the velocity around a closed path is zero for an irrotational velocity field. The line integral around any contour not enclosing the center of the potential vortex is zero, as indicated in Fig. 14-6. The center, however, is a singularity,[7] and the line integral around it has a constant, nonzero value independent of the path:

$$\oint \vec{v} \cdot \vec{dr} = \oint v_\theta r \, d\theta = -\int_0^{2\pi} \frac{C_2}{r} r \, d\theta = -2\pi C_2 \qquad (14\text{-}31)$$

This mathematically defined quantity is the circulation Γ:

$$\Gamma = \oint \vec{v} \cdot \vec{dr} \qquad (14\text{-}32)$$

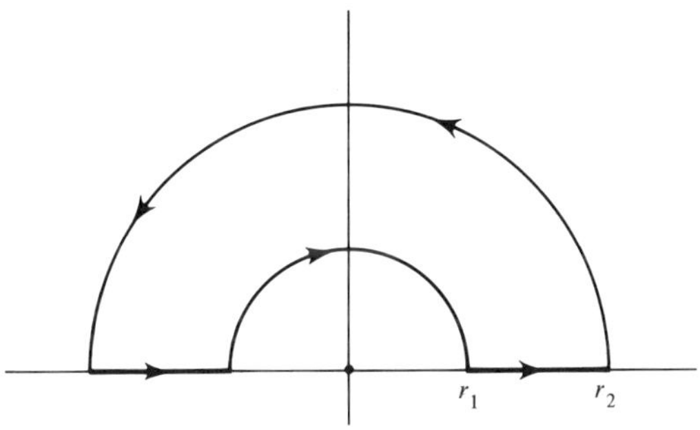

Fig. 14-6. Line contour in the neighborhood of, but not enclosing, a singularity at the origin. The integral around this contour is zero.

The constant in (14-30) and (14-31) is then $C_2 = -\Gamma/2\pi$, or, for the potential vortex field,

$$\Psi(r, \theta) = -\frac{\Gamma}{2\pi} \ln r \qquad \Phi(r, \theta) = \frac{\Gamma}{2\pi} \theta \qquad (14\text{-}33)$$

This flow field is termed a potential or irrotational vortex because the rotation is zero everywhere except at the singular point $r = 0$.

The magnetic field around a current-carrying wire, which varies as $1/r$, is the electric analog for the line vortex. This analogy can be employed with the law of Biot and Savart to calculate the "induced velocity" field for any system of vortex lines (or filaments) or to determine such a field experimentally.

[7] The stream function and the velocity are infinity at the center of the vortex, $r = 0$.

14.7 COMBINED FLOWS: Theory of Images

TWO SOURCES: SOURCE NEAR A WALL

It was stated that any linear combination of potential (or stream) functions is also a satisfactory potential (or stream) function. For example, consider two sources of equal strength a distance $2a$ apart, as illustrated in Fig. 14-7. The velocity potential and stream function are

$$\Phi = \frac{Q/b}{2\pi}(\ln r_1 + \ln r_2) \qquad \Psi = \frac{Q/b}{2\pi}(\theta_1 + \theta_2) \tag{14-34}$$

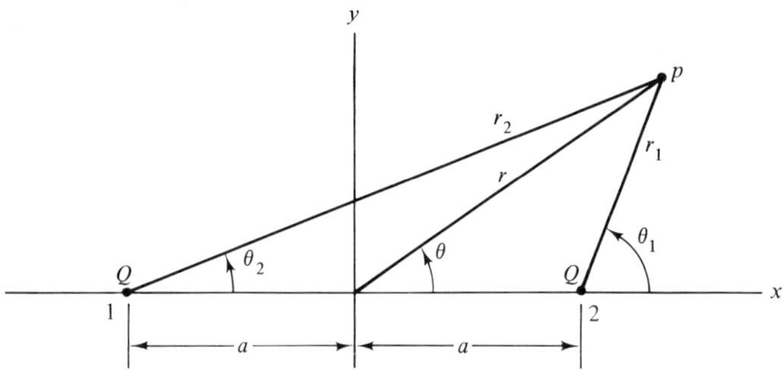

Fig. 14-7. Position coordinates for two sources of equal strength Q separated by a distance $2a$.

The x and y axes are axes of symmetry for this potential flow. Since both axes will be shown to be streamlines, they may be considered to be solid boundaries. The y axis situated between the two sources may be considered as a "reflecting wall" for either source, so one source may also be considered as mirroring the other, or as the image source with respect to the wall along the y axis.

One quadrant of the field of Fig. 14-8 can also be considered to represent the flow from a source in the floor at $x = 1$ on the x axis bounded by a wall on the left (the y axis). It was previously stated that a constant could arbitrarily be added to Φ (or Ψ) without making any change in the field representation. If

$$\frac{Q/b}{\pi} \ln a$$

is added to the potential in (14-34), Φ can be written in nondimensional parameters as

$$\Phi' = \ln r_1' r_2'$$

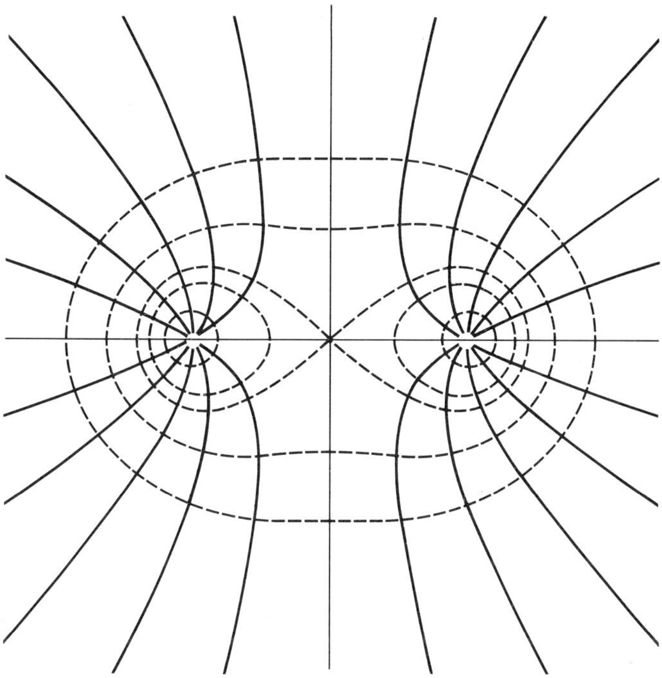

Fig. 14-8. Potential and streamlines for the flow field produced by two sources of equal strength separated by a distance $2a$ ($a = 1$). This field may also be considered to represent a source near a wall (the y axis) or a source at $x = a = 1$ on the lower wall of a 90° corner.

where $\Phi' \equiv 2\pi b\Phi/Q$ and $r' \equiv r/a$. The sources are then located at $r' = 1$, $\theta_1 = 0$, and $r_2' = 1$, $\theta_2 = \pi$ (or $r_2' = -1$, $\theta_2 = 0$). The lines of constant potential are determined by the relation

$$r_1' r_2' = e^{\Phi'} \tag{14-34a}$$

The stream function can be represented dimensionlessly as

$$\Psi' = \theta_1 + \theta_2 \tag{14-34b}$$

where $\Psi' \equiv 2\pi b\Psi/Q$. The equipotential lines for the indicated values of Φ' are shown in Fig. 14-8 together with the streamlines. For very large values of Φ' the equipotential contour approaches a circle and the streamlines approach a radial configuration. From distances much greater than $2a$, the source pair then "looks like" a single source of strength $2Q$.

The constant-potential lines of Fig. 14-8 were plotted by choosing a value of Φ' in (14-34a) and graphically constructing the contour for this value. For example, $\Phi' = 0$ gives $r_1'r_2' = 1$, where r_1' is a circle around the right-hand source point at $r' = 1$ and r_2' is a circle around the left-hand source point at $r_2' = -1$. The intersections of the circles $r_1' = \frac{1}{2}$ and $r_2' = 2$ would then be two points on the $\Phi' = 0$ contour. Other points are found by determining the intersections of circles around these origins satisfying the relation $r_1'r_2' = 1$.

The streamlines are obtained from (14-34b). For example, the horizontal axis is the streamline $\Psi' = 0$ or π. To the right of the right-hand source point, θ_1 and θ_2 are both zero. To the left of the left-hand source point, θ_1 and $\theta_2 = \pi$. Between the sources, $\theta_1 = \pi$, $\theta_2 = 0$. The vertical axis (or image plane) is the streamline $\Psi' = \pi$. A simple method exists for constructing the rest of the streamlines. Consider, for example, the streamline $\Psi' = \pi/2$ or $\theta_1 + \theta_2 = \pi/2$. At Q_1, $\theta_2^0 = 0$ (the superscript on θ will imply an incremental condition), so $\theta_1^0 = \pi/2$ or the streamline starts tangent to the vertical at Q_1. Rays can be laid out from each Q in equal angular increments. The next point on the $\pi/2$ streamline can be obtained by considering $\theta_2 = \theta_2^0 + \Delta\theta$; the corresponding value of θ_1 must be $\theta_1 - \Delta\theta$, so this next point is at the intersection of the first ray at an angle $+\Delta\theta$ (counterclockwise) from Q_1 with the first ray at an angle $-\Delta\theta$ (clockwise) from $\theta_1^0 = \pi/2$. The next point is at the intersection of the next counterclockwise ray from Q_1 with the next clockwise ray from Q_2. Once this construction is started, this and the remaining streamlines are easily completed.

14.8 ILLUSTRATIVE EXAMPLE 14-1: Stream and Potential Lines for a Vortex Near a Wall

The potential field of a line or two-dimensional vortex at a distance h from a wall can be constructed by placing an image vortex at the image point a distance h on the other side of the wall, as illustrated in Fig. 14-9.

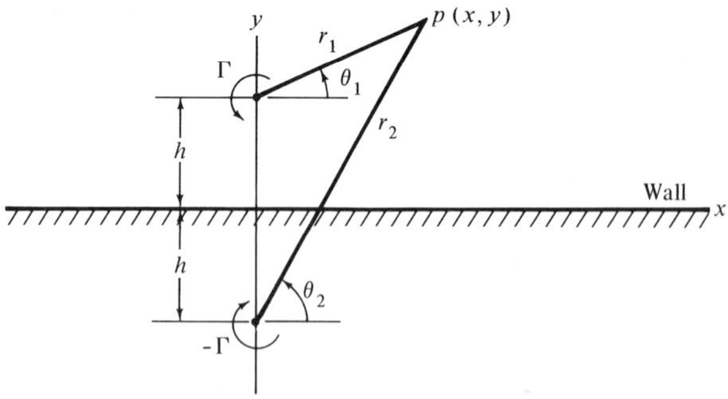

Fig. 14-9. Image-system representation for the position coordinates and the flow field of a vortex near a plane wall.

Symmetry requires that the image vortex be of opposite sign to the original vortex. This condition can also be determined mathematically by requiring that the wall (x axis) be a streamline.

The stream and potential functions at an arbitrary point P are

$$\Psi = -\frac{\Gamma}{2\pi}\ln r_1 + \frac{\Gamma}{2\pi}\ln r_2 \qquad \Phi = \frac{\Gamma}{2\pi}\theta_1 - \frac{\Gamma}{2\pi}\theta_2 \tag{14-35}$$

The equations for the stream and equipotential lines are then

$$\frac{r_2}{r_1} = e^{2\pi\Psi/\Gamma} \equiv \Psi' \qquad \text{and} \qquad \theta_1 - \theta_2 = \frac{2\pi\Phi}{\Gamma} = \Phi' \tag{14-35a}$$

which can be shown to be families of circles with "family parameters" $e^{2\pi\Psi/\Gamma}$ for the streamlines and $2\pi\Phi/\Gamma$ for the equipotential lines. The dimensionless Ψ' and Φ' are defined to simplify further calculations. To show that the equations represent circles, they can be written in terms of x,y coordinates by noting from the geometry that

$$r_1^2 = x^2 + (y-h)^2 \qquad r_2^2 = x^2 + (y+h)^2$$

$$\tan\theta_1 = \frac{y-h}{x} \qquad \tan\theta_2 = \frac{y+h}{x}$$

The stream-function equation can then be written

$$x^2 + (y+h)^2 = [x^2 + (y-h)^2]\Psi'$$

which can be arranged to give the form

$$x^2 + \left(y - h\frac{\Psi'+1}{\Psi'-1}\right)^2 = \frac{2h^2}{\Psi'-1}$$

which is an equation for circles with centers at $x = 0$, $y = h[(\Psi'+1)/(\Psi'-1)]$ (on the y axis) with radii $h\sqrt{2/(\Psi'-1)}$. The streamline $\Psi' \equiv e^{2\pi\Psi/\Gamma} = 1$ or $\Psi = 0$ is the x axis, which does satisfy the condition that the wall is a streamline.

The potential-function equation can be modified with the use of the trigonometric identity,

$$\tan(\theta_1 - \theta_2) \equiv \frac{\tan\theta_1 - \tan\theta_2}{1 + \tan\theta_1\tan\theta_2}$$

to give

$$\tan(\theta_1 - \theta_2) = \tan\Phi' = \frac{(y-h)/x - (y+h)/x}{1 + [(y^2-h^2)/x^2]} = \frac{-2hx}{x^2+y^2-h^2}$$

or

$$x^2 + \frac{2h}{\tan \Phi'} x + \left(\frac{h}{\tan \Phi'}\right)^2 + y^2 = h^2\left(1 + \frac{1}{\tan^2 \Phi'}\right)$$

or

$$\left(x + h \cot \frac{2\pi\Phi}{\Gamma}\right)^2 + y^2 = \left(h \csc \frac{2\pi\Phi}{\Gamma}\right)^2$$

The equipotential lines are then circles (orthogonal to the streamlines) with centers at $x = h \cot 2\pi\Phi/\Gamma$ on the x axis and radii $h \csc 2\pi\Phi/\Gamma$. Figure 14-10 shows the velocity distribution along the y axis produced by the *image* vortex at $y = -h$ according to the relation

$$u_2 = \frac{\partial \Phi_2}{\partial x} = +\frac{\partial \Psi_2}{\partial y} = +\frac{\Gamma}{2\pi y_2}$$

or

$$v_{\theta 2} = \frac{\partial \Phi}{r_2\, \partial \theta} = -\frac{\partial \Psi}{\partial r} = -\frac{\Gamma}{2\pi r_2}$$

On the y axis, $r_2 = y_2 = y + h$. At the point $y = +h$ where the original vortex is located, there is a velocity $u = \Gamma/4\pi h$, which implies that this vortex is in a field that causes it to be propagated at the speed $\Gamma/4\pi h$ to the right. Similarly, the vortex at $y = +h$ produces (or "induces") a velocity $u = \Gamma/4\pi h$ to the right on the vortex at $y = -h$. The vortex near a wall propagates itself along the wall with a constant speed at a fixed height above the wall.

A smoke ring is similarly self-propagating. Two small arc segments diametrically opposite each other produce velocity fields similar to those

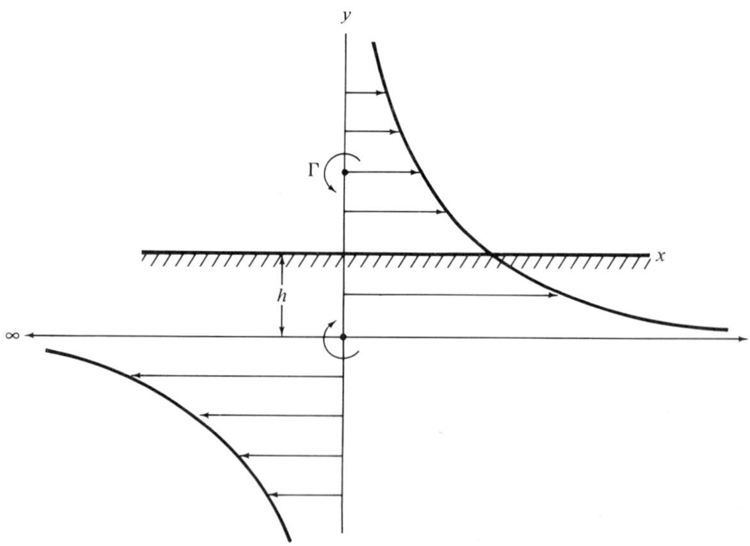

Fig. 14-10. Velocity distribution along the y axis produced by the image vortex at $y = -h$.

shown in Fig. 14-10. Actually each segment has an "induced" effect on the entire ring, which must be calculated. It is interesting that the velocity of propagation for a truly irrotational ring vortex is infinite, so a direct similarity to the smoke ring does not exist. However, the calculation can be made for a "combined vortex" with a viscous core condition added so that the velocity on the centerline of the ring is not infinite. The formulation and solution of this problem are quite complicated.

It has been assumed that the vortex is free to move or propagate. The vortex (and its singularity) is then defined as a "free vortex." It may also be specified that the vortex cannot move. A flow from the other vortex still exists. A force is then required to hold the vortex at the fixed point $x = 0$, $y = h$. The vortex is then defined as "bound."

14.9 POTENTIAL FLOW AROUND BODIES

By definition there is no flow across a streamline or stream surface. Any streamline can then be replaced by a solid boundary. If there exists a closed streamline for a particular potential flow, this streamline can be considered as a body in the surrounding flow. The particular case of a uniform approach flow will be considered with velocity parallel to the x axis, $u = -U_\infty$ (from right to left). The stream and potential functions for this flow are then given by (14-27) through (14-30). If the body is to be represented by a closed streamline, it is an obvious condition that the net source strength within the body must be zero.

FLOW PAST A SOURCE AND SINK: RANKINE OVAL

The simplest combination satisfying the condition of zero net source flow is a source at $x = s$ and sink at $x = -s$ of equal strength, as shown in Fig. 14-11. The approach flow far from the "singularity"[8] is parallel to

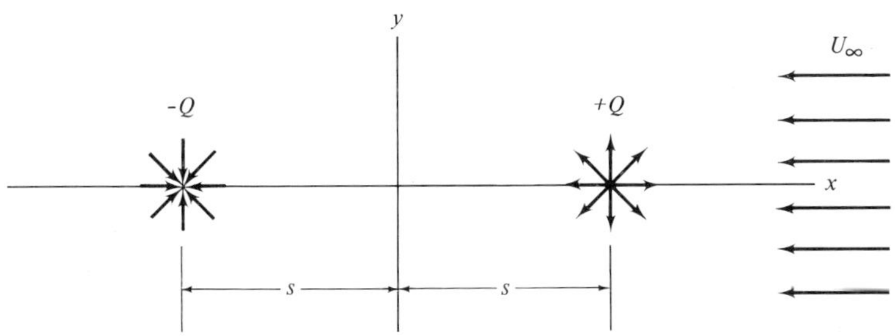

Fig. 14-11. Source, sink, and free stream approach flow for the flow field past a Rankine oval, cylindrical body.

[8] The source and sink are termed singularities, since the velocity at the source point is infinite.

the x axis. In the immediate neighborhood of the singularities, the flow field is determined by the singularity, radial in this case.

The approach flow and source flow are opposed along the x axis. Since the source flow velocity decreases with increasing radial distance from the source point, at some location $x = a$ the net effects of the source, sink, and approach flow will be expected to give a zero velocity or *stagnation point*. This point can be located in terms of U_∞, Q, and s.

The stream function is obtained by combining an approach flow with a source and sink of equal strength:

$$\Psi = -U_\infty y + \frac{Q}{2\pi}(\theta_1 - \theta_2) \tag{14-36}$$

where θ_1 and θ_2 are the angular coordinates of a point measured from the source and sink [see Section 14.7, (14-27), and Fig. 14-9]. This form of $\Psi(y, \theta_1, \theta_2)$ is not amenable to analytic attack but is directly usable for graphically plotting the streamlines. For example, it is seen that the y axis to the right of the source is the streamline $\Psi = 0$. The next streamline can be determined by giving Ψ an increment,

$$\Delta\Psi = -U_\infty \Delta y + \frac{Q}{2\pi}(\Delta\theta_1 - \Delta\theta_2)$$

For plotting, U_∞ and $Q/2\pi$ can arbitrarily be taken as unity. The radial streamlines at equal angular increments are easily drawn for the source and sink, and a family of equally spaced horizontal lines represents the approach streamlines. The easiest way to generate the streamlines is to first construct the streamlines for the source–sink combination and then add the free stream. The streamlines for the source and sink are circles of radius $r_\Psi = s/\sin \Psi''$, where $\Psi'' \equiv 2\pi\Psi/Q$. The radii for equal angular increments of Ψ'' can be calculated and the circle center located by striking an arc from either $+s$ or $-s$ and intersecting the y axis (Fig. 14-12). This y-axis intersection is the center of the source–sink streamline, since the circular streamlines of the source–sink flow must pass through both $+s$ and $-s$ (see Problem 14-15). Then $\Psi = \Psi_{\text{F.S.}} + \Psi_{\pm Q} = -U_\infty y + \Psi_{\pm Q}$; for $\Psi = \text{const.}$, the streamlines are generated by starting with any intersection of a horizontal ($y = \text{const.}$) free-stream streamline such as point 1 and a circle. The next point 2 on the same streamline will be at the next smaller circle and the next horizontal increment in y, and so on. The body streamline will be the streamline through the first y increment line (above $y = 0$) and the largest circle (point 3, Fig. 14-12). Note that the point on the x axis will be undetermined by this construction procedure since the circle through this point is of infinite radius (the x axis). The circular streamlines of the source and sink can also be located by a simple geometric procedure: (1) Lay off

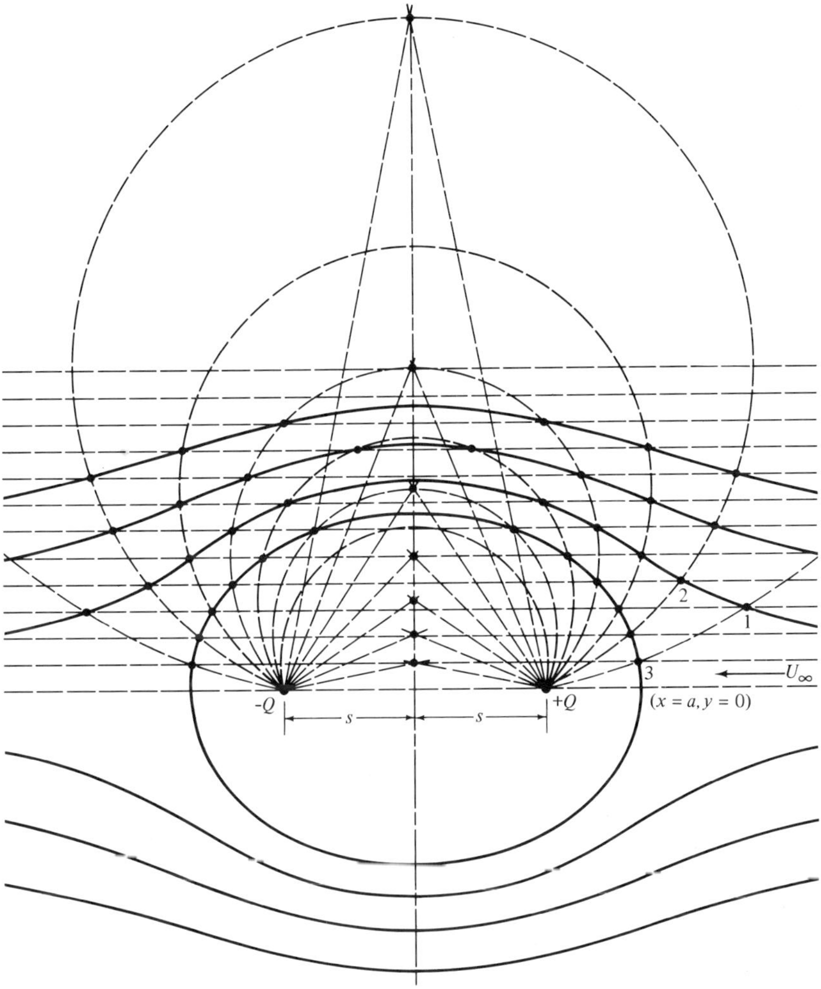

Fig. 14-12. Construction pattern for determining the streamlines for the flow field of a Rankine oval. By taking more closely spaced increments of the U_∞ streamlines, the oval shape can be flattened. Only every other streamline has been constructed.

positive radial streamlines for Q_2 in equal increments of θ_2. (2) The intersection of these rays with the y axis (also the intersection of the complementary θ_1 ray with the y axis, for this point $\theta_1 + \theta_2 = \pi$) always determines the top point on a circle through this point and $+s$ and $-s$, so the circular streamline is completely determined. The streamline pattern is illustrated in Fig. 14-12 along with some of these construction procedures. The initial angular increments chosen were $45°/4 = 11\frac{1}{4}°$.

The equation of the body streamline through the stagnation point may be determined by calculating the value of Ψ for the stagnation condition $u = 0$, $v = 0$. The stream function may be written

$$\Psi(x, y) = -U_\infty y + \frac{Q}{2\pi} \tan^{-1} \frac{2ys}{x^2 + y^2 - s^2} \tag{14-37}$$

Calculating $u = \partial\Psi/\partial y = 0$ at the stagnation point, and also noting that $y = 0$ there, gives

$$a^2 = s^2 + \frac{2Qs}{\pi U_\infty}$$

where $x = a$, $y = 0$ is the stagnation point. This result can also be considered to be a relation for the quantity Q in terms of quantities that have a physically direct connection with the geometry and flow. The value of $\Psi_B(a, s, U_\infty)$ for the body can be determined by calculating

$$\Psi(x = a, y = 0) = 0$$

It is seen that the x axis and the oval-shaped streamline are both represented by $\Psi = 0$. The semiwidth b may be calculated in terms of a and s (or Q, U_∞, and s) by setting $\Psi = \Psi_B = 0$ at $x = 0$, $y = b$ in the above equation for Ψ. The result is a transcendental function that can be solved graphically or by trial and error (see Problem 14-20).

FLOW PAST A CIRCULAR CYLINDER

If a source and sink located on the x axis are brought together at the origin with the condition that the product of the source strength and distance between them remains constant, the result is defined as a dipole or doublet. Mathematically this statement can be written

$$\lim_{\substack{s \to 0 \\ 2sQ \to \mu}} (\Psi_{+Q} + \Psi_{-Q}) = \Psi_\mu$$

where μ is the strength of the doublet. By performing this limiting calculation,

$$\Psi_\mu(x, y) = \frac{\mu}{2\pi} \frac{y}{x^2 + y^2}$$

or

$$\Psi_\mu(r, \theta) = \frac{\mu}{2\pi} \frac{\cos\theta}{r} \tag{14-38}$$

The streamlines can be determined as circles passing through the origin with centers on the y axis (see Problem 14-21).

If the stream and potential functions for the doublet are combined with that for a free stream,

$$\Psi = \Psi_\infty + \Psi_D = -U_\infty y + \frac{\mu}{2\pi}\frac{y}{x^2 + y^2}$$

$$\text{or} \qquad \Psi = -\left(U_\infty - \frac{\mu}{2\pi r^2}\right) r \sin\theta \tag{14-39a}$$

The body streamline is obtained by determining the streamline through the stagnation point $u = 0$, $v = 0$. This condition gives

$$r_0 = a = \sqrt{\frac{\mu}{2\pi U_\infty}}$$

The stream function can then be expressed as

$$\Psi(x, y) = -U_\infty y\left(1 - \frac{a^2}{x^2 - y^2}\right)$$

$$\text{or} \qquad \Psi(r, \theta) = -U_\infty r \sin\theta\left(1 - \frac{a^2}{r^2}\right) \tag{14-39b}$$

At the stagnation point, $r_0^2 = x^2 + y^2 = a^2$, so $\Psi_B = 0$ and the body streamline is a circle with radius a; the x-axis $y = 0$ or $\theta = 0$ also makes up the zero streamline. On this circle, the radial and tangential velocity components are $u_r = 0$ (a necessary condition for $r = a$ to be a streamline) and $v_\theta = 2U_\infty \sin\theta$. The magnitude of the velocity $\text{v} = \sqrt{u^2 + v^2} = \sqrt{u_r^2 + v_\theta^2}$ is then just $\text{v} = v_\theta$.

The solutions are graphically constructed by adding the stream functions as indicated in (14-39) for the condition that $\Psi = \text{const}$. This procedure is facilitated by writing the doublet stream function as

$$x^2 + \left(y - \frac{1}{\Psi'_D}\right)^2 = \frac{1}{\Psi'^2_D}$$

where $\Psi'_D \equiv 4\pi\Psi_D/\mu$, and for the free stream $\Psi'_\infty \equiv \Psi_\infty/U_\infty = -y$. The streamline pattern for this flow is illustrated in Fig. 14-13.

CYLINDER WITH CIRCULATION IN A FREE STREAM

The previous example of a free stream and doublet yielded a circle as a closed streamline that can then be considered to be a solid body streamline. Any other potential flow with a circular streamline with radius a can be superimposed on this flow, and its potential and stream functions added to that for the flow about a circle. About the only flow that satisfies this

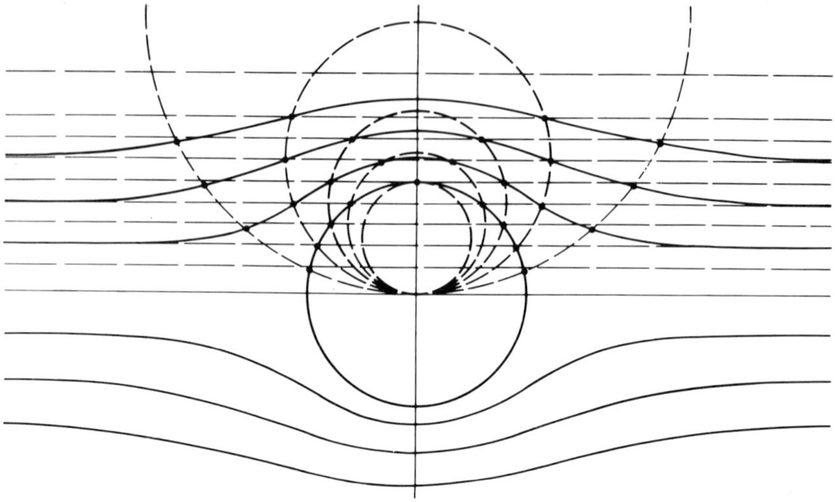

Fig. 14-13. Construction pattern for determining the streamlines for the flow field of a doublet in a free stream to give the flow around a circular cylinder. Only every other streamline has been constructed.

condition is a potential vortex. The stream and potential functions for the free stream, doublet, vortex combination for flow about a circular cylinder with circulation Γ are given by

$$\Psi(r, \theta) = -U_\infty r \sin\theta\left(1 - \frac{a^2}{r^2}\right) - \frac{\Gamma}{2\pi}\ln r$$

$$\Phi(r, \theta) = -U_\infty r \cos\theta\left(1 + \frac{a^2}{r^2}\right) + \frac{\Gamma}{2\pi}\theta \tag{14-40}$$

The streamlines can be obtained by adding circles concentric about the origin with increasing increments of $\ln r$ representing increments of Ψ_Γ to the streamlines of Fig. 14-13. The resulting streamlines are illustrated in Fig. 14-14.

The stagnation points are obtained by setting the velocity components $u_r = 0$, $v_\theta = 0$. This procedure determines the relation between Γ and the location of the stagnation points $r = a$, $\theta = \theta_s$,

$$\Gamma = -4\pi a U_\infty \sin\theta_s \tag{14-41}$$

Since Γ is a positive number, the stagnation points on the cylinder are located in the third and fourth quadrants. Equation (14-41) then relates the mathematically defined quantity Γ to the physical quantity θ_s, the geometric location of the stagnation point.

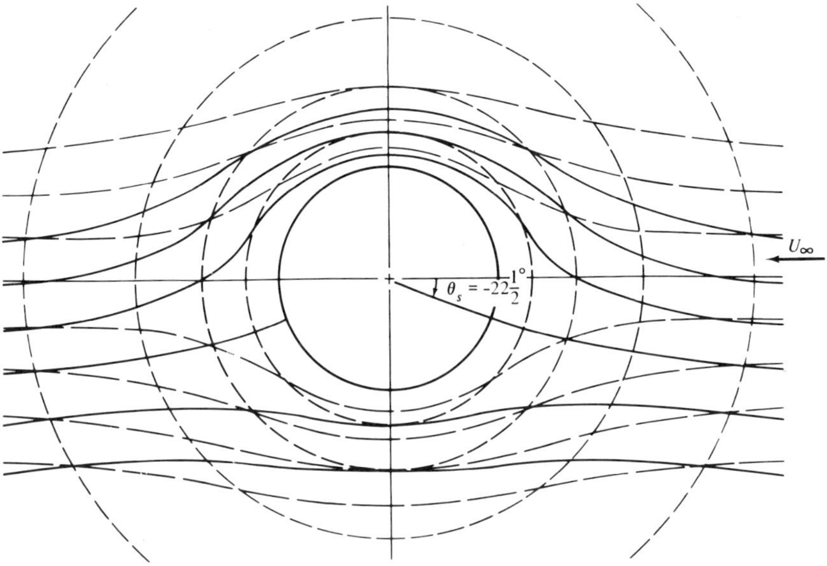

Fig. 14-14. Potential flow streamlines for the flow around a circular cylinder with circulation.

Using (14-41), (14-40), may be written

$$\Psi(r, \theta) = -U_\infty\left[r \sin\theta\left(1 - \frac{a^2}{r^2}\right) - 2a \sin\theta_s \ln r\right] \qquad (14\text{-}42)$$

A unique and limiting case exists for $\theta_s = \pi/2$ or $\Gamma = 4\pi a U_\infty$, for which the stagnation points coincide at the bottom of the cylinder. If the circulation Γ is increased beyond this value, the stagnation point moves off the cylinder into the fluid along the negative y axis.

This potential flow configuration can be considered as a first approximation to flow about a spinning cylinder. The larger than free stream velocity on the top will result in a lower pressure than on the bottom, resulting in the development of a lift force. This aerodynamic force was first studied and described by G. Magnus in connection with the deflection of musket and cannon balls.

14.10 PRESSURE AND FORCE IN POTENTIAL FLOW

The pressure distribution for frictionless incompressible flow in a potential flow field is determined directly from Bernoulli's equation. For steady flow

$$p + \frac{\rho \mathrm{v}^2}{2g_c} = p_\infty + \frac{\rho U_\infty^2}{2g_c} = \text{const.}$$

where $\mathrm{v}^2 = \vec{\mathrm{v}} \cdot \vec{\mathrm{v}}$, $\mathrm{v}^2 = u^2 + v^2$, or $\mathrm{v}^2 = u_r^2 + v_\theta^2$.

For the pressure distribution on a closed or "body" streamline, the velocity magnitude is determined by the boundary conditions on that streamline. For example, the velocity on a circular cylinder is just $\mathrm{v}(r = a) = v_\theta(r = a)$, $u_r = 0$.

The force components of primary importance are usually those parallel and perpendicular to the free stream velocity U_∞. It is therefore convenient to take the x axis parallel to U_∞, as previously considered. The force component Y in the y direction is then the lift force, or $L = Y$. The force component in the negative x direction is the drag force $D = -X$. With no friction force, the resultant force on the body is

$$\vec{F} = \oiint p \, d\vec{A}$$

For two-dimensional flow, the components for a cylindrical body of length b (perpendicular to the x,y or r,θ plane) are (Fig. 14-15)

$$X = -b \oint p \, dy \qquad \text{or} \qquad D = b \oint p \, dy$$

$$Y = L = b \oint p \, dx$$

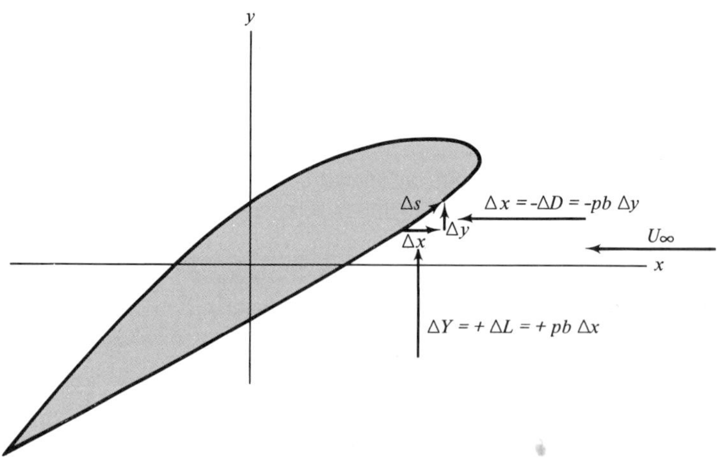

Fig. 14-15. Pressure–force components on a body contour.

In Chapter 7 it was seen that the pressure coefficient and force coefficients are more general quantities and therefore more satisfactory ones with which to deal. For a steady, incompressible, frictionless flow the pressure coefficient is

$$c_p \equiv \frac{p - p_\infty}{\rho U_\infty^2/2g_c} = 1 - \left(\frac{\mathrm{v}}{U_\infty}\right)^2 \tag{14-43}$$

The quantity v/U_∞ is determined directly from the potential flow field relationships.

The force coefficients are

$$C_L = L\bigg/\frac{\rho U_\infty^2}{2g_c}A = L\bigg/\frac{\rho U_\infty^2}{2g_c}bc \qquad C_D = D\bigg/\frac{\rho U_\infty^2}{2g_c}bc$$

where $A = bc$ is the reference area taken equal to the span times the chord length. The coefficients are then calculated as

$$C_L = \frac{b\oint p\,dx}{(\rho U_\infty^2/2g_c)bc} = \oint \frac{pd(x/c)}{\rho U_\infty^2/2g_c} = \oint c_p\,dx' \tag{14-44a}$$

where $x' \equiv x/c$, and the condition that $\oint p_\infty\,dx' = p_\infty \oint dx' = 0$ has been used. Similarly,

$$C_D = \oint c_p\,dy' \tag{14-44b}$$

14.11 ILLUSTRATIVE EXAMPLE 14-2: Tornado

The tornado that moves with a storm is a vortex produced by meteorological wind-current interactions. As an approximation, it can be assumed to behave as a line vortex. The vortex velocity distribution was obtained (14-30) as $\mathrm{v} = v_\theta = \Gamma/2\pi r$ or $rv_\theta = \text{const.} = C$. If the tangential velocity can be measured at some reference radius (for example, $v_{\theta 1}$ and r_1), the rest of the irrotational field can be calculated. Since the storm travels at a nearly uniform velocity, the velocity and pressure distributions can be calculated with respect to the uniformly moving center of the vortex. The vortex streamlines are circles, so the cylindrical coordinate system will correspond to the streamline or natural coordinate system with $\hat{\theta} = \hat{s}$ and $\hat{r} = \hat{R}_s = -\hat{n}$. The pressure distribution is then obtained from (6-28) as

$$\frac{dp}{dr} = \frac{\rho \mathrm{v}^2}{g_c r} \qquad \left(\frac{\partial p}{r\,\partial\theta} = \frac{\partial p}{\partial s} = 0\right) \tag{14-45}$$

Since the flow is irrotational, integration of (14-45) must yield Bernoulli's equation

$$p - p_1 = \frac{\rho}{2g_c}(\mathrm{v}^2 - \mathrm{v}_1^2) = \frac{\rho \mathrm{v}_1^2}{2g_c}\left[\left(\frac{r_1}{r}\right)^2 - 1\right] \tag{14-46}$$

which is obtained by integrating (14-45) with $r\mathrm{v} = \text{const.} = r_1\mathrm{v}_1$ (see Problem 14-29).

Consider the example case of a recording station that indicated a barometric reading of 28.9 in. Hg and a minimum wind speed of 1 ft/sec at a distance of 200 yards from the path of the eye of a tornado being carried by a storm with a mean speed of 7.5 mph (11 ft/sec). Relative to the moving tornado, the tangential wind speed at a 600-ft radius is 10 ft/sec. At this radius, the wind speed is relatively small, so the barometric pressure remained at nearly 730 mm Hg. At a radius of 50 ft, the tangential velocity is 120 ft/sec. The flow obviously cannot remain irrotational to the center of a real vortex, since an infinite velocity and negatively infinite pressure are predicted there. A "core" where the viscous shear becomes significant forms as a result of the high velocity gradients and therefore high shear strain rates at small radii. Consequently, as $r \to 0$, the flow approaches a solid body rotational condition described in Chapter 3. This flow transition in the flow behavior with change in radius is a complicated phenomenon. For this example the flow is approximately irrotational to a core diameter of about 50 ft. The velocity at the edge of the core is 240 ft/sec. The pressure at the edge of the core is determined from (14-46) as 67.5 lb_f/ft^2 below ambient.

The destructive force of a tornado is a consequence of this high velocity and difference in pressure. A relatively small object (automobile or garage) can be considered as a body in the wind current of the vortex. From the discussion and results of Chapter 11, the C_D for such a body can be estimated to be of the order of unity, giving a force of the order of 1500 lb_f on an automobile as the center of the eye passes within 50 ft of it. Obviously there will also be a significant torque. Another interesting phenomenon is evident. As the tornado passes over or very near a house, the pressure suddenly decreases. The pressure at a 100-ft diameter in the above example is about 17 psf below ambient. The house is then at ambient pressure and the surrounding pressure is suddenly dropped as the tornado passes. The 17-psf pressure difference across an 8- by 15-ft wall gives a 2000-lb force; the house literally explodes. The moral to this calculation might be to open all the doors and windows when a tornado warning is posted.

The inverted funneling effect is perhaps the most interesting from a fluid mechanics point of view. It is a secondary flow phenomena produced by the presence of the boundary layer at the ground. In the irrotational

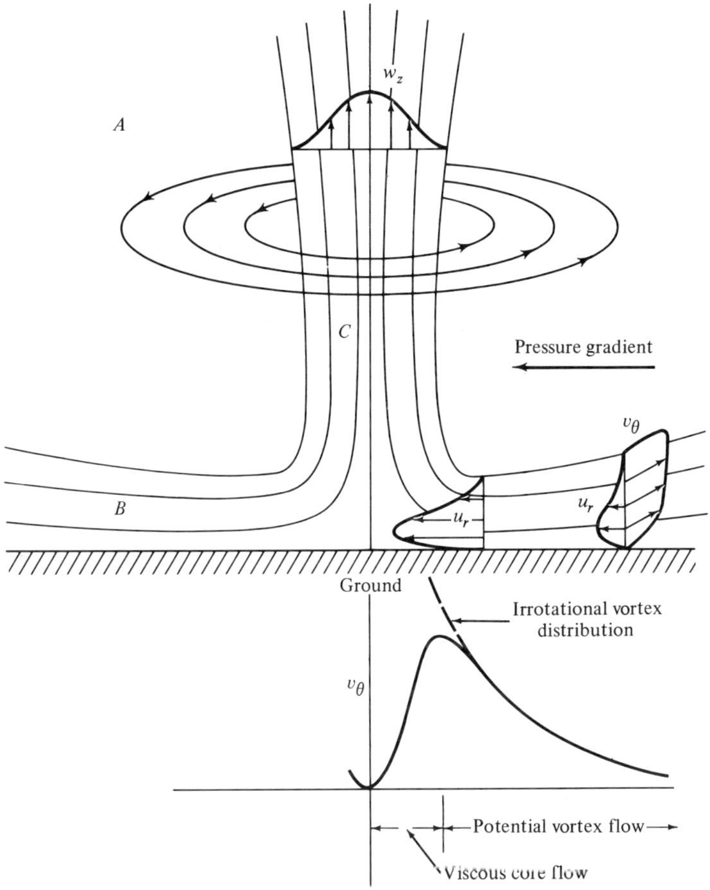

Fig. 14-16. Flow field representation for a tornado showing flow regions and velocity distributions. Regions: *A*, irrotational field; *B*, secondary boundary-layer inflow to the suction funnel *C*.

vortex flow, there exists a dynamic balance between the centrifugal inertia and the radial inwardly directed pressure gradient (see Fig. 14-16). On the ground surface the velocity must be zero; there is no inertia there. There is also no significant curvature of the flow in the plane of the ground, so the pressure gradient perpendicular to the ground must be negligible. Near the ground the pressure gradient is then the same as in the irrotational field (region A of Fig. 14-16), but the velocity and therefore the inertia at a given radius are less than those in the irrotational field. The excess of the magnitude of the radial pressure gradient over the inertia at any altitude z above the ground in the boundary-layer region B produces a radial inward flow component superimposed on the tangential boundary-layer flow. The boundary layer is then a three-dimensional skewed flow, with radial and tangential components, or a spiral helical flow. The rate at which the flow spirals into the center depends on the height z from the ground.

The dot-dash line of Fig. 14-16 indicates the edge of the boundary layer. The radial u_r and tangential v_θ components of the boundary-layer flow are indicated in the figure. The updraft w_z component in the funnel tube (region C) is also indicated. The velocity here has two components, w_z and v_θ. The v_θ distribution as a function of radius is indicated below the twister diagram.

The radial flow is interesting but complicated to evaluate for such a configuration. The average $\bar{u}_r$ must increase as the flow approaches the core because of the continuity condition and because the boundary-layer mass flow must increase as a result of the diffusion of vorticity (the viscous "pickup" of more fluid) in the flow direction.

The boundary-layer flow in region C is quite complicated, with three-dimensional and pressure-gradient effects. The inward pressure-gradient flow component and the upward draft are strong enough to lift cars, trucks, and houses, and in some cases has sucked ponds and lakes dry.

The strength of a tornado is measured by the value of $C = vr$. In this example, $C = 6000$ ft^2/sec. C may be as large as 100,000 ft^2/sec for a strong tornado. For the large dimensions and velocities involved, the boundary-layer flow is turbulent, as determined by the condition of the irrotational "external" flow. In this case, the external local reference Reynolds number, $N_R = vr/\nu = c/\nu$, is constant.

14.12 ILLUSTRATIVE EXAMPLE 14-3: The Magnus Effect

From Fig. 14-14 and (14-40) or (14-42) the velocity over the top of the cylinder is seen to be greater than that over the bottom. Therefore, the velocity and pressure distributions are not symmetric with respect to the horizontal diameter (x axis). A force in the y direction (lift force) is then expected. The lift coefficient C_L is determined directly from (14-44), with c_p determined from (14-43) and (14-40) as

$$c_p = 1 - \left(2 \sin \theta + \frac{\Gamma}{2\pi a U_\infty}\right)^2$$

to give

$$C_L = \frac{\Gamma}{aU_\infty}$$

Note that the "chord" $c = 2a$ in (14-44). The lift force per unit width of span is then

$$\frac{L}{b} = \frac{\rho U_\infty \Gamma}{g_c} \tag{14-47}$$

This result is universal for any cylindrical (two-dimensional) body in a uniform stream U_∞, and is known as the Kutta–Joukowski lift relationship.[9] The value of Γ is determined from the sum of all potential vortices within the body contour. The lift of an airfoil section is given by this relation. The problem is to calculate Γ for an arbitrary body shape subject to the required physical flow conditions. Airfoil characteristics will be discussed in Section 14.13.

In terms of the tangential "vortex speed" v_Γ on the cylinder (at $r = a$) $\Gamma = 2\pi a v_\Gamma$, the lift coefficient is $C_L = 2\pi(v_\Gamma/U_\infty)$. In terms of the stagnation angle θ_s, $C_L = -4\pi \sin \theta_s$.

The idealized case of the cylinder with circulation can be used to approximate the flow about a rotating cylinder. The cylinder surface speed $a\omega$ is assumed to be equivalent to the circulation speed v_Γ; $v_\Gamma = a\omega = 2\pi a n$, where ω is the rotational speed in rad/sec, n in rev/sec. The above expression gives C_L directly proportional to the ratio $v_\Gamma/U_\infty = a\omega/U_\infty = \alpha$, where α is a dimensionless speed ratio. A plot of L or C_L as a function of v_Γ or α for a constant approach velocity U_∞ would be a straight line with slope 2π. The Magnus effect has appeared attractive as a high-lift device on many occasions, but none has ever been proved to be practical because of mechanical complications. Near the beginning of the nineteenth century two ships were built with rotors to replace sails. Although they were able to cross the Atlantic Ocean with speed and maneuverability superior to that of a sailing ship, their origin coincided with the advent of ship screw propulsion. Even though the drive power required for the rotors was a small fraction of that for screw propulsion, the rotor ship could not compete on a speed basis. There were attempts to use the Magnus effect as a lift-augmentation device on airfoils, but the mechanical complexity was too great. One of the most recent attempts was a recovery device on large rocket boosters. The booster would be rotated by small tangential rockets so that its path would "flare out" as it dropped. The idea did not prove feasible.

The L/D performance parameter for the rotating cylinder is not good enough for it to compete with an airfoil. The experimental C_L and C_D curves for a two-dimensional cylinder are shown in Fig. 14-17. In the shaded portion of the curves, the behavior is Reynolds-number-dependent.

[9] Wilhelm Kutta, Russian hydrodynamicist (1847–1921); Nicolai Joukowski, German (1867–1944). This lift relationship was determined by Joukowski for the idealized case of a rotating cylinder. Kutta independently obtained the relation for cylindrical shapes (airfoils).

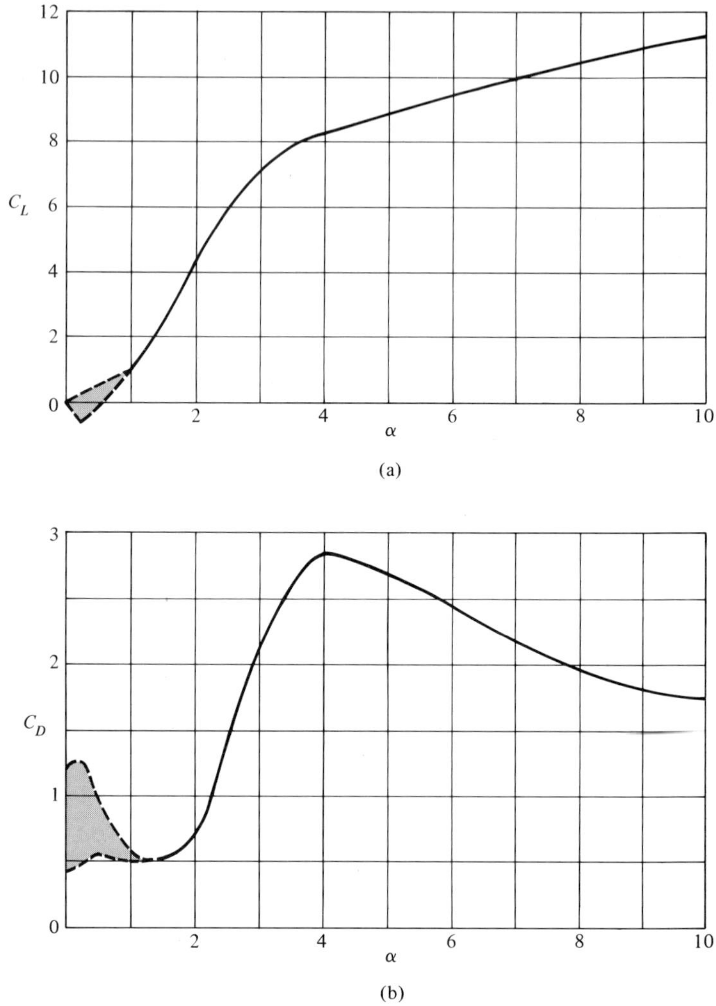

Fig. 14-17. Lift and drag curves for a rotating cylinder in an airstream: (a) lift coefficient (C_L) versus velocity ratio (α); (b) drag coefficient (C_D) versus velocity ratio (α). The characteristics are Reynolds-number dependent in the shaded areas. Note the existence of the negative C_L region. [From W. M. Swanson, The Magnus Effect: A Summary of Investigations to Date, *Trans. ASME,* **D83,** 3 (1961).]

In a critical range of Reynolds numbers and α's, a negative C_L is obtained. As in the case of a nonrotating cylinder, the flow pattern is complicated and determined primarily by boundary-layer behavior. The Magnus effect for the real flow case is a direct consequence of the effect of the rotating surface on the boundary-layer separation behavior, from which the circulation and lift result (see the reference in the Fig. 14-17 legend for a discussion of this behavior). The Magnus effect is the reason for the curve on baseballs, golf balls, and so on. The maximum deviation that can be put on a pitched baseball is about 10 in. The flight path of the baseball is a smooth, nearly circular arc curve; there is no "break" on a curved ball.

14.13 AIRFOILS

The flow about a circular cylinder with circulation is of interest in connection with the flow about an airfoil. It is possible to apply a suitable mathematical transformation (conformal mapping function) to a circle that will map it into an airfoil. The simplest such transformation is the Joukowski transformation, wherein the potential flow field about the circle is mapped into that about an airfoil. The velocity and pressure distributions can then be calculated. The potential flow field compares much more favorably with the flow actually obtained for the airfoil than for the cylinder, since the airfoil is much less blunt (more streamlined), so the separation condition is less severe. If the angle of attack is small, the adverse pressure gradient is small and separation is delayed. Under very favorable conditions (well-designed airfoil, use of boundary-layer control methods such as suction), separation may not occur. The velocity and pressure distributions can then be calculated and the C_L determined. These calculated results agree very closely (within a few percent) of those measured experimentally. If there is no separation, the C_D can be approximated by calculating the skin friction for an equivalent flat-plate length and area.

The lift for an airfoil is given by (14-47). The section C_L for small angles of attack is $C_L = 2\pi\alpha_0$, which is seen to be the same as for the circular cylinder as given in Section 14.13 (α_0 is the angle of attack measured from the zero-lift orientation). Figure 14-18 shows an example of the characteristics for a typical airfoil section. The maximum L/D ratio for this airfoil section is about 110. The number designation indicates an NACA 6-series airfoil with minimum pressure at the 6/10 chord position, a design lift coefficient $C_L = 0.2$, and a thickness ratio (thickness/chord) of 16 percent. The (215) is a chord profile designation. Airfoil section L/D ratios as high as 180 can be obtained for a smooth airfoil. Roughness conditions can decrease this value by 40 to 70 percent, depending on the magnitude of $C_{L,\max}$, the roughness conditions, and the Reynolds number. Three-dimensional effects give a further reduction, so a wing with an aspect ratio of 8 would have a maximum C_L/C_D of about 40. The overall aircraft C_L/C_D is on the order of 15 to 20. These data were obtained from I. H. Abbott, A. E. von Doenhoff, and L. S. Stivers, Jr., Summary of Airfoil Data, *NACA Report 824*, 1945.

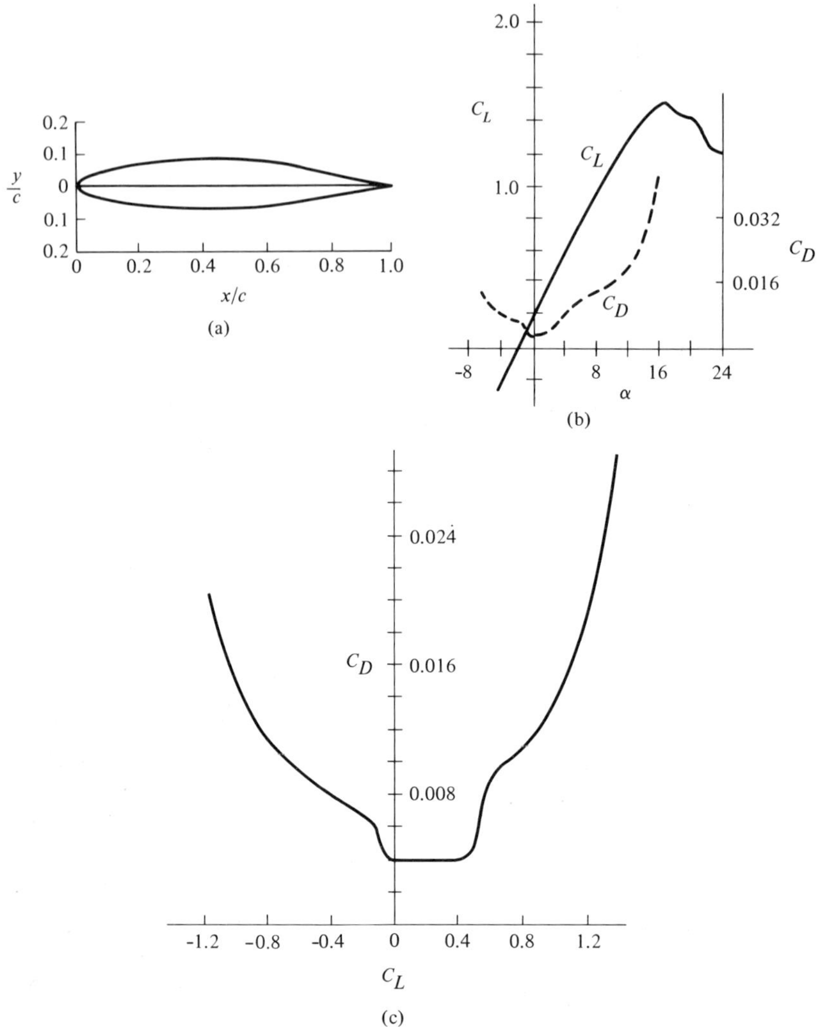

Fig. 14-18. Section characteristics for a 66(215)-216 NACA airfoil section: (a) section profile; (b) lift coefficient (C_L) as a function of angle of attack (α); (c) "polar" lift-drag plot.

14.14 GRAPHICAL POTENTIAL FIELD CONSTRUCTION

Orthogonality of the stream and potential functions implies that lines of constant Φ and lines of constant Ψ are mutually orthogonal or perpendicular. For boundary geometries that are nonuniform or irregular, an analytic solution is quite difficult. The above orthogonality property allows a method for obtaining an approximate flow field. Consider the straight-walled nozzle geometry with curved inlet and outlet as illustrated

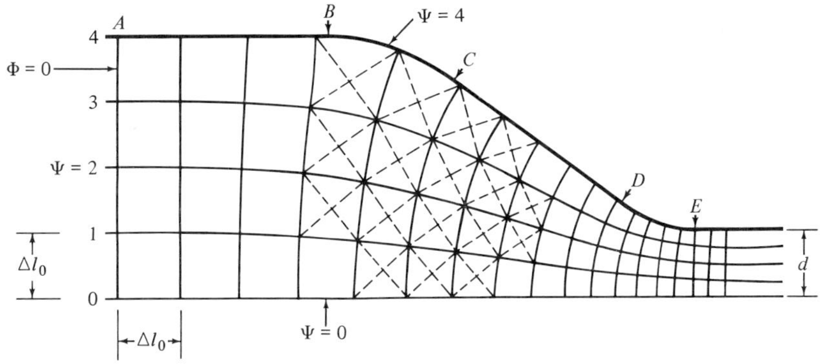

Fig. 14-19. Φ–Ψ grid or "flow net" for the potential flow in a simple contraction nozzle.

in Fig. 14-19. Upstream from the beginning of the boundary transition (at B), the flow is uniform. However, since this transition will affect the flow for some distance upstream, it is necessary to assume that the uniform conditions are valid at some arbitrary upstream station. In this example a distance of nominally ½ width has been taken (that is, $OA = AB$). Since lines of constant Φ and Ψ must be everywhere perpendicular, they will form a *net* of *curvilinear squares* (for equal initial spacing), the diagonals of which must be: (1) equal in length, (2) perpendicular to each other, and (3) bisect the corner angles. With liberal use of a pencil, eraser, and patience, the *flow net* having these properties may be constructed.

The velocity is given by

$$\mathrm{v} = \frac{\partial \Phi}{\partial s} = -\frac{\partial \Psi}{\partial n} \approx \frac{\Delta \Phi}{\Delta s} = -\frac{\Delta \Psi}{\Delta n}$$

Since $\Delta s = \Delta n = \Delta l$ by choice, $\mathrm{v}/\mathrm{v}_0 = \Delta l_0/\Delta l$ gives the velocity between adjacent streamlines anywhere in the field. Some important features of the flow pattern can be noted that aid in the construction of the flow nets. Since the walls are streamlines, the equipotential lines must meet the wall perpendicularly. The normal pressure gradient component is in a direction away from the center of curvature of the streamlines; the pressure at the wall between B and C is therefore greater than at the centerline. The velocity in the region of the wall is then less than that at the centerline, so the potential and streamline spacing is greater where the wall curvature is concave. In the convex region, the converse is true. Between D and E the curvature is significant and the velocity is high. Here $\partial p/\partial n \propto \mathrm{v}^2/R_s$; v^2 is large in the throat region, so $\partial p/\partial n$ is relatively large here. A significant variation in v is therefore expected in the throat if the curvature here is large. The velocity in the throat is a maximum at the wall, as indicated by the close streamline spacing there. (See the figure in Section 4.7 and Fig. 5-6.)

14.15 NUMERICAL SOLUTION OF POTENTIAL FLOW FIELDS

The description of a potential flow field requires solution of the governing differential equations subject to given boundary conditions. The simplest type of problem formulation is for steady potential flow governed by Laplace's equation and given boundary conditions on Φ (or Ψ). Consider the nozzle geometry of Fig. 14-19 as an example. For a numerical procedure to be programmed on a computer, a cartesian coordinate system is usually the most direct in which to formulate the problem. The best coordinate system for this problem is the one indicated by formulation of the boundary conditions. The field can be determined by solution for either the potential or the stream function. The problem formulated in terms of the potential function is:

Differential equation:

$$\nabla^2\Phi(x, y) = \frac{\partial^2\Phi}{\partial x^2} + \frac{\partial^2\Phi}{\partial y^2} = 0$$

Boundary conditions:

$$\text{(1) On } x = 0: \quad u = \frac{\partial\Phi}{\partial x} = U, \quad v = \frac{\partial\Phi}{\partial y} = 0; \quad \Phi = Ux$$

$$\text{(2) On } y = 0: \quad v = \frac{\partial\Phi}{\partial y} = 0, \quad u = \frac{\partial\Phi}{\partial x}$$

$$\text{(3) On wall:} \quad y = g(x), \quad \frac{v}{u} = \frac{dg}{dx} \quad \text{(wall is a streamline)}$$

$$\text{or} \quad \frac{\partial\Phi}{\partial y} = \frac{\partial\Phi}{\partial x}\frac{dg}{dx}$$

In terms of the stream function:

Differential equation:

$$\nabla^2\Psi(x, y) = \frac{\partial^2\Psi}{\partial x^2} + \frac{\partial^2\Psi}{\partial y^2} = 0$$

Boundary conditions:

$$\text{(1) On } x = 0: \quad u = \frac{\partial\Psi}{\partial y} = U, \quad v = -\frac{\partial\Psi}{\partial x} = 0; \quad \Psi = Uy$$

$$\text{(2) On } y = 0: \quad \Psi = \text{const.} = 0 \text{ (arbitrarily)}$$

$$\text{(3) On wall:} \quad \Psi = \text{const.} = 5 \text{ (arbitrarily)}$$

By definition of Ψ as proportional to the flow rate, $\Psi = \int u\,dy = \sum U\,\Delta y = 5U\,\Delta l_0$; U and Δl_0 are arbitrarily chosen to be unity as "normalized" quantities. In boundary condition (1) $\Delta\Psi$ increments are also unity. Because the boundary conditions are most easily stated in terms of the stream function, its solution will be undertaken. It is desirable to size the mesh (Δx and Δy) such that enough streamlines can be determined in the throat to indicate the velocity distribution $u(y)$ there (for example, the number of streamlines illustrated is $N_1 = 5$). The mesh size at the throat must be at least d/N_1. If the nozzle contraction ratio is N_C, the number of mesh lines across the exit is $N_1 N_C$ spaced at $d/N_1 N_C$. This spacing is most satisfactory if N_C is an even number. If it is not, a suitable choice is to take the nearest integer value greater than N_C.

It is now necessary to calculate Ψ at the grid intersection points, which involves numerical solution of the differential equation. The derivative $\partial\Psi/\partial x$ at a point, say, x_n of Fig. 14-20, can be approximated "from the left"[10] by

$$\left(\frac{\partial\Psi}{\partial x}\right)_L \approx \frac{\Delta_-\Psi}{\Delta x} = \frac{\Psi(x_n) - \Psi(x_{n-1})}{x_n - x_{n-1}} = \frac{\Psi_n - \Psi_{n-1}}{\Delta x} \qquad (14\text{-}48a)$$

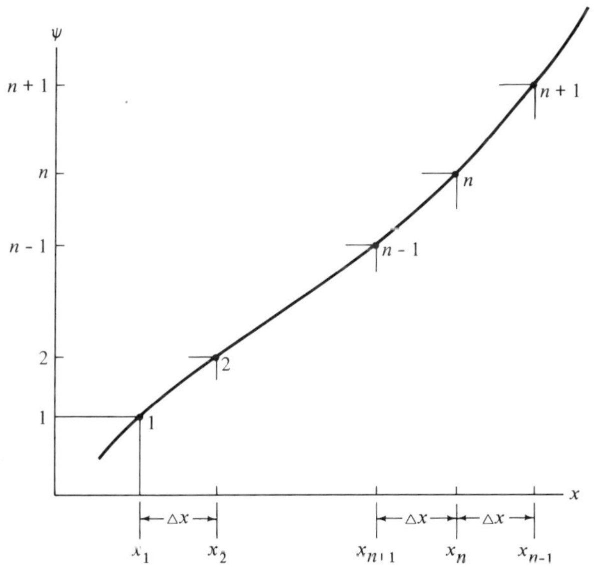

Fig. 14-20. Finite-difference relations for $\Delta\Psi$ intervals in terms of Δx intervals.

[10] With respect to the "central point" x_n, x_{n-1} is a "backward" direction, therefore sometimes noted as minus or $\Delta_-\Psi/\Delta x$, which is equivalent to the "left-derivative" approximation. Similarly, the right approximation is in the forward or plus direction, or $\Delta_+\Psi/\Delta x$.

and from the right by

$$\left(\frac{\partial \Psi}{\partial x}\right)_R \approx \frac{\Delta_+ \Psi}{\Delta x} = \frac{\Psi_{n+1} - \Psi_n}{\Delta x} = \frac{\Psi(x + \Delta x) - \Psi(x)}{\Delta x} \quad (14\text{-}48b)$$

where the Δx intervals are taken as equal increments ($\Delta x = x_n - x_{n-1} = x_{n+1} - x_n$). In the approximation with Δx finite, these derivatives will not be the same, but will approach each other as Δx is made smaller. In the limit as $\Delta x \to 0$ for the derivative, they will be equal (assuming of course, that the function is continuous at x_n).

The second derivative implies a curvature. Three points on the curve are required to determine its curvature or radius of curvature. The second derivative approximation at x_n must use both the right and left derivatives. Indicating now that Ψ is a function of both x and y, and that the derivatives are to be taken at a point x_n (which is a general point x, y),

$$\frac{\partial^2 \Psi}{\partial x^2} = \frac{\Delta}{\Delta x}\frac{\Delta \Psi}{\Delta x} = \frac{\Delta_+ \Psi - \Delta_- \Psi}{(\Delta x)^2} = \frac{\Psi_{n+1} - 2\Psi_n + \Psi_{n-1}}{(\Delta x)^2}$$

$$= \frac{\Psi(x + \Delta x, y) - 2\Psi(x, y) + \Psi(x - \Delta x, y)}{(\Delta x)^2} \quad (14\text{-}49)$$

Similarly,

$$\frac{\partial^2 \Psi}{\partial y^2} = \frac{\Psi_{m+1} - 2\Psi_m + \Psi_{m-1}}{(\Delta y)^2}$$

$$= \frac{\Psi(x, y + \Delta y) - 2\Psi(x, y) + \Psi(x, y - \Delta y)}{(\Delta y)^2}$$

giving

$$\frac{\partial^2 \Psi}{\partial x^2} + \frac{\partial^2 \Psi}{\partial y^2} \approx \frac{\Psi(x + \Delta x, y) - 2\Psi(x, y) + \Psi(x - \Delta x, y)}{(\Delta x)^2}$$

$$+ \frac{\Psi(x, y + \Delta y) - 2\Psi(x, y) + \Psi(x, y - \Delta y)}{\Delta y^2}$$

It is now expedient to make the grid a square one with $\Delta x = \Delta y = \Delta l$, as indicated in Fig. 14-21. Then

$$\nabla^2 \Psi \approx \frac{\Psi(x + \Delta l, y) + \Psi(x - \Delta l, y) + \Psi(x, y + \Delta l) + \Psi(x, y - \Delta l) - 4\Psi(x, y)}{(\Delta l)^2}$$

$$= 0$$

The value of $\Psi(x, y)$ is then the average of the four surrounding values,

$$\Psi(x, y) = \frac{\Psi(x + \Delta l, y) + \Psi(x - \Delta l, y) + \Psi(x, y + \Delta l) + \Psi(x, y - \Delta l)}{4} \quad (14\text{-}50)$$

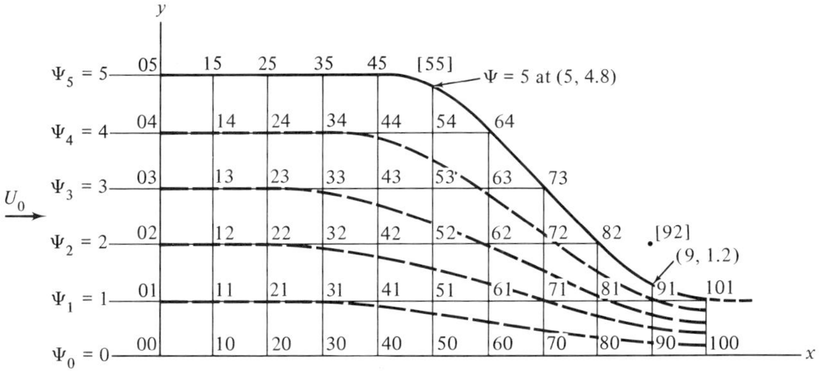

Fig. 14-21. Grid layout for numerical solution of the stream function Ψ for flow through a nozzle. The nozzle is symmetrical about the centerline (x axis), so the solution is necessary for only half of the field. The streamlines shown are those for an initial estimate for starting values at the grid points.

ITERATION METHOD

It is desired to determine $\Psi(x, y)$. The field is divided into a grid network as indicated in Fig. 14-21 and Ψ is determined for each grid intersection point using (14-50). The value of Ψ is known on the centerline ($y = 0$), on the wall, and at the upstream station ($x = 0$). The grid points of Fig. 14-21 are labeled with their x, y values in the order indicated, and subscripts are used to indicate the point. For example, the value of Ψ_{11} is [from (14-50)]

$$\Psi_{11} = \frac{\Psi_{01} + \Psi_{10} + \Psi_{12} + \Psi_{21}}{4} \tag{14-50a}$$

For this point, Ψ_{01} and Ψ_{10} are known and are fixed from the boundary conditions, but Ψ_{12} and Ψ_{21} are not known. For point 52,

$$\Psi_{52} = \frac{\Psi_{42} + \Psi_{51} + \Psi_{62} + \Psi_{53}}{4} \tag{14-50b}$$

For the problem illustrated, there are 23 grid points for which Ψ is not known; therefore, the solution is required to the set of 23 equations in 23 unknowns. There is no direct analytical procedure for the solution of this many equations, so the iteration method can be used. This method involves a step by step convergence procedure. Starting with a boundary (preferably a straight one, or a known uniform field condition), approximate values of $\Psi_{x,y}$ are tabulated at the grid points. For the example of Fig. 14-21, the solution may be started along the boundaries chosen as the

x or y axes. It is more logical in this case to start with the upstream boundary (y axis) and proceed in the flow direction. To obtain rapid convergence (and therefore decrease computation time), an initial set of values ("first estimate") can be obtained by sketching in an approximate set of streamlines, as indicated in Fig. 14-21. The iteration procedure is carried out by "solving" (14-50) as indicated in (14-50a) and (14-50b). For example, the first iteration at point 11 is

$$\Psi_{11} = \frac{\Psi_{10} + \Psi_{12} + \Psi_{01} + \Psi_{21}}{4} = \frac{0 + 2 + 1 + 1.05}{4} = 1.01$$

as indicated. The determination of the first iteration value of Ψ_{12} now uses this value of Ψ_{11}:

$$\Psi_{12} = \frac{1.01 + 3.00 + 2 + 2.05}{4} = 2.015 \qquad \text{(listed as 2.01)}$$

This procedure must be continued row by row until there is no change in the value of Ψ at any point. (It may also be carried out column by column.) With only four iterations carried out for Table 14-1, it is indicated that complete convergence has not yet been obtained at all points.

Special treatment is required for points 54 and 91, as indicated by the * in, and the square brackets around, points [55] and [91]. The increment in Δy at these two points is short because the bounding streamline $\Psi = 5$ is only 0.8 of a chosen standard increment from 54 and 0.2 from 91. These shortened increments must be considered in calculating the $+y$ derivatives at these points. For example, consider a general point with two short increments, as indicated in Fig. 14-22. Ψ_{mj} is the value to be determined

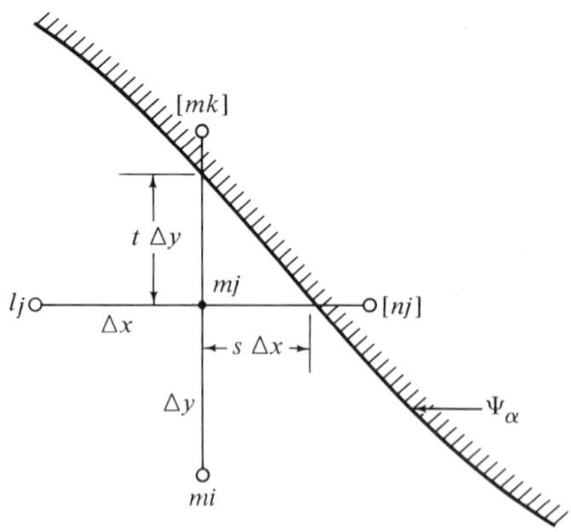

Fig. 14-22. Determination of a grid point adjacent to a boundary value.

TABLE 14-1

ITERATIVE CALCULATION PROCEDURE FOR SOLUTION OF THE FLOW FIELD OF FIG. 14-21

Grid pt.: (00)	(01)	(02)	(03)	(04)	(05)	
0	2	2	3	4	5	(Known values)
(10)	(11)	(12)	(13)	(14)	(15)	
0	1.00	2.00	3.00	4.00	5	(1st est.)
	1.01	2.01	3.00	4.00		(1st iteration)
	1.01	2.01	3.01	4.00		
	1.01	2.02	3.01	4.01		
(20)	(21)	(22)	(23)	(24)	(25)	
0	1.05	2.05	3.00	4.00	5	(1st est.)
	1.04	2.04	3.02	4.00		
	1.03	2.04	3.03	4.02		
	1.03	2.04	3.04	4.03		
(30)	(31)	(32)	(33)	(34)	(35)	
0	1.10	2.10	3.05	4.00	5	(1st est.)
	1.06	2.08	3.08	4.04		
	1.06	2.10	3.10	4.07		
	1.06	2.09	3.12	4.09		
(40)	(41)	(42)	(43)	(44)	(45)	
0	1.10	2.15	3.20	4.10	5	(1st est.)
	1.10	2.22	3.23	4.17		
	1.14	2.25	3.28	4.20		
	1.16	2.27	3.30	4.22		
(50)	(51)	(52)	(53)	(54)	[55₋] ($t = 0.8$)	
0	1.20	2.50	3.50	4.40	5	(1st est.)
	1.30	2.52	3.58	4.47*		
	1.31	2.54	3.60	4.48		
	1.32	2.55	3.62	4.49		
(60)	(61)	(62)	(63)	(64)		
0	1.60	3.00	4.10	5		(1st est.)
	1.58	2.98	4.14			
	1.59	3.01	4.15			
	1.60	3.02	4.16			
(70)	(71)	(72)	(73)			
0	2.0	3.70	5			(1st est.)
	2.07	3.76				
	2.05	3.76				
	2.04	3.76				
(80)	(81)	(82)				
0	3.0	5				
	2.84					
	2.79					
	2.78					

(continued)

TABLE 14-1 *(continued)*

Grid pt.: (00)	(01)	(02)	(03)	(04)	(05)
(90)	(91)	[91.2] ($t = 0.2$)			
0	4.30*	5			
	4.11				
	4.10				
	4.10				
(100)	(101)				
0	5				

and Ψ_α is the boundary value (for example, $\Psi_\alpha = 5.0$ in Fig. 14-21). The plus and minus derivatives are

$$\frac{\partial_+\Psi}{\partial x} \approx \frac{\Psi_\alpha - \Psi_{mj}}{s\,\Delta x} \qquad \frac{\partial_-\Psi}{\partial x} \approx \frac{\Psi_{mj} - \Psi_{lj}}{\Delta x}$$

$$\frac{\partial^2\Psi}{\partial x^2} \approx \frac{1}{\Delta x}\left[\frac{\Psi_\alpha - \Psi_{mj}}{s} - \frac{\Psi_{mj} - \Psi_{lj}}{1}\right]$$

The y derivatives are similarly determined to give

$$\Psi_{mj} = \frac{\Psi_\alpha[(1/s) + (1/t)] + \Psi_{lj} + \Psi_{mj}}{2 + (1/s) + (1/t)} \tag{14-51}$$

Consider, for example, point 54, for which $s = 1$ and $t = 0.8$ for the first iteration value:

$$\Psi_{54} = \frac{5 + 5/0.8 + 3.58 + 4.17}{3 + 1/0.8} = 4.47$$

A plot of the streamlines may be made from the data from Table 14-1. Usually the velocities, velocity distributions, and pressure distribution are of more interest and consequence. The velocity components are determined from

$$u = \frac{\partial\Psi}{\partial y} \approx \frac{\Delta\Psi}{\Delta y} \qquad v = -\frac{\partial\Psi}{\partial x} \approx -\frac{\Delta\Psi}{\Delta x}$$

At any grid point, components can be calculated from either the plus or minus derivatives, or more accurately, from the average of the two; that is,

$$u \approx \frac{1}{2}\left(\frac{\Delta_+\Psi}{\Delta y} + \frac{\Delta_-\Psi}{\Delta y}\right)$$

For example, at point 53,

$$\frac{u}{U_0} = \tfrac{1}{2}[(\Psi_{54} - \Psi_{53}) - (\Psi_{53} - \Psi_{52})] = \tfrac{1}{2}(\Psi_{54} - \Psi_{52}) = 0.97$$

$$\frac{v}{U_0} = \tfrac{1}{2}(\Psi_{43} - \Psi_{63}) = -0.43 \qquad \frac{V}{U_0} = 1.01$$

The inclination of the streamline to the initial flow direction (x axis) is

$$\theta_{53} = \tan^{-1}\frac{v}{u} = -24.9^\circ$$

The boundary velocity and pressure are quantities of interest. For example, the velocity components at the wall above 54 are

$$\frac{u}{U_0} \approx \frac{\Delta_+\Psi}{\Delta y} = \frac{5 - 4.49}{0.8} = 0.64 \qquad \frac{v}{U_0} \approx 0.51 \qquad \frac{V}{U_0} = 0.82$$

In the concave corner, the flow is seen to decelerate. The pressure at any point may be calculated from Bernoulli's equation.

All the above calculation procedures are easily programmed on a computer. The program can be set up to calculate and read out velocity or pressure data or to go through a plotter and plot the desired data directly.

The most interesting station in this example flow field is the nozzle throat. As stated above, a velocity variation across the throat is to be expected because of the curvature there. However, the size of the increment chosen for this example does not allow this distribution to be resolved or defined, since the throat spans a single increment in Δy. At least five increments would be necessary to allow definition of the velocity distribution there. To obtain this resolution would require the problem to be reworked with equal Δx and Δy increments $\tfrac{1}{5}$ the size originally chosen. The number of grid points and calculations would then increase by a factor of about 25, definitely indicating that the problem should be set up on a computer.

Another numerical technique applicable to potential flow problems is the relaxation method.[11] It is similar to the iterative method in that the same basic increment equations are used. This method is sometimes more rapidly convergent than the iteration method.

14.16 ANALOG METHODS

In dealing with potential flow fields, any physical phenomenon that exhibits a "potential behavior" can be used to "solve" potential problems experimentally. Electric potential (voltage) fields are the most direct

[11] See, for example, R. V. Southwell, *Relaxation Methods*, vols. 1 and 2, Oxford University Press, New York, 1956.

with which to deal. Stationary magnetic, thermal, and other fields are potential or conservative. "Stationary" implies steady state, or not varying with time. If the field is stationary (and with no sources), there is no mechanism for dissipation and the field is conservative. For a flow field to be conservative, no dissipation is allowable, so the condition of no viscous action must also be specified. The single defining condition for a field to be conservative is that the curl of the flux quantity must be zero.

Fluid flow field:

$$\nabla \times \vec{v} = 0 \text{ (irrotationality)}$$
$$\vec{v} = \nabla\Phi \text{ (defines velocity potential)}$$
$$\nabla\cdot\vec{v} = 0;\ \nabla^2\Phi = 0 \text{ (continuity; } \vec{v} \text{ and } \Phi \text{ fields are continuous)}$$

Electric field:

$$\nabla \times \vec{E} = 0 \text{ (} \vec{E} \equiv \text{electric field strength; no circulating currents)}$$
$$\vec{E} = -\nabla V \text{ (} V \equiv \text{voltage)}$$
$$\nabla\cdot\vec{E} = 0;\ \nabla^2 V = 0 \text{ (} \vec{E} \text{ and } V \text{ field are continuous with no sources)}$$

Magnetic field:

$$\nabla \times \vec{H} = 0 \text{ (} \vec{\mathrm{H}} \equiv \text{magnetic field intensity; no electromagnetic currents are generated)}$$
$$\vec{H} = -\nabla\mathscr{F} \text{ (} \mathscr{F} \equiv \text{magnetic potential or magnetomotive force)}$$
$$\nabla\cdot\vec{H} = 0;\ \nabla^2\mathscr{F} = 0 \text{ (} \vec{H} \text{ and } \mathscr{F} \text{ fields are continuous with no poles)}$$

Thermal field:

$$\vec{q} = -\kappa\nabla T \text{ (energy transferred per unit area} \propto T \text{ gradient; } \kappa \text{ is thermal conductivity)}$$
$$\nabla\cdot\vec{q} = 0;\ \nabla^2 T = 0 \text{ (no thermal-energy sources)}$$

The electric field is the easiest to use for simulation or analog purposes because of the availability and ease of control of potential (voltage) sources, and because of the case and accuracy of instrumentation.

A two-dimensional field plotter is illustrated in Fig. 14-23. The voltage divider is set to an integer increment of the supply voltage. The increment size is determined by the number of potential lines desired for the field geometry to be investigated. The probe is used to explore the field to determine the locus of the line of constant voltage as indicated by a zero reading on the null indicator. The field is constructed of a relatively high

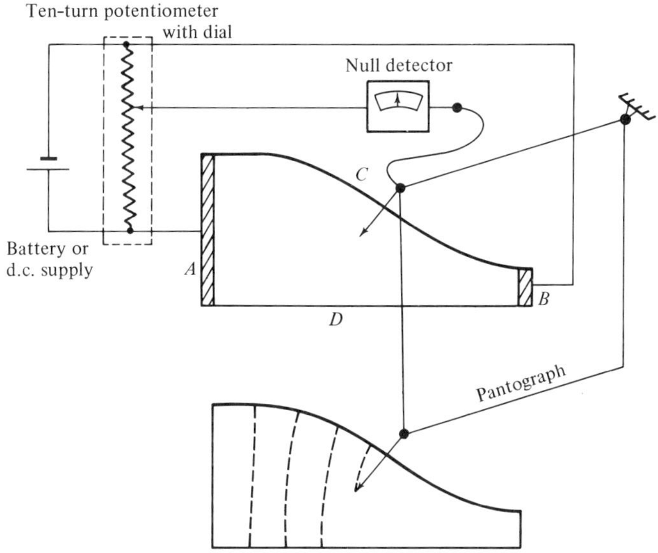

Fig. 14-23. Electric potential analog field plotter.

resistance material with a uniform surface resistivity. The constant-potential end leads are good conductors. One satisfactory material developed for this field-plotting technique is Teledeltos paper, a paper coated uniformly with a carbon compound. The paper is cut to the desired field boundary shape. The constant-potential boundary-condition leads can be painted on with a conducting paint. The accuracy is determined by the uniformity of the voltage source, the linearity of the voltage divider, and the uniformity of the resistivity of the field material. A pantograph can be arranged as illustrated to mark the field lines (a scale-magnification factor can also be obtained).

The streamlines and potential lines are orthogonal and bounding surfaces are lines of constant stream function. From these facts a procedure can be deduced for determining streamlines. The open boundaries perpendicular to solid boundaries at suitable stations are chosen for bounding potential lines. The solid boundaries (C and D of Fig. 14-23) can be connected across the voltage source and the streamlines determined in the same manner as were the potential lines. The model would have conducting boundaries along C and D and the A and B conducting boundaries would have to be removed (or another model used). Note that in general it would not be possible to obtain a "square" mesh because of the extreme difficulty of determining $\Delta\Phi$ and $\Delta\Psi$ increments before starting. An advantage of this method is that any subregion can be explored for greater resolution. For example, a finer potential increment can be obtained at the nozzle exit when determining streamlines to give a more detailed velocity distribution determination there. Also, if it is desired to know only the velocity distribution at the exit, the supply voltage may be divided into a predetermined number of increments there (say 10) and the streamline spacing determined only at the exit to give the velocity distribution there. In addition, any subregion may be explored to obtain greater resolution simply by further subdividing the voltage in that region. These advantages are not strictly possible in the numerical techniques. The overall accuracy of determination of the stream and potential lines depends on the size of increment of Ψ or Φ and the mesh size originally chosen. Subdivision of a region into a finer mesh will then depend on the accuracy of results of the initial large mesh size. It is also obviously not possible to obtain numerical results only for a subregion of the field without solving for the entire field.

A simpler version of the above electric analog method can be set up using only a battery and voltmeter with the conducting model.

The electric-field analog method is also applicable to three-dimensional models. A three-dimensional conducting field model is built with constant-voltage surfaces at known equipotential surfaces. The model is then explored with a probe insulated except at the tip. The models are usually liquids in a tank constructed to model the geometry. Semigelled or quite viscous liquids with suitable conductivity are used to prevent unwanted fluid and electric currents during probing.

14.17 SUMMARY

1. Potential flow calculations can be used to determine velocity and pressure distributions in two- and three-dimensional flow fields. If the actual or real fluid flow fields correspond to those calculated from potential flow, the method is useful. If the viscous boundary layer is of appreciable thickness or produces separation, the potential flow field will not be similar to the real one. For blunt bodies on which separation will occur, the potential flow results are then only approximately valid in certain limited regions such as in the neighborhood of the stagnation point of a sphere or cylinder.
2. The general subject of potential flow deals with partial differential equations and their solutions. The subject is therefore one for an intermediate level of study. Two-dimensional flow problems are handled most expediently with complex variable mathematics and conformal mapping and therefore are also beyond the elementary level. The potential flow covered in this chapter is included as an introduction to two- and three-dimensional problems and their methods of solution.
3. The determination of the potential flow field around a body (or an internal flow) is the first step necessary in the determination of boundary-

layer behavior. As indicated in Chapter 9, the external or potential flow pressure or velocity distribution is necessary before the solution procedure for the boundary-layer problem can be started.

4. Numerical procedures (such as the iterative method) with the aid of a computer provide a powerful method of solving potential flow geometries for arbitrary geometries. The method can be extended to three-dimensional flow fields but with a significant increase in setup procedure and calculation time.

REFERENCES

1. W. E. Rogers, *Introduction to Electric Fields*, McGraw-Hill, Inc., New York, 1954.
   An undergraduate-level book on potential fields. It contains some interesting fluid and membrane analogies to electric field configurations.
2. J. Artley, *Fields and Configurations*, Holt, Rinehart and Winston, New York, 1965.
   An intermediate-level book covering the general subject of fields and methods of solutions to field problems.
3. H. Rouse, *Fluid Mechanics for Hydraulic Engineers*, McGraw-Hill, Inc., New York, 1938.
   Chapter III gives further information on flow nets and their graphical construction.
4. M. C. Salvadori and M. L. Baron, *Numerical Methods in Engineering*, Prentice-Hall, Inc., Englewood Cliffs, N.J., 1961.
   References 4, 5, and 6 can be used for further study on numerical methods.
5. J. B. Scarborough, *Numerical Mathematical Analysis*, Johns Hopkins University Press, Baltimore, 1955.
6. R. V. Southwell, *Relaxation Methods*, Oxford University Press, Vol. 1, 1946, Vol. II, 1956.
7. L. M. Milne-Thompson, *Theoretical Hydrodynamics*, The Macmillan Company, New York, 1960.
   An intermediate-level text with much material on potential flow.

PROBLEMS

14-1 What is the effective Reynolds number for a "ideal" flow?

14-2 It is desired to examine a class of solutions $\Phi(x, y) = x^n + f(x, y)$ (n is any number, integer, or fraction). Find $f(x, y)$.

14-3 Show that a possible solution to (14-7) is $\Phi(x, y) = Ae^{nx} \cos my$ and therefore that $\Phi(x, y) = \sum_{n=-\infty}^{\infty} A_n e^{nx} \cos my$ is a solution.

14-4 Show that $\Phi(x, y, z) = Ae^{inx}e^{ky+mz}$ is a solution to (14-7) under a particular condition and determine the condition.

14-5 Show that $A \sin x \cosh y = \Phi(x, y)$ is a solution to (14-7).

14-6 Show that the velocity potential for a three-dimensional source (or sink), as given by $\Phi = -Q/4\pi r$, satisfies Laplace's equation in spherical coordinates.

14-7 Show that the velocity potential for a two-dimensional or line source (or sink), as given by $\Phi = (Q/2\pi) \ln r$, satisfies Laplace's equation in polar or cylindrical coordinates.

14-8 Consider the mathematical statements for the differential of Φ and that of Ψ:

$$d\Phi = \frac{\partial \Phi}{\partial x} dx + \frac{\partial \Phi}{\partial y} dy$$

$$d\Psi = \frac{\partial \Psi}{\partial x} dx + \frac{\partial \Psi}{\partial y} dy$$

and show that lines of constant Φ are perpendicular to lines of constant Ψ. What are the limitations on this mathematical "proof" of the orthogonality of the potential and stream functions?

14-9 (a) For a line source at the origin as indicated in the figure, show that div $\vec{v} = Q/b$ for any cylindrical volume coaxial with and surrounding a length b of the line. (b) For the volume element of length b not enclosing the origin, integrate the net flow out of this volume and show that it is zero. (c) Integrate the volume flow rate through any cylindrical surface of length b and show that it is equal to Q.

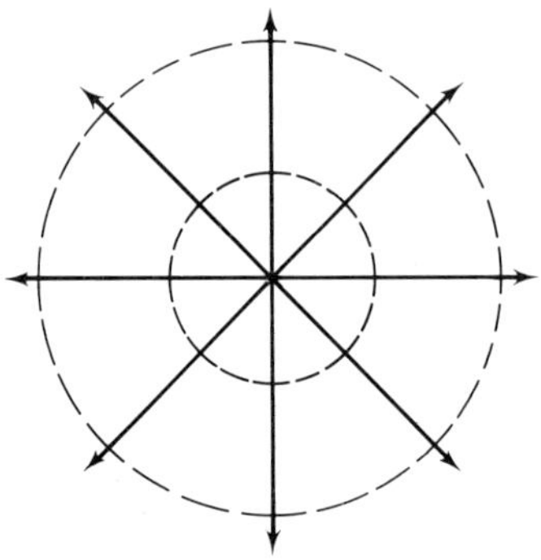

Fig. P14-9.

14-10 Derive the expressions for the velocity components in terms of Φ and Ψ in a plane polar coordinate system (14-20b) from a procedure similar to that by which the x,y velocity components u and v were developed leading to (14-19b).

14-11 Determine the stream function for a line source by taking the potential function from (14-17) and using the orthogonality relations (14-20a).

14-12 The radial component of a uniform flow parallel to the polar axis was used to determine $\Phi(r, \varphi, \theta) = -U_\infty r \cos \varphi$. Show that the same result is obtained by noting that the tangential or φ component is $U_\infty \sin \varphi$. What is the azimuthal, or θ, component?

14-13 Show that the contour integrals around the closed contours of Fig. 14-6 are zero for the velocity field obtained from the stream function for a potential vortex $\Psi(r, \theta) = C_2 \ln r$.

14-14 Examine the continuity condition for a potential vortex field as given by (14-33).

14-15 (a) Determine the equations of the streamlines and equipotential lines for a source of strength Q at $x = s$ and a sink of strength Q at $x = -s$. (b) Plot the streamlines and equipotential lines by geometric construction in a manner similar to that by which the two-source characteristics were plotted for Fig. 14-8. Use equation (14-34) for this construction. (c) Show that the equations of the streamlines are circles with radii $r_\Psi = s/\sin \Psi'$, where $\Psi' \equiv 2\pi\Psi/Q$. (d) Can this combination be considered as an image field? Explain.

14-16 For equal vortices at $x = +a$ and $x = -a$, show that the streamlines are the equivalent of the equipotential lines and the equipotential lines are the equivalent of the streamlines of the two source examples worked out in Section 14.7.

14-17 For the two sources or source near a wall illustrated in Fig. 14-7, find the instantaneous velocity of the source with respect to the wall. Can this problem be solved easily? Compare this problem with that of the vortex near the wall and discuss the formulation and solution.

14-18 Construct the potential field for a source and a vortex at the origin and find the equations of the streamlines.

14-19 Verify that (14-36) correctly represents the potential and stream functions for the specified approach flow. Sketch a plot of the equipotential and streamlines.

14-20 For the example of the flow past a source and sink, show that the half-width is determined from the transcendental equation

$$\frac{s}{b} = \tan \frac{2bs}{(a^2 - s^2)}$$

14-21 Show that the equations of the streamlines for a doublet are circles of radius $\mu/4\pi\Psi$ at $x = 0$, $y = \mu/4\pi\Psi$.

14-22 For a doublet, find the equations of the equipotential lines and specify their geometric shape.

14-23 Determine the potential function for flow past a doublet corresponding to the stream function given in (14-39).

14-24 Consider the potential flow past a cylinder of radius a and a streamline that is $a/10$ above the x axis at $x = \infty$. Calculate the spacing between the cylinder and this streamline at the top of the cylinder. From the relation between the flow rate and stream function, determine the average velocity between this streamline and the cylinder at the top of the cylinder.

14-25 Determine the velocity potential function $\Phi(x, y)$ and the equations of the equipotential lines for the flow of Fig. 14-9 and (14-35).

14-26 Determine the velocity potential $\Phi(x, y)$ and $\Phi(r, \theta)$ for flow around a cylinder. Graphically construction the equipotential contours for this potential function.

14-27 For a circular cylinder with circulation in a free stream such that the forward stagnation point is at $\theta_s = -30°$, find the magnitude of (a) the circulation Γ and (b) the velocity at the top and bottom of the cylinder.

14-28 For the case of a cylinder with circulation in a free stream where the stagnation points coincide at $\theta_s = -\pi/2$, find the velocity on the top of the cylinder.

14-29 Integrate (14-45) with the condition $r\mathrm{v}$ = const. to obtain (14-46).

14-30 Determine an equation for the pressure difference $\Delta p \equiv p_1 - p_2$ in an irrotational vortex as a function of density ρ, $C = \mathrm{v}r$, r_2, and the radius ratio r_2/r_1 ($r_2 < r_1$). Show that for $(r_2/r_1)^2 \ll 1$, $\Delta p \approx \rho C^2/2g_c r_2^2$ and that if $p = p_{\text{amb}}$, where $r_1 = \infty$, that this relation is exact.

14-31 Plot the velocity, pressure difference, and strain rate versus radius for the tornado vortex of Example 14-2 (a) assuming irrotational flow, and (b) assuming irrotational flow to the core and that the core is one of constant rotation (ω) ("solid body").

14-32 A tornado passed two weather-recording stations located 200 ft apart. The path was perpendicular to the 200-ft line between the stations. Station 1 recorded a maximum wind speed of 20 ft/sec and station 2, 30 ft/sec, both velocities in the same direction. The propagation speed of the entire storm was 10 mph in the direction of the recorded velocities. (a) Locate the center of the tornado with respect to the two stations. (b) Find the pressure at station 2 for an atmospheric pressure of 28.8 in. Hg at a considerable distance from the twister path.

14-33 A twister moving in the neighborhood of a weather station produced a peak velocity reading of 300 ft/sec. The center of its path was measured as 500 ft from the station. The storm speed was 18 mph in that vicinity. What were the velocity and pressure (below local atmospheric) at the core, the radius of which was about 100 ft?

14-34 A recording barometer reassured a maximum pressure drop of 8 in. Hg below the local ambient atmospheric pressure as a tornado passed. (a) Estimate the maximum wind speed at the barometer location. (b) Estimate the variation of atmospheric density. (c) On what factors does the error in these estimates depend?

14-35 Determine the expression for c_p for a cylinder with circulation as given in Example 14-3 and calculate L/b and C_L.

14-36 Referring to the section on numerical methods, list other reasons why it is more satisfactory to solve for the stream function than the potential function.

14-37 Discuss reasons why Ψ is more suitable to determine by the process of iteration than Φ.

14-38 Complete the indicated calculations to determine (14-51).

14-39 Complete the iteration procedure for Table 14-1.

14-40 Program on a digital computer the iterative solution to the geometry of Fig. 14-21 with the objective of determining the velocity and pressure distribution on the wall and across the throat. Use increments one fifth the size of those in the figure. Note that the number and design of initial streamlines is arbitrary. They are only used for the first estimate, so that the six shown are sufficient.

14-41 Program the solution for the streamlines and velocity components for the potential flow through the sudden contraction shown in the figure. Determine and plot the velocity components as a function of y at $x = 2.5h$ and $x = 3.5h$.

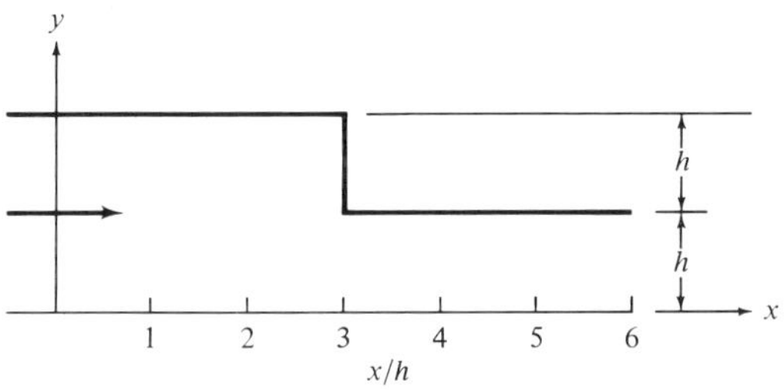

Fig. P14-41.

14-42 For the two-dimensional elliptical nozzle shown, find the velocity and pressure distribution at the throat by programming the iterative method on a computer. Take increments at least $d/10$.

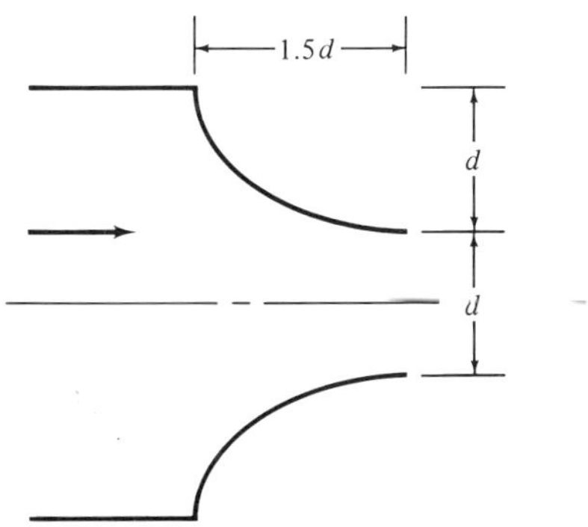

Fig. P14-42.

14-43 The Laplace equation in cylindrical coordinates for axisymmetric flow is

$$\frac{\partial^2 \Psi}{\partial r^2} + \frac{1}{r}\frac{\partial \Psi}{\partial r} + \frac{\partial^2 \Psi}{\partial z^2} = \frac{1}{r}\frac{\partial}{\partial r}\left(r\frac{\partial \Psi}{\partial r}\right) + \frac{\partial^2 \Psi}{\partial z^2} = 0$$

Show that the numerical equation is

$$\Psi_n(r, z) \approx \overline{\Psi} + \frac{1}{4}\frac{\Delta r}{r_n}[\Psi(r_n + \Delta r, z) - \Psi(r_n - \Delta r)z]$$

where $\overline{\Psi}$ is the average of the surrounding four values as in the cartesian-coordinate-system development presented. Using this result, calculate the throat velocity and pressure distribution for the geometry of Problem 14-42, where the cross section is circular. Note that this geometry is the one specified for the ASME flow nozzles. From your results, discuss and criticize this geometry. How could it be improved?

14-44 Construct an electric-field analog of the geometry of Figs. 14-21 and 14-23 to determine the potential and streamlines. Determine the velocity distribution at the exit. Compare the results with those obtained by the numerical procedure (Problem 14-40).

14-45 Sketch the circuit for an electric-field analog using only a battery source and voltmeter.

14-46 Discuss the possibility of using the electric field method to obtain streamlines in a three-dimensional flow field.

CHAPTER 15

CONTENTS

UNSTEADY FLOW

The two principal types of unsteady flow involve transients, arising during the establishment of a steady-state flow, and continual periodic oscillations. Neither of these types can be reduced to a steady frame of reference as described in Chapter 6. Except for a few very simple cases, unsteady flows are complicated by the introduction of the additional variable, time, with the introduction of phase differences in the physical system. Analytical solutions of unsteady flow problems are therefore difficult, as is a physically intuitive evaluation. This chapter will deal with only a few simple examples of unsteady flow.

Transient flow behavior occurs when starting, stopping, or changing normally steady operating conditions. Weather is a good example of a system in a combined transient and periodic state. Locally damaging transients are produced in liquids during the occurrence of water hammer and cavitation. Traveling waves such as shock, blast, and acoustic waves can be both transient and oscillatory. The flow behind a turbomachine rotor can be considered as an oscillating flow superimposed on a mean flow.

15.1 FLUCTUATIONS IN MANOMETERS

Most flow systems possess some fluctuation, so pressure measurements (or flow measurements using differential pressures) may be affected by the fluctuations. The observed pressure-measuring device may or may not record or indicate the correct pressure as a function of time, or it may fluctuate wildly, depending on the dynamic characteristics of the measuring system and the oscillating signal input. If the measuring system has a resonant frequency that coincides with a strong mode of an oscillation in the flow system, a resonant vibration will be excited. For example, if a hydraulic pump or turbine system causes a flow or pressure disturbance as a result of interactions between blade row wakes (or blade pulsations and the casing), these pulsations may cause considerable difficulty in obtaining flow or pressure measurements if the system has a natural frequency that coincides with (or is a harmonic of) the natural frequency of the measuring system. For this reason it is sometimes helpful to be able to calculate the expected resonant frequency of the measuring system.

U-TUBE MANOMETER

Consider a liquid-filled, U-tube manometer with relatively short lines connected to a Venturi tube, Pitot tube, or other air- or gas-flow system where pulsations may exist (Fig. 15-1). The mass (and therefore the inertia or momentum) of the gas is much smaller than that of the liquid in the manometer, so the momentum effects of the connecting lines will be neglected. If there is a mean pressure across the manometer giving a reading of R from initial zero as indicated in Fig. 15-1, the problem is to determine the natural frequency of the manometer under the given conditions.

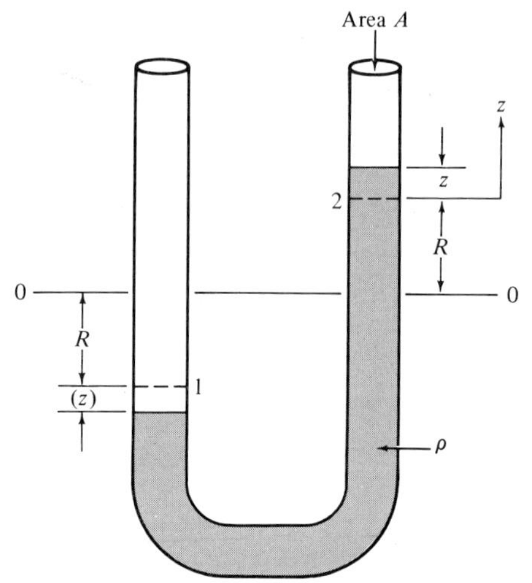

Fig. 15-1. U-tube-manometer representation for determination of oscillation frequency.

It is possible to write the equations for the solution of the problem by using either energy or force and momentum considerations. For this simple example, both developments will be made and some conclusions drawn about the preference of the one to be used as a consequence of the procedures involved.

For problems with curved geometries, the scalar energy principle is sometimes easier to use than the component (or vector) equations of force and momentum. If viscous dissipation effects are neglected for an initial approximation, only gravity-force work is done on the fluid, so only gravity potential and kinetic energy must be considered. If the column is displaced slightly and released, it will oscillate about the equilibrium reading at R with an amplitude z_0. The oscillation will be similar to that of a pendulum with constant total energy equal to the sum of the potential and kinetic energy at any state during the cycle. For a positive z, the potential energy in the right leg is $\rho Agz^2/2g_c$, with lowering of the left leg also contributing an equal amount. Since only the change in potential energy with respect to the initial equilibrium point must be considered, R does not enter into the potential energy for this particular example. The kinetic energy involves the entire mass of liquid: K.E. $= \rho ALv^2/2g_c$, where L is the length of the liquid column. Conservation of energy states that the sum of these two energy contributions is a constant. The constant can be evaluated at the end of the oscillation, where $z = z_0$ and $v = 0$: P.E.$_{max}$ $= \rho Agz_0^2/g_c$. Therefore,

$$\frac{\rho Agz^2}{g_c} + \frac{\rho ALv^2}{2g_c} = \frac{\rho Agz_0^2}{g_c} \tag{15-1a}$$

Now $v = dz/dt$, so (15-1a) gives

$$\frac{dz}{dt} = \sqrt{\frac{2g}{L}}\sqrt{z_0^2 - z^2} \tag{15-1b}$$

Integrating this equation and evaluating the constant of integration gives

$$z = z_0 \sin \sqrt{\frac{2g}{L}}\, t$$

The time required for the column to go from $z = 0$ to $z = z_0$ is one quarter of the period of vibration, or $T/4$. Equation (15-1b) can be evaluated at these limits to give

$$z_0 = z_0 \sin \sqrt{\frac{2g}{L}}\, \frac{T}{4}$$

which gives

$$T = 2\pi\sqrt{\frac{L}{2g}} \tag{15-2a}$$

as the period of vibration. The natural or resonant frequency is

$$f_n = \frac{1}{T} = \frac{1}{2\pi}\sqrt{\frac{2g}{L}} \tag{15-2b}$$

Note that the result is independent of both the area A of the tube and the density of the fluid.

The force–momentum equation may be applied along a streamline, which is the path of motion of the fluid. The problem is *not* a steady-state one, and the fluid control volume is not fixed in space. Since the control volume is a well-defined mass, it seems appropriate to use a lumped or integral approach to the problem. Indeed, if dissipation (viscous) effects are neglected, the column can be considered to be a solid mass slipping back and forth in the tube. The net unbalanced force in the z direction for a small displacement from the equilibrium position is (for both legs)

$$F_z = -\frac{2\rho A g z}{g_c}$$

The acceleration involves the entire mass and is

$$a_z = \frac{dv}{dt} = \frac{d^2 z}{dt^2}$$

The equation of motion is then

$$\frac{-z\rho A g z}{g_c} = \frac{\rho A L}{g_c}\frac{d^2 z}{dt^2} \tag{15-3a}$$

or

$$\frac{d^2 z}{dt^2} + \frac{2g}{L} z = 0 \tag{15-3b}$$

with the condition that at $t = 0$, $z = z_0$, and $dz/dt = 0$. Equation (15-3b) is a harmonic equation with solutions $z = C_1 \cos \omega t + C_2 \sin \omega t$. The first boundary (or initial) condition gives $C_1 = z_0$, the second gives $C_2 = 0$, and substitution into (15-3b) indicates that $\omega = \sqrt{2g/L} = 2\pi f_n$.

In this development, the forces and acceleration (and rate of change of momentum) in the z direction were considered. The force–momentum relationship is a vector one and the force and acceleration (also velocity and displacement) around the bend of the U tube are not in the z direction, so it is not obvious that the above procedure is correct. If one considers the details of the motion around the bend with the column broken up into small elemental lengths Δs (or ds) along the direction of motion s, the forces at the interfaces of the elements will cancel when the total effect is obtained by adding them all together for the integrated effect. The wall forces are perpendicular to the s direction of motion and so produce no acceleration (or change in s momentum). For more complicated geometries these arguments are more difficult to formulate, so the energy principle is more direct and specific in its use.

Equations (15-3) can be obtained directly from (6-11),

$$-\frac{1}{\rho}\frac{\partial p}{\partial s} - \frac{1}{g_c}\frac{\partial gz}{\partial z} = \frac{1}{g_c}\frac{\partial \mathrm{v}}{\partial t} + \mathrm{v}\frac{\partial \mathrm{v}}{\partial s}$$

where the friction has again been neglected. Assuming one-dimensional flow, $\partial \mathrm{v}/\partial n = 0$, and with $A = \text{const.}$, $\partial \mathrm{v}/\partial s = 0$ (v does not vary along s at any instant); therefore, $\partial \mathrm{v}/\partial t = d\mathrm{v}/dt$. Multiplying the equation by $dV = ds\, dA$ and integrating gives

$$-\frac{1}{\rho}\iint p\, dA - \frac{1}{g_c}\iint g_z\, dA = \frac{1}{g_c}\iiint \frac{d\mathrm{v}}{dt}\, ds\, dA$$

Let the left end of the U tube be station 1 and the right end station 2; the above equation can then be integrated to give

$$(p_1 - p_2)\frac{A}{\rho} - \frac{g}{g_c}(2R + 2z)A = \frac{LA}{g_c}\frac{d\mathrm{v}}{dt}$$

The Δp and $2R$ terms are the equilibrium manometer conditions and therefore cancel. The remaining equation is (15-3). This equation could also be developed starting with the unsteady form of (6-45) for a control volume with nonstationary boundaries. The details of and arguments for reducing it to the form of (15-3) are excessively complicated.

The energy method is the most direct but is limited when friction effects must be considered. The energy "constant" cannot be evaluated simply and directly as in the above examples, because it would include both a gravitational potential energy and an internal energy. The friction would affect a conversion from the available maximum potential energy at the end of each cycle to thermal internal energy during each cycle.

15.2 MANOMETER WITH LONG CONNECTING LINES

Consider a manometer with a column length L and area A_m connected across a Venturi meter as illustrated in Fig. 15-2. The legs are a length l above the equilibrium O-O line, and the manometer is connected by two long lines, L_1 and L_2, of area A_f to the Venturi taps (or any two taps in a fluid system that may not be at the same altitude z). Because (1) A_f is usually smaller than the manometer-tube area A_m, (2) the relatively long length and small diameter of the lines indicates a small mass or inertia proportional to D^2L, and (3) large friction proportional to L, friction effects should be included in the analysis. Unfortunately, it is not possible to include the friction effect in an energy analysis (as was done in the case of the simple U-tube manometer) because the energy constant cannot be

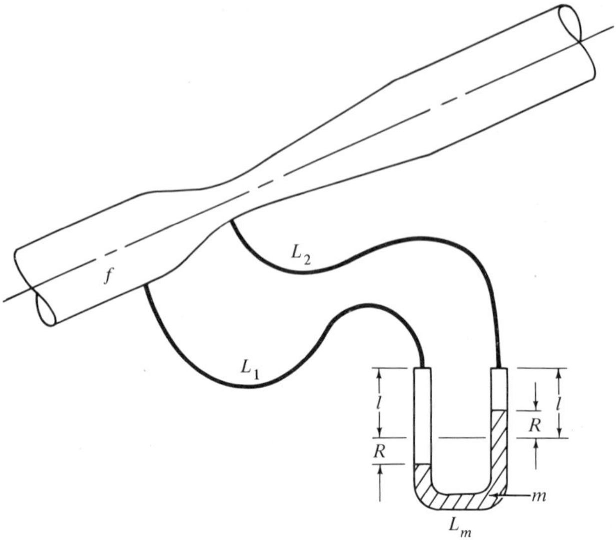

Fig. 15-2. Oscillation of a manometer connected with a Venturi meter. The line fluid is denoted by *f* and the manometer fluid by *m*.

evaluated. Use of the *s*-momentum equation has been shown to be valid. Further simplification is possible if quasi-one-dimensional flow is assumed ($\beta \approx 1$). Neglecting transition "loss" affects at the tubing connection points and noting that the mass of each fluid in the manometer remains constant, the force–momentum equation may be developed. The momentum rate or inertia term is

$$\rho_f(L_1 + L_2)A_f \frac{d\mathrm{v}_f}{dt} + \rho_f(2l)A_m \frac{d\mathrm{v}_m}{dt} + \rho_m L_m A_m \frac{d\mathrm{v}_m}{dt}$$

where the subscript f denotes fluid line and m denotes manometer-tube variables. The continuity equation gives $A_f\mathrm{v}_f = A_m\mathrm{v}_m$ for the line fluid. The inertia then becomes

$$[\rho_f(L_1 + L_2 + 2l) + \rho_m L_n]A_m \frac{d\mathrm{v}_m}{dt}$$

The force corresponding to the displacement of the manometer from its static equilibrium position is

$$(\rho_m - \rho_f)\frac{2g}{g_c} A_m z$$

The friction term can only be approximated from the steady-state, fully-developed result obtained in Chapter 7:

$$\tau = \frac{8\mu \mathrm{v}}{D} \qquad \tau A = \frac{8\mu \mathrm{v}}{D}\pi DL = 8\pi\mu \mathrm{v} L$$

The flow is not steady or fully developed. It may be laminar. The periods will be relatively long; therefore, the velocity is relatively small. The connecting lines are usually relatively small in diameter. During the flow acceleration, the boundary layer is thin and the acceleration provides a stabilizing influence as described in Chapter 9. The wall velocity gradient will be greater than that for fully developed flow during the accelerating half-cycle. For an idea of the maximum N_R, consider $z_0 = 2$ in., $\omega = 1$/sec, giving a maximum velocity of 1/6 ft/sec. For water at room temperature in a 1/4-in.-I.D. line, the maximum N_R is then ~ 350. The deceleration is destabilizing, but the mean velocity (and therefore the kinetic energy) is decreasing, so it is possible that laminar flow may be obtained. The velocity gradient for the decelerating flow is less than that for fully developed flow, so if turbulence does not occur, the above equation may give an approximation for an averaged friction effect.

Adding the friction force for each length of fluid gives

$$F_\tau = \sum \tau A = -8\pi[\mu_f(L_1 + L_2)\mathrm{v}_f + \mu_f(2l)\mathrm{v}_m + \mu_m L_m \mathrm{v}_m]$$

$$F_\tau \approx -8\pi\left[\mu \frac{A_m}{fA_f}(L_1 + L_2) + \mu_m L_m\right]\mathrm{v}_m$$

where the friction in the top part of the manometer has been neglected with respect to that in the longer, smaller-diameter lines.

The manom ter displacement force measured from the equilibrium position is

$$F_g = \frac{-(\rho_m - \rho_f)A_m g 2z}{g_c}$$

where z is the oscillation displacement from the equilibrium position in the manometer tube. The oscillation velocity v_m in the manometer tube is then dz/dt. The force–momentum equation for the free vibration motion is then

$$M\frac{d^2z}{dt^2} + N\frac{dz}{dt} + Gz = 0 \tag{15-4a}$$

where

$$M \equiv \frac{[\rho_f(L_1 + L_2 + 2l) + \rho_m L_m]A_m}{g_c}$$

$$N \equiv 8\pi\left[\mu_f \frac{A_m}{A_f}(L_1 + L_2) + \mu_m L_m\right]$$

$$G \equiv \frac{2(\rho_m - \rho_f)A_m g}{g_c}$$

Equation (15-4a) can be divided by M and written in further shorthand notation as

$$\frac{d^2z}{dt^2} + 2b\frac{dz}{dt} + \omega_0^2 z = 0 \tag{15-4b}$$

Note that $\omega_0^2 \equiv G/M$ is the undamped natural frequency and essentially has the form $\omega_0^2 = 2g'/L'$, similar to the case of the simple U tube; $b \equiv N/2M$ is the damping parameter. Since (15-4b) is a linear, homogeneous, second-order differential equation, the solution is of the form

$$z = Ce^{\lambda t}$$

where λ is to be determined as a function of the equation parameters, and C from the boundary or initial conditions. These conditions can be taken as: at $t = 0$, $z = z_0$, $dz/dt = 0$. Solution of the assumed form gives a quadratic in λ with two roots

$$\lambda^2 + 2b\lambda + \omega_0^2 = 0$$

$$\lambda_{1,2} = -\text{b} \pm \sqrt{b^2 - \omega_0^2} \equiv -b \pm r \quad \text{if } b > \omega_0 \quad (\text{note } b > r) \tag{15-5a}$$

$$\lambda_{1,2} = -b \pm i\sqrt{\omega_0^2 - b^2} \equiv -b \pm iq \quad \text{if } \omega_0 > b \tag{15-5b}$$

If $b > \omega_0$, the solution to (15-5a) is

$$z = C_1 e^{-(b-r)t} + C_2 e^{-(b+r)t} \tag{15-6a}$$

Evaluating the constants from the initial conditions yields

$$z = \frac{z_0}{2r}\left[(b + r)e^{-(b-r)t} - (b - r)e^{-(b+r)t}\right] \tag{15-6b}$$

Equation (15-6b) has the general form

$$z \sim \alpha e^{-\beta t} - \beta e^{-\alpha t}$$

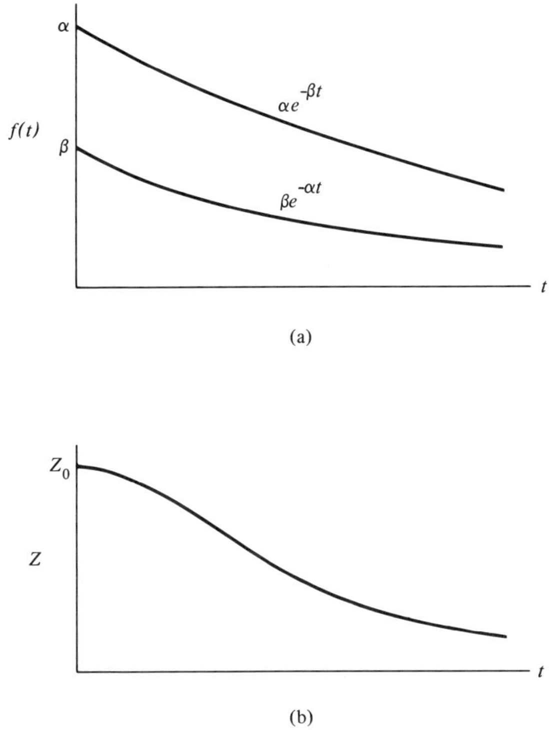

Fig. 15-3. Representation of the results indicated by (15-6) for the oscillation of a manometer with long lines.

where $\alpha \equiv b + r$ is greater than $\beta \equiv b - r$. These decreasing exponential functions have equal initial slopes at $t = 0$ and are sketched in Fig. 15-3. The curve representing the first, or α, term has the greater initial value and falls off more slowly than the second, or β, curve. The difference is a positive exponential function that remains positive (Fig. 15-3b). The behavior is termed *overdamped.* If an initial displacement is given to the system, it will never return to equilibrium but will approach it exponentially. This overdamped condition is unacceptable for a manometer, because an excessively long time would be required to attain equilibrium for an applied pressure signal. It will be shown in the next section that an overdamped manometer system is highly unlikely.

If $b < \omega_0$, the solution of (15-5b) is a damped periodic motion:

$$z = e^{-bt}(C_1 e^{iqt} + C_2 e^{-iqt})$$

or

$$z = e^{-bt}(A \cos qt + B \sin qt) \qquad (15\text{-}7a)$$

where $A \equiv C_1 + C_2$, $B \equiv i(C_1 - C_2)$, and $q = \sqrt{\omega_0^2 - b^2}$. Evaluating the constants for the given initial conditions gives

$$z = z_0 e^{-bt}\left(\cos qt + \frac{b}{q}\sin qt\right) \tag{15-7b}$$

The natural frequency of the damped system is q, and the damping factor is b. The damping increases with increasing values of b. The free vibration response to the initial conditions is indicated in Fig. 15-4.

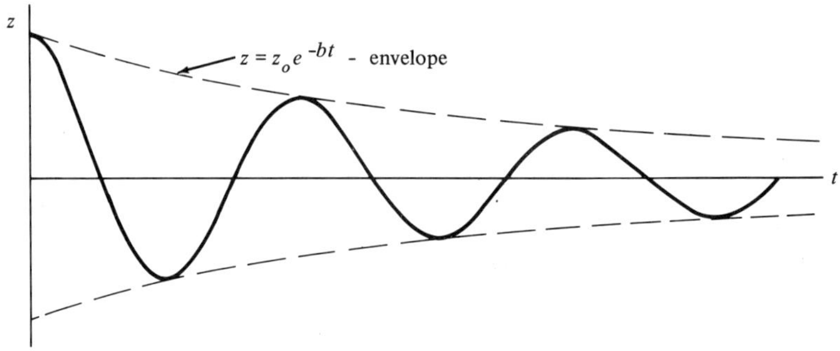

Fig. 15-4. Damped periodic motion.

A unique solution is obtained for $b = \omega_0$; the two λ roots are then identical. The solution must still contain two constants, so its form is

$$z = e^{-bt}(C_1 + C_2 t)$$

or

$$z = z_0 e^{-bt}(1 + bt) \tag{15-8}$$

The behavior of this function is indicated in Fig. 15-5. It takes a time corresponding to $bt = 6.65$ for the system to reach 99 percent of its equilibrium value (that is, $z/z_0 = 0.01$).

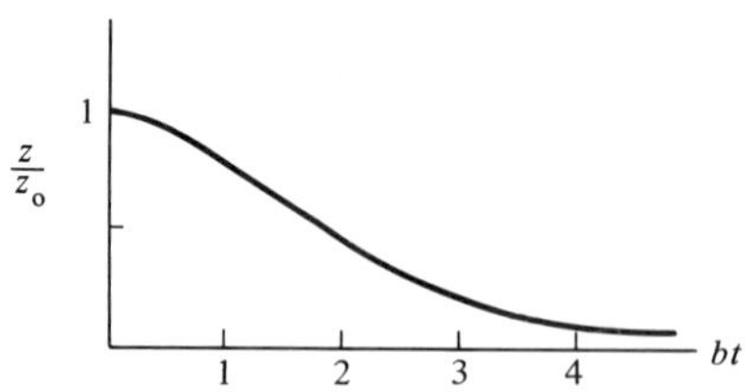

Fig. 15-5. Motion for the unique case $b = \omega_0$.

A slightly underdamped system is desirable. A damping of $b^2 = \omega_0^2/2$ is sometimes considered to be near optimum for this type of vibrating system, in which a signal output reading is wanted. This particular case is also easily calculated (see Problem 15-5).

15.3 ILLUSTRATIVE EXAMPLE 15-1: Mercury Manometer Oscillation

For the manometer system of Fig. 15-2, consider a case where L_1 and L_2 are each 10 ft of ⅛-in. tubing, the manometer-tube diameter is ¼ in., $L_m = 2$ ft, $2l = 2$ ft., $\mu_f = 2 \times 10^{-5}$ $\text{lb}_\text{f}\text{-sec/ft}^2$, $\mu_m = 3.2 \times 10^{-5}$ $\text{lb}_\text{f}\text{-sec/ft}^2$, and the specific gravity of mercury is 13.6. The undamped natural frequency is

$$\omega_0^2 = \frac{G}{M} = \frac{2(\rho_m - \rho_f)}{\rho_f(L_1 + L_2 + 2l) + \rho_m L_m} = \frac{2(S_m - 1)g}{L_1 + L_2 + 2l + S_m L_m} \tag{15-9}$$

where S_m is the manometer-fluid specific gravity. The above numerical data give $\omega_0 = 3.7/\text{sec}$, or a period of $T = 1.7$ sec. The damping parameter is

$$b = \frac{N}{2M} = \frac{4\pi[\mu(A_m/fA_f)(L_1 + L_2) + \mu_m L_m]}{[\rho_f(L_1 + L_2 + 2l) + \rho_m L_m]A_m/g_c} = 0.70 \tag{15-10}$$

With ω_0^2 much larger than b^2, the motion is only slightly damped. The periodic frequency $q = \sqrt{\omega_0^2 - b^2} = \omega_0\sqrt{1 - (b/\omega_0)^2}$ is only slightly smaller than ω_0, and therefore may be approximated by ω_0.

This relatively long period causes no particular difficulty in obtaining manometer readings. Periodic oscillations with periods on the order of 1 or 2 sec are not usual in gas or liquid systems. It does indicate that a period of a few seconds must be allowed for transients to die out before making readings. If random oscillations exist in the system with occasional long-period (1- to 2-sec) oscillations, it will be difficult to obtain an equilibrium reading.

Equation (15-10) indicates that increasing the line length has little effect on damping. Moreover, (15-9) shows that increasing the line length yields an adverse effect of increasing the period, thereby decreasing the response. Therefore, it is advantageous to make the lines as short as possible and provide damping by other means, which do not involve increasing the fluid mass. Damping may be obtained by utilizing damping needle valves in the lines, or a friction damper such as a porous plug.

This example indicates that the natural frequency of oscillation of a manometer system (without added damping devices) can be closely approximated by the undamped natural frequency. From the previous material, the frequency is most easily determined from application of the principle of conservation of energy.

15.4 RESERVOIR OSCILLATIONS

The problem of transient oscillations between two large reservoirs connected by a relatively long length of pipe yields valuable information in connection with other less obvious problems to be considered later. Consider two reservoirs, as indicated in Fig. 15-6, with an initial difference in level that is significantly smaller than the depth of either reservoir. At some initial time ($t = 0$), the valve is opened relatively rapidly (assume the time of opening is less than half the period of the resulting oscillation)

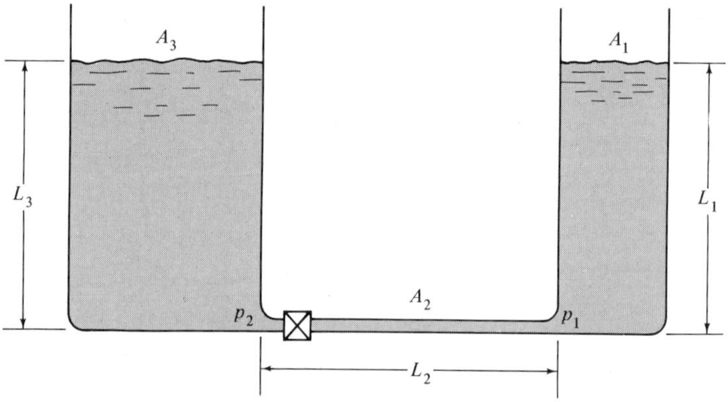

Fig. 15-6. Notation for the oscillation between two reservoirs.

and a transient flow occurs until equilibrium is reached. The initial elevation of the right reservoir above the final equilibrium position is defined as z_1. The dynamic equation is obtained by the same method used to derive (15-5),

$$\frac{\rho A_3 L_3}{g_c}\frac{d\text{v}_3}{dt} + \frac{\rho A_2 L_2}{g_c}\frac{d\text{v}_2}{dt} + \frac{\rho A_1 L_1}{g_c}\frac{d\text{v}_1}{dt} + \sum F\text{v}_2 + \frac{\rho A_1 g z_1}{g_c} + \frac{\rho A_3 g z_3}{g_c} = 0$$

The friction coefficient $\sum F$ is determined from the sum of the dissipation ("loss") coefficients for the valve, line L_2, and open discharge into the right reservoir. For simplicity of development, consider the line "dissipation" to predominate. The flow in this case will be considered to be turbulent. The friction force term is τA_{per} (where A_{per} is the peripheral area), which will be approximated by

$$\tau A_{\text{per}} = (p_2 - p_1)A_2 = f\frac{L_2}{D_2}\frac{\rho|\bar{\text{v}}_2|}{2g_c}\text{v}_2 A_2 \qquad \text{(15-11)}$$

which gives

$$N = f\frac{L_2}{D_2}\frac{\rho|\overline{\mathrm{v}_2}|}{2g_c}A_2 \tag{15-12}$$

as an average value. $|\overline{\mathrm{v}_2}|$ is the time-average absolute value of v_2, or, even more approximately, one quarter of the initial value v_0 determined from the initial reservoir-level differential Δz_0:

$$4|\overline{\mathrm{v}_2}| \equiv \mathrm{v}_0 = \sqrt{\frac{2g\,\Delta z_0}{fL/D}}$$

F in the dynamic equation is a positive scalar number and therefore is taken proportional to the average value of $\overline{\mathrm{v}}_2$. The friction force is then proportional to $|\overline{\mathrm{v}}_2|\mathrm{v}_2$ and always opposes the motion. The friction factor f can be based on $N_R = \mathrm{v}_0 D_2/4\nu$. The factor ¼ is used to account for both the cyclic time average and the overall decay average. These approximations are necessary because the turbulent friction term $\propto \mathrm{v}_2^2$ would make the equation nonlinear and quite difficult (or tedious) to solve.[1]

Introducing the continuity equation

$$A_1\mathrm{v}_1 = A_2\mathrm{v}_2 = A_3\mathrm{v}_3$$

which also gives

$$A_1 z_1 = A_2 z_2 = A_3 z_3$$

for reservoirs with vertical walls, the above dynamic equation can be written

$$\frac{\rho_1 A_1}{g_c}(L_1 + L_2 + L_3)\frac{d\mathrm{v}_1}{dt} + f\frac{L_2}{D_2}\frac{\rho|\overline{\mathrm{v}_2}|}{2g_c}A_1\mathrm{v}_1 + \frac{2\rho g A_1}{g_c}z_1 = 0$$

or

$$\frac{d^2 z_1}{dt^2} + \frac{(fL_2/D)|\mathrm{v}_2|}{2L}\frac{dz_1}{dt} + \frac{2g}{L}z_1 = 0 \tag{15-13}$$

where $L = L_1 + L_2 + L_3$. Assuming the motion to be oscillatory, the undamped natural frequency is $\omega_0 = \sqrt{2g/\sum L}$ and the oscillation frequency is $\omega = \sqrt{\omega_0^2 - b^2}$, where $b = (fL_2/D)|\overline{\mathrm{v}_2}|/4\sum L$. It is interesting to note that if $L_2 \gg L_1$ or L_3, $b \approx f|\overline{\mathrm{v}_2}|/4D$, which is independent of L, and $\omega_0 \approx \sqrt{2g/L_2}$. If $b > \omega_0$, the motion will be overdamped, as described previously.

[1] The approximate method indicated here is similar to the first step made in a numerical iterative method for the solution of the nonlinear equation.

15.5 CAVITY RESONATOR

The volume in a transducer such as illustrated in Fig. 3-15 and the connecting line can form another type of vibrating system. With reference to Fig. 15-7, assume that there exists an oscillating pressure signal $p_a(t)$ at the end of the line where a pressure measurement is desired. The magnitude of the oscillating pressure is usually much less than the mean pressure.

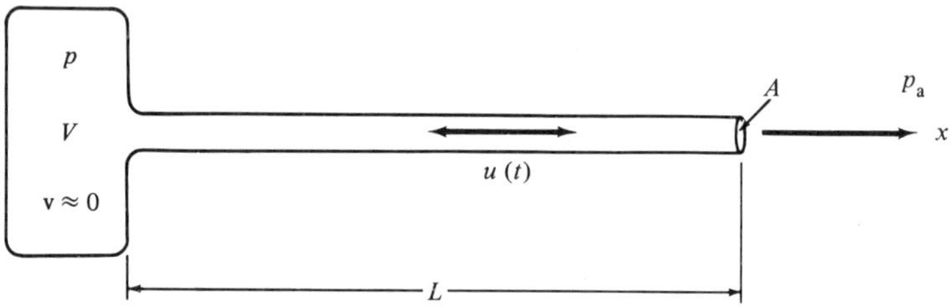

Fig. 15-7. Notation for the cavity resonator.

If the volume of the connecting line is less than the cavity volume, most of the compression takes place in the cavity, allowing the simplified assumption that the gas motion in the tube is incompressible. The mass in the tube then acts like an oscillating mass on a gas spring made up of the compressible gas in the cavity. Friction can be neglected, as indicated from previous examples.

The time-dependent variables involved are (1) the pressure (or $p - p_a$), (2) the density ρ in the cavity, and (3) the gas velocity or displacement in the tube. Since the gas column in the tube is considered to be incompressible, the cavity pressure and p_a are in phase (the signal transmission through the column is instantaneous), so $p - p_a$ is also a function of time. The problem is considered with respect to force and acceleration of the gas in the tube. The pressure oscillations will be assumed to be of sufficiently small magnitude (with respect to the mean pressure) that the volume flow in and out of the tube is much less than the volume in the tube.

If all the above assumptions are valid, the dynamic equation for the gas in the tube is

$$(p - p_a)A = \frac{\rho A L}{g_c}\frac{du}{dt} \qquad (15\text{-}14)$$

Since $p - p_a$, ρ, and u are interrelated functions of time t, more information is needed to solve this equation. It will be seen that p is the best choice for the dependent variable. The continuity relation can be used to obtain a relation between ρ and u. The variation of mass with time in the cavity must be a consequence of the flow out of the tube,

$$\frac{\partial}{\partial t}(m)_{\text{cav}} + \rho A \text{u} = 0$$

or

$$V\frac{\partial \rho}{\partial t} = -\rho A \text{u} \tag{15-15}$$

where V is the cavity volume. A relation between p and ρ can be obtained by specifying the process. If friction and heat transfer are negligible (a satisfactory assumption for $V_{\text{tube}} < V_{\text{cav}}$) the p–ρ process can be considered to be isentropic. From Chapter 12,

$$\frac{dp}{d\rho} = \frac{a^2}{g_c} \quad \text{or} \quad \frac{d\rho}{dt} = \frac{g_c}{a^2}\frac{dp}{dt}$$

which can be combined with (15-15) to give

$$\text{u} = \frac{-g_c V}{\rho a^2 A}\frac{dp}{dt} \tag{15-16}$$

Taking the derivative of (15-16) with respect to t, neglecting second-order variations involving the product of $(d\rho/dt)(dp/dt)$ and combining the result with (15-14) gives

$$\frac{d^2P}{dt^2} + \frac{a^2A}{LV}P = 0 \tag{15-17}$$

where $P \equiv p - p_a$. P varies harmonically with a period $\omega_0 = \sqrt{a^2A/LV}$. The displacement x of the gas in the column and the velocity u vary with the same period. For initial conditions of $p - p_a \equiv P_0$ and $dP/dt = 0$ at $t = 0$, the solution to (15-17) is

$$P = P_0 \cos \omega t \tag{15-18}$$

The velocity is obtained from (15-16) as

$$\text{u} = \frac{-g_c P_0}{\rho L \omega_0} \sin \omega t = -\text{u}_0 \sin \omega t \tag{15-19}$$

The displacement is obtained from $\text{u} = dx/dt$,

$$x = \frac{g_c P_0 V}{\rho a^2 A} \cos \omega t = x_0 \cos \omega t \tag{15-20}$$

As a typical numerical example consider a transducer cavity 2 cm in diameter and 2 mm in length with a connecting tube 10 cm long and 1-mm I.D. The gas is air. The transducer range is 0.1 atm and the mean signal pressure is at 1 atm. The natural frequency is $f_0 = \omega_0/2\pi = 188$ cycles/sec. If the magnitude of the pressure oscillation is nominally assumed to be 1/10 the transducer range, the velocity and displacement amplitudes can be determined as $u_0 = 7$ m/sec and $x_0 = 6$ mm. The maximum velocity amplitude is $2u_0 = 14$ m/sec, which is significantly less than $a = 343$ m/sec (for atmospheric conditions) and $2x_0 = 1.2$ cm, which is less than $L = 10$ cm, so the results are apparently within the order-of-magnitude assumptions.

ACOUSTIC OSCILLATIONS

If the volume of the cavity is much less than the tube volume, the tube will behave as if it were closed at one end. There will be no significant flow within the tube (for maximum external signal variations significantly less than 1 atm in amplitude) and the vibrations will be acoustic in nature. An impressed pressure signal at the open end will travel into the tube as a compression wave, as sketched in Fig. 15-8. At time $t = 0$ the ambient pressure is suddenly increased to $p_a + \Delta p$. This pressure signal is propagated into the gas in the tube with a velocity a. At time t, the wave has traveled a distance $x = at$. Consider the pressure at a fixed point in the tube, for example, at the closed end, as indicated in Fig. 15-8b. A time $T = L/a$ will be required before the signal is felt there. The gas behind the pressure wave front is moving in the direction of the front with a small velocity. The pressure, and therefore the density, of the gas initially at rest is increased. Therefore, the gas behind the wave is compressed into a smaller volume than it originally occupied. This condition can only be maintained if gas is supplied from the left to allow for an increase in density.

When the wave "hits" the end, the motion must be stopped. Therefore, a force is required to bring the gas to rest, which results in a further pressure increase of magnitude Δp. The pressure at the end of the tube is then $2\Delta p$, as indicated in Fig. 15-7b.

When the reflected pressure wave goes out the open end, the pressure and density in the tube are greater than ambient, so that a flow is initiated out of the tube, resulting in a pressure decrease in the tube as indicated. The round trip of this rarefaction wave also requires a time corresponding to a traverse of two lengths of the tube. A complete period then requires an equivalent length of $4L$, or the period is $T = 4L/a$, or the frequency is $f_0 = a/4L$. This frequency is sometimes called the organ pipe frequency (closed end). For the above example of the 10-cm tube, this frequency is $f_0 = 343/(4 \times 0.1) = 860$ cycles/sec, which is greater than the cavity resonant frequency. Determination of the organ pipe frequency serves as a

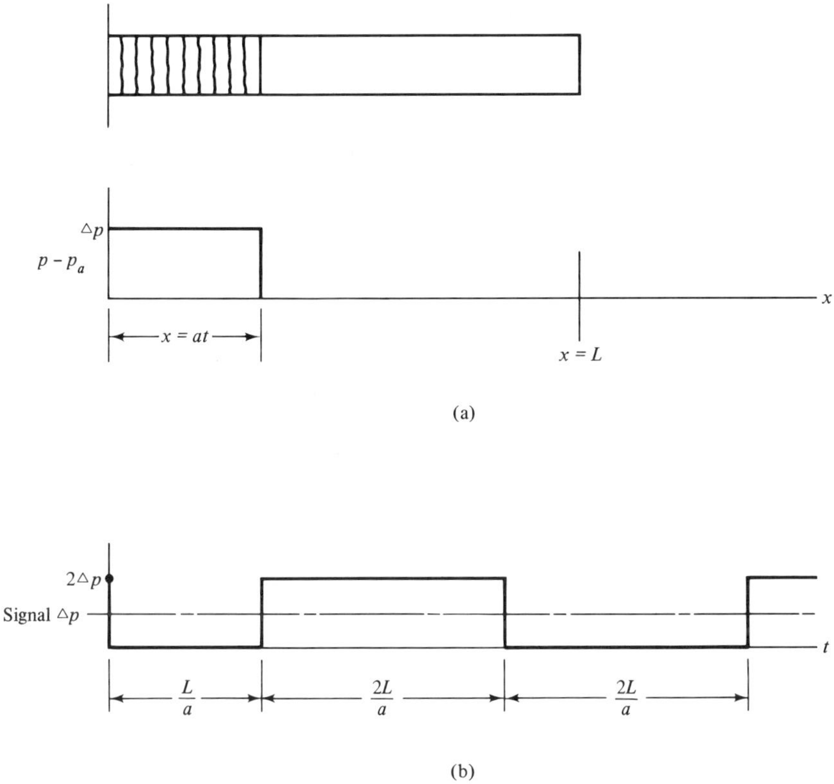

Fig. 15-8. Acoustic wave propagation in a tube: (a) compression wave propagating into the tube at time t, (b) pressure at the end of the tube as a function of time.

quick check on the validity of the assumptions of the cavity resonance frequency. If the cavity resonance frequency is approximately on the order of, or greater than, the organ pipe frequency, the assumptions for its calculation are not valid.

It is obvious that a serious difficulty can be encountered if the pressure oscillations $P(t)$ are in resonance with the natural frequency of the tube and cavity of a transducer. The oscillation signal output of the transducer may indicate a magnitude much greater than the maximum amplitude $2P_0$. The situation may be complicated further if the transducer diaphragm has the same resonant frequency. It is wise to know the natural frequency of a transducer and to calculate the cavity and organ pipe frequencies if a very large oscillation signal is being picked up and cannot be explained otherwise.

The material and physical dimensions of the connecting lines may tend to alter the above results. If the lines are relatively "soft" (thin wall or material with a low elastic modulus), the frequency and peak magnitudes may be less than calculated above for both types of oscillation. This modification will occur if the modulus of elasticity of the wall material is approximately on the order of the bulk modulus of the fluid. An example of this effect will next be illustrated for a dangerous transient condition that exists in liquid flow.

15.6 WATER HAMMER

For a transient gas wave in a duct or tube, the pressures and forces depend on the mass and velocity (or inertia) of the moving fluid column. The acoustic variations considered previously are usually not destructive in nature (with such notable exceptions as the sonic boom). In liquids with densities on the order of thousands of times those of a gas and with much larger sonic speeds, extremely large pressures may develop.

Water hammer is a transient produced in a piping system when the flow is suddenly stopped by rapidly closing a valve. Such transients occur in starting or stopping pumps or turbines or when a check valve closes as a consequence of a supply failure. If the valve in the part of a system as illustrated in Fig. 15-9 were closed suddenly, and if the fluid were completely incompressible, the rate of change of momentum of the fluid

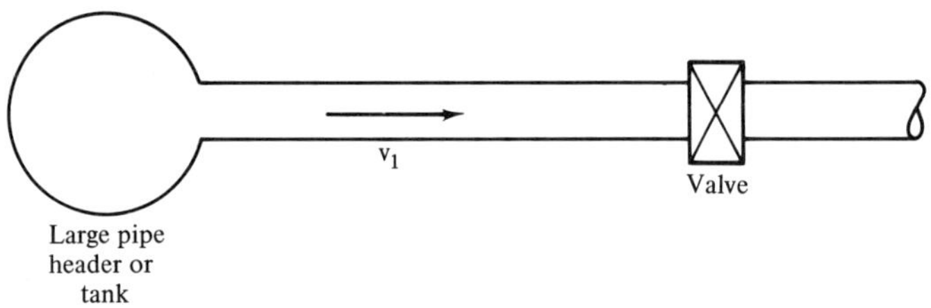

Fig. 15-9. Example system where water hammer can occur.

within the pipe would be infinite, giving an infinite force on the valve and piping. However, because of the finite compressibility of the liquid, the signal that the valve has closed is propagated back into the fluid with approximately the sonic wave speed. This wave speed (as given in Chapter 2) is

$$a = \sqrt{\frac{g_c E_B}{\rho}} = \sqrt{\frac{g_c}{\rho \kappa}}$$

where E_B is the bulk modulus of elasticity,

$$E_B = -\frac{\Delta p}{\Delta V/V} \qquad \text{and} \qquad \kappa \equiv \frac{1}{E_B}$$

and κ is the compressibility. The overpressure developed by the wave formation and propagation can be determined from the equation of motion. Figure 15-9 shows the wave traveling into the fluid. The velocity is reduced from v_1 (the flow velocity) to zero by the compression wave. In liquids, $v_w = a$ is quite large (about 3000 to 4000 ft/sec), while flow velocities are usually relatively small (of the order of 100 ft/sec or less), so the "Mach number" is relatively small. The approximation that $v_w = a$ is then satisfactory. The friction force on the control volume is relatively small compared with the pressure force $(p_1 - p_2)A$ and can be neglected.

In Fig. 15-10a the control volume is considered to be moving with the constant wave speed v_w. The force–momentum principle can be applied to the relative motion of this moving control volume as indicated in

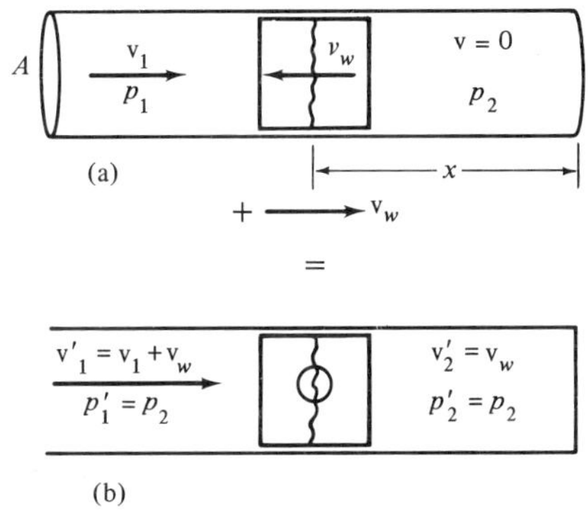

Fig. 15-10. Control volume for water-hammer analysis: (a) absolute or ground-fixed coordinate system; (b) coordinate system moving with uniform velocity v_w.

Chapter 6. A uniform motion $+v_w$ is added to the flow field to make the wave stationary as indicated in Fig. 15-10b. The fact that the pipe and valve are now moving to the right need not cause concern, since only the specified control volume is to be considered. The motion has been reduced to one of steady flow with respect to the control volume, so the combined momentum and continuity statement is

$$\sum F_x = (p_1 - p_2)A = \frac{\dot{m}}{g_c}[v_w - (v_1 + v_w)] = \frac{\rho_2}{g_c} A v_w v_1$$

or

$$(p_2 - p_1) \equiv \Delta p = \frac{\rho_2}{g_c} v_w v_1 \approx \frac{\rho}{g_c} v_w v_1 \tag{15-21a}$$

where $\rho_1 \approx \rho_2 = \rho$ has been assumed as a consequence of the small compressibility of liquids. It is also noted that $\dot{m}/A = \rho_2 \mathrm{v_w} = \rho_1(\mathrm{v}_1 + \mathrm{v}_w)$, which, with $\mathrm{v}_1 \ll \mathrm{v}_w$, also gives $\rho_1 \approx \rho_2$. With $\mathrm{v_w} = a$ in (15-21a),

$$\Delta p = \frac{\rho \mathrm{v}_1 a}{g_c} = \frac{\rho \mathrm{v}_1}{g_c}\sqrt{\frac{g_c E_B}{\rho}} = \mathrm{v}_1\sqrt{\frac{\rho E_B}{g_c}} \tag{15-21b}$$

FLEXIBLE PIPE

If the pipe elasticity is considered, the effective wave speed will be less than that in a rigid pipe because some of the energy is used to expand the pipe. The total change in volume will then be the result of the pipe strain plus the liquid compressibility. The effective compressibility is then

$$\kappa_{\text{eff}} = -\frac{\Delta V_{\text{pipe}} + \Delta V_{\text{liq}}}{V\,\Delta p} = -\left(\frac{\Delta V_{\text{pipe}}}{V\,\Delta p} + \frac{\Delta V_{\text{liq}}}{V\,\Delta p}\right)$$

or

$$\kappa_{\text{eff}} = -\frac{\Delta V_{\text{pipe}}}{V\,\Delta p} + \kappa_{\text{liq}} \tag{15-22}$$

κ_{liq} is usually known; $\Delta V_{\text{pipe}}/V$ is given by

$$\frac{\Delta V}{V} = \frac{\Delta(\pi R^2 x)}{\pi R^2 x} \approx \frac{2\,\Delta R}{R}$$

assuming that the valve position (closed pipe end) remains fixed.

The strain $\Delta R/R$ can be related to the pressure and wall thickness of the pipe (w). For a relatively thin-walled pipe ($w/R \ll 1$), the strain is primarily due to tangential ("hoop") stresses, given by

$$s = \frac{\Delta p R}{w}$$

For an elastic material, the relation between stress and strain gives

$$s = eE_P = \frac{\Delta l}{l} E_P = \frac{\Delta R}{R} E_P$$

where E_P is the linear modulus of the pipe material. These last two relations yield the strain,

$$\frac{\Delta R}{R} = \frac{\Delta p R}{w E_P}$$

The effective compressibility is then

$$\kappa_{\text{eff}} = \frac{2R}{wE_P} + \kappa_{\text{liq}} = \kappa_{\text{liq}}\left(1 + \frac{2R}{wE_P\kappa_{\text{liq}}}\right) \tag{15-23}$$

or the effective bulk modulus is given by

$$E_{\text{eff}} = \frac{1}{(2R/wE_P) + (1/E_B)} = \frac{E_B}{1 + (2RE_B/wE_P)} \tag{15-24}$$

The speed of propagation of the wave is then

$$\text{v}_{w,\text{eff}} = \sqrt{\frac{g_c E_{\text{eff}}}{\rho}} = \sqrt{\frac{g_c E_B}{\rho[1 + (2RE_B/wE_P)]}} \tag{15-25}$$

which is seen to be less than that in an ideal, rigid tube. For water in a steel pipe, $E_B \approx 3 \times 10^5$ psi and $E_P = 3 \times 10^7$ psi, so the added factor will not be very large. For copper tubing with $E_P \approx 10^7$ psi, the wave speed will be somewhat slower. The speed as given by (15-25) may be approximated by

$$\text{v}_{w,\text{eff}} \approx \left(1 - \frac{RE_B}{wE_P}\right)\sqrt{\frac{g_c E_B}{\rho}} \tag{12-25a}$$

for cases where $RE_B/wE_P < 1/10$.

The $\Delta\rho$ may be determined simply by substituting v_w into (15-21),

$$\Delta p = \frac{\rho \text{v}_1}{g_c}\sqrt{\frac{g_c E_B}{\rho[1 + (2RE_B/wE_P)]}} \tag{15-26}$$

The pressure at the valve will remain constant at $p + \Delta p$ as the wave travels to the end of the pipe and is dispersed within the connecting tank, reservoir, or large pipe (Fig. 15-9).

As the pressure wave leaves the open end of the pipe, it is reflected as an expansion wave. The pressure behind the expansion wave is limited in its drop by the vapor pressure of the liquid. If the wave is such that a pressure less than the vapor pressure would occur, the fluid boils locally ("cavitates"). The growth and collapse of the cavitating region causes further transients in the system. If there is an abrupt turn (tee or el) in the line, or another valve with a small orifice (for example, globe valve), the pressure wave will be reflected from it. The reflection, dispersion, diffusion, and cancellation of the waves in this last case is a quite complex phenomenon. As is often heard in plumbing systems without pulsation dampers, there may be several wave reflections before the energy is dissipated. The time history of the pressure at a fixed point in the pipe is similar to that shown in Fig. 15-7b.

The above development is approximate for nearly instantaneous valve closure. A sudden closure is one where closing is completed in a time less than $2L/\text{v}_w$ (that is, before the reflected wave reaches the valve). The development of the relations for a gradual closure are more complicated, requiring a step-by-step procedure for solution. More information on water hammer, and solutions to water-hammer problems are given in reference 1 at the end of the chapter.

It is interesting to consider the concept of impedance and admittance (or, together, immittance) as related to the problem of the flexible pipe. *Impedance* is defined as the change in potential per change in flow or current, and *admittance* is its reciprocal. The simplest form of impedance is resistance. Mass inertia or "inertance" is analogous to inductive reactance in alternating-current circuits. In (15-22) and (15-24), the potential Δp is common in both compressibility or inertial effects (or responses) to the traveling wave. Parallel circuits are simple circuits where the potential is common. In parallel circuits, the admittances add, and in series circuits, the impedances add. The compressibility is then similar to admittance and the modulus to impedance. Also in parallel circuits, the flow or change in quantity (in this case ΔV) of the circuit branches add, as indicated in (15-22). Such concepts of analogous circuit behavior are useful for the construction and use of electric analogs, where the parameters are easier to obtain and manipulate than for the fluid counterpart. Under certain conditions, exact electric analog circuits can be constructed for the fluid circuits. There have recently been introduced such electric elements as nonlinear resistances to model square (or other) power-law fluid resistance.

15.7 SURGE CHAMBERS

In order to prevent high transient peak pressures, such as occur under water-hammer conditions or starting and stopping "high-head" turbomachinery, a surge capacitance can be provided to absorb and store some of the energy, thereby reducing the magnitude of the pressure surge. In a pumping plant, a surge tank can be used to smooth out flow-demand variations and to provide a controlled (slowed-down) pressure buildup across the pump during starting. On shutdown the surge tank prevents water-hammer-type high-pressure differences across the pump.

Figure 15-11 illustrates a surge chamber placed upstream of a fluid device (valve, turbine, and so on), in which the surge transients are to be controlled or minimized. If the device were a booster pump in a long fluid transmission system, the pump would directly follow the surge chamber junction, with a shutoff and check valve downstream of the pump. On shutdown, the valve would be closed with the pump still running in order to decrease the magnitude of the transient pressure (and flow) surges in the entire system. A centrifugal pump can have its discharge closed off without overloading the pump drive motor; the shutoff power is less than the normal operating power for this type of pump (and about equal to normal operating power for a propeller-type pump). It is not desirable to

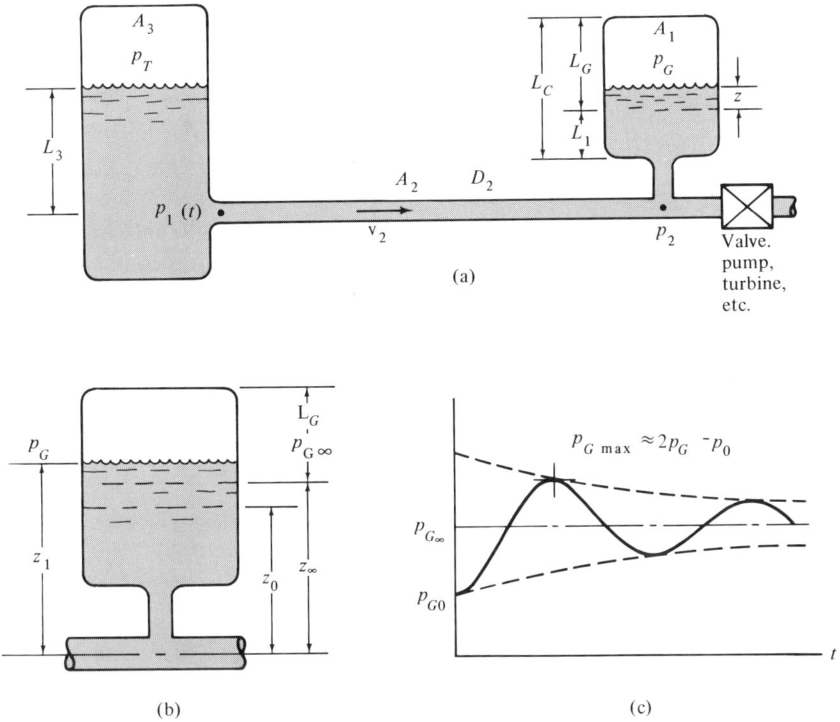

Fig. 15-11. (a) Surgechamber in a line adjacent to a transient-producing device; (b) notation for variables in the surge chamber; (c) pressure oscillation history in the chamber.

throttle the suction of a pump, since damaging cavitation and cavitation surging can be produced. The check and shutoff valve for a turbine are usually placed upstream of the turbine to protect the turbine from high surge pressures.

While the flow is steady, the level in the surge chamber is constant. The pressure in the chamber is the static pressure ($\sim p_2$) in the line ahead of the valve. This pressure is less than the static pressure p_1 at the supply inlet by the amount of the pressure drop in the system. For the simple system of Fig. 15-11, the steady-flow pressure drop is just that due to friction in the length of the pipe, L_2.

If the flow is stopped by the valve, a transient period will ensue. The water-hammer pressure wave from the valve will be diffused and dispersed into the surge chamber. The inertia of the mass in the line L_1 will produce a flow into this chamber. The energy of the mass is converted into potential energy of compression of the gas above the liquid surface and gravitational potential energy of the level rise. This combination energy storage then acts as an energy-storage "spring" for the traveling mass, also analogous to an electric capacitance for the storage of electric charge. When the liquid column is brought to rest, the potential energy in the capacitance chamber may be greater than that represented by the reservoir supply pressure p_T if the line or system frictional resistance is not too great. For most systems of the type illustrated in Fig. 15-11, the mainline resistance or "battery potential" p_T is essentially constant. The surge chamber capacitance will then discharge back into the line as the stored potential energy is converted into kinetic energy. The inertia corresponding to the kinetic energy in the line will produce an overdischarge and the system will oscillate (again depending on the magnitude of the resistive dissipation in the line). A damped oscillating motion will occur similar to that in the manometer systems.

It is of value to be able to analyze such a system to determine the peak system pressures. Also if an analytical method can be formulated, it is possible to synthesize the design parameters for a surge capacitance required to limit the maximum value of p_2. The representative system of Fig. 15-11 will be analyzed with respect to these objectives.

Since the oscillating system of Fig. 15-11 is similar to that of Fig. 15-6, valuable information can be obtained by comparing the two. The supply at reservoir pressure p_T can be considered to be equivalent to a reservoir with height $H_T = g_c p_T/\rho g$, which is equivalent to L_3 in (15-13). The objective of utilizing this similarity and substitution is to eliminate the necessity of having $p_1(t)$ a time-dependent variable that cannot be a given or specified function of time in the equation.

If the surge tank is closed (as indicated in Fig. 15-11), a vertical displacement z of the fluid from equilibrium will produce a downward gravitational force and a downward pressure force from the gas compression in the confined volume. The gravitational force is the same as that developed with respect to the manometer and reservoir oscillations. If the pulsations are relatively rapid, no appreciable heat transfer from the gas will occur as it is compressed (or to the gas as it expands), and the compression may be assumed to be adiabatic with the process relation $pV^\gamma = \text{const}$. The variation of gas pressure from the equilibrium value is then obtained from the relation

$$\frac{\Delta p}{p} = \frac{\gamma \, \Delta V}{V} = \frac{\gamma \, \Delta z}{L} \tag{15-27}$$

which can be obtained by taking the differential of the above equation, since small variations from the mean or initial values are assumed, and since the tank cross-sectional area is constant.

The equilibrium level that the surge or capacitance chamber will reach after the transient has died out (as $t \to \infty$) will be z_∞ (Fig. 15-11b). There will be no flow at this condition, so a static equilibrium equation can be written

$$p_T + \frac{\rho g L_T}{g_c} = p_{G\infty} + \frac{\rho g z_\infty}{g_c} \tag{15-28}$$

Before the transient begins [until $t = 0$ and at $t = 0(-)$]

$$p_T + \frac{\rho g L_T}{g_c} = p_{G0} + \frac{\rho g z_0}{g_c} + \left(1 + f\frac{L_2}{D}\right)\frac{\rho V_{20}^2}{2g_c} \tag{15-29}$$

where steady, quasi-one-dimensional flow has been assumed, and v_{20} is the initial average velocity in the pipe, and the reservoir pressure p_T and L_T are assumed to remain constant. If the valve is closed rapidly, the transient period will be as illustrated in Fig. 15-11c, assuming that the system is underdamped. The inertia of the fluid will produce a flow into the chamber, raising the level and the gas pressure p_G therein. An overshoot of the equilibrium value (z_∞ and $p_{G\infty}$) will result in a reverse flow from the chamber into the open system. The system will continue to oscillate until final static equilibrium has been attained. The maximum pressure $p_{G,\max}$ (and the maximum pressure in the system) will then be approximately equal to (or less than) $p_{G\infty} + p_{G\infty} - p_{G0} = 2p_{G\infty} - p_{G0}$. If the change in level $z_\infty - z_0$ is small with respect to L_G (the height of the gas space above the equilibrium level) the liquid-level oscillations in the tank will exhibit the same behavior as the pressure oscillations. This similarity of behavior of p_G and z oscillations, $\Delta p_G \propto \Delta z$, is mathematically indicated by (15-27), which can be written

$$\frac{\Delta p}{p_{G\infty}} = \frac{p_G - p_{G\infty}}{p_{G\infty}} = \frac{\gamma(z - z_\omega)}{L_G} = \frac{\gamma\,\Delta z}{L_G} \tag{15-27a}$$

The reservoir or main tank is represented in Fig. 15-11a as having a large area ($A_3 \gg A_1$) with a level above the surge chamber L_3 and a possible charge pressure p_T above the free surface. The dynamic equation for the fluid system up to the valve, written for a system displacement from initial equilibrium, is

$$\frac{-\rho_3 L_3 A_3}{g_c}\frac{dv_3}{dt} + \frac{\rho_2 L_2 A_2}{g_c}\frac{dv_2}{dt} + \frac{\rho_1 L_1 A_1}{g_c}\frac{dv_1}{dt} + Nv_2 + \left(-\Delta p_T - \frac{Rg\,\Delta z_3}{g_c}\right)A_3 + \left(p_1 + \frac{\rho g\,\Delta z}{g_c}\right)A_1 = 0 \tag{15-30a}$$

where Δp and Δz are the displacements from equilibrium (or from steady-state values). Application of the continuity condition gives

$$A_1 v_1 = A_2 v_2 = -A_3 v_3 \quad \text{and} \quad A_1\,\Delta z_1 = -A_3\,\Delta z_3 = A_2\,\Delta x \tag{15-31}$$

The minus signs appear on the reservoir terms because z is measured upward, but a reference positive displacement in the chamber (and positive v_1) gives a downward displacement in the reservoir, therefore a negative v_3. Combining (15-27a), (15-30a), and (15-31) gives

$$\frac{\rho A_1}{g_c}(L_1 + L_2 + L_3)\frac{dv_1}{dt} + N\frac{A_1}{A_2}v_1 + \left[\gamma\left(-\frac{p_{T\infty}}{L_{T\infty}} + \frac{p_{G\infty}}{L_{G\infty}}\right) + \frac{2\rho g}{g_c}\right]A_1\,\Delta z_1 = 0$$

or (15-30b)

$$(L_1 + L_2 + L_3)\frac{dv_1}{dt} + \frac{g_c N}{\rho A_2}v_1 + \left[\frac{g_c\gamma}{\rho}\left(\frac{p_{G\infty}}{L_{G\infty}} - \frac{p_{T\infty}}{L_{T\infty}}\right) + 2g\right]\Delta z_1 = 0$$

For the friction term,

$$\frac{g_c N}{\rho A_2} = \begin{cases} \dfrac{8\pi g_c \mu L}{\rho A_2} = \dfrac{8\pi\nu L}{A_2} & \text{laminar [from (15-4)]} \\[2ex] \dfrac{fL_2}{2D_2}\,|\overline{v_2}| & \text{turbulent [from (15-12)]} \end{cases}$$

In terms of the variable z_1, (15-28b) can be written

$$(L_1 + L_2 + L_3)\frac{d^2z_1}{dt^2} + \frac{g_c N}{\rho A_2}\frac{dz_1}{dt} + \left[\frac{g_c\gamma}{\rho}\left(\frac{p_{G\infty}}{L_{G\infty}} - \frac{p_{T\infty}}{L_{T\infty}}\right) + 2g\right](z_1 - z_\infty) = 0 \quad \text{(15-30c)}$$

The z_∞ term with its coefficient can be written on the right side of (15-28c) to give an inhomogeneous equation, or it can be left as it is, and the transformation of variable $z' \equiv z - z_0$, $dz' = dz$ can be made. The latter procedure is the most direct for the solving the equation and applying the initial conditions to obtain the constants of integration. In terms of z', and with $L_1 + L_2 + L_3 \equiv L$, (15-28c) is

$$\frac{d^2z'}{dt^2} + \frac{g_c N}{\rho A_2 L}\frac{dz'}{dt} + \left[\frac{g_c\gamma}{\rho L}\left(\frac{p_{G\infty}}{L_{G\infty}} - \frac{p_{T\infty}}{L_{T\infty}}\right) + \frac{2g}{L}\right]z' = 0 \quad \text{(15-30d)}$$

or

$$\frac{d^2z'}{dt^2} + 2b\frac{dz'}{dt} + \omega_0^2 z' = 0 \quad \text{(15-30e)}$$

where

$$\omega_0^2 = \sqrt{\frac{g_c\gamma}{\rho L}\left(\frac{p_{G\infty}}{L_{G\infty}} - \frac{p_{T\infty}}{L_{T\infty}}\right) + \frac{g}{L}} \quad \text{(15-32)}$$

is the undamped radial frequency,

$$b = \frac{g_c N}{2\rho A_2 L}$$

is the damping factor, and the radial oscillation frequency is

$$\omega = \sqrt{\omega_0^2 - b^2} \qquad \text{if } \omega_0 > b$$

If $b > \omega_0$, the motion is overdamped, as previously described. The initial conditions at $t = 0$ are (1) $z(0) = z_0$ or $z'(0) = z'_0$, and (2) $dz/dt = 0 = dz'/dt$. The solution of (15-28e) with these conditions is

$$(z - z_\infty) = (z_0 - z_\infty)e^{-bt}\left(\cos \omega t + \frac{b}{\omega_0} \sin \omega t\right) \tag{15-33}$$

The condition $t \to \infty$ gives $z = z_\infty$.

Whether ω_0 is greater or less than b depends on the initial flow condition (laminar or turbulent) and the parameters of the surge chamber (that is, the surge chamber size L_G and initial charge pressure $p_{G\infty}$ in terms of the known system factors L_2, $p_{T\infty}$, and $g_c N/\rho A_2$). Note that L_1 is usually so small with respect to H_0 or L_2 that it can be neglected. The most obvious and logical design criterion is the maximum pressure p_2 that can be tolerated at the valve. Neglecting $L_2 = z_\infty$ with respect to L_1 and H_0 implies $p_2 \approx p_G$. The maximum p_2 would then be on the order of $p_{G\infty}$ or, if the damping is small, would be the pressure corresponding to the maximum amplitude of z. With reference to Fig. 15-15c for "mild" damping, this pressure is seen to correspond to $z \approx 2z_\infty - z_0$. With this value of z in (15-27a),

$$p_{G\max} - p_{G\infty} = \frac{\gamma p_{G\infty}}{L_G}(z_\infty - z_0)$$

If the problem is to design the chamber to obtain specifically L_G and A_1, this information may be obtained from (15-27) to (15-31) in terms of specified reservoir conditions, L_T and p_T, maximum allowable pressure, line size (A_2 and L_2), and flow rate Q or line velocity v_2. Equations (15-28) and (15-29) can be combined to give

$$\Delta p = p_{G\infty} - p_{G0} = \frac{\rho g}{g_c}(z_0 - z_\infty) + \left(1 + f\frac{L_2}{D_2}\right)\frac{\rho v_{20}^2}{2g_c} \tag{15-34}$$

With (15-27a), Δz can be eliminated to give

$$\Delta p = \left(1 + \frac{\rho g L_G}{g_c \gamma p_{G\infty}}\right) = \left(1 + f\frac{L_2}{D_2}\right)\frac{\rho v_{20}^2}{2g_c} \tag{15-35}$$

Equation (15-33) is a relation between v_{20} and Δp. It can be noted that Δp or Δp_{max} cannot be specified independently of v_{20}. However, if $p_{G\infty}$ (or p_T), Δp, and v_{20} were given (along with L_2 and D_2), (15-36) could be regarded as a final equation from which L_G could be determined. Equation (15-33) cannot be used for determining Δp as a function of v_{20} even if the parameters $p_{G\infty}$, L_G, and fL_2/D_2 are given, since the initial flow velocity v_{20} is determined by the initial flow conditions in the entire system. It may, therefore, be noted that

$$v_{20}^2 > \frac{2g_c\,\Delta p}{1 + f(L_2/D_2)}$$

is always valid if the problem is to have any practical significance.

Although v_{20} is a system parameter, it may be eliminated from determination of the design variables in terms of these variables by noting that a relation for the dynamics of the problem for $t > 0$ can also be obtained from (15-33), (15-30), and (15-31). After the valve closes, $v_1 = dz/dt$ and $v_2 = (A_1/A_2)v_1$. The transient velocity is then

$$v_{20} = \frac{A_1}{A_1}\,|\overline{v_1}| = \frac{A_1}{A_2}\left|\frac{dz}{dt}\right|$$

which can be obtained from (15-33) as

$$|v_2| = \frac{A_1}{A_2}\left|(z_\infty - z_0)\,\frac{b^2 + \omega^2}{\omega}\,e^{-bt}\sin\omega t\right| \tag{15-36}$$

The 1 subscript will be dropped from the Δz term with the understanding that z and Δz then refer to the chamber. Comparison of (15-36) with (15-33) indicates that v_2 is a damped sine wave similar to, but 90° out of phase with, the pressure (or displacement) shown in Fig. 15-11c. At first this result appears paradoxical: at $t = 0$, $dz/dt = 0$, but v_2 is $v_{20} = v_{2max}$, which is obvious, since it is a given initial condition on the problem. However, only the amplitudes are of interest. As the pressure changes from p_{G0} to $p_{G\infty}$ through Δp, the line velocity is decelerated from v_{20} to zero. The v_{20} and Δp values are then the same as those appearing in (15-35), if v_2 in (15-36) is taken as the initial modulus or magnitude of (15-36); that is,

$$|v_2| = v_{20} = \frac{A_1}{A_2}\,(z_\infty - z_0)\,\frac{\omega^2 + b^2}{\omega} = \frac{A_1}{A_2}\,\omega\,(z_\infty - z_0)\left(1 + \frac{b^2}{\omega^2}\right)$$

$$= \frac{A_1}{A_2}\,\omega(z_\infty - z_0)\left(\frac{\omega_0}{\omega}\right)^2 \tag{15-37}$$

After eliminating $z_\infty - z_0$ in terms of Δp from (15-27a), (15-37) can be combined with (15-35) to give (with $\omega^2 + b^2 \equiv \omega_0^2$)

$$\mathrm{v}_{20} = \frac{A_1}{A_2}\Delta z\,\frac{\omega_0^2}{\omega} = \frac{A_1}{A_2}\,\omega_0\,\Delta z\,\frac{\omega_0}{\omega} \approx \frac{A_1}{A_2}\,\frac{\Delta p_G L_G}{\gamma p_{G\infty}}\,\omega_0 \tag{15-38}$$

In the last approximate form, the ratio ω_0/ω has been taken approximately equal to unity from the following considerations:

$$\frac{\omega_0}{\omega} = \sqrt{\frac{\omega_0^2}{\omega_0^2 + b^2}} = \sqrt{\frac{1}{1 + (b/\omega_0)^2}} = (1 - \zeta)^{-1/2} \approx 1 + \frac{\zeta}{2}$$

if $\zeta^2 \ll 1$ or $\zeta < 1$, where $\zeta \equiv b/\omega_0$ is the damping ratio. If $\zeta^2 \ll 1$, the system is only slightly damped. For this problem

$$\zeta^2 = \frac{[(fL_2/2D_2)(\bar{\mathrm{v}}_2/L)]^2}{(g_c\gamma/\rho L)(p_{G\infty}/L_G - p_{T\infty}/L_T) + 2g/L}$$

which will be less than unity for most practical values of the parameters encountered. The approximate form of (15-38) combined with (15-35) gives

$$1 + \frac{\rho g L_G}{g_c \gamma p_{G\infty}} = \left(1 + f\frac{L_2}{D_2}\right)\frac{\rho}{2g_c}\left(\frac{A_1}{A_2}\right)^2 \frac{\Delta p L_G^2}{\gamma^2 p_{G\infty}}\left[\frac{g_c\gamma}{\rho L}\left(\frac{p_{G\infty}}{L_G} - \frac{p_{T\infty}}{L_T}\right) + \frac{2g}{L}\right]$$

which can be factored to give

$$\left(\frac{A_1}{A_2}\right)^2 = \frac{2\gamma p_{G\infty} L/L_G\,\Delta p}{1 + f(L_2/D_2)}\,\frac{1 + (\rho g L_G/\gamma g_c p_{G\infty})}{1 + (2\rho g L_G/g_c\gamma p_{G\infty}) - (p_{T\infty}/p_{G\infty})(L_G/L_T)} \tag{15-39a}$$

or

$$\left(\frac{A_1}{A_2}\right)^2 = \frac{1}{1 + f(L_2/D_2)}\,\frac{2\gamma p_{G\infty}}{\Delta p L_G}\,\frac{1 + r'}{1 + 2r' - s'} \approx \frac{1}{1 + f(L_2/D)}\,\frac{2\gamma p_{G\infty}}{\Delta p L_G} \tag{15-39b}$$

where

$$r' \equiv \frac{\rho g L_G}{g_c \gamma p_{G\infty}} \qquad \text{and} \qquad s' \equiv \frac{p_{T\infty}}{p_{G\infty}}\,\frac{L_G}{L_T}$$

In nearly all practical cases, $r' < 1$ and $s' < 1$, so $(1 + r')/(1 + 2r' - s') \approx 1 - r' + s' \approx 1$. For most cases, z_∞ will be small enough that $p_{G\infty} \approx p_{T\infty}$. The equivalent overall length $L = H_T + L_2$ and fL_2/D_2 are also known input information. $\Delta p = p_{G\infty} - p_{G0}$ can be determined in terms of the specified maximum overpressure $\Delta p_{\max} = p_{\max} - p_\infty = p_{G\infty} - p_{G0}$ from

Fig. 15-11c. The ratio $\Delta p_{max}/p_\infty = \Delta p/p_T$ can be defined as the design pressure factor P.F. and (15-39b) can be written

$$\frac{D_1}{D_2} = \left\{\frac{2\gamma L/L_G}{[1 + f(L_2/D_2)]\text{P.F.}}\right\}^{1/4} \tag{15-40}$$

This final result relates D_1 and L_G in terms of known or specified information; it does not give either quantity explicitly. However, (15-40) can be used to determine both L_G and D_1. Design of the surge tank requires that its L_G/D_1 ratio is a practical one, nominally $1 \leq L_G/D_1 < 5$; also, practical considerations indicate that $2 < D_1/D_2 < 10$. It is, therefore, possible to choose an initial value of L_G and calculate D_1 (or D_1/D_2) and examine L_G/D_1 to see if the result is practical. The variables have been plotted in Fig. 15-12 to aid in such a procedure. The line for $\Delta p/p$ = P.F. = 1.0 is shown dashed, because the approximation $\Delta z/L_G = \Delta p/\gamma p$ is good only for $\Delta z/L_G \ll 1$, and therefore for $\Delta p/p \ll 1$.

L_G is the height of the gas column and not the overall surge-chamber height L_C, which should be approximately twice L_G.

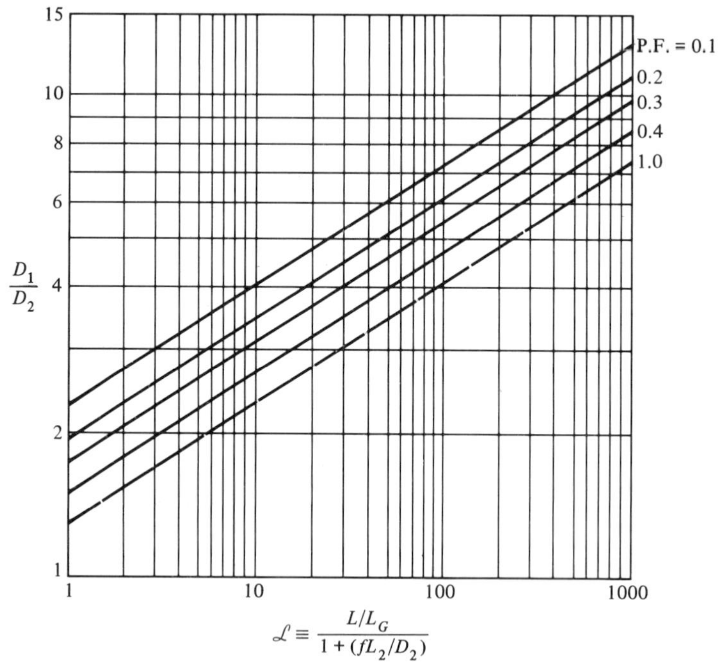

Fig. 15-12. Design-parameter functions for pulsation-damper design.

15.8 ILLUSTRATIVE EXAMPLE 15-2: Surge-Tank Design

A chemical reactor pumping system draws a reagent from a tank in which the level is maintained at 6 ft above the transfer pipe. A 50-ft length of ½-in.-I.D. tubing leads to a high-pressure reactor injection pump. A solenoid-actuated ball valve provides rapid shutoff in case of emergency. The flow rate is 5 gpm of fluid with the same properties as water. Because of the quick-closing valve, a surge chamber is necessary. The size and volume of the chamber will be determined.

The equilibrium p_∞ corresponds to 6 ft of fluid or about 2.6 psig (17.3 psia). The flow rate of 5 gpm through a ½-in.-diameter tube gives $v_{20} = 8.2$ ft/sec. The N_R for a fluid with properties the same as water is then $\sim 3 \times 10^4$, giving $f = 0.023$ for smooth tubing.

The overall equivalent L is 56 ft; $1 + fL/D = 28.6$. Since p_∞ is not large, a relatively large overpressure can be tolerated. The maximum $\Delta p/p_\infty$ that can be used with the above development is unity, which is actually too large for the approximations to be valid. For this case, however, assume $\Delta p/p_\infty = 1.0$. Since D_2 is ½ in., L_G can be taken as approximately 6 in. ($L_C = 12$ in.). Using Fig. 15-12, a first approximation is obtained:

$$\frac{L/L_G}{1 + f(L_2/D_2)} = \frac{112}{28.6} \approx 4 \qquad \frac{D_1}{D_2} = 1.82 \qquad \text{or} \qquad D_1 = 0.91 \text{ in.}$$

The surge chamber is then essentially a capped standpipe about 12 in. high and about 1 in. in I.D. Taking $L_G = 4$ in. gives $D_1/D_2 = 2$ or $D_1 = 1$ in., so the chamber would be 8 in. long by 1 in. in diameter. The Δz variation from the midpoint z_∞ is $\Delta z/L_G = \text{P.F.}/\gamma \approx 0.7$. More accurately from

$$\frac{p_{\max}}{p_{G\infty}} = \left(\frac{L_G}{L_G - \Delta z_{\max}}\right)^\gamma$$

or, equivalently,

$$\frac{\Delta z_{\max}}{L_G} = 1 - \left(1 + \frac{\Delta p}{p_{G\infty}}\right)^{-1/\gamma} = 1 - (1 + \text{P.F.})^{1/\gamma} = 1 - (0.5)^{1/1.4} = 0.39$$

and

$$\frac{p_{\min}}{p_{G\infty}} = \frac{p_{G0}}{p_{G\infty}} = \left(\frac{L_G}{L_G + \Delta z_{\min}}\right)^\gamma$$

or

$$\frac{\Delta z_{\min}}{L_G} = (1 - \text{P.F.})^{-1/\gamma} - 1$$

which does not give a realistic result for P.F. = 1.

The chamber diameter is relatively small. It would also be possible to use a larger one, which would reduce the magnitude of the surge pressure (and therefore reduce P.F.). For example, if a 2-in.-diameter, 4-in.-long tank were arbitrarily specified as a compromise after making the above calculation, the P.F. can then be calculated from (15-40),

$$\text{P.F.} = \frac{2\gamma L/G}{[1 + f(L_2/D_2)](D_1/D_2)^4} \tag{15-40a}$$

For $L = 56$ ft, $L_G = \frac{1}{6}$ ft, the P.F. is 0.135. Since P.F. $= \Delta p/p$ is small, the $\Delta z/z_G$ is also small, $\Delta z/L_G \approx 0.1$, so the approximations are valid. This magnitude of $\Delta z/\Delta_G$ is also more desirable. This second calculation shows the sensitivity of the design to the chamber diameter due to the fourth-power relation.

The natural frequency, ω_0, is determined from (15-30e) as

$$\omega_0 = \sqrt{\frac{g_c\gamma p_G}{\rho L L_G} + \frac{2g}{L}} = \sqrt{\frac{2g}{L}\left[1 + \frac{\gamma}{2}\left(\frac{L_3}{L_G} - 1\right)\right]} \approx \sqrt{\frac{\gamma g L_3}{L L_G}}$$

($\Delta p_T = 0$; therefore, p_T does not appear). Since $p_{G\infty} = \rho g(L_3 - L_G)/g_c$,

$$\omega_0 = \sqrt{\frac{64.4}{56}\left[1 + 0.7\left(\frac{6}{1/6} - 1\right)\right]} = 5.4/\text{sec}$$

or

$$\omega_0 \approx \sqrt{\frac{\gamma g L_3}{L L_G}} = 5.4/\text{sec}$$

giving $T \approx 1.2$ sec. (Note that ω_0^2 is nearly proportional to $1/L_G$.) The damping ratio is

$$\zeta = \frac{b}{\omega_0} = \frac{(fL_2/2D_2)(\bar{v}_2/L)}{\omega_0}$$

which with $\bar{v}_2 = v_2/2$ gives

$$\zeta = \frac{(28.6/2)[8.2/(2)(56)]}{5.4} = 0.194$$

Since $\zeta^2 \ll 1$, the system is lightly damped and $\omega \approx \omega_0$ (within 2 percent).

Note that some means must be provided for charging (or bleeding) the surge tanks to give the correct initial L_G/L_C ratio, so the minimum level z_0 is not less than zero. For many applications it is not desirable to have the gas and liquid in contact; the liquid may absorb the gas, which may be undesirable from process considerations, as well as removing the gas from the chamber. For these applications, commercial surge chambers are available with corrugated or rolling flexible diaphragms and charging

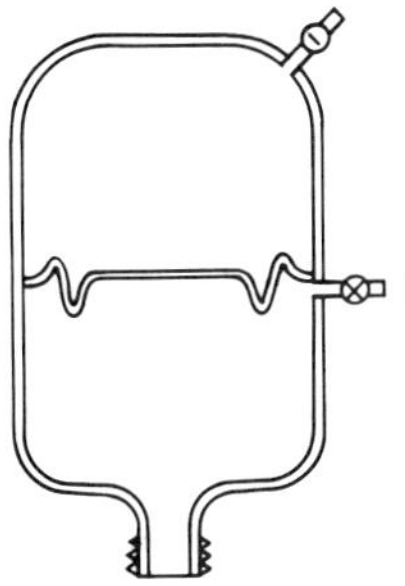

Fig. 15-13. Schematic of the construction of surge tanks.

valves as illustrated in Fig. 15-13. Where the steady-state operating pressures are relatively large (for example, greater than 1 atm), this type of chamber may be used in any orientation.

15.9 PULSATION DAMPER: Circuit Analogs

A pulsation damper is similar to a surge chamber but has a different objective. In many cases it is necessary to obtain a smooth flow from a pulsating or alternating source. This objective is the same as that required for a D.C. power supply. The pulsations must be damped, filtered, or smoothed out. The jolts in a mechanical system are smoothed or "absorbed" by shock absorbers. The parameters involved are analogous. The damping and resonant characteristics of the circuits are similar, and the similarities can be used in a number of ways. Most engineers are familiar with or study the characteristics of transient mechanical and electrical systems. It is advantageous to be able to consider the similarities and use the best, easiest, and most obvious points and manipulation methods from each. Characteristics of pulsation dampers will be developed with this objective.

Consider the simple series circuits of Fig. 15-14. When the mass is at a displacement x from equilibrium, Newton's second law applied to its motion in the x direction gives

$$\frac{m}{g_c}\frac{d^2x}{dt^2} = -c\frac{dx}{dt} - kx + F(t) \tag{15-41}$$

The inertial action corresponding to a change of motion is the $ma/g_c = (m/g_c)(d^2x/dt^2)$. The laminar or linear frictional resistance is $cv = c(dx/dt)$, where c is the Coulomb damping coefficient, assuming that the frictional resistance is of the Coulomb type, that is, directly proportional to v. Some of the work done by the external force in displacing the

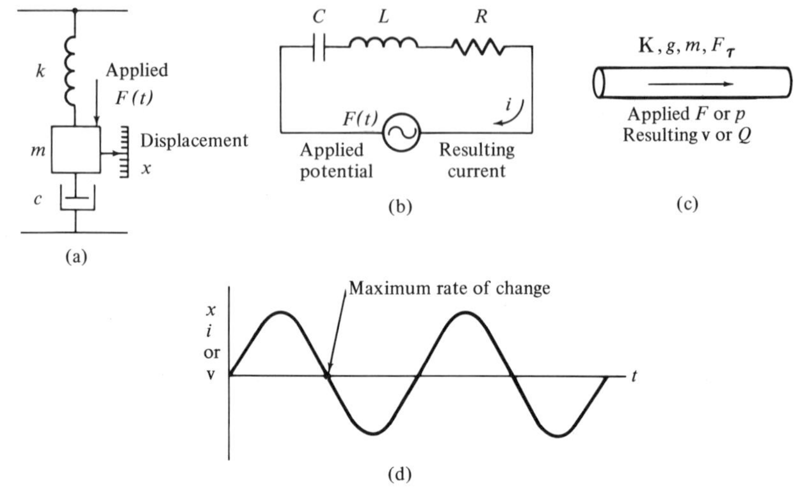

Fig. 15-14. Simple series (a) mechanical, (b) electrical, and (c) hydraulic circuits. (d) Sinusoidal displacement (x or Q) or rate (v or i) as a function of time.

system from its equilibrium position is stored in the spring as potential energy of displacement. This work is reversible. The spring constant k (in units of force/displacement) represents the storage capacity of the spring. $F(t)$ is the externally applied force that produces the motion.

In the electric circuit, the flux is the electric current i, which is the rate of flow of electric charge[2]:

$$i = \frac{dQ_e}{dt}$$

The applied driving force or potential is the electric potential (or voltage) e. An equation for (15-41) can be written for the sum of the "forces" or potentials for each element of the circuit. As a varying flux is produced in the inductor coil, a varying magnetic field is produced which generates the potential. The potential for the coil is then proportional to the rate of change of flux:

$$e_L = L\frac{di}{dt}$$

where L is the inertial or inductive coefficient, the *inductance*.

[2] The subscript e will be used for electric quantities where the notation is otherwise the same, as for electrical or fluid quantities or properties.

As the charge flows into the capacitor (when a current or flux is initiated), a potential charge is built up that depends on the amount of charge accumulated or stored. The potential for the capacitor is then proportional to its charge,

$$e_C = \frac{1}{C} Q_e = \frac{1}{C} \int i \, dt$$

The capacitance C is then defined as the change in charge obtained by a change in potential,

$$C = \frac{\Delta Q_e}{\Delta e_C}$$

The potential required to drive a flux i through a resistance is

$$e_R = iR$$

A potential $e(t)$ applied to the circuit of Fig. 15-14 produces a flux i through the circuit according to the relation

$$e(t) = e_L + e_R + e_C \equiv L \frac{di}{dt} + Ri + \frac{1}{C} i \, dt \qquad \text{(15-42)}$$

Using the relation between charge and flux, (15-42) can be written

$$L \frac{d^2 Q_e}{dt^2} + R \frac{dQ_e}{dt} + \frac{1}{C} Q_e = e(t) \qquad \text{(15-42a)}$$

Equations (15-41) and (15-42a) are identical in form. The sets m/g_c and L, c and R, k and $1/C$ are then analogous. Any solution to a series electric circuit will then apply to a mechanical circuit with equivalent parameters. If the motion is "free," that is, if there is no driving or forcing function $F(t)$ or $e(t)$, the solutions to the above equations are similar to those to (15-4a), as given by (15-6), (15-7), and (15-8). The undamped natural or resonant frequency is $\omega_0 = 1/\sqrt{LC}$, the damped natural frequency is $\omega_n = \sqrt{\omega_0^2 - b^2}$, where $2b = R/L$, or $\omega_n = \omega_0\sqrt{1 - \zeta^2}$ and $\zeta^2 \equiv (\omega_0/b)^2 \equiv R^2C/4L$. If the driving functions are periodic, the response of the system [that is, the solution $i(t)$ or $x(t)$] will depend on the driving frequency or, more specifically, on the ratio of driving frequency to resonant frequency, ω/ω_0. If $\omega = \omega_0$ and the system is driven at its natural frequency, the excursion amplitude (i_{max} or x_{max}) may become extremely large, depending on the magnitude of ζ. The analysis of oscillating systems is treated in more detail in books on vibrations and circuit analysis. Resonance is a consequence of the fact that the inductive and capacitive components are in opposite phase, so that during each half-cycle one enforces the other. In the case of simple U-tube oscillations, the "flux" of liquid (and therefore the velocity, inertia, and kinetic energy) is a maximum when the two legs are at the same level *O-O*. When the flux is zero, the potential energy stored as a consequence of the maximum height difference between the two columns is a maximum. This simple *free* oscillation occurs at the *natural* frequency of the system. If a slight push were applied to the top of one of the columns each time it reached its maximum, it would start down with a greater velocity, as determined by the effect of the impulse of the forcing function on increasing the inertia. Maintaining such a forcing function with only a small amount of work per impulse at this point of each cycle would soon result in an extremely large *resonant* buildup. The oscillation of a mass on a spring as discussed in the introduction of Chapter 10 is a similar example. For these examples it can be observed that the maximum potential energy occurs one quarter of a cycle (or 90°) *after* the maximum kinetic energy. The occurrence of the maximum kinetic energy or flux can then be said to "lead" the occurrence of maximum potential energy (or displacement) by 90°. This behavior can also be observed in the A.C. electric circuit. When the voltage is just starting to increase, there is no charge in the capacitor; the current can then flow in easily. As the voltage reaches its peak, the capacitance is charged and there is little current flow. During periodic variations of e, the current then runs ahead of, or "leads," the voltage in the capacitance by one quarter of a period or 90°. In the inductor, the maximum potential occurs when the maximum rate of change of current occurs (at $i = 0$, Fig. 15-14d). The voltage in the inductance then lags the current by one quarter of a period (90° for periodic motion). The resistance has no effect on the phase, so e_R is in phase with i.

For a series circuit, the same flux must pass through all elements; the flux is therefore a common reference base and is the same for all elements. The voltages (force or pressures) across each element are not in phase and therefore add vectorially rather than linearly. If the common current is considered as a base reference, the relation among the voltage components may be indicated by a vector diagram, as indicated in Fig. 15-15a. The variation of the potential components with time is obtained by rotating the vector diagram ("phasor") about its origin at the radial frequency ω and tracing the ordinate component of the vector along the time axis, as indicated in Fig. 15-15b. The subscript m on the vectors implies maximum positive amplitude. Since the i is common, the resultant circuit voltage is the sum of the vector components,

$$e_m = \sqrt{e_R^2 + (e_L - e_C)^2} = i_m\sqrt{R^2 + (X_L - X_C)^2} \qquad (15\text{-}43)$$

which is, of course, just the amplitude of the applied voltage $e(t)$ across the

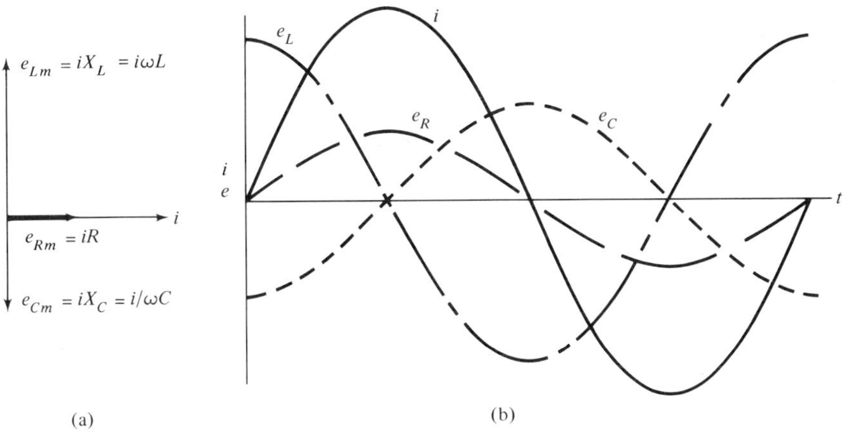

Fig. 15-15. (a) Potential-vector-phase relations; (b) components as a function of time.

circuit. From the vector diagram and the above equation it is obvious that, for a given e, the current (displacement, flow) will be a maximum when $e_L = e_C$ or when the capacitive reactance $X_C = 1/\omega C$ is just equal to the inductive reactance $X_L = \omega L$, or when $\omega = \omega_0 = 1/\sqrt{C}$.

Consider a mass hanging on a spring: If the mass is pulled down and released, it will oscillate with a natural frequency $\omega_0 = \sqrt{g_c k/m}$ if the damping ζ is small. If a small oscillating force with period ω_0 is applied to the mass, quite large resonant oscillations will occur. The inertia of a mass undergoing a sinusoidal periodic motion is a maximum at maximum acceleration from zero velocity. The inertial action I_m then leads the velocity. For the spring, there is no force at maximum velocity (zero displacement) and F_s develops subsequently as the displacement increases and velocity decreases, so the F_s can be represented as following the velocity. The phasor vector diagram for the series mechanical system is then similar to the potential phasor diagram Fig. 15-15a.

The series fluid circuits considered have been represented by the equation

$$\frac{\rho AL}{g_c}\frac{dv}{dt} + Nv + \left(\frac{\gamma p}{L} + \frac{\rho g}{g_c}\right)A\,\Delta z = F(t) = p(t)A \qquad (15\text{-}44)$$

where $\Delta z = \int v\,dt$. This equation resembles (15-43), but it is not quite equivalent. First, the current i at any instant is the same in any element of the circuit, while v depends on the cross-sectional area. However, $Q = A$v does possess the same flux equivalence as i. Second, the force also is a function of area and is therefore not similar to the potential e. The pressure, however, can be considered as equivalent to e. Inserting $\text{v} = Q/A$ and dividing by A, (15-44) becomes

$$\Delta p_I + \Delta p_R + \Delta p_C \equiv \frac{\rho L}{g_c A}\frac{dQ}{dt} + \frac{NQ}{A^2} + \left(\frac{\gamma p}{LA^2} + \frac{\rho g}{g_c A}\right)\int Q\,dt = p(t) \tag{15-44a}$$

where Δp_I represents the Δp due to the inertia of the fluid mass, Δp_R is resistive, and Δp_C is capacitive. Analogous inertance $I = \rho L/g_c A$, resistance $F = N/A^2$, and capacitance $C_\kappa = LA^2/\gamma p = V/\gamma p$ and $C_g = g_c A/\rho g$ quantities can be defined for the fluid circuit and the methods of electric-circuit analysis can be applied directly to fluid circuits.

All the fluid circuits previously considered in this chapter have been series circuits. The form of the final equations indicates that the equivalent inertance for series circuits is just the sum of the individual inertances, which also applies to addition of inductances in an electric circuit. The resistances also add linearly. The inductive reactance in an electric circuit element is a function of the frequency and is just $X_L = \omega L$; the inertial reactance of a fluid element is $X_I = \omega I$. The R and X_L reactances are in general termed *impedances*; they impede the flux (or flow of charge) produced by an applied potential. On the other hand, a capacitance admits a charge and so is termed an admittance.[3]

The compressive C_κ and gravitational C_g capacitances add according to

$$\frac{1}{C} = \frac{1}{C_\kappa} + \frac{1}{C_g}$$

For a D.C. potential, the capacitance is an open circuit (no flux possible), or it is an infinite impedance (zero admittance). As the frequency is increased, the impedance to an A.C. flux decreases or the admittance increases, so capacitive admittance is proportional to the frequency ω. In a series circuit, each capacitance must be charged before the next, so the admittance decreases with the number of capacitances in series. On the other hand, for capacitances in parallel, the admitting capacity is obviously increased, so capacitances and capacitive admittances in parallel add. The capacitive admittance is therefore $Y_C = \omega C$. Reactance X and admittance Y are conceptually reciprocal quantities, as they are mathematically $X = 1/Y$. In general, reactances add vectorially in series circuits, and admittances add vectorially in parallel circuits. The vector addition is represented in Fig. 15-15a; the current i is common to each potential component vector and may be removed to give only a reactance diagram. In parallel circuits, the potential (or p) is common with the flux dividing

[3] Impedance and admittance are included in a general term immittance.

into the various branches according to the admittance of each, so like admittances add linearly, and unlike add vectorally. The phasing of admittances is also inverse to that of reactances, since they are reciprocal quantities. These concepts are particularly important for use in combining series-parallel circuits.

The compressive capacitance for a compressible fluid system is defined as C = gas volume charge/change in potential; therefore, for given capacitor volume, $C = \Delta V/\Delta p$. In terms of the compressibility of the fluid,[4]

$$\kappa \equiv \frac{\Delta p}{\Delta V/V} \qquad C_\kappa = \frac{V}{\kappa}$$

If the compressed fluid is a perfect gas, and if the compression process is assumed to be isentropic, (15-27) can be used for relatively small $\Delta V/V$ to give

$$C_\kappa = \frac{\Delta V}{\Delta p} = \frac{V}{\gamma p}$$

TABLE 15-1

CIRCUIT PARAMETERS

	Mechanical	*Electrical*	*Fluid*
Applied potential	Force F	Electrical potential or voltage e	Pressure p
Flux	Velocity $v = dx/dt$	Current i	Flow rate Q
Resistance	Friction C, Coulomb damping coefficient	Resistance R	Friction $R = N/A^2$ $R_{lam} = 8\pi\mu L/A^2$ $R_{turb} = (\rho fL/2AD)(\bar{v}/g_c)$
Inertance	Mass m	Inductance L	Mass inertance $I = \rho L/g_c A$
Potential-energy (storage capacity)	Spring (k spring constant)	Capacitance $C = \Delta Q_e/\Delta e$	Capacitance $C = \Delta V/\Delta p$ $C_\kappa = V/\kappa,\ g_c V/\rho a^2,\ V/\gamma p$ $C_g = g_c A/\rho g$

Reactance	
Capacitive	$X_C = 1/\omega C$
Inductive	$X_L = \omega L$
Admittance	
Capacitive	$Y_C = \omega C$
Inductive	$Y_L = 1/\omega L$

[4] As used here, a positive Δp yields a positive charge ΔV, so κ requires a positive sign for a capacitor charge.

The gravitational capacitance due to the liquid-level-rise potential is

$$C_g = \frac{\Delta V}{\Delta p} = \frac{\Delta V}{\rho g \, \Delta z / g_c}$$

For a cylindrical chamber, $\Delta V = A \, \Delta z$, giving

$$C_g = \frac{g_c A}{\rho g}$$

These last two capacity quantities are those which appear in (15-44a). These capacitances and the other circuit parameters are tabulated in Table 15-1.

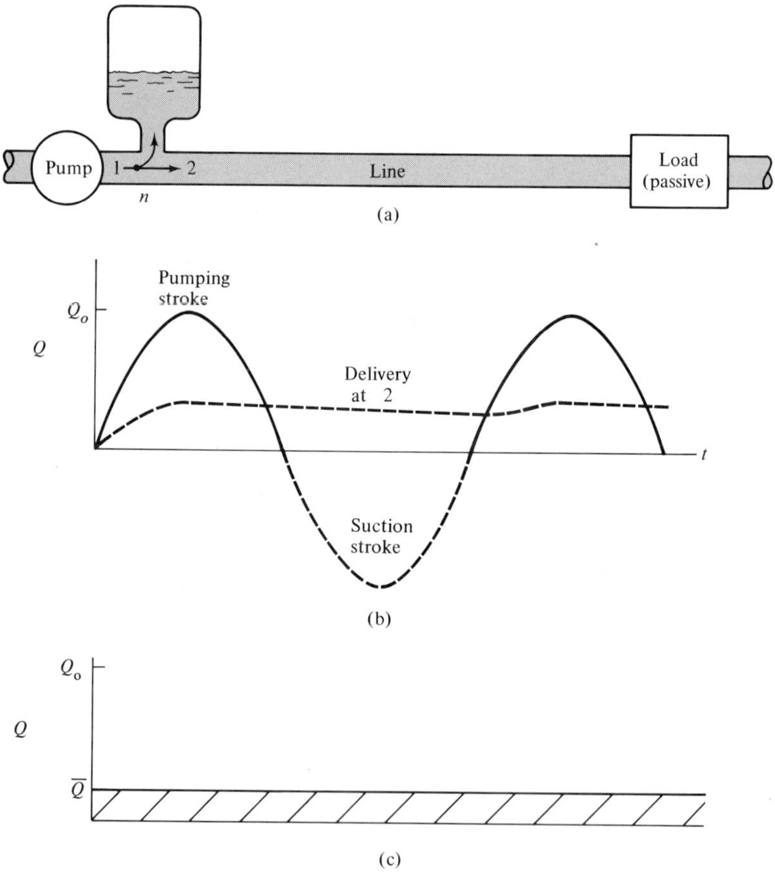

Fig. 15-16. (a) Pulsation damper installed following an oscillation source. (b) Pump output flow rate at 1; the dashed line is the flow rate at 2. (c) Desired output into the line at 2.

PULSATION DAMPER

A system consisting of a pulsing pump output into a delivery line and load is sketched in Fig. 15-16a. The worst case of a pulsating pump delivery is presented by a single-cylinder, single-acting pump with one delivery stroke and one suction stroke per revolution. The flow output of such a pump is indicated in Fig. 15-16b. For this analysis, the time between each delivery stroke will be taken as one period $T = 2\pi/\omega$. The flow rate can then be represented by the relation

$$Q = Q_0 \sin \omega t \tag{15-45}$$

The average flow rate during one period is

$$\bar{Q} \equiv \frac{1}{T}\int_0^{T/2} Q\,dt = \frac{Q_0}{T}\int_0^{T/2} \sin \omega t\,dt = \frac{Q_0}{\pi}$$

or the pulsation magnitude Q_0 is π times the average delivery rate $\bar{Q}$.

The desired outflow output at 2, with a minimum of pulsating or A.C. ripple, is indicated in Fig. 15-16c. The object of the addition of the pulsation damper is then to smooth out the pulsations to a desired or predicted value by the addition of the capacitance. The smoothed performance can be specified in terms of a *ripple factor* defined as $\alpha \equiv Q_r/\bar{Q}$, where Q_r is the magnitude of the A.C., which is superimposed on the average flow rate $\bar{Q}$. The design problem is then to determine the capacitance for a specified value of α.

Without the pulsation damper, the circuit can be treated by methods developed previously in this chapter. The equivalent electrical circuit for the line and load without the capacitance is shown in Fig. 15-17a. Addition of the pulsation damper adds a significant complexity. With it, the circuit has a parallel branch with a branch point or "node" at n (Fig. 15-16a). The pump is then feeding one branch with a series inertance and resistance and one branch with a capacitance and inertance. The inertance of the pulsation damper liquid mass can be neglected with respect to other circuits reactances, as was demonstrated in the case of the surge chamber. Similarly, the liquid capacitance C_g can be neglected with respect to the gas compression capacitance $C_\kappa = V/\gamma p$. The simplified equivalent electrical circuit is then as shown in Fig. 15-17b. The line and load resistance and inertance (or inductance) are shown lumped together, which will be shown to be a satisfactory procedure. It has also previously been shown that resistances and inductances or inertances (or resistances and reactances) add in a series branch. The impedance of the resistance–inertance branch is $Z_0 = \sqrt{R^2 + X_I^2}$. The current through the circuit is $Q_0(t) = Q(t)$, as given in (15-45). The pressure $p_0(t) = Q_0 Z_0$. In the modified circuit of Fig. 15-17b, the flow rate $Q(t) = Q_0$ delivered by the pump is the same.

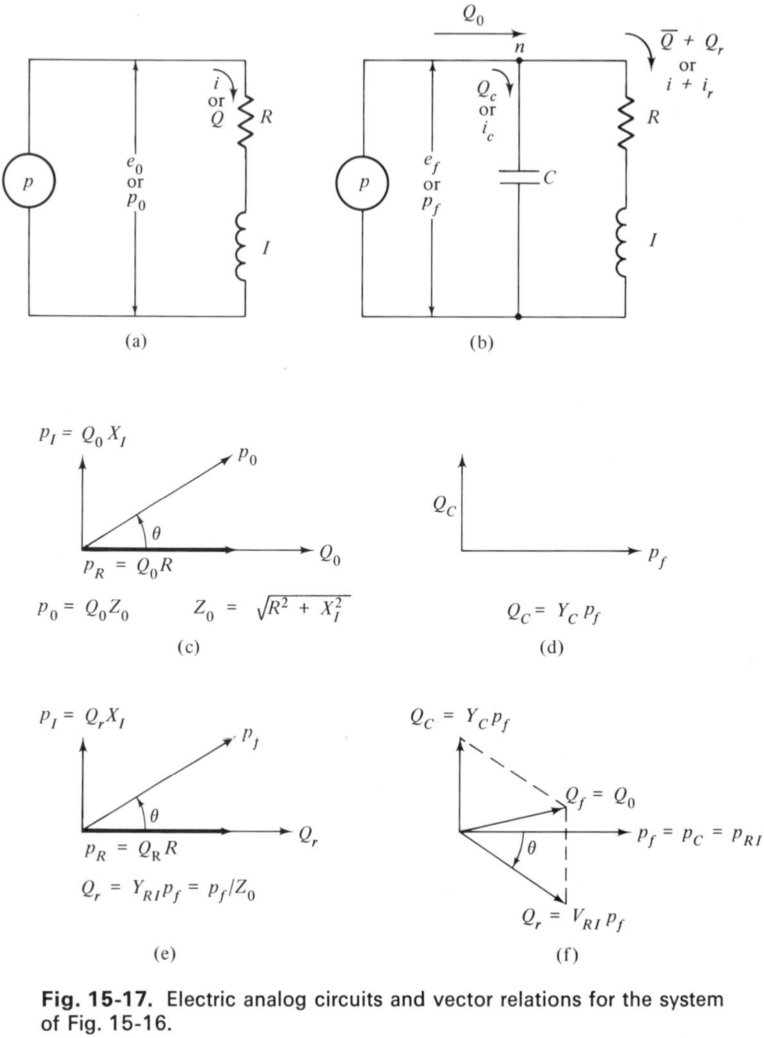

Fig. 15-17. Electric analog circuits and vector relations for the system of Fig. 15-16.

The object of the capacitance is to absorb the energy (and volume or charge) of the pump delivery stroke and then to discharge smoothly into the load circuit as the delivery decreases (on the downhill portion of the curve of Fig. 15-16b) and during the suction stroke. Some of the flow will also go into the load circuit during the delivery stroke. The net flow at 2 is then indicated by the dashed line in Fig. 15-16b, for a relatively small amount of smoothing capacitance.

At this point, two important facts are readily apparent: The flow (current) into the capacitance is not in phase with that into the load, but leads it, and the admittance Y_C of the capacitive branch of the parallel circuit must be much greater than that of the load branch to allow easy charging during the pumping surge to the extent necessary for discharge during the rest of the cycle to provide the required smoothing effect. It may also be noted that, although the supply flow (Q_0) to both circuits is the same, the magnitude of the pressure (voltage) of the original circuit p_0 is not the same as the final p_f. Also, p_f is common or the same for both branches of the parallel circuit. In the final configuration (Fig. 15-17b) the flow is shown as the D.C. value $\bar{Q}$ plus a ripple component Q_r. The load branch is the same for both circuits, or its impedance is Z_0 for both cases.

The phase diagram for the pressures (voltages) in the two elements of the original series load circuit is shown in Fig. 15-17c. In the parallel circuit, the flow splits into the two branches with the pressure (voltage) remaining common to each. The phase diagram for the capacitance branch is shown in Fig. 15-17d. The load branch has the same configuration as the original circuit, but the magnitudes are different; $p_f \neq p_0$, as shown in Fig. 15-17e. In fact, since $Y_C \gg Y_0$ or $Z_C \ll Z_0$, the final p_f that must be developed by the pump for the smooth circuit is much less than p_0,

$$\frac{p_f}{p_0} \approx \frac{Z_C}{Z_0} = \frac{1}{Y_C Z_0} \tag{15-46}$$

The p_f for Fig. 15-17d and *e* are the same common value, so d and e can be superimposed to show the relation between the flow (current) components Q_C and Q_r. The final flow Q_0 is then the vector sum of Q_C and Q_r and indicated on Fig. 15-17f. (In reality, Q_r is much smaller than Q_C.)

The ripple factor α was defined as

$$\alpha \equiv \frac{Q_r}{\bar{Q}} = \frac{Q_r}{Q_0}\frac{Q_0}{\bar{Q}} \tag{15-47}$$

In the parallel circuit Q_r is the A.C. through the resistance–inductance branch. In the original circuit, Q_0 is the A.C. through the resistance–inductance branch $Q_0 = p_0 Y_0 = p_0 Z_0$, and $Q_r = p_f Y_{\mathrm{RI}} = p_f/Z_0$ since $Z_{\mathrm{RI}} = Z_0$. Therefore,

$$\alpha = \frac{Q_0}{\bar{Q}}\frac{Q_r}{Q_0} = \frac{Q_0}{\bar{Q}}\frac{p_f}{p_0}$$

Defining $Q_0/\bar{Q} = q'$ ($q' - r$ for the example case considered) and noting that $p_f = Q_0 Z_f = Q_0/Y_f$ and $p_0 = Q_0/Z_0 = Q_0 Y_0$ gives

$$\alpha = q'\frac{Q_r}{Q_0} = q'\frac{p_f}{p_0} = q'\frac{Z_f}{Z_0} = q'\frac{Y_0}{Y_f} = \frac{q'}{Y_f Z_0} \tag{15-48}$$

or

$$\frac{p_f}{p_0} = \frac{Z_f}{Z_0} = \frac{Q_f}{Q_0}$$

The last form for α is the most satisfactory because Z_0 is directly evaluated for the original series circuit, and Y_f is directly evaluated for the final parallel circuit. As seen from Fig. 15-17f, since p_f is common, the admittances Y_C and Y_{RI} will add vectorially in the same manner as the flows. This addition will be shown directly by determining $Q_f \equiv Q_0$ for Fig. 15-17f:

$$\begin{aligned} Q_0^2 &= (Q_C - Q_r \sin\theta)^2 + (Q_r \cos\theta)^2 \\ &= p_f^2(Y_C - Y_{RI}\sin\theta)^2 + p_f^2(Y_{RI}\cos\theta)^2 \\ &= p_f^2 Y_f^2 \end{aligned} \tag{15-49a}$$

or, with $Y_{RI} = 1/Z_0$ and using the relations for θ given in Fig. 15-17,

$$Y_f^2 = \left(Y_C - \frac{X_I}{Z_0}\right)^2 + \left(\frac{R}{Z_0}\right)^2 \tag{15-49b}$$

The product $Y_f Z_0$ appearing in α (15-48) is then

$$Y_f Z_0 = \left(Y_C Z_0 - \frac{X_I}{Z_0}\right)^2 + \left(\frac{R}{Z_0}\right)^2 \approx Y_C Z_0 \tag{15-50}$$

The approximation is added here as a consequence of a comparison with (15-46) and (15-48), which could have been made earlier. This approximation is seen to be valid at this point by comparing orders of magnitude of the quantities within the radical. The design problem is to determine C or $Y_C = \omega C$ for a given α and initial circuit parameters. Using (15-50) with (15-48) to solve for the dimensionless quantity $Y_C Z_0$,

$$\left(Y_C Z_0 - \frac{X_L}{Z_0}\right)^2 + \left(\frac{R}{Z_0}\right)^2 = \left(\frac{q'}{\alpha}\right)^2$$

or

$$Y_C Z_0 = \frac{q'}{\alpha}\left[1 - \left(\frac{\alpha R}{q' Z_0}\right)\right] + \frac{X_I}{Z_0} \approx \frac{q'}{\alpha} \tag{15-51}$$

because $q' > 1$, $\alpha \ll 1$, and R/Z_0 and X_I/Z_0 are < 1. The approximation is quite good even for a rather large value of α of approximately 0.1 (the error is only on the order of 1 or 2 percent). The value of the capacitance is then

$$C = \frac{q'}{\alpha\omega Z_0} \tag{15-52}$$

The gas volume of the pulsation damper is then obtained from the approximation $C = V/\gamma p$ from Table 15-1,

$$V = \frac{q'\gamma\bar{p}_f}{\alpha\omega Z_0} \tag{15-53}$$

The pressure $\overline{p_f}$ is the D.C. pressure in the circuit with insertion of the pulsation damper. The pump delivery pressure is $\overline{p_f} + p_r \approx \overline{p_f}$, which is the D.C. pressure drop through the circuit resistance only $(\overline{p_f} = Q_0 R)$. In an actual circuit, Q_0 and R would be known, or R would be calculated for the known load configuration or specification. Otherwise, R could be determined from $R = p_f/Q_0$, if the D.C. pressure could be determined for a given load Q_0.

In connection with the above system it is interesting to note that if $R \ll X_I$, $Y_C Z_0 = \omega^2 LC = (\omega/\omega_0)^2$, where ω_0 is the resonant frequency of the I-C loop "tank." If $R \neq 0$, $Y_C Z_0 = \omega^2 IC(1 - \zeta^2) = (\omega/\omega_n)^2$.

For a gas system with a pulsating delivery, a cavity resonator can be used as a branch capacitance chamber to dampen the pulsations and acoustic waves from the pulsating compressor. The storage tank on a reciprocating compressor system usually provides this added service.

15.10 ILLUSTRATIVE EXAMPLE 15-3: Pulsation-Damper Design

A two-stroke, single-acting metering pump driven at 400 rpm supplies fuel oil with a specific gravity of 0.86 and a viscosity of 6 cP to a burner at a rate of 10 liters/hr. The supply line is 5 m long and 5 mm in I.D. It is desired to limit the pulsations to 1 percent of the mean flow at the burner inlet. The burner is a 0.5-mm nozzle with the stream atomized against an impingement plate. What is the minimum size of pulsation damper for this job?

The two-stroke, single-acting pump has two delivery strokes per revolution. Its discharge curve, $Q(t)$, resembles that of Fig. 15-16b with an additional delivery pulse where the suction gaps occur. The fluctuation parameter is then $q' \equiv Q_0/\bar{Q} = \pi/2$. The inertance is[6]

$$I = \frac{\rho L}{g_c A} = \frac{(860)(5)}{(\pi/4)(0.005)^2} = 2.19 \times 10^8 \, \frac{\text{N-sec}^2}{\text{m}^5}$$

and the inertial reactance is

$$X_I = \omega I = 2\,\frac{400}{120\pi} = 4.64 \times 10^8 \, \frac{\text{N-sec}}{\text{m}^5}$$

The factor 2 enters because there are two delivery cycles per revolution. The load resistance is the pipe friction resistance plus the dynamic pressure corresponding to the nozzle velocity. The Reynolds number in

[6] All quantities have been converted to the mks system for calculations.

the pipe is $N_R = \rho v D/g_c\mu$. The pipe velocity is 14.2 cm/sec (and the nozzle velocity is 1420 cm/sec). The pipe Reynolds number is then

$$N_R = \frac{(860)(0.142)(0.005)}{(0.006)} = 102$$

so the pipe flow is laminar. The pipe resistance is then

$$R_P = \frac{8\pi\mu L}{A^2} = \frac{128\mu L}{\pi D^4} = \frac{(128)(0.006)(5)}{\pi(0.005)^4} = 1.96 \times 10^9 \frac{\text{N-sec}}{\text{m}^5}$$

The nozzle Reynolds number is 10,200, so the nozzle load resistance can be taken as the dynamic pressure. The nozzle resistance, with $v_{noz} = 100 \cdot v_{pipe}$ and $Q = 10^{-5}/3.6\ \text{m}^3/\text{sec}$, is

$$R_{noz} = \frac{\rho v_n^2}{2g_c Q} = \frac{(860)(0.142)^2(3.6)}{2} 10^5 = 3.1 \times 10^6 \frac{\text{N-sec}}{\text{m}^5}$$

The total series load resistance is then essentially that of the pipe, $R = 1.96 \times 10^9$ N-sec/m^5. The load impedance is

$$Z_0 = 10^9\sqrt{(1.96)^2 + (0.464)^2} = 2.02 \times 10^9\ \text{N-sec/m}^5$$

which indicates that $Z_0 \approx R$ (+3 percent). Note that the load resistance is primarily pipe friction resistance, which may be expected for a passive load when the pipe friction is laminar.[7] Also $Z_0 \approx R \approx 2 \times 10^9$ N-sec/m^5.

The capacitance gas volume can now be calculated from (15-13):

$$V = \frac{q' r p_f}{\alpha Z_0} = \frac{q'\gamma \bar{Q} R}{\alpha Z_0}$$

and since $R \approx Z_0$,

$$V \equiv \frac{q'\gamma\bar{Q}}{\alpha\omega} = \frac{(\pi/2)(1.4)(10^{-4})}{(0.01)(2.12)(3.6)} = 2.9 \times 10^{-3}\ \text{m}^3 \qquad (2.9\ \text{liters})$$

The value of γ has been taken as 1.4 for a representative diatomic gas (air, nitrogen, and so on).

If a cylindrical tank is chosen with the L and D of the *gas* volume equal,

$$L_G = D = \left(\frac{4}{\pi} V\right)^{1/3} = 15.5\ \text{cm}$$

[7] Passive implies fixed resistive, inductive, and capacitive elements. An *active* element would be a turbine, motor, or hydraulic piston, for which the load characteristics would have to be known to determine R, I, and C at the operating flow rate.

which is a satisfactory size. If a diaphragm type of pulsation damper is chosen, the diaphragm should be close to the bottom of the capacitance to minimize the liquid volume therein. The volume of the pipe is only about 1 liter. If the fluid volume in the capacitance chamber is on the same order, its inertance must be considered. The effect would be to add a series inertance in the parallel capacitance branch of the circuit and thereby to decrease the damping. The effects of this and other damping circuits are considered further in Problems 15-27 through 15-30.

Methods of handling circuits and damper or filter design can be found in many reference texts on electric-circuit analysis and in particular in reference 3.

15.11 SUMMARY

1. Simplified or approximate analyses of oscillating systems usually require linearization of motions or parameters which are functions of displacements or fluxes (velocities). The validity of these approximating assumptions should be checked with the solutions obtained.
2. In the material presented on resonant oscillations and surge and pulsation damping, only the primary or fundamental frequency has been considered. However, if the wave form is not a perfect sine wave, it can be considered as a Fourier series of sine and cosine components with various amplitudes and frequencies, the largest amplitude and lowest frequency of which is the fundamental. Since the propagation and damping actions are inversely proportional to frequency, the higher frequencies will cause the least trouble, so it is usually sufficient to design for the fundamental.
3. Methods of generalized circuit analysis are useful and helpful in handling problems where transient oscillations occur.
4. Important and interesting studies of fluid transients is involved in the design of fluid amplifiers. These units allow complete power control and actuation with only a single fluid power source; electric and electronic secondary control systems and power sources are eliminated. Fluidic devices also have the advantage that no moving control parts are necessary. All the control signals, and so on, can be accomplished with fluid jets. The fundamental fluid mechanics of jets, entrainment vortex flows, and separation is involved in the design and analysis of these jet devices. Oscillators, computer logic devices, rectifiers, and amplifiers are some of the circuit operations possible with these pure fluid systems.

   The methods of circuit analysis presented in this chapter, and the fundamental fluid mechanics in the preceding chapters, provide the background required for study and investigations in these and other current and future fields of interest.

REFERENCES

1. J. Parmakian, *Waterhammer Analysis*, Prentice-Hall, Inc., Englewood Cliffs, N.J., 1955.
2. R. S. Sandford, *Physical Networks*, Prentice-Hall, Inc., Englewood Cliffs, N.J., 1965.
   Methods of deriving circuit parameters for electric, mechanical (vibrational and rotational), fluid acoustic, and coupled systems are given, and methods of analysis are developed.
3. E. M. Sabbagh, *Circuit Analysis*, The Ronald Press Company, New York, 1961.

PROBLEMS

15-1 Discuss the physical reasons for the fact that the period of vibration of the U-tube manometer fluid is independent of both tube cross-sectional area and fluid density.

15-2 Calculate the period and natural frequency of a U-tube water manometer with a water column 2 ft long.

15-3 Using the principle of conservation of energy, formulate the equation for the undamped oscillation of the inclined manometer illustrated. L and H_0 are the zero-pressure equilibrium positions and R is the scale reading. The tube and well areas are A_1 and A_2 and the angle of inclination is α. Show that the period of oscillation is

$$T = 2\pi\sqrt{\frac{L + R}{g[(A_1/A_2) + \sin\alpha]}}$$

(Consider the manometer in two parts, m_1 and s_1, the mass and oscillation displacement in the inclined tube, and m_2 and z, the mass and vertical oscillation displacement in the well.) Find the period for a water manometer with $A_1/A_2 = 100$, $\sin\alpha = 1/5$, $L = 6$ cm, and $R = 10$ cm.

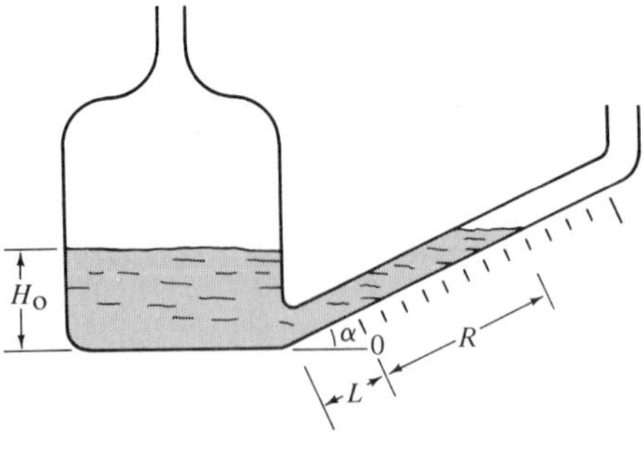

Fig. P15-3.

15-4 For a damped oscillatory motion as defined in (15-7b), the logarithm of the ratio of two successive amplitudes separated by one period is defined as the log decrement. Show that this log decrement is

$$\ln \frac{z_{n+1}}{z_n} = \frac{2\pi b}{q}$$

15-5 For the undamped case for $b^2 = \omega_0^2/2$, (a) find q/ω_0 (the damped to undamped frequency ratio), (b) plot the response curve z/z_0 versus either qt or bt, and (c) determine the time for the oscillation amplitude to drop 1 percent of its initial value. (d) Find the ratio of the first amplitude peak to the initial value; that is, find z_1/z_0 for $bt = 2\pi$. (e) Find the first overshoot amplitude ratio, that is, the amplitude ratio at $bt = \pi$, and comment on the result.

15-6 Find the percentage difference between the natural frequency ω_n and the resonant frequency ω_0 for the mercury manometer of Example 15-1.

15-7 Given two reservoirs as illustrated, find the undamped natural frequency and the damping parameter. Describe the oscillating motion.

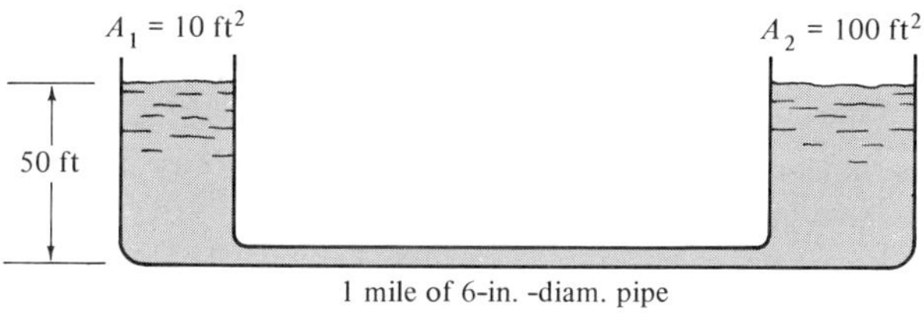

Fig. P15-7.

15-8 For a valve with a dissipation coefficient K in the line between two reservoirs, show that the damping parameter is

$$b = \frac{|\bar{V}_2|}{4\Sigma L}\left(f\frac{L_2}{D_2} + K + 1\right)$$

where the exit kinetic energy dissipation is also taken into account.

15-9 For the oscillation between reservoirs which do not have vertical walls (see sketch) show that the results are the same as those given in (15-13). (For any reservoir, the assumption that $\Delta z/L \ll 1$ must also imply $\Delta A/A \ll 1$ as z varies during the transient condition.)

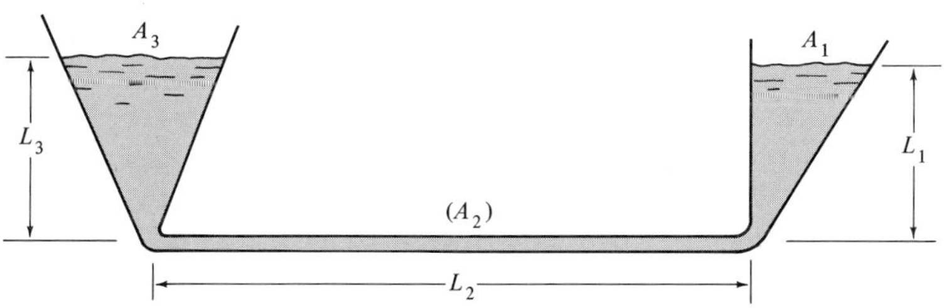

Fig. P15-9.

15-10 For the oscillation between reservoirs sketched in the figure, show that the results are the same as given in (15-13).

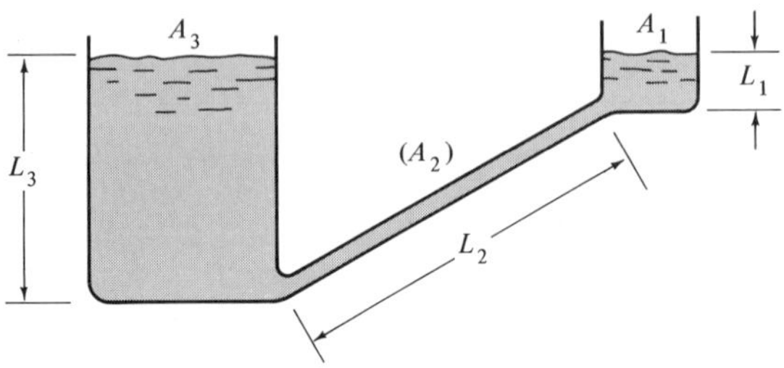

Fig. P15-10.

15-11 From the results of the above two problems, show that the results of (15-13) are applicable to a general reservoir configuration as sketched.

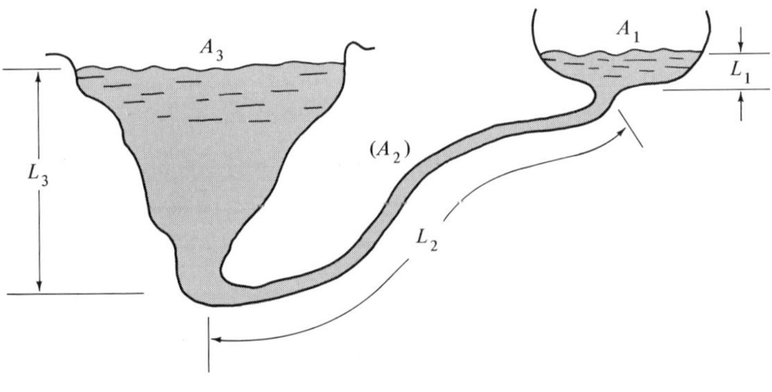

Fig. P15-11.

15-12 Calculate the natural cavity resonant frequency for a 1/16-in.-diameter 1-ft-long tube connected to a 1-in.-diameter ⅛-in.-long cavity. Assume that the mean pressure is 10 psia and that p_0 corresponds to 5 in. of water. Also calculate x_0 and u_0. Examine and discuss the validity of the assumptions made in the derivation and calculations. What is the organ pipe frequency?

15-13 For water (E_B = 320,000 psi), find the wave-propagation speed. For water flowing in a pipe at 50 ft/sec, what is the pressure rise when a valve is suddenly closed (assuming the pipe is rigid)? What is the force on the valve in a 1-in.-I.D. pipe?

15-14 Equation (15-21a) can also be determined as

$$p_2 - p_1 = \Delta p = \rho_1(\mathrm{v}_1 + \mathrm{v}_w)\mathrm{v}_1 = \rho \mathrm{v}_1 \mathrm{v}_w\left(1 + \frac{\mathrm{v}_1}{\mathrm{v}_w}\right)$$

Calculate Δp for Problem 15-13 and compare the results. What is the relative density increase?

15-15 Find the velocity of propagation of an acoustic wave in water contained in a steel and a copper pipe ½-in. in I.D. by 1/16-in. wall. What is the increase in pipe stress in each case if the velocity is initially 20 ft/sec?

15-16 Show that for the system of Fig. 15-15a, ω_0 in (15-30e) can be expressed

$$\omega_0^2 = \frac{g_c \gamma p_T}{\rho L}\left(\frac{1}{L_G} - \frac{1}{L_T}\right) + \frac{2g}{L}\left[1 + \frac{\gamma}{2}\left(\frac{L_3}{L_G} - 1\right)\right]$$

15-17 In Table 15-1, the units of fluid resistance R are determined by $R = \Delta p/\Delta Q$. What are the dimensions of R in terms of the fundamental dimensions F, M, L, and t? Do the units of R_{lam} and T_{turb} check with $\Delta p/\Delta Q$?

15-18 What are the units of fluid inertance and capacitance as given in Table 15-1?

15-19 Determine the inertance and acoustic capacitance for a hydraulic system ($E_B = 3 \times 10^5$ psi) with a ½-in.-I.D. line 10 ft long which then narrows to a ¼-in.-I.D. line 20 ft long. Tell why the capacitance and inertance (or the corresponding admittances or impedances) must be considered to be in parallel rather than in series.

15-20 Determine the inertance and compressive capacitance (C_κ) for a ½-in.-I.D. hydraulic line 10 ft long which then branches into two ¼-in.-I.D lines 10 ft long ($E_B = 3 \times 10^5$ psi.)

15-21 Draw the equivalent circuit and phase diagrams for the simple U-tube manometer of Fig. 15-1. Determine the equations for its resonant and natural frequencies.

15-22 Using the electric-circuit analog method, sketch the circuit and determine the resonant and natural frequencies of the mercury-manometer problem in Example 15-1.

15-23 Using the electric-circuit analogy, sketch the circuit and determine the relation for the resonant frequency of the cavity resonator of Section 15-5.

15-24 In Example 15-3, what is the maximum pressure developed by the pump before and after installation of the pulsation damper?

15-25 The simplest type of pulsation damper for the problem in Example 15-3 would be a cylindrical tank, say, 16.5 cm in diameter, with a pipe connection screwed into a tee in the line. Oil absorbs very little air, so a diaphragm is not necessary for this application. If an *open* pulsation damper tank were connected to the line in this manner, calculate the necessary overall length of the tank and the volume of oil in the tank corresponding to the average steady-flow conditions.

15-26 For the pulsation damper tank diameter of 16.5 cm in Problem 15-25, what would be the level variation in the tank during the pumping cycle? (What is $\Delta z/L_G$?)

15-27 For a pulsation damper where the liquid volume in the tank is such that its inertance is not small compared with the load inertance, sketch the equivalent electrical circuit diagram and the phasor diagram corresponding to Fig. 15-17c through f.

15-28 It is possible that a circuit such as that of Example 15-3 would yield a capacitance volume that is impractically large. If the pump configuration is given (ω and q'), consider the most practical steps that can be taken to decrease the volume, assuming that the ripple factor α must be maintained at its prescribed value.

15-29 A pumping circuit with a single-cylinder, single-acting metering pump delivers methanol catalyst to a continuous-process chemical reactor. The flow rate is 2 liters/hr through 8 m of 5-mm-I.D. tubing. The pressure in the reactor is maintained at 3.5×10^6 N/m^2. (a) Specify a suitable pulsation damper to maintain the flow rate constant within ½ percent. (b) Comment on the size and pressure in the gas volume. (c) Specify the wall thickness of the pulsation damper.

15-30 A two-cylinder, two-stage (one delivery stroke per revolution) air compressor running at 600 rpm delivers air into an accumulator tank with a volume of 10 ft^3 which is maintained at 150 psig by a pressure controller. The compressor supplies air at the rate of $\dot{m} = 0.1$ lb$_m$/sec to a turbine through a pressure regulator just ahead of the turbine, which maintains the turbine inlet pressure at 100 psig. The temperature in the air supply is constant and uniform at 90°F. The supply line is 100 ft of 1-in. pipe, which accounts for most of the volume between the accumulator and the turbine. Estimate the ripple at the turbine inlet when the compressor is running.

CHAPTER 16

CONTENTS

ONE-DIMENSIONAL MAGNETOHYDRODYNAMICS

A qualitative and semiquantitative understanding and evaluation of the effects of electromagnetic fields on fluid motions can be obtained by considering the principal effects of electric and magnetic fields coupled with the force and motion fields studied previously. These electromagnetic effects are the same as those which occur in a conventional electric generator or motor. The motion of a conducting fluid element in a magnetic field will result in the generation of an electric field, just as the motion of the copper conductors (windings) of a generator rotor through the stator fields results in the generation of an electric field, or electric potential or voltage. If a voltage is applied to a conductor in a magnetic field such that a current flows in it, a magnetomotive force (mmf) is exerted on the conductor. Consideration of these effects will indicate that they usually occur together in any machine. The resultant action is determined by the effect that predominates.

In equilibrium flow conditions such as exist in steady-state operating magnetohydrodynamic generators, accelerators, electromagnetic pumps, and so on, the fluid starts through the apparatus with no net or gross charge. In an ionized or conducting fluid, all the original elements are neutral, so even though the molecules may be dissociated or partially ionized the net number of positively charged particles is equal to the number of negatively charged ones. This state of net neutral charge affords a simplification in dealing with the electromagnetic effects. The state of dissociation and ionization is a function of the temperature and density of the fluid. The random molecular kinetic energy (temperature) tends to break the molecules apart and knock the electrons off the atoms; the lower the density, the smaller is the possibility of the particles to reassociate. Since the state of the fluid (or its components) is determined by pressure, density, and temperature, it is easier to specify the degree of dissociation and ionization as functions of pressure and temperature rather than density and temperature, since pressure is a directly measurable or observable quantity. As energy is added to, or work is done on the fluid, its temperature will change, as will its state of dissociation and ionization, so that its conductivity will change. For the following simplified development, such changes as these will be neglected.

16.1 ELECTROMAGNETIC FUNDAMENTALS

First, consider the generator effect with respect to the constant-area flow channel of Fig. 16-1. For the following discussion the electromagnetic and flow fields will be assumed to be uniform, so that all the developments will be one-dimensional. In some cases, the quantities will be on a per unit volume basis, and can be applied to the more general equations

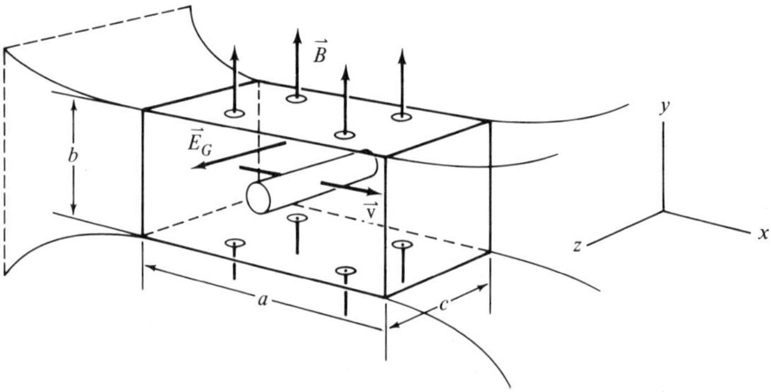

Fig. 16-1. Generation of an electric field $\vec{E}_G$ by the motion $\vec{v}$ of a conducting fluid in a magnetic field $\vec{B}$.

previously developed. Figure 16-1 indicates a fluid element conductor moving with a velocity $\vec{v}$ in a magnetic field of strength $\vec{B}$ (webers per square meter, Wb/m^2; weber = volt-sec; gauss = weber per square centimeter). According to Faraday's law, an electric field $\vec{E}_G$ (volts per meter, V/m) will be generated equal to the product of the perpendicular components of $\vec{v}$ and $\vec{B}$

$$\vec{E}_G = \vec{v} \times \vec{B} \tag{16-1}$$

The relation between the electric field strength E and the scalar voltage or electrical potential Φ is

$$\vec{E} = -\nabla\Phi \tag{16-2}$$

The voltage (or potential difference) is then obtained by integration,

$$\Delta\Phi = -\int \vec{E}\cdot\vec{ds} \tag{16-3}$$

For the one-dimensional configuration of Fig. 16-1, $\Phi_G = E_G \cdot c = \mathrm{v}Bc$ would be measured on a voltmeter. If the two sides of area *ab* are conducting plates, a load can be connected across them.

Next consider the case illustrated in Fig. 16-2, where a voltage Φ_A is applied across the plates producing an electric field E_A ($E_A = \Phi_A/c$). A current I through the fluid from one plate to another will result. The current density $\vec{j}$ (amperes per square meter, A/m²) in the magnetic field $\vec{B}$ will produce a force $\vec{f}$ equal to the product of the perpendicular components of $\vec{j}$ and $\vec{B}$ according to Ampere's law,

$$\vec{f} = \vec{j} \times \vec{B} \tag{16-4}$$

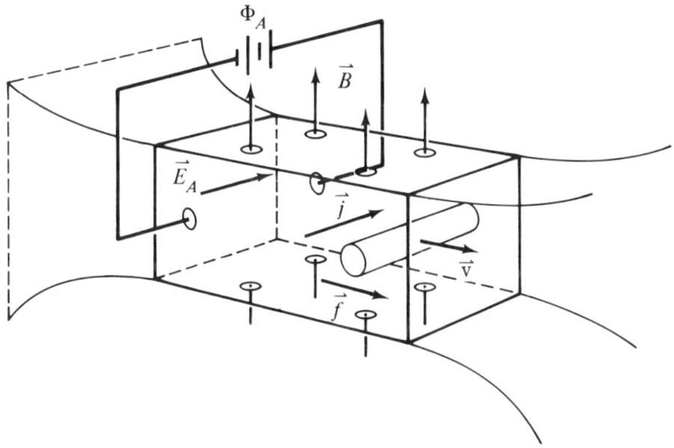

Fig. 16-2. Force $\vec{f}$ produced on a conducting fluid element by the interaction of a current $\vec{j}$ with a magnetic field $\vec{B}$.

with the units of

$$\frac{\mathrm{A}}{\mathrm{m}^2}\frac{\mathrm{Wb}}{\mathrm{m}^2} = \frac{\mathrm{A}}{\mathrm{m}^2}\frac{\mathrm{V\text{-}sec}}{\mathrm{m}^2} = \frac{\mathrm{W\text{-}sec}}{\mathrm{m}^4} = \frac{(\mathrm{J/sec})\mathrm{sec}}{\mathrm{m}^4} = \frac{\mathrm{N\text{-}m}}{\mathrm{m}^4} = \frac{\mathrm{N}}{\mathrm{m}^3}$$

so $\vec{f}$ is a magnetomotive body force per unit volume, called the Lorentz force.[1]

16.2 MAGNETOHYDRODYNAMIC GENERATOR

If a conducting fluid flow is maintained through a duct by an applied pressure differential as indicated in Fig. 16-1, the electric field $\vec{E}_G$ or

[1] The units used in this chapter are the mks international system as adopted by the National Bureau of Standards and available in E. A. Mechtly, The International System of Units, Physical Constants and Conversion Factors, *NASA Special Publication 7012*, 1964.

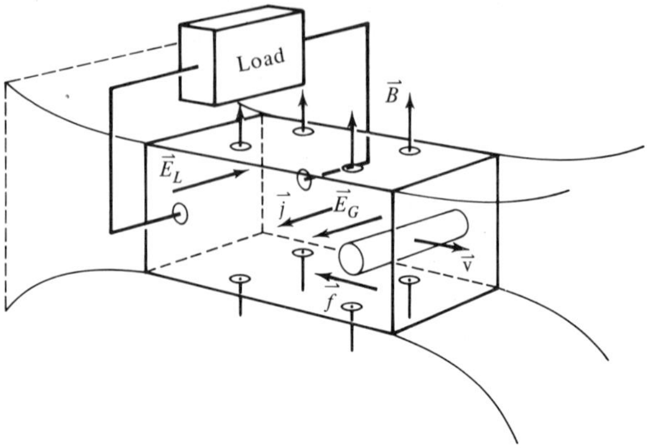

Fig. 16-3. Generated and load electric fields in a magnetohydrodynamic channel section.

$\Phi_G = E_G c$ is generated. The generator system of Fig. 16-3 has both effects: An electric field $\vec{E}_G$ is generated, so there is a current around the series circuit through the load resistance R_L and the fluid resistance R_F. The voltage drop across the load $IR_L = \Phi_L = \Phi_A$ is now equivalent to an applied voltage, and therefore to an applied field $E_L = E_A$, as shown. For purposes of generality, E_A rather than E_L will be used for the field in the fluid resulting from the load voltage. The load field E_A will always be opposite to E_G; however, since these are vector quantities the net field is $\vec{E}_G + \vec{E}_A$. Ohm's law can be written for the voltage drop across the fluid resistance of the circuit,

$$\Phi_G - \Phi_L = IR_F \qquad \text{or} \qquad I = \frac{\Phi_G - \Phi_L}{R_F} \tag{16-5}$$

Rather than the resistance R, or the resistivity of the fluid, the conductivity σ (reciprocal of resistivity) is usually used. The conduction $G = 1/R$ is proportional to area and inversely proportional to path length with the conductivity σ as the proportionality function

$$G = \frac{1}{R} = \frac{\sigma A}{L} = \frac{\sigma ab}{c} \quad (1/\text{ohms})$$

The reciprocal ohm is termed the mho, so the units of σ are mhos per meter. The magnitude of the current density $\vec{j}$ is $j = I/A = I/ab$. Equation (16-5) as a scalar equation is

$$(E_G - E_A)c\frac{\sigma ab}{c} = jab \tag{16-6a}$$

or, in vector form, Ohm's law is

$$\vec{j} = \sigma(\vec{E}_G + \vec{E}_A) \tag{16-6b}$$

GENERATOR OPERATING CHARACTERISTICS

The generated field $\vec{E}_G = \vec{v} \times \vec{B}$, or the corresponding voltage Φ_G, is the no-load or open-circuit voltage. From (16-5), the load voltage can be determined as

$$\Phi_L = \Phi_G - IR_F$$

The generated voltage $\Phi_G = vBc$ is assumed to be constant because $\vec{v}$ and $\vec{B}$ are applied quantities, and are assumed to be held constant for operation as a generator. The voltage characteristic for the generator is then shown in Fig. 16-4.

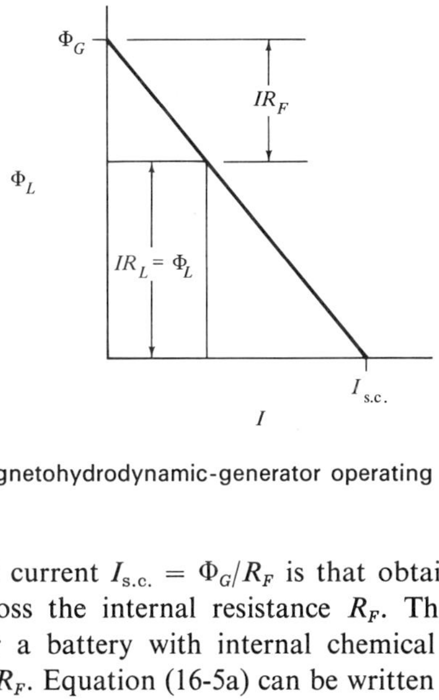

Fig. 16-4. Magnetohydrodynamic-generator operating characteristics.

The short-circuit current $I_{s.c.} = \Phi_G/R_F$ is that obtained from the generated voltage across the internal resistance R_F. This characteristic is similar to that for a battery with internal chemical potential Φ_G and internal resistance R_F. Equation (16-5a) can be written in nondimensional form by dividing through by the fixed reference voltage Φ_G and using the relation $\Phi_G = I_{s.c.}R_F$:

$$\Phi' \equiv \frac{\Phi_L}{\Phi_G} = 1 - \frac{I}{I_{s.c.}} \equiv 1 - I' \tag{16-7}$$

with the dimensionless voltage ratio Φ' and current ratio I' as indicated. The dimensionless characteristic plot is shown in Fig. 16-5.

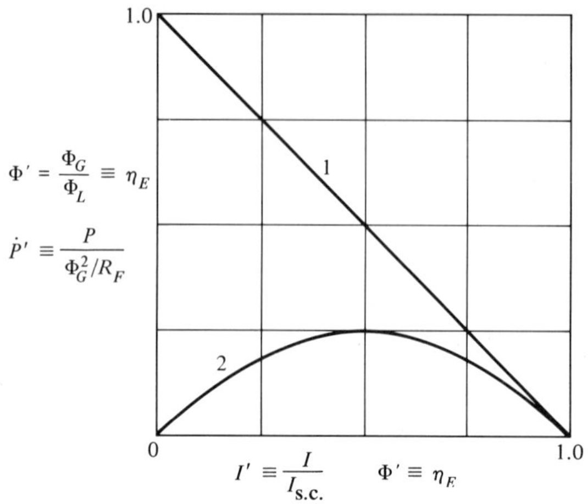

Fig. 16-5. Dimensionless voltage–current characteristic versus I' and the dimensionless power characteristic P′ versus Φ'. Note that since $\Phi' \equiv \eta_E$, the voltage–current characteristic is also the efficiency versus current characteristic. Curve 1, Φ' versus I' (or η_E versus I'); curve 2, P' versus η_E (equivalent to P' versus Φ').

The electric power output is

$$P_E = \Phi_L I = \frac{\Phi_L(\Phi_G - \Phi_L)}{R_F}$$

A dimensionless power P' can be defined and determined in terms of a dimensionless voltage ratio Φ' as

$$P' = \frac{P}{\Phi_G^2/R_F} = \frac{\Phi_L}{\Phi_G}\left(1 - \frac{\Phi_L}{\Phi_G}\right) = \Phi'(1 - \Phi') \tag{16-8}$$

The power characteristic, P' versus Φ', is also shown in Fig. 16-5.

The power input is the work rate required to overcome the magnetomotive force

$$\dot{W} = \vec{F}\cdot\vec{\text{v}} = (\vec{j} \times \vec{B}abc)\cdot\vec{\text{v}} \qquad \text{W} \tag{16-9a}$$

Manipulation of the vector quantities allows (16-9a) to be written

$$\dot{W} = -(\vec{j}\cdot\vec{\text{v}} \times \vec{B})abc = -(\vec{j}ab)\cdot(\vec{E}_G c) \tag{16-9b}$$

or as a scalar for the assumed one-dimensional fields,

$$\dot{W} = I\Phi_G \tag{16-9c}$$

The power efficiency is defined as the output power divided by the input work rate,

$$\eta_E \equiv \frac{P_E}{\dot{W}} = \frac{I\Phi_L}{I\Phi_G} = \Phi' \tag{16-10}$$

yielding the interesting and important result that the efficiency is equal to the voltage ratio, as indicated in Fig. 16-5. The magnetic field B should be as large as possible to maximize the output power to size ratio and maintain the required velocities at optimum values. An electromagnet is therefore required. The power to operate the magnet is not included in the above input power and efficiency calculations, because it is possible that a magnetohydrodynamic device could work using a permanent magnet, which requires no power to maintain $\vec{B}$.

The generated voltage depends only on the fluid velocity and the magnetic field. The magnetic field strengths required are quite large. To obtain these fields, electromagnets with superconducting coil windings operating at cryogenic temperatures are required. The power also depends on the current, which, for a given load resistance and generated potential Φ_G, depends only on the fluid conductivity. The conductivity of a gas depends on the degree of ionization. Gases do not ionize easily, so high temperatures are required. The temperatures necessary for ionization of air and fuel combustion products are higher than can be handled by most materials. Other means have been investigated to increase the conductivity. Small amounts of potassium or cesium help to catalyze or "seed" the ionization and greatly increase the conductivity. Ultraviolet and radio-frequency excitation energy will also produce ionization. The conductivities of some representative substances are given in Table 16-1.

The maximum power is determined from (16-8) as

$$P'_{\max} = 0.25 \qquad \text{at} \qquad I' = 0.5 = \eta_E$$

This is a rather low operating efficiency. At $\eta_E = 0.75 = I'$, $P' = 0.75P'_{\max}$ which may be a more desirable operating condition.

The body force $\vec{f} = \vec{j} \times \vec{B}$ is in a direction opposite to the fluid motion as shown in Fig. 16-3,

$$\vec{f} = +\hat{x}j_z \times \hat{y}B_y = -\hat{z}j_zB_y$$

For a fluid element as illustrated in Fig. 16-6, the equation of motion for steady, one-dimensional flow reduces to a force balance between surface pressure forces and the body magnetomotive force, giving

$$-\frac{\partial p}{\partial x} + f_x = 0 \tag{16-11a}$$

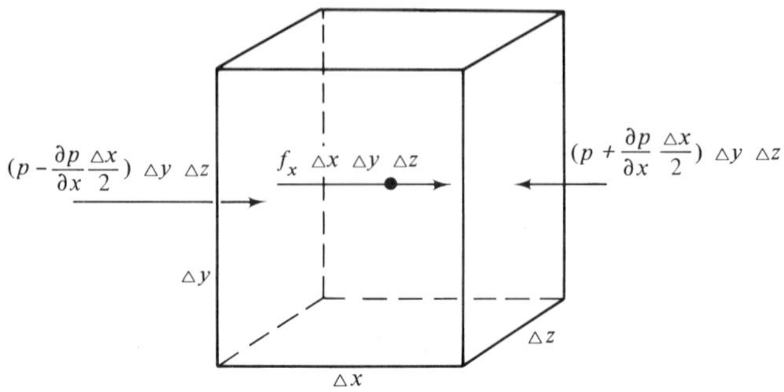

Fig. 16-6. Equilibrium of a fluid element under an electromagnetic body force per unit volume and pressure forces.

TABLE 16-1

CONDUCTIVITIES OF SOME REPRESENTATIVE SUBSTANCES

Substance	*Conductivity* (σ, mhos/m)
Copper (25°C)	6×10^7
Mercury (25°C)	10.8×10^5
Saturated salt water (25°C)	25
Distilled water (25°C)	2×10^{-5}
Air (2500°K, 1 atm)	10^{-7}
Air with 1 percent potassium (2000°K, 1 atm)	1.5
Air with 1 percent potassium (2500°K, 1 atm)	15
Helium with 1 percent cesium (2000°K, 1 atm)	6.6
Helium with 1 percent cesium (2500°K, 1 atm)	62
Hydrocarbon combustion products with 1 percent cesium (2500°K, 1 atm)	2–20

Note: Cesium seeding yields about twice the conductivity as that obtained with potassium.

where f_x is the x component of the magnetomotive force per unit volume. In the above example, $\partial p/\partial x$ is the only component of the pressure gradient, and $f_x \equiv f$ is the only component of the body force. Equation (16-11a) can be generalized for two- or three-dimensional flows as

$$\nabla p - \vec{f} = 0 \qquad (16\text{-}11b)$$

For the one-dimensional example case above, (16-9a) gives

$$\Delta p = f_x \Delta x = fa = -jBa \qquad \text{or} \qquad p_1 - p_2 = jBa \qquad (16\text{-}11c)$$

The pressure force per unit area in the $+x$ direction, which results from the electromagnetic action, is termed the "magnetic pressure," just as $\rho gz/g_c$ is called the hydrostatic pressure and $\rho v^2/2g_c$ is the dynamic pressure.

16.3 CROSSED-FIELD ACCELERATOR

If a voltage source is connected across the side-plate electrodes as shown in Fig. 16-7, there will be a current $\vec{j}$ through the fluid. As described in connection with Fig. 16-2, the magnetomotive force $\vec{f} = \vec{j} \times \vec{B}$ will be produced as indicated in Figs. 16-2 and 16-7. This force will tend to

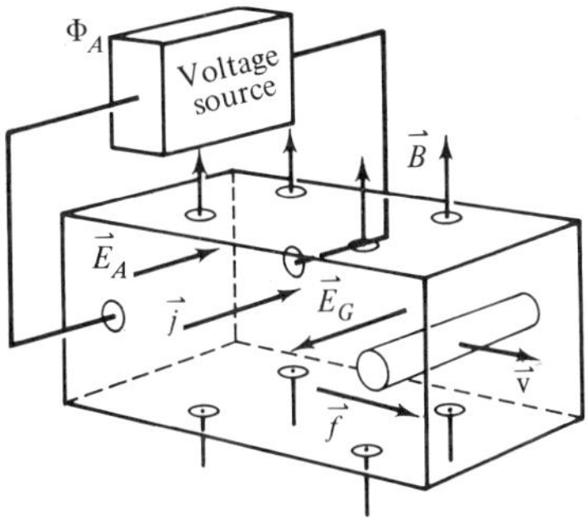

Fig. 16-7. Voltage applied to a conducting fluid sets up a current $\vec{j}$ that interacts with the magnetic field $\vec{B}$ to produce a magnetomotive force and a motion $\vec{v}$.

accelerate the fluid to the right. Under steady-state conditions of equilibrium among the magnetic, pressure, and inertia forces, a generated field $\vec{E}_G = \vec{v} \times \vec{B}$ in a direction opposite to $\vec{E}_A$ will be generated as shown. As before, the current I or the current density $\vec{j}$ will be

$$\vec{j} = \sigma(\vec{E}_A + \vec{E}_G)$$

It is obvious that when $|E_A| = |E_G|$ (scalar values), $\vec{j}$ and therefore $\vec{f}$ are zero.

The equilibrium or steady flow condition maintained above will exist when the fluid is accelerated to a velocity where the magnetic force overcomes the pressure and friction forces in the ducting system, as determined by the velocity v. The thrust efficiency can be defined as the ratio of the thrust power in the fluid to the electric power input,[2]

$$\eta_T = \frac{\vec{f}\cdot\vec{v}(abc)}{(\vec{E}_A c)\cdot(\vec{j}ab)} = \frac{\vec{f}\cdot\vec{v}}{\vec{E}_A\cdot\vec{j}} = \frac{\vec{j}\times\vec{B}\cdot\vec{v}}{\vec{E}_A\cdot\vec{j}} = -\frac{\vec{E}_G\cdot\vec{j}}{\vec{E}_A\cdot\vec{j}} \qquad (16\text{-}12a)$$

[2] From Fig. 16-7, $\vec{E}_G\cdot\vec{j}$ is seen to be negative, so η_T is a positive quantity.

For the parallel fields as illustrated,

$$\eta_T = \frac{E_G}{E_A} = \frac{1}{\Phi'} = \frac{1}{\eta_E} \tag{16-12b}$$

This electromagnetic thrust can be used to accelerate the conducting fluid or ionized gas as an electromagnetic pump or rocket augmenter. This type of pump has the advantages that it has no moving parts or shaft seals and can be constructed to be simple and reliable.

16.4 ELECTROMAGNETIC FLOW METER

If the load of Fig. 16-3 is replaced by a voltmeter (with a high internal resistance, $R_{\text{meter}} > R_F$), the current drawn will be quite small, so the meter will measure essentially Φ_G, which is directly proportional to the average or integrated velocity through the stream for a given applied magnetic field B. For a one-dimensional approximation, the velocity is then given by $\text{v}_{\text{meter}} = \Phi/Bc$.

16.5 ILLUSTRATIVE EXAMPLE 16-1: Magnetohydrodynamic Generator

Consider a generator with a configuration shown in Fig. 16-3, with a field of $B = 4$ Wb/m^2 (40,000 G), 1 percent potassium-seeded combustion gases (for which $\gamma = 1.25$, $\overline{\mathscr{M}} = 20$) at $T = 2500$°K ($\sim$4000°F) and a channel with a volume of 1 m^3. The ratio $a{:}b{:}c$ of the configuration is to be examined. Assume a representative value of $\sigma = 10$ mhos/m (see Fig. 16-8) and an operating channel Mach number $M = 0.3$, which will give a relatively high velocity but still allow assumption of incompressible flow. Initially, an average pressure of 1 atm will be assumed. It is desired to determine the maximum power, the load voltage possible for these conditions, and the pressure drop across the generator. The relation between the configuration and the operating characteristics is determined by the governing equations:

$$P_{\text{max}} = \frac{\sigma E_G^2}{4}\,\text{vol} = \frac{\sigma \text{v}^2 B^2\,\text{vol}}{4}$$

$$\Phi_G = \text{v}Bc \qquad \Phi_L = \eta\Phi_G \qquad \eta \equiv \Phi'$$

$$I = jab = \sigma \text{v}B(1 - \eta)ab$$

$$\Delta p = jBc = \sigma \text{v}B^2(1 - \eta)c$$

The maximum power depends on the given operating conditions, v, B, and the volume, but is independent of the configuration ($a{:}b{:}c$ ratios). It is

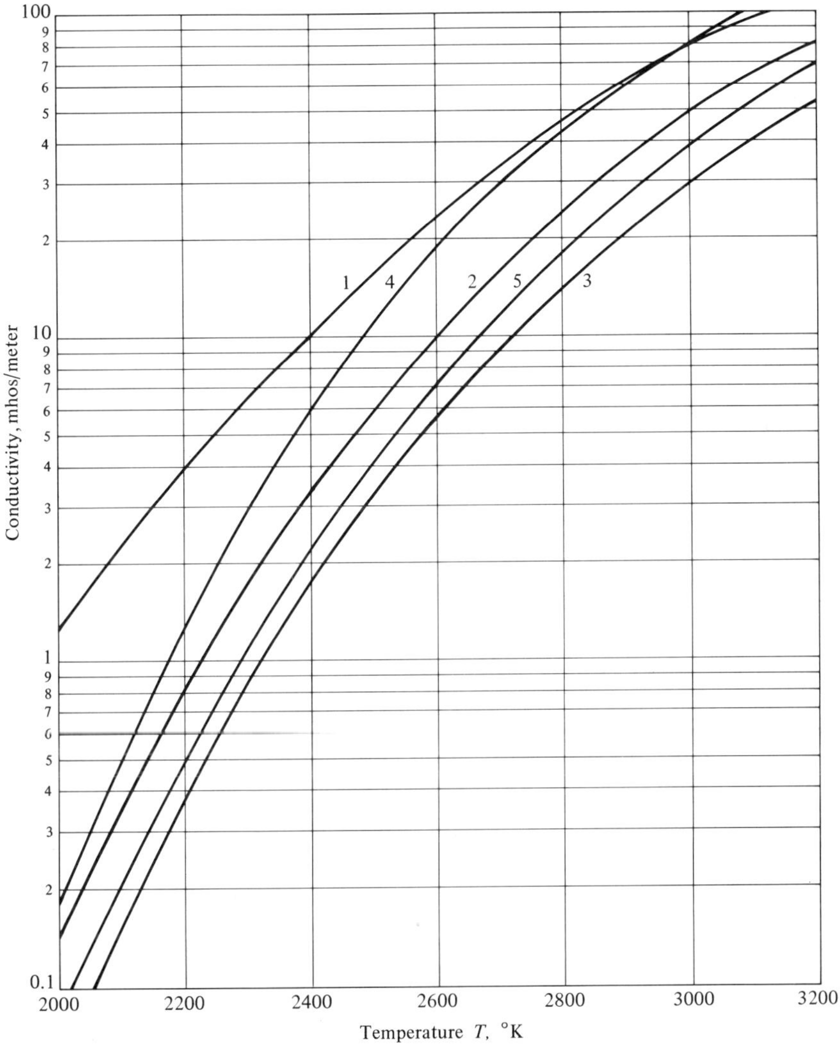

Fig. 16-8. Conductivity as a function of temperature and pressure for some selected gases. Curve 1: air at 1 atm with 1 percent cesium; curve 2: natural gas and air, stoichiometric mixture, 1 atm, 1 percent potassium; curve 3: natural gas and air, stoichiometric mixture, 2 atm, 1 percent potassium; curve 4: natural gas and air, stoichiometric mixture, 1 atm, 1 percent cesium; curve 5: ethyl alcohol and stoichiometric oxygen with 1:1 N_2:O_2, 1 atm, 1 perçent potassium.

desirable to operate such that $\eta = 0.70$ to 0.85. A specified design voltage Φ_L can be obtained by adjusting the width c and by varying v and B. (There are also other optimizing criteria possible, but they cannot be considered within the restrictions of one-dimensional flow.) The current depends on σ, v, B, and the side electrode plate area ab. If the voltage is prescribed, v and B are determined as above. The conductivity is a function of temperature and pressure, as indicated in Fig. 16-8. It is advantageous to operate at the maximum temperature allowable by the materials available. The above value of 2500°K is somewhat optimistic.

For a given or otherwise optimized v and B, the Δp is directly proportional to the length. For a generator the v and Δp will be determined somewhat by the combustion processes and operating temperature. For this example, it will be considered desirable to operate with a small Δp. For a nominal volume of 1 m^3 a cubic configuration will be chosen with $a = b = c = 1$ m^3.

The velocity is $\text{v} = Ma$, where $a = \sqrt{\gamma g_c R_0 T/\mathscr{M}}$ is the velocity of sound in the gas; R_0 is the universal gas constant (8314 J/kg-mole-°K) and $\mathscr{M}$ is the molecular mass number (in this case the average particle mass number in kg/mole). In the mks system $g_c = 1$ kg-m/N-sec^2, so

$$\text{v} = 0.3\sqrt{\frac{(1.25)(8314)(2500)}{20}} = 342 \text{ m/sec}$$

Since $c = 1$, the magnitude of the generated voltage Φ_G is equal to the generated field strength E_G:

$$\Phi_G = \text{v}Bc = (342)(4)(1) = 1368 \text{ V}$$

If the efficiency $\eta = \Phi'$ is nominally taken as 0.75, the operating voltage is $\Phi_L = 1026$ V.

The maximum power is

$$P_{\max} = \frac{\sigma E_G^2}{4} abc = \frac{10}{4}(1368)^2(1) = 4.7 \times 10^6 \text{ W} \quad \text{or} \quad 4.7 \text{ MW}$$

which is a sizable power per volume ratio. For $\eta = 0.75$ the peak generated power available would be ~3.5 MW. The corresponding load current would be

$$I = \sigma E_G(1 - \eta)ab = (10)(1368)(0.25)(0.1) = 3420 \text{ A}$$

The current at maximum power is twice this quantity.

The pressure drop is

$$\Delta p = jBa = (3420)(4)(1) = 13{,}680 \text{ N/m}^2 \quad \text{or} \quad 0.135 \text{ atm}$$

The relative pressure drop $\Delta p/p = 13.5$ percent for an assumed inlet pressure of 1 atm.

It is also of particular interest to calculate the ΔT for two reasons: σ is a rapidly varying function of temperature and pressure (Fig. 16-8), and the velocity will vary as determined by both p and T (or if the velocity is to be maintained constant, a modification of the channel configuration would be required). Assuming the velocity (and therefore the kinetic energy) will be held constant, the temperature difference across the channel section can be determined approximately from the energy balance (assuming $\Delta T/T \ll 1$, an assumption that must be checked to validate the assumptions of nearly constant v and σ):

$$mc_v \, \Delta T = -P$$

(The negative sign is necessary since P is an output.) The mass flow rate $\dot{m}$ can be written in terms of the given quantities as (see Chapter 12)

$$\dot{m} = \rho A \mathrm{v} = \frac{p}{RT} AM\sqrt{\gamma g_c RT}$$

and c_v can be determined from the perfect gas relationships,

$$c_p - c_v = R \qquad \text{and} \qquad \frac{c_p}{c_v} = \gamma$$

which give $c_v = R/(\gamma - 1)$. The product $\dot{m}c_v$ is then

$$\dot{m}c_v = \frac{pAM}{\gamma - 1}\sqrt{\frac{\gamma g_c R}{T}} \ \frac{\text{J}}{\text{sec-°K}}$$

ΔT can be calculated as

$$\Delta T = -\frac{\gamma - 1}{pAM}\sqrt{\frac{T}{\gamma g_c R}}\, P$$

$$= -\frac{0.25}{(101{,}325)(1)(0.3)}\,\frac{2500}{(1.25)(416)}\, 3.5 \times 10^6 = -63\text{°K}$$

As energy is taken from the fluid, internal energy (as measured by T) and availability to do work (as indicated by p) both decrease, or the combined effect decreases as indicated by a decrease in the enthalpy.

For steady flow (constant $\dot{m}$) and constant area, the relative change in velocity can be obtained from the continuity equation and the equation of state, assuming relatively small changes in p and T and neglecting changes in gas composition (that is, neglecting effective changes in molecular mass number),

$$\frac{\Delta \mathrm{v}}{\mathrm{v}} = -\frac{\Delta \rho}{\rho} = \frac{\Delta T}{T} - \frac{\Delta p}{p} = -\frac{63}{2500} + 0.135 = +0.11$$

This result indicates that the channel area must increase in the direction of the flow if the velocity is to be maintained constant. It is not desirable to have a variation of velocity with channel distance because the generated voltage depends on velocity, and the duct losses depend on v^2.

The gases are still quite hot and can be used to generate additional power in a gas turbine. In this case of a binary power cycle, the inlet gas pressure must be greater than 1 atm.

16.6 DIFFERENTIAL EQUATIONS FOR STEADY PLASMA FLOW

CONTINUITY EQUATION

If all the mass is considered, the overall mass must remain constant independent of the dissociation and ionization taking place due to local or overall changes in temperature and pressure. Therefore, the continuity equation is still $\dot{m} = \rho A\mathrm{v} = \text{const.}$, or the differential equation is the same as (14-19) with $\partial\rho/\partial t = 0$.

ENERGY EQUATION

The principles developed in Chapter 5 can be extended to include electromagnetic effects. A statement of the energy equation for a defined control volume (C.V.) is

$$\begin{bmatrix}\text{RATE OF CHANGE}\\ \text{OF INTERNAL ENERGY}\\ \text{IN C.V. AT ANY}\\ \text{INSTANT}\end{bmatrix} \rightarrow \begin{bmatrix}\text{TRANSPORT RATE OF MASS}\\ \text{ASSOCIATED ENERGY } \textit{in}\\ \text{ACROSS C.V. BOUNDARIES}\\ \text{BY FLOW FIELD}\end{bmatrix} + \begin{bmatrix}\text{WORK}\\ \text{RATE}\\ \textit{on}\text{ FLUID}\\ \text{IN C.V.}\end{bmatrix}$$
$$+ \begin{bmatrix}\text{ENERGY RATE } \textit{to}\\ \text{FLUID } \textit{in}\text{ C.V.:}\\ \text{BY HEAT TRANSFER}\end{bmatrix} + \begin{bmatrix}\text{ELEC. POWER}\\ \textit{to}\text{ FLUID}\\ \textit{in}\text{ C.V.}\end{bmatrix}$$

For steady flow, the left-hand term is zero. The first term on the right side can be transposed to the left and written as positive energy *out*. The force-times-displacement work rate $(\vec{F}\cdot\vec{\mathrm{v}})$ is due to pressure and shear surface forces and body (mass-acting) forces.

An energy equation for these conditions can be written

$$\oiint e\rho\vec{\mathrm{v}}\cdot d\vec{A} = \sum \dot{W}_{\text{surf}} + \sum \dot{W}_{\text{body}} + \dot{Q}_{H.T.} + P_E$$

where e is the internal energy per unit mass, $\dot{Q}_{H.T.}$ the heat transfer to the fluid in the control volume, and P_E the *net* electromagnetic energy rate or power added to the fluid in the control volume from an external source. For steady flow of a neutral fluid, there will be no net charge energy stored or transported, so

$$e \equiv \tilde{u} + \frac{\mathrm{v}^2}{2g_c}$$

The divergence theorem can be applied to the left side to obtain

$$\oiiint \nabla \cdot [(e)(\rho\vec{\mathrm{v}})]\, dV$$

which can be expanded in terms of the indicated product to give

$$[(\rho\vec{\mathrm{v}}) \cdot \nabla e + e\nabla \cdot (\rho\vec{\mathrm{v}})]\, dV$$

The second term within the brackets is zero for the steady flow of a neutral fluid.

The surface work rate is produced by pressure and shear forces:

$$\sum \dot{W}_{\mathrm{surf}} = \dot{W}_p + \dot{W}_\tau$$

The pressure work rate is

$$\dot{W}_p = \int \vec{\mathrm{v}} \cdot d\vec{F}_p = \oiint -\vec{\mathrm{v}} \cdot (p\, d\vec{A}) = \oiint -p\vec{\mathrm{v}} \cdot d\vec{A}$$

The shear work rate is

$$\dot{W}_\tau = \int \vec{\mathrm{v}} \cdot d\vec{F}_\tau = \oiint \vec{\mathrm{v}} \cdot (d\vec{A} \cdot \vec{\tau}) = \oiint d\vec{A} \cdot \tau \cdot \vec{\mathrm{v}} = \oiint \vec{\mathrm{v}} \cdot \tau \cdot d\vec{A}$$

where τ is the tensor shear stress (a symmetric tensor) expressed as a dyadic.

The body-force work rate is considered to be produced by gravity-type forces ($\vec{f}_g$ force per unit mass) and the electromagnetic body force per unit volume $\vec{f}_{\mathrm{em}} = \vec{j} \times \vec{B}$. The net electromagnetic energy addition or power P_E is the integral over the control volume of the net power per unit volume, which is the product of the net electric field times the current. The arrangement of Fig. 16-6 will be used as a general case, since $\vec{E}_A \cdot \vec{j}$ is a positive input quantity and $\vec{j} \times \vec{B}$ is a positively directed force. Φ_A

is an external force potential *applied to* the fluid in the control volume.

$$P_E = \oiiint \vec{j}\cdot(\vec{E}_A + \vec{E}_G)\,dV$$

The steady-flow energy equation is then

$$\oiiint \rho\vec{\mathrm{v}}\cdot\nabla\left(\frac{\mathrm{v}^2}{2g_c} + \tilde{u}\right)dV = \oiint [-p\vec{\mathrm{v}}\cdot d\vec{A} + \vec{\mathrm{v}}\cdot\tau\cdot d\vec{A}]$$

$$+ \oiiint (\rho\vec{f}_g + \vec{j}\times\vec{B})\cdot\vec{\mathrm{v}}\,dV$$

$$+ \dot{Q}_{H.T.} + \oiiint \vec{j}\cdot(\vec{E}_A + \vec{E}_G)\,dV \quad (16\text{-}13a)$$

For the present development, the flow is assumed to be frictionless and adiabatic: $\dot{W}_\tau = 0$, $\dot{Q}_{H.T.} = 0$. The divergence theorem can be applied to the area integral term, and the electromagnetic body force integrand can be manipulated to give

$$\vec{j}\times\vec{B}\cdot\vec{\mathrm{v}} = \vec{j}\cdot\vec{B}\times\vec{\mathrm{v}} = -\vec{j}\cdot(\vec{\mathrm{v}}\times\vec{B}) = -\vec{j}\cdot\vec{E}_G$$

Applying these assumptions and manipulations to (16-13a) gives

$$\oiiint \rho\vec{\mathrm{v}}\cdot\nabla\left(\frac{\mathrm{v}^2}{2g_c} + \tilde{u}\right)dV = \oiiint [-\nabla\cdot(p\vec{\mathrm{v}}) + \rho\vec{f}_g\cdot\vec{\mathrm{v}} + \vec{j}\cdot\vec{E}_A]\,dV \quad (16\text{-}13b)$$

The equation is now entirely in terms of volume integrals, so a differential equation can be dedûced,

$$\rho\vec{\mathrm{v}}\cdot\nabla\left(\frac{\mathrm{v}^2}{2g_c} + \tilde{u}\right) = -p\nabla\cdot\vec{\mathrm{v}} - \vec{\mathrm{v}}\cdot\nabla p + \rho\vec{f}_g\cdot\vec{\mathrm{v}} + \vec{j}\cdot\vec{E}_A \quad (16\text{-}14)$$

MOMENTUM EQUATION

The momentum principle applied to a fluid control volume such as illustrated in Fig. 16-7 can be stated as

$$\begin{bmatrix}\text{RATE OF CHANGE}\\ \text{OF MOMENTUM OF}\\ \text{FLUID IN C.V.}\end{bmatrix} = \begin{bmatrix}\text{TRANSPORT RATE}\\ \text{OF MOMENTUM } \textit{in}\\ \text{ACROSS C.V. BOUNDARIES}\end{bmatrix} + \begin{bmatrix}\text{FORCES ON}\\ \text{FLUID IN}\\ \text{C.V.}\end{bmatrix}$$

For steady flow, the left-hand term is zero. The first term on the right side can be transposed to the left side and written as positive momentum transport *out*. The forces are the same as those specified in the development of the energy equation. The momentum equation can be written and developed using methods similar to those of Chapter 6:

$$\frac{1}{g_c}\oiint (\vec{v})(\rho\vec{v})\cdot d\vec{A} = \sum \vec{F}_{\text{surf}} + \sum \vec{F}_{\text{body}}$$

$$= \oiint d\vec{A}\cdot\tau - \oiint p\, d\vec{A} + \oiiint (\rho\vec{f}_g + \vec{f}_{\text{em}})\, dV \qquad (16\text{-}15a)$$

Again neglecting shear stress and applying the divergence theorem gives

$$\frac{1}{g_c}\oiiint \nabla\cdot[(\vec{v})(\rho\vec{v})]\, dV = \oiiint (-\nabla p + \rho\vec{f}_g + \vec{j}\times\vec{B})\, dV \qquad (16\text{-}15b)$$

The left side can be expanded as $\vec{v}\nabla\cdot(\rho\vec{v}) + \rho\vec{v}\cdot\nabla\vec{v}$. The first term contains the steady-state continuity equation term, which is consequently zero. The second term can be expanded by using the vector identity

$$\vec{v}\cdot\nabla\vec{v} = \nabla\left(\frac{v^2}{2}\right) - \vec{v}\times\nabla\times\vec{v} = \nabla\left(\frac{v^2}{2}\right) - \vec{v}\times 2\vec{\omega}$$

Since $\vec{\omega}$ is assumed zero, no tangential shear force term can be applied to a fluid element, so it cannot be rotated. If the fluid is initially irrotational, it must remain irrotational, and $\vec{\omega} = 0$ for the entire field.[3] A differential equation can now be deduced from (16-15b),

$$\rho\nabla\left(\frac{v^2}{2g_c}\right) = -\nabla p + \rho\vec{f}_g + \vec{j}\times\vec{B} \qquad (16\text{-}16)$$

COMBINED EQUATION

Equation (16-16) can be multiplied by $\vec{v}\cdot$ and combined with (subtracted from) (16-14) to give

$$\rho\vec{v}\cdot\nabla\tilde{u} = -p\nabla\cdot\vec{v} + \vec{j}\cdot\vec{E}_A - \vec{v}\cdot\vec{j}\times\vec{B} \qquad (16\text{-}17a)$$

The last terms can be combined to give $\vec{j}\cdot(\vec{E}_A + \vec{E}_G)$ and with Ohm's law (16-6b), (16-17a) becomes

$$\rho\vec{v}\cdot\nabla\tilde{u} = -p\nabla\cdot\vec{v} + \frac{j^2}{\sigma} \qquad (16\text{-}17b)$$

[3] More detailed analyses of magnetofluiddynamic fields indicate that vorticity may be produced by a nonuniform $\vec{j}\times\vec{B}$ field.

Using the continuity equation in the form $\nabla \cdot \vec{v} = -\vec{v} \cdot \nabla\rho/\rho$, (16-17b) can also be written

$$\rho\vec{v} \cdot \nabla\tilde{u} = \frac{p}{\rho}\vec{v} \cdot \nabla\rho + \frac{j^2}{\sigma} \tag{16-17c}$$

This equation states that the transport rate of internal energy is a result of the compression work on the fluid and the Joule or Ohmic conversion. For a one-dimensional flow, (16-17b) is easily written for the flow or streamline direction as

$$\rho v \frac{d\tilde{u}}{ds} = -\frac{pv}{\rho}\frac{d\rho}{ds} + \frac{j^2}{\sigma} \tag{16-17d}$$

With $d\tilde{u} = c_v\, dT$ and $p/\rho = (R_0/\mathscr{M})T$ (where $\mathscr{M}$ is the average molecular mass number). Equation (16-17d) can also be written

$$c_v \frac{dT}{ds} = \frac{R_0}{\mathscr{M}}\frac{T}{\rho}\frac{d\rho}{ds} + \frac{j^2}{\rho v\sigma} \tag{16-17e}$$

If the flow is incompressible, $d\tilde{u}/ds = j^2/\rho v\sigma$. For a uniform one-dimensional flow (constant j, ρ, v, and σ) this result can be integrated over a length Δs to get $\Delta\tilde{u} \equiv \Delta\tilde{u}_{em} = (j^2/\rho v\sigma)\,\Delta s$. Therefore, in the interpretive context in which (6-23b) was written for a quasi-one-dimensional, steady incompressible flow including friction and heat-transfer effects,

$$\Delta\tilde{u} = \Delta\tilde{u}_q + \Delta\tilde{u}_\tau + \Delta\tilde{u}_{em} = \tilde{q} + \frac{\overline{4\tau\,\Delta s}}{\rho D_H} + \frac{\overline{j^2\,\Delta s}}{\rho v\sigma} \tag{16-18}$$

16.7 SUMMARY

1. The much-simplified considerations of magnetohydrodynamic channel flows presented here allow only idealized or first approximation determination of performance characteristics of magnetohydrodynamic devices. Boundary-layer effects, end effects, and other generated fields and currents (Hall effect) and other two- and three-dimensional effects have been neglected. The study of magnetofluiddynamics requires further and advanced study of electricity and magnetism, statistical mechanics, and fluid dynamics. A current comprehensive advanced or intermediate presentation that brings these subjects together is listed in reference 1.
2. Magnetohydrodynamic thrust for interplanetary flight has been considered with the power supplied by solar-energy-conversion cells. Magnetohydrodynamic fields have also been suggested for slowing

reentry vehicles with magnetohydrodynamic fields in the ionized flow about the vehicle acting as reverse thrustors (opposite to the accelerator effect discussed above). The field interaction also has the effect of thickening the boundary layer and decreasing the heat transfer to the vehicle.

REFERENCES

1. G. W. Sutton and A. Sherman, *Engineering Magnetohydrodynamics*, McGraw-Hill, Inc., New York, 1965.
2. S. I. Pai, *Magnetogasdynamics and Plasma Dynamics*, Prentice-Hall, Inc., Englewood Cliffs, N.J., 1962.
3. A. B. Cambel, *Plasma Physics and Magnetofluidmechanics*, McGraw-Hill, Inc., New York, 1963.
4. R. G. Deissler, One-Dimensional Analysis of MHD Energy Conversion, *NASA Technical Note D-680*, 1961.

PROBLEMS

16-1 Show that the dimensions of (16-8a) are watts = J/sec = N-m/sec.

16-2 What is the ratio of power/max. power at which the product of power and efficiency is maximized, and what is the corresponding efficiency?

16-3 Check the units of the velocity in Example 16-1 as given by $v = Ma = M\sqrt{\gamma g_c R_0 T/\mathscr{M}}$.

16-4 Calculate the mass flow rate of the combustion gas through the magnetohydrodynamic generator of Example 16-1.

16-5 In Example 16-1, determine approximately the conductivity of the gas leaving the generator, and discuss the significance of the determination.

16-6 In Example 16-1, assume that the inlet pressure is 3 atm (instead of 1 atm) and that the remaining conditions are the same, and determine the power available if the exhaust gas from the magnetohydrodynamic generator is passed through a gas turbine with an efficiency of 85 percent. What is the ratio of magnetohydrodynamic power to generator power?

16-7 Check the units of each term of (16-16). Is this equation homogeneous in units of the terms? What are the units in the mks or international system?

16-8 Permanent magnets are available with field strengths of 2000 G in air, 4-cm^2 poles, and 2-cm gaps. Determine the head-flow rate (or Δp flow rate) characteristics for an experimental laboratory magnetohydrodynamic pump with the applied field supplied by a 12-V battery for (a) mercury and (b) salt water.

16-9 Determine the Δp versus Q performance characteristics for a magnetohydrodynamic pump with a configuration as shown in Fig. 16-2. Discuss the optimum relative configuration, that is, $a:b:c$ ratios.

16-10 Sketch and make initial design calculations for an electromagnetic flowmeter to measure the flow rate of water in a piping system where the expected rate is 0.2 to 20 gpm. What features are necessary such that the meter indication is independent of fluid conductivity?

16-11 Sketch the field orientations for a magnetohydrodynamic thrust-augmented solid-propellant rocket with a 10- by 10-cm exhaust area, and find the additional thrust for a potassium-seeded exhaust that gives $\sigma = 20$ mhos/m. The operating characteristics are $v_e = 2000$ m/sec, $\gamma = 1.3$, $B = 1$ Wb, $\Phi_A = 10{,}000$ V. Discuss the effects of the magnetohydrodynamic thrustor section on nozzle geometry.

16-12 An electromagnetic flowmeter is constructed as shown. The maximum water flow rate is to be 0.01 m^3/sec ($\sigma = 0.1$ mho/m). The meter section is 4 by 2 by 4 cm. An electromagnet produces a B field of 0.5 Wb. Find (a) the maximum voltage generated, (b) the fluid path resistance, and (c) the meter pressure drop. (d) Discuss the meter characteristics for this application.

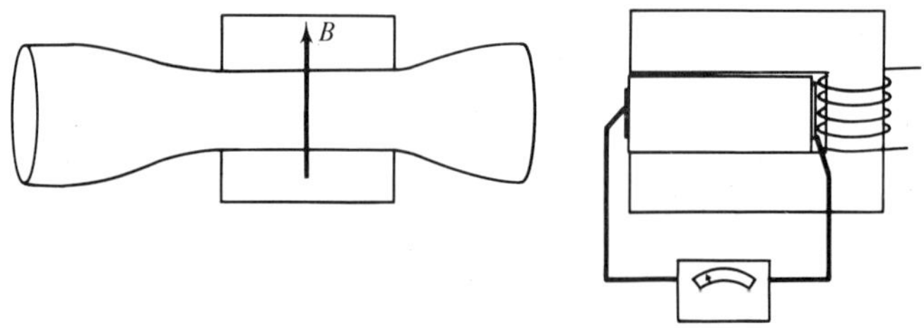

Fig. P16-12.

APPENDIX A

FORCE AND MASS UNITS

All units of measurement[1] of any quantity are arbitrary. There is no possible or logical starting point for the establishment of any prime unit for measurement. The establishment of reference standard units therefore depends on the language and the person who first decided that a reference standard was necessary. As a result of this arbitrariness, there are numerous systems of measurement in use today.

The most logical basis would be to define a unit of mass as the mass of a fixed number of molecules, say, 10^{20} of a reference substance, for example, oxygen (since the periodic table is based on O_2 as the reference standard). *Mass* is a basic quantity, but it is interesting that there is no way of determining or measuring mass uniquely. The standard methods of measuring the mass of a substance involve the determination of the effect of a force field on the mass. One such method is "weighing" the mass, or determining the relative effect of the earth's attraction for this mass as compared with an established standard mass, or with the aid of a calibrated spring. In either case, there must be a standard reference unit of mass. The absolute standard kilogram mass is a block of platinum (containing about 3.1×10^{24} molecules). Since the molecules cannot be counted, secondary standards can only be established against this primary standard by a comparative procedure ("weighing" on a very accurate balance).

There are many ways in which a unit of force could be established: by the effect of gravitational or inertial forces, by a relative deflection of a beam or spring, and so on. Once the mass unit is established, the most simple and apparently reproducible method is to determine the action of the earth's gravitational field on the standard unit mass and define the resultant action as the unit of force. To obtain the maximum possible accuracy and reproducibility in the establishment of the unit force, a definite reference location for this determination must be established, since the gravity attraction varies with latitude, longitude, and altitude.

[1] A unit of measurement is more specifically defined by the terminology "reference dimensional unit of measurement," or *unit* for short.

Force and mass can *exist independently* and, therefore, are *independent* quantities in nature. The force in a compressed spring has no *direct* relationship with any mass. On the other hand, an indirect effect or relation may be defined arbitrarily. The weight force-to-mass relationship that exists on earth is completely different from the one that would exist on other planets. It is, therefore, logical to consider force and mass as independent quantities with independent units of measurement.

Under standard gravitational conditions ($g_0 = 32.1740$ ft/sec^2, 980.7 m/sec^2) 1 pound-force is arbitrarily defined as the force exerted by the earth's gravitational field on the standard pound-mass. Note that the pound-mass (lb_m) is the actual and logical fundamental established quantity in this system.

The relationship among mass, force, and acceleration in four systems can be stated:

1. One dyne is the force required to accelerate 1 gram at a rate of 1 cm/sec^2.
2. One pound-force (lb_f) is the force required to accelerate 1 pound-mass at the rate of 32.1740 ft/sec^2 $= g_0$.
3. One newton (N) is the force required to accelerate 1 kilogram (kg) at the rate of 1 m/sec^2. (1 N = 1,000,000 dynes.)
4. One pound-force is the force required to accelerate 1 slug at the rate of 1 ft/sec^2.

The mks system is being more widely used and has been adopted as the international standard. In the English (lb_m-lb_f-ft-sec) system, 1 lb_m and 1 lb_f are numerically equal. When using force and mass units in an equation, one must be careful. First, any equation written to express some physical phenomenon must be homogeneous in both dimensions and units of measurement to have any meaning. (This requirement is the basis for dimensional analysis.) For example, consider Newton's second law in the form

$$F \propto ma$$

If three fundamental units of measurement (mks or ft-lb_m-sec) are specified in the English system,

$$F(\text{lb}_f) \propto m(\text{lb}_m)\cdot a(\text{ft/sec}^2)$$

But this equation is not homogeneous in the three basic units, so a proportionality sign has been used instead of an equality. For an equality, a proportionality constant must be added that will allow a balance of units on both sides of the equation. This procedure is required since it cannot logically be said that "force *equals* mass times acceleration," because fundamentally different physical concepts are involved. To eliminate this difficulty, the above equation is written with a constant of proportionality as

$$F(\text{lb}_f) = k(?)\cdot m(\text{lb}_m)\cdot a(\text{ft/sec}^2)$$

Solving the dimensional system to determine the units of k gives

$$[k] = \frac{\text{lb}_f}{\text{lb}_m\text{-ft/sec}^2}$$

and the dimensions of both sides of the equations are the same. Now using the gravitational relationship for the action of the earth's gravitational field on a mass, which is the action defined as "weight" ("W" = F_g),

$$1\ (\text{lb}_f) = 1\ (\text{lb}_m)\cdot k\,\frac{\text{lb}_f}{\text{lb}_m\text{-ft/sec}^2}\cdot g_e\ (\text{ft/sec}^2)$$

where $a = g_e$ is not a constant but is the local acceleration produced by the earth's gravitational field. If $g_e = g_0 = 32.1740$ ft/sec^2 is chosen as the *standard* value of gravitational acceleration, 1 lb$_m$ will be acted upon by the local *standard* gravity field ($g_0 = 32.1740$ ft/sec^2) to give an indicated measurement for the *standardized force* unit of 1 lb$_f$,

$$1\ (\text{lb}_f) = 1\ (\text{lb}_m)\cdot k\,\frac{\text{lb}_f}{\text{lb}_m\text{-ft/sec}^2}\cdot 32.1740\ (\text{ft/sec}^2)$$

Solving for the numerical value of k gives (taking $g_e = 32.2$ ft/sec^2)

$$k\,\frac{\text{lb}_f}{\text{lb}_m\text{-ft/sec}^2} = \frac{1\ (\text{lb}_f)}{1\ (\text{lb}_m)}\,\frac{1}{32.2\ (\text{ft/sec}^2)}$$

Instead of k, $1/k$ is usually used and defined as a proportionality constant or conversion factor for this system of units and denoted g_c. From the above equation,

$$g_c = \frac{1}{k} = 32.2\,\frac{\text{lb}_m\text{-ft}}{\text{lb}_f\text{-sec}^2}$$

so that g_c is numerically equal to g_0 but has different units. With respect to gravitational force or weight and mass,

$$F_g\ (\text{lb}_f) = \frac{m\ (\text{lb}_m)\cdot g_e\ (\text{local acceleration, ft/sec})}{g_c\ (\text{lb}_m\text{-ft/lb}_f\text{-sec}^2)}$$

which may also be expressed as the ratio

$$\frac{F_g}{g} = \frac{W}{g} = \frac{m}{g_c} \qquad \frac{\text{lb}_f}{g} = \frac{\text{lb}_m}{g_c}$$

For a given object, the mass is fixed; also g_c has a fixed value, but F_g (or W) may vary *depending* on the local g, with the ratio W/g remaining constant.[2]

The above system was derived using Newton's second law to attempt to make a three-unit system rigorously correct. In an absolute system four basic units are necessary. In other systems (the cgs, mks, slug-ft-sec: 1, 3, and 4 above) the derived force–mass–acceleration relationship is expressed as a dimensional *equivalent*. For example, 1 slug *is equivalent* to 1 pound (force) divided by g (local); 1 gram is equivalent to 1 dyne/g; 1 newton is equivalent to 1 kg/g.[3] In these systems, the equivalence is understood as a consequence of Newton's second law and one unit is replaced by a combination of the others.

Another justification for the use of the *absolute system of units* is that most tables currently using the English system (steam tables, air tables, and so on) give values of the variables *per pound-mass* (lb_m) of the substance (or on a lb_m-mole basis).

CONVERSIONS

The previous section indicates another interesting point: It is not strictly correct to write 1 ft = 0.3048 m ("equals"), since the units are not the same on both sides of the equating sign. What is meant is that 1 ft $\equiv$ 0.3048 m ("equivalent to"), implying that a conversion factor exists between the two methods of specifying a length measurement. This conversion factor is $K_{\text{ft-m}} = 0.3048$ m/ft, so that to convert feet to meters: 1 (ft) $\times$ K (m/ft) = 0.3048 (m). Both sides of the equation are now equal in magnitude and units.

FORCE AND MASS UNITS IN THE ENGLISH SYSTEM

A pound-mass is absolute. It is the mass of 2.73×10^{26} molecules (where $\mathscr{M}$ is the "reference molar mass number" of the substance, sometimes unfortunately called molecular weight). It is, therefore, a property of a given number of molecules of a substance.

[2] It is interesting to note that the illegitimate unit, the "slug," is defined as this ratio:

$$s = \frac{W}{g} = \frac{m}{g_c} \frac{\text{lb}_f\text{-sec}^2}{\text{ft}}$$

and that it does not possess a mass unit in its strange combinational unit.

[3] Or $1\text{ N} = k \dfrac{1\text{ kg}}{1\text{ m/sec}^2} \qquad k = 1 \dfrac{\text{N-sec}^2}{\text{kg-m}}$

A pound-force is absolute. It is established as and defined by the action of the earth's *gravitational force field* (expressed as an equivalent accelerational field) *at a given, specified location on the equator* on 1 pound-mass. A pound-force is a basic, defined, constant (or absolute) unit of "push."

A pound-weight is nonsense. It is defined as the action of the *local* gravitational field (which is expressed as an equivalent accelerational field) on a pound-mass. The *weight* is then a function of the local "*gravity value*" (not a constant!), which is a function of longitude, latitude, and altitude, or, in general, the location or local gravity conditions, which are in turn functions of the masses of the bodies in the surroundings and the distances between them. Weight is *not* a property of a given volume or mass substance.

Consider a given mass (n molecules) of a substance in the shape of a cube. This cube then has a definite mass ($m = nm_i$), where m_i is the mass/molecule. At a particular standard reference point on the equator it will experience a gravitational force $F_{g0} = mg_0/g_c$. "m" is a fixed, absolute quantity, g_c is a conversion constant, g_0 is the measure of the local *standardized* gravity field. At some other point, $F_g = mg_e/g_c$, where g_e is the local acceleration (depending on location). On the moon $F_{gm} = mg_m/g_c$, where g_m is the acceleration at some point on the moon (depending on moon altitude, longitude, and latitude). Obviously F_{g0}, F_g, and F_{gm} are *not* the same. But this is the quantity called "weight." The difficulty occurs because of the association of the variable quantity "weight" with absolute quantities of mass and force as used in everyday experiences.

The statement "I weigh 160 pounds" means: At the point where I stand, the *local gravity* field acts on my mass producing a force of 160 lb_f. On the moon I would *weigh* 27 lb_f, but it is still the same mass. The confusion exists because of the tendency to associate "weight" with mass. To eliminate this ambiguity, the terms weight and specific weight are omitted from this book.

CONCLUSION

The absolute or *F-L-M-t* dimension system is necessary to maintain rigorous accounting of dimensions in equations and to specifically account for variations of local gravity.

PROBLEMS

A-1 Starting with Newton's fundamental law of mutual mass attraction stating that the attracting force between two masses is proportional to the products of the masses and inversely proportional to the distance between their centers, show that g varies with altitude (z) according to

$$g = g_0\left(1 + \frac{z}{r_0}\right)^{-2}$$

where r_0 is the radius of the earth.

A-2 Find the altitudes at which g varies by 1 percent and by 5 percent from g_0.

Answer: 10 and 50 miles

A-3 What speed would a satellite have to have to orbit the earth in a circular path at an altitude of 200 miles?

A-4 Deduce the escape velocity $v_e = \sqrt{2g_0r_0}$ and show that

$$v_z = \frac{v_e}{\sqrt{z[1 + (z/r_0)]}}$$

where v_z is the orbiting speed at altitude z.

A-5 The net force produced by the mutual mass attraction and centrifugal force at the earth's surface is the net "weight" of a body. What portion of g_0 is made up by each of these effects?

APPENDIX B

VECTORS

A vector may be interpreted as a graphical or mathematical representation of a physical quantity that has both magnitude and direction. Examples of vector quantities in mechanics are straight-line distance or displacement between two points, velocity, acceleration, momentum, and so on.

A number of advantages are obtained by using vectors in the study of mechanics. The vector, vector notation, and mathematical operations allow a more direct physical *interpretation* of mathematical equations and their derivations. Mathematical manipulations are simplified by the use of vectors and may be made without specifying a coordinate system. It is much easier to work with a single vector equation in general notation than with an array of equations in a specific coordinate system.

The distance between two points 1 and 2 has a magnitude and a direction. Its magnitude in a cartesian coordinate system is

$$s = \sqrt{(x_2 - x_1)^2 + (y_2 - y_1)^2 + (z_2 - z_1)^2}$$

and its direction cosines are

$$\cos\alpha = \frac{x_2 - x_1}{s} \qquad \cos\beta = \frac{y_2 - y_1}{s} \qquad \cos\gamma = \frac{z_2 - z_1}{s}$$

The quantities s, α, β, and γ are four quantities needed to specify the magnitude and direction of s, and they are all functions of the x, y, and z coordinates of points 1 and 2. Until it is necessary to make actual numerical calculations of point locations, it is much easier to represent this line distance simply as $\vec{s}_{1-2}$. It may be possible that the result would be specified better in a cylindrical coordinate system than in a cartesian coordinate system. The necessary manipulations can be carried out using the general vector notation, and the translation into specific coordinates left until it is necessary to determine specific results. The coordinate system that best fits the geometric characteristics of the problem and the solution can then be chosen so that the determination of a quantity or numerical answer will be made with the least amount of effort and error.

FIELDS

A field is any region of interest where physical or mathematical quantities can be specified, measured, or otherwise determined. As an example of a field in a flowing fluid, the pressure distribution in any flow system or subregion of that system can be considered as a *pressure field*. Pressure is a scalar quantity; therefore, the pressure field is a scalar field. In the same flow region (for example, Fig. 2-1) the velocity distribution can be represented as *velocity vector field*.

COORDINATE SYSTEMS

In locating the point in a field wherein a quantity is to be specified, it is necessary to have some base reference point, or *coordinate system origin*. In connection with a physical flow field within or around a piece of machinery, the coordinate system origin is chosen as the most logical place from which specific points in the field can be easily measured. Any point in the field is then located by a *location* or *position* vector $\vec{r}$ from the origin to the specific point.

In a three-dimensional field, the location vector will have three independent components. These three independent *components* are *not* unique but depend on the orientation and type of coordinate system chosen. *Orthogonal coordinate systems*, where the independent coordinate system component directions are mutually perpendicular, are the most common. Examples of this type of coordinate system are (1) cartesian (x, y, z) (orthogonal, rectilinear), (2) cylindrical polar (r, θ, z), and (3) spherical polar (r, φ, θ) (both orthogonal curvilinear). More complex and not as usual are elliptical, bipolar, and other, coordinate systems, which are also orthogonal and curvilinear. The following material and developments will be concerned only with cartesian, cylindrical, and polar coordinate systems. Position, location, or radius vectors may be specified in any one of these coordinate systems (Fig. 4-2). Each component is a vector, since each has a magnitude and the coordinate direction. The coordinate system orientation is arbitrary but is chosen with the most logical orientation for specifying the points and quantities to be determined for the field under investigation.

The local coordinate directions are indicated by *unit vectors*, which have unity (therefore constant) magnitude, and the local coordinate directions. In a rectilinear coordinate system, these directions are constant; they do not change from one point to any other point in the field. In curvilinear coordinate systems, the magnitudes are always unity, but the directions can possibly change, so the unit vectors are not necessarily constant vectors. A *constant vector* is then one whose *direction and magnitude* are both always constant. In curvilinear coordinate systems the directions of the unit vectors are seen to depend on the field point P in question. The position vector $\vec{r}_P$ expressed in the three example coordinate systems is shown in Fig. 4-2.

A pressure field is specified by giving a value of the pressure at every point in the field, or $p(\vec{r})$, for example, $p(x, y, z)$ or $p(r, \theta, z)$, and so on. If p may also vary with time at a given point, then $p(\vec{r}, t)$ indicates that the p field may be time-varying or nonstationary.

In most physical problems, the coordinate system origin is a point fixed in space and the coordinate system orientation is fixed (Eulerian frame of reference and noninertial system).

A vector (velocity) field would be represented by $\vec{v}(\vec{r})$. The velocity and each of its components are functions of all the space coordinates. If the velocity at a given point also varies with time, $\vec{v}(\vec{r}, t)$, and the flow is unsteady.

VECTOR ALGEBRA

The geometric properties of vectors provide greater meaning to the mathematics of geometry and trigonometry.

If two vectors (say $\vec{A}$ and $\vec{B}$) are equal in magnitude and have the same direction, they are equal (Fig. B-1). Since the vector $\vec{A}_{1-2} = \vec{B}_{3-4}$, a plane in space can be drawn that will contain $\vec{A}$ and $\vec{B}$.

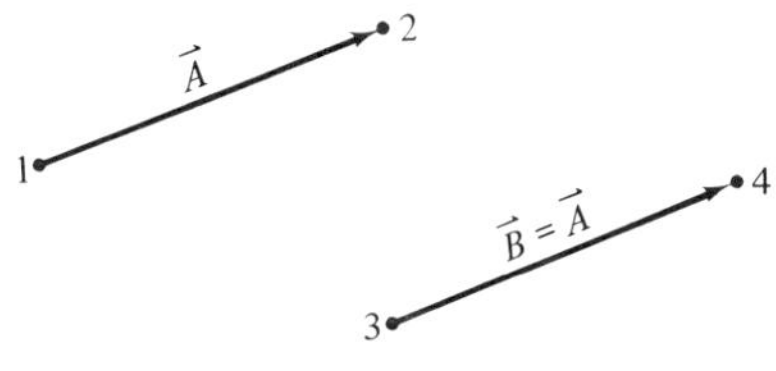

Fig. B-1.

If $\vec{B}$ is equal in magnitude to $\vec{A}$ but opposite in direction, $\vec{B}$ is defined as the negative of $\vec{A}$ (Fig. B-2).

VECTOR ADDITION

The sum of two vectors is obtained by geometrically adding one vector onto the other (Fig. B-3). This addition is possible only if the two vectors

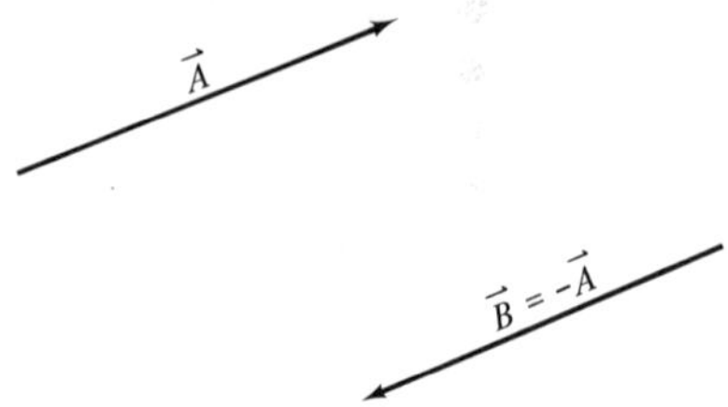

Fig. B-2.

are coplanar. If two vectors intersect, they define a plane. If $\vec{A}$ and $\vec{B}$ are two coplanar vectors in space, addition is accomplished graphically by adding them at their point of intersection, as shown in Fig. B-3. The resultant is the vector $\vec{C} = \vec{A} + \vec{B}$. $\vec{C}$ is seen to be the same if obtained by adding $\vec{B}$ to $\vec{A}$ in the above manner. $\vec{B}$ and $\vec{A}$ can then be *commuted* with respect to the order of addition; that is, vector addition is commutative:

$\vec{A} + \vec{B} = \vec{B} + \vec{A}$ COMMUTATIVE PROPERTY OF VECTOR ADDITION

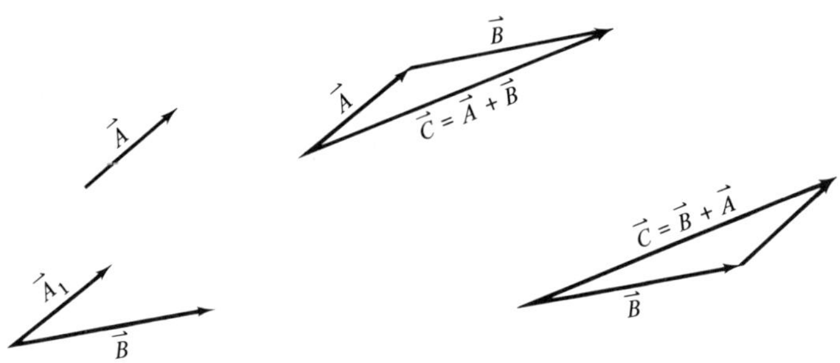

Fig. B-3.

VECTOR SUBTRACTION

The difference of two vectors $\vec{A}$ and $\vec{B}$ would be $\vec{A} - \vec{B}$, which is obtained graphically by first determining $-\vec{B}$ and adding it to $\vec{A}$ (Fig. B-4). Vector subtraction (like scalar subtraction) is seen to be noncommutative: $\vec{C} = \vec{A} - \vec{B}$, $\vec{D} = \vec{B} - \vec{A}$, $\vec{D} = -\vec{C}$.

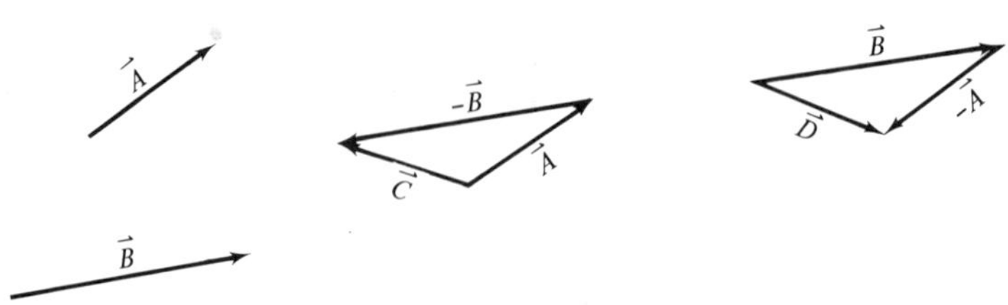

Fig. B-4.

If three or more vectors are combined by addition (and/or subtraction) the resultant can be obtained by combining any two (as indicated by the + or − operation) and combining this resultant with the third, and so on. Geometric constructions similar to those above will show that the order of combination does not affect the resultant vector, or the resultant is independent of the manner in which these steps are *associated*:

$$\vec{A} + (\vec{B} - \vec{C}) = (\vec{A} - \vec{C}) + \vec{B} = (\vec{A} + \vec{B}) - \vec{C} \quad \text{and so on}$$

ASSOCIATIVE PROPERTY OF VECTOR COMBINATION

VECTOR MULTIPLICATION: MULTIPLICATION OF A VECTOR BY A SCALAR

Multiplication of a vector by a scalar can affect only the magnitude of the vector (Fig. B-5). It therefore follows that

$$\alpha\vec{A} = \vec{A}\alpha \qquad \text{COMMUTATIVE PROPERTY OF MULTIPLICATION}$$

$$\alpha(\vec{A} + \vec{B}) = \alpha\vec{A} + \alpha\vec{B} \qquad \text{DISTRIBUTIVE PROPERTY OF MULTIPLICATION}$$

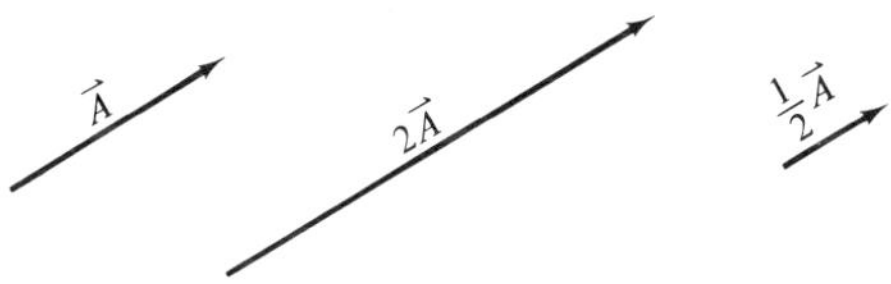

Fig. B-5.

MULTIPLICATION OF A VECTOR BY A VECTOR: SCALAR MULTIPLICATION, SCALAR OR "DOT" PRODUCT

The first type of product, the scalar or "*dot*" product of two vectors, is best considered by means of a physical example. The work done by a force $\vec{F}$ in moving through a linear distance $\vec{s}$, not collinear with the direction of $\vec{F}$, is determined only by the product of the collinear components of $\vec{F}$ and $\vec{s}$. The component of $\vec{F}$ perpendicular to $\vec{s}$ is not found in the definition of the quantity work; that is (Fig. B-6),

$$\text{W} = (\vec{F} \cdot \vec{s})_{\text{collin}}$$

Force and distance are vector quantities, but the resultant work is a scalar quantity. This type of scalar product of $\vec{F}$ and $\vec{s}$ is designated by the dot between the vectors and is therefore termed the *dot product*. The collinear component of $\vec{F}$ with $\vec{s}$ is $F \cos \alpha$; the resultant scalar product of $\vec{F}$ and $\vec{s}$ is then

$$W = \vec{F} \cdot \vec{s} = (F \cos \alpha)s = F(s \cos \alpha) = Fs \cos \alpha$$

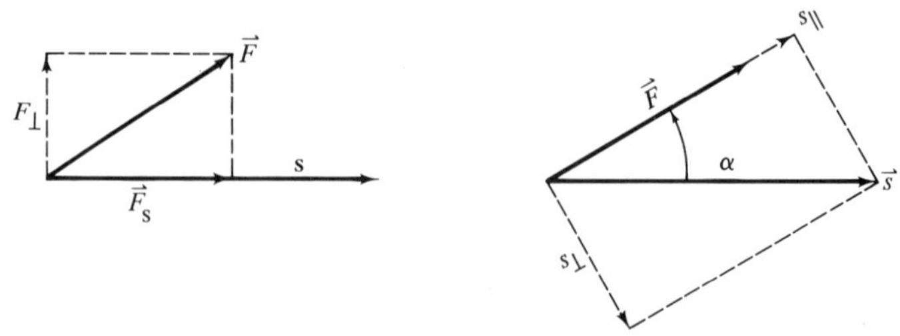

Fig. B-6.

The scalar product is commutative. It may also be represented as

$$\vec{F}\cdot\vec{s} = Fs\cos(F, s)$$

where the notation $(\vec{F}, \vec{s})$ means the angle between $\vec{F}$ and $\vec{s}$ measured in a positive (counterclockwise) sense from $\vec{F}$ toward $\vec{s}$. For example, $(\vec{F}, \vec{s}) = \alpha$, as indicated in Fig. B-6.

Some interesting properties of the vectors and their defined operations are often quite useful. For example, consider the equation

$$\vec{F}\cdot\vec{s} = 0$$

There are three possible solutions to this equation: (1) $\vec{F} = 0$, (2) $\vec{s} = 0$, and (3) $\vec{F}$ is perpendicular to $\vec{s}$. This and other properties of vectors and vector operations allow many simplifications of problems in analytic geometry.

VECTOR OR CROSS PRODUCT

The vector or "cross" product of two vectors is best illustrated with the relationship among force, radius, and torque. If a force acts through a point P at a distance $\vec{r}$ from the point O, a torque about O will be produced. Let the distance OP be the vector $\vec{r}$. The torque is then produced about an axis through O perpendicular to the plane of $\vec{F}$ and $\vec{r}$. Since this torque has an orientation which can be specified with respect to the axis through 0, it has an orientation and magnitude. The positive direction of the torque axis is defined by the "right-hand screw" rotation of the force about the axis through O perpendicular to the plane of $\vec{r}$ and $\vec{F}$, as shown in Fig. B-7. The magnitude of the torque is

$$|\vec{T}| = (r)(F\sin\theta) = (r\sin\theta)(F) = rF\sin\theta$$

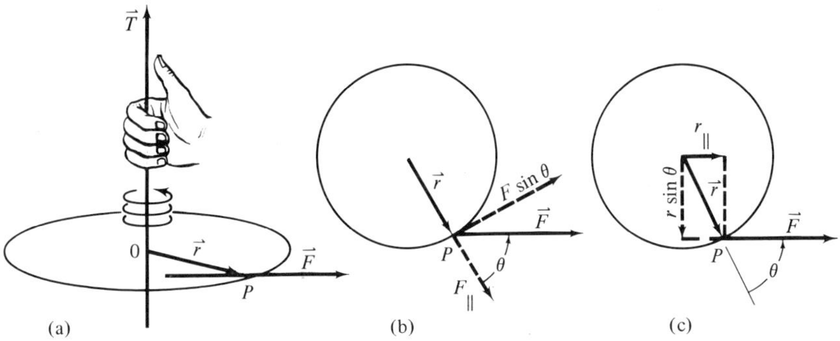

Fig. B-7.

The direction of θ is defined as positive from $\vec{r}$ into $\vec{F}$, as indicated on Fig. B-7b. The direction of $\vec{T}$ is defined as perpendicular to the plane of $\vec{r}$ and $\vec{F}$ and from $\vec{r}$ into $\vec{F}$. Since torque is a vector quantity, the product is termed the vector product and is defined by the relation $\vec{T} = \vec{r} \times \vec{F}$, with the magnitude given by the above relation. The vector defined by $\vec{F} \times \vec{r}$ (from $\vec{F}$ into $\vec{r}$) would have the same magnitude but the opposite direction, or

$$\vec{r} \times \vec{F} = -\vec{F} \times \vec{r}$$

so the vector product is *not* commutative.

The cross product also represents the area of the parallelogram between the two vectors $\vec{B}$ and $\vec{C}$ of Fig. B-8. The direction of the product vector $\vec{A}$ is perpendicular to $\vec{B}$ and $\vec{C}$. A plane area $\vec{A}$ (or elemental area $\overrightarrow{\Delta A}$) is then represented by a vector perpendicular to its surface plane.

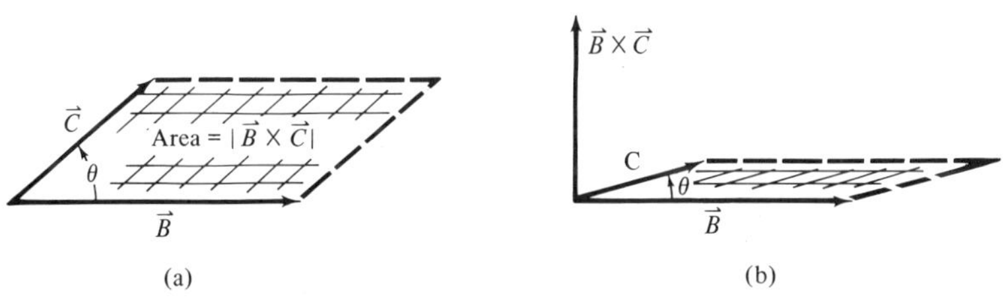

Fig. B 8.

TRIPLE PRODUCTS: TRIPLE SCALAR PRODUCT

The product $\vec{A} \cdot (\vec{B} \times \vec{C})$ represents the volume of a parallelopiped of base area $\vec{B} \times \vec{C}$ times the perpendicular altitude, as illustrated in Fig. B-9. Since volume has no unique direction, the resultant of this product is a

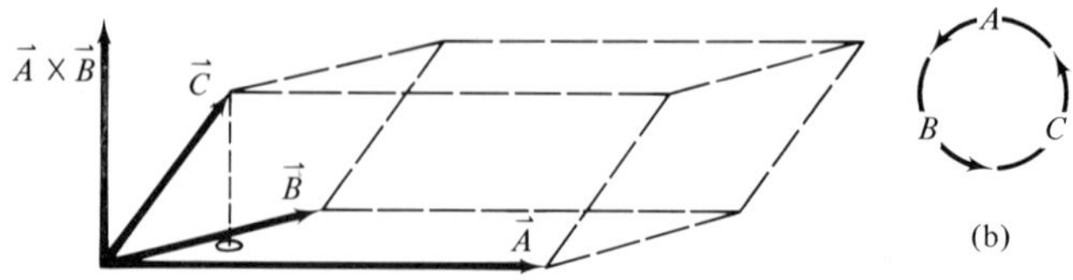

Fig. B-9.

scalar quantity. The same volume will be obtained by any combination of the vectors $\vec{A}$, $\vec{B}$, and $\vec{C}$,

$$V = \vec{A}\cdot(\vec{B} \times \vec{C}) = \vec{B}\cdot(\vec{C} \times \vec{A}) = \vec{C}\cdot(\vec{A} \times \vec{B})$$

but any other combination will produce a resultant opposite in sign to the above. A multiple operation for which this property holds is termed *cyclicly permutative* (Fig. B-9b). If the direction of the cycle is reversed (acyclic), the sense or sign of the result is reversed.

Note that the order of the operation is indicated by the parentheses in the above equation for V. It can be seen that no other possibility of performing the operation exists [for example, $(\vec{A}\cdot\vec{B}) \times \vec{C}$ has no meaning].

TRIPLE VECTOR PRODUCT

The triple vector product, $\vec{A} \times \vec{B} \times \vec{C}$, must give a vector resultant. The order in which the products are taken is now important. By considering a few geometric examples it can be seen that

$$\vec{A} \times (\vec{B} \times \vec{C}) \neq (\vec{A} \times \vec{B}) \times \vec{C}$$

or the triple cross product is *not associative.*

VECTOR-COORDINATE SYSTEM RELATIONSHIPS

To determine numerical results with respect to a given coordinate system origin and orientation, it is necessary to consider the coordinate components of the vector. A vector can be decomposed into coordinate components by using a procedure inverse to that by which three vectors are added to produce a resultant. Figure B-10 shows the vector $\vec{A}$, which is composed of the three components, $\vec{A}_1$, $\vec{A}_2$, $\vec{A}_3$; the three components can be considered to have been obtained by decomposing $\vec{A}$: $A_1 = A \cos(A, x_1)$, $A_2 = A \cos(A, x_2)$, $A_3 = A \cos(A, x_3)$. The angles (A, x_1),

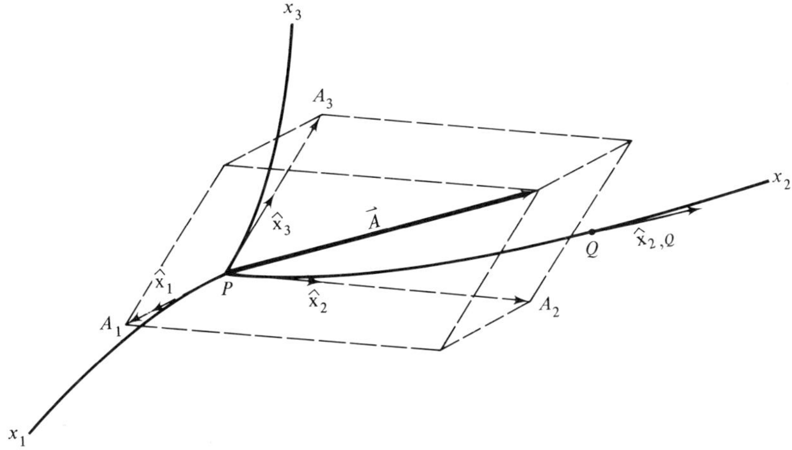

Fig. B-10.

and so on, are the angles between $\vec{A}$ and the coordinate axes. The x_1,x_2,x_3 coordinate system shown is a general one, with no well-defined properties such as orthogonality. It is a nonorthogonal, curvilinear coordinate system: nonorthogonal because the coordinates are not mutually perpendicular and curvilinear because the coordinate directions are not straight lines from P.

UNIT VECTORS

The *directions* of the coordinates at P are given by the *unit vectors* at P. These are vectors of unit magnitude and direction determined by the direction of the coordinate system at the point under consideration. In this text these unit vectors are denoted by a circumflex to avoid confusion and ambiguity with the i,j,k subscript notation. For example, the set of unit vectors can be written in abbreviated form as $\hat{x}_i$, where the i subscript is defined as indicating all possible values of i for the system; in this case, i equals 1, 2, 3. $\vec{A}_1$, $\vec{A}_2$, and $\vec{A}_3$ are the vectors in the $\hat{x}_1$, $\hat{x}_2$, and $\hat{x}_3$ directions that can be combined to give $\vec{A}$,[1]

$$\vec{A} = \vec{A}_1 + \vec{A}_2 + \vec{A}_3$$

Since the A_i components are in the corresponding $\hat{x}_i$ direction, they can also be expressed as $\hat{x}_i A_i$, or

$$\vec{A} = \hat{x}_1 A_1 + \hat{x}_2 A_2 + \hat{x}_3 A_3 = \sum_{i=1}^{3} \hat{x}_i A_i \equiv \hat{x}_i A_i$$

[1] By use of the subscript notation, this statement could be abbreviated to read "A_i are the vectors in the $\hat{x}_i$ directions . . . ."

The summation indication on the last term of the right side is dropped as an *abbreviation convention*: Whenever the same subscript notation is repeated on a single term, the summation over all possibilities for that subscript is assumed.

As previously stated, unit vectors may not be *constant vectors* since their direction may change from one point to another in a curvilinear coordinate system. For example, the direction of $\hat{x}_2$ at Q is different from that of $\hat{x}_2$ at P in Fig. B-10. The unit vectors are (in general) functions of the coordinates, that is, $\hat{x}_i(x_1, x_2, x_3)$, and so on, or $\hat{x}_i(x_j)$.

In an orthogonal rectilinear (cartesian) coordinate system, the unit vectors $\hat{x}$, $\hat{y}$, and $\hat{z}$ are constant vectors: Their magnitudes are unity and at every point in space the coordinate directions are constant.

In the cylindrical polar coordinate system, the unit vector $\hat{r}$ varies from one point to another. For example, at a point with a θ coordinate different from the point P shown in Fig. 4-2b, $\hat{r}$ will have a different direction. However, if θ and z are fixed and only r varies, $\hat{r}$ is constant along r. Similarly, $\hat{r}$ does not vary with z; therefore, $\hat{r}$ (θ only). Similar consideration indicates that $\hat{z}$ is a constant vector in both the cartesian and cylindrical polar coordinate systems. For the case of the spherical polar coordinate system, all the unit vectors are seen to change direction with an arbitrary change of position from the point P to any other point Q.

For plane or two-dimensional coordinate systems (where all field variables are independent of the z coordinate), the field in any plane parallel to the base plane is identical to that in the base plane. The entire flow field can then be represented by the base-plane section, for example, the flow field around a cylinder or air foil of infinite span with its axis parallel to the z axis. In every plane perpendicular to this z axis the flow field is the same. A more physically realizable example is that of the cylinder or air foil spanning a wind tunnel. The flow is essentially independent of the end conditions and therefore can be considered as two-dimensional or "plane."

CURVES IN SPACE: LINE ELEMENT

Consider an arbitrary curve in space (for example, a streamline). This curve can be considered as a locus of points specified by their locations along s, by the position vectors $\vec{r}$ of these points (Fig. B-11). A line element is then represented by $\overrightarrow{\Delta s}$ along s, which is identical to $\overrightarrow{\Delta r}$. The line element ($\overrightarrow{\Delta s}$ or $\overrightarrow{\Delta r}$) can be considered to be composed of its coordinate vector components, for example,

$$\begin{aligned}\overrightarrow{\Delta s} = \overrightarrow{\Delta r} &= \hat{x}\,\Delta x + \hat{y}\,\Delta y + \hat{z}\,\Delta z = \hat{r}\,\Delta r + \hat{\theta} r\,\Delta\theta + \hat{z}\,\Delta z \\ &= \hat{r}\,\Delta r + \hat{\varphi} r\,\Delta\varphi + \hat{\theta} r \sin\varphi\,\Delta\theta = \hat{x}_i\,\Delta x_i\end{aligned}$$

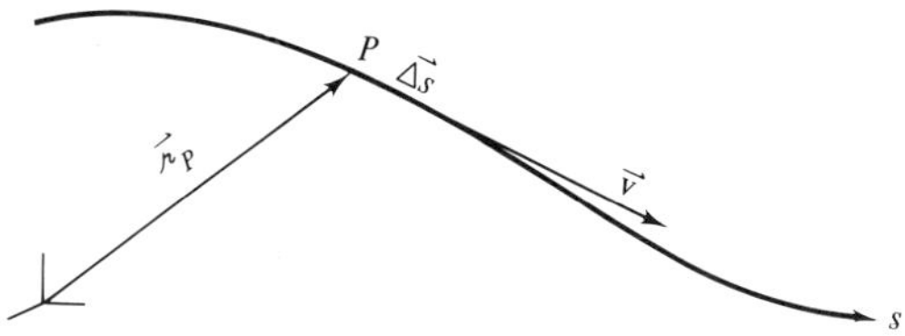

Fig. B-11.

where again $\hat{x}_i$ and Δx_i are used to denote the coordinate unit vectors and components of Δs in any arbitrary coordinate system. Note that the Δx_i are defined as coordinate *length* increments, and not just coordinate increments. For example, $\Delta r_\varphi = \Delta x_\varphi = r\,\Delta\varphi$.

In the limit for a differential line element, $\vec{\Delta s}$ approaches $\vec{ds}$ ($\vec{\Delta s} \to \vec{ds}$):

	$\hat{x}_i$	dx_i
Cartesian	$\hat{x}, \hat{y}, \hat{z}$	dx, dy, dz
Cylindrical polar	$\hat{r}, \hat{\theta}, \hat{z}$	$dr, r\,d\theta, dz$
Spherical polar	$\hat{r}, \hat{\varphi}, \hat{\theta}$	$dr, r\,d\varphi, r\sin\varphi\,d\theta$

DERIVATIVE OF A VECTOR

CHANGE IN POSITION WITH RESPECT TO TIME

The change in position of a point or particle as it moves along a path s with respect to time (t), or the derivative of position along s with respect to time, is defined by the operation

$$\lim_{\delta t \to 0} \frac{\vec{\delta s}}{\delta t} = \frac{\vec{ds}}{dt} = \frac{\vec{dr}}{dt} = \vec{v}$$

Since $\vec{\Delta s}$ and $\vec{\Delta r}$ are vectors and t is a scalar, the result is the vector velocity in the direction of s at the point P (Fig. B-12).

Since $\vec{r} = \hat{x}_i x_i$, $\vec{dr}/dt$ or $\vec{ds}/dt$ implies the derivations of a product

$$\vec{v} = \frac{\vec{ds}}{dt} = \frac{d}{dt}(\hat{x}_i x_i) = \hat{x}_i \frac{dx_i}{dt} + x_i \frac{d\hat{x}_i}{dt}$$

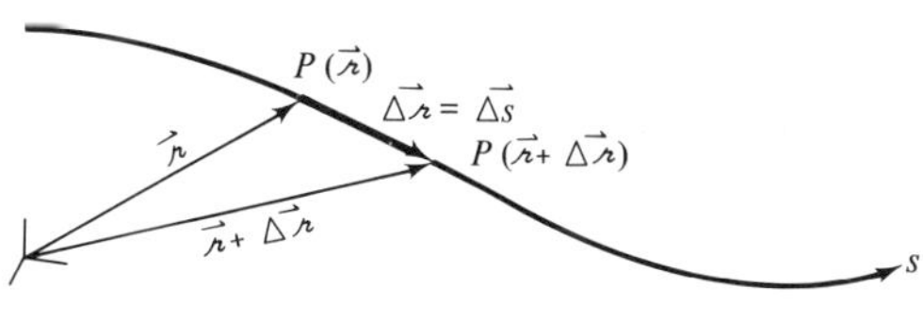

Fig. B-12.

but $d\hat{x}_i/dt$ is zero for any coordinate system, since the unit vectors are functions of space only and therefore are not time-varying (where the coordinate system origin is fixed in space). The component speeds dx_i/dt can be denoted by v_i; then

$$\vec{v} = \frac{\vec{ds}}{dt} = \hat{x}_i \frac{dx_i}{dt} = \hat{x}_i v_i$$

For example,

$$\vec{v} = \hat{x}u + \hat{y}v + \hat{z}w = \hat{r}v_r + \hat{\theta}v_\theta + \hat{z}w_z = \hat{r}v_r + \hat{\varphi}v_\varphi + \hat{\theta}v_\theta$$

In the above case, where s is any arbitrary line in space, $\vec{ds}/dt$ gives the rate of change of s along that line. If s is specifically a streamline (denoted by ψ), which is a line defined as drawn everywhere tangent to the velocity, $\vec{ds}/dt$ is the field velocity $\vec{v}$.

DERIVATIVE OF A VECTOR

Next consider $\vec{dv}/dt$. $\vec{v}$ is a function of the position P where it is measured or to be determined, and it may be a function of time: $\vec{v}(\vec{r}, t)$, or $\vec{v}(x_i, t)$. $\vec{v}$ may then be a function of four independent coordinates, three space coordinates and time. By taking the literal definition of the derivative of the velocity $\vec{v}$:

$$\frac{\vec{dv}}{dt} = \frac{d}{dt}[\vec{v}(\hat{x}_i, t)] = \frac{\partial \vec{v}}{\partial t} + \frac{\partial \vec{v}}{\partial x_i}\frac{\partial x_i}{\partial t} = \frac{\partial \vec{v}}{\partial t} + v_i \frac{\partial \vec{v}}{\partial x_i} = \frac{\partial(\hat{x}_i v_i)}{\partial t} + v_i \frac{\partial(\hat{x}_j v_j)}{\partial x_i}$$

The quantity $\partial x_i/\partial t$ is just the x_i component of the velocity, or v_i. Only in the case of a cartesian coordinate system are the unit vectors ($\hat{x}_i$ or $\hat{x}_j$) independent (or nonconstant), so the above derivative can be written in expanded form as

$$\begin{aligned}\frac{\vec{dv}}{dt} &= \hat{x}\left(\frac{\partial u}{\partial t} + u\frac{\partial u}{\partial x} + v\frac{\partial u}{\partial y} + w\frac{\partial u}{\partial z}\right)\\ &+ \hat{y}\left(\frac{\partial v}{\partial t} + u\frac{\partial v}{\partial x} + v\frac{\partial v}{\partial y} + w\frac{\partial v}{\partial z}\right)\\ &+ \hat{z}\left(\frac{\partial w}{\partial t} + u\frac{\partial w}{\partial x} + v\frac{\partial w}{\partial y} + w\frac{\partial w}{\partial z}\right) = \hat{x}a_x + \hat{y}a_y + \hat{z}a_z = \vec{a}\end{aligned}$$

The vector $\vec{a}$ is the acceleration, which is made up of two parts, $\partial\vec{v}/\partial t$ (the variation of the velocity at a fixed point in space as the field varies with time) and the variation of velocity with respect to direction at a fixed instant of time. The significance of the second part is easier to see if it is considered in scalar form as

$$\frac{d\mathrm{v}}{dt} = \frac{\partial \mathrm{v}}{\partial t} + \frac{\partial \mathrm{v}}{\partial s}\frac{\partial s}{\partial t} = \frac{\partial \mathrm{v}}{\partial t} + \mathrm{v}\frac{\partial \mathrm{v}}{\partial s}$$

where s is a streamline curve, so that v is only in the s direction. The term $\mathrm{v}(\partial \mathrm{v}/\partial s)$ is termed the "*convective acceleration*" or the acceleration due to the variation in velocity from one point to another when measured with respect to a coordinate system, the origin of which is fixed in space. The term $\mathrm{v}(\partial \mathrm{v}/\partial s)$ is the simple scalar from of the vector quantity $v_i[\partial(\hat{\mathrm{x}}_i \mathrm{v}_i)/\partial x_i]$, which is usually written in the form $\vec{\mathrm{v}} \cdot \nabla \vec{\mathrm{v}}$, where the symbol ∇ ("del") is used to denote the vector differential operator $\nabla \equiv \hat{\mathrm{x}}_i(\partial/\partial x_i)$.

As an example, consider the acceleration in a two-dimensional plane polar coordinate system:

$$\vec{a}(\vec{r}, t) = \vec{a}(r, \theta, t) = \frac{d}{dt}\vec{\mathrm{v}}(r, \theta, t)$$

$$= \frac{d}{dt}[\hat{\mathrm{r}}(r, \theta)u_r(r, \theta, t) + \hat{\theta}(r, \theta)v_\theta(r, \theta, t)]$$

$$= \hat{\mathrm{r}}\frac{\partial u_r}{\partial t} + \frac{\partial(\hat{\mathrm{r}}u_r)}{\partial r}\frac{\partial r}{\partial t} + \hat{\theta}\frac{\partial v_\theta}{\partial t} + \frac{\partial(\hat{\theta}v_\theta)}{r\,\partial\theta}\frac{r\,\partial\theta}{\partial t}$$

(Note that $\Delta x_\theta = r\,\Delta\theta$ or $\partial x_\theta = r\,\partial\theta$ and $r(\partial\theta/\partial t)$ is then v_θ, the tangential or "θ-velocity" component; the radial velocity component is $u_r = \partial r/\partial t$.)

Expanding the above gives

$$\vec{a}(r, \theta, t) = \hat{\mathrm{r}}\frac{\partial u_r}{\partial t} + u_r\left(\frac{\partial \hat{\mathrm{r}}}{\partial r}u_r + \hat{\mathrm{r}}\frac{\partial u_r}{\partial r}\right) + \hat{\theta}\frac{\partial v_\theta}{\partial t} + v_\theta\left(\frac{\partial\hat{\theta}}{r\,\partial\theta}v_\theta + \hat{\theta}\frac{\partial v_\theta}{r\,\partial\theta}\right)$$

At this point it is required that a method be available for evaluating the derivatives of the unit vectors in a curvilinear coordinate system before the radial and tangential components of the acceleration can be determined.

DERIVATIVES OF THE UNIT VECTORS

The most direct way to determine the derivatives of the unit vectors is by a graphical evaluation of the rates of change of the unit vectors with respect to the directions indicated. It was previously stated that the derivatives of all nine unit vectors in the cartesian coordinate system are zero because the vectors $\hat{\mathrm{x}}$, $\hat{\mathrm{y}}$, and $\hat{\mathrm{z}}$ are constant vectors; that is,

$$\frac{\partial\hat{\mathrm{x}}}{\partial x} = 0, \quad \frac{\partial\hat{\mathrm{x}}}{\partial y} = 0, \quad \frac{\partial\hat{\mathrm{x}}}{\partial z} = 0, \quad \ldots \qquad \text{or} \qquad \frac{\partial\hat{\mathrm{x}}_i}{\partial x_j} = 0$$

To determine the partial derivatives of the unit vectors in the plane or cylindrical polar coordinate system, their rates of change with each of the coordinate directions will be determined. The derivatives of the unit vectors to be determined in the cylindrical polar coordinate system are

$$\frac{\partial \hat{\mathbf{r}}}{\partial r} \qquad \frac{\partial \hat{\mathbf{r}}}{r\,\partial \theta} \qquad \frac{\partial \hat{\mathbf{r}}}{\partial z}$$

$$\frac{\partial \hat{\boldsymbol{\theta}}}{\partial r} \qquad \frac{\partial \hat{\boldsymbol{\theta}}}{r\,\partial \theta} \qquad \frac{\partial \hat{\boldsymbol{\theta}}}{\partial z}$$

$$\frac{\partial \hat{\mathbf{z}}}{\partial r} \qquad \frac{\partial \hat{\mathbf{z}}}{r\,\partial \theta} \qquad \frac{\partial \hat{\mathbf{z}}}{\partial z}$$

In order to determine $\partial \hat{\mathbf{r}}/\partial r$, consider the unit vector $\hat{\mathbf{r}}$ at some point $P(r, \theta)$, as shown in Fig. B-13. The change in $\hat{\mathbf{r}}$ with r only can be obtained by determining $\hat{\mathbf{r}}(r + \Delta r)$ at $P(r + \Delta r, \theta)$. The change in $\hat{\mathbf{r}}$ is then

$$\hat{\mathbf{r}}(r + \Delta r, \theta) - \hat{\mathbf{r}}(r, \theta)$$

and

$$\frac{\partial \hat{\mathbf{r}}}{\partial r} = \lim_{\Delta r \to 0} \frac{\hat{\mathbf{r}}(r + \Delta r, \theta) - \hat{\mathbf{r}}(r, \theta)}{\Delta r}$$

The unit vector in the r direction at P' is a vector of magnitude unity and direction determined by the coordinate r direction at P'. The *change* in the unit vector is then

$$\hat{\mathbf{r}}(r + \Delta r, \theta) - \hat{\mathbf{r}}(r, \theta) = \left(\hat{\mathbf{r}} + \frac{\partial \hat{\mathbf{r}}}{\partial r}\Delta r\right) - \hat{\mathbf{r}} = \frac{\partial \hat{\mathbf{r}}}{\partial r}\Delta r$$

By performing the graphical vector substitution subtraction indicated, the vector $(\partial \hat{\mathbf{r}}/\partial r)\,\Delta r$ is zero, since there is no change in magnitude and no change in direction of $\hat{\mathbf{r}}$ in going from P to P'.

The direction of $\hat{\mathbf{r}}$ *does* change with the change in angle θ, as seen in Fig. B-13b. The direction of the derivative of $\hat{\mathbf{r}}$ with respect to the direction of θ can be evaluated from

$$\frac{\partial \hat{\mathbf{r}}}{\partial \theta} = \lim_{\Delta \theta \to 0} \frac{\hat{\mathbf{r}}(r, \theta + \Delta\theta) - \hat{\mathbf{r}}(r, \theta)}{\Delta \theta} = \lim_{\Delta \theta \to 0} \frac{[\hat{\mathbf{r}} + (\partial \hat{\mathbf{r}}/\partial \theta)\,\Delta\theta] - \hat{\mathbf{r}}}{\Delta \theta}$$

The process of the vector subtraction indicated shows that the direction of $(\partial \hat{\mathbf{r}}/\partial \theta)\,\Delta\theta$ is $+\hat{\boldsymbol{\theta}}$. The magnitude is determined from the vector triangle of Fig. B-13b. Since $\Delta\theta$ is a very small angle (and considering magnitudes only)

$$|\hat{\mathbf{r}}\,\Delta\theta| = \left|\frac{\partial \hat{\mathbf{r}}}{\partial \theta}\Delta\theta\right| \qquad \text{or} \qquad \left|\frac{\partial \hat{\mathbf{r}}}{\partial \theta}\right| = |\hat{\mathbf{r}}| = 1$$

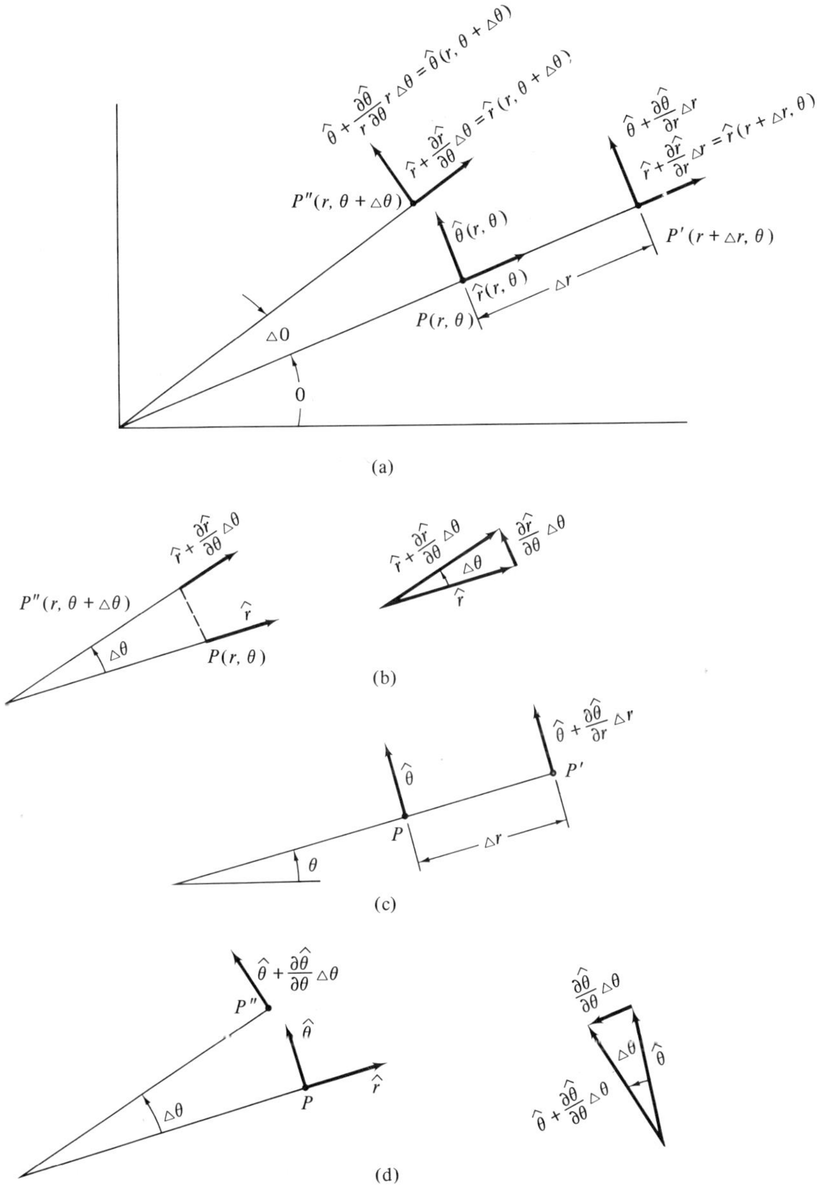
$\hat{\theta} + \frac{\partial\hat{\theta}}{r\,\partial\theta} r\,\Delta\theta = \hat{\theta}(r, \theta + \Delta\theta)$
$\hat{r} + \frac{\partial\hat{r}}{\partial\theta}\Delta\theta = \hat{r}(r, \theta + \Delta\theta)$
$\hat{\theta} + \frac{\partial\hat{\theta}}{\partial r}\Delta r$
$\hat{r} + \frac{\partial\hat{r}}{\partial r}\Delta r = \hat{r}(r + \Delta r, \theta)$
$P''(r, \theta + \Delta\theta)$
$\hat{\theta}(r, \theta)$
$P'(r + \Delta r, \theta)$
$\hat{r}(r, \theta)$
Δr
$P(r, \theta)$
$\Delta\theta$
θ
(a)
$\hat{r} + \frac{\partial\hat{r}}{\partial\theta}\Delta\theta$
$P''(r, \theta + \Delta\theta)$
$\hat{r}$
$\Delta\theta$
$P(r, \theta)$
$\frac{\partial\hat{r}}{\partial\theta}\Delta\theta$
(b)
$\hat{\theta} + \frac{\partial\hat{\theta}}{\partial r}\Delta r$
$\hat{\theta}$
P'
P
Δr
θ
(c)
$\hat{\theta} + \frac{\partial\hat{\theta}}{\partial\theta}\Delta\theta$
P''
$\hat{\theta}$
$\hat{r}$
P
$\Delta\theta$
$\frac{\partial\hat{\theta}}{\partial\theta}\Delta\theta$
(d)

Fig. B-13.

The result is obtained that $\partial\hat{\mathbf{r}}/\partial\theta$ is a unit vector in the $+\theta$ direction, or $\partial\hat{\mathbf{r}}/\partial\theta = +\hat{\boldsymbol{\theta}}$.

The derivatives of the unit vector $\hat{\boldsymbol{\theta}}$ are determined in the same manner (Fig. B-13c and d). The value of $\partial\hat{\boldsymbol{\theta}}/\partial r$ is zero, since there is no change in direction involved. A nonzero value of the derivative of the unit vector $\hat{\boldsymbol{\theta}}$ in the θ direction is expected, since θ is a curvilinear coordinate. Consider

$$\frac{\partial\hat{\boldsymbol{\theta}}}{\partial\theta} = \lim_{\Delta\theta\to 0} \frac{\hat{\boldsymbol{\theta}}(r, \theta + \Delta\theta) - \hat{\boldsymbol{\theta}}(r, \theta)}{\Delta\theta} = \frac{(\partial\hat{\boldsymbol{\theta}}/\partial\theta)\,\Delta\theta}{\Delta\theta}$$

The vector quantity $(\partial\hat{\boldsymbol{\theta}}/\partial\theta)\,\Delta\theta$ has the direction $-\hat{\mathbf{r}}$. The magnitude is determined from the vector triangle of Fig. B-13d,

$$|\hat{\boldsymbol{\theta}}\,\Delta\theta| = \left|\frac{\partial\hat{\boldsymbol{\theta}}}{\partial\theta}\Delta\theta\right| \qquad \text{or} \qquad \left|\frac{\partial\hat{\boldsymbol{\theta}}}{\partial\theta}\right| = |\hat{\boldsymbol{\theta}}| = 1$$

which gives $\partial\hat{\boldsymbol{\theta}}/\partial\theta = -\hat{\mathbf{r}}$.

All partial derivatives of all unit vectors with respect to the z direction are zero, since the change in z (that is, Δz) will produce no change in direction of any of the other unit vectors. Also, the change in $\hat{\mathbf{z}}$ with respect to the remainder of the coordinates is zero. In particular, $\partial\hat{\mathbf{z}}/\partial\theta$ is seen to be zero, since $\hat{\mathbf{z}}$ is always perpendicular to the plane in which θ changes (while r is held constant), and therefore no change in the direction of $\hat{\mathbf{z}}$ is produced with a change of θ only. The derivatives of the unit vectors for the cylindrical polar coordinate system can be presented in tabular (matrix) form:

$$\begin{array}{ccc}
\dfrac{\partial\hat{\mathbf{r}}}{\partial r} = 0 & \dfrac{\partial\hat{\mathbf{r}}}{\partial\theta} = \hat{\boldsymbol{\theta}} & \dfrac{\partial\hat{\mathbf{r}}}{\partial z} = 0 \\[2ex]
\dfrac{\partial\hat{\boldsymbol{\theta}}}{\partial r} = 0 & \dfrac{\partial\hat{\boldsymbol{\theta}}}{\partial\theta} = -\hat{\mathbf{r}} & \dfrac{\partial\hat{\boldsymbol{\theta}}}{\partial z} = 0 \\[2ex]
\dfrac{\partial\hat{\mathbf{z}}}{\partial r} = 0 & \dfrac{\partial\hat{\mathbf{z}}}{\partial\theta} = 0 & \dfrac{\partial\hat{\mathbf{z}}}{\partial z} = 0
\end{array}$$

which can be abbreviated in matrix form as

$$\frac{\partial(\hat{\mathbf{r}}, \hat{\boldsymbol{\theta}}, \hat{\mathbf{z}})}{\partial(r, \theta, z)} = \begin{pmatrix} 0 & \hat{\boldsymbol{\theta}} & 0 \\ 0 & -\hat{\mathbf{r}} & 0 \\ 0 & 0 & 0 \end{pmatrix}$$

The derivatives of the unit vectors in the spherical polar coordinate system are more difficult to evaluate. The method of graphical evaluation again gives the easiest and most direct method of evaluating these derivatives.

GENERAL UNIT-VECTOR-DERIVATIVE RULES

In the above derivation methods, two general procedural rules are evident. A third rule is available for orthogonal coordinate systems as a consequence of the cross-product relation

$$\hat{\mathrm{x}}_1 \times \hat{\mathrm{x}}_2 = \hat{\mathrm{x}}_3 \qquad \text{or} \qquad \hat{\mathrm{x}}_i \times \hat{\mathrm{x}}_j = \hat{\mathrm{x}}_k$$

where *ijk* is any right-hand cyclic permutation of the coordinates. These rules are:

1. The derivative of any unit vector with respect to a rectilinear coordinate (for example, r) is zero.
2. The derivative of any unit vector with respect to a curvilinear coordinate, the plane of variation of which is perpendicular to the unit vector, is zero (for example, $\partial\hat{\mathrm{z}}/\partial\theta = 0$ in the cylindrical polar coordinate system).
3. If the derivatives of two unit vectors with respect to a coordinate are known, the derivative of the third with respect to that coordinate may be obtained from the derivative of the cross product:

$$\frac{\partial \hat{\mathrm{x}}_3}{\partial x_1} = \frac{\partial}{\partial x_1}(\hat{\mathrm{x}}_1 \times \hat{\mathrm{x}}_2) = \frac{\partial \hat{\mathrm{x}}_1}{\partial x_1} \times \hat{\mathrm{x}}_2 + \hat{\mathrm{x}}_1 \times \frac{\partial \hat{\mathrm{x}}_2}{\partial x_1}$$

This relation may also be used as a check on the graphical determinations.

UNIT-VECTOR DERIVATIVES IN THE SPHERICAL POLAR COORDINATE SYSTEM

The relations among the unit vectors and the coordinates are shown in Fig. 4-2c. The derivatives to be found, and the results, are

$$\frac{\partial(\hat{\mathrm{r}}, \hat{\varphi}, \hat{\theta})}{\partial(r, \varphi, \theta)} = \begin{pmatrix} \dfrac{\partial\hat{\mathrm{r}}}{\partial r} & \dfrac{\partial\hat{\mathrm{r}}}{\partial\varphi} & \dfrac{\partial\hat{\mathrm{r}}}{\partial\theta} \\ \dfrac{\partial\hat{\varphi}}{\partial r} & \dfrac{\partial\hat{\varphi}}{\partial\varphi} & \dfrac{\partial\hat{\varphi}}{\partial\theta} \\ \dfrac{\partial\hat{\theta}}{\partial r} & \dfrac{\partial\hat{\theta}}{\partial\varphi} & \dfrac{\partial\hat{\theta}}{\partial\theta} \end{pmatrix} = \begin{pmatrix} 0 & \hat{\varphi} & \hat{\theta}\sin\varphi \\ 0 & -\hat{\mathrm{r}} & \hat{\theta}\cos\varphi \\ 0 & 0 & -\hat{\mathrm{r}}\sin\varphi - \hat{\varphi}\cos\varphi \end{pmatrix}$$

By rule 1, the left-hand column elements are all zero. By rule 2, $\partial\hat{\theta}/\partial\varphi$ is zero. The remaining five derivatives must be calculated. The determination of $\partial\hat{\mathrm{r}}/\partial\varphi$ is identical to the determination of $\partial\hat{\mathrm{r}}/\partial\theta$ in the cylindrical coordinate system, so that $\partial\hat{\mathrm{r}}/\partial\varphi = \hat{\varphi}$; likewise, $\partial\hat{\varphi}/\partial\varphi = -\hat{\mathrm{r}}$.

The derivative $\partial\hat{\varphi}/\partial\theta$ is evaluated from the geometry of Fig. B-14. The vector variation $(\partial\hat{\varphi}/\partial\theta)\,\Delta\theta$ at P' has the same magnitude and direction when projected on the base plane. The direction of the derivative is $+\hat{\theta}$. Its magnitude is determined in the plane of variation of the independent variable (θ), the base plane in this case. The magnitude is obtained from the base-plane triangle from the condition that the arc length $(\partial\hat{\varphi}/\partial\theta)\,\Delta\theta$ is equal to the radius-angle product $|\hat{\varphi}\,\Delta\theta\cos\varphi|$. The result is $\partial\hat{\varphi}/\partial\theta = \hat{\theta}\cos\varphi$.

The derivative $\partial\hat{\mathrm{r}}/\partial\theta$ is determined in exactly the same manner. The projection of $\hat{\mathrm{r}}$ on the base plane is $\hat{\mathrm{r}}\sin\varphi$, which indicates that $\partial\hat{\mathrm{r}}/\partial\theta = \hat{\theta}\sin\varphi$.

The remaining derivative $\partial\hat{\theta}/\partial\theta$ is more difficult to determine by the geometric procedure. Its variation is again determined in the base plane, but its direction of variation is radially inward *in the base plane*. This direction has both $-\hat{\mathrm{r}}$ and $-\hat{\varphi}$ components. Since all other derivatives have been found, $\partial\hat{\theta}/\partial\theta$ may also be obtained by using rule 3:

$$\frac{\partial\hat{\theta}}{\partial\theta} = \frac{\partial(\hat{\mathrm{r}}\times\hat{\varphi})}{\partial\theta} = \frac{\partial\hat{\mathrm{r}}}{\partial\theta}\times\hat{\varphi} + \hat{\mathrm{r}}\times\frac{\partial\hat{\varphi}}{\partial\theta} = (\hat{\theta}\sin\varphi)\times\hat{\varphi} + \hat{\mathrm{r}}\times(\hat{\theta}\cos\varphi)$$

$$\frac{\partial\hat{\theta}}{\partial\theta} = -\hat{\mathrm{r}}\sin\varphi - \hat{\varphi}\cos\varphi$$

ILLUSTRATIVE EXAMPLE: **div $\vec{\mathrm{v}}\,(r, \varphi, \theta)$**

The mathematical manipulation implied is

$$\nabla\cdot\vec{\mathrm{v}} = \left(\hat{\mathrm{r}}\frac{\partial}{\partial r} + \hat{\varphi}\frac{\partial}{r\,\partial\varphi} + \frac{\hat{\theta}}{r\sin\varphi\,\partial\theta}\right)\cdot(\hat{\mathrm{r}}u_r + \hat{\varphi}v_\varphi + \hat{\theta}w_\theta)$$

There are three derivatives of three two-term products, indicating the possibility of 18 terms:

$$\nabla\cdot\vec{\mathrm{v}} = \hat{\mathrm{r}}\cdot\frac{\partial}{\partial r}(\hat{\mathrm{r}}u_r + \hat{\varphi}v_\varphi + \hat{\theta}w_\theta) + \frac{\hat{\varphi}}{r}\cdot\frac{\partial}{\partial\varphi}(\hat{\mathrm{r}}u_r + \hat{\varphi}v_\varphi + \hat{\theta}w_\theta)$$
$$+ \frac{\hat{\theta}}{r\sin\dot{\varphi}\,\partial\theta}\cdot(\hat{\mathrm{r}}u_r + \hat{\varphi}v_\varphi + \hat{\theta}w_\theta)$$

Since all $\partial\hat{\mathrm{x}}_i/\partial r$ are zero, the r derivatives give only $\partial u_r/\partial r$.

The φ-derivative term gives $\partial v_\varphi/r\,\partial\varphi$ plus $\partial\hat{\mathrm{x}}_i/\partial\varphi$ terms with φ components. From the partial differential array, the only one is $\partial\hat{\mathrm{r}}/\partial\varphi = \hat{\varphi}$. The only additional term in the $\partial/\partial\varphi$ derivative is then

$$\frac{\hat{\varphi}}{r}\frac{\partial}{\partial\varphi}(\hat{\mathrm{r}}u_r) = \frac{\hat{\varphi}}{r}\cdot\hat{\varphi}u_r = \frac{u_r}{r}$$

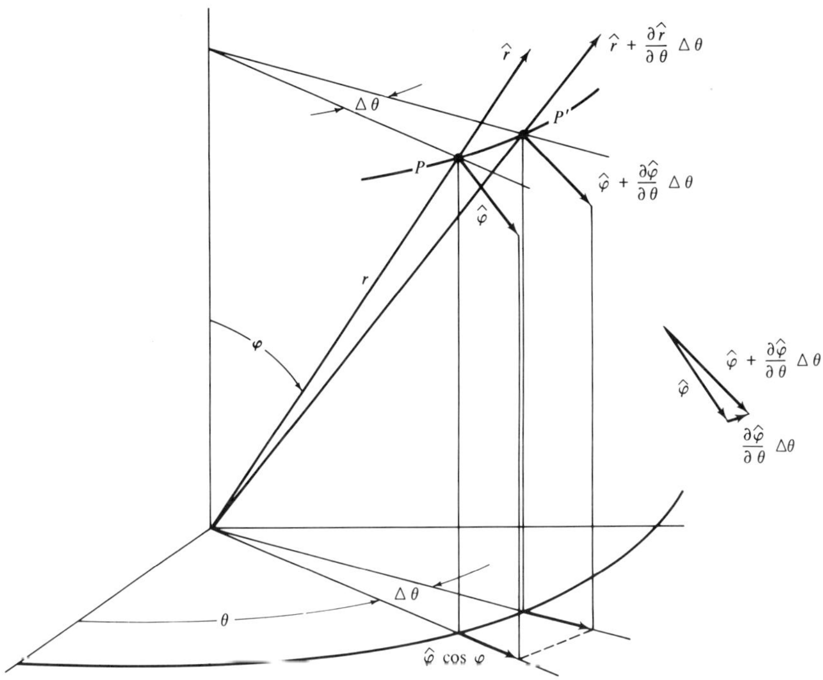

(a)

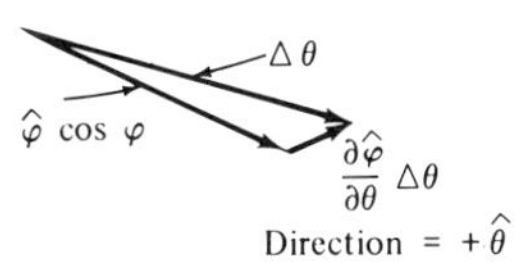

$$\left| \frac{\partial \hat{\varphi}}{\partial \theta} \Delta \theta \right| = \left| \hat{\varphi} \cos \varphi \, \Delta \theta \right|$$

$$\frac{\partial \hat{\varphi}}{\partial \theta} = \hat{\theta} \cos \varphi$$

(b)

Fig. B-14.

The θ-derivative term gives $\partial w_\theta / r \sin\varphi\, \partial\theta$ plus the two derivatives of unit vectors with θ components: $\partial\hat{r}/\partial\theta$ and $\partial\hat{\varphi}/\partial\theta$,

$$\frac{\hat{\theta}}{r \sin\theta} \frac{\partial}{\partial\theta} (\hat{r} u_r + \hat{\varphi} v_\varphi) = \frac{u_r}{r} + \frac{v_\varphi}{r} \frac{\cos\varphi}{\sin\varphi}$$

The result is then

$$\nabla \cdot \vec{v}(r, \varphi, \theta) = \frac{\partial u_r}{\partial r} + \frac{2u_r}{r} + \frac{\partial v_\varphi}{r\, \partial\varphi} + \frac{v_\varphi}{r} \frac{\cos\varphi}{\sin\varphi} + \frac{\partial w_\theta}{r \sin\varphi\, \partial\theta}$$

which may be contracted to

$$\nabla \cdot \vec{v}(r, \varphi, \theta) = \frac{1}{r} \frac{\partial(r^2 u_r)}{\partial r} + \frac{\partial(v_\varphi \sin\varphi)}{r\, \partial\varphi} + \frac{\partial w_\theta}{r \sin\varphi\, \partial\theta}$$

Other operations involving the curl of a vector and the divergence of a dyadic are evaluated in a similar manner.

It is much easier and more accurate to obtain such expansions in various coordinate systems by this direct method than by making a coordinate transformation from a cartesian coordinate system expansion of the quantity.

APPENDIX C

SIMPLIFIED ELEMENTARY KINETIC THEORY OF A PERFECT GAS

The classical method of the study of thermodynamics considers the relationships between work and energy and methods for calculating states and processes for various substances as functions of the physical properties of the substance. No detailed consideration of the physical (atomic) structure of the substances is assumed or used in these methods. In many cases, certain simplifying hypotheses are made to make calculations and predictions possible, the most notable of which is the assumption of a perfect gas. Fortunately, perfect gas behavior is approached by most gases under certain "favorable" conditions of pressures below the critical pressure and temperatures above the critical temperature.[1]

Additional information can be obtained if the structure and behavior of the "working substance" is examined with respect to its atomic and molecular structure and composition. Insight into the reasons for the characteristic behavior of many substances can be obtained by making certain idealizing assumptions regarding the microstructure (molecular configuration) of the substance.

An idealized perfect gas will be considered from the microscopic viewpoint. The primary perfect gas assumptions are: (1) there is a very large number of molecules, uniformly distributed in space at any instant, or that any position for any molecule is equally likely, and (2) the motion of the molecules is completely random with respect to the speed and direction of travel.

A given number of molecules has a definite mass equal to the mass of one molecule times the number of molecules:

$$m = N m_i$$

where m is the mass, N the number of molecules of the particular mass under consideration, and m_i the mass of one molecule.

[1] An upper limit on temperature is imposed as that at which the molecules start to dissociate. A lower limit on density exists when the mean free path approaches the characteristic dimension of the physical system in which the gas is in contact.

The *molecular mass* or *molar mass*[2] of any substance is the mass corresponding to the molecular mass number $\mathcal{M}$ of that substance. For example, a molar mass of O_2 is 32 kg in the mks system, or 32 lb_m in the English (lb_m-lb_f-ft-sec) system. Since the mass of the molecule is also proportional to this number (by definition), it follows that each molar mass of any substance must contain the same number of molecules. If $\mathcal{M}$ denotes the molecular mass number and m_i is the mass of one molecule, then for any substance the ratio

$$\frac{\mathcal{M}}{m_i} = \frac{\text{molar mass}}{\text{mass/molecule}} = \text{number of molecules in a molar mass} \equiv N_A$$

TABLE C-1

SOME COMPARATIVE MAGNITUDES OF MOLECULAR NUMBER DENSITIES FOR A PERFECT GAS

Metric measurement	*English measurement*
6.023×10^{26} molecules in 1 kg-mole, or	2.73×10^{26} molecules in 1 lb_m-mole, or
6.025×10^{23} molecules in 22.414 liters at standard temperature and pressure, or	2.73×10^{26} molecules in 385 ft^3 at standard temperature and pressure, or
2.69×10^{19} molecules/cm^3 at s.t.p., or	4.10×10^{14} molecules in one millionth of a cubic inch at s.t.p., or
2.69×10^{16} molecules/mm^3 at s.t.p., or	
$\sim 3.7 \times 10^{-17}$ mm^3 is the volume around one molecule at s.t.p., or	$\sim 2.44 \times 10^{-21}$ $in.^3$ is the volume around one molecule at s.t.p., or
the average distance between molecules at s.t.p. is $\sim 3.3 \times 10^{-6}$ mm	the average distance between molecules at s.t.p. is $\sim 1.31 \times 10^{-7}$ in.
The average "diameter" of a molecule is ~ 2 or 3×10^{-7} mm, or	The average diameter of a molecule is $\sim 10^{-8}$ in.
the ratio of intermolecular distance to molecular "diameter" is about 10:1.	
Visible light has a wavelength of ~ 0.6 micron, or	
6×10^{-4} mm, or the distance between molecules at s.t.p. is $\sim 1/2000$ of a wavelength of light.	

Note: The number density of water is about 1100 times that of a gas at s.t.p.

[2] A molar mass is termed a *mole*.

is a constant. This fixed number of molecules per mole (known as Avogadro's number) is 6.023×10^{26} molecules/kg-mole or 2.73×10^{26} molecules/lb_m-mole. Table C-1 indicates the number of molecules in a given volume at standard conditions (1 atm, 20°C). The assumption of randomness of distribution of molecules is valid since there are 440 molecules (a medium-sized sample) in a cube one millionth of an inch on a side (a volume of 10^{-18} in.^{3}), which is a nearly immeasurable size since one millionth of an inch is less than one half of a wavelength of visible light.

AN APPROXIMATE ANALYSIS OF MOLECULAR MOTION

By analyzing the random motion of the molecules within an arbitrary volume, the effects of the motion will be obtained and interpreted.

Consider a small subvolume of gas within a container with one of the walls of the container as one of its sides. This subvolume can be taken to be about 0.001 in. on a side in the form of a cube as shown in Fig. C-1. (This volume would contain approximately 10^{10} molecules.)

In order to make the analysis as simple as possible, the following assumptions will be made:

1. The volume is large enough that complete randomness of molecule position, speed, and direction of travel is obtained, so statistical averages of speed and direction of travel can be taken.
2. The gas consists of perfectly elastic spherical molecules as point masses with zero, or negligible volume (with respect to the volume of the surroundings). The assumption of perfectly elastic spheres implying equal incidence and reflection angles, and no energy loss on collision, is an interesting one with respect to its implications and will be discussed later. In addition, the mass of the molecule will be considered to be negligible with respect to that of the container, so the container walls can be considered to be entirely rigid.

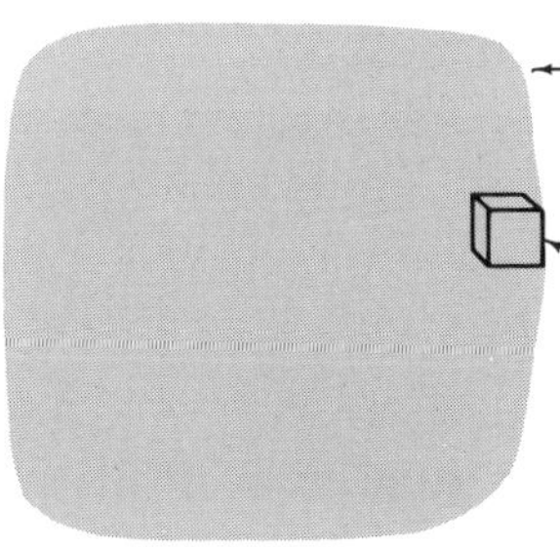

Fig. C-1.

3. Intermolecular forces will be neglected with respect to the inertial forces of the molecules.
4. Newton's laws of motion apply to the microscopic motion.
5. The macroscopic principles pertaining to a perfect gas apply:

$$\frac{p}{\rho} = RT, \quad c_p - c_v = R, \quad \ldots$$

where c_p, c_v, R, and $\gamma \equiv c_p/c_v$ are constants.

Consider one molecule traveling in the x direction as it hits and bounces off the right-hand wall (Fig. C-2). At the time of collision, the molecule experiences an impulse of

$$F\,dt = \frac{m(\Delta \text{v})}{g_c} \tag{C-1a}$$

or

$$F_i\,dt = \frac{m_i[-u_i - (-u_i)]}{g_c} = \frac{-2m_i u_i}{g_c} \tag{C-1b}$$

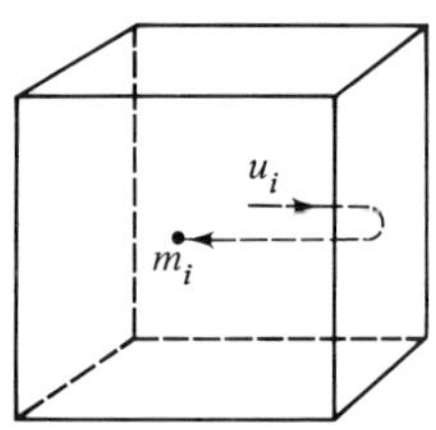

Fig. C-2.

where the subscript i is used to denote the conditions for one molecule and u_i is the x component of velocity. The impulse imparted to the wall is equal and opposite to that on the molecule, or

$$\underset{\text{ON WALL}}{F_i\,dt} = \frac{2m_i u_i}{g_c} \tag{C-2}$$

Since the velocities (speeds and directions) of the molecules are random, an average speed is considered as characteristic of the average motion of all molecules; this speed will be denoted by $\tilde{\text{v}}$.

On the average, there will be N molecules in the subvolume V. Since the motion and distribution are random, $N/3$ of these molecules will be traveling in the x direction (+ and −) or $N/6$ in the positive x direction.

Over some time period τ a sufficiently large number of collisions will occur that the net effect will be a force on the wall. For simplicity, the period τ will be taken such that all the $N/6$ molecules traveling in the x direction within the volume V will make just one collision with the right-hand wall. Then

$$\bar{F}\tau = \int_\tau F_i \, dt$$

or

$$\bar{F}\tau = \int_\tau \frac{2m_i\tilde{v}}{g_c} = \sum \frac{2m_i\tilde{v}}{g_c} \tag{C-3}$$

where the summation is taken over all the $N/6$ collisions occurring in the period τ. This summation procedure can be written

$$\bar{F}\tau = \frac{2m_i}{g_c}\tilde{v} \times (\text{number of collisions in } \tau) \tag{C-4}$$

The number of collisions in τ seconds is equal to (number of collisions/sec) $\times$ (τ sec), where the number of collisions/sec is the collision frequency f_c. The length of the volume V in the x direction is $L_1 = \tau\tilde{v}$, or $\tau = L_1/\tilde{v}$. The collision frequency of the $N/6$ molecules in τ sec is

$$f_c = \frac{N}{6}\frac{1}{\tau} = \frac{N\tilde{v}}{6L_1} \tag{C-5}$$

From (C-4), the average force is

$$\bar{F}\tau = \frac{2m_i\tilde{v}}{g_c}(f_c\tau)$$

Inserting f_c from (C-5) gives

$$\bar{F} = \frac{2m_i\tilde{v}}{g_c}\frac{N}{6}\frac{\tilde{v}}{L_1} = \frac{Nm_i(\tilde{v})^2}{3g_cL_1} \tag{C-6}$$

Of more importance than the force is the specific force, or the force per unit area, $p \equiv F/A = F/L_2L_3$, or

$$p = \frac{1}{3}\frac{Nm_i\tilde{v}^2}{L_1L_2L_3g_c}$$

With $L_1L_2L_3 \equiv V$,

$$pV = \frac{1}{3}\frac{Nm_i}{g_c}\tilde{v}^2 \qquad \text{or} \qquad pV = \frac{2}{3}\frac{Nm_j\tilde{v}^2}{2g_c} \tag{C-7}$$

Noting that $Nm_i/V = m/V = \rho$, (C-7) can be written

$$\frac{p}{\rho} = \frac{2}{3}\left(\frac{\tilde{v}^2}{2g_c}\right) \tag{C-8}$$

where $\tilde{v}^2/2g_c$ is the kinetic energy per unit mass of the random motion of the molecules.

Next consider the *energy* of the molecules in the volume. For these idealized molecules (point masses without any gravitational- or electrostatic-type force fields) the only energy mode is kinetic. The average kinetic energy of the gas in V is[3]

$$\frac{m\langle v^2\rangle}{2g_c} = \frac{m\langle u^2\rangle}{2g_c} + \frac{m\langle v^2\rangle}{2g_c} + \frac{m\langle w^2\rangle}{2g_c} = \frac{m}{2g_c}[\langle u^2\rangle + \langle v^2\rangle + \langle w^2\rangle] \tag{C-9}$$

Since the velocity is random, each kinetic-energy part must be equal to the other, or the K.E. is *equipartitioned* among the three parts. The random translational K.E. of this perfect gas represents the only possible constituent of the *internal energy*. The internal energy of a perfect gas is defined macroscopically as a function of the temperature only. This defining relation is a basic concept of physics,

$$\frac{m_i\langle u_i^2\rangle}{2g_c} \equiv \frac{1}{2}\hbar T \tag{C-10}$$

where $\hbar$ is the *Boltzmann constant* relating T to the mean K.E. per molecule ($\hbar = 1.38 \times 10^{-23}$ J/°K). For the three parts of (C-9),

$$\frac{m_i\langle v^2\rangle}{2g_c} = \frac{3}{2}\hbar T \tag{C-11}$$

The K.E. is averaged over all possible molecule speeds. Since $\langle v^2\rangle \neq \langle v\rangle^2$, ($\langle v\rangle^2 \equiv \tilde{v}^2$), $\langle v^2\rangle/2g_c$ in (C-9) is then not equivalent to $\tilde{v}^2/2g_c$ in (C-8); however, for simplicity, it will be assumed that they are equal; that is, $\langle v^2\rangle \approx \tilde{v}^2$. Comparison of (C-8) with (C-11) gives

$$\frac{p}{\rho} = \frac{\hbar}{m_i}T \tag{C-12}$$

[3] The pointed brackets imply time averages: $\langle v\rangle \equiv 1/\tau \int_\tau v\,dt$; $\langle v^2\rangle \equiv 1/\tau \int_\tau v^2\,dt$, where τ is some time period much larger than the time between molecular "collisions." $\langle v\rangle^2$ and $\langle v^2\rangle$ would only be equal if v were uniform (constant) independent of time. Since $v(t)$ is a random (for example, Gaussian) function, the two are not equal, but differ only by a constant that is a function of the distribution $v(t)$. Because of the convenience of notation $\langle v\rangle \equiv \tilde{v}$ is used.

Equation (C-12) may be written

$$\frac{p}{\rho} = \frac{N_A k}{N_A m_i} T = \frac{R_0}{\mathscr{M}} T \tag{C-13}$$

where $R_0 = N_A k$ is the universal gas constant relating the internal energy of a mole of gas to its translational kinetic energy. Note that k, R_0, and N_A are universal constants. R_0 is the gas constant for 1 mole ($\mathscr{M}$) containing N_A molecules and k is the gas constant for one molecule (m_i), or $R_0/\mathscr{M} = k/m_i$.

With $m = \mathscr{M}n$ (mass = mass/mole × number of moles in m), (C-13) may be written

$$\frac{p}{\rho} = \frac{nR_0}{m} \qquad \text{or} \qquad pV = nR_0T \tag{C-14}$$

The ratio $R_0/\mathscr{M} \equiv R$ is often used where R is the specific gas constant. Equation (C-13) is then

$$\frac{p}{\rho} = RT \tag{C-15}$$

Equations (6-13) through (6-15) are the familiar macroscopic equations of state for a perfect gas.

The principle of "equipartition of energy" can be extended to other modes of motion that are possible for "real" molecules (polyatomic molecules) that do not satisfy all the original assumptions for the perfect ("billiard-ball") gas model. From the above development, an amount of energy equal to $\frac{1}{2}kT$ is associated with each degree of freedom of motion of the molecule.

Consider other kinds of energy possible for the ideal billiard-ball gas model. The only possible energy that can be associated with the random motion of this gas molecule is the *thermal internal energy*. This equivalence of random-motion kinetic energy to *thermal* internal energy gives a physical meaning to the quantity "internal energy."

The *specific energy at constant volume* c_v is defined as the rate of change of internal energy with respect to temperature at constant volume, or

$$c_v \equiv \frac{\partial}{\partial T}\text{(internal energy)}$$

For the monatomic, billiard-ball perfect gas,

$$c_{vi} \frac{\partial}{\partial T}\left[\left(\frac{3}{2}\right)kT\right] = \left(\frac{3}{2}\right)k \text{ const.} \tag{C-16a}$$

For N_A molecules,

$$c_{v0} = \left(\frac{3}{2}\right) N_A k = \left(\frac{3}{2}\right) R_0 \tag{C-16b}$$

is the specific energy per mole. For a unit mass of a substance (for example, 1 lb_m) the specific heat at constant volume is

$$c_v = \frac{3}{2} \frac{R_0}{\mathcal{M}} = \left(\frac{3}{2}\right) R \tag{C-16c}$$

Also, for a perfect gas

$$c_p - c_v = R \qquad \text{or} \qquad c_p = \left(\frac{5}{2}\right) R \tag{C-17}$$

The ratio of specific energies, $\gamma \equiv c_p/c_v$, is then

$$\gamma = \frac{5/2}{3/2} = \left(\frac{5}{3}\right) = 1.67$$

for a monatomic gas as it approaches perfect gas conditions. The above values for c_v, c_p, and γ are amazingly accurate for many monatomic gases in states sufficiently far from their vapor–liquid line states.

It is interesting to consider the nature of molecular "collisions." Perfectly elastic collisions (without energy dissipation) are not usual in nature. When two "almost elastic" bodies collide, the final kinetic energy of the two bodies is not quite equal to the initial sum of their *kinetic* energies. The law of energy conservation states that the difference in the initial and final K.E.'s appears as an increase of the internal energy of the bodies; their temperature is increased. If any of the kinetic energy of the molecular collisions is dissipated, it can be converted only to internal energy, but the internal energy is identical with the kinetic energy. This "collision" process is then conservative on a molecular basis, where the molecular motion is completely random. Such a state of complete randomness is defined as an equilibrium state.

It is also of interest that molecular "collisions" do not occur in a literal sense. Real molecules have force fields of a complicated nature. An example of the behavior of a molecular force field as a function of the distance between molecules is shown in Fig. 2-9. When the molecules are very far apart, these intermolecular forces are very small and the substance behaves as a perfect gas. As this gas is compressed, gravity-type mass-attraction forces come into play. The behavioral state of the substance is seen to be determined by the ratio of the forces; for example, a highly compressed gas could still behave as a nearly perfect gas if the temperature is high enough that the translational momentum or inertia forces of the molecules are still much greater than the intermolecular field forces of attraction. The liquid state is obtained when the intermolecular forces are large enough to hold the molecules together, but where the inertial forces are large enough to maintain the random "thermal" (Brownian) motion. In the solid state there is no random translational motion of the molecules. They are bound together by the forces of attraction but are in a state of oscillatory dynamic equilibrium, between the attraction and repulsion conditions.

The curve of Fig. 2-9 rises extremely rapidly in the repulsion region, as evidenced by the great resistance of liquids and solids to compressive stress (very small compressibility).

The idealized perfect gas was not considered to have any intermolecular force fields. The "collisions" of a real gas molecule according to the force field behavior of Fig. 2-9 indicates a complicated and not a true collision. As one molecule makes a collision-line approach relative to another, it first feels an attraction. As the distance decreases, the field becomes repulsive, which increases extremely rapidly as the distance continues to decrease. This repulsion "barrier" provides the "bounce" for the collision.

BEHAVIOR OF REAL GASES

Equation (C-15) was obtained for perfect gas behavior. With $v_s \equiv 1/\rho$, where v_s is the specific volume (volume per unit mass),

$$pv_s = RT \tag{C-15a}$$

The average displacement volume of the molecules will be a function of the characteristic force field behavior of the specific substance, its temperature, and the number of molecules present in a given volume. If this displacement volume is b, the above relation can be modified to give

$$p(v_s - b) = RT \tag{C-18}$$

The attraction force field is initially inversely proportional to intermolecular distance. The molecule will then feel an attractive force toward the main mass of the gas as it approaches the immediate proximity of the wall. This pull away from the wall will reduce the momentum imparted to the wall. Since the force falls off very fast, only the molecules in the immediate vicinity of the molecule "hitting" the wall will provide any significant retardation. Assuming this force to be proportional to r^{-6}, and since v_s is proportional to r^3 (where r is the intermolecular distance, or an average distance) the retarding force that subtracts from the pressure is proportional to v_s^{-2}. Including this correction gives the van der Waals equation of state:

$$p = \frac{RT}{v_s - b} - \frac{a}{v_s^2} \qquad \text{or} \qquad \left(p + \frac{a}{v_s^2}\right)(v_s - b) = RT \tag{C-19}$$

The constants a and b depend on the substance.

Next, consider the possible energy components or degrees of freedom associated with a slightly more complicated or diatomic molecule. Assume this ideal diatomic molecule to be made up of two point mass atoms joined by a spring that represents the attraction–repulsion field similar to that illustrated in Fig. 2-9. The possible modes of motion with their respective degrees of freedom (°f) are

1. K.E. of translation: 3 °f
2. Rotational K.E.: 2 °f
3. Oscillatory: kinetic energy of motion + potential energy of the spring stretch $\propto r^2$ – 2 °f

The total possible °f is then 7. Such a molecule is sketched in Fig. C-3. There are only two significant rotational degrees of freedom since the moment of inertia about x axis is negligible with respect to the other two.

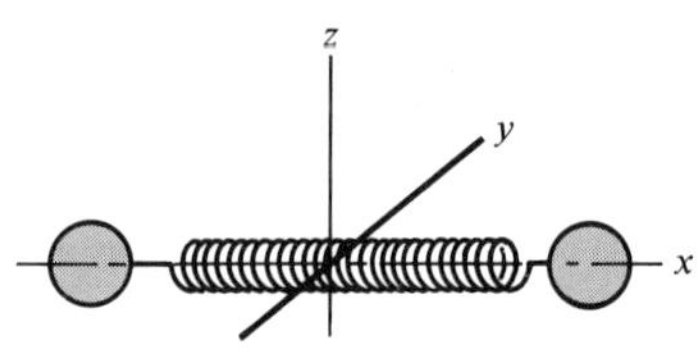

Fig. C-3.

Using the principle of equipartition of energy, the total energy for a molecule is equal to the total number of degrees of freedom × ½kT,

$$(\text{K.E.})_i = {}^\circ\text{f}_{\text{tot}} \times \tfrac{1}{2}kT$$

Therefore (C-16) gives

$$c_{vi} = \frac{^\circ\text{f}}{2}k \qquad c_{v0} = \frac{^\circ\text{f}}{2}R_0 \qquad \text{or} \qquad c_v = \frac{^\circ\text{f}}{2}R \tag{C-20}$$

For this diatomic gas, the following values are expected:

$$c_v = \frac{7}{2}R \qquad c_p = \frac{9}{2}R \qquad \text{and} \qquad \gamma = \frac{9}{7} = 1.3$$

However, for common diatomic gases near room temperature, $\gamma = 7/5 = (5 + 2)/5$, or $\gamma = (°f + 2)/°f$, so the °f for these diatomic gases in this state is °f = 5 instead of 7. The discrepancy occurs because the oscillatory mode is not excited near room temperature. At higher temperatures the oscillatory mode is excited and the other two °f occur. The value of γ then approaches 9/7 or 1.3. The converse is also true and the rotation mode is "frozen out" at very low temperatures, so that the value of γ approaches 5/3 as the temperature approaches zero. The above description does not allow for fractions of a degree of freedom. This paradox existed until classical statistical mechanics was combined with the quantum mechanics to give predicted analytical behaviors that correspond closely with those obtained by experimental observation.

Note that *temperature* for polyatomic gases is associated with the *translational kinetic energy* only. Consideration of "collisions" with the wall for a diatomic gas will indicate that, on the average, no net momentum exchange can take place through the spin or vibration modes. If the molecule is spinning very slowly and hits at an angle with respect to its major axis, some spin may be imparted to it as it rebounds. The converse is also possible. Since all possible orientations are statistically possible, the average effect is zero. The same reasoning holds true for the vibrational mode.

Although only gases have been considered, the above development can be applied to solids. In a solid there is no translational mode and no significant molecular spin mode. The only motion is an oscillatory motion of the atoms in the crystal lattice (or amorphous continuum). The oscillations occur in all three mutually perpendicular directions, and in each direction there is 1 °f for the kinetic energy and 1 °f for the potential energy. The total is 6 °f, which indicates that the specific heat should be $c_v = (6/2)R = 3R$, which is close to experimental values near room temperatures.

For liquids, the analysis becomes very difficult, because of the complicated intermolecular force behavior.

ATTEMPTS TO MODIFY THE PERFECT GAS FOR REAL GAS OR REAL MOLECULE EFFECTS

Instead of a point mass, a molecule occupies some finite volume, as was discussed above. Imagine that the effective radius r_c of this volume is determined by the collision path with another molecule, as illustrated in Fig. C-4. The effective volume is $(4/3)\pi r_c^3 = V_c$, or the cross-sectional area $\pi r_c^2 \equiv A_c$ is defined as the collision cross section. This effective volume or the collision cross section depends on the relative velocity (angle and speed of approach) and the mass of the molecules. However, since these quantities are random, a *mean* cross section can be considered. Considering

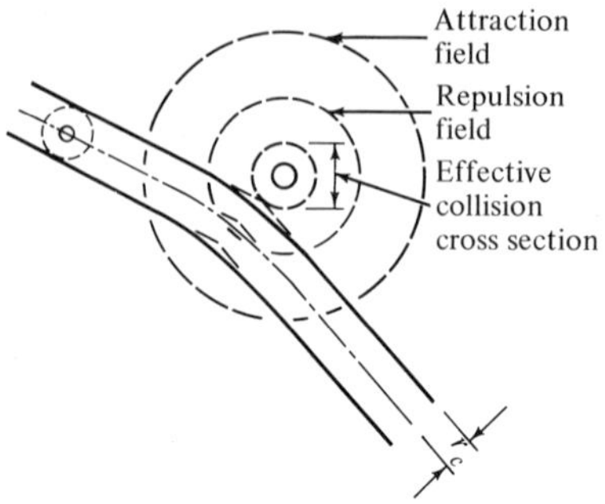

Fig. C-4.

only this displacement volume effect on the equation of state would give

$$p(v_s - b) = RT$$

as previously obtained. This is the Clausius equation of state. Further considerations of the intermolecular force fields led to the van der Waals equation, as was shown.

MEAN FREE PATH (λ)

The mean free path λ is the mean distance a molecule travels between collisions. This distance can be expected to be a function of the "size" of the molecule, or its collision cross section, and the number of molecules per unit volume, that is, the number density. It is also expected that the mean free path will be larger than the mean distance between the molecules. To visualize the collision process better, consider a molecule moving through a volume occupied by molecules that are at rest relative to the molecule under observation, Fig. C-5. In a period of time, δt, this molecule will travel a distance $\tilde{v}\,\delta t$ and "sweep out" a volume $A_c\tilde{v}\,\delta t$. Multiplying this volume by the number of molecules per unit volume gives the number of collisions occurring in the time δt:

$$\text{Number of collisions} = \text{volume} \times \frac{\text{number of molecules}}{\text{unit volume}}$$

or $$A_c\tilde{v}\,\delta t n' = \text{number of collisions in } \delta t \qquad (n' \equiv N/V)$$

or $$A_c\tilde{v}n' = \frac{\text{number of collisions}}{\text{unit time}} = \text{collision frequency}$$

The mean free path is the distance covered in the time δt divided by the number of collisions in δt,

$$\lambda = \frac{\tilde{v}\,\delta t}{A_c\tilde{v}n'\,\delta t} = \frac{1}{A_c n'} \tag{C-21}$$

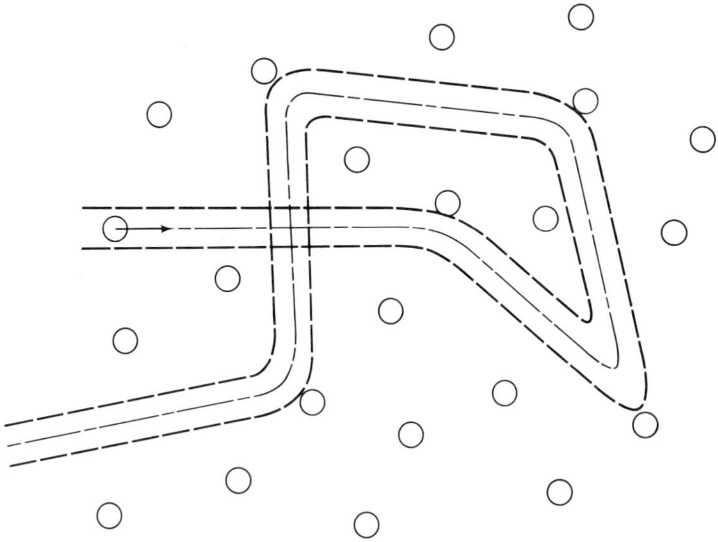

Fig. C-5.

The simplifying assumption that the motion of the molecules in the field is zero yields some error. Taking into account the relative motion of the molecules and their statistical distribution of speed gives

$$\lambda = \frac{1}{\sqrt{2}A_c n'} \tag{C-22}$$

COEFFICIENT OF VISCOSITY (μ)

For a perfect gas, the viscosity coefficient is a molecular momentum exchange coefficient; therefore, it is expected to depend on the mass and velocity of the molecules.

Consider a shear flow where a velocity profile exists, such as in the flow of a fluid near a solid boundary as indicated in Fig. C-6. Next, consider

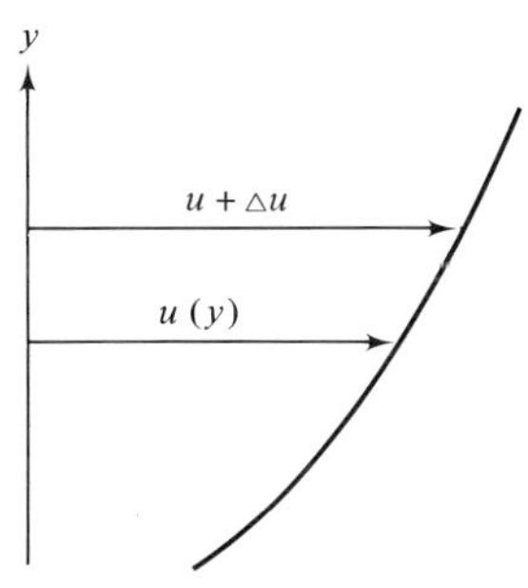

Fig. C-6.

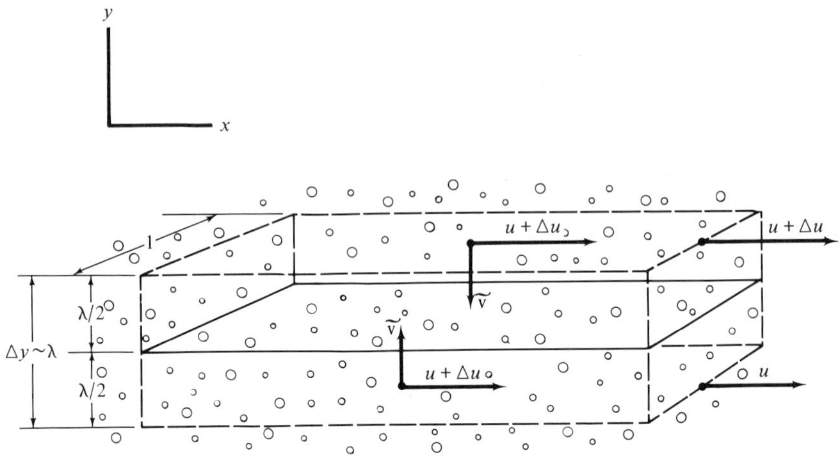

Fig. C-7.

an imaginary plane parallel to the direction of flow, and a volume circumscribed around this plane a distance $\lambda/2$ above and below the reference plane (Fig. C-7). The ordered flow velocity is $u(y)$. The velocities of the molecules consist of this ordered velocity plus the random molecular volume, $N/6$ will have a component of their random velocity in the $+y$ direction; all molecules can be considered to have the random speed $\tilde{v}$. The average directed or ordered velocity of these molecules is u. Now consider the x component of the force–momentum relationship for the transfer of momentum from the molecules in the lower half volume as they move up with velocity $\tilde{v}$ and cross the plane with an original momentum $m_i u$ (per molecule) and exchange momentum with the molecules in the upper half volume. For this momentum exchange, Newton's second law can be written

$$F_x = \frac{1}{g_c} \text{(mass flow rate across plane)} \times (\Delta u)$$

The force F_x is the accelerating force produced on the slower moving lower molecules by the upper molecules.

The mass flow rate across the plane is the density times the area times the average speed. The density is

$$\rho = n m_i \qquad \text{(number of molecules/unit volume)} \times \text{(mass/molecule)}$$

Since only $N/6$ of the molecules are involved, the mass flow rate is

$$\dot{m} = \frac{nm_i}{6}\,\tilde{v}A$$

and

$$F_x = \frac{nm_i\tilde{v}A}{6g_c}\,\Delta u$$

for the $+y$ molecules. An equal number will be traveling in the $-y$ direction. The force–momentum balance will be similar to that derived above. The net effect will be to produce a net force twice that given, or

$$F_x = \frac{nm_i\tilde{v}A}{3g_c}\,\Delta u$$

From the defining relationship for Newtonian viscosity,

$$\tau_{yx} = \mu\,\frac{du}{dy} = \mu\,\frac{\Delta u}{\Delta y}$$

The value of Δy for the above derivation is just λ. Solving the above equation for τ or F_x/A and with $\Delta y = \lambda$ gives

$$\tau = \frac{F_x}{A} = \frac{nm_i\tilde{v}\lambda}{3g_c}\,\frac{\Delta u}{\Delta y}$$

With $\rho = nm_i$,

$$\tau = \mu\,\frac{\Delta u}{\Delta y} = \frac{\rho\tilde{v}\lambda}{3g_c}\,\frac{\Delta u}{\Delta y}$$

or

$$\mu = \frac{\rho\tilde{v}\lambda}{3g_c} = \frac{nm_i\tilde{v}\lambda}{3g_c}$$

With $\lambda = 1/(\sqrt{2}\,A_c n)$,

$$\mu = \frac{m_i\tilde{v}}{3\sqrt{2}\,g_c A_c} \qquad \text{(C-23)}$$

A more exact analysis gives

$$\mu = \frac{0.493\rho\tilde{v}\lambda}{g_c} = \frac{0.493nm_i\tilde{v}\lambda}{g_c} \qquad \text{(C-24)}$$

If the relationship between the mean speed, $\tilde{v}$, and the temperature is now used,

$$\tilde{v} = \sqrt{\frac{3kT}{m_i}}$$

and also substituting $\lambda = 1/(\sqrt{2}\, A_c n)$,

$$\mu = \frac{0.493}{\sqrt{2}} \frac{\sqrt{3m_i \bar{k} T/g_c}}{A_c} \text{ (units of force·time/length}^2)^4 \qquad \text{(C-25)}$$

The only state variable in this relationship for μ is T. The viscosity of a perfect gas is then directly proportional to the square root of the absolute temperature. (The effect of T on the collision cross section is neglected, and the molecular displacement volume is considered to be negligible.)

Viscosity is easily measured in the laboratory. The only quantity that cannot be determined easily in (C-25) is A_c. The above equation then gives a method for determining the collision cross sections of gases under conditions at which they approach perfect gas behavior.

KINEMATIC VISCOSITY (ν)

The kinematic viscosity is defined as

$$\nu \equiv \frac{\mu}{\rho/g_c} = 0.493\tilde{v}\lambda \text{ (m}^2\text{/sec, cm}^2\text{/sec} \equiv \text{stokes, or ft}^2\text{/sec)} \qquad \text{(C-26)}$$

It is significant as a momentum diffusion coefficient or "momentum diffusivity," indicating the rate at which viscous momentum (or its effect, the shear) will "diffuse" into a flowing fluid, as in a jet or in the boundary layer near a solid boundary.

THERMAL CONDUCTIVITY (k)

The thermal conductivity is a transport or diffusion coefficient indicating the rate of diffusion of energy by molecular collision similar to the diffusion of momentum. The thermal conductivity k is the internal energy exchange coefficient. From the macroscopic viewpoint,

$$\dot{Q}_H = -\iint k \text{ grad } T \cdot d\vec{A}$$

for internal energy or "heat" transfer. The internal, or molecular, kinetic energy was determined as $(°\text{f}/2)\bar{k}T$. A simplified analysis for the transport of energy across an area by random molecular motion gives

$$k = \frac{°\text{f}}{2} \frac{\bar{k} n \tilde{v} \lambda}{3} \qquad \text{(C-27)}$$

For a monatomic, perfect gas, $°\text{f} = 3$ and $k \approx n\tilde{v}\lambda/2$.

[4] For units of fluid properties, and conversion constants, see Appendix D.

Because of the similarity between the thermal conductivity as an energy diffusivity and the viscosity as a momentum diffusivity, it is of interest to examine the ratio of these two quantities. Taking the value of μ from the simplified analysis,

$$\mu = \frac{nm_i \tilde{\text{v}} \lambda}{3g_c}$$

the ratio is

$$\frac{k}{\mu} = \frac{{}^\circ\text{f} \not{k} g_c}{2m_i}$$

Substituting molar values,

$$c_{v0} = \frac{{}^\circ\text{f}}{2} R_0 \qquad \not{k} = \frac{R_0}{n_0} \qquad m_i = \frac{\mathscr{M}}{N_A}$$

gives

$$\frac{k\mathscr{M}}{g_c \mu c_{v0}} = 1 \tag{C-28}$$

The agreement with this ratio and that obtained from experimental observations is not good. Better agreement is obtained if intermolecular force field effects on both the force–momentum and work–energy relations are considered. The ratio of (C-28) is then about 2.5 for a monatomic gas, about 2 for a diatomic gas, and about 5/3 for a triatomic gas.

THERMAL DIFFUSIVITY (α)

The thermal diffusivity is defined as

$$\alpha \equiv \frac{k}{\rho c_p} \tag{C-29}$$

It is the internal energy diffusion coefficient and has the same units as the kinematic viscosity. With the above value for k, and with

$$c_p = c_v + R = \left(1 + \frac{{}^\circ\text{f}}{2}\right) R$$

$$\alpha = \frac{({}^\circ\text{f}/2)(\tilde{\text{v}}\lambda/3)}{1 + ({}^\circ\text{f}/2)} \tag{C-30}$$

The ratio of momentum diffusivity to thermal diffusivity (ν/α) occurs in convective heat-transfer analyses and is known as the Prandtl number.

For the simplified analyses ($\nu = \tilde{v}\lambda/3$),

$$N_{\text{Pr}} \equiv \frac{\nu}{\alpha} = \frac{c_p \mu g_c}{k} = \frac{\tilde{v}\lambda/3}{(°f/2)(\tilde{v}\lambda/3)/(1 + °f/2)} = \frac{(1 + °f/2)}{°f/2} = \gamma$$

which is a somewhat surprising (but inaccurate) result. With the ratio (from C-28)

$$\frac{k\mathcal{M}}{g_c \mu c_{v0}} \neq 1$$

and with the appropriate value from a more exact analysis as a modification for the Prandtl number (for air, as an example, with $°f = 5$),

$$\frac{\nu}{\alpha} = \frac{\gamma}{2} = 0.7$$

which is quite close to the experimental value obtained near room temperature.

NO SLIP CONDITION AT THE WALL

If a billiard-ball type of "collision" of ideal molecules is considered at a solid boundary, it would be expected that the angle of reflection would be equal to the angle of incidence. Therefore, no flow retardation would result, and there would be no wall shear action.

The resolution to this paradox is that a *smooth wall* does not exist on a molecular scale. A wall surface has random roughness distribution. The randomness of the molecular velocity distribution is therefore maintained in collisions with the wall if there is no ordered flow velocity of the fluid.

Consider a representative molecule at a distance of one mean free path (λ) or less above the wall indicated in Fig. C-8. Assume the velocity of this molecule (or all molecules within a distance λ from the wall) to have a mean ordered flow velocity U plus a random rms speed $\tilde{v}$ equally likely in any direction. Only random motion with a downward component is significant. The addition of U will produce more collisions with the back faces of the wall irregularities than with the front faces. Consequently, there will be a greater backward reflected motion. This back reflection will just cancel U, so that within one λ from the wall, there is no mean flow velocity; therefore, U must be zero within the distance λ from the wall. For distances greater than λ, the viscous momentum exchange conditions of Chapter 2 and this appendix exist.

If the *length* of the flow passage is on the order of λ, many molecules will traverse the length without colliding with the wall. For example,

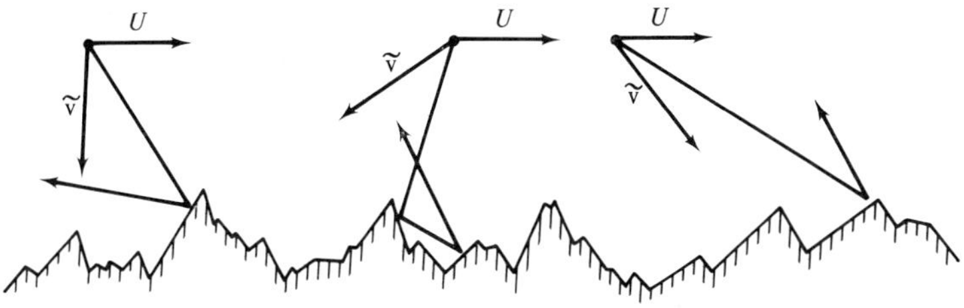

(a)

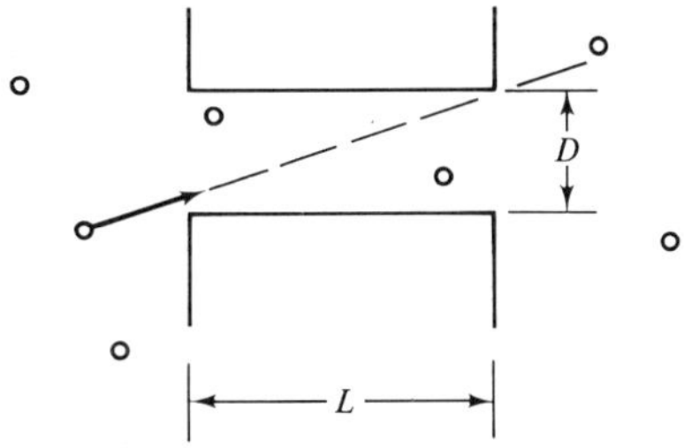

(b)

Fig. C-8.

Fig. C-8b, for $L > \lambda$, indicates the possibility of molecules within a cone angle $\alpha = \tan^{-1}(D/L)$ traversing the passage with no collision with the wall or with other molecules. For such a physical case where $L > \lambda$, the density must be quite small. A significant dimensionless parameter for rarefied gas flow is then the ratio L/λ or D/λ. This parameter is known as the Knudsen number, N_{Kn}. When it is very large, the flow may be treated with continuum mechanics. When it is of the order of unity, statistical mechanics is used.

NOMENCLATURE

a	constant in the van der Waals equation of state
A_c	collision cross section (area)
b	constant in the van der Waals equation
c_p	constant-pressure specific energy
c_v	constant-volume specific energy
c_{v0}	molar constant-volume specific energy
f	frequency (events/sec)
f_c	collision frequency

°f	degrees of freedom
F	force
g	local gravitational acceleration
g_c	conversion constant
g_0	standard reference value of g
$\hbar$	Boltzmann constant (R_0/n_0)
k	thermal conductivity
L	length
m	mass
m_i	mass of one molecule
$\mathscr{M}$	molar mass number (mass of 1 mole)
n	number of molecules per unit volume (number density)
N_A	number of molecules/mole (Avogadro's number)
N	total number of molecules in any given volume V (or mass m)
p	pressure
R	gas "constant"
R_0	universal gas constant
t	time
T	temperature
u	x component of velocity
$\tilde{u}$	internal energy
v	y component of velocity
$\vec{\mathrm{v}}$	vector velocity
$\tilde{\mathrm{v}}$	rms speed
v_s	specific volume ($1/\rho$)
V	volume
x, y, z	coordinates
α	thermal diffusivity
γ	specific heat ratio (c_p/c_v)
λ	mean free path
μ	coefficient of viscosity (absolute viscosity)
ν	kinematic viscosity
ρ	density
τ	time period
τ_{ij}	shear stress (also used without subscripts to denote shear stress)

PROBLEMS

C-1 If the average molecular mass of air is 29 and all the elements except O_2 and N_2 are in negligible quantities, find (a) the mass fractions of O_2 and N_2, and (b) the volume fractions.

C-2 Find the mean speed ($\tilde{\mathrm{v}}$) for the O_2 and N_2 molecules in air at 68°F.

C-3 Find the speed for the traces of H_2 in air at 68°F.

C-4 Assuming the average molar mass of air to be 29, calculate c_p, c_v, R (in units of Btu/lb$_m$), and γ and compare with tabulated values.

C-5 From tabulated molar values, calculate c_v, c_p, R, and γ and compare the results with those of Problem A-4: (a) at 100°R, (b) at 68°F (528°R), (c) at 1000°R, (d) at 2000°R, and (e) at 3500°R.

C-6 Calculate the degrees of freedom for the molecules from Problem C-5.

C-7 Compute $\tilde{v}$, λ, and A_c for air and for H_2 (a) at 68°F and 1 atm, and (b) at 68°F and 1 mm Hg absolute pressure, where the kinematic viscosity at 68°F (ft^2/sec) is given as air, 1.63×10^{-4}; H_2, 1.01×10^{-8}.

APPENDIX D

TABLES

Tables 2-1 and 2-2 list characteristic properties of some liquid and gases.

Figure 2-5 presents absolute viscosity as a function of temperature for various fluids.

Figure 2-8 presents kinematic viscosity as a function of temperature for various fluids.

Drag coefficients for cylinders, spheres, and various shapes are presented in the figures of Chapter 11 and in Table 11-1.

TABLE D-1

PROPERTIES OF THE ATMOSPHERE[a]

Altitude, ft	T, °F	p, lb_f/ft^2	p/p_0	ρ, lb_m/ft^3	ρ/ρ_0	μ, $lb_f\text{-}sec/ft^2$	a, ft/sec
0	59.00	2116.2	1.0000	7.6475×10^{-2}	1.0000	3.737×10^{-7}	1116.4
500	57.22	2078.3	0.9821	7.536	0.9855	3.727	1114.5
1,000	55.43	2040.9	0.9644	7.246	0.9711	3.717	1112.6
5,000	41.17	1760.9	0.8321	6.590	0.8617	3.637	1097.1
10,000	23.36	1455.6	0.6878	5.648	0.736	3.534	1077.4
20,000	−12.26	973.3	0.4599	4.077	0.5332	3.325	1036.9
30,000	−47.83	629.7	0.2975	2.866	0.3747	3.107	994
36,152[b]	−69.70	472.7	0.2234	2.272	0.2971	2.960	984.5
40,000	−69.70	393.1	0.1858	1.890	0.2471	2.969	968.1
50,000	−69.70	243.6	0.1151	1.171	0.1531	2.969	968.1
60,000	−69.70	151.0	7.137×10^{-2}	7.259×10^{-3}	9.492×10^{-2}	2.969	968.1
70,000	−69.70	93.67	4.437	4.502	5.887	2.969	968.1
80,000	−69.70	58.13	2.747	2.794	3.613	2.969	968.1
82,345[c]	−69.70	51.98	2.456	2.497	3.268	2.969	968.1
90,000	−57.20	36.29	1.715	1.690	2.210	3.048	983.4
100,000	−40.89	23.09	1.091	1.033	1.351	3.150	1003.2
150,000	−40.43	3.060	1.446×10^{-3}	1.147×10^{-4}	1.500×10^{-3}	3.532	1096.3
200,000	−2.71	0.4754	2.246×10^{-4}	1.950×10^{-5}	2.550×10^{-4}	3.381	1047.9
250,000	−105.3	4.864×10^{-2}	2.295×10^{-5}	2.573×10^{-6}	3.364×10^{-5}	2.737	922.7
300,000	−105.3	3.698×10^{-3}	1.748×10^{-6}	1.951×10^{-7}	2.552×10^{-6}	2.743	923.9
400,000	2.76	6.096×10^{-5}	2.881×10^{-8}	2.112×10^{-9}	2.762×10^{-8}		

(continued)

TABLE D-1 *(continued)*

Altitude, ft	T, °F	p, lb_f/ft^2	p/p_0	ρ, lb_m/ft^3	ρ/ρ_0	μ, lb_f-sec/ft^2	a, ft/sec
500,000	365.6	4.875×10^{-6}	2.304×10^{-9}	9.210×10^{-11}	1.204×10^{-9}		
600,000	755.7	1.152×10^{-6}	5.444×10^{-10}	1.447×10^{-11}	1.893×10^{-10}		
700,000	875.4	3.925×10^{-7}	1.855×10^{-10}	4.107×10^{-12}	5.371×10^{-11}		
800,000	999.2	1.592×10^{-7}	7.524×10^{-11}	1.430×10^{-12}	1.869×10^{-11}		
900,000	1124	7.332×10^{-8}	3.465×10^{-11}	5.771×10^{-13}	7.546×10^{-11}		
1,000,000	1251	3.718×10^{-8}	1.757×10^{-11}	2.607×10^{-13}	3.409×10^{-12}		

[a] Adapted from R. H. Minzner, W. S. Ripley, and T. P. Condron, *U.S. Extension to the ICAO Standard Atmosphere*, U.S. Department of Commerce, Weather Bureau, Washington, D.C., 1958. See also Standard Atmosphere Tables and Data for Altitudes up to 65,800 Feet, *NACA Report 1235*, 1955.

[b] Tropopause: the altitude at which the temperature becomes constant at $T = 216.66°K$. This altitude is specified as 11,000 geopotential meters.

[c] Upper limit of isothermal layer; specified as 25,000 geopotential meters.

TABLE D-2

PROPERTIES OF DRY AIR AT 1 ATM

T, °F	ρ, lb_m/ft^3	$\mu \times 10^7$, lb_f-sec/ft^2	$\nu \times 10^4$, ft^2/sec	k, Btu/(hr-ft-°F)	$\beta \times 10^3$, 1/°F
−200	0.1538	2.09	0.463	0.0080	3.90
−100	0.1104	2.80	0.814	0.0108	2.82
−50	0.1020	3.08	0.972	0.0121	2.46
0	0.0867	3.41	1.267	0.0133	2.21
32	0.0807	3.59	1.433	0.0140	2.03
40	0.0794	3.63	1.472	0.0143	2.00
50	0.0780	3.68	1.520	0.0143	1.96
60	0.0765	3.74	1.575	0.0145	1.92
80	0.0735	3.86	1.689	0.0151	1.85
100	0.0709	3.96	1.800	0.0156	1.79
120	0.0684	4.08	1.917	0.0161	1.72
150	0.0651	4.23	2.092	0.0169	1.64
200	0.0601	4.49	2.403	0.0178	1.52
250	0.0560	4.75	2.733	0.0192	1.41
300	0.0524	5.00	3.075	0.0203	1.32
400	0.0461	5.45	3.800	0.0225	1.16
500	0.0413	5.86	4.567	0.0247	1.04
600	0.0374	6.26	5.383	0.0267	0.94
700	0.0342	6.67	6.272	0.0288	0.86
800	0.0315	7.02	7.175	0.0306	0.79
900	0.0292	7.39	8.156	0.0324	0.74
1000	0.0272	7.71	9.128	0.0342	0.68
1200	0.0239	8.35	11.24	0.0373	0.60
1400	0.0213	8.92	13.45	0.0399	0.54
1600	0.0193	9.47	15.83	0.0423	0.49
1800	0.0176	9.98	18.30	0.0443	0.44

TABLE D-3

PROPERTIES OF WATER[a]

T, °F	p_{sat}, $lb_f/in.^2$	ρ, lb_m/ft^3	$\mu \times 10^6$, lb_f-ft/sec^2	$\nu \times 10^5$, ft^2/sec	$\beta \times 10^4$, 1/°F	$\sigma \times 10^3$, lb_f/ft
32	—	62.43	37.43	1.929	−0.38	5.18
40	0.122	62.43	32.06	1.652	+0.15	5.14
50	0.173	62.41	27.65	1.40	0.43	5.03
60	0.256	62.36	23.83	1.23	0.85	5.02
70	0.363	62.30	20.44	1.06	1.12	5.08
80	0.507	62.20	17.76	0.919	1.51	4.90
90	0.698	62.12	15.71	0.825	1.79	4.84
100	0.949	62.00	14.22	0.738	2.03	4.28
120	1.692	61.71	11.65	0.607	2.50	4.65
140	2.889	61.38	9.256	0.511	2.91	4.52

(continued)

TABLE D-3 *(continued)*

T, °F	p_{sat}, $lb_f/in.^2$	ρ, lb_m/ft^3	$\mu \times 10^6$, lb_f-ft/sec^2	$\nu \times 10^5$, ft/sec^2	$\beta \times 10^4$, 1/°F	$\sigma \times 10^3$, lb_f/ft
160	4.741	61.00	8.314	0.439	3.28	4.39
180	7.510	60.58	7.175	0.381	3.63	4.26
200	11.53	60.12	6.328	0.340	3.98	4.12
212	14.696	59.83	5.905	0.317	4.18	4.03
220	17.19	59.63	5.603	0.302	4.33	3.97
240	24.97	59.10	5.059	0.276	4.67	3.82
260	34.43	58.86	4.600	0.252	5.00	3.66
280	49.20	57.95	4.205	0.234	5.34	3.49
300	67.01	57.32	3.882	0.220	5.70	3.33
320	89.66	56.65	3.583	0.203	6.10	3.16
340	118.0	55.94	3.325	0.191	6.50	3.00
360	153.0	55.21	3.125	0.182	6.94	2.83
380	195.8	54.38	2.939	0.174	7.43	2.67
400	247.3	53.64	2.780	0.168	7.97	2.50
450	422.6	51.5	2.474	0.155	9.72	2.10
500	680.8	49.0	2.237	0.145	12.1	1.65
550	1045	45.87	20.34	0.139	16.4	1.29
600	1543	42.37	1.830	0.137	23.6	0.75
650	2208	37.33	1.579	0.136	50	0.30
705.4	3206	19.91	1.053	0.170	—	0

[a] All properties for temperatures between 32 and 212°F are at 1-atm pressure; properties at temperatures above 212°F are at the saturation pressure.

TABLE D-4

QUANTITY DEFINITIONS

International (standard) kilogram (kg): mass of platinum–irridium. Equivalent to mass of 1.000027×10^{-3} m^3 of pure water at 4°C and 1 atm.
Pound-mass (lb_m): 0.05359237 kg.
Pound-force (lb_f): the force required to accelerate 1 lb_m at the rate of 32.1740 ft/sec^2 or the force of gravity on 1 lb_m at standard sea level and 45° latitude.
Liter: volume of 1 kg of pure water at 4°C and 1 atm.
Dyne: the force required to accelerate 1 g at the rate of 1 cm/sec^2.
Newton (N): the force required to accelerate 1 kg-mass at the rate of 1 m/sec^2.
Erg = dyne-cm.
Absolute joule: 1 N-m (10^7 ergs).
Watt (W): 1 J/sec (also 1 V-A).
British thermal unit (Btu): amount of energy required to raise 1 pound-mass of water at 39.1°F (temperature of maximum density) and 1 atm through a temperature difference of 1°F.
Calorie (cal): Amount of energy required to raise 1 g of water at 4°C and 1 atm through a temperature difference of 1°C.
Poise (P): Absolute (or "dynamic") viscosity unit; 1 P is equivalent to 1 dyne-sec/cm^2.
Centipoise (cP): one one-hundredth of a poise (10^{-2} P).
Stoke: kinematic viscosity unit; 1 stoke is equivalent to 1 cm^2/sec.
Centistoke: one one-hundredth of a stoke (10^{-2} stoke).

TABLE D-5

PHYSICAL CONSTANTS

1. Standard sea-level atmospheric pressure
   14.696 $lb_f/in.^2$
   101,325 N/m^2
   1013.25 millibars
   760.00 mm Hg (density 13.595 g/cm^3 at $g_0 = 9.80665$ m/sec^2)
2. Temperature
   0°F 459.688°R
   0°C 273.16°K
3. Lapse rate (λ) of the troposphere
   0.00356616°F/ft
4. Acceleration of gravity (standard, sea level, 45° latitude)
   32.1740 ft/sec^2
   9.80665 m/sec^2
   980.665 cm/sec^2
5. Gas constant (R_0)
   1544 $ft\text{-}lb_f/lb_m\text{-}°R$
   8.31438×10^7 ergs/g-mole-°K
   8314.38 abs. J/kg-mole-°K
6. Boltzmann constant
   $k = 1.3805 \times 10^{-16}$ erg/molecule-°K
   $k = 1.38032 \times 10^{-23}$ J/molecule-°K
7. Mechanical equivalent of heat (Joule's constant)
   $J = 4.1840 \times 10^7$ ergs/cal
   $J = 778.16$ $ft\text{-}lb_f/Btu$

TABLE D-6

CONVERSION FACTORS

The units presented in these tables are those encountered in the absolute English, cgs, and mks systems, as discussed in Appendix A.

1. LENGTH (L)

Multiply ↓ by / to get →	*Inches* (in.)	*Feet* (ft)	*Centimeters* (cm)	*Meters* (m)
Inches (in.)	1	12	2.5400	0.02540
Feet (ft)	1/12	1	30.480	0.30480
Centimeters (cm)	0.3937	0.03281	1	0.0100
Meters (m)	39.370	3.2808	100	1

(continued)

TABLE D-6 *(continued)*

2. AREA (A)

Multiply ↓ by / to get→	Square inches ($in.^2$)	Square feet (ft^2)	Square centimeters (cm^2)	Square meters (m^2)
Square inches ($in.^2$)	1	6.9444×10^{-3}	6.4516	6.4516×10^{-4}
Square feet (ft^2)	144	1	929.03	9.2903×10^{-2}
Square centimeters (cm^2)	0.15500	1.0764×10^{-3}	1	10^{-4}
Square meters (m^2)	1550.0	10.764	10^4	1

3. VOLUME (V)

Multiply ↓ by / to get→	Cubic inches ($in.^3$)	Cubic feet (ft^3)	Gallons (gal)	Cubic centimeters (cm^3)	Liters
Cubic inches ($in.^3$)	1	5.7870×10^{-4}	4.3290×10^{-3}	16.387	1.6387×10^{-2}
Cubic feet (ft^3)	1728	1	7.4805	28317	28.316
Gallons (gal)	231.00	0.13368	1	3785.4	3784.3
Cubic centimeters[a] (cm^3)	6.1023×10^{-2}	3.5314×10^{-5}	2.6417×10^{-4}	.1	0.99997×10^{-3}
Liters	61.025	3.5315×10^{-2}	0.26418	1000.03	1

[a] One milliliter (ml) and one cubic centimeter (cm^3) are nearly the same. By definition, 1 ml is equivalent to 1.000028 cm^3, or 1 liter is equivalent to 10000.028 cm^3 (or 1.000028×10^{-3} m^3, since the meter is the fundamental established standard of length measurement).

4. MASS (m)

Multiply ↓ by / to get→	Pounds (lb_m)	Grams (g)	Kilograms (kg)
Pounds (lb_m)	1	453.59	0.45359
Grams (g)	2.2046×10^{-3}	1	10^{-3}
Kilograms (kg)	2.2046	10^3	1

5. DENSITY (ρ)

Multiply ↓ by / to get→	lb_m/ft^3	$lb_m/in.^3$	g/cm^3	kg/m^3
lb_m/ft^3	1	5.7870×10^{-3}	1.6018×10^{-2}	16.018
$lb_m/in.^3$	1728	1	27.680	27680
g/cm^3	62.428	3.6127×10^{-2}	1	1000
kg/m^3	6.2428×10^{-2}	3.6127×10^{-5}	10^{-3}	1

(continued)

TABLE D-6 *(continued)*

6. *FORCE (F)*

Multiply by ↓ / to get →	*Pounds force* (lb_f)	*Dynes*	*Newtons* (N)
Pounds force (lb_f)	1	4.4482×10^5	4.4482
Dynes	2.2481×10^{-6}	1	10^{-5}
Newtons (N)	0.22481	10^{-5}	1

7. *PRESSURE (p)*

Multiply by ↓ / to get →	lb_f/ft^2	$lb_f/in.^2$	atm	$dynes/cm^2$	N/m^2
lb_f/ft^2	1	6.9444×10^{-3}	4.7254×10^{-4}	487.80	47.880
$lb_f/in.^2$	144	1	6.8046×10^{-2}	68947	6894.7
atm	2116.2	14.696	1	1.0133×10^6	1.0133×10^5
$dynes/cm^2$	2.0886×10^{-3}	1.4505×10^{-5}	9.8692×10^{-7}	1	0.10000
N/m^2	2.0886×10^{-2}	1.4505×10^{-4}	9.8692×10^{-6}	10	1

The barometric pressure unit, the "bar," is defined as 100 $dynes/m^2$ or 10^6 $dynes/cm^2$. A torr (from Torricelli) is a hydrostatic pressure head of 1 mm Hg.

8. *ENERGY (Work, Heat)*

Multiply by ↓ / to get →	$ft\text{-}lb_f$	Btu	abs. J	cal	erg
$ft\text{-}lb_f$	1	1.285×10^{-0}	1.3558	0.32405	1.3558×10^7
Btu	778.16	1	1055.0	252.16	1.0550×10^{10}
abs. J	0.73756	9.4783×10^{-4}	1	0.23901	10^7
cal	3.0860	3.9657×10^{-3}	4.1840	1	4.1840×10^7
erg	7.3756×10^{-8}	9.4783×10^{-11}	10^{-7}	2.3901×10^{-8}	1

1 J is defined as 1 N-m; 1 erg is 1 dyne-cm (1 N = m is 10^7 ergs)

1 international joule is equivalent to 1.000165 absolute joules

1 international steam table calorie (I.T. cal) is equivalent to 1.000654 cal (thermochemical calorie)

9. *SPECIFIC ENERGY*

Multiply by ↓ / to get →	Btu/lb_m	cal/g	abs. J/kg
Btu/lb_m	1	0.55592	2326.0
cal/g	1.7988	1	4184.0
abs. J/kg	4.2993×10^{-4}	2.3901×10^{-4}	1

(continued)

TABLE D-6 *(continued)*

10. ABSOLUTE VISCOSITY (μ)

Multiply by ↓ / to get→	lb_f-sec/ft^2	P	N-sec/m^2
lb_f-sec/ft^2	1	478.8	4788
P	2.0885×10^{-3}	1	10^{-1}
N-sec/m^2	2.0885×10^{-2}	10	1

1 cP = 1/100 P

11. KINEMATIC VISCOSITY (ν)

Multiply by ↓ / to get→	ft^2/sec	stoke	m^2/sec
ft^2/sec	1	929.03	9.2903×10^{-2}
stoke	1.076×10^{-3}	1	10^{-4}
m^2/sec	10.764	10^4	1

1 cS = 1/100 Stoke

12. POWER (*P*)

Multiply by ↓ / to get→	hp	ft-lb_f/sec	Btu/hr	W	ergs/sec
hp	1	550	2544	745.7	7.457×10^9
ft-lb_f/sec	1.818×10^{-3}	1	4.6254	1.356	1.356×10^7
Btu/hr	3.929×10^{-3}	0.2162	1	0.2930	2.930×10^6
W	1.341×10^{-3}	0.7378	3.413	1	10^7
ergs/sec	1.341×10^{-10}	7.738×10^{-7}	3.431×10^{-7}	10^{-7}	1

1 W is 1 J/sec or 1 N-m/sec

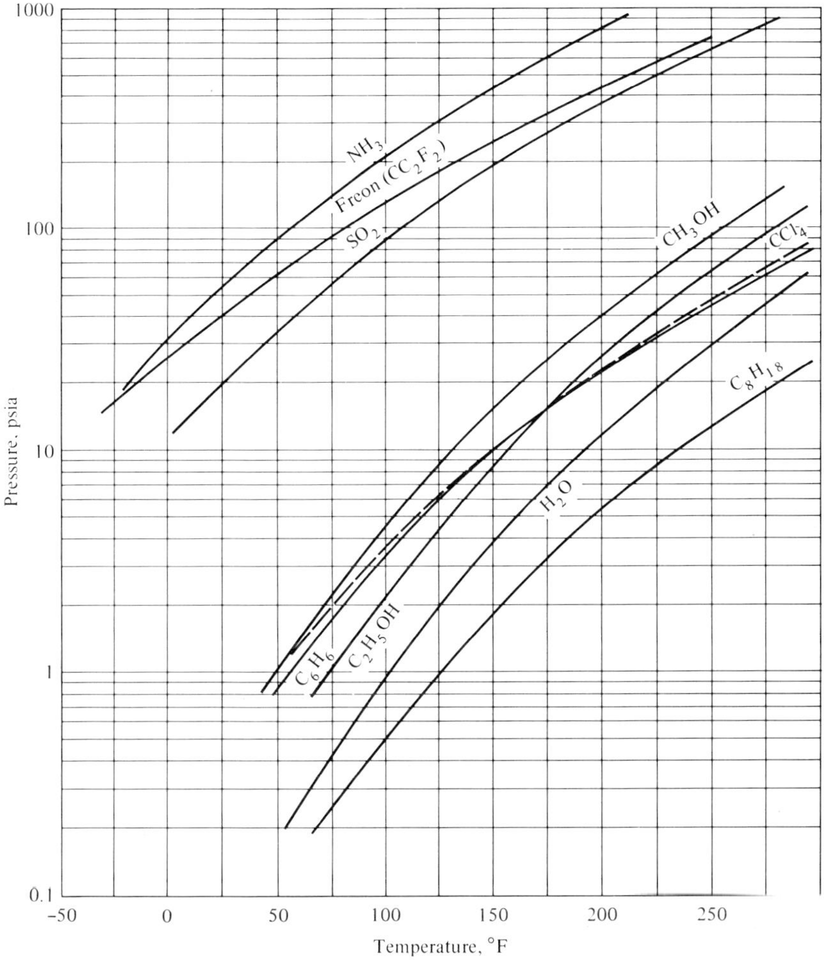

Fig. D-1. Vapor pressure versus temperature for various fluids.

INDEX

INDEX